AF905426

Signals and Communication Technology

More information about this series at http://www.springer.com/series/4748

Jens-Rainer Ohm

Multimedia Signal Coding and Transmission

Springer

Jens-Rainer Ohm
Institute of Communication Engineering
RWTH Aachen University
Aachen
Germany

ISSN 1860-4862 ISSN 1860-4870 (electronic)
Signals and Communication Technology
ISBN 978-3-662-46690-2 ISBN 978-3-662-46691-9 (eBook)
DOI 10.1007/978-3-662-46691-9

Library of Congress Control Number: 2015934224

Springer Heidelberg New York Dordrecht London
© Springer-Verlag Berlin Heidelberg 2015
This work is subject to copyright. All rights are reserved by the Publisher, whether the whole or part of the material is concerned, specifically the rights of translation, reprinting, reuse of illustrations, recitation, broadcasting, reproduction on microfilms or in any other physical way, and transmission or information storage and retrieval, electronic adaptation, computer software, or by similar or dissimilar methodology now known or hereafter developed.
The use of general descriptive names, registered names, trademarks, service marks, etc. in this publication does not imply, even in the absence of a specific statement, that such names are exempt from the relevant protective laws and regulations and therefore free for general use.
The publisher, the authors and the editors are safe to assume that the advice and information in this book are believed to be true and accurate at the date of publication. Neither the publisher nor the authors or the editors give a warranty, express or implied, with respect to the material contained herein or for any errors or omissions that may have been made.

Printed on acid-free paper

Springer-Verlag GmbH Berlin Heidelberg is part of Springer Science+Business Media
(www.springer.com)

Preface

During the last decades, digital audiovisual information has become omnipresent in professional and private life. Audiovisual media have changed our ways of communication and collaboration, information acquisition, learning, working and interacting. Simultaneously, Internet, mobile networks and digital storage have superseded traditional ways of distribution and access. In this context, multimedia communication establishes novel forms of communication between people and/or machines, which are characterized by ubiquity, multimodality, interactivity and increasing intelligence.

Based on my lectures on topics of multimedia communication systems held over many years at RWTH Aachen University, this book, along with another release to appear soon, represents a substantially upgraded version of the textbook "Multimedia Communication Technology" of 2004. The topic of the current book is multimedia signal compression and transmission, the second book's topic is identification and recognition of multimedia signals entitled as "Multimedia Content Analysis" (MCA). Both books (as well as the two lectures they are based upon) are self-contained and therefore not to be understood as volume 1 and 2 of a package. However, due to commonalities between coding and content analysis (both based on concepts from signal processing and information theory), it is not surprising that the reader will frequently find cross references (including pointers to section/chapter numbers) between one and the other book. Fundamental knowledge about both SP and IT is helpful for understanding, but chapters 2 and 4 target to summarize the most relevant underlying concepts.

Since the above-mentioned 2004 release, the progress that was made in compression of audiovisual data has again been breath-taking. Consequently, newest developments are reflected, including the deeper understanding of concepts that enabled the emergence of the High Efficiency Video Coding (HEVC) standard, as well as methods to compress 3D video and audio data, and possible future trends are sketched. However, rather than explaining the precise implementation of existing standards, the book intends to provide the reader with deep understanding for the underlying concepts, to eventually give support becoming a designer of novel technology in the field.

Most chapters are again supplemented by homework problems, for which solutions are made available from **http://www.ient.rwth-aachen.de**.

The book would not have been possible without contributions of numerous students, scientific staff, colleagues in academia and standardization bodies and many other people who have worked with me on topics of image, video and audio processing, encoding and recognition over more than 25 years. For the sake of brevity, I would like to express my thanks globally without naming anybody explicitly.

Aachen, December 21, 2014 Jens-Rainer Ohm

Table of Contents

1 Introduction

Multimedia communication means combination of multiple information types, which are typically perceptible by human senses. In this context, audiovisual information (speech/audio/sound/image/video/graphics) is of particular importance, and a challenge w.r.t. the complexity of signal processing. Digital representation provides new degrees of freedom in exchange, distribution and acquisition of data, where communication includes exchange of information between humans, between humans and machines, or between machines only. Sufficient quality is desirable in the reproduction of audiovisual data, whereas the amount of bits for data transmission is often limited, such that compression is a key technology. This chapter introduces basic concepts and terminology of multimedia signal compression and transmission, and gives an overview about acquisition, common formats and data amounts in digital representation of audiovisual signal sources.

1.1 Concepts and terminology

A basic paradigm of communication system design is about conveying information at lowest possible cost, while providing sufficiently high quality and maximum benefit to the user. In multimedia communication, types of information which require high transmission rates and high processing speed due to their large volume, i.e. *audio*, *image*, *video* and *graphics* sources are most critical; communication systems (networks and devices) are often reaching their limits for these types of data. Therefore, processing, storage and transmission of audiovisual signals has become one of the most important driving factors in developing devices with ever increasing processing power and memory, and networks with higher capacity. Multimedia sources are either generated by sensor capture (camera, microphone, with analog to digital conversion), or they are generated synthetically, such as graphics, synthesized speech or sounds. For reproduction and consumption by humans, another conversion into the corresponding physical medium (e.g. generating a sound wave through a loudspeaker) is often necessary.

© Springer-Verlag Berlin Heidelberg 2015

J.-R. Ohm, *Multimedia Signal Coding and Transmission*,
Signals and Communication Technology, DOI 10.1007/978-3-662-46691-9_1

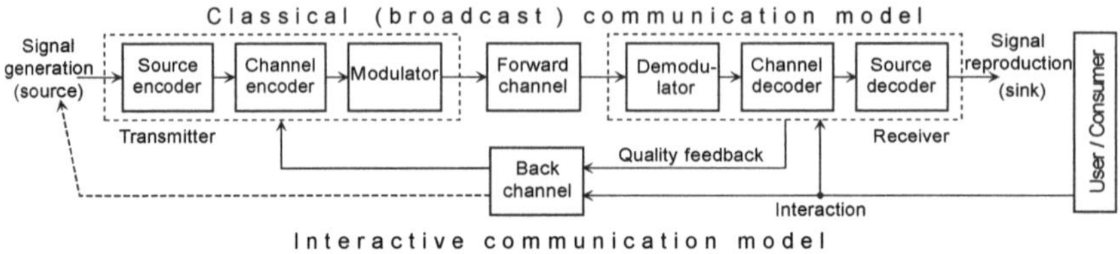

Fig. 1.1. Concept of a classical communication system and feedback/interactive paths

The classical concept of a uni-directional communication system is illustrated in the upper part of Fig. 1.1. This model is suitable for applications where no back channel exists, such as broadcast. Devices on the left-hand side (transmitter) have their complementary parts on the right-hand side (receiver). Source coder and decoder, channel coder and decoder cooperate by using negotiated codes, trying to compensate for errors and capacity limitations imposed by the channel. The goal of this system is to convey as much information as possible with a quality as high as possible. This is achieved by optimization of the source encoder, which compresses the source to lowest possible number of bits, the channel encoder, which protects the bit stream from possible losses, and the modulator which sends symbol carriers in the most suitable way over a physical channel. Joint optimization of these building blocks is necessary, taking into consideration characteristics of the channel, such as losses caused by interference or noise, and available physical bandwidth. The channel has a *capacity* which expresses the maximum information rate that can be conveyed without errors between transmitter and receiver. In this context, the main task of source encoding is to *remove redundancy* that typically is present in the original capture of multimedia sources, and compress them to a lower rate. Channel encoding *adds redundancy* for the purpose of error protection in case of losses. The rate produced by the source shall not exceed the channel capacity, and if that were the case, either information losses would occur, or a more aggressive source compression has to be used (see Sec. 1.1.1). The latter may be the better choice, as it is under control of the encoder and, even when introducing more distortion, would provide a more graceful degradation than random channel losses.

In the classical communication concept, the receiver is passive and deterministically decodes whatever is received via the channel. This model is suitable in broadcast, where it would not make much sense to react on a single user's intervention, however in point-to-point communication this is different. The lower part of Fig. 1.1 illustrates how the classical model can be augmented by a back channel and interactivity components. The user himself can interact, and can take influence on any part of the chain, including the signal generation. Further, automatic feedback can be generated by any part of the receiver, e.g. to request retransmission of lost information, or to inform the transmitter about current channel conditions. More 'intelligent' systems could even include elements of signal content analysis to assist the user in interaction and search for multimedia data.

Beyond traditional broadcast channels, multimedia traffic nowadays is often transferred over *heterogeneous channels* (e.g. the physical transmission underlying in the internet) which can have largely varying characteristics. In particular in wireless and mobile transmission, it is necessary to perform frequent and proper adaptation due to instantaneous fluctuation of channel characteristics. In a generalized view, channels can be networks or storage devices, where storage elements not only reside at the ends of the transmission chain (server and client), temporary storage is done in proxy servers for quick access to content that is likely to be retrieved multiple times. In case of erroneous transmission, recovery at the client side often goes far beyond traditional channel and source decoding, e.g. by using advanced error concealment methods including content analysis.

Ultimately, a multimedia communication system will be judged by the value it provides to the user, which is interpreted in terms of quality over cost. The term *Quality of Service* (QoS) is widely used, which shall be interpreted here as the quality perceived, being influenced by the network parameters such transmission rate, delay and losses, as well as the compression performance of source encoding/decoding, and the inter-relationship with the channel characteristics. The cost charged from the user also depends on the cost of devices used to fulfill these tasks. Therefore, a less complex system is better serving the user's needs.

This book concentrates on the topics of *coded representation* and *transmission* of multimedia signals. The remaining part of this section will give a high-level overview about concepts and terminology related to each of these aspects.

1.1.1 Signal representation by source coding

Through *multimedia signal compression*, the most compact representation providing highest possible perceptual quality shall be achieved. After capturing, the signal is converted into a digital representation with finite number of samples and amplitude levels. This step already influences the final quality. If the range of rates that a candidate channel can convey, or the resolution required by an application are not known by the time of acquisition, it is advisable to capture the signal by highest possible quality, and if necessary scale it down to lower quality later.

Properties of the signal itself which enable reduction of the rate by the *source coder* without information loss can be expressed in terms of *redundancy* (e.g. expected similarity between samples of the signal, or more frequent occurrence of certain source states). If consumption by a human is intended, it may further be useful to adapt the source coding method considering *perceptual properties*, as it would be useless to convey a finer granularity of quality than the user can (or would desire to) perceive. Some degradations occurring in the process of source coding and decoding may not be perceivable, or be minor such that they can be tolerated, or a modification may even be desirable (such as removal of sensor noise). This is the *irrelevance* part of the original signal, which may very well be omitted in encoding[1]. If however only a low capacity channel is available, just exploitation of redundancy or irrelevance may not be sufficient to reduce the rate. Then, it is unavoidable to accept additional distortions. The nature of these distortions can be different, e.g. reduction of spatial or temporal resolution, or coding/quantization noise. *Content-related* properties can also be taken into consideration. This can e.g. be done by assigning better quality to parts or pieces of the signal which are likely to be more important to the user's attention.

The encoded information is usually represented in form of binary digits (*bits*). The bit rate is measured either in *bit/sample*[2], or *bit per second* (*bit/s*), where the latter results from the *bit/sample* ratio, multiplied by the number of *samples/s* (the sampling rate). An important criterion to judge the performance of a source coding scheme is the *compression ratio*. This is the ratio between the bit rate necessary for representation of the uncompressed source and its compressed representation. If e.g. for digital high-definition TV the uncompressed source requires at least 800 *Mb/s*[3], and if the rate after compression shall be 4 *Mb/s*, the necessary compression ratio is 800:4=200. If compressed signal streams are stored as files on computer discs, the file size can be evaluated to judge the compression performance. When translating into bit rates, it must be observed that file sizes are often measured in *KByte*, *MByte* etc., where one *Byte* consists of 8 *bit*, 1 *KByte*=1,024 *Byte*, 1 *MByte*=1,024 *KByte* etc.

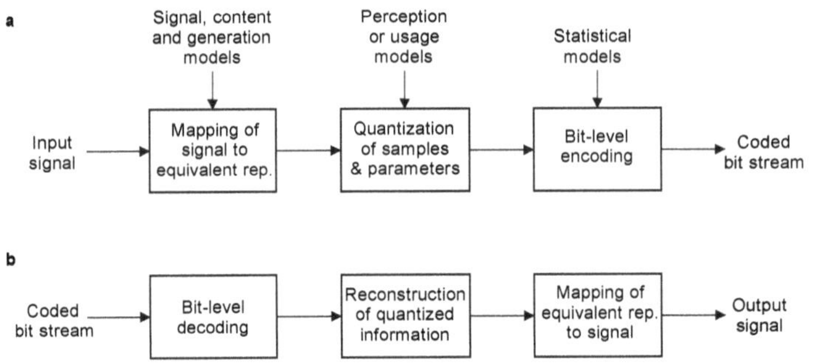

Fig. 1.2. Basic principle of a coding system for multimedia signals: **a** Encoder **b** Decoder

Fig. 1.2 shows the basic block diagram of typical multimedia signal coding and decoding systems:

---

[1] In the case where consumption is made by a machine, the definition of irrelevance would also need to consider the intended purpose of using the signal. For example, distortions by source coding which would make a content analysis fail may be unacceptable.

[2] For image signals also *bit/pixel* (*b/p*) is used, where 'pixel' is a widely-used acronym for *picture element*, an image sample. To be more neutral about the type of signals, we prefer the term sample also in case of images and video.

[3] *Mb/s*: 10^6 bit per second, also *kb/s*: 10^3 bit per second etc.

- The first step is a mapping of signal samples into an equivalent representation, which in ideal case allows a reverse mapping for perfect recovery. Usually, the representation in the equivalent representation space is sparser than in the original signal space, such that the subsequent steps become simpler and can be better optimized. Sparseness means that the equivalent representation requires less samples, or the values of the samples are concentrated at fewer amplitudes. Important principles for this mapping are *prediction* and *transforms*. In case of linear operations employed for this purpose, the mapping is also denoted as *de-correlation*, as it usually removes the linear statistical dependencies between samples in the equivalent representation space. To optimize such systems, statistical models are used, but the generation (e.g. properties of the acquisition process) and the semantic content could be considered as well. If the mapping function is varying (as usually necessary for multimedia signals with locally variable properties), additional *side information* parameters may need to be generated, which are required to perform the reverse mapping during decoding.
- The second step is *quantization*, which maps the signal, its equivalent representation or side information parameters into values from discrete alphabets (if that was not the case before, or if more aggressive lossy compression shall be performed). Perceptual constraints of the observer or other criteria for irrelevance (related to intended usage) should be considered during quantization such that as much as possible of the relevant information is retained with a given number of discrete (quantized value) states. Typically, a lower number of quantized value states naturally leads to lower number of bits per sample, but also increases the distortion.
- The final step is *bit-level encoding*, which has the goal to represent the discrete information states by even lower rate, usually without changing them. The optimization of encoding is mostly performed on basis of statistical criteria, such as frequency of occurrence of the different states. The minimum rate that is needed for the unique representation of the states, i.e. letters of a finite discrete alphabet, is denoted as *entropy*.

Beyond *rate* and *distortion*, other parameters to tune a source coding algorithm are *latency* and *complexity*. All four parameters have mutual influence on each other. The relation between rate and distortion is determined by the *rate distortion function* (see Sec. 2.5.5 and 4.2.1) which gives a theoretical lower bound of the rate if a certain maximum distortion limit is required. Improved rate/distortion performance (which means improved compression ratio while keeping distortion constant) can often be achieved by increasing the complexity of the encoding/decoding algorithm, e.g. adapting it in a more sophisticated way to varying source statistics. Alternatively, increased latency also helps to increase compression performance; if for example an encoder is able to look ahead on effects of current decisions for future encoding steps, this can provide an advantage.

Whereas rate and latency can straightforwardly be quantified by a set of values such as mean, peak and variation, quantification of distortion would require more sophisticated metrics that take into account the user perception, whereas criteria such as squared error or signal to noise ratio (SNR) are often used and are simple to compute (see Sec. 3.3). Quantification of complexity is again challenging, as building blocks of a codec may be simpler or more difficult to implement on different platforms, and judgement of complexity requires to take into account multiple modalities such as gate count, chip size, power consumption, memory usage, software / hardware friendliness, capability for parallelization of operations etc.

1.1.2 Optimization of transmission

The interface between the source coder and the channel is also of high importance for the overall Perceptual QoS. While the source encoder removes redundancy from the signal, channel coding *adds redundancy* to the bit stream for the purpose of protection and recovery in case of losses. At the receiver end, the channel decoder removes the redundancy inserted by the channel encoder, using it to perform error correction, while the source decoder supplements the redundancy which was removed by the source encoder, reconstructing the source as good as possible from the information received. From this point of view, the operations of source encoding/channel decoding and channel encoding/source decoding are similar. Actually, the more complex part is usually on the side where redundancy is removed, as this needs to target estimating the *relevant information* within an over-complete representation. In fact, source and channel encoding should be optimized jointly for optimum performance. For example, it is not useful to add redundancy by channel encoding for parts of bit streams which are less relevant from a user perspective. The transmission over the channel also includes modulation, where the combination of channel coding principles and modulation at the physical level is commonly used in communication technology to approach the limits of channel capacity [Proakis, Salehi 2007].

In the context of designing multimedia systems it is often advantageous to interpret the channel as a 'black box' using a model. This in particular concerns error/loss characteristics, transmission rate, delay (latency) etc., which are the most important parameters of network quality. When a certain minimum of quality is guaranteed by the network, adaptation between source coding and the network transmission can be made in an almost optimum way. This is usually done by negotiation protocols. If no network quality is guaranteed, specific mechanisms can be introduced for adaptation at the server and client sides. This includes application-specific error protection based on estimated network quality or usage of retransmission protocols. Introduction of latency is also a viable method to improve the transmission quality, e.g. by optimization of transmission schedules, temporary buffering of information at the receiver side before presentation is started, or scrambling/interleaving of streams when burst losses in transmission are expected (cf. Sec. 9.3). In extreme case, the receiver could inquire re-transmission of lost or corrupted information until it finally arrives, but this would increase the transmission delay. Therefore, also for the channel coding part, improved quality can be expected when *higher latency* or *higher complexity* of the systems are allowed.

However, this may not be suitable for all applications. For example, real-time conversational services (like video telephony) only allow for low latency; the same is true for any interactive applications where a quick reaction is expected. For mobile devices, where battery capacity is critical, the overall complexity of systems should generally be kept low.

Physically, the abstract channel will often consist of a chain of several network paths having different characteristics. Herein, the entire chain can never be stronger than its weakest element. In heterogeneous networks, capabilities for easy adaptation of media streams under changing network characteristics plays an important role. In some cases, it may be necessary to *transcode* the streams into a different format that is better suitable for the specific network characteristics. From this point of view, source coding methods producing *scalable streams*, which are adaptable and can be truncated independently of the encoding process, are highly advantageous (Sec. 5.4.2). Another approach is the usage of *multiple descriptions* of the content (Sec. 5.4.3), which are conveyed to the receiver via parallel transmission paths.

1.2 Signal sources and acquisition

Multimedia systems usually process digital representations of signals, while the acquisition and generation of *natural signals* may not directly be performed by a digital device; e.g. electro-magnetic (microphone), optical (lens), chemical (film) media can be involved in the conversion. The analog-to-digital conversion itself consists of a sampling step which maps a spatio-temporally continuous signal into discrete samples, and a quantization step which maps an amplitude-continuous signal into numerical values. Acquisition devices like digital cameras equipped by CCD or CMOS sensors inherently provide sampled signals and quantized representation at the output. Multimedia signals can also be synthetically generated, like graphics, synthetic sounds or artificial audio effects, or mixtures of natural and synthetic content can be produced. If natural signals are captured, part of the information originally available in the outside (three-dimensional) world is lost due to

- limited frequency bandwidth or resolution of the acquisition device;
- 'Non-pervasiveness' of the acquisition device, which resides at a singular position in the 3D exterior world, such that the properties of the signal are available only for this specific view or listening point; even when using multiple cameras or microphones, the acquisition of 3D spatial information will be *incomplete* due to the fact that sensor positions cannot be arbitrary dense.

This relationship between exterior world and the incomplete projection into the signal shall now be studied for the example of a camera system. The 'world coordinate system' denotes locations $\mathbf{W}=[W_1,W_2,W_3]^{\mathrm{T}}$, where an imaging property (also denoted as *plenoptic function*) $I(\mathbf{W},\mathbf{\Theta},t)$ at a specific location $\mathbf{W}$ and time t

is given by the color and intensity of light arriving from an angular orientation $\mathbf{\Theta}=[\theta_1\ \theta_2\ \theta_3]^T$ [McMillan, Bishop 1995]. This could be captured by a camera positioned at **W** with an optical axis ('ray of view') orientation towards $\mathbf{\Theta}$. For this 'light field' representation, a single camera delivers only one sample in $(\mathbf{W},\mathbf{\Theta})$, and it will become more complete with additional cameras; optionally, additional views can be generated by interpolation or extrapolation (MCA, Ch. 7).

A *pin-hole camera* model is shown in Fig. 1.3. Here, the world coordinate system is defined with origin at the position of the camera, where the optical axis of the camera is on the W_3 (depth) axis, and perpendicular to the image plane. The 2-dimensional image plane (e.g. film or electronic sensor) of the camera receives a projection from the 3D exterior world into the image plane, captured at time t. In the pin-hole model, it is guaranteed that only *one light ray* from the exterior world would arrive at a point of the image plane, such that focus of the picture is guaranteed[4]. The relationship between a point P at exterior world coordinate $\mathbf{W}_P=[W_{1,P}\ W_{2,P}\ W_{3,P}]^T$ and its projection into the point $\mathbf{t}_P=[t_{1,P}\ t_{2,P}]^T$ of the image plane is characterized by the *central projection equation*, where F is the focal length of the camera,

$$t_{i,P} = F\frac{W_{i,P}}{W_{3,P}} \quad \text{for } i=1,2. \tag{1.1}$$

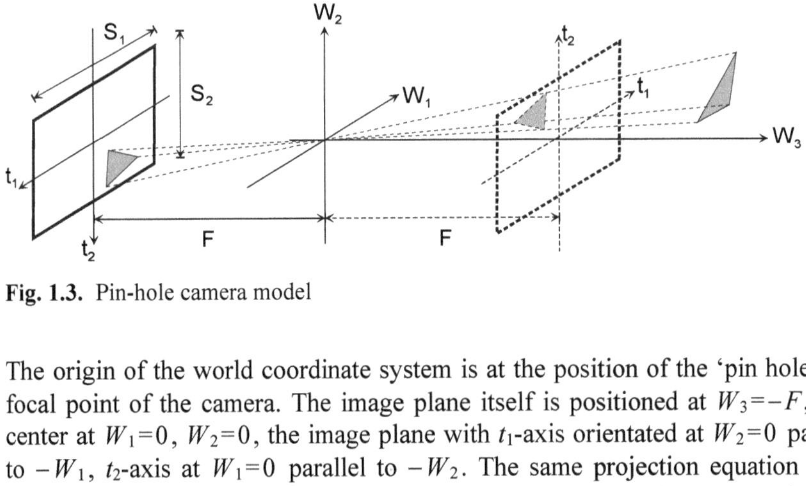

Fig. 1.3. Pin-hole camera model

The origin of the world coordinate system is at the position of the 'pin hole', the focal point of the camera. The image plane itself is positioned at $W_3=-F$, with center at $W_1=0$, $W_2=0$, the image plane with t_1-axis orientated at $W_2=0$ parallel to $-W_1$, t_2-axis at $W_1=0$ parallel to $-W_2$. The same projection equation holds when the rays pass through a mirrored image plane positioned at $W_3=F$, then the two axes t_i will not need to be mirrored relative to the corresponding W_i. The

[4] Physically, this is not possible, as the hole itself must have some minimum diameter. In cameras with lenses, a number of rays originating from the same point in the exterior world are intended to converge towards one point in the image plane, which is only achieved if the lens is in focus.

image plane itself has a width S_1 and a height S_2. The horizontal and vertical view angles spanning symmetrically around the optical axis are

$$\varphi_i = \pm \arctan \frac{S_i}{2F} \quad ; \quad i = 1,2 \,. \tag{1.2}$$

In electronic cameras and scanners, the 2D image plane is often scanned and sampled sequentially left-right / top-down row by row. The image plane of width S_1 and height S_2 is then mapped into $N_1 \mathrm{x} N_2$ discrete sampling locations with distances $T_i = S_i / N_i$. Such sampled and spatially bounded images can be expressed as matrices. Usually, the top left sample of the image is assigned with coordinate (0,0) and is the top left element of the matrix as well[5].

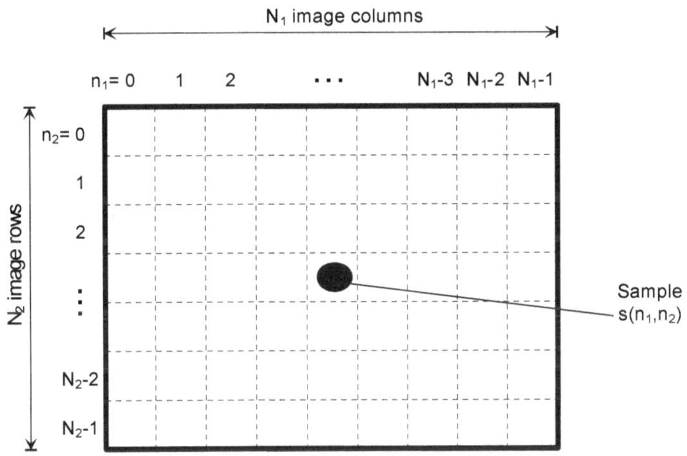

Fig. 1.4. Definition of the discrete image matrix

This is illustrated in Fig. 1.4 for the matrix containing $N_1 \times N_2$ samples,

$$\mathbf{S} = \begin{bmatrix} s(0,0) & s(1,0) & s(2,0) & \cdots & \cdots & s(N_1-1,0) \\ s(0,1) & s(1,1) & & & & s(N_1-1,1) \\ s(0,2) & & \ddots & \ddots & & \vdots \\ \vdots & & & & & s(N_1-1,N_2-2) \\ s(0,N_2-1) & s(1,N_2-1) & \cdots & \cdots & s(N_1-2,N_2-1) & s(N_1-1,N_2-1) \end{bmatrix} . \tag{1.3}$$

In some cases (e.g. if values from different rows of the matrix shall be included in one single arithmetic operation) it is more convenient to re-arrange all samples

[5] Observe that this definition of a coordinate system has an offset compared to the 2D coordinate system used in the central projection, which had its origin where the optical axis traverses the image plane (which could be between existing sample locations).

from the image matrix into one single vector, again sequentially row-wise scanned, starting at the top left sample of the image:

$$\mathbf{s} = \big[s(0,0) \quad s(1,0) \quad \cdots \quad s(N_1-1,0) \quad s(0,1) \quad s(1,1) \quad \cdots \\ \cdots \quad s(0,N_2-1) \quad \cdots \quad s(N_1-1,N_2-1)\big]^{\mathrm{T}} . \tag{1.4}$$

Historically, image sampling was first performed in the temporal dimension, either with a single sample in case of a still picture camera, or with a sequence of picture samples in movie film capture. If the time distance between the captured pictures is T_3, the number of pictures per second (also denoted as *frame rate*) is $1/T_3$. In electronic (analog) video cameras, vertical spatial sampling was then introduced using a line structure, while the signal within each line was still continuous over the horizontal direction[6] and in amplitude (not quantized). In digital imaging the signal is finally also sampled in the horizontal dimension and quantized into numerically represented amplitude values.

In analog video technology (and still in some digital video cameras), *interlaced* acquisition has been used (Sec. 2.3.3). In this case, a *frame* consists of *two field pictures*, containing the even and odd lines, respectively. The two fields are captured with a time offset of half a frame period. When the entire frame picture is captured simultaneously (as typically done in movie cameras), the acquisition is denoted as *progressive*.

In stereo and multi-view displays, the 3D spatial sensation is achieved by exposing different pictures (with stereoscopic parallax shift as per capture from two slightly distant cameras) to the observer's left and right eyes. Multi-view displays go beyond stereo in that they improve the sensation by changing the views when the observer moves in front of the display (motion parallax). To generate content for stereo and multi-view displays, it is therefore necessary to capture a scene by two or more cameras, or provide depth information to generate additional views synthetically, which further increases the data rate necessary for storage and transmission (see Sec. 7.6 and MCA, CH. 7).

1.3 Digital representation of multimedia signals

The process of analog to digital conversion of a signal consists of sampling (Sec. 2.3) and quantization (ch. 4). The resulting 'raw' digital format is denoted as *Pulse Code Modulation* (PCM) representation. Raw formats are often regarded as the original references in digital multimedia signal processing applications; specific characteristics related to such formats for visual and audio signals are explained in this section.

---

[6] Note that in analog video capture and display the lines are scanned sequentially over time, such that in principle there is no unique time of capture associated with the entire images. This fact is of course only relevant in cases where motion occurs.

1.3.1 Image and video signals

When image and video signals are represented digitally by B bit per component[7] sample, 2^B different levels can be represented by PCM. Currently, B=8 (256 levels) is still mainly used for consumer-quality photo and video cameras and scanners, with tendency going towards B=10 or higher. In professional and scientific applications, up to B=16 are used. Electronic cameras usually perform a nonlinear mapping of light intensity to amplitude values, providing a finer scale of amplitudes for darker values. This is denoted as *gamma transfer characteristic* of the acquisition device, described by the (approximate) transfer equation

$$V = c_1 \cdot \Phi^\gamma + c_2 \,, \tag{1.5}$$

where Φ is the luminous flux normalized to a maximum level, V the amplitude of the signal, c_1 the sensitivity of the camera and c_2 an offset value[8].

These nonlinear characteristics are typically ignored when processing digital representations, but are important either for the purpose of fidelity presentation (printing, display) or for precise analysis. When image signals are mapped into a linear amplitude scale (as e.g. often done in computer graphics applications), 16 bit or more are useful for representation of each component. The maximum number of bits that is required for representing data acquired by image sensors mainly depends on the amount of noise. When the sensor noise level exceeds the expected quantization noise (Sec. 4.1), the less significant bits become random. As newer generations of camera sensors (e.g. CMOS) tend to be less noisy, the usage of larger bit-depths becomes more and more beneficial, and higher dynamic ranges of amplitude can be realized from the sensor itself. Note that even the amplitude range that can be represented by 16 bits is still relatively low as compared to the available scale of visually distinguishable natural illumination levels, which covers a range over nine decimal powers in light intensity. Hence, human eyes and also cameras are equipped with regulation mechanisms, which are the iris, the diaphragm and the sensitivity of the receptor medium; in cameras also the shutter time is adapted to the intensity of light. In *high dynamic range* (HDR) imaging, several captures of the scene are made using different exposures, and depending on the brightness of a certain part of the scene these are combined into one image with artificially extended bit depth. This however only works perfectly for static scenes and fixed camera position; otherwise, registration is necessary which may cause artifacts when not correct. HDR imaging is an example where the store

---

[7] Components are colors, luminance or chrominance, see below.

[8] For example, the ITU-R Rec. BT.709 defines the transfer characteristic in two ranges, where the lower range has a linear characteristic:

$c_1 = 1.099$; $c_2 = -0.099$; $\gamma = 0.45$ for $1 \geq \Phi \geq 0.018$

$c_1 = 4.500$; $c_2 = 0.000$; $\gamma = 1.00$ for $0 \leq \Phi < 0.018$.

picture is not directly gained from sensor data, but additional signal processing steps are applied. In other types, e.g. medical image data where the image amplitudes often represent the density and absorption properties of tissue or bones, processing before a useful image can be released for storing is even more common.

Beyond brightness/color related data, other modalities of images that can be captured such as depth data (e.g. by range sensors or by depth estimation from a stereo camera setup) or so-called alpha (transparency) maps which are often used in blending/composing scenes from different images (SEE MCA, CH. 7).
To capture and represent color images, the most common representation consists of three primary components of active light, red (*R*), green (*G*) and blue (*B*), which is in coincidence with the color sensitivity of the human eye (Sec. 3.1.3). These components are separately acquired and sampled. In case of three separate sensors for color acquisition (as used in high-end cameras), the count of samples is higher by a factor of three as compared to monochrome images, but the color triplet can be interpreted as a vector $\mathbf{s}(\mathbf{n})=[s_R(\mathbf{n})\ s_G(\mathbf{n})\ s_B(\mathbf{n})]^T$ at every sample position $\mathbf{n}$. True representation of color may even require more components in a multi-spectral representation, or other data such as depth or transparency associated with the same coordinate $\mathbf{n}$ can be additional components of the vector.

Cameras with one sensor chip use a spatial multiplex with an optical color filter array, where often the *Bayer pattern* is used which uses double number of green samples compared to red and blue (see Fig. 1.5). This is justified by the fact that the spectrum range of green color is approximately matching the maximum sensitivity of the human eye[9]. It should be noted that full-resolution RGB images generated by such cameras are produced by interpolation post processing rather than native acquisition.

B	G	B	G	B	G	B	G
G	R	G	R	G	R	G	R
B	G	B	G	B	G	B	G
G	R	G	R	G	R	G	R
B	G	B	G	B	G	B	G
G	R	G	R	G	R	G	R

Fig. 1.5. Bayer pattern for one-chip sensor RGB sampling

Similar spatial multiplex arrangements as the Bayer pattern are also used in displays. Some higher-quality displays artificially generate yellow samples (by interpolating mainly from red and green samples), and replace half of the green samples with the intent of a more natural reproduction of color mixtures.

The native *RGB* representation is not directly usable for interpreting the physical light spectrum and associated color mapping. For this purpose, the artificial *XYZ* color space which is based on the so-called *standard observer* was defined in

[9] Also note that in the Bayer pattern the green component uses a quincunx sampling (Sec. 2.3.2) which allows for purely horizontal and vertical orientations 2x the maximum frequency compared to the rectangular sampling grids of red and blue.

[CIE 1931]. It allows specifying reference illumination and display color, and can be mapped into the *chromaticity diagram*, which gives interpretation about the range of color tones which can be mixed from primaries in terms of hue and saturation.[10]

Color images and video are often represented by a *luminance* component Y and two *chrominance* (color difference) components. For the transformation between RGB and luminance/chrominance representations, different definitions exist, depending on the particular application domain. For example, in standard TV resolution video, the following mapping is usual (as defined in ITU-R rec. BT.601):

$$Y = 0.299 \cdot R + 0.587 \cdot G + 0.114 \cdot B;$$
$$C_\mathrm{b} = \frac{0.5}{0.866} \cdot (B - Y) \quad ; \quad C_\mathrm{r} = \frac{0.5}{0.701} \cdot (R - Y). \tag{1.6}$$

For high definition (HD) video, the commonly used mapping is defined in part 2 of ITU-R rec. BT.709 as[11]

$$Y = 0.2126 \cdot R + 0.7152 \cdot G + 0.0722 \cdot B;$$
$$C_\mathrm{b} = \frac{0.5}{0.9278} \cdot (B - Y) \quad ; \quad C_\mathrm{r} = \frac{0.5}{0.7874} \cdot (R - Y) \tag{1.7}$$

For Ultra HD (UHD) display devices, even more extension of the color gamut range is foreseen. This is reflected in ITU-R rec. BT.2020, which is defining raw UHD formats, by the following mapping:

$$Y = 0.2627 \cdot R + 0.6780 \cdot G + 0.0593 \cdot B;$$
$$C_\mathrm{b} = \frac{0.5}{0.9407} \cdot (B - Y) \quad ; \quad C_\mathrm{r} = \frac{0.5}{0.7373} \cdot (R - Y) \tag{1.8}$$

The component Y covers the same amplitude range $0 \ldots A$ as the original components RGB, and the chrominance components cover a range $-A/2 \ldots +A/2$. It is useful to avoid overload by clipping the ranges to $A_{\min} \ldots A_{\max}$, where $A_{\min}$ is the black level and $A_{\max}$ the nominal peak level in case of luminance (or correspondingly $A_{\min} = -A_{\max}$ the minimum negative and $A_{\max}$ the maximum positive values in case of chrominance). This is achieved by the following modification:

---

[10] see also [MCA, SEC. 4.1.1]. Representing digital pictures by XYZ values might be advantageous in terms of enabling a more generic color mapping for various output device characteristics.

[11] One of the goals here is extending the color gamut, i.e. the possible range of chromaticities which can be generated out of the YC_bC_r representation. Whereas the mapping (1.6) was mostly optimized for CRT with a more restricted range of displayable colors through the properties of illuminant phosphor, today's displays such as LCD provide a wider gamut which is better supported by definitions (1.7) and (1.8).

$$Y' = Y\frac{A_{\max} - A_{\min}}{A} + A_{\min}; \quad C'_{\mathrm{b|r}} = C_{\mathrm{b|r}}\frac{A_{\max} - A_{\min}}{A} \tag{1.9}$$

Finally, since all considerations so far are still assuming continuous amplitude levels (though restricted by value range), it is necessary to perform quantization and achieve a representation by a finite number of bits. A uniform quantization is used here (cf. Sec. 4.1). For simplicity, assuming normalization A=1, the following operation provides as unsigned integer values of B bit precision[12]:

$$DY' = \lfloor Y' \cdot 2^B \rfloor; \quad DC'_{\mathrm{b|r}} = \lfloor (C'_{\mathrm{b|r}} + 0.5) \cdot 2^B \rfloor. \tag{1.10}$$

If the RGB color space is used, the same clipping and rounding as for Y in (1.9)/(1.10) can be applied to the three color primaries separately. It should be observed that there is no guarantee of lossless recovery of a quantized B bit representation in RGB after transformation into a quantized YC_bC_r representation (and vice versa). Nevertheless, the transformation into YC_bC_r is useful under compression aspects, since the possible color variations in the RGB are restricted such that perceptually and statistically more important colors are represented more accurately. The YC_bC_r also reduces the correlation between the three RGB color primary components, where structural detail (texture) is concentrated in Y. Further compression is achieved by sub-sampling the chrominance components, targeting the irrelevance of the human visual system which is less sensitive w.r.t. spatial resolution in chroma than for luma brightness and contrast (see Sec. 3.1.3). In some extreme cases of color patterns with sharp edges, this subsampling can however be critical. This is mostly the case with synthetic content, less with camera captures where edges tend to be slightly blurred anyway through the influence of the optical system.

For interlaced video, sub-sampling of the chrominance components is usually performed only in horizontal direction, while with progressive sampling, both horizontal and vertical chrominance sub-sampling is often used. Component sampling ratios are often expressed in a notation 'N_Y:N_{C1}:N_{C2}' to express the relative numbers of samples[13]. For example,

- when the same number of samples is used for all three components, the expression is '4:4:4';

---

[12] Typical values to be used in (1.9) are $A_{\min}$=1/16 and $A_{\max}$=15/16 for luminance, and $A_{\max}$=7/16=$-A_{\min}$ for chrominance.

[13] The rationale behind these notations is going back to analog video, but they are still commonly used. N_Y expresses the number of luma samples per chroma sample numbers in the two subsequent values, where it is assumed that luma is present in each line. N_{C1} und N_{C2} express the number of chroma samples (identical for both chroma components) in even and odd lines, respectively (with first even line being line zero). Actually, the expression by only three numbers is not complete enough to cover all possible sampling ratios of three components independently both in horizontal and vertical directions, e.g. disallowing vertical subsampling by ratios larger than 2. For RGB sources, N_Y is used equivalently for the green component, whereas red and blue are replacing the two chroma components.

- a sampling structure with horizontal-only sub-sampling of the two chrominance components by factors of two is expressed by the notation '4:2:2', while '4:1:1' indicates horizontal-only sub-sampling by a factor 4;
- if sub-sampling of chrominance components is performed in both directions, i.e. half number of samples along both horizontal and vertical directions, the notation is '4:2:0'.

The respective source format standards also specify the sub-sampled component sample positions in relation to the luminance sample positions.

Table 1.1. Resolution formats of digital still images

	640x480 (VGA)	1024x768 (XGA)	1536x1024 (1.5 MPix)	2048x1536 (3 MPix)	3072x2048 (6 MPix)	4256x2848 (12 MPix)	5472x3648 (20 MPix)
Number of samples	307,200	786,432	1.572,864	3.145,728	6.291,456	12.121,088	19.961,856
Picture aspect ratio	4:3	4:3	3:2	4:3	3:2	3:2	3:2
Number of bits per component, 8 bit precision	2.457,600	6.291,456	12.582,912	25.165,824	50.331,648	96.968,704	159.694,848

Table 1.1 shows example resolutions typically used for still image acquisition by digital cameras, even though many variations exist here in the market. Picture aspect ratio *width*:*height* of 3:2 is mostly used, which is identical to traditional landscape photo formats; some formats are also derived from the 4:3 aspect ratio of early computer displays; nowadays, since most displays and projectors adopt a 16:9 ratio, picture aspect ratios and resolutions of HD and UHD video are emerging for still images as well (cf. table 1.2). Often, formats are aligned such that picture width and height in number of samples is a multiple of 16, 32 or 64, which is advantageous for block based compression algorithms, e.g. transform coding. Therefore, formats denoted as '6 Megapixels' do not necessarily have exactly 6.000,000 samples. Total number of samples, number of bits per component image when using B=8 bit precision are also listed in the examples.

Generally, the required data volume for a raw data representation can be computed by multiplication of the numbers from up to five contributing dimensions: width, height, number of components, bit depth per component, chrominance subsampling (if applicable), and number of pictures (e.g. in case of image series or multi-view video acquisition).

For video representation, besides the total number of bits e.g. required to store a movie, the *number of bits per second* is of interest which results by multiplying the number of bits per picture by the number of pictures per second. Table 1.2 describes parameters of commonly used video formats. In some cases, numbers of samples, frame rates etc. allow variations.

Table 1.2. Digital video source formats and their properties

	Format						
	QVGA/ QWVGA	*CIF/ SIF*	*VGA/ WVGA*	*ITU-R BT.601 (SDTV)*	*ITU-R BT.709*[14] *(1080x)*	*SMPTE 296M (HD 720p)*	*ITU-R BT.2020 (4Kx2K/ 8Kx4K)*
Sampling frequency (*Y*) [*MHz*][15]	--	--	--	13.55	74.25 … 148.5	74.25	--
Sampling structure*)	P	P	P	I	P/I	P	P
Samples/line (*Y*)	320 (432)	352	640 (864)	720	1920	1280	3840/ 7680
Number of lines (*Y*)	240	288 (240)	480	576 (480)	1080	720	2160/ 4320
Color sampling	4:2:0	4:2:0	4:2:0	4:2:2	4:2:2	4:2:2	4:4:4/ 4:2:2/ 4:2:0
Picture aspect ratio	4:3 (16:9)	4:3	4:3 (16:9)	4:3	16:9	16:9	16:9
Frame rate [*Hz*]	10-30	10-30	24-60	25 (30)	24-60	24-60	24-120
PCM bit depth [*b*]	8	8	8	8	8-10	8-10	10-12
	Data rates (uncompressed)						
per frame [*kbyte*] 1 *kbyte* = 1024 x 8 *b*	112.5 (151.8)	148.5 (123.8)	450 (607.5)	810 (675)	4,050 … 5,063	1,800 … 2,250	30,375 … $145 \cdot 10^3$
per second [*Mb/s*]	9.22 … 37.3	10.13 … 36.5	88.44 … 298.6	165.9	796.2 … 2,488	353.9 … 1,106	2,986 … $143 \cdot 10^3$

*) P/I: Progressive/Interlaced

For standard TV resolution, the origin of the digital TV formats was the analog TV signal of 625 lines in Europe (or 525 lines in Japan or US). These analog signals are sampled by a rate of 13.55 MHz for the luminance. After removal of vertical blanking intervals, 575 (480) active lines remain. The digital formats

[14] The specifications refer to the 'common image format' defined in part 2 of BT.709, which is widely deployed in digital HD video.

[15] The actual sampling frequency cannot be listed for all-digital format specifications such as BT.2020.

listed in Tab. 1.2 are storing only those active samples with a very small overhead of few surplus samples from the horizontal blanking intervals. Analog Japanese and US formats are traditionally using roughly 60 fields per second (30 frames per second)[16], while in Europe, 50 fields per second (25 frames per second) is used in analog TV (PAL, SECAM). The digital standards defining HD formats are more flexible in terms of frame and field rates, allowing ranges of 24, 25, 29.97, 30, 50 or 60 frames/second progressive, 50 or 60 fields/second interlaced. For computer display or mobile devices, also formats such as Common Intermediate Format (*CIF*), *VGA*, quarter VGA (*QVGA*) and wide-screen VGA (WVGA) are used, but it can be expected that they will more and more be replaced by common HD formats. Ultra HD (UHD) formats are currently emerging. These usually have double number of samples horizontally/vertically as compared to HD1080 (then called '*4Kx2K*'), or even quadrupled ('*8Kx4K*'). Though using progressive sampling, frame rates may further increase in the future (72 frames per second and beyond).

Fig. 1.6 illustrates the sampled image areas (when assuming same size/density of samples) supported in formats between QVGA and HDTV. An increased number of samples can either be used to increase the resolution (spatial detail), or to display scenes over an extended area. For example, in a cinema movie close-up views of human faces are rarely shown. Movies displayed on a cinema screen allow the observer's eye to explore the scene at different locations on the screen, while on standard definition TV screens and even more for the smaller formats, this capability is very limited, and mostly close-ups are shown.

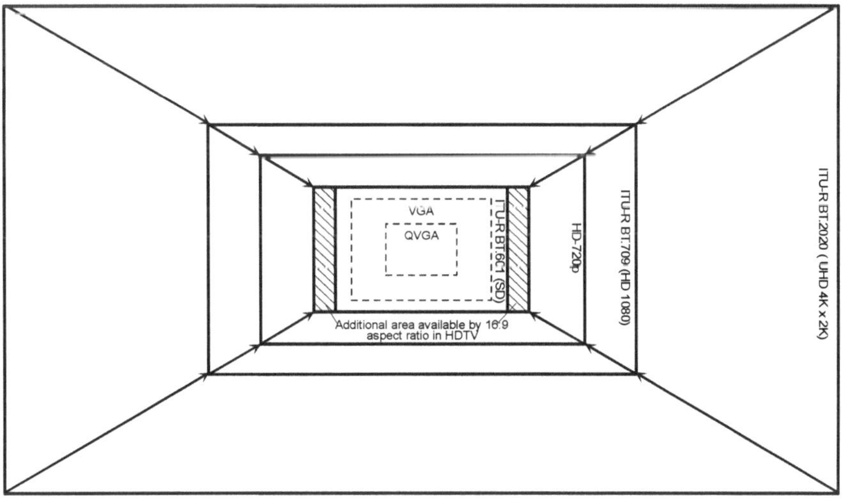

Fig. 1.6. Sizes and aspect ratios of digital video/image formats[17]

---

[16] These are commonly referred to as NTSC formats. The precise frame rate of analog NTSC formats is $30\times(1000/1001)\approx 29.97$ frames/s.

[17] 8Kx4K is not shown due to space limitation.

For medical and scientific purposes, digital images with much higher resolution than in movie production are used, resolutions of up to 10,000x10,000= 100.000,000 samples are quite common.

1.3.2 Speech and audio signals

For audio signals, parameters such as sampling rate, precision (bit depth) and number of channels take most influence on the resulting data rates of the digital representation. These parameters highly depend on the properties of the signals, and on the requirements for quality. In speech signal quantization, nonlinear mappings using logarithmic amplitude compression are used[18], which for the case of low amplitudes provides an equivalently low quantization noise as in 12 bit quantization, even though only 8 bit/sample are used. For music signals to be acquired with Audio CD quality, linear 16 bit representation is at least necessary. For some specialized applications, even higher bit-depths and higher sampling rates are used than those common for CD. Table 1.3 shows typical parameters in the areas of speech and broadband-audio digital representation. The required bit rates (per second) are calculated by multiplication of the sampling rate by the bit depths and the number of audio channels. The number of bits required for storage is the rate per second multiplied by the duration of the audio track.

1.3.3 Need for compression technology

Due to the tremendous amount of rates necessary for representation of the original un-coded formats, the requirement for data compression by application of image, video and audio coding is permanently present, even though the available transmission bandwidth is further increasing by advances in communication technology. In general, the past experience has shown that multimedia traffic increases faster than new capacity is becoming available, and compressed transmission of data is inherently cheaper. If sufficient capacity is available, it is more efficiently used in terms of quality serving the user, if the resolution of the signal is increased. With improved compression, this could even be achieved without increasing the data rate. Further, certain types of communication channels (in particular in mobile transmission) exist where the bandwidth is inherently scarce and expensive, and more rigid limits exist in terms of bandwidth efficiency (number of bit/s that can be reliably transported per unit of channel bandwidth), due to physical limitations. This must however be balanced against the complexity that is necessary for the implementation of a compression algorithm, which may lead to higher cost of the device and higher power consumption, which is in particular critical for mobile devices.

---

[18] Denoted as μ-law and A-law quantization characteristic, see sec. 4.1.

Table 1.3. Sampling and raw bit rates for audio signals

	Frequency range [Hz]	Sampling rate [KHz]	Number of channels and PCM resolution [bits/sample]	Max. PCM rate [kb/s]
Telephony speech	300 – 3,400	8	8-16	64-128
Broadband speech	50 – 7,000	16	8-16	128-256
CD audio	10 – 20,000	44.1	2 x 16 (stereo)	1410
DVD multichannel	10 – 22,000	48	(5+1) x 16	$4.6 \cdot 10^3$
DVD audio	10 – 44,000	Up to 96	2 x 24	Up to $4.6 \cdot 10^3$
Multichannel audio (5+1, 7+2, *M*+*N*)	10 – 44,000	Up to 96	(*M*+*N*) x 24	Up to $18.4 \cdot 10^3$
Surround sound audio	10 – 44,000	48	(up to 200) x 16	Up to $153.6 \cdot 10^3$

1.4 Problems

Problem 1.1.

A movie of 150 minutes duration is digitally represented in progressive HD format (1920x1080 samples luminance, 2x960x1080 samples chrominance per frame, 60 frames/*s*, 8 bit/sample PCM).

a) Which hard disc capacity is required to store the entire movie in original format?
b) The movie shall be encoded using HEVC compression, achieving a compression factor of 200:1 over PCM. Which file size is now required to store the stream on the hard disc?
c) To transmit the movie over a satellite channel, error protection by channel coding is provided. This increases the rate by 10 %. Which overall transmission rate (*Mb/s*) over the channel is necessary?

Problem 1.2.

Video from a smartphone shall be transmitted via an LTE network, providing a data rate of 720 *kb/s*. 20 *kb/s* are used for wide-band speech transmission. The temporal sampling rate shall be 30 frames/*s*.

a) As the video coder does not output a constant-rate stream, it is necessary to buffer the information before transmitting over the channel. What maximum buffer size (measured in bits) is allowable, if the maximum delay caused by buffering shall be 100 *ms*?

b) The output rate of the encoder can be changed by modification of the quantizer setting (the discrete-step 'Q factor'). As a rule of thumb, lowering the Q factor by one step may increase the rate by a factor 1.1. For one picture, it is measured that encoding has consumed 19280 bit. By how many steps should the Q factor be lowered or increased to avoid an over-run or under-run of the buffer?

2 Fundamentals

The main purpose of this chapter is to introduce notation used throughout this book, and review fundamental principles of signal processing, statistical analysis and modeling that are used in subsequent chapters. Readers who are familiar e.g. with principles of one- and multi-dimensional sampling, random signal analysis, linear prediction and linear transforms may browse quickly over these topics.

2.1 Signals and systems

2.1.1 Elementary signals

A two-dimensional cosine signal defined over continuous coordinates $\mathbf{t}=[t_1\ t_2]^{\mathrm{T}}$ is given as

$$s_{\cos}(t_1,t_2)=\cos\left[2\pi\left(F_1t_1+F_2t_2\right)\right]=\cos\left[2\pi\mathbf{f}^{\mathrm{T}}\mathbf{t}\right] \text{ with } \mathbf{f}=\left[F_1\quad F_2\right]^{\mathrm{T}}. \quad (2.1)$$

After applying a coordinate transformation

$$\begin{bmatrix}\tilde{t}_1\\ \tilde{t}_2\end{bmatrix}=\begin{bmatrix}\cos\varphi & \sin\varphi\\ -\sin\varphi & \cos\varphi\end{bmatrix}\begin{bmatrix}t_1\\ t_2\end{bmatrix} \text{ with } \varphi=\arctan\frac{F_2}{F_1} \quad \left(\text{for } F_1\geq 0\right), \quad (2.2)$$

only a one-dimensional dependency remains as

$$s_{\cos}(\tilde{t}_1,\tilde{t}_2)=\cos\left[2\pi F\tilde{t}_1\right] \text{ with } F=\sqrt{F_1^2+F_2^2}=\left\|\mathbf{f}\right\|_2=\frac{F_1}{\cos\varphi}=\frac{F_2}{\sin\varphi}. \quad (2.3)$$

(2.1) can be interpreted as a sinusoidal wave front with orientation by an angle φ relative to the t_1 axis. Sections of this wave front in parallel with one of the two axes are observed as sinusoids of frequencies F_1 or F_2, respectively. These correspond to the periods or wavelengths (measured along the coordinate axes, see Fig. 2.1a)

© Springer-Verlag Berlin Heidelberg 2015

J.-R. Ohm, *Multimedia Signal Coding and Transmission*,
Signals and Communication Technology, DOI 10.1007/978-3-662-46691-9_2

$$T_1 = \frac{1}{F_1} \quad ; \quad T_2 = \frac{1}{F_2}. \tag{2.4}$$

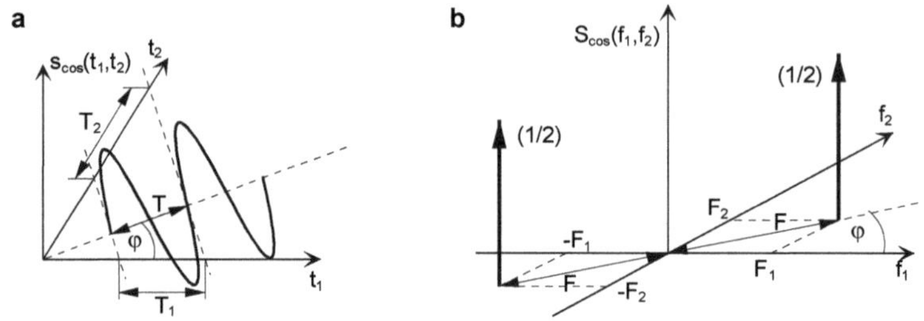

Fig. 2.1. **a** Directional orientation and wavelength of a sinusoid in a 2D plane **b** spectrum

As another interpretation, consider a cosine of period $T_1 = 1/F_1$ along the t_1 orientation, with phase shifted by $\phi(t_2)$ depending on the position in t_2,

$$s_{\cos}(t_1,t_2) = \cos\left[2\pi F_1 t_1 + \phi(t_2)\right]. \tag{2.5}$$

With linear dependency $\phi(t_2) = 2\pi F_2 t_2$, this is identical to (2.1). Then, for any $t_2 = k / F_2$ ($k \in \mathbb{Z}$), $\phi(t_2) = 2\pi k$. This determines distances where the signal has equal amplitude for a fixed t_1, i.e. $T_2 = 1/F_2$ is the period length along the t_2 orientation. Thus, the 2-dimensional cosine can also be interpreted as a sinusoid over one dimension which has a linear phase shift depending on the other dimension. This is illustrated in Fig. 2.2. Alternative formulations of the same signal would be

$$\begin{aligned} s_{\cos}(t_1,t_2) &= \cos\left[2\pi F_2 t_2 + \phi(t_1)\right] \text{ with } \phi(t_1) = 2\pi F_1 t_1 \text{ or} \\ s_{\cos}(t_1,t_2) &= \cos\left[2\pi F_1 t_1 + \phi(t_2)\right] \text{ with } \phi(t_2) = 2\pi F_2 t_2, \end{aligned} \tag{2.6}$$

such that any horizontal or vertical section over the different phase-shifted versions will also give a sinusoid of period T_2. Whereas T_1 and T_2 are the periods that can be measured w.r.t. to the coordinate axis orientations, the effective period of the 2-dimensional sinusoid, measured by the direction of wave front propagation, can be determined from (2.3) and (2.4) as

$$T = \frac{1}{F} = \frac{T_1 T_2}{\sqrt{T_1^2 + T_2^2}}. \tag{2.7}$$

Even though the example given here is based on a cosine function, a similar principle can be applied for any sinusoid. Likewise, it can be extended to a one- or multidimensional complex periodic exponential function

$$s_{\text{exp}}(\mathbf{t}) = \mathrm{e}^{\mathrm{j}2\pi F_1 t_1}\,\mathrm{e}^{\mathrm{j}2\pi F_2 t_2}\cdots\mathrm{e}^{\mathrm{j}2\pi F_\kappa t_\kappa} = \mathrm{e}^{\mathrm{j}2\pi\left[F_1 t_1 + F_2 t_2 + \ldots + F_\kappa t_\kappa\right]} = \mathrm{e}^{\mathrm{j}2\pi\breve{\mathbf{f}}^{\mathrm{T}}\mathbf{t}}$$
$$= \cos\left(2\pi\breve{\mathbf{f}}^{\mathrm{T}}\mathbf{t}\right) + \mathrm{j}\sin\left(2\pi\breve{\mathbf{f}}^{\mathrm{T}}\mathbf{t}\right) \text{ with } \breve{\mathbf{f}} = \begin{bmatrix}F_1 & \cdots & F_\kappa\end{bmatrix}^{\mathrm{T}}, \mathbf{t} = \begin{bmatrix}t_1 & \cdots & t_\kappa\end{bmatrix}^{\mathrm{T}} \qquad (2.8)$$

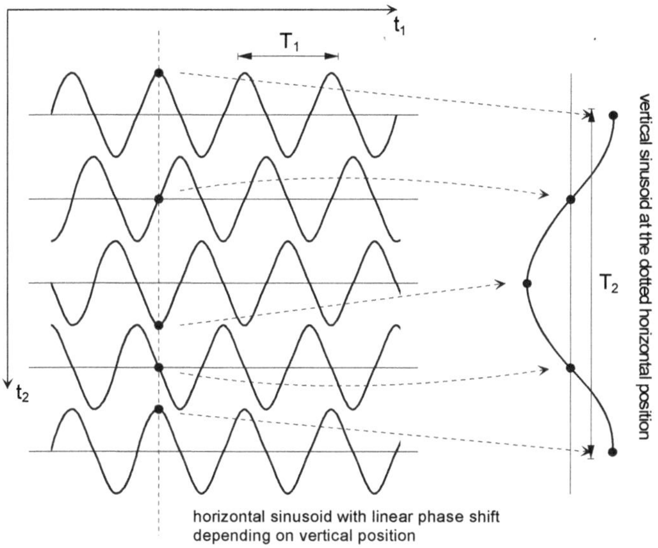

Fig. 2.2. Interpretation of a two-dimensional sinusoid: Linear phase shift depending on vertical position results in vertical wavelength and frequency

For the case κ=2, (2.1) is the real part of $s_{\text{exp}}(\mathbf{t})$. For expression as a one-dimensional signal as in (2.3), $\kappa-1$ rotations are necessary. If a signal can be defined by independent one-dimensional functions (as is the case with (2.8)), it is called *separable*, i.e.

$$s_{\text{sep}}(\mathbf{t}) = s(t_1)\cdot s(t_2)\cdot\cdots\cdot s(t_\kappa). \qquad (2.9)$$

Some examples of aperiodic elementary 1D signals (which could be used to construct corresponding separable multi-dimensional signals) are the *sinc function*[1]

$$s(t) = \frac{\sin(\pi t)}{\pi t} = \mathrm{si}(\pi t)\,, \qquad (2.10)$$

the *rectangular impulse*

$$\mathrm{rect}(t) = \begin{cases} 1, & |t| \le 1/2 \\ 0, & |t| > 1/2, \end{cases} \qquad (2.11)$$

the *unit step function*

[1] sinc= *sin*us *c*ardinalis, si(x)=sin(x)/x with si(1)=1.

$$\varepsilon(t) = \begin{cases} 1, & t \geq 0 \\ 0, & t < 0, \end{cases} \tag{2.12}$$

and the *Gaussian impulse*

$$s(t) = e^{-\pi t^2}. \tag{2.13}$$

2.1.2 Systems operations

A system generally performs a mapping (transfer) of an input $s(t)$ into an output $g(t)=\mathrm{Tr}\{s(t)\}$. A system is *linear*, if superposition using a weighted combination with constants a_i can be applied either at the input or at the output,

$$\mathrm{Tr}\left\{\sum_i a_i s_i(t)\right\} \overset{!}{=} \sum_i a_i \,\mathrm{Tr}\{s_i(t)\} = \sum_i a_i g_i(t). \tag{2.14}$$

Further, the system is *time invariant*, if for any shift t_0 of the input the output is shifted equally,

$$\mathrm{Tr}\{s(t-t_0)\} = g(t-t_0). \tag{2.15}$$

If a system fulfills both (2.14) and (2.15), it is called *linear time invariant* (LTI). The output signal of an LTI system fed by a *Dirac impulse* $\delta(t)$ as input is the *impulse response* $h(t)$. The transfer between input and output is given by the *convolution integrals*

$$s(t) = \int_{-\infty}^{\infty} s(\tau)\delta(t-\tau)\,\mathrm{d}\tau = s(t) * \delta(t), \tag{2.16}$$

$$g(t) = \int_{-\infty}^{\infty} s(\tau)h(t-\tau)\,\mathrm{d}\tau = s(t) * h(t). \tag{2.17}$$

The most important rules of convolution algebra are

a) The Dirac impulse is the *unity element* of convolution, according to (2.16).

b) *Commutative property*,

$$s(t) * h(t) = g(t) = \int_{+\infty}^{-\infty} s(t-\theta)h(\theta)(-\mathrm{d}\theta) = \int_{-\infty}^{+\infty} h(\theta)s(t-\theta)\,\mathrm{d}\theta = h(t) * s(t). \tag{2.18}$$

c) *Associative property*[2],

$$f(t) * s(t) * h(t) = [f(t) * s(t)] * h(t) = f(t) * [s(t) * h(t)]. \tag{2.19}$$

d) *Distributive property*,

$$f(t) * [s(t) + h(t)] = [f(t) * s(t)] + [f(t) * h(t)]. \tag{2.20}$$

Convolution can straightforwardly be extended to signals with multi-dimensional dependencies, e.g. image signals where an amplitude is defined for positions with horizontal/vertical coordinates (t_1,t_2). An example of a two-dimensional convolu-

---

[2] For combinations of convolution with other operations, in particular multiplication of functions, this is not true; the sequence of processing needs to be observed.

tion integral, with both the signal and the impulse response having two-dimensional dependencies, is defined as

$$g(t_1,t_2)=\int_{-\infty}^{\infty}\int_{-\infty}^{\infty} s(\tau_1,\tau_2)h(t_1-\tau_1,t_2-\tau_2)\,\mathrm{d}\tau_1\,\mathrm{d}\tau_2 = s(t_1,t_2)**h(t_1,t_2). \qquad (2.21)$$

If κ dimensions are combined into a vector $\mathbf{t}=[t_1,\ldots,t_\kappa]^\mathrm{T}$, same with the variables of the convolution integral $\boldsymbol{\tau}=[\tau_1,\ldots.,\tau_\kappa]^\mathrm{T}$, multi-dimensional convolution is defined by[3]

$$s(\mathbf{t})=\int_{-\infty}^{\infty}\cdots\int_{-\infty}^{\infty} s(\boldsymbol{\tau})\delta(\mathbf{t}-\boldsymbol{\tau})\,\mathrm{d}^\kappa\boldsymbol{\tau} = s(\mathbf{t})\ast\delta(\mathbf{t}). \qquad (2.22)$$

$$g(\mathbf{t})=\int_{-\infty}^{\infty}\cdots\int_{-\infty}^{\infty} s(\boldsymbol{\tau})h(\mathbf{t}-\boldsymbol{\tau})\,\mathrm{d}^\kappa\boldsymbol{\tau} = s(\mathbf{t})\ast h(\mathbf{t}). \qquad (2.23)$$

(2.22) can be interpreted via the *sifting property* of the Dirac impulse, which contributes only the signal value $\boldsymbol{\tau}=\mathbf{t}$ to the result of the integration. The multi-dimensional Dirac impulse can be described as a separable combination of a series of 1D Dirac impulses[4], each of which performs sifting in one dimension. Therefore,

$$\delta(\mathbf{t})=\delta(t_1)\cdot\delta(t_2)\cdot\ldots \quad \text{with} \quad \int_{-\infty}^{\infty}\cdots\int_{-\infty}^{\infty}\delta(\mathbf{t})\,\mathrm{d}^\kappa\mathbf{t}=\int_{-\infty}^{\infty}\delta(t_1)\,\mathrm{d}t_1\int_{-\infty}^{\infty}\delta(t_2)\,\mathrm{d}t_2\cdots=1 \qquad (2.24)$$

Properties (2.18)-(2.20) still hold for multi-dimensional convolution. An interesting class of two- and multi-dimensional LTI[5] systems are the *separable systems* with an impulse response that can be written as a multiplication of two or more functions, e.g. in the two-dimensional case[6]

$$h(t_1,t_2)=h_1(t_1)\cdot h_2(t_2)=\left[h_1(t_1)\cdot\delta(t_2)\right]**\left[\delta(t_1)\cdot h_2(t_2)\right]. \qquad (2.25)$$

Inserting (2.25) into (2.21) unveils that convolution in case of a 2D separable system can be implemented as a concatenation of two 1D convolutions to be performed at any position of the respective other dimension,

---

[3] The bold star symbol ($\ast$) expresses convolution over vector variables, to be performed by nested integrations.

[4] A 1D Dirac impulse $\delta(t_1)$ in a two- or multi-dimensional coordinate system can be interpreted as a line impulse, plane impulse or hyper-plane impulse (depending on the number of dimensions). It is zero for any $t_1\neq 0$, but can be interpreted as an infinite-amplitude slice positioned at $t_1=0$ with infinite extension over the remaining dimension(s), with volume integration over the entire multi-dimensional space giving a value of 1.

[5] For sake of simplicity, the denotation *time invariant* is not changed, even though typically at most one of the dependencies in a multi-dimensional system is along the time axis.

[6] In the expression by 2D convolution the line impulses are needed to indicate the presence of the impulse response at any position of the other dimension(s).

$$g(t_1,t_2) = s(t_1,t_2)**h(t_1,t_2) = \int_{-\infty}^{\infty}\int_{-\infty}^{\infty} s(\tau_1,\tau_2)h_1(t_1-\tau_1)h_2(t_2-\tau_2)\,\mathrm{d}\tau_1\,\mathrm{d}\tau_2$$
$$= \int_{-\infty}^{\infty} h_2(t_2-\tau_2)\underbrace{\int_{-\infty}^{\infty} s(\tau_1,\tau_2)h_1(t_1-\tau_1)\,\mathrm{d}\tau_1}_{g_1(t_1,\tau_2)}\,\mathrm{d}\tau_2 \tag{2.26}$$
$$= \underbrace{s(t_1,t_2)**[h_1(t_1)\cdot\delta(t_2)]}_{g_1(t_1,t_2)}**[\delta(t_1)\cdot h_2(t_2)].$$

Due to the associative property (2.19), the sequence of processing the dimensions is irrelevant in case of separable systems.

Eigenfunctions have the property that their shape is not changed when they are transmitted over an LTI system; the output can be computed by multiplication with a complex amplitude factor H, the related *eigenvalue*. A periodic 1D eigenfunction can be defined as a special case of (2.8),

$$s_{\mathrm{E}}(\mathbf{t}) = \mathrm{e}^{\mathrm{j}2\pi\mathbf{f}^{\mathrm{T}}\mathbf{t}} = \cos\left(2\pi\mathbf{f}^{\mathrm{T}}\mathbf{t}\right) + \mathrm{j}\sin\left(2\pi\mathbf{f}^{\mathrm{T}}\mathbf{t}\right). \tag{2.27}$$

Transmission over an LTI system gives

$$s_{\mathrm{E}}(\mathbf{t}) * h(\mathbf{t}) = \int_{-\infty}^{\infty}\cdots\int_{-\infty}^{\infty} h(\boldsymbol{\tau})\mathrm{e}^{\mathrm{j}2\pi\mathbf{f}^{\mathrm{T}}(\mathbf{t}-\boldsymbol{\tau})}\,\mathrm{d}^{K}\boldsymbol{\tau}$$
$$= \mathrm{e}^{\mathrm{j}2\pi\mathbf{f}^{\mathrm{T}}\mathbf{t}}\underbrace{\int_{-\infty}^{\infty}\cdots\int_{-\infty}^{\infty} h(\boldsymbol{\tau})\mathrm{e}^{-\mathrm{j}2\pi\mathbf{f}^{\mathrm{T}}\boldsymbol{\tau}}\,\mathrm{d}^{K}\boldsymbol{\tau}}_{H(\mathbf{f})} = H(\mathbf{f})\mathrm{e}^{\mathrm{j}2\pi\mathbf{f}^{\mathrm{T}}\mathbf{t}}. \tag{2.28}$$

The type of complex periodic eigenfunctions plays an important role in *Fourier analysis*, establishing a relation between signal (**t**) and Fourier (**f**) domains. Herein,

$$H(\mathbf{f}) = \int_{-\infty}^{\infty}\cdots\int_{-\infty}^{\infty} h(\mathbf{t})\mathrm{e}^{-\mathrm{j}2\pi\mathbf{f}^{\mathrm{T}}\mathbf{t}}\,\mathrm{d}^{\kappa}\mathbf{t} \tag{2.29}$$

is the relation of Fourier transform of the impulse response $h(\mathrm{t})$, giving the frequency-dependent Fourier transfer function $H(\mathbf{f})$ of an LTI system. Feeding an eigenfunction into a series of two LTI systems with impulse responses $h_{\mathrm{A}}(\mathbf{t})$ and $h_{\mathrm{B}}(\mathbf{t})$ gives the result

$$s_{\mathrm{E}}(\mathbf{t}) * h_{\mathrm{A}}(\mathbf{t}) * h_{\mathrm{B}}(\mathbf{t}) = \left[H_{\mathrm{A}}(\mathbf{f})s_{\mathrm{E}}(\mathbf{t})\right] * h_{\mathrm{B}}(\mathbf{t}) = H_{\mathrm{A}}(\mathbf{f})\cdot H_{\mathrm{B}}(\mathbf{f})\cdot s_{\mathrm{E}}(\mathbf{t}) \tag{2.30}$$

It can be concluded that the convolution product in the time domain is mapped to an algebraic product in the frequency domain.

The Fourier transform is applicable not only for impulse responses $h(\mathbf{t})$, but for any signals $s(\mathbf{t})$, $g(\mathbf{t})$ etc. into their corresponding Fourier spectra

$$S(\mathbf{f}) = \int_{-\infty}^{\infty}\cdots\int_{-\infty}^{\infty} s(\mathbf{t})\mathrm{e}^{-\mathrm{j}2\pi\mathbf{f}^{\mathrm{T}}\mathbf{t}}\,\mathrm{d}^{\kappa}\mathbf{t}.$$

Two- and multi-dimensional extensions of eigenfunctions are straightforward and establish the basis of multi-dimensional Fourier spectra that are discussed in the subsequent sections. Due to the separable property of the multi-dimensional complex eigenfunctions, the multi-dimensional Fourier transform can be computed sequentially over the different dimensions, but the final result still provides an interpretation by directional orientation.

2.2 Signals and Fourier spectra

2.2.1 Spectra over two- and multi-dimensional coordinates

Rectangular coordinate systems. The amplitude of an image signal is dependent on the spatial position in two dimensions t_1 and t_2 – horizontally and vertically. Related frequency axis orientations shall be f_1 (horizontally) and f_2 (vertically). The two-dimensional Fourier transform of a spatially continuous signal is

$$S(f_1,f_2)=\int_{-\infty}^{\infty}\int_{-\infty}^{\infty} s(t_1,t_2)\,\mathrm{e}^{-\mathrm{j}2\pi f_1 t_1}\,\mathrm{e}^{-\mathrm{j}2\pi f_2 t_2}\,\mathrm{d}t_1\,\mathrm{d}t_2. \tag{2.31}$$

(2.31) can be extended into a generic definition of κ-dimensional spectra associated with a κ-dimensional signal, where all frequency coordinates $\mathbf{f}=[f_1\ f_2\ \dots\ f_\kappa]^{\mathrm{T}}$ and signal coordinates in space and time $\mathbf{t}=[t_1\ t_2\ \dots\ t_\kappa]^{\mathrm{T}}$ are expressed as vectors. This gives

$$S(\mathbf{f})=\int_{-\infty}^{\infty}..\int_{-\infty}^{\infty} s(\mathbf{t})\,\mathrm{e}^{-\mathrm{j}2\pi\mathbf{f}^{\mathrm{T}}\mathbf{t}}\,\mathrm{d}^{\kappa}\,\mathbf{t}. \tag{2.32}$$

The complex spectrum can be interpreted by *magnitude* and *phase* of a contributing oscillation at a given frequency **f**,

$$\begin{aligned}
|S(\mathbf{f})| &= \sqrt{\left[\mathrm{Re}\{S(\mathbf{f})\}\right]^2+\left[\mathrm{Im}\{S(\mathbf{f})\}\right]^2}=\sqrt{S(\mathbf{f})S^*(\mathbf{f})}\\
\varphi(\mathbf{f}) &= \arctan\frac{\mathrm{Im}\{S(\mathbf{f})\}}{\mathrm{Re}\{S(\mathbf{f})\}}\pm\pi\cdot k(\mathbf{f})\ \text{ with }\ k(\mathbf{f})=\begin{cases}1 & \text{for }\ \mathrm{Re}\{S(\mathbf{f})\}<0\\ 0 & \text{else.}\end{cases}
\end{aligned} \tag{2.33}$$

By *inverse Fourier transform*, the signal can be reconstructed from the Fourier spectrum:

$$s(\mathbf{t})=\int_{-\infty}^{\infty}\dots\int_{-\infty}^{\infty} S(\mathbf{f})\,\mathrm{e}^{\mathrm{j}2\pi\mathbf{f}^{\mathrm{T}}\mathbf{t}}\,\mathrm{d}^{\kappa}\,\mathbf{f}. \tag{2.34}$$

Coordinate system mapping. Rectangular (orthogonal) coordinate systems are only a special case for the description of two- and multidimensional signals. They allow expressing the multi-dimensional Fourier transform through eigenfunctions which are also orthogonal (i.e. independent in terms of signal analysis properties) between the different dimensions. Two unity vectors $\mathbf{e}_1 = [1\ 0]^T$ and $\mathbf{e}_2 = [0\ 1]^T$ define the orientations of the axes. Any coordinate pair (t_1,t_2) can then be expressed as a vector $\mathbf{t} = t_1\mathbf{e}_1 + t_2\mathbf{e}_2$. The relationship with frequency vectors $\mathbf{f} = f_1\mathbf{e}_1 + f_2\mathbf{e}_2$ is given by (2.32), using the same orientation. Now, a linear coordinate mapping $\tilde{\mathbf{t}} = t_1\mathbf{t}_1 + t_2\mathbf{t}_2 = \mathbf{Tt}$ shall be applied to the signal (leaving the coordinate origin unchanged), which can be expressed through the mapping matrix[7]

$$\mathbf{T} = \begin{bmatrix} t_{11} & t_{12} \\ t_{21} & t_{22} \end{bmatrix} = [\mathbf{t}_1 \quad \mathbf{t}_2]. \tag{2.35}$$

The vectors $\mathbf{t}_1$ und $\mathbf{t}_2$ are the *basis vectors* of this mapping. A complementary mapping of frequency coordinates shall exist, expressed similarly as $\tilde{\mathbf{f}} = \mathbf{Ff}$ by using a mapping matrix

$$\mathbf{F} = \begin{bmatrix} f_{11} & f_{12} \\ f_{21} & f_{22} \end{bmatrix} = [\mathbf{f}_1 \quad \mathbf{f}_2]. \tag{2.36}$$

Unless the determinants of matrices **T** or **F** are zero, the mappings must be reversible, such that $\mathbf{t} = \mathbf{T}^{-1}\tilde{\mathbf{t}}$ and $\mathbf{f} = \mathbf{F}^{-1}\tilde{\mathbf{f}}$. The relations are given by biorthogonality (A.25) of **T** and **F** [see e.g. OHM 2004],

$$\mathbf{T}^{-1} = \mathbf{F}^T; \quad \mathbf{F}^{-1} = \mathbf{T}^T \quad \Rightarrow \quad \mathbf{F} = \left[\mathbf{T}^{-1}\right]^T; \quad \mathbf{T} = \left[\mathbf{F}^{-1}\right]^T. \tag{2.37}$$

The Fourier transform in the mapped coordinate system can then be expressed as follows, assuming amplitude invariance of the mapped samples,

$$\tilde{S}(\tilde{\mathbf{f}}) = \int_{-\infty}^{\infty} \cdots \int_{-\infty}^{\infty} s(\tilde{\mathbf{t}})\, \mathrm{e}^{-\mathrm{j}2\pi\tilde{\mathbf{f}}^T\tilde{\mathbf{t}}}\, \mathrm{d}^\kappa \tilde{\mathbf{t}} = |\mathbf{T}|\, S(\mathbf{f}). \tag{2.38}$$

2.2.2 Spatio-temporal signals

In a video signal, two-dimensional pictures vary over time. The time dependency t is mapped into a 'temporal' frequency f_3, where the Fourier spectrum is

$$S(f_1, f_2, f_3) = \int_{-\infty}^{\infty}\int_{-\infty}^{\infty}\int_{-\infty}^{\infty} s(t_1,t_2,t_3)\, \mathrm{e}^{-\mathrm{j}2\pi f_1 t_1}\, \mathrm{e}^{-\mathrm{j}2\pi f_2 t_2}\, \mathrm{e}^{-\mathrm{j}2\pi f_3 t_3}\, \mathrm{d}t_1\, \mathrm{d}t_2\, \mathrm{d}t_3. \tag{2.39}$$

For the case of sinusoids, the spectral property resulting by temporal changes can

---

[7] It is assumed here that the origin of the coordinate transform is not changed by the mapping. A more general form is the mapping $\tilde{\mathbf{t}} = \mathbf{Tt} + \boldsymbol{\tau}$, where $\boldsymbol{\tau}$ expresses a shift of the origin. This is also denoted as *affine mapping*. Regarding the Fourier spectrum, the additional translation only effects a linear phase shift $\mathrm{e}^{\mathrm{j}2\pi\mathbf{f}^T\boldsymbol{\tau}}$.

be interpreted similarly to Fig. 2.3. In particular, if motion is constant (without local variations of shift and without acceleration) and the amplitude of the signal is only changing by motion, the behavior of the signal can be expressed by a linear phase shift in t_1 and t_2, depending on time t_3. Consider first the case of zero motion, $s(t_1,t_2,t_3) = s(t_1,t_2,0)$. Then, the three-dimensional Fourier spectrum (2.39) is

$$\begin{aligned} S(f_1,f_2,f_3) &= \int_{-\infty}^{\infty}\int_{-\infty}^{\infty} s(t_1,t_2,0)\,\mathrm{e}^{-\mathrm{j}2\pi f_1 t_1}\,\mathrm{e}^{-\mathrm{j}2\pi f_2 t_2}\,\mathrm{d}t_1\,\mathrm{d}t_2 \cdot \int_{-\infty}^{\infty} \mathrm{e}^{-\mathrm{j}2\pi f_3 t_3}\,\mathrm{d}t_3 \\ &= S(f_1,f_2)\big|_{t_3=0} \cdot \delta(f_3). \end{aligned} \tag{2.40}$$

The Dirac impulse $\delta(f_3)$ indicates that the 3D spectrum in case of unchanged signals is a sampled plane, with non-zero components only at f_3=0:

$$S(f_1,f_2,f) \sim \begin{cases} S(f_1,f_2)\big|_{t_3=0} & \text{when } f_3 = 0 \\ 0 & \text{when } f_3 \neq 0. \end{cases} \tag{2.41}$$

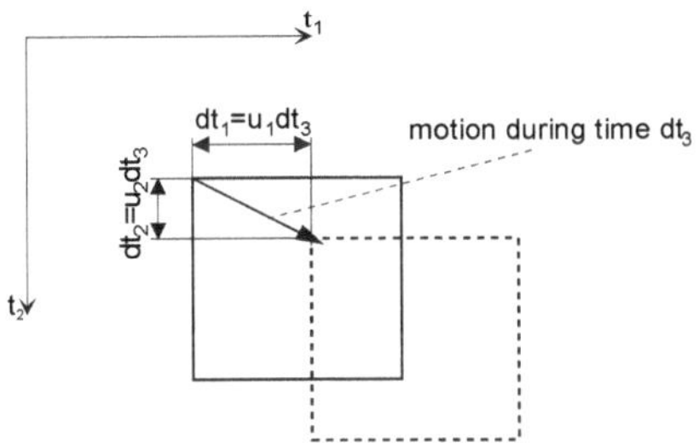

Fig. 2.3. Spatial shift caused by translational motion of velocity **u**

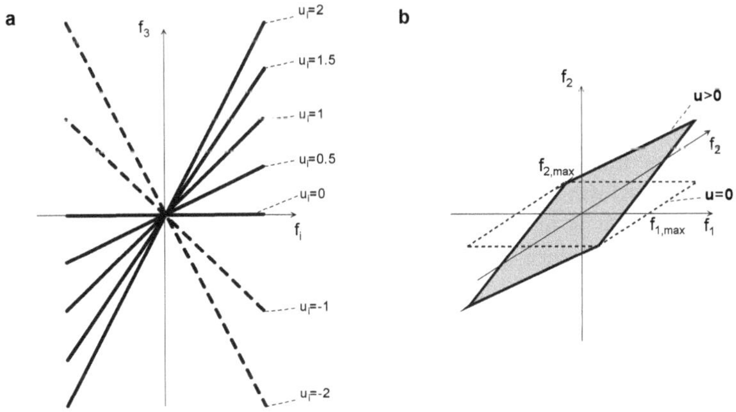

Fig. 2.4. **a** Shear of the non-zero spectral components by different translational motion velocities, shown in an (f_i,f_3) section (i=1,2) of the 3D frequency domain **b** Position of the non-zero spectral components in cases of zero and non-zero 2D translational motion

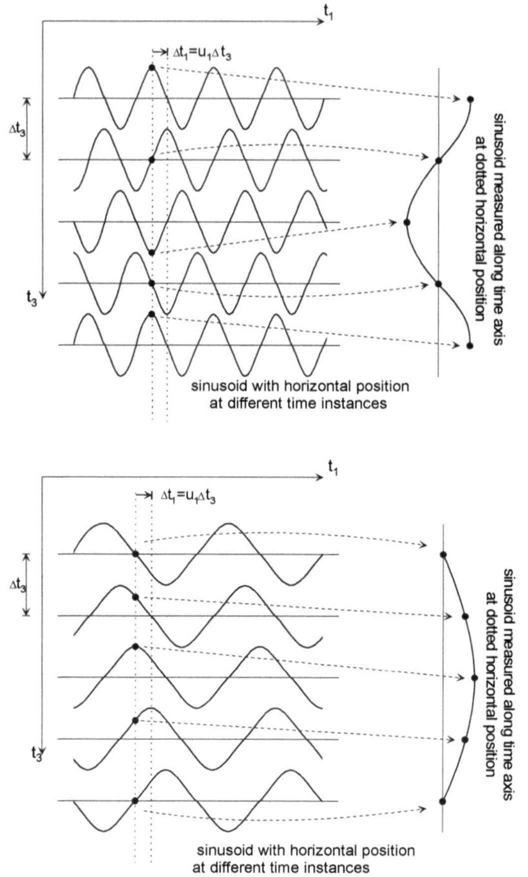

Fig. 2.5. Interpretation of the frequency in f_3 for two sinusoids of different spatial frequencies, which are subject to the same translational motion

If constant-velocity translation motion is present in the signal, a spatial shift $\mathrm{d}t_1=u_1\mathrm{d}t_3$ in horizontal direction and $\mathrm{d}t_2=u_2\mathrm{d}t_3$ in vertical direction occurs within time interval $\mathrm{d}t_3$, which is linearly dependent on the velocity vector $\mathbf{u}=[u_1,u_2]^{\mathrm{T}}$ (see Fig. 2.3). Taking reference to the signal for $t_3=0$, this gives

$$s(t_1,t_2,t_3)=s(t_1+u_1t_3,t_2+u_2t_3,0) \tag{2.42}$$

and

$$S(f_1,f_2,f_3)=\int_{-\infty}^{\infty}\int_{-\infty}^{\infty}\int_{-\infty}^{\infty} s(t_1+u_1t_3,t_2+u_2t_3,0)\,\mathrm{e}^{-\mathrm{j}2\pi f_1t_1}\,\mathrm{e}^{-\mathrm{j}2\pi f_2t_2}\,\mathrm{e}^{-\mathrm{j}2\pi f_3t_3}\,\mathrm{d}t_1\,\mathrm{d}t_2\,\mathrm{d}t_3. \tag{2.43}$$

By replacing $\tau_i=t_i+u_it_3 \Rightarrow \mathrm{d}\tau_i=\mathrm{d}t_i, t_i=\tau_i-u_it_3$, the temporal dependency can be separated in the Fourier integration

$$S(f_1,f_2,f_3)=\int_{-\infty}^{\infty}\int_{-\infty}^{\infty} s(\tau_1,\tau_2)\mathrm{e}^{-\mathrm{j}2\pi f_1\tau_1}\,\mathrm{e}^{-\mathrm{j}2\pi f_2\tau_2}\,\mathrm{d}\tau_1\,\mathrm{d}\tau_2\cdot\int_{-\infty}^{\infty} e^{-\mathrm{j}2\pi(f_3-f_1u_1-f_2u_2)t_3}\,d\tau_3$$
$$=S(f_1,f_2)\big|_{t_3=0}\cdot\delta(f_3-f_1u_1-f_2u_2). \tag{2.44}$$

Thus,

$$S(f_1,f_2,f_3)\sim\begin{cases} S(f_1,f_2)\big|_{t=0} & \text{when } f_3=f_1u_1+f_2u_2 \\ 0 & \text{when } f_3\neq f_1u_1+f_2u_2. \end{cases} \tag{2.45}$$

The spectrum $S(f_1,f_2)$ is now sampled on a plane $f_3=f_1u_1+f_2u_2$ in the 3D frequency domain. Fig. 2.4a shows positions of non-zero spectrum planes for different normalized velocities u_i, where the (f_i,f_3) section is shown for $f_j=0$, $i\neq j,(i,j)\in[1,2]$. Fig. 2.4b shows qualitatively the behavior in the full (f_1,f_2,f_3) space for the zero-motion case and for motion by one constant velocity $\mathbf{u}>\mathbf{0}$, where further the spectrum is assumed to be band-limited in f_1 and f_2.

Fig. 2.4 illustrates that the positions of non-zero spectral values in case of constant velocity are found via a linear relationship between f_3 and the frequencies relating to the spatial coordinates. This effect can also be interpreted in the signal domain. Fig. 2.5 shows two sinusoids of different frequencies f_1, both moving by the same velocity. The phase shift occurring due to the constant-velocity motion linearly depends on the given spatial frequency.

2.3 Sampling of multimedia signals

Ideal sampling describes the multiplication (modulation) of a signal by a regular (equidistant) train of Dirac impulses. In the 1D case, this gives

$$s_{\delta_T}(t)=s(t)\sum_{n=-\infty}^{\infty}\delta\left(t-nT\right)=\sum_{n=-\infty}^{\infty}s(nT)\delta\left(t-nT\right). \tag{2.46}$$

An example is shown in Fig. 2.6.

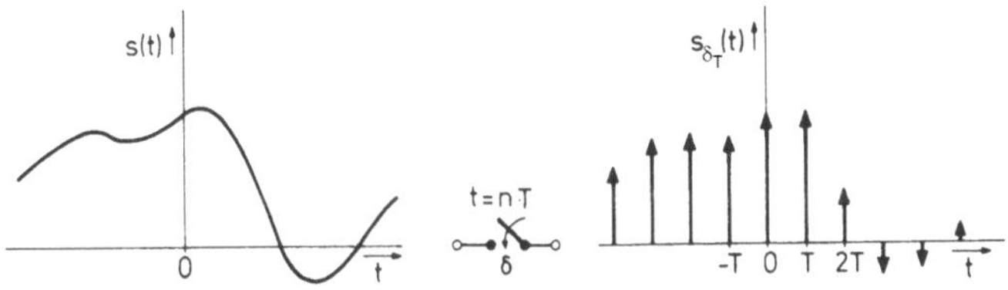

Fig. 2.6. Output $s_{\delta_T}(t)$ of an ideal sampling unit

The *ideal sampler* generates a discrete-time, equidistant series of weighted Dirac impulses from the continuous-time signal $s(t)$. The weights are the *samples* $s(nT)$. The Fourier spectrum of the sampled signal is

$$\begin{array}{ccccccc} s_{\delta_T}(t) & = & s(t) & \cdot & \sum_{n=-\infty}^{\infty} \delta(t-nT) \\ \circ\!\!-\!\!\bullet & & \circ\!\!-\!\!\bullet & \circ\!\!-\!\!\bullet & \circ\!\!-\!\!\bullet \\ S_{\delta_T}(f) & = & S(f) & * & \frac{1}{|T|}\sum_{n=-\infty}^{\infty} \delta\left(f-\frac{n}{T}\right) \end{array} \tag{2.47}$$

$S_{\delta_T}(f)$ is periodic by the *sampling rate* $1/T$, with spectral copies scaled in amplitude,

$$S_{\delta_T}(f) = \frac{1}{|T|}\sum_{k=-\infty}^{\infty} S\left[f-k/T\right]. \tag{2.48}$$

Fig. 2.7 shows this relation for a real-valued band-limited *lowpass signal* with zero-valued spectrum at any $|f| \geq f_c$ with cut-off frequency f_c.

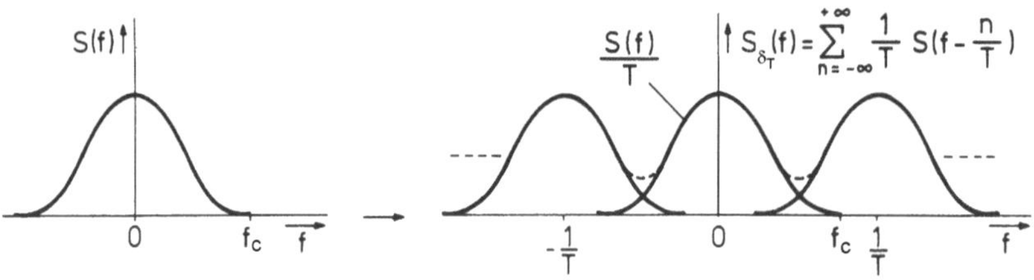

Fig. 2.7. Periodic components in the Fourier spectrum of a sampled signal $s_{\delta_T}(t)$

When sampling is performed using a sampling period

$$T \leq \frac{1}{2f_c}, \tag{2.49}$$

the periodic copies of the spectrum $S(f)$ in $S_{\delta_T}(f)$ are not overlapping, such that the original $S(f)$ can be perfectly reconstructed by suitable lowpass filtering from $S_{\delta_T}(f)$. This basic idea of sampling is shown in Fig. 2.8. If (2.49) is violated, frequency components from the periodic copies may appear in the baseband after the lowpass filtering, which is denoted as *aliasing*.

The lowpass shall have a transfer function which is flat in the range $|f| < f_c$ of the pass band and shall perfectly discard frequencies $|f| > 1/T - f_c$ from $S_{\delta_T}(f)$. Assuming that an *ideal lowpass* is used, the reconstruction of the continuous-time signal from the sampled signal can be formulated in the frequency and time domains as (see Fig. 2.8)

$$\begin{array}{ccccccc} S(f) & = & S_{\delta_T}(f) & \cdot & T\,\mathrm{rect}\left(\dfrac{f}{2f_c}\right) & & \\ \bullet\!\!-\!\!\circ & & \bullet\!\!-\!\!\circ & & \bullet\!\!-\!\!\circ & & (2.50) \\ s(t) & = & s_{\delta_T}(t) & * & 2f_c T\,\mathrm{si}\left(\pi 2 f_c t\right). & & \end{array}$$

If the largest possible sampling period $T=1/(2f_c)$ is used, (2.50) gives

$$s(t)=\left[\sum_{n=-\infty}^{\infty} s(nT)\delta\left(t-nT\right)\right]*\mathrm{si}\left(\pi\frac{t}{T}\right)=\sum_{n=-\infty}^{\infty} s(nT)\,\mathrm{si}\left(\pi\frac{t-nT}{T}\right). \quad (2.51)$$

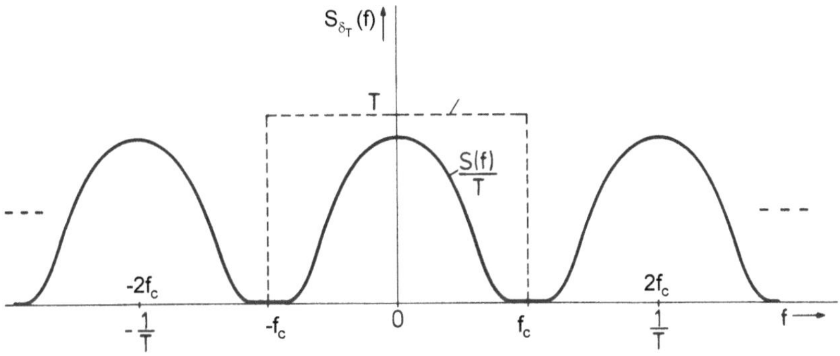

Fig. 2.8. Reconstruction of the Fourier spectrum $S(f)$ from $S_{\delta_T}(f)$ by using an ideal lowpass filter of cut-off frequency f_c

This formulation of the *sampling theorem* shows, that a real-valued signal which is band limited within a given lowpass range limited by f_c can be described without any errors by an equidistant series of weighted sinc functions. This is also denoted as the *cardinal series* of $s(t)$. The weights are equal to the samples of the signal as extracted with distances $T=1/(2f_c)$ from the signal. Fig 2.9 shows this principle.

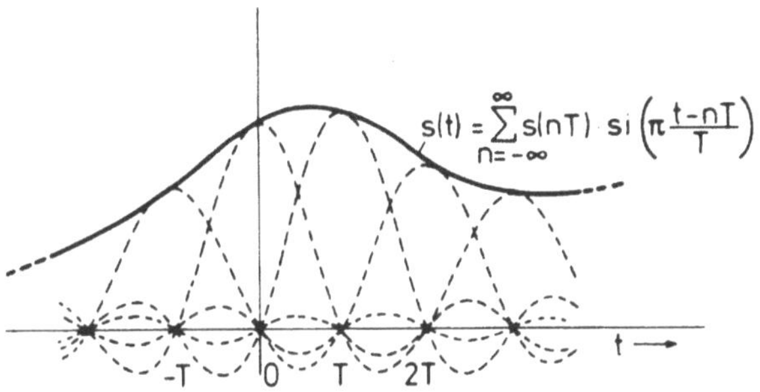

Fig. 2.9. Band limited real-valued lowpass signal $s(t)$ reconstructed by superposition of weighted sinc functions with distances $T=1/(2f_g)$

2.3.1 Separable two-dimensional sampling

Separable two- or multidimensional sampling is independent in the respective dimensions. This can be expressed from 1D *Dirac impulse trains* (refer to (2.47))

$$\delta_{T_i}(t_i)=\sum_{n_i=-\infty}^{\infty}\delta(t_i-n_iT_i)\,. \tag{2.52}$$

By multiplication of two impulse trains, which are separable on a rectangular grid, a two-dimensional ideal sampling function is defined as

$$\delta_{T_1,T_2}(t_1,t_2)=\delta_{T_1}(t_1)\cdot\delta_{T_2}(t_2)=\sum_{n_1=-\infty}^{\infty}\sum_{n_2=-\infty}^{\infty}\delta(t_1-n_1T_1,t_2-n_2T_2)\,. \tag{2.53}$$

Due to separable property, the rectangular impulse grid has a 2D spectrum

$$\begin{aligned}\frac{1}{|T_1|}\delta_{1/T_1}(f_1)\cdot\frac{1}{|T_2|}\delta_{1/T_2}(f_2)&=\frac{1}{|T_1T_2|}\delta_{1/T_1,1/T_2}(f_1,f_2)\\&=\frac{1}{|T_1T_2|}\sum_{k_1=-\infty}^{\infty}\sum_{k_2=-\infty}^{\infty}\delta(f_1-k_1/T_1,f_2-k_2/T_2).\end{aligned} \tag{2.54}$$

The operation of ideal rectangular-grid sampling of a spatially-continuous 2D signal $s(t_1,t_2)$ is expressed by multiplication with $\delta_{T_1,T_2}(t_1,t_2)$. The sample aspect ratio is defined as T_1/T_2. The discrete signal $s(n_1,n_2)$ consists of amplitude samples $s(n_1T_1,n_2T_2)$. Its spectrum is

$$\begin{aligned}S_{\delta_{T_1T_2}}(f_1,f_2)&=\frac{1}{|T_1T_2|}S(f_1,f_2)**\delta_{1/T_1,1/T_2}(f_1,f_2)\\&=\frac{1}{|T_1T_2|}\sum_{k_1=-\infty}^{\infty}\sum_{k_2=-\infty}^{\infty}S\big(f_1-k_1/T_1,f_2-k_2/T_2\big).\end{aligned} \tag{2.55}$$

It is also possible to compute the periodic spectrum directly from the discrete series of samples:

$$\begin{aligned}S_{\delta_{T_1T_2}}(f_1,f_2)&=\int_{-\infty}^{\infty}\int_{-\infty}^{\infty}s_{\delta_{T_1,T_2}}(t_1,t_2)\,\mathrm{e}^{-\mathrm{j}2\pi f_1t_1}\,\mathrm{e}^{-\mathrm{j}2\pi f_2t_2}\,\mathrm{d}t_1\,\mathrm{d}t_2\\&=\int_{-\infty}^{\infty}\int_{-\infty}^{\infty}\sum_{n_1=-\infty}^{\infty}\sum_{n_2=-\infty}^{\infty}s(n_1T_1,n_2T_2)\delta(t_1-n_1T_1,t_2-n_2T_2)\,\mathrm{e}^{-\mathrm{j}2\pi f_1t_1}\,\mathrm{e}^{-\mathrm{j}2\pi f_2t_2}\,\mathrm{d}t_1\,\mathrm{d}t_2\\&=\sum_{n_1=-\infty}^{\infty}\sum_{n_2=-\infty}^{\infty}s(n_1T_1,n_2T_2)\int_{-\infty}^{\infty}\int_{-\infty}^{\infty}\delta(t_1-n_1T_1,t_2-n_2T_2)\,\mathrm{e}^{-\mathrm{j}2\pi f_1t_1}\,\mathrm{e}^{-\mathrm{j}2\pi f_2t_2}\,\mathrm{d}t_1\,\mathrm{d}t_2\\&=\sum_{n_1=-\infty}^{\infty}\sum_{n_2=-\infty}^{\infty}s(n_1T_1,n_2T_2)\,\mathrm{e}^{-\mathrm{j}2\pi f_1n_1T_1}\,\mathrm{e}^{-\mathrm{j}2\pi f_2n_2T_2},\end{aligned} \tag{2.56}$$

or by performing normalization by setting $T_1=T_2=1$,

$$S_\delta(f_1,f_2)=\sum_{n_1=-\infty}^{\infty}\sum_{n_2=-\infty}^{\infty}s(n_1,n_2)\mathrm{e}^{-\mathrm{j}2\pi f_1 n_1}\mathrm{e}^{-\mathrm{j}2\pi f_2 n_2}. \tag{2.57}$$

2D pulse grid sampling generates periodic copies of the spectrum along *both directions*. Examples of Fourier amplitude spectra $|S(f_1,f_2)|$ and $|S_\delta(f_1,f_2)|$ in case of rectangular sampling are shown in Fig. 2.10.

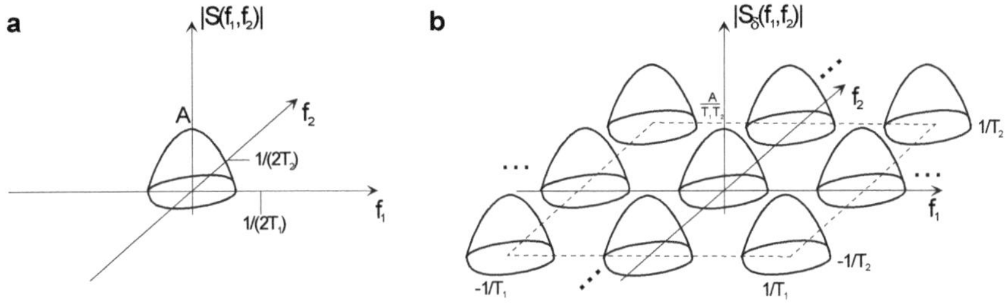

Fig. 2.10. Spectra of 2D image signals: **a** Continuous signal **b** sampled signal

To allow reconstruction by a 2D lowpass filter, $s(t_1,t_2)$ has to be band limited before sampling. 2D separable sampling allows perfect reconstruction by a lowpass interpolation filter if

$$S(f_1,f_2)\overset{!}{=}0 \quad \text{for} \quad |f_1|\geq\frac{1}{2T_1} \quad \text{or} \quad |f_2|\geq\frac{1}{2T_2}, \tag{2.58}$$

such that

$$S(f_1,f_2)=T_1T_2S_{\delta_{T_1T_2}}(f_1,f_2)\cdot\mathrm{rect}(T_1f_1)\cdot\mathrm{rect}(T_2f_2)$$

$$\bullet\!\!-\!\!\circ$$

$$\begin{aligned}s(t_1,t_2)&=s_{\delta_{T_1T_2}}(t_1,t_2)**\mathrm{si}\left(\pi\frac{t_1}{T_1}\right)\cdot\delta(t_2)**\mathrm{si}\left(\pi\frac{t_2}{T_2}\right)\cdot\delta(t_1)\\&=\sum_{n_1=-\infty}^{\infty}\sum_{n_2=-\infty}^{\infty}s(n_1T_1,n_2T_2)\,\mathrm{si}\left[\pi\left(\frac{t_1}{T_1}-n_1\right)\right]\mathrm{si}\left[\pi\left(\frac{t_2}{T_2}-n_2\right)\right].\end{aligned} \tag{2.59}$$

This method of separable sampling can be straightforwardly extended to an arbitrary number of dimensions.

2.3.2 Non-separable two-dimensional sampling

Equidistant one-dimensional sampling and separable multi-dimensional sampling have only one degree of freedom (per dimension) in varying the sampling distance T. In case of non-separable sampling, sampling positions are still following a regular pattern, but need to be formulated with mutual dependency. Different regular grids of 2D sampling are illustrated in Fig. 2.11. Regularity means a systematic periodicity of a basic structure, which can be expressed by a system of *basis vectors* $\mathbf{t}_1=[t_{11}\ t_{21}]^{\mathrm{T}}$, $\mathbf{t}_2=[t_{12}\ t_{22}]^{\mathrm{T}}$. Linear combinations of these vectors, when multiplied by the integer vector index $\mathbf{n} = [n_1,n_2]^{\mathrm{T}}$, point to the effective positions $\mathbf{t}(\mathbf{n})=n_1\mathbf{t}_1+n_2\mathbf{t}_2$, which could be interpreted as 'centers of sampling cells'. The basis vectors are the columns of a coordinate transformation matrix $\mathbf{T}$, which in this context is also denoted as *sampling matrix*:

$$\underbrace{\begin{bmatrix} t_1(n_1,n_2) \\ t_2(n_1,n_2) \end{bmatrix}}_{\mathbf{t}(\mathbf{n})} = \underbrace{\begin{bmatrix} t_{11} & t_{12} \\ t_{21} & t_{22} \end{bmatrix}}_{\mathbf{T}} \cdot \underbrace{\begin{bmatrix} n_1 \\ n_2 \end{bmatrix}}_{\mathbf{n}} . \tag{2.60}$$

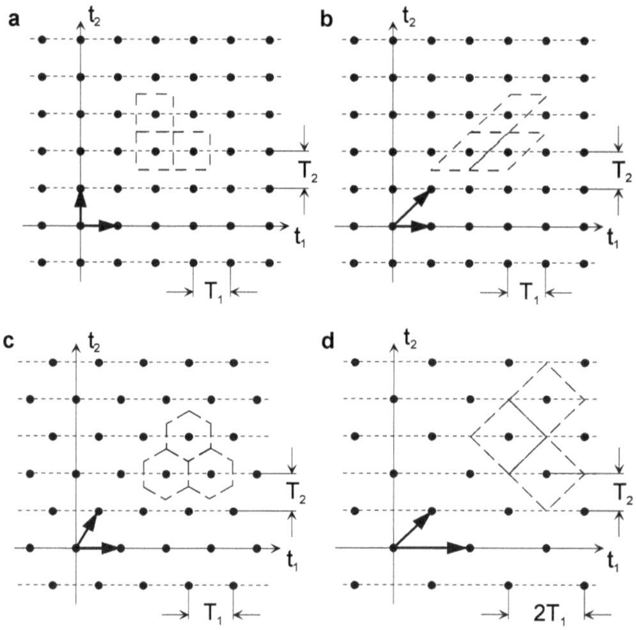

Fig. 2.11. 2D sampling grids: **a** rectangular **b** horizontal shear, v=1 **c** hexagonal **d** quincunx

For the separable case, the sampling distances T_1 in horizontal and T_2 in vertical direction are independent of each other. The corresponding sampling matrix is diagonal, with a frequency matrix according to (2.37),

$$\mathbf{T}_{\text{rect}} = \begin{bmatrix} T_1 & 0 \\ 0 & T_2 \end{bmatrix} \Rightarrow \mathbf{F}_{\text{rect}} = \begin{bmatrix} \frac{1}{T_1} & 0 \\ 0 & \frac{1}{T_2} \end{bmatrix}. \tag{2.61}$$

For the case of *shear sampling* (horizontal or vertical shear alternatively)

$$\mathbf{T}_{\text{shear}} = \begin{bmatrix} T_1 & v \cdot T_1 \mid 0 \\ 0 \mid v \cdot T_2 & T_2 \end{bmatrix} \Rightarrow \mathbf{F}_{\text{shear}} = \begin{bmatrix} \frac{1}{T_1} & 0 \mid -\frac{v}{T_1} \\ -\frac{v}{T_2} \mid 0 & \frac{1}{T_2} \end{bmatrix}, \tag{2.62}$$

the effective sampling grid would still appear as rectangular when v is an integer value (see Fig. 2.11b). Shear sampling can be interpreted as an alternative approach of adapting the sampling process by directional signal characteristics, where one axis of the coordinate system is tilted by the propagation direction of the signal. Such an approach may be useful when in a multidimensional sampling process the sampling positions in some dimensions cannot be changed due to system restrictions, e.g. when an image is scanned line-wise, or with fixed temporal sampling positions in case of video sampling.

Two other cases which can be interpreted as special cases of shear sampling (using non-integer shear factors v) are the hexagonal sampling scheme (Fig. 2.11c) and the quincunx sampling scheme (Fig. 2.11d). The basis vectors are tilted such that each sample has same distances towards its six or four nearest neighbors, respectively. To achieve this, a common scaling of sampling distance T (equal to the vertical distance between lines in these two cases) is used for both basis vectors.

$$\mathbf{T}_{\text{hex}} = T \cdot \begin{bmatrix} \frac{2}{\sqrt{3}} & \frac{1}{\sqrt{3}} \\ 0 & 1 \end{bmatrix} \Rightarrow \mathbf{F}_{\text{hex}} = \frac{1}{T} \begin{bmatrix} \frac{\sqrt{3}}{2} & 0 \\ -\frac{1}{2} & 1 \end{bmatrix};$$

$$\mathbf{T}_{\text{quin}} = T \cdot \begin{bmatrix} 2 & 1 \\ 0 & 1 \end{bmatrix} \Rightarrow \mathbf{F}_{\text{quin}} = \frac{1}{T} \begin{bmatrix} \frac{1}{2} & 0 \\ -\frac{1}{2} & 1 \end{bmatrix}. \tag{2.63}$$

To determine the positions of periodic spectral copies in the case of non-separable sampling, a non-separable 2D Dirac impulse grid with sampling positions defined by $\mathbf{T}$ is mapped by a coordinate transformation into a separable, unity-distance

Dirac impulse grid[8] $\text{Ш}(\mathbf{t}) \equiv \delta_{\mathbf{I}}(\mathbf{t}) \circ\!\!-\!\!\bullet\, \delta_{\mathbf{I}}(\mathbf{f}) \equiv \text{Ш}(\mathbf{f})$, with $T_1=T_2=...=1$. This gives, using (2.38):

$$\text{Ш}(\mathbf{T}^{-1}\mathbf{t}) \circ\!\!-\!\!\bullet \left|\mathbf{T}\right| \text{Ш}(\mathbf{F}^{-1}\mathbf{f}) \;\; \text{with } \mathbf{F} = \left[\mathbf{T}^{-1}\right]^{\mathrm{T}} \tag{2.64}$$

It should however be observed, that by applying the coordinate transformation to the 'sheh' function, the Dirac impulses are scaled reciprocally, following the determinant of the respective coordinate transformation matrix (in both **t** and **f** domains). Therefore, explicitly expressed by sums of non-scaled Dirac impulses,

$$\text{Ш}(\mathbf{T}^{-1}\mathbf{t}) = \sum_{\mathbf{n}} \left|\mathbf{T}\right| \delta(\mathbf{t}-\mathbf{Tn}) \;\; \text{and} \;\; \text{Ш}(\mathbf{T}^{-1}\mathbf{f}) = \sum_{\mathbf{k}} \left|\mathbf{F}\right| \delta(\mathbf{f}-\mathbf{Fk}), \tag{2.65}$$

and finally

$$\underbrace{\sum_{\mathbf{n}} \delta(\mathbf{t}-\mathbf{Tn})}_{\delta_{\mathbf{T}}(\mathbf{t})} \circ\!\!-\!\!\bullet \frac{1}{\left|\mathbf{T}\right|} \underbrace{\sum_{\mathbf{k}} \delta(\mathbf{f}-\mathbf{Fk})}_{\delta_{\mathbf{F}}(\mathbf{f})} \tag{2.66}$$

The spectrum of a multi-dimensional signal being ideally sampled with the scheme defined by the sampling matrix **T** then is

$$s_{\delta_{\mathbf{T}}}(\mathbf{t}) = s(\mathbf{t}) \cdot \delta_{\mathbf{T}}(\mathbf{t}) \circ\!\!-\!\!\bullet\, S_{\delta_{\mathbf{T}}}(\mathbf{f}) = S(\mathbf{f}) * \frac{1}{\left|\mathbf{T}\right|} \delta_{\mathbf{F}}(\mathbf{f}) = \frac{1}{\left|\mathbf{T}\right|} \sum_{\mathbf{k}} S(\mathbf{f}-\mathbf{Fk}), \tag{2.67}$$

which gives specifically for the 2-dimensional case

$$S_{\delta_{\mathbf{T}}}(f_1, f_2) = \frac{1}{\left|\mathbf{T}\right|} \sum_{k=-\infty}^{\infty} \sum_{l=-\infty}^{\infty} S(f_1 - k_1 f_{11} - k_2 f_{12}, f_2 - k_1 f_{21} - k_2 f_{22}). \tag{2.68}$$

(2.67) and (2.68) can be interpreted in a way that each κ-tuple of integer values in **k** points to one copy of the spectrum by the corresponding linear combination of the basis vectors, **Fk.** Again, direct computation of $S_\delta(\mathbf{f})$ as in (2.56) would be possible from the series of samples,

$$S_{\delta_{\mathbf{T}}}(\mathbf{f}) = \sum_{\mathbf{n}} s(\mathbf{Tn}) \mathrm{e}^{-\mathrm{j}2\pi \mathbf{f}^{\mathrm{T}}\mathbf{Tn}} = \sum_{\mathbf{n}} s(\mathbf{Tn}) \mathrm{e}^{-\mathrm{j}2\pi \left[\mathbf{F}^{-1}\mathbf{f}\right]^{\mathrm{T}}\mathbf{n}}. \tag{2.69}$$

$\mathbf{F}^{-1}\mathbf{f}$ in (2.64) could be interpreted as normalized frequency where spectral copies are at integer vector positions **k**, and $\mathbf{n}=\mathbf{T}^{-1}\mathbf{t}$ would describe a discrete signal over integer vector indices **n**, corresponding to a normalization of the **t** coordinates by **T**. Then, a separable Fourier sum over a normalized frequency could be computed directly from the signal samples $s(\mathbf{n})$, regardless of the actual sampling structure, as a generalization of (2.57),

$$S_\delta(\mathbf{f}) = \sum_{\mathbf{n}} s(\mathbf{n}) \mathrm{e}^{-\mathrm{j}2\pi \mathbf{f}^{\mathrm{T}}\mathbf{n}}. \tag{2.70}$$

---

[8] Using the letter 'sheh' from the Russian alphabet as a symbolic expression of a unity distance Dirac impulse grid

It must however be noted that the normalization assumed in (2.70) may be misleading because reasonable conditions about band limitation, as necessary for alias-free sampling and reconstruction, are not fully reflected here. A mapping of conditions in (2.58) would unnecessarily restrict the degrees of freedom in defining band limitation in the non-rectangular sampling case, because the matrix **F** only allows describing a linear coordinate transformation of the baseband boundaries from a square-shaped lowpass. Two examples of such a mapping for the case of quincunx sampling (2.63) are shown in Fig. 2.12. The resulting limitation of the base band in Fig. 2.12b would be asymmetric, giving different preference to orientations. Moreover, other grids with identical sampling points (though differently indexed in **k**) can be defined using alternative sampling matrices **T**. For example, two possible definitions of a position-wise identical quincunx grid as in (2.63) would be (case II is based on a vertical shear, case III is a rotation of coordinates)

$$\mathbf{T}_{\text{quin-II}} = T\begin{bmatrix} 1 & 0 \\ -1 & 2 \end{bmatrix} \quad ; \quad \mathbf{T}_{\text{quin-III}} = T\begin{bmatrix} 1 & 1 \\ -1 & 1 \end{bmatrix}. \tag{2.71}$$

These different definitions would have a significant impact on the band limitation, when back-projection from the unity grid (2.64) is used as reference. In the specific case of quincunx, the rotation would give the somewhat optimum omnidirectional packing as discussed below, but it may not be possible to make such definition from the sampling matrix for any sampling structure, particularly in higher dimensions.

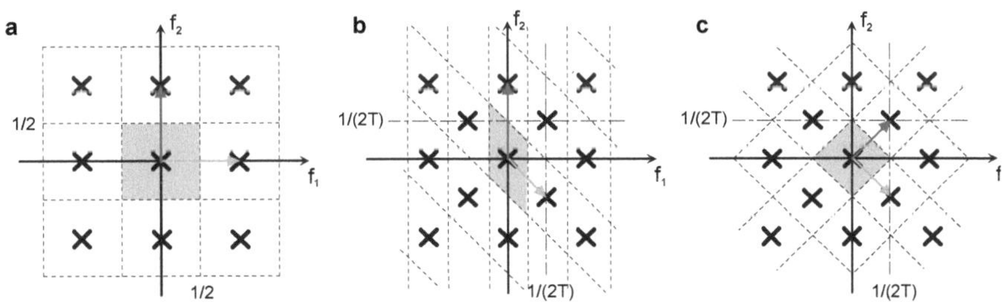

Fig. 2.12. **a** Base band and its periodic copies for rectangular-grid sampling in the normalized frequency plane **b/c** corresponding reverse mapping $\mathbf{F}^{-1}\mathbf{f}$ for the case of quincunx sampling according to (2.63) (**b**) and version III from (2.71) (**c**).

Best omnidirectional lowpass band limitation for non-rectangular sampling can be derived from the theory of dense packing of identically shaped cells (areas or volumes) in multiple dimensions. For this, the position of the center of the base band at the origin of the frequency plane is regarded in relation to the positions of centers of directly neighbored spectral copies. In order to make the shapes identical and symmetric for different directional orientations, the cut-off frequency should be at half distance between the zero frequency and the centers of closest spectral copies. This can be determined by drawing interconnection lines (called

Delaunay lines) from **f**=**0** to those center points. In the 2D case, *Voronoi lines* (which would become *planes* or *hyper planes* in higher dimensions) are intersecting at the mid position of the respective Delaunay line with perpendicular orientation (i.e. the orientation of the Delaunay line could be interpreted as normal vector of the Voronoi boundary). The connection of all Voronoi lines closest to frequency zero establishes the boundary of the base band. For the examples of quincunx and hexagonal sampling matrices in (2.63), the shapes of base bands and the periodic copies thereof are illustrated in Fig. 2.13a/b, and conditions are explicitly determined in the following paragraphs. In the quincunx case, this becomes indeed identical to the mapping of conditions (2.58) by $\mathbf{F}_{\text{quin-III}}$ of (2.71); in the hexagonal case, no such direct mapping is possible.

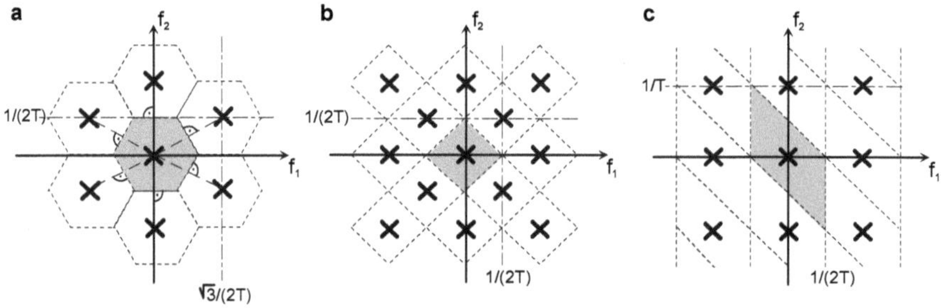

Fig. 2.13. Positions of base band and spectral copies for different 2D sampling grids. **a** Hexagonal **b** Quincunx **c** Shear, $v = 1$ [– – – Delaunay line ----- Voronoi line]

Hexagonal sampling. The hexagonal shape of the base band requires piecewise definition, but is symmetric over all four quadrants. When $|f_1| \leq \sqrt{3}/6T$, the boundary of the base band is parallel with the f_1 axis, while for higher frequencies $|f_1|$, four lines with slopes $a = \pm\sqrt{3}$ and intercepts $b = \pm 1/T$ define the boundary. This results in sampling conditions (see problem 2.1)

$$S(f_1, f_2) \overset{!}{=} 0 \quad \text{for} \quad |f_2| \geq \frac{1}{2T} \quad \text{or} \quad |f_1| + \frac{|f_2|}{\sqrt{3}} \geq \frac{1}{\sqrt{3}T}. \tag{2.72}$$

Quincunx sampling. The boundary of the base band is described by four lines of slopes $a = \pm 1$ and intercepts $b = \pm 1/(2T)$. This gives the condition

$$S(f_1, f_2) \overset{!}{=} 0 \quad \text{for} \quad |f_1| + |f_2| \geq \frac{1}{2T}. \tag{2.73}$$

In quincunx sampling, pure horizontal or vertical sinusoids can be reconstructed up to the same frequency as with quadrangular (i.e. rectangular where $T = T_1 = T_2$) sampling, though the number of samples is reduced by a factor of two. For sinusoids of diagonal orientation, the maximum allowable frequency is however lower by a factor $\sqrt{2}$. Quincunx sampling better matches human perception which is less sensitive to fine detail in the diagonal directions. In the Bayer pattern (Fig. 1.5), a quincunx sampling grid is therefore applied to the G (green) component. For interpolation into full resolution, a 2D sinc function rotated by 45° (or an approximation thereof) can be applied:

$$h(t_1,t_2)=\mathrm{si}\left(\pi\frac{t_1-t_2}{4T}\right)\mathrm{si}\left(\pi\frac{t_1+t_2}{4T}\right). \tag{2.74}$$

For any two- and multi-dimensional sampling system, the allowable bandwidth of the signal (area or volume covered by the base bands in Fig. 2.13) is identical to the determinant of the frequency sampling matrix **F**. Likewise, the area or volume of each 'sampling cell' is the determinant of the sampling matrix **T**. Due to (2.37), the density of samples and the alias-free signal bandwidth are mutually reciprocal. The definition of the base band allows certain degrees of freedom, in trading the resolution ranges between the different dimensions. As an example for this, alias-free quincunx sampling could also be realized using a separable reconstruction filter of horizontal pass-band cut-off $\pm 1/(4T)$, vertical cut-off $\pm 1/(2T)$ or vice versa. The question whether this makes sense can only be answered by an analysis of signal characteristics, and by the actual goal of sampling, e.g. the effective bandwidth of signals along each of the dimensions.

The theory of densest packing as mentioned above can not only be used for determining the boundaries of the baseband (respectively the cut-off characteristics of the lowpass interpolation filter), but also to determine the best two- or multi-dimensional sampling grid. Assume that a goal would be to represent directional sinusoids such as (2.1),(2.3) with highest possible frequency F regardless of the orientation. From that point of view, the optimum shape of the baseband would be a circle, in higher dimensions, it becomes a sphere or hyper sphere. If a circle or sphere of given radius r (e.g. $r=1/2$) is fitted with the minimum baseband cut-off, the determinant $|\mathbf{F}|$ of the related matrix is a criterion for the necessary number of samples per unit to allow a cut-off at $f=1/2$ at minimum. For example, in the case of separable 2D sampling (2.61) (see Fig. 2.12a), dense packing of circles with $r=1/2$ is possible when $T_1=T_2=1$, $|\mathbf{F}|=1$. In hexagonal sampling (2.63) (see Fig. 2.12c), this is possible using $T=1$, which gives $|\mathbf{F}|=\sqrt{3}/2\approx 0.866$. This is denoted as *sphere packing advantage* of the hexagonal structure, meaning that sinusoids of arbitrary directional orientation with a given maximum frequency can be sampled using less than 87% of the samples that would be necessary for the separable case. Alternatively, using the same number of samples, the cut-off frequency can be increased by the reciprocal square root of that factor. In 2D, the hexagonal scheme provides the best possible sampling in that sense. The quincunx scheme does not provide a sphere packing advantage.

Two- and multi-dimensional sampling structures can also be constructed by superimposing different systems of basis vectors. For example, a quincunx scheme as in Fig. 2.11c can be interpreted as a superposition of two rectangular schemes (see problem 2.2). Similarly, a grid of equal-sized triangle cells can be formed by a superposition of two hexagonal grids of Fig. 2.11c, where the second is vertically offset by $2T_2/3$. However in this case, the cells corresponding to the two sub-grids have different orientation, and each point has only three nearest neighbors with equal distance, which indicates that the packing would be less dense than with a single hexagonal grid.

2.3.3 Sampling of video signals

A video sequence of pictures can be interpreted as a three-dimensional (2D spatial+temporal) signal (see Fig. 2.14). Let the time distance between subsequent sampled pictures be T_3. An extension of separable sampling (2.60) to the third dimension then leads to the following mapping of sampling positions in the spatio-temporal continuum:

$$\begin{bmatrix} t_1(n_1,n_2,n_3) \\ t_2(n_1,n_2,n_3) \\ t_3(n_1,n_2,n_3) \end{bmatrix} = \begin{bmatrix} n_1T_1 \\ n_2T_2 \\ n_3T_3 \end{bmatrix} = \mathbf{T}_{\text{prog}} \begin{bmatrix} n_1 \\ n_2 \\ n_3 \end{bmatrix} . \tag{2.75}$$

For the example of Fig. 2.14a, samples have identical spatial positions in any picture. Such a configuration is denoted as *progressive sampling*, which is shown in Fig. 2.14b over the vertical and temporal directions. The sampling matrix related to fully-separable progressive sampling is given as

$$\mathbf{T}_{\text{prog}} = \begin{bmatrix} T_1 & 0 & 0 \\ 0 & T_2 & 0 \\ 0 & 0 & T_3 \end{bmatrix} \Rightarrow \mathbf{F}_{\text{prog}} = \begin{bmatrix} 1/T_1 & 0 & 0 \\ 0 & 1/T_2 & 0 \\ 0 & 0 & 1/T_3 \end{bmatrix} , \tag{2.76}$$

however, any sampling that handles the temporal dimension independent of the two spatial dimensions could also be entitled as progressive (e.g. quincunx or hexagonal only in the two spatial dimensions). In analog video, *interlaced sampling* was typically used, and interlaced formats still exist in some digital video cameras. Even and odd lines are sampled in a time-interleaved fashion, such that for each time instance, only half of the lines is sampled and available for subsequent processing. The resulting pictures consisting of either even or odd lines are called the even and odd *fields*, respectively (see. Fig. 2.14c). The sampling matrix in this case can be defined as[9]

---

[9] In Fig. 2.14c and in the sampling matrix (2.77) a configuration is shown where the top field (lines 0,2,4,..) is the field which is sampled first within the frame. In NTSC TV and digital 60 Hz interlaced video derived thereof, the bottom field is sampled first. This is however only relevant if field pictures are grouped together as a 'frame', e.g. when absolute timing information is assigned.

$$\mathbf{T}_{\text{inter}} = \begin{bmatrix} T_1 & 0 & 0 \\ 0 & 2T_2 & T_2 \\ 0 & 0 & T_3 \end{bmatrix} \Rightarrow \mathbf{F}_{\text{inter}} = \begin{bmatrix} 1/T_1 & 0 & 0 \\ 0 & 1/(2T_2) & 0 \\ 0 & -1/(2T_3) & 1/T_3 \end{bmatrix}. \qquad (2.77)$$

This could be interpreted as a quincunx sampling grid[10] applied to the vertical/temporal continuum in 3D. By this, higher vertical frequencies can be supported only when no significant temporal changes (e.g. caused by motion) are present.

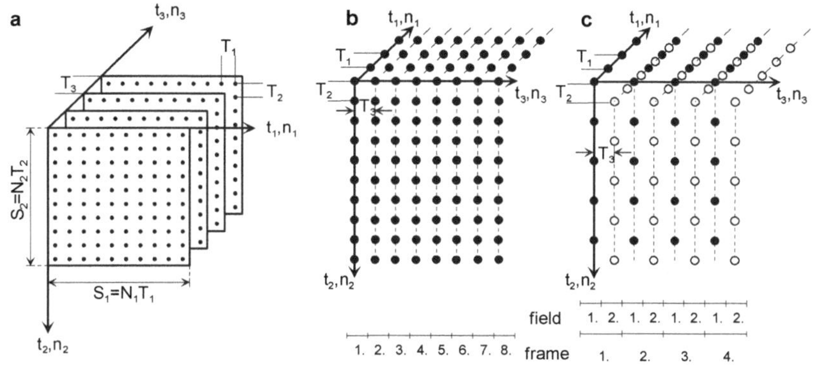

Fig. 2.14. **a** Progressively sampled image sequence **b/c** Video sampling in vertical/temporal directions: Progressive (**b**) and interlaced (**c**) schemes.

In progressive sampling – which is the 3D version of separable sampling – the conditions of the sampling theorem for avoiding alias can be formulated independently in each dimension. In this case the sampling matrix is diagonal, such that no interrelationships occur:

$$S(f_1,f_2,f_3)\overset{!}{=}0 \quad \text{when } |f_1| \geq \frac{1}{2T_1} \text{ or } |f_2| \geq \frac{1}{2T_2} \text{ or } |f_3| > \frac{1}{2T_3}. \qquad (2.78)$$

In interlaced sampling, only the condition for the first dimension can be separated, since the horizontal sampling positions are independent,

$$S(f_1,f_2,f_3)\overset{!}{=}0 \quad \text{when } |f_1| \geq \frac{1}{2T_1} \quad \text{or} \quad \frac{|f_2|}{T_3} + \frac{|f_3|}{T_2} \geq \frac{1}{2T_2T_3}. \qquad (2.79)$$

In video acquisition, spatial sampling is often assumed to be alias free, as the elements of the acquisition system (lenses etc.) naturally have a lowpass effect. As was shown in (2.45), the frequency f_3 depends on spatial frequency and the strength of motion. Assume that the signal could contain sinusoids of almost the

[10] The bottom-right 2x2 sub-matrix in (2.77) is indeed similar to (2.63) except for the fact that T_2 and T_3 actually express different physical units (space and time). Therefore, setting $T_1=T_2$ (as it was done in the 2D case) is not meaningful here.

maximum allowed spatial frequencies ($F_1 \approx 1/(2T_1)$, $F_2 \approx 1/(2T_2)$). Substituting the condition for f_3 from (2.78) into (2.45), the following limiting condition must then be imposed on the velocity to achieve alias-free sampling:

$$|u_1| \cdot \frac{T_3}{T_1} + |u_2| \cdot \frac{T_3}{T_2} \overset{!}{<} 1 \text{ resp. } |k_1| + |k_2| \overset{!}{<} 1 \quad \text{with} \quad k_i = u_i \cdot \frac{T_3}{T_i} . \tag{2.80}$$

Herein, the k_i express the horizontal/vertical displacements in units of samples from one picture to the next, if the velocity u_i is observed in the continuous signal in the respective dimension. The strict limitation, disallowing shifts larger than one spatial sample per time unit, appears surprising at first sight, as humans usually are capable to watch moving pictures of much higher motion without any problem. However, the limitation in (2.80) assumes that only *one* sinusoid of close-to-highest allowable spatial frequency would be sampled. Spectra of natural video signals are non-sparse with high energy in low-frequency ranges, which allow perceiving the motion reliably and alias-free. Particularly, the observer's eyes can track the motion which compensates alias by projecting the spectrum towards frequency f_3=0 (or alternatively could be interpreted as using a shear of the reconstruction filter pass-band).

To illustrate the effects of alias occurring in the case of progressive sampling, Fig. 2.15 shows a vertical/temporal section (f_2,f_3) of the 3D frequency space. A spatial sinusoid of close to half vertical sampling frequency is assumed which has a spectrum consisting of two Dirac impulses (⦿). Centers of periodic spectral copies are marked by '**x**'. Fig. 2.15a shows the spectrum of the signal without motion. Fig. 2.15b indicates skewing of the position in direction of f_3, when the signal is moved by half a special sampling unit per time unit ($u_2=-0.5T_2/T_3$) upwards, Fig. 2.15c illustrates the case of motion by 1.5 units ($u_2=-1.5T_2/T_3$). In the latter case, alias components appear in the base band, such that a viewer could interpret this as a motion by half a unit downwards ($u_2=0.5T_2/T_3$). The spatial frequency of the signal remains unchanged in any case, i.e. aliasing in f_3 only causes wrong interpretation of motion in the case of progressive sampling. In cinema, this is known as the 'stage coach effect', where rotating wheels equipped with periodic spokes seem to move slower, stand still or turn backwards, depending on the combined effect of temporal sampling distance, angular distance between the spokes and the speed of the wheel.

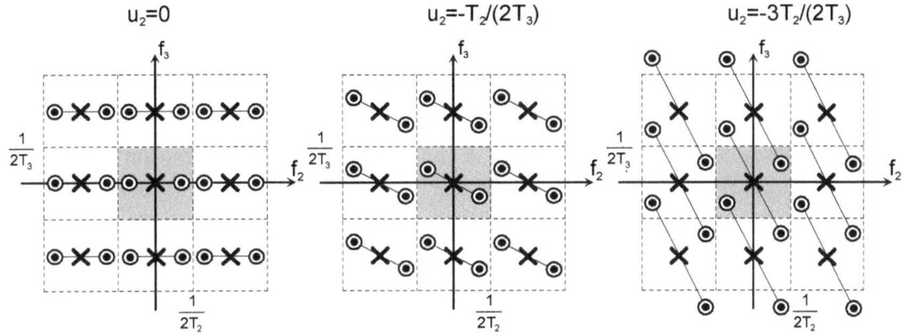

Fig. 2.15. Effect of alias, vertical motion of a progressively-sampled sinusoid.

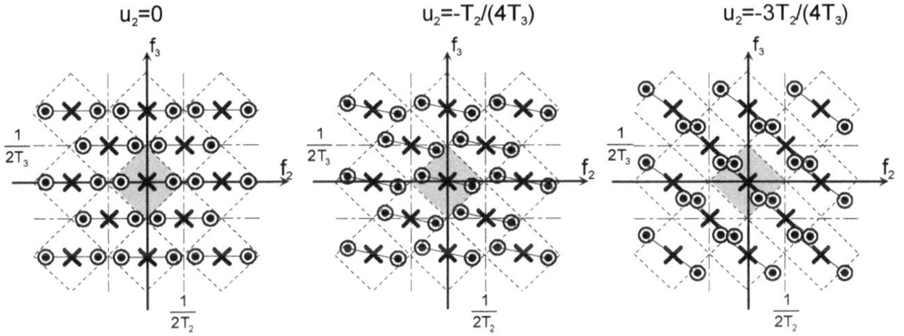

Fig. 2.16. Effect of alias, vertical motion of an interlaced-sampled sinusoid.

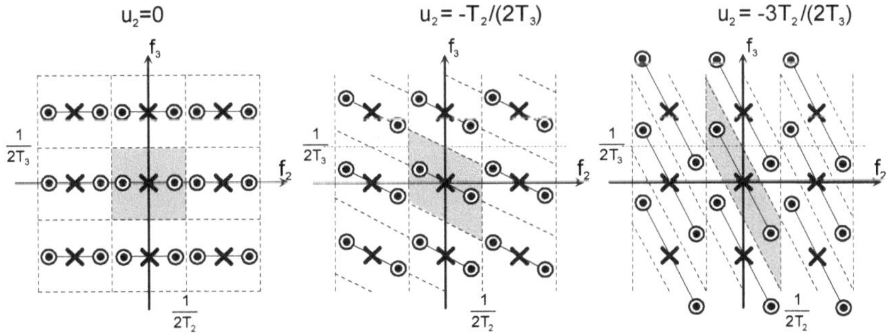

Fig. 2.17. Avoidance of alias by adaptation of the human visual system; tracking by the eyes effects correct reconstruction in sheared sampling.

Fig. 2.16 shows the effect for the case of interlaced sampling of the same signal[11]. First, it is obvious that aliasing already occurs with lower motion than in the pro-

---

[11] Note that the temporal sampling distance T_3 according to the sampling matrix in (2.77) refers to frame units, i.e. the sampling distance between fields is $T_3/2$. Likewise, the vertical sampling distance between the adjacent lines of each field is $2T_2$ by this definition.

gressive sampling case. Second, if alias spectra originate from diagonally-adjacent spectral copies and with vertical frequency of the sinusoid as F_2, an alias component of frequency $\tilde{F}_2 = 1/(2T_2) - F_2$ appears in the base band. In particular when highly-detailed periodic stripes are present in the scene and moving, this can result in appearance of strange sinusoidal components, typically also having different orientations than the original, as the horizontal frequency component F_1 would not be affected and orientation follows from (2.2).

As motion causes a tilt of spectra towards positions $f_3 \neq 0$, but does not cause a spreading of spectra, perfect reconstruction and correct perception would in principle be possible when the motion is known to the observer. This can either be interpreted to relate to the case of shear sampling (where the spectral shape of the reconstruction filter is aligned towards $f_3 = u_1 f_1 + u_2 f_2$), or as motion compensation (where the observer 'transforms' the reference coordinate system according to the motion). Fig. 2.17 illustrates that a single sinusoid moving by higher velocity can still be interpreted correctly; however, from a *single* sinusoid it is usually not possible to determine the actual motion, as the signal is periodic and multiple correspondences are detected between the subsequent pictures (a typical observer would assume the lowest possible velocity, which means that the displacement should not be larger than half a period in any direction). However, for structured signals which contain salient points, edges etc., the true motion can be tracked accordingly, as all frequency components are identically shifted (consistent linear phase shift). Motion-compensated processing in video compression performs a similar task, allowing to compress signals based on their actual redundancy along the temporal axis, thus avoiding alias components.

2.4 Discrete signal processing

2.4.1 LSI systems

The one- or multidimensional operation[12]

$$g(\mathbf{n}) = \sum_{\mathbf{m} \in \mathbf{Z}_\kappa} s(\mathbf{m}) h(\mathbf{n} - \mathbf{m}) = s(\mathbf{n}) * h(\mathbf{n}) \tag{2.81}$$

is denoted as *discrete convolution*. Its properties are similar to the continuous-time convolution integral, e.g. the associative, commutative and distributive properties apply. The *unit impulse*

---

[12] The Z-lattice $\mathbf{Z}_\kappa$ is an infinite set of vectors consisting of all possible integer number combinations over κ dimensions.

$$\delta(\mathbf{n}) = \begin{cases} 1 & \text{für } \mathbf{n} = \mathbf{0} \\ 0 & \text{für } \mathbf{n} \neq \mathbf{0}, \end{cases} \tag{2.82}$$

also denoted as *Kronecker impulse*, is the unity element,

$$s(\mathbf{n}) = \delta(\mathbf{n}) * s(\mathbf{n}) = \sum_{\mathbf{m}\in\mathbf{Z}_K} s(\mathbf{m})\delta(\mathbf{n}-\mathbf{m})\,. \tag{2.83}$$

Discrete convolution (2.81) is *linear* (2.14) and *shift invariant*, the latter property being equivalent with time invariance (2.15). Therefore, a system performing the discrete convolution operation is denoted as LSI system, for which (2.81) provides the unique mapping between input and output, with behaviour fully described by the impulse response $h(\mathbf{n})$. The operation of certain classes of LSI systems can be interpreted by finite order *difference equations*, for which a causal form[13] is

$$\sum_{\mathbf{p}\in\mathcal{N}_{\mathbf{p}}^{0+}} \tilde{b}_{\mathbf{p}} g(\mathbf{n}-\mathbf{p}) = \sum_{\mathbf{q}\in\mathcal{N}_{\mathbf{q}}^{0+}} \tilde{a}_{\mathbf{q}} s(\mathbf{n}-\mathbf{q}). \tag{2.84}$$

This gives the input/output relation (simplified when normalizing $\tilde{b}_0 = 1$)

$$g(\mathbf{n}) = \underbrace{\sum_{\mathbf{q}\in\mathcal{N}_{\mathbf{q}}^{0+}} a_{\mathbf{q}} s(\mathbf{n}-\mathbf{q})}_{\text{FIR part}} + \underbrace{\sum_{\mathbf{p}\in\mathcal{N}_{\mathbf{p}}^{+}} b_{\mathbf{p}} g(\mathbf{n}-\mathbf{p})}_{\text{IIR part}} \quad \text{with } a_{\mathbf{q}} = -\frac{\tilde{a}_{\mathbf{q}}}{\tilde{b}_{\mathbf{0}}},\ b_{\mathbf{p}} = -\frac{\tilde{b}_{\mathbf{p}}}{\tilde{b}_{\mathbf{0}}}. \tag{2.85}$$

The corresponding digital filters consist of an FIR (*Finite Impulse Response*) part taking reference to $|\mathcal{N}_{\mathbf{q}}^{+}|$ previous samples of the input, and an IIR (*Infinite Impulse Response*) part using feedback from $|\mathcal{N}_{\mathbf{p}}^{+}|$ previously processed output samples.

2.4.2 Discrete Fourier transform

Similar to (2.47), a spectrum $S(\mathbf{f})$ shall be represented by samples which have distances that are expressed by a separable (diagonal) sampling matrix $\mathbf{F}$ on the frequency axis[14]:

$$S_{\mathrm{p}}(\mathbf{f}) = \sum_{\mathbf{k}\in\mathbf{Z}_K} S(\mathbf{Fk})\delta(\mathbf{f}-\mathbf{Fk}) = S(\mathbf{f})\sum_{\mathbf{k}\in\mathbf{Z}_K} \delta(\mathbf{f}-\mathbf{Fk})\,. \tag{2.86}$$

Applying the inverse Fourier transform gives

---

[13] Herein, $\mathcal{N}^{0+}$ is a finite set of integer index vectors $\mathbf{p}|\mathbf{q}$ corresponding to a neighbourhood of previously available input samples, including the current sample with $\mathbf{p}|\mathbf{q}=\mathbf{0}$. For example, in 1D, the range of values is q=0…Q. Similarly, $\mathcal{N}^{+}$ is excluding the current sample, e.g. in 1D, with range of values p=1…P

[14] In principle, the following considerations are extensible to non-separable spectrum sampling, which for simplicity is omitted here.

$$\begin{array}{ccccc} S_{\mathrm{p}}(\mathbf{f}) & = & S(\mathbf{f}) & \cdot & \sum_{\mathbf{k}\in\mathbf{Z}_\kappa} \delta(\mathbf{f}-\mathbf{F}\mathbf{k}) \\ \bullet\!\!-\!\!\circ & & \bullet\!\!-\!\!\circ & \bullet\!\!-\!\!\circ & \bullet\!\!-\!\!\circ \\ s_{\mathrm{p}}(\mathbf{t}) & = & s(\mathbf{t}) & * & \frac{1}{|\mathbf{F}|}\sum_{\mathbf{n}\in\mathbf{Z}_\kappa} \delta(\mathbf{t}-\mathbf{T}\mathbf{n}) \end{array} \quad \text{with } \mathbf{T}=\left[\mathbf{F}^{-1}\right]^{\mathrm{T}}. \tag{2.87}$$

Spectrum sampling described by $\mathbf{F}$ effects a periodic repetition of the $\mathbf{t}$-domain function described by $\mathbf{T}$. If the duration of $s(\mathbf{t})$ fits into one 'periodic cell' of $\mathbf{T}$, it can be reconstructed from $s_{\mathrm{p}}(\mathbf{t})$ by multiplying it with a separable rectangular window function that has the shape of the cell, which in the frequency domain corresponds to a separable sinc function:

$$\begin{array}{ccccc} s(\mathbf{t}) & = & s_{\mathrm{p}}(\mathbf{t}) & \cdot & |\mathbf{F}|\operatorname{rect}(\mathbf{T}\mathbf{t}) \\ \circ\!\!-\!\!\bullet & & \circ\!\!-\!\!\bullet & \circ\!\!-\!\!\bullet & \circ\!\!-\!\!\bullet \\ S(\mathbf{f}) & = & S_{\mathrm{p}}(\mathbf{f}) & * & \operatorname{si}(\pi\mathbf{F}\mathbf{f}). \end{array} \tag{2.88}$$

From these considerations, periodic signals possess discrete spectra, but also signals that are time limited to a range that is equivalent to one period of $\mathbf{T}$ are completely represented by spectral samples over $\mathbf{F}$.

As band-limited signals can be described from a series of samples over time, it can further be concluded that a signal which is considered as limited and could therefore be equivalently periodic *in both time and frequency domains* can also be perfectly represented by *finite series of samples* in any of the two domains. A signal $s_{\mathrm{d}}(\mathbf{n})$ shall be nonzero only in ranges $[0;M_i-1]$ within all of its κ dimensions ($i=1\ldots\kappa$), or equivalently be periodic over M_i samples. Then, samples of the periodic Fourier spectrum taken at distances $F_i=1/M_i$ are giving a unique representation, where in the two subsequent equations $\mathbf{F}=[\mathbf{M}^{-1}]^{\mathrm{T}}$ is a diagonal matrix with the F_i values of the different dimensions as entries (and $\mathbf{M}$ similarly holding the M_i values). This gives the *Discrete Fourier Transform* (DFT) over κ dimensions,

$$S_{\mathrm{a}}(\mathbf{F}\mathbf{k}) = S_{\mathrm{d}}(\mathbf{k}) = \sum_{n_1=0}^{M_1-1}\cdots\sum_{n_\kappa=0}^{M_\kappa-1} s_{\mathrm{d}}(\mathbf{n})\mathrm{e}^{-\mathrm{j}2\pi\mathbf{n}^{\mathrm{T}}\mathbf{F}\mathbf{k}};\quad k_i=0,\ldots,M_i-1, \tag{2.89}$$

with the inverse DFT allowing reconstruction of all $|\mathbf{M}|$ samples,

$$s_{\mathrm{d}}(\mathbf{n}) = \sum_{n_1=0}^{M_1-1}\cdots\sum_{n_\kappa=0}^{M_\kappa-1} S_{\mathrm{d}}(\mathbf{k})\mathrm{e}^{\mathrm{j}2\pi\mathbf{n}^{\mathrm{T}}\mathbf{F}\mathbf{k}};\quad n_i=0,\ldots,M_i-1. \tag{2.90}$$

2.4.3 *z* transform

A condition for existence of the Fourier sum (2.57) of a discrete-time signal is finite absolute summation

$$\sum_{\mathbf{n}\in\mathbf{Z}_\kappa} |s(\mathbf{n})| < \infty. \tag{2.91}$$

An exception is established for periodic signals which have Fourier spectra $S_\delta(f)$

containing Dirac impulses. Otherwise, for a larger class of signals that do not grow stronger than exponentially on at most one side, convergence can be achieved by an exponential weighting $\mathrm{e}^{-\boldsymbol{\sigma}^{\mathrm{T}}\mathbf{n}} = \mathrm{e}^{-\sigma_1 n_1} \cdots \mathrm{e}^{-\sigma_\kappa n_\kappa}$ (σ_i values real),

$$e^{-\boldsymbol{\sigma}^{\mathrm{T}}\mathbf{n}} s(\mathbf{n}) \circ\!\!-\!\!\bullet \sum_{\mathbf{n}\in\mathbf{Z}_\kappa} \left(s(\mathbf{n})\mathrm{e}^{-\boldsymbol{\sigma}^{\mathrm{T}}\mathbf{n}} \right) \mathrm{e}^{-\mathrm{j}2\pi\mathbf{f}^{\mathrm{T}}\mathbf{n}} = \sum_{\mathbf{n}\in\mathbf{Z}_\kappa} s(\mathbf{n})\mathrm{e}^{-(\boldsymbol{\sigma}+\mathrm{j}2\pi\mathbf{f})^{\mathrm{T}}\mathbf{n}}. \tag{2.92}$$

Substituting $z_i = \mathrm{e}^{(\sigma_i + \mathrm{j}2\pi f_i)}$ by polar coordinates $z_i = \rho_i \mathrm{e}^{\mathrm{j}2\pi f_i}$ with $\rho_i = \mathrm{e}^{\sigma_i} \geq 0$ ($\rho_i > 0$ and σ_i real valued, $\rho_i \to 0$ for $\sigma_i \to -\infty$) and defining

$${}^{\kappa}\mathbf{z}^{(\mathbf{l})} = \prod_{i=1}^{\kappa} z_i^{l_i}, \tag{2.93}$$

the two-sided κ-dimensional z-transform of the signal $s(\mathbf{n})$ is

$$S(\mathbf{z}) = \sum_{\mathbf{n}\in\mathbf{Z}_\kappa} s(\mathbf{n})\, {}^{\kappa}\mathbf{z}^{(-\mathbf{n})}. \tag{2.94}$$

Values of $\mathbf{z}$ where a solution exists are contained within the *region of convergence* (RoC) of the complex z hyperspace. The z-transform is particularly useful in LSI system analysis and synthesis. Convolution in the time domain can again be expressed by multiplication in the z domain,

$$g(\mathbf{n}) = s(\mathbf{n}) * h(\mathbf{n}) \overset{z}{\circ\!\!-\!\!\bullet} G(\mathbf{z}) = S(\mathbf{z}) \cdot H(\mathbf{z}) \quad \text{with } \mathrm{RoC}\{G\} = \mathrm{RoC}\{S\} \cap \mathrm{RoC}\{H\}, \tag{2.95}$$

and a delay by $\mathbf{k}$ samples can be expressed as

$$s(\mathbf{n}-\mathbf{k}) = s(\mathbf{n}) * \delta(\mathbf{n}-\mathbf{k}) \overset{z}{\circ\!\!-\!\!\bullet} S(\mathbf{z}) \cdot \sum_{\mathbf{n}\in\mathbf{Z}_\kappa} \delta(\mathbf{n}-\mathbf{k})\, {}^{\kappa}\mathbf{z}^{(-\mathbf{n})} = S(\mathbf{z})\, {}^{\kappa}\mathbf{z}^{(-\mathbf{k})}. \tag{2.96}$$

A causal FIR/IIR filter with difference equation (2.84), where the z-transform is separately applied to the left and right sides, gives

$$\begin{aligned} &\sum_{\mathbf{q}\in\mathcal{N}_{\mathbf{q}}^{0+}} a_{\mathbf{q}} s(\mathbf{n}-\mathbf{q}) \overset{z}{\circ\!\!-\!\!\bullet} S(\mathbf{z}) \cdot A(\mathbf{z}) \ \text{ with } A(\mathbf{z}) = \sum_{\mathbf{q}\in\mathcal{N}_{\mathbf{q}}^{0+}} a_{\mathbf{q}}\, {}^{\kappa}\mathbf{z}^{(-\mathbf{q})} \\ &\sum_{\mathbf{p}\in\mathcal{N}_{\mathbf{p}}^{+}} b_{\mathbf{q}} \cdot g(\mathbf{n}-\mathbf{p}) \overset{z}{\circ\!\!-\!\!\bullet} G(\mathbf{z}) \cdot B(\mathbf{z}) \ \text{ with } B(\mathbf{z}) = \sum_{\mathbf{p}\in\mathcal{N}_{\mathbf{p}}^{+}} b_{\mathbf{p}}\, {}^{\kappa}\mathbf{z}^{(-\mathbf{p})} \end{aligned} \tag{2.97}$$

and therefore

$$G(\mathbf{z}) \cdot [1 - B(\mathbf{z})] = S(\mathbf{z}) \cdot A(\mathbf{z}) \quad \Rightarrow H(\mathbf{z}) = \frac{G(\mathbf{z})}{S(\mathbf{z})} = \frac{A(\mathbf{z})}{1 - B(\mathbf{z})} = \frac{\sum_{\mathbf{q}\in\mathcal{N}_{\mathbf{q}}^{0+}} a_{\mathbf{q}}\, {}^{\kappa}\mathbf{z}^{(-\mathbf{q})}}{1 - \sum_{\mathbf{p}\in\mathcal{N}_{\mathbf{p}}^{+}} b_{\mathbf{p}}\, {}^{\kappa}\mathbf{z}^{(-\mathbf{p})}}. \tag{2.98}$$

The FIR part of the filter corresponds to the numerator polynomial and the zero locations of the z transform, whereas the IIR part relates to the denominator and its singularities (poles). From (2.98) it is straightforward to design an inverse filter which performs *de-convolution*, i.e. reproduces $s(\mathbf{n})$ from $g(\mathbf{n})$,

$$S(\mathbf{z}) = \frac{G(\mathbf{z})}{H(\mathbf{z})} = G(\mathbf{z}) \cdot H^{(-1)}(\mathbf{z})$$

$$\Rightarrow H^{(-1)}(\mathbf{z}) = \frac{S(\mathbf{z})}{G(\mathbf{z})} = \frac{1 - B(\mathbf{z})}{A(\mathbf{z})} = \frac{1 - \sum_{\mathbf{p} \in \mathcal{N}_\mathbf{p}^+} b_\mathbf{p} \, {}^\kappa\mathbf{z}^{(-\mathbf{p})}}{\sum_{\mathbf{q} \in \mathcal{N}_\mathbf{q}^{0+}} a_\mathbf{q} \, {}^\kappa\mathbf{z}^{(-\mathbf{q})}} = \frac{\frac{1}{a_0} - \sum_{\mathbf{p} \in \mathcal{N}_\mathbf{p}^+} \frac{b_\mathbf{p}}{a_0} \, {}^\kappa\mathbf{z}^{(-\mathbf{p})}}{1 - \sum_{\mathbf{q} \in \mathcal{N}_\mathbf{q}^+} \frac{a_\mathbf{q}}{a_0} \, {}^\kappa\mathbf{z}^{(-\mathbf{q})}}. \tag{2.99}$$

Properties of the multi-dimensional *z* transform. Properties of the multidimensional *z* transform are very similar to those of the Fourier transform:

Linearity: $$\sum_i a_i s_i(\mathbf{n}) \overset{z}{\circ\!\!-\!\!\bullet} \sum_i a_i S_i(\mathbf{z}) \tag{2.100}$$

Shift: $$s(\mathbf{n} - \mathbf{k}) \overset{z}{\circ\!\!-\!\!\bullet} {}^\kappa\mathbf{z}^{(-\mathbf{k})} S(\mathbf{z}) \tag{2.101}$$

Convolution: $$g(\mathbf{n}) = s(\mathbf{n}) * h(\mathbf{n}) \overset{z}{\circ\!\!-\!\!\bullet} G(\mathbf{z}) = S(\mathbf{z}) \cdot H(\mathbf{z}) \tag{2.102}$$

Inversion[15]: $$S(-\mathbf{n}) \overset{z}{\circ\!\!-\!\!\bullet} S(\mathbf{z}^{(-\mathbf{I})}) \tag{2.103}$$

Scaling[16]: $$s_{\mathbf{U}\downarrow}(\mathbf{n}) = s(\mathbf{U}\mathbf{n}) \overset{z}{\circ\!\!-\!\!\bullet} S\left(\mathbf{z}^{(\mathbf{U}^{-1})}\right) \tag{2.104}$$

Expansion: $$s_{\mathbf{U}\uparrow}(\mathbf{n}) = \begin{cases} s(\mathbf{m}), & \mathbf{n} = \mathbf{U}\mathbf{m} \\ 0, & \text{else} \end{cases} \overset{z}{\circ\!\!-\!\!\bullet} S_{\mathbf{U}\uparrow}(\mathbf{z}) = S(\mathbf{z}^{(\mathbf{U})}) \tag{2.105}$$

Modulation: $$s(\mathbf{n}) \cdot e^{j2\pi\mathbf{F}\mathbf{n}} \overset{z}{\circ\!\!-\!\!\bullet} S(\mathbf{z}e^{-j2\pi\mathbf{F}}). \tag{2.106}$$

2.4.4 Multi-dimensional LSI systems

The set of samples accessed by a two- and multi-dimensional system is entitled as 'support region' or neighborhood $\mathcal{N}$. An interesting class of symmetric 2D support regions is established by a *homogeneous neighborhood*, where signal samples at positions (m_1,m_2) belong to the neighborhood of a sample at position $\mathbf{n}=[n_1\ n_2]^\mathrm{T}$ according to a maximum distance norm of order P[17]:

---

15 $\mathbf{z}^{(\mathbf{A})}$ expresses a coordinate mapping in the multi-dimensional *z* domain such that in the ith dimension $z_i^{(\mathbf{A})}=\prod z_j^{a_{ji}}$. With $z_i=e^{j2\pi f_i}$, the equivalent mapping in the Fourier domain is $\mathbf{Af}$.

16 Scaling is a sub-sampling operation with integer values $U > 1$. The *z* transform mapping as expressed in (2.104) is strictly valid when no information loss occurs, i.e. where only samples in $s(n_1,n_2,...)$ which are at positions n_iU_i were non-zero.

17 Homogeneous neighbourhood systems are *symmetric* in terms of shape, but also in terms of mutual relationship of samples, which means that the current sample at position (n_1,n_2) is

$$\mathcal{N}_C^{(P)}(\mathbf{n}) = \left\{ \mathbf{m} \ : \ 0 < \sum_i |m_i - n_i|^P \leq C \right\}. \tag{2.107}$$

The parameter $C \geq 0$ influences the size of the neighborhood support region, whereas $P \geq 0$ influences the shape. The discrete *multi-dimensional convolution* of a signal $s(\mathbf{n})$ by the impulse response $h(\mathbf{n})$ is then defined as a finite-neighborhood operation

$$g(\mathbf{n}) = \sum_{\mathbf{m} \in \mathcal{N}(\mathbf{0})} s(\mathbf{m}) \cdot h(\mathbf{n} - \mathbf{m}) = \sum_{\mathbf{m} \in \mathcal{N}(\mathbf{0})} h(\mathbf{m}) \cdot s(\mathbf{n} - \mathbf{m}). \tag{2.108}$$

The support region $\mathcal{N}$ in (2.108) can specify impulse responses which have either finite or infinite extension.

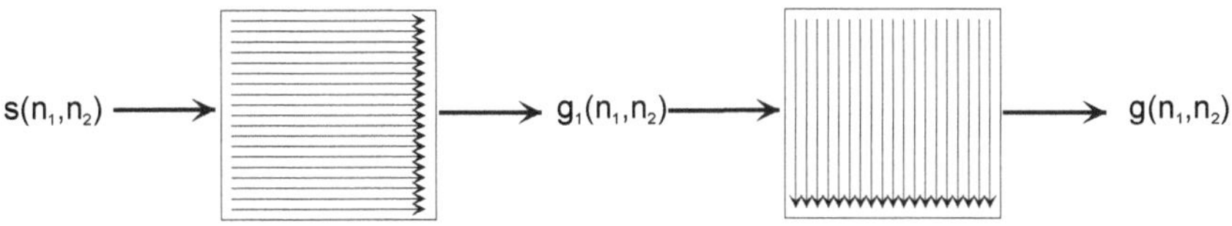

Fig. 2.18. Principle of a separable 2D LSI system with horizontal filter step first

Separable 2D LSI systems can be implemented in a similar fashion as per (2.26). Fig. 2.18 shows the principle, where first a horizontal 1D convolution is performed along each row, resulting in $g_1(n_1,n_2)$. In the next step, $g(n_1,n_2)$ is computed by convolving each column of $g_1(n_1,n_2)$. *Infinite Impulse Response* (IIR) filters are not realized by direct implementation of the convolution equation (2.108), but use feedback from previous output values $g(n_1,n_2)$. A given sequence of processing has to be obeyed due to the recursive relationship. For a 2D geometry, all positions which need to be previously processed to provide the input for the current position establish the support region $\mathcal{N}$. Fig. 2.19 shows three different causal IIR filter geometries with their respective $\mathcal{N}$ geometries: The *wedge plane filter*, the *quarter plane filter* and the *asymmetric half plane filter*. For the cases of quarter-plane and wedge-plane filter masks, either row-wise or column-wise recursion scans are possible; these filters also allow processing sequences with diagonal scans, or computation of all samples positioned on a diagonal in parallel (denoted as *wavefront processing*). For the asymmetric half plane filter, row-wise processing (starting at the top left position) is the only possible sequence of recursion. On a rectangular grid, only quarter-plane filters can be defined from separable causal 1D filters.

---

also a member of the same neighbourhood systems when applied to any of its neighbours (m_1,m_2). The neighbourhood can also be infinitely extended, e.g. for $P=0$ and $C \geq 2$.

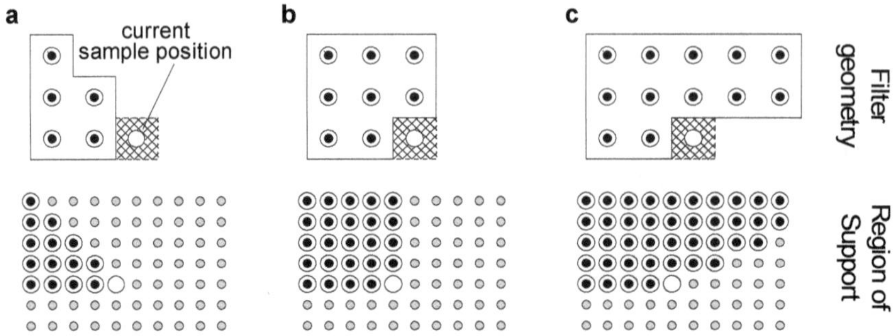

Fig. 2.19. Causal 2D filter masks and geometries of their support regions:
a Wedge plane **b** Quarter plane **c** Asymmetric half plane

A recursive 2D quarter-plane filter, where the filter geometry defines the feedback from $(P_1+1)(P_2+1)-1$ previously filtered samples, generates the output signal

$$g(n_1,n_2)=s(n_1,n_2)+\sum_{\substack{m_1=0\\(m_1,m_2)\neq(0,0)}}^{P_1}\sum_{m_2=0}^{P_2} b(m_1,m_2)\cdot g(n_1-m_1,n_2-m_2)\,. \tag{2.109}$$

In case of separable recursive filtering, lines and columns of a picture can be processed sequentially, such that the result of filtering along one of the dimensions is input to the filter along the other dimension, e.g. with horizontal processing first as

$$\begin{aligned} g_1(n_1,n_2)&=s(n_1,n_2)+\sum_{m_1=1}^{P_1} b_1(m_1)\cdot g_1(n_1-m_1,n_2)\ \ \text{(for all } n_2)\\ g(n_1,n_2)&=g_1(n_1,n_2)+\sum_{m_2=1}^{P_2} b_2(m_2)\cdot g(n_1,n_2-m_2)\ \ \text{(for all } n_1). \end{aligned} \tag{2.110}$$

The actual relation between the recursive coefficients in 1D and 2D can be determined by the difference equation (2.84) and its modification (2.85), such that

$$b(m_1,m_2)=-\tilde{b}_1(m_1)\tilde{b}(m_2)\ \text{ with } \tilde{b}_i(0)=1,\ \tilde{b}_i(m_i)=-b_i(m_i) \text{ for } 1\le m_i\le P_i\,. \tag{2.111}$$

Pictures are finite, where the output of filtering shall usually have the same size as the input e.g. for display purposes. Indices **n**−**m** in (2.108) can however have values less than zero or larger than the maximum coordinates M_1-1 or M_2-1, when samples close to the image boundaries shall be processed. Hence, it is necessary to define a signal extension beyond the boundaries of the input signal to consistently compute the convolution. Zero-setting of values is not useful, as pictures typically have non-zero mean. Three methods copying samples from the picture beyond the boundary, and therefore applicable for FIR filtering, are shown in Figs. 2.20a-c[18].

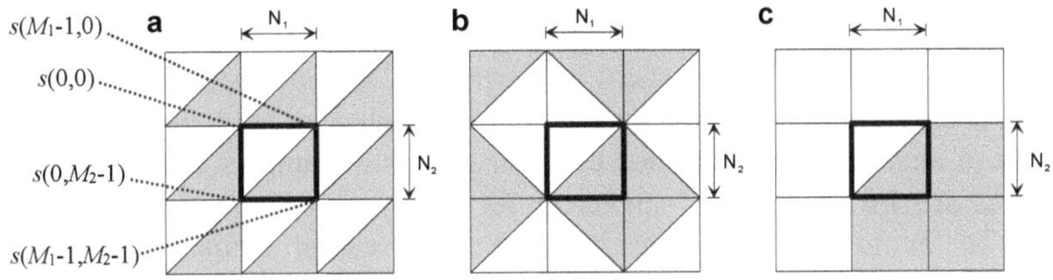

Fig. 2.20. Boundary extensions of finite image signals.
a periodic **b** symmetric (antiperiodic) **c** constant value

Fourier transfer functions of multi-dimensional filters. The multi-dimensional Fourier transform of the discrete impulse response is

$$H_\delta(\mathbf{f}) = \sum_{n_1=-\infty}^{\infty} \cdots \sum_{n_K=-\infty}^{\infty} h(\mathbf{n}) \cdot \mathrm{e}^{-\mathrm{j}2\pi\mathbf{f}^\mathrm{T}\mathbf{n}} . \tag{2.112}$$

If the system has FIR or causal IIR property, the summation limits can be bounded, such that the complex transfer function can directly be determined. For example, a 2D FIR system with a symmetric neighborhood of size $(Q_1+1)(Q_2+1)$ (Q_1 and Q_2 even) gives

$$H_\delta(f_1,f_2) = \sum_{n_1=-Q_1/2}^{Q_1/2} \sum_{n_2=-Q_2/2}^{Q_2/2} a(n_1,n_2)\, \mathrm{e}^{-\mathrm{j}2\pi n_1 f_1} \mathrm{e}^{-\mathrm{j}2\pi n_2 f_2} , \tag{2.113}$$

or for the case of a 2D quarter-plane IIR system, the Fourier transfer function is

$$H_\delta(f_1,f_2) = \frac{1}{1 - \sum\limits_{\substack{n_1=0 \\ (n_1,n_2)\neq(0,0)}}^{P_1} \sum\limits_{n_2=0}^{P_2} b(n_1,n_2)\, \mathrm{e}^{-\mathrm{j}2\pi n_1 f_1} \mathrm{e}^{-\mathrm{j}2\pi n_2 f_2}} . \tag{2.114}$$

The filter types and geometries of (2.113) and (2.114) are often used in the context of spatial prediction and interpolation of pictures.

[18] In case of IIR filters, it is necessary to define start values for the recursion from values $g(\mathbf{n}-\mathbf{m})$ which would be outside of the picture. Usually this should reflect the mean expectation, e.g. zero for audio/speech, mean gray value for pictures.

2.5 Statistical analysis

Statistical analysis is mainly discussed here for *sampled* multimedia signals $s(\mathbf{n})$, however similar properties hold for continuous signals $s(\mathbf{t})$. An ideal assumption would be *stationarity*, i.e. statistical properties not dependent on the position in time or space. For multimedia signals this does usually not hold; however, similar methods of analysis can be applied on local groups of samples assuming that the properties are invariant there, sufficient for giving reliable empirical measurements. To avoid differentiation between such cases, statistical parameters throughout this chapter are discussed as if they were independent of measurement time and place, and of the data set's size.

It should be observed that in the design of multimedia compression technology, it is normally necessary to use test data sets that exhibit all possible variety. It is even useful to augment test sets by more 'untypical' data which put challenges to the compression algorithm. Even though in adaptive methods usually local statistical properties are exploited, it is still necessary to allow possible adaptation states which give support to the whole variety of data that are expected to be fed into a coder.

2.5.1 Sample statistics

Statistical properties of samples from signals can be characterized by the *Probability Density Function* (PDF) $p_s(x)$, interpreting observed signal amplitudes as instantiations of a *random variable* x of an underlying random process $s(\mathbf{n})$. For the case of continuous amplitudes, the PDF provides information about expected occurrences of certain ranges of amplitude. The probability of a value observation $s(\mathbf{n})\leq x$ is given by the *Cumulative Distribution Function* (CDF)

$$P_s(x) \equiv \Pr\left[s(\mathbf{n}) \leq x\right] = \int_{-\infty}^{x} p_s(\xi)\,\mathrm{d}\xi\,. \tag{2.115}$$

The CDF is monotonically increasing and has a value in the range $P_s(-\infty) = 0 \leq P_s(x) \leq 1 = P_s(\infty)$. The probability of a signal amplitude to be within an interval range $[x_\mathrm{a};x_\mathrm{b}]$ is therefore

$$\Pr\left[x_\mathrm{a} < s(\mathbf{n}) \leq x_\mathrm{b}\right] = \int_{x_\mathrm{a}}^{x_\mathrm{b}} p_s(\xi)\,\mathrm{d}\xi = P_s(x_\mathrm{b}) - P_s(x_\mathrm{a}) \geq 0\,. \tag{2.116}$$

Furthermore,

$$\int_{-\infty}^{\infty} p_s(x)\,\mathrm{d}x = P_s(\infty) - P_s(-\infty) = 1\,. \tag{2.117}$$

The *expected value* $\mathcal{E}\{f[x]\}$ is the mean over a set of signal observations with a function $f[x]$ applied to the samples; it is related to the PDF by[19]

$$\mathcal{E}\{f[s(\mathbf{n})]\} = \lim_{N\to\infty} \frac{1}{N} \sum_{\mathbf{n}} f[s(\mathbf{n})] = \int_{-\infty}^{\infty} f(x) p_s(x)\,\mathrm{d}x\,. \tag{2.118}$$

From these definitions, the following important parameters are defined describing *sample statistics*:

– $f(x)=x$: Mean value $m_s = \int_{-\infty}^{\infty} x \cdot p_s(x)\,\mathrm{d}x = \mathcal{E}\{s(\mathbf{n})\}$ (2.119)

– $f(x)=x^2$: Quadratic mean (power) $Q_s = \int_{-\infty}^{\infty} x^2 \cdot p_s(x)\,\mathrm{d}x = \mathcal{E}\{s^2(\mathbf{n})\}$ (2.120)

– Variance $\sigma_s^{\,2} = \int_{-\infty}^{\infty} (x - m_s)^2 p_s(x)\,\mathrm{d}x = \mathcal{E}\{(s(\mathbf{n}) - m_s)^2\} = Q_s - m_s^{\,2}$ (2.121)

For numeric (digital) processing, signal samples are quantized, which means they are mapped into a set of discrete amplitudes (see section 4.1). The mapping function is the *quantization characteristic*, which is a staircase function (see Fig. 2.21 for a case of uniform quantization of a finite positive amplitude range using a step size Δ). The value of the discrete *probability mass function* (PMF) of the quantized process can then be determined from the areas under the PDF of the unquantized process within the respective quantization intervals j with lower boundary x_j, upper boundary x_{j+1} and reconstruction[20] y_j,

$$p_{s_Q}(y_j) = \int_{x_j}^{x_{j+1}} p_s(x)\,\mathrm{d}x\,. \tag{2.122}$$

---

[19] The terminology 'expected value' is used here both for cases of finite and infinite data sets. Only the latter is mathematically precise. If a finite set of N measurements is used, the expected value is *empirical*, but could be regarded as reliable if it is not significantly changing when N would be further increased.

[20] Typically in uniform quantization of step size Δ, the reconstruction value is placed at the center of the interval, i.e. $x_j=y_j-\Delta/2$ and $x_{j+1}=y_j+\Delta/2$ (see section 4.1). Note that in the context of quantization we will typically assume that representation (encoding) by a finite alphabet is possible. In general, a PMF can also consist of an infinite number of discrete values. This is of no harm if the probability of values converges towards zero at both ends of the amplitude range.

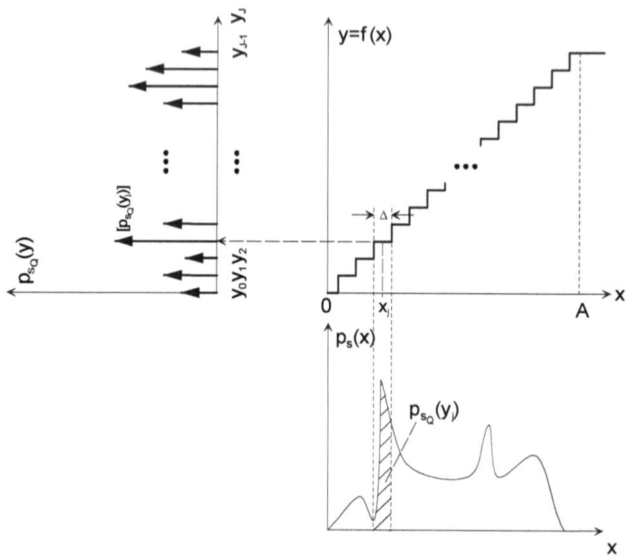

Fig. 2.21. Quantization characteristic and mapping of the PDF $p_s(x)$ of the continuous-amplitude signal to the probability mass function $p_{sQ}(y_j)$ of the quantized (discrete-amplitude) signal.

The PMF expresses the probability of the quantized (discrete) amplitude values y_j. The related PDF consists of a weighted sum of Dirac impulses[21]

$$p_{s,\delta}(x) = \sum_j p_{s_Q}(y_j)\delta(x - y_j) \tag{2.123}$$

where further from (2.117),

$$\sum_j p_{s_Q}(y_j) = 1 . \tag{2.124}$$

and

$$\mathcal{E}\left\{f\left[s_Q(\mathbf{n})\right]\right\} = \int_{-\infty}^{\infty} f(x)\sum_j p_{s_Q}(y_j)\delta(x - y_j)\,\mathrm{d}x = \sum_j p_{s_Q}(y_j)f(y_j) . \tag{2.125}$$

PDF models are useful to characterize the statistical behavior of a random process. For example, mean value and variance could be measured and used as parameters

---

[21] In the sequel, the subscript ‘Q’ is usually omitted, as the fact that the signal has been quantized is obvious from the context. Discrete probability functions (PMF) are written as $p_s(y_j)$. In the case of finite alphabets, this can also be expressed as $\Pr(S_j)$, where S_j is one discrete state with index j (without explicitly expressing an amplitude value).

under the assumption that a certain PDF shape is given. For multimedia signals, the generalized Gaussian distribution is often useful to express sample statistics[22]:

$$p_s(x) = a\mathrm{e}^{-|b(x-m_s)|^\gamma} \quad \text{with } a = \frac{b\gamma}{2\Gamma\left(\frac{1}{\gamma}\right)} \quad \text{and } b = \frac{1}{\sigma_s}\sqrt{\frac{\Gamma\left(\frac{3}{\gamma}\right)}{\Gamma\left(\frac{1}{\gamma}\right)}}. \tag{2.126}$$

For $\gamma = 2$, (2.126) gives the Gaussian normal PDF

$$p_s(x) = \frac{1}{\sqrt{2\pi\sigma_s^2}}\mathrm{e}^{-\frac{(x-m_s)^2}{2\sigma_s^2}}, \tag{2.127}$$

for which many optimization problems can be solved analytically. The normal PDF plays an important role, as according to the central limit theorem, it is the result of the superposition of a large number of statistically independent random signals. For $\gamma = 1$, (2.126) gives the Laplacian PDF:

$$p_s(x) = \frac{1}{\sqrt{2\sigma_s^2}}\mathrm{e}^{-\frac{\sqrt{2}|x-m_s|}{\sigma_s}}. \tag{2.128}$$

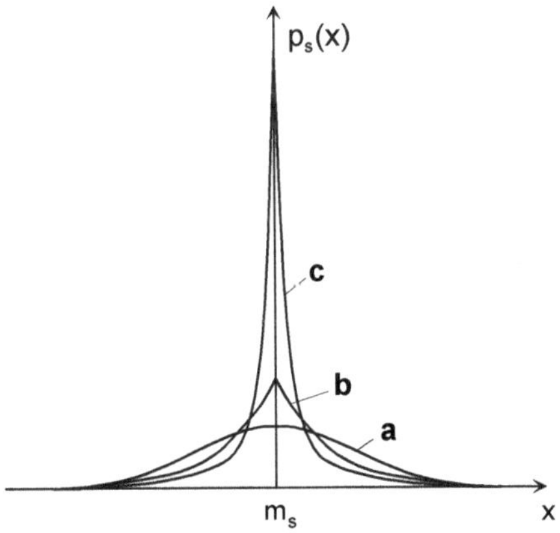

Fig. 2.22. Generalized Gaussian PDF for different values of γ: γ=2, Gaussian (**a**); γ=1, Laplacian (**b**); γ=0.5 (**c**)

Both, Gaussian and Laplacian cases are shown in Fig. 2.22, as well as a more narrow case ($\gamma = 0.5$). The Laplacian PDF has been reported to be a suitable mod-

---

22 The function $\Gamma(\cdot)$ which influences the shape of the PDF via the parameter γ, is defined as $\Gamma(u) = \int_0^\infty \mathrm{e}^{-x} x^{u-1} \mathrm{d}x$.

el for the probability distribution of DCT block transform coefficients extracted from still images [REININGER, GIBSON 1983] [LAM, GOODMAN 2000], and from motion compensated residual signals [BELLIFEMINE ET AL. 1992] as used in video coding. Finally, for $\gamma \to \infty$, (2.126) also expresses a uniform distribution (see problem 2.4),

$$p_s(x) = \frac{1}{\sqrt{12\sigma_s^2}} \operatorname{rect}\left(\frac{x - m_s}{\sqrt{12\sigma_s^2}} \right). \tag{2.129}$$

Models for discrete PMFs can be derived from analytic PDF models by applying an appropriate quantization in (2.122). Direct *sampling* of a PDF might give similar results in the case of small quantization step size Δ, but would typically lead to violation of (2.124) which would require re-normalization of the values. Another approach is representation of a continuous PDF by a mixture distribution – mixtures of Gaussians are often used for this purpose,

$$p_s(x) = \sum_i w_i \frac{1}{\sqrt{2\pi\sigma_{s_i}^2}} \mathrm{e}^{-\frac{(x-m_{s_i})^2}{2\sigma_{s_i}^2}} . \tag{2.130}$$

The parameters m_{s_i}, σ_{s_i} and the weights w_i of the different contributing Gaussian hulls, as well as the number of hulls have to be estimated. This can be achieved by initially identifying local peaks in the PDF to be described, analyze the slopes around the peaks, and then refine the match by algorithms such as expectation maximization or kernel density estimation (see MCA, CH. 5).

Models of PMFs can also be formulated directly in a finite discrete number space. As an example, the Bernoulli or binomial PMF defines probabilities of J discrete values, such that the jth value state occurs by probability

$$\Pr(S_j) = \binom{J-1}{j-1} p^{j-1} (1-p)^{J-j}; \quad 1 \le j \le J . \tag{2.131}$$

Alternatively, the probability values of the Bernoulli distribution in the J discrete states can be obtained by convolutions involving J–1 subsequent [p 1–p] FIR filter kernels. The symmetric case of the Bernoulli distribution (p=0.5) with increasing J can also be interpreted as discrete counterpart of the Gaussian normal PDF, which would be approached by iterative convolution of narrow continuous rectangular pulses.

2.5.2 Joint statistical properties

Joint probability functions (CDF, PDF or PMF) are used to express statistics about joint observations of two or multiple random values. Herein, the values can either stem from the same or from different signals, and/or from same or different locations in time and space. Therefore, joint probability functions express dependencies that exist either between the samples of only one or of different random signals. Joint probability functions have a K-dimensional dependency when K values are observed jointly. For the following paragraphs, the case K=2 is discussed, assuming $s_1(\mathbf{n})$ and $s_2(\mathbf{n}+\mathbf{k})$ are two observations with a relative shift of $\mathbf{k}$ samples. The concepts straightforwardly extend to higher K when additional observations are made.

The joint PDF $p_{s_1s_2}(x_1,x_2;\mathbf{k})$ is a 2-dimensional function (for one value of $\mathbf{k}$). The basic rules which are given in this section are applicable likewise to the discrete PMF or other discrete joint probability functions. Firstly, the joint functions are symmetric,

$$p_{s_1s_2}(x_1,x_2;\mathbf{k}) = p_{s_2s_1}(x_2,x_1;\mathbf{k}) . \tag{2.132}$$

In the hypothetical case that the observed samples were generally identical,

$$p_{s_1s_2}(x_1,x_2;\mathbf{k}) = p_{s_1}(x_1)\delta(x_2 - x_1) = p_{s_2}(x_2)\delta(x_1 - x_2) , \tag{2.133}$$

whereas for statistical independence,

$$p_{s_1s_2}(x_1,x_2;\mathbf{k}) = p_{s_1}(x_1)p_{s_2}(x_2) . \tag{2.134}$$

Conditional probabilities allow to express an expectation about the probability of random variables x_1 for the first observation, if it is already known that the other observation came as x_2, expressing the '*probability of* x_1 *given* x_2'. No uncertainty about the conditioning event exists, such that the conditional probabilities can be gained from the joint probability, normalized by the probability of the condition,

$$p_{s_1s_2}(x_1|x_2;\mathbf{k}) = \frac{p_{s_1s_2}(x_1,x_2;\mathbf{k})}{p_{s_2}(x_2)} \; ; \quad p_{s_2s_1}(x_2|x_1;\mathbf{k}) = \frac{p_{s_1s_2}(x_1,x_2;\mathbf{k})}{p_{s_1}(x_1)} . \tag{2.135}$$

For statistically independent processes, (2.134) and (2.135) give $p_{s_1s_2}(x_1|x_2;\mathbf{k})=p_{s_1}(x_1)$ and $p_{s_2s_1}(x_2|x_1;\mathbf{k})=p_{s_2}(x_2)$, i.e. the given condition does not help to decrease uncertainty.

These concepts can likewise be extended to joint statistics of more than two signals or more than two samples from one signal. If e.g. K values from one or several continuous-amplitude signal(s) are combined into a vector $\mathbf{s} = [s_1,s_2, \ldots , s_K]^{\mathrm{T}}$, the joint probability density becomes also K-dimensional and is denoted as *vector PDF*[23]

$$p_{\mathbf{s}}(\mathbf{x}) = p_{s_1s_2\ldots s_K}(x_1,x_2,\ldots,x_K) , \tag{2.136}$$

where specifically for the case of statistical independency of the vector elements

---

[23] For simplicity, it is not explicitly expressed here that the samples of the vector can stem from various locations; in principle, individual shift parameters $\mathbf{k}$ would optionally need to be specified for the elements of the vector.

$$p_{\mathbf{s}}(\mathbf{x}) = p_{s_1}(x_1) \cdot p_{s_2}(x_2) \cdot \ldots \cdot p_{s_K}(x_K) . \tag{2.137}$$

The conditional PDF of a sample $s(\mathbf{n})$, provided that a conditioning vector $\mathbf{s}$ is given (which shall not include the sample itself), is defined as

$$p_{s|\mathbf{s}}(x \mid \mathbf{x}) = \frac{p_{s\mathbf{s}}(x, \mathbf{x})}{p_{\mathbf{s}}(\mathbf{x})}, \tag{2.138}$$

which for each given $\mathbf{x}$ is a one-dimensional PDF over variable x. In the context of joint analysis, also the definition of the joint expected value has to be extended to functions over several variables which are taken from distant positions in the signal, such as

$$\begin{aligned} \mathcal{E}\{f[s_1(\mathbf{n}), s_2(\mathbf{n}+\mathbf{k}),\ldots]\} &= \lim_{N\to\infty} \frac{1}{N} \sum_{\mathbf{n}} f[s_1(\mathbf{n}), s_2(\mathbf{n}+\mathbf{k}),\ldots] \\ &= \int_{-\infty}^{\infty}\int_{-\infty}^{\infty} p_{s_1 s_2 ..}(x_1, x_2, \ldots; \mathbf{k}) f(x_1, x_2, \ldots)\, \mathrm{d}x_2\, \mathrm{d}x_1 . \end{aligned} \tag{2.139}$$

The joint PDF $p_{s_1 s_2}(x_1,x_2;\mathbf{k})$ expresses the probability of a constellation where one random sample $s_1(\mathbf{n})$ has a value x_1, while the other sample $s_2(\mathbf{n}+\mathbf{k})$ has a value x_2. From this, linear statistical dependencies between the two samples are expressed by the *correlation function*[24]:

$$\begin{aligned} \varphi_{s_1 s_2}(\mathbf{k}) = \mathcal{E}\{s_1(\mathbf{n}) s_2(\mathbf{n}+\mathbf{k})\} &= \lim_{N\to\infty} \frac{1}{N} \sum_{\mathbf{n}} s_1(\mathbf{n}) s_2(\mathbf{n}+\mathbf{k}) \\ &= \int_{-\infty}^{\infty}\int_{-\infty}^{\infty} x_1 x_2 p_{s_1 s_2}(x_1, x_2; \mathbf{k})\, \mathrm{d}x_1\, \mathrm{d}x_2 . \end{aligned} \tag{2.140}$$

For the case $s_1=s_2=s$ (samples for correlation calculation taken from the same signal $s(\mathbf{n})$), (2.140) is an *autocorrelation function* (*ACF*), otherwise a *cross correlation function* (*CCF*). The *covariance function* is similarly computed by separating the mean values:

$$\mu_{s_1 s_2}(\mathbf{k}) = \mathcal{E}\left\{\left[s_1(\mathbf{n}) - m_{s_1}\right]\left[s_2(\mathbf{n}+\mathbf{k}) - m_{s_2}\right]\right\} = \varphi_{s_1 s_2}(\mathbf{k}) - m_{s_1} m_{s_2} . \tag{2.141}$$

The autocorrelation (2.140) and autocovariance (2.141) for $\mathbf{k}=\mathbf{0}$ give the power (2.120) and the variance (2.121), respectively. These are the maximum values of these functions. When normalized by their respective maxima, the resulting *standardized autocorrelation* and *autocovariance functions* have values between -1 and $+1$:

---

24 For quantized signals, the expected value can be computed from the PMF by applying (2.125) analogously, which is used here.

$$\alpha_{ss}(\mathbf{k}) = \frac{\varphi_{ss}(\mathbf{k})}{\varphi_{ss}(\mathbf{0})} = \frac{\varphi_{ss}(\mathbf{k})}{Q_s} \quad ; \quad \rho_{ss}(\mathbf{k}) = \frac{\mu_{ss}(\mathbf{k})}{\mu_{ss}(\mathbf{0})} = \frac{\mu_{ss}(\mathbf{k})}{\sigma_s^{\ 2}}. \tag{2.142}$$

A similar normalization by the cross power and cross variance (values for $\mathbf{k}=\mathbf{0}$) is applicable to the cross correlation and covariance functions,

$$\alpha_{s_1s_2}(\mathbf{k}) = \frac{\varphi_{s_1s_2}(\mathbf{k})}{\sqrt{Q_{s_1}Q_{s_2}}} \quad ; \quad \rho_{s_1s_2}(\mathbf{k}) = \frac{\mu_{s_1s_2}(\mathbf{k})}{\sigma_{s_1}\sigma_{s_2}}. \tag{2.143}$$

Correlation and covariance functions analyze *linear statistical dependencies*. If two signals are *uncorrelated*, $\varphi_{s1s2}(\mathbf{k})=m_{s1}m_{s2}$ and $\mu_{s1s2}(\mathbf{k})=0$ over all $\mathbf{k}$. Unless periodic components are present in a signal, the following conditions hold for the ACF and covariance if $|\mathbf{k}|$ grows large[25]:

$$\lim_{|\mathbf{k}|\to\infty} \varphi_{ss}(\mathbf{k}) = m_s^{\ 2} \quad ; \quad \lim_{|\mathbf{k}|\to\infty} \mu_{ss}(\mathbf{k}) = 0. \tag{2.144}$$

It should be observed that 'uncorrelated' signals or signal samples are not necessarily statistically independent. More general *nonlinear dependencies* cannot be identified by correlation functions. Cases of such nonlinear dependencies are real-valued signals that are similar by amplitude but have random sign, or complex-valued signals that are similar in amplitude but have random phase properties compared to each other.

Two correlated or uncorrelated, zero-mean stationary Gaussian processes $s_1(\mathbf{n})$ and $s_2(\mathbf{n})$ shall be given. After normalizing their amplitudes by the standard deviations, a sum process and a difference process are established as follows:

$$\Sigma(\mathbf{n},\mathbf{k}) = \frac{s_1(\mathbf{n})}{\sigma_{s_1}} + \frac{s_2(\mathbf{n}+\mathbf{k})}{\sigma_{s_2}}; \ \Delta(\mathbf{n},\mathbf{k}) = \frac{s_1(\mathbf{n})}{\sigma_{s_1}} - \frac{s_2(\mathbf{n}+\mathbf{k})}{\sigma_{s_2}}. \tag{2.145}$$

Sum and difference processes are zero-mean Gaussian as well, having the following variances:

$$\sigma_\Sigma^2(\tau) = \mathcal{E}\left\{\left[\frac{s_1(\mathbf{n})}{\sigma_{s_1}} + \frac{s_2(\mathbf{n}+\mathbf{k})}{\sigma_{s_2}}\right]^2\right\} = 2[1+\rho_{s_1s_2}(\mathbf{k})], \tag{2.146}$$

and similarly

$$\sigma_\Delta^2(\tau) = \mathcal{E}\left\{\left[\frac{s_1(\mathbf{n})}{\sigma_{s_1}} - \frac{s_2(\mathbf{n}+\mathbf{k})}{\sigma_{s_2}}\right]^2\right\} = 2[1-\rho_{s_1s_2}(\mathbf{k})], \tag{2.147}$$

where $\rho_{s1s2}(\tau)$ is the standardized cross covariance following the principle of (2.142). The correlation between the sum and difference processes is

---

[25] In case of multi-dimensional correlation functions It is sufficient when one of the values in the vector $\mathbf{k}$ grows large.

$$\mathcal{E}\{\Sigma(\mathbf{n},\mathbf{k})\Delta(\mathbf{n},\mathbf{k})\} = \mathcal{E}\left\{\left[\frac{s_1(\mathbf{n})}{\sigma_{s_1}} + \frac{s_2(\mathbf{n}+\mathbf{k})}{\sigma_{s_2}}\right]\left[\frac{s_1(\mathbf{n})}{\sigma_{s_1}} - \frac{s_2(\mathbf{n}+\mathbf{k})}{\sigma_{s_2}}\right]\right\}$$
$$= \frac{\mathcal{E}\{s_1^2(\mathbf{n})\}}{\sigma_{s_1}^2} - \frac{\mathcal{E}\{s_2^2(\mathbf{n}+\mathbf{k})\}}{\sigma_{s_2}^2} = 0. \tag{2.148}$$

Due to Gaussian property, the uncorrelated sum and difference processes are furthermore statistically independent. The joint PDF therefore is

$$p_{\Sigma\Delta}(y_1,y_2,\mathbf{k}) = \underbrace{\frac{1}{\sqrt{4\pi[1+\rho_{s_1s_2}(\mathbf{k})]}}\mathrm{e}^{-\frac{y_1^2}{4[1+\rho_{s_1s_2}(\mathbf{k})]}}}_{p_\Sigma(y_1)} \cdot \underbrace{\frac{1}{\sqrt{4\pi[1-\rho_{s_1s_2}(\mathbf{k})]}}\mathrm{e}^{-\frac{y_2^2}{4[1-\rho_{s_1s_2}(\mathbf{k})]}}}_{p_\Delta(y_2)}$$
$$= \frac{1}{4\pi\sqrt{1-\rho_{s_1s_2}{}^2(\mathbf{k})}}\mathrm{e}^{-\frac{y_1^2[1-\rho_{s_1s_2}(\mathbf{k})]+y_2^2[1+\rho_{s_1s_2}(\mathbf{k})]}{4[1-\rho_{s_1s_2}{}^2(\mathbf{k})]}} \tag{2.149}$$

Reverse mapping from y_1 and y_2 to the random variables x_1 and x_2 of the original processes $s_1(\mathbf{n})$ and $s_2(\mathbf{n})$ gives

$$y_1 = \frac{x_1}{\sigma_{s_1}} + \frac{x_2}{\sigma_{s_2}} = \frac{\sigma_{s_2}x_1 + \sigma_{s_1}x_2}{\sigma_{s_1}\sigma_{s_2}};\; y_2 = \frac{x_1}{\sigma_{s_1}} - \frac{x_2}{\sigma_{s_2}} = \frac{\sigma_{s_2}x_1 - \sigma_{s_1}x_2}{\sigma_{s_1}\sigma_{s_2}} \tag{2.150}$$

such that

$$p_{s_1s_2}(x_1,x_2;\mathbf{k}) = \frac{1}{2\pi\sigma_{s_1}\sigma_{s_2}\sqrt{1-\rho_{s_1s_2}{}^2(\mathbf{k})}}\mathrm{e}^{-\frac{\sigma_{s_2}^2x_1^2+\sigma_{s_1}^2x_2^2-2\sigma_{s_1}\sigma_{s_2}\rho_{s_1s_2}(\mathbf{k})x_1x_2}{2\sigma_{s_1}^2\sigma_{s_2}^2(1-\rho_{s_1s_2}{}^2(\mathbf{k}))}}. \tag{2.151}$$

Generalization to the case of non-zero mean processes further gives

$$p_{s_1s_2}(x_1,x_2;\mathbf{k}) = \frac{1}{2\pi\sigma_{s_1}\sigma_{s_2}\sqrt{1-\rho_{s_1s_2}{}^2(\mathbf{k})}}\mathrm{e}^{-\frac{\sigma_{s_2}^2(x_1-m_{s_1})^2+\sigma_{s_1}^2(x_2-m_{s_2})^2-2\sigma_{s_1}\sigma_{s_2}\rho_{s_1s_2}(\mathbf{k})(x_1-m_{s_1})(x_2-m_{s_2})}{2\sigma_{s_1}^2\sigma_{s_2}^2(1-\rho_{s_1s_2}{}^2(\mathbf{k}))}}. \tag{2.152}$$

A more compact expression of (2.152) is possible by the following matrix notation using a *covariance matrix* $\mathbf{C}_{s_1s_2}$,

$$p_{s_1s_2}(x_1,x_2;\mathbf{k}) = \frac{1}{\sqrt{(2\pi)^2\cdot|\mathbf{C}_{s_1s_2}(\mathbf{k})|}}\cdot e^{-\frac{1}{2}\xi^{\mathrm{T}}\mathbf{C}_{s_1s_2}(\mathbf{k})^{-1}\xi} \quad \text{with } \xi = \begin{bmatrix} x_1 - m_{s_1} \\ x_2 - m_{s_2} \end{bmatrix}$$
$$\text{and } \mathbf{C}_{s_1s_2}(\mathbf{k}) = \mathcal{E}\{\xi\cdot\xi^{\mathrm{T}}\} = \begin{bmatrix} \sigma_{s_1}^2 & \mu_{s_1s_2}(\mathbf{k}) \\ \mu_{s_1s_2}(\mathbf{k}) & \sigma_{s_2}^2 \end{bmatrix} \tag{2.153}$$
$$\Rightarrow \mathbf{C}_{s_1s_2}(\mathbf{k})^{-1} = \frac{1}{\underbrace{\sigma_{s_1}^2\sigma_{s_2}^2\left(1-\rho_{s_1s_2}{}^2(\mathbf{k})\right)}_{|\mathbf{C}_{s_1s_2}(\mathbf{k})|}}\begin{bmatrix} \sigma_{s_2}^2 & -\sigma_{s_1}\sigma_{s_2}\rho_{s_1s_2}(\mathbf{k}) \\ -\sigma_{s_1}\sigma_{s_2}\rho_{s_1s_2}(\mathbf{k}) & \sigma_{s_1}^2 \end{bmatrix}.$$

The transformation (2.145) into sum and difference processes can be interpreted as a coordinate transformation from a Cartesian (x_1,x_2) coordinate space into the rotated (y_1,y_2) coordinate space, where the axes y_1 and y_2 are still orthogonal. Equal values of the PDF, according to the exponent in (2.149), can be found on ellipses with principal axes along the y_1 and y_2 axes[26], scaled by $\sqrt{1+\rho_{s_1s_2}(\mathbf{k})}$ and $\sqrt{1-\rho_{s_1s_2}(\mathbf{k})}$, respectively.

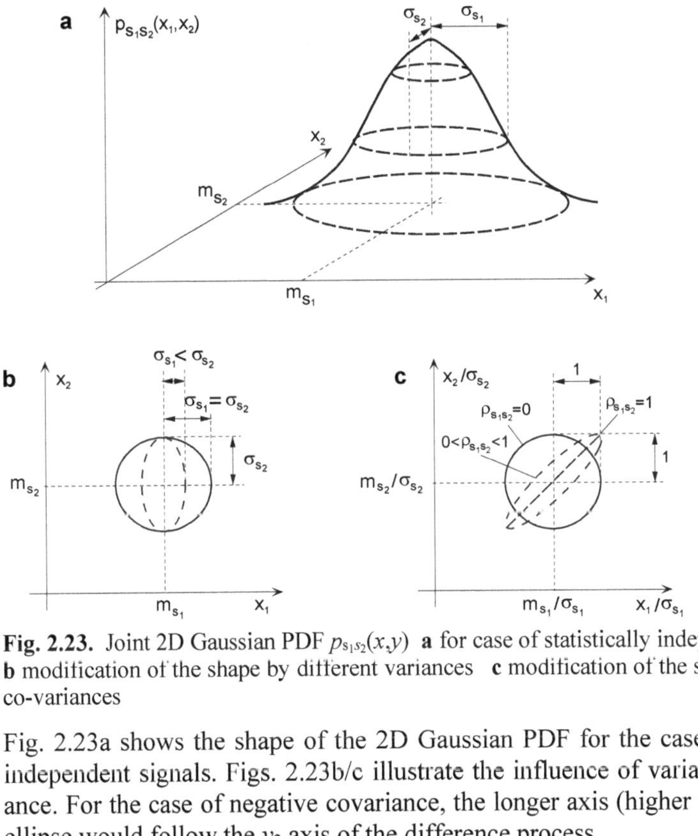

Fig. 2.23. Joint 2D Gaussian PDF $p_{s_1s_2}(x,y)$ **a** for case of statistically independent signals **b** modification of the shape by different variances **c** modification of the shape by different co-variances

Fig. 2.23a shows the shape of the 2D Gaussian PDF for the case of statistically independent signals. Figs. 2.23b/c illustrate the influence of variance and covariance. For the case of negative covariance, the longer axis (higher variance) of the ellipse would follow the y_2 axis of the difference process.

(2.153) straightforwardly extends to the general case where the correlation properties between measurements of K random values combined in a vector notation (2.136) are formulated in a covariance matrix

$$\mathbf{C}_{\mathbf{ss}} = \mathcal{E}\{\mathbf{ss}^{\mathrm{T}}\} - \mathbf{m}_{\mathbf{s}}\mathbf{m}_{\mathbf{s}}^{\mathrm{T}} = \left[\mathcal{E}\{s_i s_j\} - m_{s_i} m_{s_j}\right]. \tag{2.154}$$

using the vector of linear mean values

[26] Note that the axes y_1 and y_2 are defined after normalization of the x and y axes by the standard deviations of the respective processes. The ellipse's orientation therefore is 45 degrees relative to the normalized x and y axes.

$$\mathbf{m}_\mathrm{s} = \mathcal{E}\{\mathbf{s}\} = \left[\mathcal{E}\{s_i\}\right] \text{ with } 1 \le i \le K. \tag{2.155}$$

The joint PDF in this case can be expressed as *vector Gaussian PDF*

$$p_\mathrm{s}(\mathbf{x}) = \frac{1}{\sqrt{(2\pi)^K \cdot \left|\mathbf{C}_\mathrm{ss}\right|}} \cdot e^{-\frac{1}{2}[\mathbf{x}-\mathbf{m}_\mathrm{s}]^\mathrm{T} \mathbf{C}_\mathrm{ss}^{-1} [\mathbf{x}-\mathbf{m}_\mathrm{s}]}. \tag{2.156}$$

Again, to show its properties, it would be necessary to find an alternative representation by applying a linear transformation to the normalized combination of random samples (as in the case above by the sum and difference operations). After this, K statistically independent output processes are available, which in case of Gaussian processes means they are uncorrelated. In (2.279)-(2.282) it will be shown that this is possible by computing the set of eigenvectors of the covariance matrix, which establish a new orthogonal coordinate system, on which the amplitudes of random vectors **s** are projected. In case of a Gaussian PDF, equal values are then found on the hull of a K-dimensional hyper-ellipsoid, with principal axes having same orientations as the corresponding eigenvectors, and widths of the ellipsoid axes proportional to the square roots of the related eigenvalues.

$$\mathbf{C}_\mathrm{ss} = \begin{bmatrix} \mu_{ss}(0) & \mu_{ss}(1) & \mu_{ss}(2) & \cdots & \mu_{ss}(K-1) \\ \mu_{ss}(1) & \mu_{ss}(0) & \mu_{ss}(1) & \ddots & \vdots \\ \mu_{ss}(2) & \mu_{ss}(1) & \mu_{ss}(0) & \ddots & \\ \vdots & \ddots & \ddots & \ddots & \mu_{ss}(1) \\ \mu_{ss}(K-1) & \cdots & & \mu_{ss}(1) & \mu_{ss}(0) \end{bmatrix}$$
$$= \sigma_s^{\,2} \cdot \begin{bmatrix} 1 & \rho_{ss}(1) & \rho_{ss}(2) & \cdots & \rho_{ss}(K-1) \\ \rho_{ss}(1) & 1 & \rho_{ss}(1) & \ddots & \vdots \\ \rho_{ss}(2) & \rho_{ss}(1) & 1 & \ddots & \\ \vdots & \ddots & \ddots & \ddots & \rho_{ss}(1) \\ \rho_{ss}(K-1) & \cdots & & \rho_{ss}(1) & 1 \end{bmatrix}, \tag{2.157}$$

A special case applies, if the observations combined in the vector **s** are K samples from one single stationary Gaussian process, which are taken at equidistant time positions. In this case, the covariance matrix becomes an autocovariance matrix which has the following Toeplitz structure of (2.157)[27], where the mean vector is filled by a constant mean value,

$$\mathbf{m}_\mathrm{s} = m_s \cdot \mathbf{1} = \begin{bmatrix} m_s & m_s & \cdots & m_s \end{bmatrix}^\mathrm{T} \tag{2.158}$$

[27] In case of stationarity, variance and covariance values only depend on the distance, i.e. $\mathcal{E}\{s(0)s(1)\}=\mathcal{E}\{s(1)s(2)\}=\ldots$, which leads to this structure.

2.5.3 Spectral properties of random signals

The Fourier transform of the correlation function gives the *power density spectrum*[28]

$$\varphi_{ss}(\mathbf{k}) = \mathcal{E}\{s(\mathbf{n})s(\mathbf{n}+\mathbf{k})\} \quad \circ\!\!-\!\!\bullet \quad \Phi_{ss,\delta}(\mathbf{f}) = \mathcal{E}\left\{\left|S_\delta(\mathbf{f})\right|^2\right\} \tag{2.159}$$

The relationship between the power (quadratic mean) value and the power density spectrum is expressed by *Parseval's theorem*,

$$Q_s = \varphi_{ss}(\mathbf{0}) = \int_{-1/2}^{1/2} \cdots \int_{-1/2}^{1/2} \Phi_{ss,\delta}(\mathbf{f}) \mathrm{d}^{\kappa} \mathbf{f} \,. \tag{2.160}$$

If a random process is zero-mean, its autocorrelation and autocovariance functions are identical. Otherwise, the autocorrelation is increased by m_s^2. Likewise, for non-zero mean processes, a Dirac impulse is contained in the power density spectrum at $\mathbf{f}=\mathbf{0}$ (and at all periodic copies) with a weight m_s^2, corresponding to the power of the mean value (DC component). With presence of periodic components, Dirac impulses would be contained in the power density spectrum at the corresponding frequency locations.

Estimation of power spectra is often done via the DFT (2.89), i.e. a sampled frequency axis is used in computing the expected value in the right part of (2.159). For this purpose, blocks of M samples (ΠM_i samples for two- and multi-dimensional finite signals) are transformed into instantaneous DFT energy spectra $|S_d(k)|^2$. To minimize the effect of the inherent periodic continuation of the DFT, window functions can be used to let the signal decay towards zero at the boundaries of the analysis block. An alternative way to estimate power density spectra can be achieved via autoregressive (AR) modeling (see section 2.6.1). Both DFT-based spectral estimation, as well as AR modeling can be applied locally over a finite number of samples e.g. with the goal to adapt a compression algorithm by instantaneous (local) signal properties, or globally by computing expected values (power density spectrum or ACF) over a sufficiently large number of samples of a random process, which could be used to tune the general properties of a compression algorithm by thc typical statistics of thc givcn class of multimedia signals.

2.5.4 Markov chain models

The *state change behavior* of random processes with rather discrete appearance needs to be modeled for binary signals $b(n) \in \{0,1\}$ (e.g. two-tone images), bit streams, or for features on a more abstract level, e.g. segment transitions in space or time, where a segment relates to a semantic unit (spoken word, video scene,

[28] Note that the power spectrum of sampled random signals is periodic. Furthermore, in the formulation of the expected value over spectra from random signals in (2.159), a normalization by the time span of the Fourier transform must be performed to get an expression about the average spectral power density within a given frequency range.

region in an image). A simple model to define finite states of signals with memory is the *Markov chain*, in simplest case a 2-state (binary) model as shown in Fig. 2.24a[29]. As $b(n)$ has only two states S_0='0' and S_1='1', the model is fully defined by transition probabilities of its temporal sequence[30] $\Pr(S_0|S_1)$ (S_0 follows S_1), and $\Pr(S_1|S_0)$ (S_1 follows S_0). The remaining probabilities $\Pr(S_0|S_0)$ and $\Pr(S_1|S_1)$, which express occurrence of sequences with two equal values, can in case of the two-state chain be derived as

$$\Pr(S_i \mid S_i) = 1 - \Pr(S_j \mid S_i) . \tag{2.161}$$

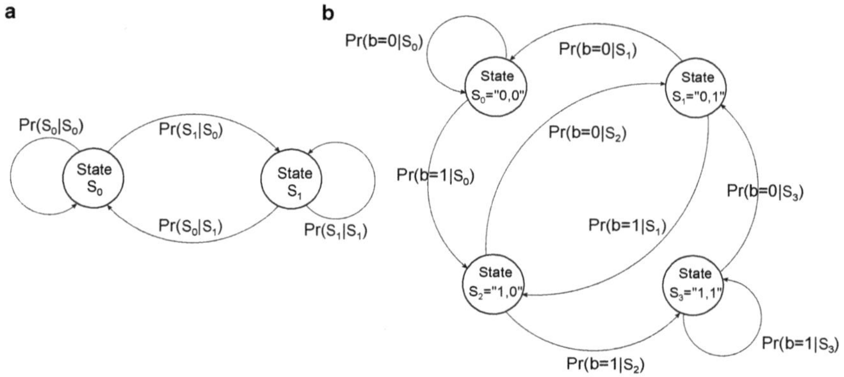

Fig. 2.24. Binary sequences modeled by Markov chain of **a** two states **b** four states giving dependency on two previous binary states

The 'Markov property' of the model process shall fulfill two conditions:

- The probability to be in a state is only dependent on the transition probabilities leading to this state, coupled with the respective probability of the state from which the transition is made;
- The model shall be stationary, the probability of states shall be independent of time or location of observation.

This can be formulated as follows for the two-state model, based on a state transition matrix **P**:

$$\begin{bmatrix} \Pr(S_0) \\ \Pr(S_1) \end{bmatrix} = \underbrace{\begin{bmatrix} \Pr(S_0 \mid S_0) & \Pr(S_0 \mid S_1) \\ \Pr(S_1 \mid S_0) & \Pr(S_1 \mid S_1) \end{bmatrix}}_{\mathbf{P}} \begin{bmatrix} \Pr(S_0) \\ \Pr(S_1) \end{bmatrix} . \tag{2.162}$$

From this, the global probabilities of two states can be determined as

---

[29] The problem will be discussed here mainly for binary sequences $b(n)$, but it can formally be extended to any sequences of discrete events $\mathfrak{s}(n) \in \{S_j; j=1,\ldots,J\}$. Extensions to continuous-amplitude signals $s(n)$ are also made by *Markov Random Fields* (cf. Sec. 6.6.2).

[30] For simple notation, $\Pr(S_i|S_j) \equiv \text{Prob}[b(n)=S_i|b(n-1)=S_j]$.

$$\Pr(S_i) = \frac{\Pr(S_i \mid S_j)}{\Pr(S_i \mid S_j) + \Pr(S_j \mid S_i)} = 1 - \Pr(S_j)\,; \quad [i,j] \in \{0,1\}. \tag{2.163}$$

Once a state is given, the probability of '0'- or '1'-sequences of remaining length l can be determined by concatenating the probabilities that the model rests in the state for another $l-1$ cycles and then changes,

$$\begin{aligned} \text{Prob}[b(n) = \{..\underbrace{00..0}_{\text{length } l}1..\}] &= \Pr(S_1 \mid S_0)\cdot\left[1-\Pr(S_1 \mid S_0)\right]^{l-1} \\ \text{Prob}[b(n) = \{..\underbrace{11..1}_{\text{length } l}0..\}] &= \Pr(S_0 \mid S_1)\cdot\left[1-\Pr(S_0 \mid S_1)\right]^{l-1}. \end{aligned} \tag{2.164}$$

These probabilities decay exponentially by increasing length l. Successive binary samples would be statistically independent for the case where $\Pr(S_0|S_1)=\Pr(S_0)$ and $\Pr(S_1|S_0)=\Pr(S_1)$. Markov chains with more than two states can be defined accordingly, where again the full set of transition probabilities between all states suffices to define the model. This general formulation of the Markov chain transitions for a model of J states can be written as an extension of (2.162)

$$\begin{bmatrix} \Pr(S_0) \\ \Pr(S_1) \\ \vdots \\ \Pr(S_{J-1}) \end{bmatrix} = \begin{bmatrix} \Pr(S_0 \mid S_0) & \Pr(S_0 \mid S_1) & \cdots & \Pr(S_0 \mid S_{J-1}) \\ \Pr(S_1 \mid S_0) & \Pr(S_1 \mid S_1) & & \vdots \\ \vdots & & \ddots & \\ \Pr(S_{J-1} \mid S_0) & & \cdots & \Pr(S_{J-1} \mid S_{J-1}) \end{bmatrix} \cdot \begin{bmatrix} \Pr(S_0) \\ \Pr(S_1) \\ \vdots \\ \Pr(S_{J-1}) \end{bmatrix}. \tag{2.165}$$

Due to the Markov property, the probability of transition into a state only depends on *one previous state*, such that for the binary 2-state model

$$\begin{aligned} &\text{Prob}\left[b(n) = S_i \mid b(n-1) = S_j, b(n-2) = S_k, b(n-3) = S_l, \ldots\right] \\ &\quad = \Pr\left(S_i \mid S_j, S_k, S_l, \ldots\right) = \Pr\left(S_i \mid S_j\right). \end{aligned} \tag{2.166}$$

If a binary sequence $b(n)$ shall be defined where the state of a sample depends on *two previous samples*, the transition probabilities have to be expressed as $\Pr(S_i|(S_j, S_k))$. It is then necessary to define a Markov chain with four states, relating to the four configurations of $[\,b(n-1)=S_j, b(n-2)=S_k\,]$. As however the current $[\,b(n), b(n-1)]$ will become $[\,b(n-1), b(n-2)]$ in the follow-up state, certain state transitions are impossible, which can be implemented by assigning zero as transition probability. A state diagram related to this given case is shown in Fig. 2.24b. This model straightforwardly extends to the case where a sample $b(\mathbf{n})$ is conditioned by a K-dimensional vector $\mathbf{b}$ of previous values, which can be established from a one- or multi-dimensional neighborhood context $\mathcal{C}(\mathbf{n})$ of K members, not including the current position. The model will then be based on 2^K different states, and is fully described by 2^{K+1} state transitions

$$\Pr\left(b(\mathbf{n}) = \beta \mid \mathbf{b}\right) \quad ; \quad \mathbf{b} = \left\{[b(\mathbf{i})] \mid \mathbf{i} \in \mathcal{C}(\mathbf{n}); \mathbf{i} \neq \mathbf{n}\right\} \quad ; \quad \beta \in \{0,1\}. \tag{2.167}$$

However, only 2^K state transitions are freely selectable, as in the example above, if the current sample would become member of **b** in the next step.

$$\Pr\left(b(\mathbf{n}) = 0 \mid \mathbf{b}\right) = 1 - \Pr\left(b(\mathbf{n}) = 1 \mid \mathbf{b}\right). \tag{2.168}$$

The follow-up state can also be constrained by zero-probability transitions, as in the example above, ruled by the fact that certain values are not independent.

Even though in the cases discussed so far the number of states is finite, the complete sequences $b(n)$ or $b(\mathbf{n})$ can be regarded as infinite. If a Markov chain model allows a transition with a non-zero probability from any state to any other state within a finite number of steps, it is said to be *irreducible*. This would not be the case for chains where one or several states S_i exist with *all* outgoing transition probabilities $\Pr(S_j|S_i) = 0$, but at least one incoming transition probability $\Pr(S_i|S_j) > 0$. This S_i will be a terminating state which once reached can never again be left. Such models can be useful in cases where finite sequences with expected termination shall be modeled.

2.5.5 Statistical foundation of information theory

Considerations about certainty and uncertainty of an information establish the foundations of *information theory*. In general, sending an information intends reducing the *uncertainty* about an event, a letter from a text, the state of a signal etc. Assume a discrete set $\boldsymbol{S}$ is given, characterizing J possible states S_j of an event. Each state shall have a probability $\Pr(S_j)$. The goal is to define a measure for the information $I(S_j)$ which is related to the knowledge that the event would be in state S_j. Consequently, the mean of information over all states will be $H(\boldsymbol{S}) = \Sigma \Pr(S_j) I(S_j) = \mathcal{E}\{I(S_j)\}$. Availability of *complete information* means that any uncertainty is removed about the state of the event. The function $H(\boldsymbol{S})$ shall retain its consistency if the amount of certainty is varied, e.g. if it stays uncertain whether the state is S_0 or S_1, while it is already certain that the state will not be $S_2 \ldots S_{J-1}$. Assuming that the $I(S_j)$ shall be related to the probabilities of the states, the following condition must be observed:

$$\begin{aligned} H(S) &= H\left\{\Pr(S_0), \Pr(S_1), \Pr(S_2), \ldots, \Pr(S_{J-1})\right\} \\ &\overset{!}{=} H\left\{\Pr(S_0) + \Pr(S_1), \Pr(S_2), \ldots, \Pr(S_{J-1})\right\} \\ &\quad + \left(\Pr(S_0) + \Pr(S_1)\right) H\left\{\frac{\Pr(S_0)}{\Pr(S_0) + \Pr(S_1)}, \frac{\Pr(S_1)}{\Pr(S_0) + \Pr(S_1)}\right\}. \end{aligned} \tag{2.169}$$

If (2.169) is valid, an arbitrary separation of the information into certainty and uncertainty about any of the states of the event is possible. It can be shown that the only function fulfilling (2.169) is the *self information* of a discrete event of state S_j from the set $\mathbf{S}$ defined as[31]

$$I(S_j) = \log_2 \frac{1}{\Pr(S_j)} = -\log_2 \Pr(S_j) . \tag{2.170}$$

The mean value of the self information over all possible states is denoted as the *entropy*

$$H(\mathbf{S}) = -\sum_{j=0}^{J-1} \Pr(S_j) \log_2 \Pr(S_j) . \tag{2.171}$$

If two distinct events defined over sets $\mathbf{S}_1$ and $\mathbf{S}_2$ occur, their *joint information* and *joint entropy* can be defined via the joint probability

$$\begin{aligned} H(\mathbf{S}_1,\mathbf{S}_2) &= \sum_{j_1=0}^{J_1-1} \sum_{j_2=0}^{J_2-1} \Pr(S_{j_1},S_{j_2}) I(S_{j_1},S_{j_2}) \\ &\text{with } I(S_{j_1},S_{j_2}) = -\log_2 \Pr(S_{j_1},S_{j_2}). \end{aligned} \tag{2.172}$$

The joint entropy is lower and upper bounded by

$$\max\{H(\mathbf{S}_1), H(\mathbf{S}_2)\} \le H(\mathbf{S}_1,\mathbf{S}_2) \le H(\mathbf{S}_1) + H(\mathbf{S}_2) , \tag{2.173}$$

where the upper bound is valid in the case of statistical independence of the two events, and the lower bound applies if they always come with identical joint occurrence of states. The concept of conditional probability defines the probability of an event in $\mathbf{S}_2$ to be in state S_{j2}, provided that the state S_{j1} of the other event in $\mathbf{S}_1$ is given. This allows reflecting the statistical dependency of state S_{j2} from S_{j1} in terms of the remaining uncertainty in the *conditional information*

$$I(S_{j_2} | S_{j_1}) = -\log_2 \Pr(S_{j_2} | S_{j_1}) = -\log_2 \frac{\Pr(S_{j_1},S_{j_2})}{\Pr(S_{j_1})} . \tag{2.174}$$

For statistically independent events, due to (2.134) and (2.135) $\Pr(S_{j2}|S_{j1})=\Pr(S_{j2})$, which makes the conditional information identical to the self information $I(S_{j2})$. The difference between the self information and the conditional information is the *mutual information.* It signifies the amount of information in state S_{j2}, which was already provided by the state S_{j1}. Likewise, this can be interpreted as the amount of information which could possibly be saved (e.g. needs not to be encoded or transmitted) when the statistical dependency is exploited:

---

[31] In (2.170), any base of the logarithm can be selected, where the unit of information is 'bit' in case of base 2 (counting the amount of binary digits). A probability $P(S_j)=0$ would lead to an infinite self information; in the subsequent definitions of entropy this is not a problem, as $\lim_{x\to 0}\left(x \cdot \log\frac{1}{x}\right) = 0$.

$$I(S_{j_2};S_{j_1}) = I(S_{j_2}) - I(S_{j_2}|S_{j_1}). \tag{2.175}$$

Combining (2.170) and (2.174) into (2.175), further considering (2.135) gives

$$I(S_{j_2};S_{j_1}) = \log_2 \frac{\Pr(S_{j_2}|S_{j_1})}{\Pr(S_{j_2})} = \log_2 \frac{\Pr(S_{j_1},S_{j_2})}{\Pr(S_{j_1})\cdot\Pr(S_{j_2})} = \log_2 \frac{\Pr(S_{j_1}|S_{j_2})}{\Pr(S_{j_1})} = I(S_{j_1};S_{j_2}). \tag{2.176}$$

This shows the symmetry property of mutual information. If two events are statistically independent, the mutual information becomes zero in all states. This is an ultimate condition for statistical independency, which even allows to test for presence or absence of nonlinear dependencies, being a more rigid criterion than the cross correlation. The mean of conditional information over all state combinations S_{j_1} and S_{j_2} is the *conditional entropy*[32],

$$\begin{aligned} H(S_2|S_1) &= -\sum_{j_1=0}^{J_1-1}\sum_{j_2=0}^{J_2-1} \Pr(S_{j_1})\Pr(S_{j_2}|S_{j_1})\log_2 \Pr(S_{j_2}|S_{j_1}) \\ &= -\sum_{j_1=0}^{J_1-1}\sum_{j_2=0}^{J_2-1} \Pr(S_{j_1},S_{j_2})\log_2 \Pr(S_{j_2}|S_{j_1}). \end{aligned} \tag{2.177}$$

The *mean of mutual information* can also be expressed from (2.175) and (2.176) as follows[33]:

$$\begin{aligned} H(S_2;S_1) = H(S_1;S_2) &= \sum_{j_1=0}^{J_1-1}\sum_{j_2=0}^{J_2-1} \Pr(S_{j_1},S_{j_2})\log_2 \frac{\Pr(S_{j_1},S_{j_2})}{\Pr(S_{j_1})\Pr(S_{j_2})} \\ &= H(S_2) - H(S_2|S_1) = H(S_1) - H(S_1|S_2). \end{aligned} \tag{2.178}$$

The general relationships between entropy, conditional entropy and mean of mutual information are shown in Fig. 2.25a by a diagram of information flow. In principle, the whole schema is invertible, i.e. the states S_{j_1} and S_{j_2} can change their roles, while the mutual information will not change. In addition, Fig. 2.25b shows an interpretation borrowed from set algebra, where the circles indicate the total amount of information from the events defined by $\boldsymbol{S}_1$ and $\boldsymbol{S}_2$. The intersection is the mean of mutual information which is shared, such that at least some statistical dependency between the two events must be in effect.

Entropy, conditional entropy and mutual information can be used to express the problem of encoding information by *discrete alphabets*. Typical examples of

---

[32] If $H(\boldsymbol{S}_2|\boldsymbol{S}_1) < H(\boldsymbol{S}_2)$ and the state of $\boldsymbol{S}_1$ is known at the decoder, it is usually possible to reduce the data rate by utilizing this prior information. This is the basis of predictive coding (see Sec. 5.2) and context-dependent entropy coding (see Sec. 4.4.5).

[33] (2.178) also is often by itself denoted as *mutual information.* Consequently, it should be called mutual entropy, but this is hardly established.

discrete alphabets are finite sets of alphanumeric letters, or sets of reconstruction values in case of signal quantization. Let a source alphabet $\mathcal{A}$ be defined, which contains all distinct letters that a discrete source could ever produce. Further, a reconstruction alphabet $\mathcal{B}$ is given. Both alphabets need not necessarily be identical (however only if $\mathcal{A}$ is identical with $\mathcal{B}$ or a subset thereof, it is possible at all to perform lossless coding and decoding). The mapping of values from $\mathcal{A}$ into values from $\mathcal{B}$ is defined by a code $\mathcal{C}$. Then,

$$H(\mathcal{A};\mathcal{B})\big|_e = H(\mathcal{A}) - H(\mathcal{A}\,|\,\mathcal{B})\big|_e \tag{2.179}$$

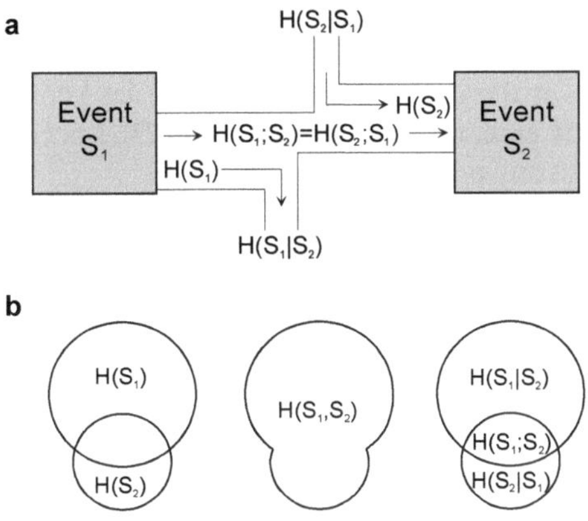

Fig. 2.25. Graphical interpretation of information-theoretic statistical parameters: **a** in terms of information flow **b** in terms of set algebra

As the mutual information cannot become negative,

$$0 \le H(\mathcal{A}\,|\,\mathcal{B})\big|_e \le H(\mathcal{A}), \tag{2.180}$$

where for $H(\mathcal{A}\,|\,\mathcal{B})\big|_e = 0$ it is possible to perform lossless decoding, while for $H(\mathcal{A}\,|\,\mathcal{B})\big|_e = H(\mathcal{A})$ *nothing is known* after decoding about the state of the source. For any values of $H(\mathcal{A}\,|\,\mathcal{B})\big|_e$ between these extremes, lossy decoding will be in effect, such that *distortion* occurs. Let $\mathcal{C}_D$ define the set of all codes, which are capable to perform the mapping from $\mathcal{A}$ onto $\mathcal{B}$ by effecting a given value of distortion D[34]. The best possible code among all $\mathcal{C}_D$ is the one which needs lowest rate for its representation, which is the code requiring least mutual information when the mapping from $\mathcal{A}$ into $\mathcal{B}$ is performed with that distortion. The lowest bound for the rate by a given distortion D will then be

[34] At this point, D shall be introduced in a quite abstract way, more concrete definitions will be used in Ch. 4.

$$R(D) = \min_{e \in e_D} H(\mathcal{A};\mathcal{B})\big|_e \,. \tag{2.181}$$

This definition, however, does not indicate a direct method to design codes, only allows judging their performance. $R(D)$ is the *rate distortion function* (RDF), which defines an interrelationship between rate R and distortion D. In this abstract form, the definition is valid for arbitrary source alphabets and arbitrary definitions of distortion. From (2.179)-(2.181) and the related reasoning, the following conclusions can be drawn:

- Lossless coding of a source generating letters from a discrete source alphabet $\mathcal{A}$, can only be achieved by investing a minimum rate $R_{min} = H(\mathcal{A})$.
- The minimum rate is zero, where at the decoder *nothing* would be known about the state of the source. In this case, a maximum distortion D_{max} occurs which should never be superseded at any positive rate.
- If the source has continuous amplitude, the number of letters in the source alphabet $\mathcal{A}$ would grow towards infinity. Hence, it is not possible to achieve zero distortion (lossless encoding) using a finite rate. If the reconstruction alphabet $\mathcal{B}$ is sufficiently large, the distortion may however become negligibly small.

Qualitative graphs of rate distortion functions for both cases of continuous-amplitude and discrete-amplitude sources are shown in Fig. 2.26[35]. Typically, the rate distortion function is convex and continuously decreasing until the maximum distortion (for rate zero) is reached.

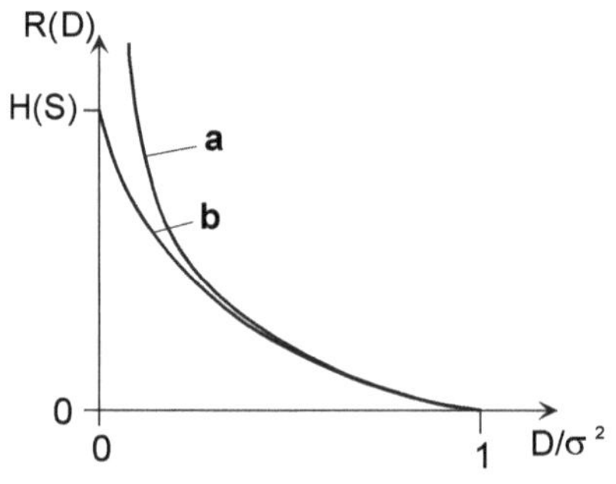

Fig. 2.26. Examples of $R(D)$ for sampled continuous (**a**) and discrete (**b**) sources

Example: Entropy of a Markov process. A Markov chain of J states is defined by the transition probability definitions in (2.165). Due to the property that the probabilities of next-state transitions are independent of history, the entropy of each

[35] Here, the distortion is expressed in terms of squared error, i.e. Euclidean distance, and normalized by $D_{max} = \sigma_s^2$, which occurs in case of zero reconstruction.

state can first be defined independently by the respective probabilities of the next-state transitions

$$H(S_j) = -\sum_{i=0}^{J-1} \Pr(S_i \mid S_j) \log_2 \Pr(S_i \mid S_j) \quad ; \quad j = 0,\ldots,J-1 . \qquad (2.182)$$

The overall entropy of the Markov process can then be computed as the probability-weighted average over all states,

$$H(S) = \sum_{j=0}^{J-1} \Pr(S_j) H(S_j) = -\sum_{i=1}^{J}\sum_{j=1}^{J} \Pr\left(S_i, S_j\right) \log_2 \Pr\left(S_i \mid S_j\right) . \qquad (2.183)$$

Differential entropy and entropy of Gaussian processes. The concept of entropy can be extended in relation to the PDF of continuous-amplitude sources. However, in principle the number of bits necessary to represent a continuous source (and therefore its entropy) would be infinite. With uniform quantization using intervals $\left[(i-1/2)\Delta ; (i+1/2)\Delta\right]$, the PMF $p_s(i)$ is according to (2.122)

$$p_s(i) = \int_{(i-1/2)\Delta}^{(i+1/2)\Delta} p_s(x)\,\mathrm{d}x \approx \Delta p_s(i\Delta) . \qquad (2.184)$$

The entropy of the discrete distribution for $\Delta \rightarrow 0$ becomes

$$\begin{aligned} H_s &= \lim_{\Delta\rightarrow 0}\left(-\sum_{i=-\infty}^{\infty} p_s(i) \log p_s(i) \right) = \lim_{\Delta\rightarrow 0}\left(-\sum_{i=-\infty}^{\infty} p_s(i) \log\left(p_s(i\Delta)\Delta\right)\right) \\ &= \lim_{\Delta\rightarrow 0}\left(-\sum_{i=-\infty}^{\infty} p_s(i\Delta)\Delta \log\left(p_s(i\Delta)\right) - \sum_{i=-\infty}^{\infty} p_s(i) \log\left(\Delta\right)\right) \qquad (2.185) \\ &= -\int_{-\infty}^{\infty} p_s(x) \log p_s(x)\,\mathrm{d}x - \lim_{\Delta\rightarrow 0} \log(\Delta). \end{aligned}$$

Whereas the term $-\log(\Delta)$ converges towards infinity for $\Delta \rightarrow 0$ and is independent of the PDF, the left term is denoted as the *differential entropy*[36],

$$H_s = -\int_{-\infty}^{\infty} p_s(x) \log p_s(x)\,\mathrm{d}x . \qquad (2.186)$$

(2.186) cannot be used as an information-theoretic criterion about the quantitative amount of information contained in a source, and the value could even become negative. It is however useful for comparing properties of PDFs or in optimization. All other variants such as joint, vector and conditional entropies can be defined similarly.

Specifically taking the natural logarithm (base e), the differential entropy of a zero-mean Gaussian process is (using the unit '*nat*' which refers to a symbol count based on Euler's number instead of the binary number system)

---

[36] Note that H_s as defined in (2.186) cannot quantitatively be compared against (2.171).

$$
\begin{aligned}
H_s &= -\int_{-\infty}^{\infty} p_s(x)\left[-\frac{x^2}{2\sigma_s^2} - \log\sqrt{2\pi\sigma_s^2}\right]\mathrm{d}x \\
&= \mathcal{E}\left\{\frac{x^2}{2\sigma_s^2}\right\} + \frac{1}{2}\log\left(2\pi\sigma_s^2\right) = \frac{1}{2}\log\left(2\pi\mathrm{e}\,\sigma_s^2\right) \quad \text{[nats]}
\end{aligned}
\quad . \tag{2.187}
$$

Similarly, the extension to a K-dimensional vector Gaussian process gives

$$
H_{\mathbf{s}} = \frac{1}{2}\log\left(\left(2\pi\mathrm{e}\right)^K \left|\mathbf{C}_{\mathbf{ss}}\right|\right) \leq KH_s \quad \text{[nats]}, \tag{2.188}
$$

which provides a quantitative expectation that the entropy for the correlated source will be lower than the K-fold entropy of single samples.

2.6 Linear prediction

2.6.1 Autoregressive models

Algorithms of multimedia signal processing often require a model about the statistical properties of sources for analytic optimization (cf. Sec. 3.4). If statistical assumptions are made which go beyond sample statistics, modeling of statistical dependencies between samples is required. The autocovariance is usually sufficient to optimize *linear systems* for a given purpose, as it characterizes *linear statistical dependencies* between samples.

A random signal of an *autoregressive (AR) process* (Fig. 2.27) is generated by a recursive filter with z transfer function $B(\mathbf{z})=1/(1-H(\mathbf{z}))$ from a stationary white Gaussian noise process $v(\mathbf{n})$ as input. The process $s(\mathbf{n})$ at the filter output possesses spectral distribution properties which are only determined by the amplitude transfer function of the filter. The PDF of this stationary process is also Gaussian.

The property of stationarity does not usually apply to multimedia sources. Moreover, a high degree of variation is observed in the local properties of image, speech and audio signals, such that a local adaptation of model parameters is typically necessary. Even then, the AR model helps to simplify problems of optimization due to its simple analytic properties. If the AR model generates a stationary Gaussian process, it is indeed fully described by a covariance matrix. In this case, an AR process would perfectly follow the vector Gaussian PDF (2.156).

The autoregressive model of first order [AR(1)] is often used to model the global statistics of image signals. For the 1D case, a Gaussian zero-mean white-noise process $v(n)$ (*innovation*) of variance σ_v^2 is fed into a recursive filter with z transfer function

$$B(z) = \frac{1}{1-\rho z^{-1}} . \tag{2.189}$$

The computation of the output is

$$s(n) = \rho s(n-1) + v(n) . \tag{2.190}$$

The AR(1) process has an autocovariance function[37]

$$\mu_{ss}(k) = \sigma_s^{\ 2} \rho^{|k|} \quad ; \quad \sigma_s^{\ 2} = \frac{\sigma_v^{\ 2}}{1-\rho^2} , \tag{2.191}$$

and a power density spectrum

$$\phi_{ss}(f) = \sigma_s^{\ 2} \sum_{k=-\infty}^{\infty} \rho^{|k|} e^{-j2\pi fk} = \frac{\sigma_s^{\ 2}(1-\rho^2)}{1-2\rho\cos(2\pi f)+\rho^2} . \tag{2.192}$$

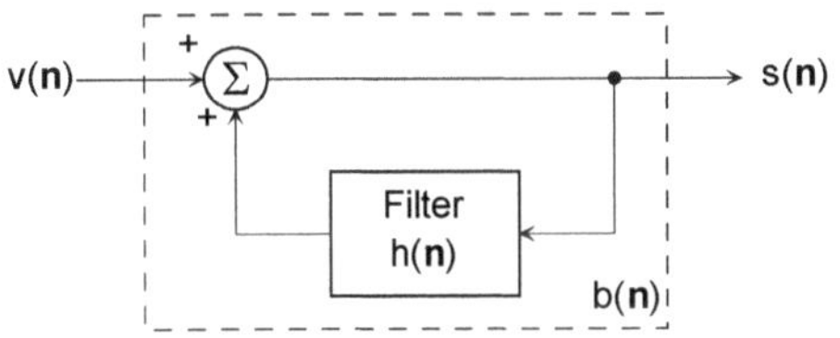

Fig. 2.27. System for generating samples from an autoregressive process

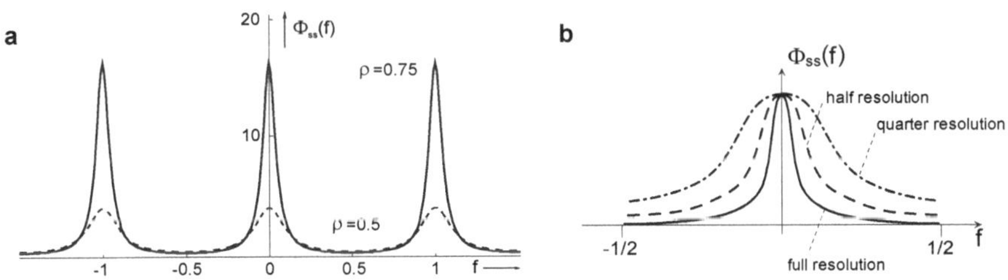

Fig. 2.28. a Power density spectra of AR(1) processes with $\sigma_s^2=1$, for two different values of $\rho(1)$ **b** Effect of decreasing sampling resolution by factors U=2 and U=4.

With zero-mean input, also the autoregressive output of the filter has zero-mean property[38]. Obviously, the AR(1) model is fully characterized by the filter parameter ρ, which is identical to the standardized autocovariance coefficient $\rho_{ss}(1)$, and one of the variances, σ_v^2 or σ_s^2. Typical values of $\rho(1)$ for natural images are between .85 and .99, which exhibit extreme concentration of spectral power

---

[37] For a proof on (2.191) and (2.192), see Problem 2.9.

[38] Optionally, a mean value can be added either at the input or at the output, with $m_s = m_v \cdot B(\mathbf{z})\big|_{\mathbf{z}=1}$.

around the zero frequency. Examples with lower values of $\rho(1)$=.75 and .5 are shown in Fig. 2.28a. It should be observed that the measurement of the correlation parameter ρ which is used to adapt an AR(1) process also depends on the sample density (resolution of the signal). If possible alias effects are ignored, downsampling the ACF by a factor of U leads to a modification into $\rho_U(1)=\rho(U)=\rho(1)^U$. The effect of increasing high-frequency components in the power density spectrum is illustrated in Fig. 2.28b.

For simple extensions of the AR(1) model into two and multiple dimensions, expression by separate standardized autocovariance coefficients $\rho_1 \equiv \rho_1(1)$ and $\rho_2 \equiv \rho_2(1)$ for horizontal and vertical directions can be used. Properties of three 2D methods are illustrated in Fig. 2.29, showing lines of constant autocovariance in the (m_1,m_2) plane (only positive values of ρ are assumed here).

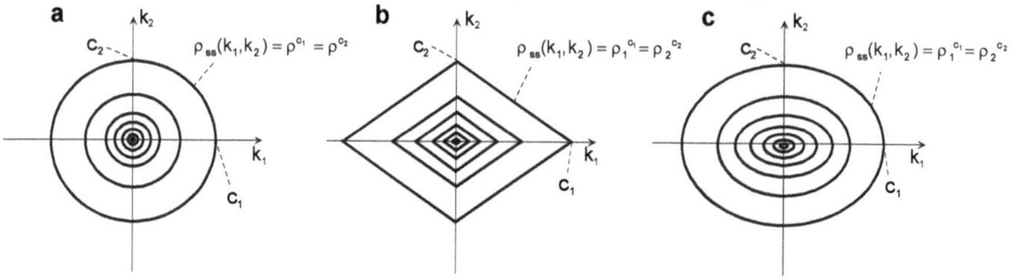

Fig. 2.29. Lines of constant autocovariance in 2D AR(1) models.
a isotropic **b** separable **c** elliptic

The *isotropic* model has an autocovariance function

$$\varphi_{ss}(m_1,m_2) = \sigma_s^{\,2} \rho^{\sqrt{m_1^{\,2}+m_2^{\,2}}} , \tag{2.193}$$

expressing circular-symmetric values independent of the direction, $\rho_1 = \rho_2$ is inherently assumed. Constant values appear on circles of radius $|\mathbf{m}| = \sqrt{m_1^{\,2}+m_2^{\,2}}$ (see Fig. 2.29a). The two-dimensional power density spectrum of the isotropic model is then also circular-symmetric[39],

$$\phi_{ss}(f_1,f_2) = \frac{\sigma_s^{\,2}(1-\rho^2)}{1-2\rho\cos\left(2\pi\sqrt{f_1^{\,2}+f_2^{\,2}}\right)+\rho^2} . \tag{2.194}$$

For the remaining models, autocovariance values are defined differently for the horizontal and vertical directions. In natural images, it can be observed that autocovariance statistics sometimes differ per orientation. It is often found that the

---

[39] Note that this is not fully precise due to the fact that the nearest periodic copies of the spectrum are only present at some angular orientations. The best coincidence would be found for the case of hexagonal sampling, or for $\rho \rightarrow 1$.

covariance along the vertical axis is lower than along the horizontal axis. The *separable model* with autocovariance function

$$\varphi_{ss}(m_1,m_2)=\sigma_s^{\ 2}\rho_1^{|m_1|}\rho_2^{|m_2|} \quad ; \quad \sigma_s^{\ 2}=\frac{\sigma_v^{\ 2}}{(1-\rho_1^{\ 2})(1-\rho_2^{\ 2})}\,, \tag{2.195}$$

shows straight lines of constant autocovariance[40]. These lines intersect with axes m_1 and m_2 at positions m_1' and m_2' where $\rho_1^{|m_1'|}=\rho_2^{|m_2'|}$ (see Fig. 2.29b). The generation of the discrete 2D signal can be implemented by a separable recursive filter, whose output is expressed by the equation

$$s(n_1,n_2)=\rho_1 s(n_1-1,n_2)+\rho_2 s(n_1,n_2-1)-\rho_1\rho_2 s(n_1-1,n_2-1)+v(n_1,n_2)\,. \tag{2.196}$$

The related power density spectrum is

$$\phi_{ss}(f_1,f_2)=\sigma_s^{\ 2}\frac{1-\rho_1^{\ 2}}{1-2\rho_1\left(\cos 2\pi f_1\right)+\rho_1^2}\cdot\frac{1-\rho_2^{\ 2}}{1-2\rho_2\left(\cos 2\pi f_2\right)+\rho_2^2}\,. \tag{2.197}$$

The *elliptic model* has an autocovariance function (for $0<\rho_i<1$)

$$\varphi_{ss}(m_1,m_2)=\sigma_s^{\ 2}\mathrm{e}^{-\sqrt{(\beta_1 m_1)^2+(\beta_2 m_2)^2}} \quad \text{with } \beta_i=-\ln\rho_i\,. \tag{2.198}$$

It shows constant autocovariance values on elliptic shapes due to the ellipse equation in the exponent (Fig. 2.29c). This model can also be interpreted as an extension of the isotropic model, which would be a special case $\beta_1=\beta_2$. Intersections of constant autocovariance graphs with the coordinate axes are identical to the case of the separable model. All models introduced so far can be interpreted as special cases of a generalized 2D AR(1) model with autocovariance[41]

$$\varphi_{ss}(m_1,m_2)=\sigma_s^{\ 2}\cdot e^{-\left[\left(\beta_1\cdot|m_1|\right)^\gamma+\left(\beta_2\cdot|m_2|\right)^\gamma\right]^{\frac{1}{\gamma}}}\,. \tag{2.199}$$

For the isotropic and for the elliptic model, γ=2; for the separable model γ=1. When $\gamma>1$, lines of constant autocovariance are convex over all the four quadrants of m_i coordinates. The intersections with the axes remain identical as in the case of separable and elliptic models, irrespective of γ. The relation of factors β_i with the horizontal/vertical autocovariance coefficients is equal to (2.198).

Autoregressive models of higher order (more than one autocovariance value per dimension) are frequently applied in speech analysis and for texture analysis of images. As those signals typically do not have the property of stationarity, the

---

[40] If the two exponential expressions in (2.195) are modified for a common basis, a line equation over absolute values appears in the exponent, see (2.199) with γ=1.

[41] This model is, like the elliptic model, only applicable for positive ρ_1 and ρ_2 values. In both cases, a precise analytic expression of the power density spectrum is difficult to define due to the directional dependent alias effect.

autocovariance function needs to be estimated over segments (finite time windows or 2D regions) of samples. A generic synthesis equation expressing AR filtering over a finite causal neighborhood $\mathcal{N}_{\mathbf{p}}^{+}$ is[42]

$$s(\mathbf{n}) = \sum_{\mathbf{p}\in\mathcal{N}_{\mathbf{p}}^{+}} a(\mathbf{p})s(\mathbf{n}-\mathbf{p}) + v(\mathbf{n}). \tag{2.200}$$

With white noise input, the output process has a power density spectrum

$$\phi_{ss}(\mathbf{f}) = \frac{\sigma_v^{\,2}}{\left|1 - \sum_{\mathbf{p}\in\mathcal{N}_{\mathbf{p}}^{+}} a(\mathbf{p})\mathrm{e}^{-j2\pi\mathbf{f}^{\mathrm{T}}\mathbf{p}}\right|^2}. \tag{2.201}$$

Next, a causal model shall be optimized under the assumption that the white-noise signal $v(n)$ which is to be fed into the AR synthesis filter shall have lowest possible variance:

$$\begin{aligned}\sigma_v^{\,2} &= \mathcal{E}\left\{v^2(\mathbf{n})\right\} = \mathcal{E}\left\{\left[s(\mathbf{n}) - \sum_{\mathbf{p}\in\mathcal{N}_{\mathbf{p}}^{+}} a(\mathbf{p})s(\mathbf{v}-\mathbf{p})\right]^2\right\} \\ &= \mathcal{E}\left\{s^2(\mathbf{n})\right\} - 2\mathcal{E}\left\{\left[s(\mathbf{n})\sum_{\mathbf{p}\in\mathcal{N}_{\mathbf{p}}^{+}} a(\mathbf{p})s(\mathbf{n}-\mathbf{p})\right]\right\} + \mathcal{E}\left\{\left[\sum_{\mathbf{p}\in\mathcal{N}_{\mathbf{p}}^{+}} a(\mathbf{p})s(\mathbf{n}-\mathbf{p})\right]^2\right\} \overset{!}{=} \min.\end{aligned} \tag{2.202}$$

The minimization is achieved by computing partial derivatives over each filter coefficient:

$$\frac{\partial\, \sigma_v^{\,2}}{\partial\, a(\mathbf{k})} \overset{!}{=} 0 \Rightarrow \mathcal{E}\left\{s(\mathbf{n})s(\mathbf{n}-\mathbf{k})\right\} = \sum_{\mathbf{p}\in\mathcal{N}_{\mathbf{p}}^{+}} a(\mathbf{p})\mathcal{E}\left\{s(\mathbf{n}-\mathbf{p})s(\mathbf{n}-\mathbf{k})\right\}. \tag{2.203}$$

This the linear *Wiener-Hopf equation system*, where the optimum filter coefficients fulfill the condition

$$\mu_{ss}(\mathbf{k}) = \sum_{\mathbf{p}\in\mathcal{N}_{\mathbf{p}}^{+}} a(\mathbf{p})\mu_{ss}(\mathbf{k}-\mathbf{p}), \tag{2.204}$$

or specifically for the 1D case with order P:

---

[42] Observe that the definitions (2.201) and (2.202) do not implicitly postulate the *causality* of the AR synthesis filters. In fact, all following deductions can likewise be made for non-causal filter sets without any limitation. Non-causal recursive filtering is indeed practically applicable for signals of finite extension, e.g. image signals. Only the current position **n** must be excluded, which means that $a(\mathbf{0})=0$. For more detail on non-causal AR modeling of images, see e.g. [JAIN 1989].

$$\mu_{ss}(k)=\sum_{p=1}^{P}a(p)\mu_{ss}(k-p) \quad ; \quad 1\le k\le P. \tag{2.205}$$

Due to the symmetry of the autocovariance, $\mu_{ss}(k-p)=\mu_{ss}(p-k)$, the problem can be simplified in a more regular matrix structure, and the Wiener-Hopf equation can be written as follows, where $\mathbf{C}_{ss}$ is the autocovariance matrix (2.157):

$$\underbrace{\begin{bmatrix}\mu_{ss}(1)\\ \mu_{ss}(2)\\ \vdots\\ \vdots\\ \mu_{ss}(P)\end{bmatrix}}_{\mathbf{c}_{ss}}=\underbrace{\begin{bmatrix}\mu_{ss}(0) & \mu_{ss}(1) & \cdots & \cdots & \mu_{ss}(P-1)\\ \mu_{ss}(1) & \mu_{ss}(0) & \mu_{ss}(1) & \cdots & \mu_{ss}(P-2)\\ \vdots & \mu_{ss}(1) & \mu_{ss}(0) & \ddots & \vdots\\ \vdots & \vdots & \ddots & \ddots & \mu_{ss}(1)\\ \mu_{ss}(P-1) & \mu_{ss}(P-2) & \cdots & \mu_{ss}(1) & \mu_{ss}(0)\end{bmatrix}}_{\mathbf{C}_{ss}}\underbrace{\begin{bmatrix}a(1)\\ a(2)\\ \vdots\\ \vdots\\ a(P)\end{bmatrix}}_{\mathbf{a}}. \tag{2.206}$$

The solution is obtained when the vector $\mathbf{c}_{ss}$ and the matrix $\mathbf{C}_{ss}$ are filled by the autocovariance estimates (computed either locally from a given signal or globally from an ensemble)[43], and then the matrix is inverted:

$$\mathbf{a}=\hat{\mathbf{C}}_{ss}^{-1}\hat{\mathbf{c}}_{ss}. \tag{2.207}$$

In the 1D case, $\mathbf{C}_{ss}$ has a *Toeplitz structure*, which means that it is diagonally symmetric (identical to its transpose), and values on each one of the diagonals (main diagonal and its off-diagonals) are identical. Even though the matrix is not sparse, it is highly regular, such that the problem of matrix inversion is simplified. This is mainly due to the fact that the lower and upper triangular matrices, as well as many sub-matrices are identical, such that sub-matrix inversions need to be computed only once. Furthermore, the matrix has full rank and is positive-definite (except for degenerate cases). Therefore, Cholesky decomposition or Levinson-Durbin recursion can be used for computationally more efficient solutions. Furthermore, the latter can be used to map the predictor into a ladder structure of subsequent first-order filters, where PARCOR (partial correlation) coefficients are determined step by step, guaranteeing stability of the synthesis filter under the simple rule that the absolute value of each coefficient in the structure must be smaller than one (see [RABINER, SCHAFER 1978] as reference for more detail on these approaches).

---

43 When computing an autocovariance estimate for a finite segment, boundary conditions need to be observed. When using samples from the previous segment (as far as they are used to predict the samples from the current segment), this may lead to better prediction results, but also may cause instability of the resulting synthesis filter (see [RABINER, SCHAFER 1978], [MARAGOS, SCHAFER 1984] and the notes about positive-definiteness of autocovariance matrices under (2.210)).

The variance of the innovation signal for the case of a 1D AR(P) model is[44]

$$\sigma_v^2 = \mathcal{E}\left\{v^2(n)\right\} = \mathcal{E}\left\{\left(s(n) - \sum_{p=1}^{P} a(p)s(n-p)\right)^2\right\} = \underbrace{\mathcal{E}\left\{\left(s(n)\right)^2\right\}}_{=\sigma_s^2}$$
$$-2\sum_{p=1}^{P} a(p)\underbrace{\mathcal{E}\left\{s(n)s(n-p)\right\}}_{=\mu_{ss}(-p)} + \sum_{p=1}^{P}\sum_{q=1}^{P} a(p)a(q)\underbrace{\mathcal{E}\left\{s(n-p)s(n-q)\right\}}_{=\mu_{ss}(p-q)} \tag{2.208}$$
$$= \sigma_s^2 - 2\sum_{p=1}^{P} a(p)\mu_{ss}(p) + \sum_{p=1}^{P} a(p)\underbrace{\sum_{q=1}^{P} a(q)\mu_{ss}(p-q)}_{=\mu_{ss}(p) \text{ acc. to W-H eq.}} = \sigma_s^2 - \sum_{p=1}^{P} a(p)\mu_{ss}(p).$$

This leads to an alternative formulation of the Wiener-Hopf equation, where the computation of the innovation signal variance is included by the first row of the matrix:

$$\sigma_v^2\delta(k) = \mu_{ss}(k) - \sum_{p=0}^{P} a(p)\cdot\mu_{ss}(k-p) \quad ; \quad 0 \le k \le P$$
$$\underbrace{\begin{bmatrix} \sigma_v^2 \\ 0 \\ \vdots \\ \vdots \\ 0 \end{bmatrix}}_{\mathbf{c}_{ss}} = \underbrace{\begin{bmatrix} \mu_{ss}(0) & \mu_{ss}(1) & \mu_{ss}(2) & \cdots & \mu_{ss}(P) \\ \mu_{ss}(1) & \mu_{ss}(0) & \mu_{ss}(1) & \cdots & \mu_{ss}(P-1) \\ \mu_{ss}(2) & \mu_{ss}(1) & \ddots & & \vdots \\ \vdots & \vdots & & \ddots & \vdots \\ \mu_{ss}(P) & \mu_{ss}(P-1) & \cdots & \cdots & \mu_{ss}(0) \end{bmatrix}}_{\mathbf{C}_{ss}} \underbrace{\begin{bmatrix} 1 \\ -a(1) \\ \vdots \\ \vdots \\ -a(P) \end{bmatrix}}_{\mathbf{a}}. \tag{2.209}$$

Furthermore, it can be concluded that

$$\sigma_v^2 = \mathbf{a}^{\mathrm{T}}\mathbf{c}_{ss} = \mathbf{a}^{\mathrm{T}}\mathbf{C}_{ss}\mathbf{a}\,, \tag{2.210}$$

which means that the autocovariance matrix with Toeplitz structure has to be positive-definite, or at least positive semi-definite for the degenerate case of $\sigma_v^2 = 0$. This is guaranteed by the property $\mu_{ss}(0) \ge |\mu_{ss}(k)|$, which will also guarantee stability of the filter. When samples from a finite window are used to estimate the autocovariance values, the property of positive-definiteness would by guarantee be fulfilled if either a periodic extension of the window is used, or if zero values are padded outside of the window.

For a separable two- or multi-dimensional model, filter coefficients can be optimized independently by solving 1D Wiener-Hopf equations, using 1D autocovariance measurements over the different coordinate axes. However, separable models do not allow optimum adaptation for all properties of multidimensional signals. For example, in the two-dimensional case, characteristic diagonal orientations of autocovariance cannot be considered explicitly. When non-separable autocovariance functions are used, non-separable IIR filters must also be defined

---

44 This is a generalization of (2.191) and also covers the AR(1) case.

as AR generator (or predictor) filters. Using the 2D autocovariance function, a 2D Wiener-Hopf equation can be defined as an extension of (2.209). For the case of a quarter-plane 2D filter, optimization gives

$$\sigma_v^2\delta(k_1,k_2)=\mu_{ss}(k_1,k_2)-\sum_{\substack{p_1=0\\(p_1,p_2)\neq(0,0)}}^{P_1}\sum_{p_2=0}^{P_2}a(p_1,p_2)\mu_{ss}(k_1-p_1,k_2-p_2)\,, \tag{2.211}$$

which can again be written as $\mathbf{c}_{ss}=\mathbf{C}_{ss}\mathbf{a}$. $\mathbf{C}_{ss}$ here is a *block Toeplitz matrix* [DUDGEON/MERSEREAU 1984]

$$\mathbf{C}_{ss}=\begin{bmatrix}\phi_0 & \phi_{-1} & \cdots & \cdots & \phi_{-P_2}\\ \phi_1 & \phi_0 & \cdots & \cdots & \phi_{1-P_2}\\ \vdots & \vdots & \ddots & & \vdots\\ \vdots & \vdots & & \ddots & \vdots\\ \phi_{P_2} & \phi_{P_2-1} & \cdots & \cdots & \phi_0\end{bmatrix} \tag{2.212}$$

with associated sub-matrices

$$\phi_p=\begin{bmatrix}\mu_{ss}(0,p) & \mu_{ss}(-1,p) & \cdots & \cdots & \mu_{ss}(-P_1,p)\\ \mu_{ss}(1,p) & \mu_{ss}(0,p) & \cdots & \cdots & \mu_{ss}(-P_1+1,p)\\ \vdots & \vdots & \ddots & & \vdots\\ \vdots & \vdots & & \ddots & \vdots\\ \mu_{ss}(P_1,p & \mu_{ss}(P_1-1,p) & \cdots & \cdots & \mu_{ss}(0,p)\end{bmatrix}. \tag{2.213}$$

The vector of coefficients is arranged by row-wise order

$$\mathbf{a}=\left[1,-a(1,0),..,-a(P_1,0),-a(0,1),..,-a(P_1,P_2)\right]^{\mathrm{T}}, \tag{2.214}$$

and the 'autocovariance vector' on the left side is

$$\mathbf{c}_{ss}=\left[\sigma_v^2,0,0,...,0\right]^{\mathrm{T}}. \tag{2.215}$$

The lengths of the vectors and the row/column lengths of the quadratic matrix are $(P_1+1)(P_2+1)$. The task is to determine the $(P_1+1)(P_2+1)-1$ unknown coefficients in $\mathbf{a}$. This is achieved as in (2.208), inverting the autocovariance matrix $\mathbf{C}_{ss}$. The full matrix of the 2D formulation does however no longer have a Toeplitz structure, because the sub-matrices (2.213) are not diagonally symmetric, since $\mu_{ss}(k,p)\neq\mu_{ss}(-k,p)$. As a consequence, the inversion cannot use the same efficient decomposition as in the 1D case, and also the number of covariance values to be used in the optimization is larger than the number of filter coefficients to be determined; therefore, a unique revertible mapping between model parameters and autocovariance does not longer exist. If positive-definiteness is violated, this may also lead to unstable synthesis filters. As an alternative, a 2D PARCOR structure was proposed in [MARZETTA 1980]. However, it is reported that in the non-separable 2D case this does not guarantee stability, either.

2.6.2 Linear prediction

Autoregressive modeling of signals is closely related to *linear prediction*, where a *predictor filter* computes an estimate $\hat{s}(\mathbf{n})$ for the signal value $s(\mathbf{n})$. The difference is the *prediction error*

$$e(\mathbf{n}) = s(\mathbf{n}) - \hat{s}(\mathbf{n}) . \tag{2.216}$$

The signal can be reconstructed by using the prediction error and the estimate,

$$\tilde{s}(\mathbf{n}) \overset{!}{=} s(\mathbf{n}) = e(\mathbf{n}) + \hat{s}(\mathbf{n}) . \tag{2.217}$$

If estimates $\hat{s}(\mathbf{n})$ are exclusively computed by past values of the signal, the prediction error $e(\mathbf{n})$ also is a unique equivalent of $s(\mathbf{n})$[45]. The prediction is typically performed by an FIR filter with transfer function $H(\mathbf{z})$ (Fig. 2.30a); the *prediction error filter* (Fig. 2.30b), performing the operation described in (2.216), has a transfer function

$$A(\mathbf{z}) = 1 - H(\mathbf{z}) . \tag{2.218}$$

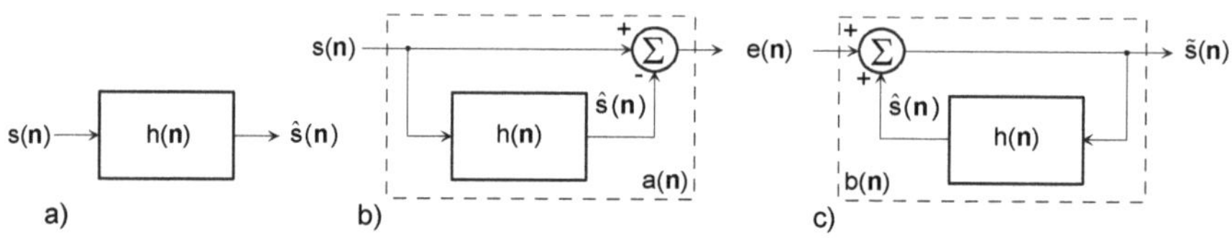

Fig. 2.30. System elements in linear prediction: **a** Predictor filter $h(\mathbf{n})$ **b** Prediction error filter (analysis filter) $a(\mathbf{n})$ **c** inverse prediction error filter (synthesis filter) $b(\mathbf{n})$

The *inverse prediction error filter* (synthesis filter, Fig. 2.30c) performs the operation (2.217). It is a recursive filter with transfer function

$$B(\mathbf{z}) = \frac{1}{A(\mathbf{z})} = \frac{1}{1 - H(\mathbf{z})} . \tag{2.219}$$

The filter (2.219) can be regarded equivalent to the synthesis filter of an AR model. Therefore, the prediction error signal would actually be Gaussian white noise if an AR process is optimally predicted (i.e. using a predictor which inverts the synthesis filter by which the process was generated). In the context of linear pre-

[45] In practical implementations, the equivalence may not be up to mathematical precision, when rounding errors occur. This can be avoided by performing systematic rounding as part of the prediction, which however would introduce a nonlinear element that can no longer be described as an LSI operation.

diction, the ratio of signal variance and prediction error variance is denoted as the *prediction gain*

$$G = \frac{\sigma_s^2}{\sigma_e^2}, \tag{2.220}$$

which can be determined from (2.208) for the case of an AR model.

Backward-adaptive prediction. Whereas the solution of the Wiener-Hopf equation assumes that autocovariance statistics either globally or of the current local segment is known, *backward-adaptive* methods of predictor filter adaptation use analysis of *past* samples under the assumption that the statistical properties are only slowly changing. The *least mean squares* algorithm (LMS) is often applied in this context [ALEXANDER, RAJALA 1985]. Predictor filter coefficients $a_{\mathbf{n}}(\mathbf{p})$ shall be used at the current position $\mathbf{n}$ in the prediction equation

$$\hat{s}(\mathbf{n}) = \sum_{\mathbf{p}} a_{\mathbf{n}}(\mathbf{p}) \cdot s(\mathbf{n}-\mathbf{p}) \quad \text{and} \quad e(\mathbf{n}) = s(\mathbf{n}) - \hat{s}(\mathbf{n}). \tag{2.221}$$

After computing the prediction error, it is evaluated how each filter coefficient would need to be modified to achieve a lower prediction error. The partial derivative of $e^2(\mathbf{n})$ over $a_{\mathbf{n}}(\mathbf{p})$ is

$$\frac{\partial e^2(\mathbf{n})}{\partial a_{\mathbf{n}}(\mathbf{p})} = -2e(\mathbf{n}) \cdot s(\mathbf{n}-\mathbf{p}), \tag{2.222}$$

such that an LMS update of coefficients to be used at the next position[46] reduces the prediction error by optimizing with regard to its negative gradient,

$$a_{\mathbf{n}+1}(p) = a_{\mathbf{n}}(\mathbf{p}) + \alpha\, e(\mathbf{n}) s(\mathbf{n}-\mathbf{p}). \tag{2.223}$$

The step size factor α influences the adaptation speed.

Two-dimensional prediction. The prediction equation in the case of a 2D quarter-plane predictor filter of order $(P_1+1)(P_2+1)-1$ is

$$\hat{s}(n_1,n_2) = \sum_{\substack{p_1=0 \\ (p_1,p_2)\neq(0,0)}}^{P_1} \sum_{p_2=0}^{P_2} a(p_1,p_2) s(n_1-p_1, n_2-p_2). \tag{2.224}$$

The z transfer function of this filter is

$$\begin{aligned} H(z_1,z_2) = {} & a(1,0)z_1^{-1} + \ldots + a(P_1,0)z_1^{-P_1} + a(0,1)z_2^{-1} + \ldots + a(P_1,1)z_1^{-P_1}z_2^{-1} \\ & + \ldots + a(0,P_2)z_2^{-P_2} + \ldots + a(P_1,P_2)z_1^{-P_1}z_2^{-P_2}. \end{aligned} \tag{2.225}$$

[46] The 'next' position at which the updated coefficient is used in 2D and multi-dimensional can be determined from the prediction direction, e.g. vertical down for a coefficient that performs vertical prediction from the sample above.

For the case of 2D signals, 2D prediction can be expected to better minimize the variance of the prediction error signal, as compared to 1D (horizontal or vertical) prediction.

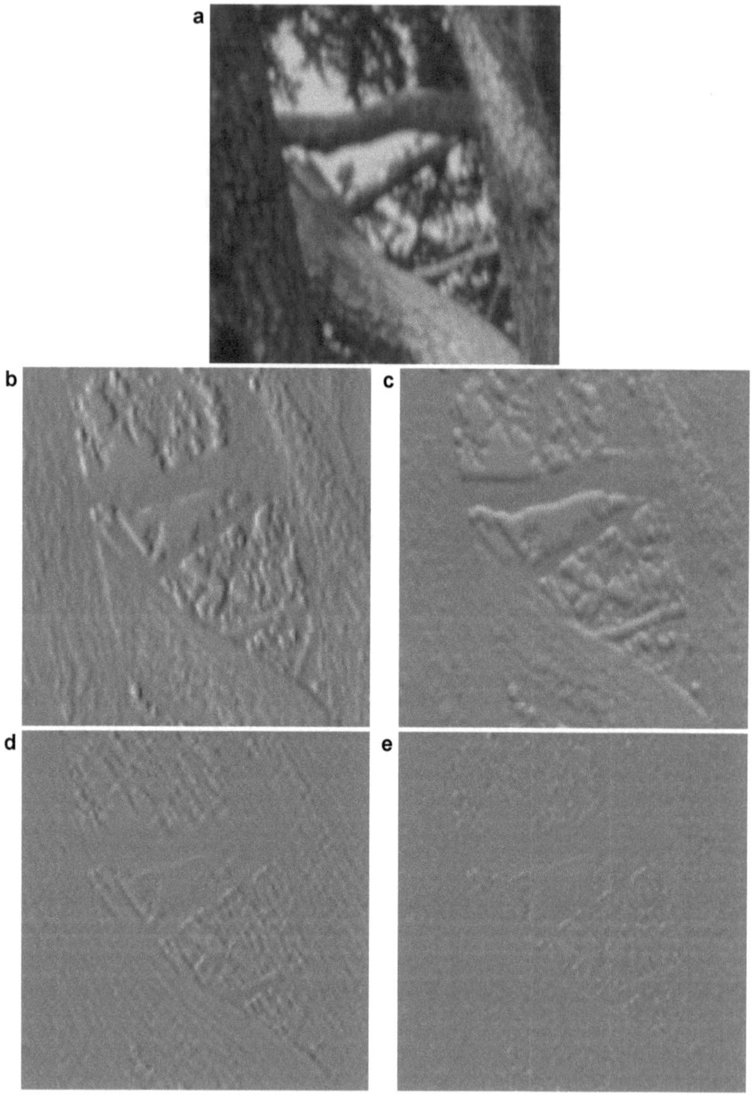

Fig. 2.31. Original image (**a**) and prediction error images: **b-c** 1D prediction row-wise, P_1=1, ρ_1=0.95 (b) 1D column-wise, P_2=1, ρ_2=0.95 (*c*) **d** 2D separable, fixed coefficients P_1=P_2=1, ρ_1=ρ_2=0.95, **e** 2D non-separable with local adaptation, quarter-plane P_1=P_2=2

Assume that 2D prediction is applied to a separable 2D AR(1) model, where the same prediction filter $H(z_1,z_2)$ is used as in the recursive loop of the model generator. Hence, the prediction error filter $A(z_1,z_2) = 1 - H(z_1,z_2)$ will exactly reproduce the Gaussian white noise fed into the generator of the AR process. For the 2D separable AR(1) model, the optimum predictor filter is constructed from the two (horizontal and vertical) 1D filters as follows:

$$\begin{aligned} &H(z_1) = \rho_1 z_1^{-1} \Rightarrow A(z_1) = 1 - \rho_1 z_1^{-1}, \\ &H(z_2) = \rho_2 z_2^{-1} \Rightarrow A(z_2) = 1 - \rho_2 z_2^{-1}, \\ &A(z_1,z_2) = A(z_1)A(z_2) = 1 - \rho_1 z_1^{-1} - \rho_2 z_2^{-1} + \rho_1\rho_2 z_1^{-1} z_2^{-1} \\ &\Rightarrow H(z_1,z_2) = 1 - A(z_1,z_2) = \rho_1 z_1^{-1} + \rho_2 z_2^{-1} - \rho_1\rho_2 z_1^{-1} z_2^{-1}. \end{aligned} \tag{2.226}$$

Fig. 2.31 shows an original image (*a*), prediction error images obtained by horizontal (*b*) and vertical (*c*) 1D prediction, and by separable 2D prediction (*d*). While the horizontal prediction is not capable to predict vertical edges, the vertical filter fails at horizontal edges, but the 2D prediction performs reasonably well in both cases. Specifically in areas showing regular textured structures (e.g. grass, water, hairs, etc.), the usage of higher-order 2D predictors can be advantageous, if adapted properly to the signal (Fig. 2.31e). In the given example, the adaptation block size was 16x16 samples, quarter-plane prediction filters of size 3x3 were optimized by solving the Wiener-Hopf equation system (2.211).

Motion compensated prediction. When temporal prediction from previous picture(s) of a video signal shall be performed, an autoregressive model cannot reasonably capture the temporal changes occurring by object or camera motion, as it is not efficiently considering the sparseness of a moving video signal's spectrum from (2.44). In *motion compensated prediction*, predictor adaptation is rather performed by *motion estimation.* Samples in picture n_3 shall be predicted, and the best-matching position in a prediction reference picture (e.g. the previous picture n_3-1) is found to be displaced by k_1 samples horizontally and k_2 samples vertically. Then, the prediction equation is

$$\begin{aligned} &e(n_1,n_2,n_3) = s(n_1,n_2,n_3) - \hat{s}(n_1,n_2,n_3) \\ &\qquad\qquad \text{with } \hat{s}(n_1,n_2,n_3) = s(n_1+k_1,n_2+k_2,n_3-1). \end{aligned} \tag{2.227}$$

This motion compensated predictor filter can be characterized by the 3D z transfer function[47]

$$H(z_1,z_2,z_3) = z_1^{k} z_2^{l} z_3^{-1}, \tag{2.228}$$

which describes a multi-dimensional shift (or delay); motion-compensated prediction therefore is a specific type of linear prediction. This simple type of filter uses a copy of samples from one previous picture and shifts them by an integer number of sample positions to generate the estimate. If the brightness of the signal changes, it could be more appropriate to multiply the amplitude by an additional factor, or shift it by an offset; in a more generalized approach, values from different ref-

[47] Basically, a random motion shift could also be included in an AR synthesis filter to obtain a reasonable video model.

erence pictures can be superimposed for prediction: Each of them may be weighted individually by a weighting factor $a(p_3)$, an offset c can optionally be added. If prediction shall further support sub-sample displacements, a spatial interpolation filter with impulse response $h_{\text{int}}(\mathbf{n})$ has to be included, with coefficients $a(p_1,p_2)$ in the convolution equation pending on the sub-sample phase $d_{1|2}$. The estimate is then computed by using up to P_3 reference pictures[48]

$$\hat{s}(n_1,n_2,n_3) = c(n_1,n_2,n_3) + \sum_{p_3=1}^{P_3} a(p_3)\cdot \\ \cdot \sum_{p_1=-Q_1/2}^{Q_1/2-1} \sum_{p_2=-Q_2/2}^{Q_2/2-1} a_{\text{int}}^{[d_1(p_3),d_2(p_3)]}(p_1,p_2)s\left[n_1+k_1(p_3)-p_1, n_2+k_2(p_3)-p_2, n_3+k_3(p_3)\right]. \tag{2.229}$$

With an offset $c(\mathbf{n})=0$, the z transfer function of the entire predictor filter can be described by

$$H(z_1,z_2,z_3) = \sum_{p_3=1}^{P_3} a_3(p_3) A_{\text{int}}^{[d_1(p_3),d_2(p_3)]}(z_1,z_2) z_1^{k_1(p_3)} z_2^{k_2(p_3)} z_3^{k_3(p_3)} . \tag{2.230}$$

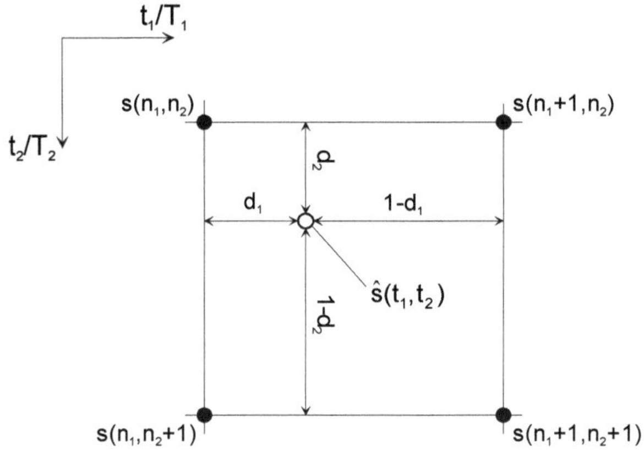

Fig. 2.32. Bilinear interpolation

The simplest approach of 2D interpolation is *bilinear interpolation*, which is separable, $h(t_1,t_2)=\Lambda(t_1)\Lambda(t_2)$ with $\Lambda(t)=\text{rect}(t)*\text{rect}(t)$. The principle is illustrated in Fig. 2.32. The value to be estimated at position (t_1,t_2) is computed from samples of

[48] 2D FIR interpolators with even impulse response lengths Q_1 and Q_2 are assumed for the horizontal and vertical sub-sample interpolations. As sample and sub-sample shifts can be different for each reference picture used in the prediction, the interpolation filter and sample motion shifts k_i are defined depending on the reference index p_3. Practically, displacements vary locally, such that the predictor filter is also a shift-variant system and the a, c and k_i parameters may also depend on n_1,n_2. In video coding, the reference pictures do not necessarily need to be ordered by their temporally sequence (see Sec. 7.2.4). This is expressed by the index $n_3+k_3(p_3)$ which defines an arbitrary list mapping.

four neighboring positions, which are weighted depending on the horizontal and vertical sub-sample phases d_1 and d_2 ($0 \le d_i < 1$):

$$\hat{s}(t_1,t_2) = s(n_1,n_2)(1-d_1)(1-d_2) + s(n_1+1,n_2)d_1(1-d_2) \\ + s(n_1,n_2+1)(1-d_1)d_2 + s(n_1+1,n_2+1)d_1 d_2. \tag{2.231}$$

However, bilinear interpolation has a relatively strong lowpass effect, and also does not provide good alias suppression[49]. Therefore, in practice, higher-order interpolation filters are used for better performance in motion compensated prediction with sub-sample accuracy (cf. Sec. 7.2.5).

In video coding, *block matching* is often used for motion estimation. Let Λ express a partition, for which a common horizontal/vertical displacement shift vector $\mathbf{k}+\mathbf{d} = [k_1+d_1, k_2+d_2]$ shall be determined for a given reference picture with distance k_3 from the current picture n_3. Π describes a set of candidate displacement shifts. *Cost functions* based on minimization of difference criteria of norm Q are often used for this purpose[50],

$$[\mathbf{k}+\mathbf{d}]_{\text{opt}}(k_3) = \\ \underset{[k_1,k_2]\in\Pi}{\arg\min} \left| \frac{1}{|\Lambda|} \sum_{(n_1,n_2)\in\Lambda}\sum \left| s(n_1,n_2,n_3) - \hat{s}(n_1+k_1+d_1, n_2+k_2+d_2, n_3+k_3) \right|^Q \right|^{\frac{1}{Q}}. \tag{2.232}$$

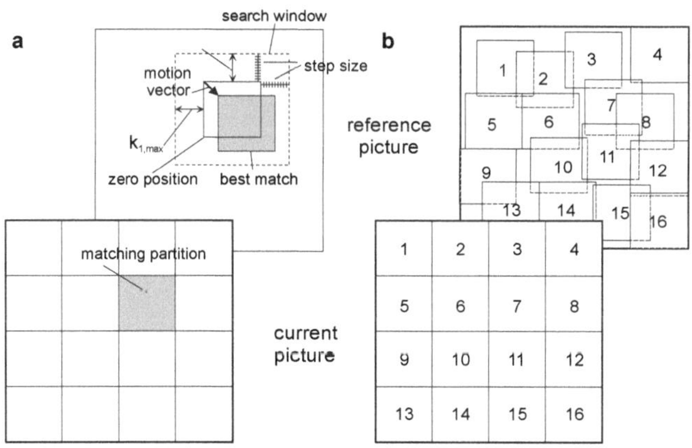

Fig. 2.33. Block matching motion estimation. **a** Definition of matching partition, search range and step size in the current picture **b** Possible overlaps of best-matching blocks in the reference picture

---

[49] Due to the triangular impulse response of the underlying 1D filter, its spectral transfer function is sinc^2, which has its first zero at $|f| = 1/2$, the first two side lobe in the first alias band $1/2 \le |f| \le 3/2$, and further side lobes in higher-frequency alias bands.

[50] Q=1 for *sum of absolute difference* (SAD), Q=2 for *sum of squared difference* (SSD). Sub-sample accurate shift parameters $l_i = k_i + d_i$ are used here, which means that for the case $d_i \neq 0$ it is necessary to compute $\hat{s}$ by interpolation filtering, cf. (2.229)/(2.230).

Fig. 2.33 illustrates the method. In Fig. 2.33a, all samples of a given partition in the current picture are subject to the same horizontal/vertical shift, and the sample pattern of the partition is compared against patterns from the set of candidate positions in the given reference picture. As a result, the displacement vector corresponding to the best pattern match is selected. Fig. 2.33b indicates an inconsistency of rigid block partitioning, as this may cause unreasonable overlaps or gaps between adjacent blocks in the reference picture, at positions where the motion is discontinuous (e.g. at object boundaries). Generally, the partitions may either be of equal size (as shown in the figure) or of variable size. As an example, with full search, scanning over all possible shift positions within a 2D search window, the total number of positions to be compared is growing linearly with the area of the search window and with the density (reciprocal squared value of step size Δ, which is the distance between adjacent shifted candidate positions).

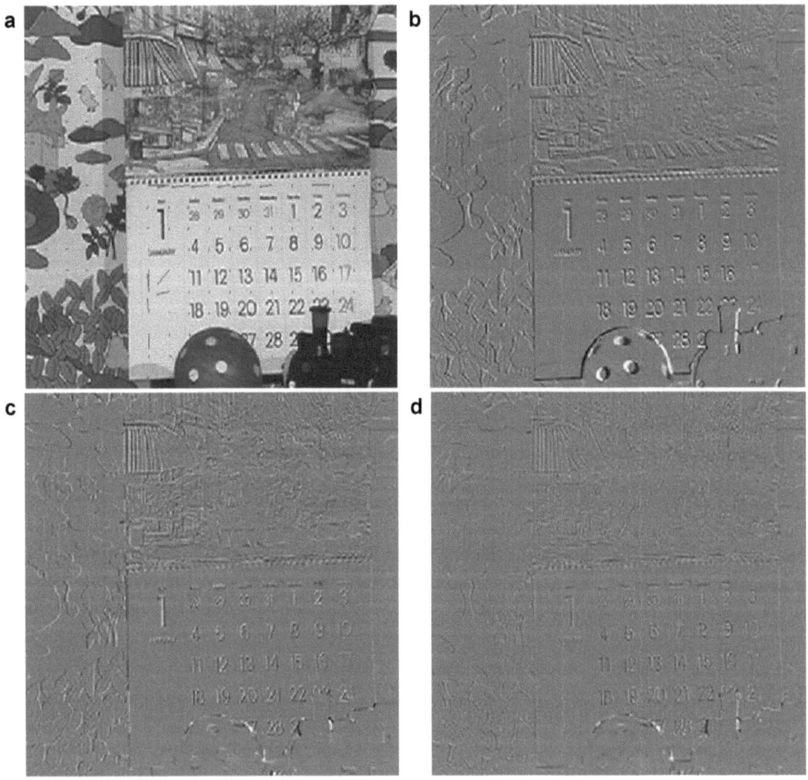

Fig. 2.34. **a** Picture (with MVs) from a video sequence, and prediction error pictures: **b** without motion compensation **c** with motion compensation, full-sample shift accuracy **d** with motion compensation, half-sample shift accuracy (both motion compensated examples with constant block grid of size 16x16, half-sample shift by bilinear interpolation)

Fig. 2.34 shows results of a picture predicted without and with motion compensation, the latter case also with bilinear interpolation filtering for half-sample accuracy (all with block grid of size 16x16). Fig. 2.35 shows corresponding results with quarter-sample accuracy, with bilinear and higher-quality (8-tap filter) interpolation, the latter as well with reduced sizes of the block grid, 8x8 and 4x4 samples (contrast enhanced 1.5x in residual pictures for better visibility).

a b c d

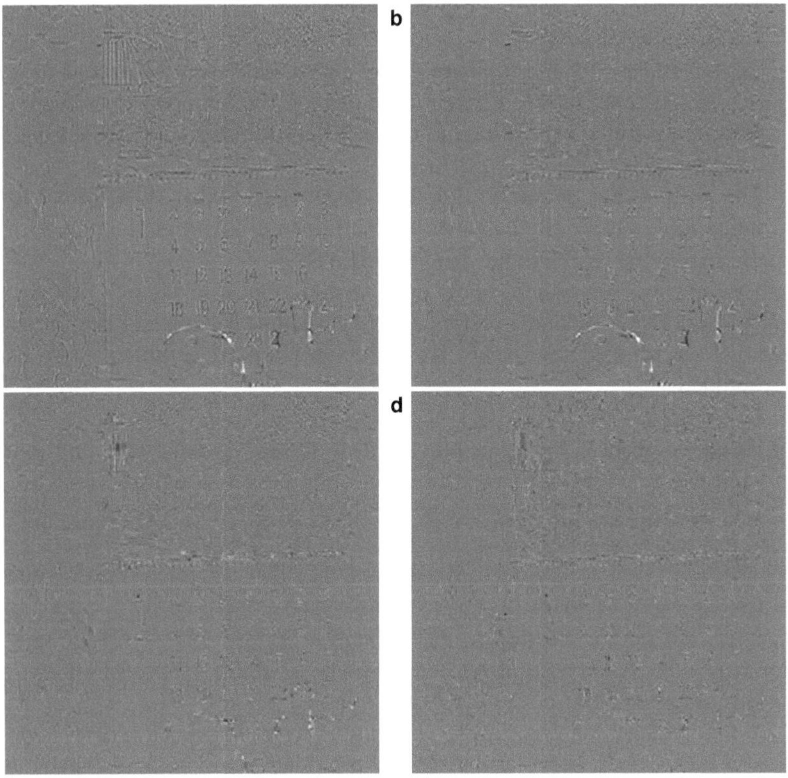

Fig. 2.35. Examples with quarter-sample accuracy in motion compensation: **a** bilinear interpolation, 16x16 block grid, and further examples with 8-tap interpolation filter: **b** 16x16 block grid **c** 8x8 block grid **d** 4x4 block grid

The true motion shift between two pictures will typically be by sub-sample units. It is however not useful to test all possible sub-sample positions over the entire search window range, as it can be expected that the cost criterion in (2.232) varies smoothly over **k**. Therefore, strategies for fast search are used which start by larger Δ values and refine the estimated motion vector into sample or sub-sample accuracy only by the last few steps.

Fast search algorithms do not test all possible candidates and therefore may no longer guarantee that the global optimum over the **k**+**d** parameter space is reached. However, with same complexity, fast algorithms can often achieve even better results compared to exhaustive (full search) approaches, since they avoid testing unreasonable candidates and can instead investigate an extended parameter space. In one or the other way, fast motion estimation algorithms inherently exploit

- the smoothness of cost functions in dependency of **k**+**d**, due to the fact that the sample patterns at adjacent candidate positions in the reference picture are almost identical when the step size Δ is small; this allows optimizing the result by iterative steps;
- the smoothness of displacement vector fields, both over the spatial coordinate (e.g. consistent motion of larger objects) and the temporal coordinate (along the motion trajectory), which allows predicting initial candidates from previous estimates;
- The joint scaling property of picture and motion vector field, where for spatially downsampled signals the number of sample-wise operations, as well as the size of the search range can be reduced[51].

Furthermore, it is possible to apply early termination of the search, when a sufficiently good displacement (in terms of cost function) has already been found. All previously mentioned approaches for search speedup are complementary and can be combined.

Multi-step search. Two principles of fast motion estimation algorithms are shown in Fig. 2.36a/b. Both are based on testing only a subset of search positions out of the entire set of parameters, where the favorable direction of changing **k** is traced for optimization of the cost criterion. In Fig. 2.36a/b, all positions tested in the particular steps are drawn as black dots, the steps are referenced by numbers, and the optimum as found in the respective step is marked by a circle. In both examples, the motion vector is finally found as $k_1=-5$, $k_2=2$. These two algorithms are typical representatives for a variety of similar approaches, one of the first was suggested in [KOGA ET AL 1981].

In the method of Fig. 2.36a, originally denoted as *three-step search* in [MUSMANN ET AL. 1985], only a small set of 9 candidate positions is evaluated in each step. Simultaneously, the search step size is decreased gradually ($\Delta=3$, 2, 1 sample width for the three iterations in the example shown). The center of the search range in iteration step r is selected from the best-matching position of the previous iteration $r-1$, such that cost criteria need to be computed only for 8 new positions in iterations 2 and 3. In the example shown, a total of $9+8+8=25$ candidate posi-

---

[51] The scaling property also imposes an interesting relationship between the picture size and the complexity of motion estimation. If the picture size is doubled horizontally and vertically, the density of samples is increased likewise. However, the size of the search range has also to be doubled horizontally and vertically, as now the related displacement maps into a motion vector **k** of double length. Considering exhaustive search, this leads to a complexity increase by a factor of 16 when doubling the picture size. It could be argued that when downsampling the step size Δ should be decreased (e.g. going from half sample to quarter sample precision of motion compensation), which would diminish this benefit. This is however not the case, if proper lowpass filtering is applied in the context of downsampling, which loses spatial detail to some extent.

tions are compared in the three iterations; the maximum range in the given example is $k_{1,\max}=k_{2,\max}=\pm 6$ samples. A full search with same range would require testing of $13^2=169$ positions. The factor of reducing computational complexity increases with larger search range (more iteration steps). Typically, a complexity dependency on $k_{\max}$ or $\log(k_{\max})$ (instead of $k_{\max}^2$ for full search) can be achieved.

In the search method shown in Fig. 2.36b, 5 different positions in $\mathcal{N}_1$ arrangement are compared in the first iteration step. In the example, a step size $\Delta=2$ is used. After finding the best match among these, only three more positions adjacent to the previous optimum need to be compared in any remaining step. This process is continued until the best-match position remains unchanged, which indicates that a local minimum over the cost function has been approached. Then, in a final step, all 8 shift positions around this optimum, or additional sub-sample positions, are checked as candidates. In the example shown, only 5+2·3+8=19 candidate positions have to be tested in total.

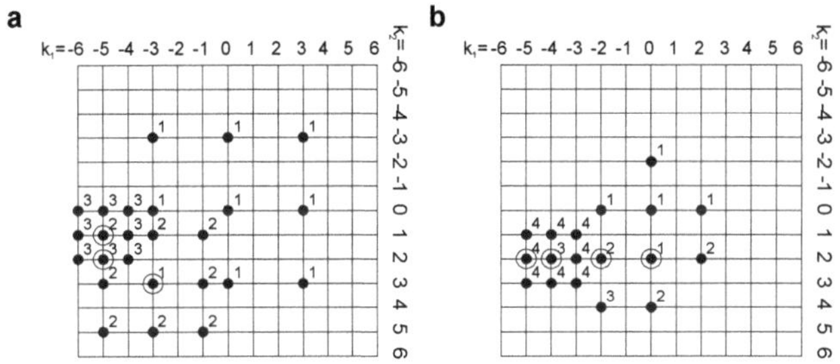

Fig. 2.36. Multi-step block-matching estimation methods – examples of **a** 'three-step search' **b** 'logarithmic search'

These concepts will approach the *globally-optimum* result (as in full search), if the cost function is truly convex and therefore improving over the motion parameter space towards the optimum position. If local extremes of the cost function exist, it is possible to get stuck in such a position. This can be the case when several similar structures (e.g. periodicities) are present.

Displacement vector predictors. As continuity of the motion vector field can be assumed, reasonable predictions for correct displacement vectors are often available from previous estimates in the spatial or temporal neighborhood. Fig. 2.37 shows possible candidate vectors from adjacent partitions (here: blocks), which can be used to predict the displacement for the current partition. The temporal predictor can be selected from the 'collocated' position in the reference picture, or as a vector which points from a location in the reference picture into the current block partition[52].

---

[52] It may be necessary to scale the candidate vectors based on the time distance between the current picture and the reference picture in comparison to the time distance that is in

Different approaches are possible to determine the final estimate for the displacement vector of the current partition:

- The mean or median from a set of previously-estimated vectors (candidates) is computed as starting point, and the new value is optimized by testing candidates within an additional search range around this initial hypothesis[53];
- The search range is determined from the range between minimum and maximum displacement values found in the set of predictor candidates;
- Different search ranges are tested around the values of several predictor candidate's displacement vectors (if they are not identical).

In this context, it is also common practice to terminate the estimation without further refinement when one of the initial candidates already provides a good estimate, which further speeds up the search on average [DE HAAN ET AL. 1993].

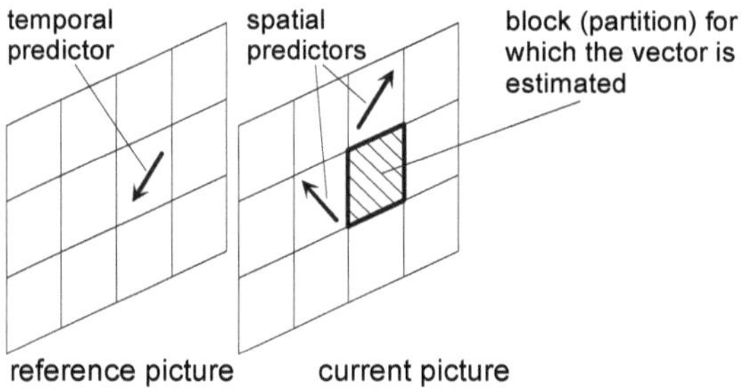

Fig. 2.37. Examples of temporally and spatially adjacent displacement candidates in predictive motion estimation for block matching

Multi-resolution motion estimation. Multi-resolution estimation determines candidates from downsampled pictures. For the same content, the displacement shift is down-scaled as well with the picture size/resolution; therefore, the search range can be down-scaled as well, whereas the motion is still captured [NAM ET AL. 1995]. In subsequent steps, the picture resolution is increased, but the estimation starts from the result of the previous step, which can be expected to be already close to the true motion, such that the search range can again be small. Furthermore, also the size of the partitions for which a common motion displacement is estimated can be decreased, in which case the spatial resolution of available displacement vectors (i.e. the density of the vector field) also increases with the picture resolution. In terms of the hierarchical representation of the pictures, such an approach can be interpreted as a Gaussian pyramid (Sec. 2.8); with decreased

effect in the motion compensation where they are used. The number of candidates can also be variable, depending on the local variation of motion.

[53] The median value can be computed separately for the horizontal and vertical displacements, or jointly for both, depending on the vector length; this may however generate a combined displacement which effectively does not exist.

partition sizes, also a *pyramid of motion vector fields* with both increasing resolution and precision is generated.

In full-search motion estimation, the size of the search range per direction $k_{i,\max}$ has the most important effect on complexity. With this regard, multi-resolution estimation has a similar benefit as multi-step search[54], but due to the additional subsampling of the pictures reduces the complexity even further and is eventually more stable due to the lowpass characteristics of downsampled pictures.

When up-scaled displacement vectors from a lower resolution level are used as starting points for estimating in several adjacent partitions in the next higher resolution level, and the differences between them are small due to using small search ranges around the candidate, hierarchical estimation implicitly can generate spatially more continuous motion vector fields. The relationships between matching areas and estimated motion vectors at two different levels are illustrated in Fig. 2.38.

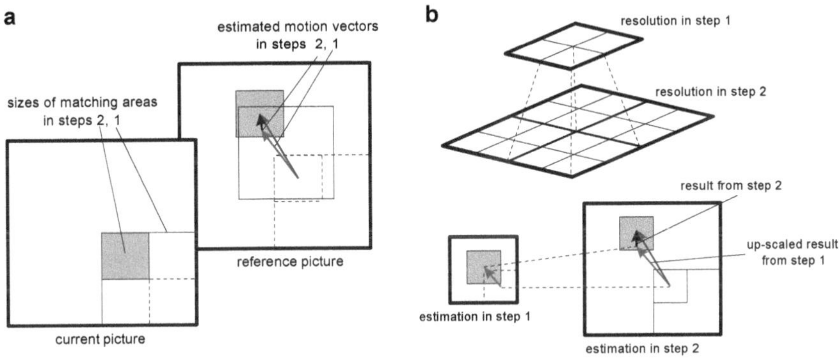

Fig. 2.38. Hierarchical motion estimation over two steps **a** Interpretation at full resolution **b** Principle of reduced resolution in the first step, and up-scaling of the resulting displacement vector

Variable block-size estimation. Displacement vector fields representing motion shifts are continuous (with only small amount of changes) within areas of background or larger moving objects, but discontinuous at object boundaries. Both properties can best be reflected when variable-size partitions are used for regions that are assigned to a common displacement vector. Typical strategies in optimizing displacement estimation for variable block sizes start from larger partitions and performs splitting into smaller partitions in cases where this has advantages w.r.t. the cost function. However, it should be observed that generally for areas with less detail estimated vectors could be ambiguous (this is denoted as *aperture problem*, see [JÄHNE 2005]). Therefore, splitting should be justified by a significant benefit in the cost function, and eventually large deviations in the displacements

54 Both multi-step and multi-resolution methods are sometimes entitled as *hierarchical motion estimation.*

of adjacent split partitions may be inhibited by additional constraints (e.g. smoothness criteria, see subsequent section). Beyond the splitting strategy, another approach would be to start with smaller partitions and merge them, if the same displacement can be applied without significant disadvantage to the cost function. Again here, a smoothness constraint can be used as part of the cost function.

Constrained estimation. *Additional constraints* are often introduced in block matching, where the cost function of a given estimate is modified by a *penalty term* $\lambda\mathcal{P}$ e.g.

- establishing interrelationships between motion vectors estimated in adjacent blocks by a smoothness constraint that takes into account motion vector differences,
- regularizing estimates in low-detail regions where no unique motion vector can be determined, by aligning them with the displacement of adjacent higher-detail regions,
- taking into account the rate that would be required to encode the displacement vector [GIROD 1994].

An example for a constrained optimization criterion in analogy with (2.232) is

$$\mathbf{k}_{\text{opt}} = \arg\min_{\mathbf{k}\in\Pi}\left[\frac{1}{|\mathbf{\Lambda}|}\sum_{\mathbf{n}\in\mathbf{\Lambda}}|s(\mathbf{n})-s(\mathbf{n}+\mathbf{k})|^{P}+\lambda\mathcal{P}(\mathbf{k})\right] \quad (2.233)$$

State of the art fast motion estimation algorithms used in video coding are often using combinations of the aforementioned approaches.

2.7 Linear block transforms

2.7.1 Orthogonal basis functions

The Discrete Fourier Transform (DFT) (2.89) is computed by multiplying M samples from a signal by an orthogonal set of complex basis functions. In general, two finite discrete 1D (real or complex) functions $t_i(n)$ and $t_j(n)$, each of length M, are *orthogonal*, if their linear combination gives zero,

$$\mathbf{t}_i^{\mathrm{T}}\mathbf{t}_j^{*} = \sum_{n=0}^{M-1} t_i(n)t_j^{*}(n) = \varphi_{t_it_j}(0) = 0 \quad \text{with} \quad \mathbf{t}_k = \left[t_k(0) \quad \cdots \quad t_k(M-1)\right]^{\mathrm{T}}. \quad (2.234)$$

If the functions $t_k(n)$ are interpreted as impulse responses of linear filters, and if the operations $c_k(n)=s(n)*t_k(n)$ are performed over all n, it can be shown that the cross correlation between any two resulting outputs $\varphi_{c_ic_j}(0)=0$ for $i\neq j$. Therefore, the usage of orthogonal basis functions can provide a *de-correlated representation* of a signal[55].

For an *orthogonal set* of basis functions, each member of the set is orthogonal with any other. The computation of the transform coefficient $c_k = \mathbf{s}^{\mathrm{T}}\mathbf{t}_k$ is a mapping from the signal domain into a transformed domain (which could be interpreted as a sampled frequency domain, provided that the basis functions have an appropriately ordered frequency transfer behaviour). If reconstruction of the signal samples is possible, the discrete set of transform coefficients establishes an equivalent representation. For processing of longer-duration signals, *local* or *short-time transforms* are often applied, in simplest case processing non-overlapping block segments (vectors) **s** of length M from the signal in a *block transform*. In the following, this problem will first be discussed for the case of one-dimensional transforms, from which two- and multidimensional transforms can straightforwardly be constructed by separable processing over the different coordinate axes. A segment from the signal $s(n)$, consisting of M subsequent samples and starting at position mM, shall be mapped into a set of U transform coefficients

$$c_k(m) = \sum_{n=0}^{M-1} s(mM+n)t_k(n) \quad ; \quad 0 \le k < U . \tag{2.235}$$

It shall be possible to reconstruct this segment of the signal by a complementary set of synthesis functions (inverse transform), such that

$$s(mM+n) = \sum_{k=0}^{U-1} c_k(m) r_k(n) \; ; \quad 0 \le n < M . \tag{2.236}$$

Substituting (2.236) into (2.235) gives

$$\sum_{n=0}^{M-1} t_k(n) \sum_{l=0}^{U-1} \frac{c_l(m)}{c_k(m)} r_l(n) = 1 \quad ; \quad 0 \le (k,l) < U . \tag{2.237}$$

This condition can only hold for all k, if the factor c_l / c_k is zero for $l \neq k$, such that

$$\sum_{n=0}^{M-1} t_k(n) r_l(n) = \begin{cases} 1 & \text{for} \quad k = l \\ 0 & \text{for} \quad k \neq l \end{cases} \quad ; \quad 0 \le (k,l) < U . \tag{2.238}$$

In the special case of an orthogonal set $\{\mathbf{t}_k\}$[56], this is fulfilled by choosing the matching analysis and synthesis bases $\mathbf{t}_k$ and $\mathbf{r}_k$ as complex conjugates, by which

---

55 If however the sequences $c_k(n)$ are subsampled, as often applied in the context of transform coding to avoid an overcomplete representation, correlation may occur partially due to aliasing.

56 Observe that the fulfillment of (2.238) does not necessarily require that the analysis basis functions $\mathbf{t}_k$ or synthesis basis functions $\mathbf{r}_l$ by themselves establish orthogonal sets; it is only necessary that function k from one set is orthogonal with function $l \neq k$ from the other set. This joint property of two sets is called *bi-orthogonality*; the choice of an orthogonal set $\{\mathbf{t}_k\}$ and $\{\mathbf{r}_k\}=\{\mathbf{t}_k^*\}$ is a special case thereof.

$\mathbf{t}_k$ is implicitly orthogonal with any other synthesis basis $\mathbf{r}_l$. The further constraint $\mathbf{t}_k^{\mathrm{T}}\mathbf{r}_k=1$ can be avoided by the generalization

$$\mathbf{r}_k = \frac{\mathbf{t}_k^*}{A_k} \quad \text{s. t.} \quad \mathbf{t}_k{}^{\mathrm{T}}\mathbf{t}_k^* = A_k \ \text{(real, positive)}. \tag{2.239}$$

By combining (2.238) and (2.239), a more general orthogonality condition for the set $\{\mathbf{t}_k\}$ is

$$\mathbf{t}_k{}^{\mathrm{T}}\mathbf{t}_l^* = \sum_{n=0}^{M-1} t_k(n)t_l^*(n) = \begin{cases} A_k & \text{for} \quad k=l \\ 0 & \text{for} \quad k \neq l \end{cases} \quad ; \quad 0 \le (k,l) < U . \tag{2.240}$$

For the example of the DFT (2.89) and IDFT (2.90), $t_k(n)=\mathrm{e}^{-\mathrm{j}2\pi nk/M}$, $r_k(n)=\mathrm{e}^{\mathrm{j}2\pi nk/M}/M$ and $A_k = M$ fulfills these conditions. A general transform from M signal values into U coefficients can also be formulated by the matrix notation

$$\underbrace{\begin{bmatrix} c_0(m) \\ c_1(m) \\ \vdots \\ \vdots \\ c_{U-1}(m) \end{bmatrix}}_{\mathbf{c}(m)} = \underbrace{\begin{bmatrix} t_0(0) & t_0(1) & \cdots & \cdots & t_0(M-1) \\ t_1(0) & t_1(1) & \cdots & \cdots & t_1(M-1) \\ \vdots & & \ddots & & \vdots \\ \vdots & & & \ddots & \vdots \\ t_{U-1}(0) & t_{U-1}(1) & \cdots & \cdots & t_{U-1}(M-1) \end{bmatrix}}_{\mathbf{T}} \cdot \underbrace{\begin{bmatrix} s(mN+N_0) \\ s(mN+N_0+1) \\ \vdots \\ \vdots \\ s((m+1)N+N_0-1) \end{bmatrix}}_{\mathbf{s}(m)}, \tag{2.241}$$

where the signal vector $\mathbf{s}$ consists of M samples, the transform matrix $\mathbf{T}$ has size $M\mathrm{x}U$ with rows establishing *basis vectors* and the result $\mathbf{c}$ includes U transform coefficients c_k. As a minimum condition for reconstruction, the transformed representation $\{\mathbf{c}(m)\}$ over all blocks shall have the same number of samples as the signal $s(n)$. This can be achieved when the starting positions of subsequent vectors $\mathbf{s}(m)$ are $N=U$ samples apart, where N_0 is an optional constant offset. In simplest case of a non-overlapping block transform, $N=M=U$ and $M_0=0$. Then, since the rows of $\mathbf{T}=[\mathbf{t}_0\ \mathbf{t}_1\ \ldots\ \mathbf{t}_{U-1}]^{\mathrm{T}}$ are the *basis functions* from an orthogonal set, they are linearly independent, $\mathbf{T}$ is a square matrix, will have full rank and is invertible. Following (2.241), the values of $s(n)$ in $\mathbf{s}(m)$ can uniquely be reconstructed from the coefficients c_k in $\mathbf{c}(m)$,

$$\mathbf{s}(m) = \mathbf{T}^{-1}\mathbf{c}(m). \tag{2.242}$$

The transform is *orthonormal*, if $A_k = 1$ in (2.240). Analysis and synthesis vectors in (2.239) are identical for a real-valued orthonormal transform basis. More generally, for a complex orthonormal transform from (A.26)[57]

$$\mathbf{T}^{-1} = \left[\mathbf{T}^*\right]^{\mathrm{T}} = \mathbf{T}^{\mathrm{H}}. \tag{2.243}$$

The synthesis functions $\mathbf{r}_l$ from (2.238) are the columns of $\mathbf{T}^{\mathrm{H}}$; in combination

[57] $\mathbf{T}^{\mathrm{H}}$ is the Hermitian matrix (conjugate transpose) of $\mathbf{T}$.

with (2.243), this gives $\mathbf{TT}^{-1}=\mathbf{I}$. In orthonormal linear transforms, the quadratic norm (energy) of signal vectors can directly be computed without any normalization from the coefficient vectors,

$$\mathbf{s}^{\mathrm{T}}\mathbf{s}=\mathbf{s}^{\mathrm{T}}\underbrace{\mathbf{T}^{\mathrm{H}}\mathbf{T}}_{\mathbf{I}}\mathbf{s}=\left[\mathbf{c}^{*}\right]^{\mathrm{T}}\mathbf{c}\quad\text{or}\quad\left\|\mathbf{s}\right\|^{2}=\left\|\mathbf{c}\right\|^{2}. \tag{2.244}$$

Otherwise, if basis vector norms are different from unity, an equivalence is still found when the values in $\|\mathbf{c}\|^2$ are scaled by the individual A_k values.

The series of transform vectors $\mathbf{c}(m)$ is computed from signal vectors $\mathbf{s}(m)$ with starting positions $n_0(m)=mN+N_0$. With hop size $N>U$, reconstruction cannot be guaranteed, with $N<U$, the result of the transform would be over-complete. For the purpose of coding, $N=U$ is most appropriate. In case of block-overlapping transforms (Sec. 2.7.4), the vectors $\mathbf{s}$ are longer than vectors $\mathbf{c}$, i.e. $M>U$. In this case, though a single $\mathbf{c}(m)$ can uniquely be computed from the corresponding $\mathbf{s}(m)$, reconstruction may require involvement of other vectors $\mathbf{c}(m)$ that also depend on samples in $\mathbf{s}(m)$, which can be achieved by a weighted *over-lap-and-add* procedure as a secondary step. In the remaining part of the current section, $N=U=M$ is assumed.

A *separable two-dimensional transform* can be expressed as concatenation of two matrix multiplications using a horizontal transform $\mathbf{T}_{\mathrm{h}}$ and a vertical transform $\mathbf{T}_{\mathrm{v}}$,

$$\underbrace{\begin{bmatrix} c_{0,0} & c_{1,0} & \cdots & c_{U_1-1,1} \\ c_{0,1} & c_{1,1} & \cdots & c_{U_1-1,1} \\ \vdots & \vdots & \ddots & \vdots \\ c_{0,U_2-1} & c_{1,U_2-1} & \cdots & c_{U_1-1,U_2-1} \end{bmatrix}}_{\mathbf{C}} = \underbrace{\begin{bmatrix} t_0(0) & t_0(1) & \cdots & t_0(M_2-1) \\ t_1(0) & t_1(1) & \cdots & t_1(M_2-1) \\ \vdots & \vdots & \ddots & \vdots \\ t_{U_2-1}(0) & t_{U_2-1}(1) & \cdots & t_{U_2-1}(M_2-1) \end{bmatrix}}_{\mathbf{T}_{\mathrm{v}}} \cdot \cdots \tag{2.245}$$

$$\cdots \cdot \underbrace{\begin{bmatrix} s(0,0) & s(1,0) & \cdots & s(M_1-1,0) \\ s(0,1) & s(1,1) & \cdots & s(M_1-1,1) \\ \vdots & \vdots & \ddots & \vdots \\ s(0,M_2-1) & s(1,M_2-1) & \cdots & s(M_1-1,M_2-1) \end{bmatrix}}_{\mathbf{S}} \cdot \underbrace{\begin{bmatrix} t_0(0) & t_1(0) & \cdots & t_{U_1-1}(0) \\ t_0(1) & t_1(1) & \cdots & t_{U_1-1}(1) \\ \vdots & \vdots & \ddots & \vdots \\ t_0(M_1-1) & t_1(M_1-1) & \cdots & t_{U_1-1}(M_1-1) \end{bmatrix}}_{\mathbf{T}_{\mathrm{h}}^{\mathrm{T}}}.$$

In a first step, all columns (length M_2) of the image matrix $\mathbf{S}$ are transformed separately giving $\mathbf{C}_{\mathrm{v}}=\mathbf{T}_{\mathrm{v}}\mathbf{S}$, the result of the vertical transform applied separately over all columns. The subsequent horizontal transform of $\mathbf{C}_{\mathrm{v}}$ is performed by using the transposed transform matrix $\mathbf{T}_{\mathrm{h}}^{\mathrm{T}}$ rather than transposing the matrix $\mathbf{C}_{\mathrm{v}}$[58]. The

58 Alternatively, the second step could be $\mathbf{C}=[\mathbf{T}_{\mathrm{h}}\mathbf{C}_{\mathrm{v}}^{\mathrm{T}}]^{\mathrm{T}}$, however the above formulation gives the output in correct (not transposed) order right away. It is of course also possible to perform the horizontal transform first, where mathematically the final result is identical.

matrix equations for the separable 2D transform and the related inverse transform are as follows[59]:

$$\mathbf{C} = \underbrace{[\mathbf{T}_\text{v}\mathbf{S}]}_{\mathbf{C}_\text{v}}\mathbf{T}_\text{h}^\text{T} \quad \Rightarrow \quad \mathbf{S} = \left[\mathbf{T}_\text{v}^{-1}\mathbf{C}\right]\left[\mathbf{T}_\text{h}^{-1}\right]^\text{T} . \tag{2.246}$$

The basis functions relating to U_1U_2 coefficients of the separable 2D transform are

$$t_{k_1,k_2}(n_1,n_2) = t_{k_1}(n_1)t_{k_2}(n_2) \quad ; \quad 0 \le k_i < U_i \quad ; \quad \mathbf{T}_{k_1,k_2} = \mathbf{t}_{k_1}\mathbf{t}_{k_2}^\text{T} . \tag{2.247}$$

The 2D basis matrices $\mathbf{T}_{k_1,k_2}$ are also denoted as *basis images*. A two- or multi-dimensional expression can generally be written as $t_\mathbf{k}(\mathbf{n})$ and $\mathbf{T}_\mathbf{k}$, where the related (scalar) transform coefficient can be expressed as the Frobenius product (A.10) of matrices or tensors,

$$c_\mathbf{k} = \mathbf{T}_\mathbf{k} : \mathbf{S} . \tag{2.248}$$

2.7.2 Types of orthogonal transforms

In this section, basis functions of some important transforms are introduced mostly by their one-dimensional versions. They extend to the case of two-dimensional separable transforms according to (2.246).

$$\mathbf{T}^\text{Haar}(8) = \frac{1}{2\sqrt{2}}\begin{bmatrix} 1 & 1 & 1 & 1 & 1 & 1 & 1 & 1 \\ 1 & 1 & 1 & 1 & -1 & -1 & -1 & -1 \\ \sqrt{2} & \sqrt{2} & -\sqrt{2} & -\sqrt{2} & 0 & 0 & 0 & 0 \\ 0 & 0 & 0 & 0 & \sqrt{2} & \sqrt{2} & -\sqrt{2} & -\sqrt{2} \\ 2 & -2 & 0 & 0 & 0 & 0 & 0 & 0 \\ 0 & 0 & 2 & -2 & 0 & 0 & 0 & 0 \\ 0 & 0 & 0 & 0 & 2 & -2 & 0 & 0 \\ 0 & 0 & 0 & 0 & 0 & 0 & 2 & -2 \end{bmatrix} = \begin{bmatrix} \mathbf{t}_0^\text{T} \\ \mathbf{t}_1^\text{T} \\ \vdots \\ \vdots \\ \\ \\ \vdots \\ \mathbf{t}_7^\text{T} \end{bmatrix} . \tag{2.249}$$

Rectangular basis functions. The analysis block lengths M of the following rectangular basis function transforms are typically dyadic ($M=2^l$, $l \in \mathbb{N}$). Typically (except for scaling necessary to achieve orthonormality), these transforms can be computed without multiplications. The Haar transform uses basis functions of non-constant length[60], where identical elementary functions (performing difference analysis over neighbored samples) are re-used at different positions of the

---

[59] In case of orthonormality, $\mathbf{S}=\mathbf{T}_\text{v}^\text{H}\mathbf{C}\mathbf{T}_\text{h}^*$.

[60] A more systematic construction of the Haar and Walsh transforms can be found in the formulation of Problem 2.13. The Haar transform can also be defined as a discrete wavelet transform (see Sec. 2.8.4) from the filter basis (2.312).

block. As an example, the transform matrix of an orthonormal Haar transform with $U=M=8$ is shown in (2.249). For the orthonormal transform, the scaling factors for the different basis types vary. Basis functions for the case $M=8$ are shown in Fig. 2.39a. The *Walsh basis* consists of $U=M$ basis functions, the set for the case $M=8$ is shown in Fig. 2.39b. The corresponding transform matrix is

$$\mathbf{T}^{\text{Walsh}}(8)=\frac{1}{2\sqrt{2}}\begin{bmatrix}1&1&1&1&1&1&1&1\\1&1&1&1&-1&-1&-1&-1\\1&1&-1&-1&-1&-1&1&1\\1&1&-1&-1&1&1&-1&-1\\1&-1&-1&1&1&-1&-1&1\\1&-1&-1&1&-1&1&1&-1\\1&-1&1&-1&-1&1&-1&1\\1&-1&1&-1&1&-1&1&-1\end{bmatrix}=\begin{bmatrix}\mathbf{t}_0^{\mathrm{T}}\\\mathbf{t}_1^{\mathrm{T}}\\\vdots\\\vdots\\ \\ \\\vdots\\\mathbf{t}_7^{\mathrm{T}}\end{bmatrix}. \quad (2.250)$$

The Walsh transform can be interpreted to be analyzing 'frequency' (based on toggling rectangles rather than oscillating sinusoids), as the number of zero crossings is steadily increasing with index k.

The *Hadamard transform* has the same set of functions as the Walsh basis, however the ordering (index numbering of basis functions) is different, not allowing interpretation by 'increasing frequency'. The rule for recursive construction implicitly guarantees orthogonality. Starting from a 1x1 identity matrix with $M'=1$, the recursion doubles the block length by each step,

$$\mathbf{T}^{\text{Had}}(1)=[1],$$
$$\mathbf{T}^{\text{Had}}(2M')=\frac{1}{\sqrt{2}}\begin{bmatrix}\mathbf{T}^{\text{Had}}(M') & \mathbf{T}^{\text{Had}}(M')\\ \mathbf{T}^{\text{Had}}(M') & -\mathbf{T}^{\text{Had}}(M')\end{bmatrix}\quad \text{for } M'=1,2,4,\ldots,M/2. \quad (2.251)$$

The Hadamard transform matrix for the case $M=8$ then is (see also Fig. 2.39c):

$$\mathbf{T}^{\text{Had}}(8)=\frac{1}{2\sqrt{2}}\begin{bmatrix}1&1&1&1&1&1&1&1\\1&-1&1&-1&1&-1&1&-1\\1&1&-1&-1&1&1&-1&-1\\1&-1&-1&1&1&-1&-1&1\\1&1&1&1&-1&-1&-1&-1\\1&-1&1&-1&-1&1&-1&1\\1&1&-1&-1&-1&-1&1&1\\1&-1&-1&1&-1&1&1&-1\end{bmatrix}=\begin{bmatrix}\mathbf{t}_0^{\mathrm{T}}\\\mathbf{t}_1^{\mathrm{T}}\\\vdots\\\vdots\\ \\ \\\vdots\\\mathbf{t}_7^{\mathrm{T}}\end{bmatrix}. \quad (2.252)$$

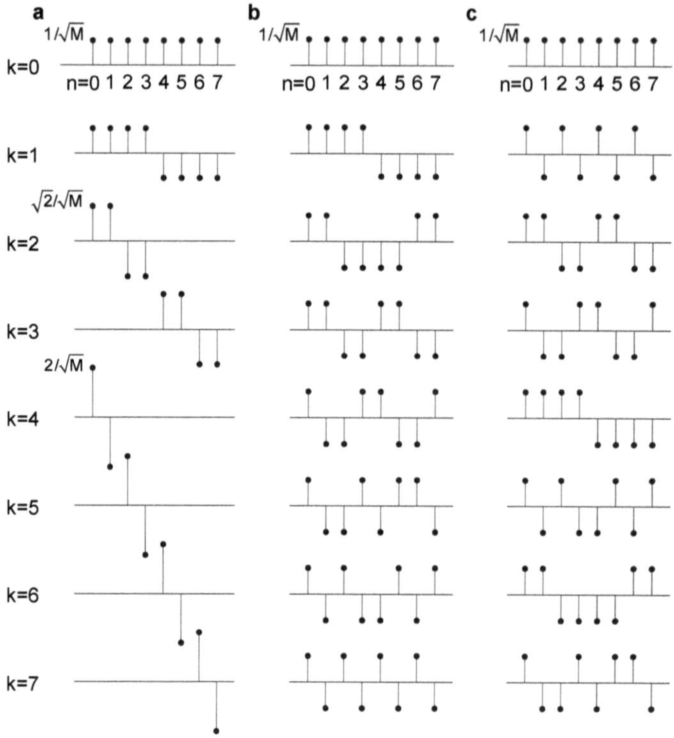

Fig. 2.39. Rectangular basis function systems **a** Haar **b** Walsh **c** Hadamard

$$\mathbf{T}^{\mathrm{DFT}} = \begin{bmatrix} 1 & 1 & 1 & 1 & 1 & \cdots & 1 \\ 1 & \mathrm{e}^{-\mathrm{j}\frac{2\pi}{M}} & \mathrm{e}^{-\mathrm{j}\frac{4\pi}{M}} & \mathrm{e}^{-\mathrm{j}\frac{6\pi}{M}} & \cdots & & \mathrm{e}^{-\mathrm{j}\frac{2\pi(M-1)}{M}} \\ 1 & \mathrm{e}^{-\mathrm{j}\frac{4\pi}{M}} & \mathrm{e}^{-\mathrm{j}\frac{8\pi}{M}} & & \ddots & & \mathrm{e}^{-\mathrm{j}\frac{4\pi(M-1)}{M}} \\ 1 & \mathrm{e}^{-\mathrm{j}\frac{6\pi}{M}} & & & & & \\ 1 & \vdots & \ddots & & \ddots & & \vdots \\ \vdots & & & & & & \\ 1 & \mathrm{e}^{-\mathrm{j}\frac{2\pi(U-1)}{M}} & \mathrm{e}^{-\mathrm{j}\frac{4\pi(U-1)}{M}} & & \cdots & & \mathrm{e}^{-\mathrm{j}\frac{2\pi(U-1)(M-1)}{M}} \end{bmatrix}. \qquad (2.253)$$

Sinusoidal basis functions. The Discrete Fourier Transform (DFT) is defined as

$$c_k = \sum_{n=0}^{M-1} s(n) W_M{}^{-mk} \quad s(n) = \sum_{k=0}^{M-1} c_k W_M{}^{mk} \quad \text{with } W_M = \mathrm{e}^{\mathrm{j}\frac{2\pi}{M}}. \qquad (2.254)$$

The complex exponential basis can be interpreted as harmonic sinusoids of specific frequency and phase. The transform matrix of the DFT is shown in (2.253). Note that this version of the DFT is not orthonormal. From (2.240), $A_k = M$ for a 1D transform and $A_{k_1k_2} = M_1M_2$ for a separable 2D transform[61]. Further, the DFT implicitly interprets a series of samples as periodic, even if they only represent a segment from a longer signal. Therefore, occasional amplitude differences between the left and right boundaries of the analysis segment are interpreted as discontinuity (see Fig. 2.40a), and spectral energy appears over broad frequency ranges. Further, when the signal is locally periodic, but the wave length (or a multiple thereof) does not match with M, energy is also spread over a certain range of the spectrum. Therefore, the DFT possesses undesirable properties with the threat of producing artifacts both in picture/video and audio compression. One approach of avoiding this is usage of window functions with roll-off towards the ends, typically used with overlap of adjacent blocks.

The amplitude discontinuity can also be avoided, if a (mirror) symmetric extension of the signal is constructed, which leads to an even symmetry and a real-valued DFT spectrum. In a first approach, even symmetry can be implemented around the points $n=0$ and $n=M-1$, with a period length of $2M-2$ (from M independent samples), as shown in Fig. 2.40b. Computing a DFT

$$c_k = \sum_{n=-M+1}^{M-2} s(n)\mathrm{e}^{-\mathrm{j}2\pi\frac{kn}{2M-2}} \quad \text{with} \quad s(n) = s(-n) \quad \text{for} \quad n<0, \tag{2.255}$$

gives real-valued coefficients as

$$\begin{aligned} c_k &= s(0) + (-1)^k s(M-1) + \sum_{n=1}^{M-2} s(n)\left[\mathrm{e}^{\mathrm{j}2\pi\frac{kn}{2M-2}} + \mathrm{e}^{-\mathrm{j}2\pi\frac{kn}{2M-2}}\right] \\ &= s(0) + (-1)^k s(M-1) + 2\sum_{n=1}^{M-2} s(n)\cos\left[\frac{\pi}{M-1}nk\right]. \end{aligned} \tag{2.256}$$

The values c_k are periodic over k with $2M-2$, and are again even symmetric around $k=0$ and $k=M-1$. Therefore, the inverse transform is identical, except for a normalization factor $1/(2M-2)$[62]. This real-valued transform is entitled as DCT type-I (for a detailed description of the different types and their implementation, see [BRITANAK ET AL. 2010] [CHEN, SMITH, FRALICK 1977]).

For applications in data compression, the DCT-I is however not best suitable due to the property that all basis functions have a maximum amplitude value at $n=0$, which can lead to high errors at the left block edge when transform coeffi-

---

61 For an orthonormal version of the 1D DFT, normalization by a factor $1/\sqrt{M}$ has to be applied both in the analysis and synthesis (IDFT). Likewise, for a 2D transform, the normalization must use a factor $1/\sqrt{M_1M_2}$.

62 Alternatively, forward and inverse transform are identical in case of orthonormality, where a normalization $1/\sqrt{1/(2M-2)}$ is applied in (2.256).

cients are discarded or heavily quantized. Second, the basis functions do not have symmetry properties themselves due to the misalignment between the length of M samples and the cosine which is periodic over $M-1$ or a multiple thereof. Third, the lowest frequency is representing approximately a full cosine period over the length of the basis function, such that signals with a slower increasing amplitude are not efficiently presented (which are often observed particularly in image signals).

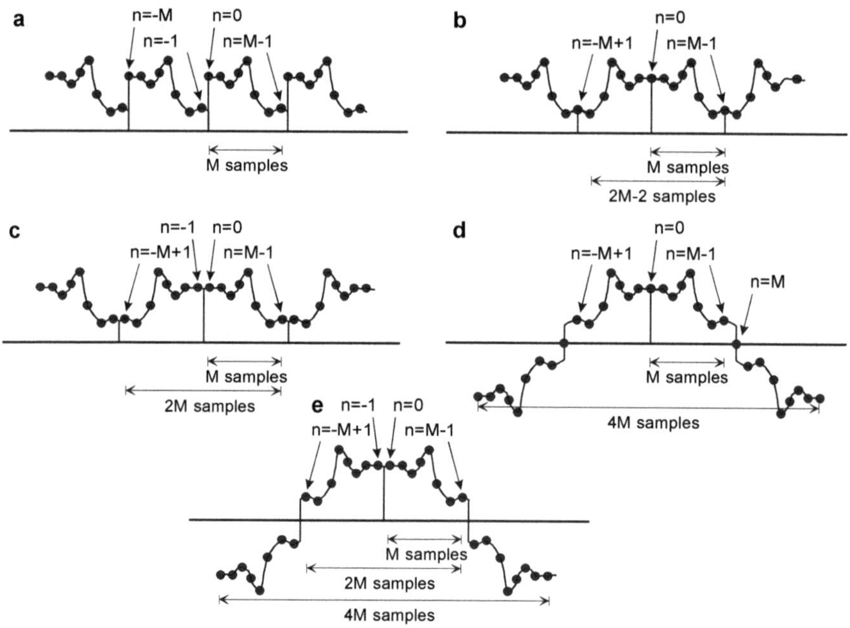

Fig. 2.40. Extension of the signal at boundaries of a finite analysis segment of length M **a** periodic (DFT case) **b** DCT type I **c** DCT type II **d** DCT type III **e** DCT type IV

To overcome these problems, the points of even symmetry could also be put at $n=-½$ and $n=M-½$, such that these two points are duplicated as well, and the even Fourier transform has to be computed over a length of $2M$, where now exactly half of the samples is redundant (see Fig. 2.40c). This can be realized by a modification of the DFT basis function, through a shift by half a sample in the complex exponent. Then,

$$c_k = \sum_{n=-M}^{M-1} s(n)\mathrm{e}^{-\mathrm{j}2\pi\frac{k}{2M}\left(n+\frac{1}{2}\right)} \quad \text{with} \quad s(n)=s(-n-1) \quad \text{for} \quad n<0, \tag{2.257}$$

which can be re-written as DCT type II:

$$c_k = \sum_{n=0}^{M-1} s(n)\left[\mathrm{e}^{-\mathrm{j}2\pi\frac{k}{2M}\left(-n-1+\frac{1}{2}\right)} + \mathrm{e}^{-\mathrm{j}2\pi\frac{k}{2M}\left(n+\frac{1}{2}\right)}\right] = 2\sum_{n=0}^{M-1} s(n)\cos\left[k\left(n+\frac{1}{2}\right)\frac{\pi}{M}\right]. \tag{2.258}$$

In the coefficient domain, the following observations are made: Coefficients c_M and c_{-M} are zero; otherwise, the following symmetries apply:

$$c_{-k} = c_k \;\text{ for }\; 0 < k < M; \quad c_{|k|} = -c_{|2M-k|} \;\text{ for }\; M < |k| < 2M\,. \tag{2.259}$$

This means that the series of coefficients has an odd symmetry around $k=\pm M$, and is periodic in k over a length of $4M$, where however still only M independent coefficients exist, all other are redundant. This can be explained by the fact that by introducing the shift by half a sample virtually the sampling rate is doubled and the block length of the DCT would also be $4M$, where however each second sample is implicitly zero. Therefore, alias spectra appear within the spectrum period (cf. Sec. 2.8.1). In terms of the inverse DCT, the following computation is necessary (formally, the sum should run over $4M$ samples with normalization by $1/(4M)$, but the coefficients for $|k| \geq M$ with corresponding complex exponentials would give exactly the same contribution and can therefore be omitted):

$$\begin{aligned} s(n) &= \frac{1}{2M}\left[\sum_{k=0}^{M-1} c_k\, \mathrm{e}^{\mathrm{j}2\pi\frac{k}{2M}\left(n+\frac{1}{2}\right)} + \sum_{k=-M+1}^{1} c_k\, \mathrm{e}^{\mathrm{j}2\pi\frac{k}{2M}\left(n+\frac{1}{2}\right)}\right] \text{ with } c_{-k} = c_k \\ &= \frac{1}{M}\left(\frac{c_0}{2} + \sum_{k=1}^{M-1} c_k \cos\left[k\left(n+\frac{1}{2}\right)\frac{\pi}{M}\right]\right). \end{aligned} \tag{2.260}$$

The transform of (2.260), when applied to a signal, is also entitled as DCT type III. Its symmetry properties (even around $n=0$ and odd around $n=M$) are shown in Fig. 2.40d. The corresponding inverse transform is the DCT-II.

Due to the shift by half a sample and usage of symmetric basis functions, the combination of DCT-II and DCT-III can also be used for linear interpolation (upsampling) of signals. In case of upsampling by a factor of 2, this can be achieved replacing the sign-inverted coefficients at positions $k=M+1\ldots 2M-1$ by zero values[63], and extending the inverse transform to the full length of $4M$ samples, such that $2M$ samples are generated by the inverse transform. Similarly, filling more zeroes and extending the block length of the inverse transform allows higher upsampling ratios.

Both DCT-II (2.258) and DCT-III (2.260) have orthogonal basis vectors, but do not fulfill (2.240), as the norm of $\mathbf{t}_0$ is different. By the following modification, orthonormality is achieved, and the basis vectors of the DCT-II can be written as

$$\begin{aligned} t_k^{\text{DCT-II}}(n) &= \sqrt{\frac{2}{M}} C_0 \cos\left[k\left(n+\frac{1}{2}\right)\frac{\pi}{M}\right] \quad \text{for } 0 \leq \{n,k\} < M \\ &\text{with } \; C_0 = \frac{1}{\sqrt{2}} \text{ for } k=0 \;\;;\;\; C_0 = 1 \text{ for } k \neq 0. \end{aligned} \tag{2.261}$$

Another concept is the DCT type IV, which somewhat combines the properties of DCT-II and DCT-III by using even symmetry around $n = -½$ and odd symmetry

[63] This is equivalent to suppression of alias spectra in interpolation filtering, cf. Sec. 2.8.1.

around $n=\pm M-½$ (see Fig. 2.40e). The period over n is $4M$ (again with only M independent samples), where however now the contributions of the values beyond the odd symmetry point contribute differently. Since $s(n)$ is even, values for $n<0$ would contribute as complex conjugate such that the DFT over length $4M$ gives

$$c_k = 2\,\mathrm{Re}\left\{\sum_{n=0}^{2M-1} s(n)\,\mathrm{e}^{-\mathrm{j}2\pi\frac{k}{2M}\left(n+\frac{1}{2}\right)}\right\} \text{with } s(n) = -s(2M-n-1) \text{ for } \quad n \geq M\,, \tag{2.262}$$

which can be re-written as

$$\begin{aligned} c_k &= 2\,\mathrm{Re}\left\{\sum_{n=0}^{M-1} s(n)\left[\mathrm{e}^{-\mathrm{j}2\pi\frac{k}{4M}\left(n+\frac{1}{2}\right)} - \mathrm{e}^{-\mathrm{j}2\pi\frac{k}{4M}\left(2M-n-1+\frac{1}{2}\right)}\right]\right\} \\ &= 2\,\mathrm{Re}\left\{\sum_{n=0}^{M-1} s(n)\left[\mathrm{e}^{-\mathrm{j}\pi\frac{k}{2M}\left(n+\frac{1}{2}\right)} - \mathrm{e}^{-\mathrm{j}\pi k}\,\mathrm{e}^{\mathrm{j}\pi\frac{k}{2M}\left(n+\frac{1}{2}\right)}\right]\right\}. \end{aligned} \tag{2.263}$$

The result is zero for even values of k. Replacing $k\rightarrow 2k+1$ for considering only the non-zero coefficients at odd positions finally gives

$$c_k = 4\sum_{n=0}^{M-1} s(n)\cos\left[\left(k+\frac{1}{2}\right)\left(n+\frac{1}{2}\right)\frac{\pi}{M}\right]. \tag{2.264}$$

In the DCT-IV, n and k have identical influences on the basis function. Furthermore, it can be concluded that virtual zero samples should exist both over n and k; including the frequency zero which would be at $k=-½$ in (2.264). Due to the symmetry of the coefficient series which is even around $k=-½$ and odd around $k=\pm M-½$, the inverse transform is identical to (2.264), except the need for amplitude scaling by $1/(8M)$[64]. The lowest discrete frequency comes with the basis function $\mathbf{t}_0$ which is a cosine with one full period over $4M$ (or a quarter of a period over M).

Due to these properties, the DCT-IV is most suitable for compression of zero-mean signals, such as audio. It is also used as a basis for block-overlapping transforms (see Sec. 2.7.4), where typically the length of the basis function is extended to an equivalent of $2M$ but multiplied by a window function that decays towards the tails and completely chops off the negative mirrored parts. By this, any undesirable effects of the odd symmetry at $n=\pm(M-½)$ are avoided, which is denoted as 'time domain alias cancellation' [PRINCEN, BRADLEY 1986].

As further variants of symmetric extensions with extended DFT basis functions, it is also possible to apply an *odd symmetry* around $n=0$ or $n=-½$. In this case, the real part of the DFT will be zero, but the values of the imaginary part can be used as if they were real valued coefficients. It should however be considered that odd symmetry requires the symmetry point itself to be zero, which has to

---

64 $8M$ is the actual period over k when the zero coefficients would be included. Alternatively, both forward and inverse transforms can be scaled by $\sqrt{2/M}$.

apply whenever n=0 or $n=\pm M$ are used as symmetry points. When there are M non-zero values, nevertheless the extended DFT length needs to include the zero values. Since odd signal components relate to the (imaginary) sine component of the DFT's complex exponential, this class of transforms is categorized as *Discrete Sine Transform* (DST). Similar to the discussion above, there are mainly four types:

- DST type I: Odd symmetries both at n=0 and $n=\pm M$, effective DFT period (including two zero samples) $2M$+2; it has symmetric basis functions.
- DST type II: Odd symmetries both at $n=-½$ and $n=M-½$, no zero samples, effective DFT period $2M$;
- DST type III: Odd symmetry at n=0, even symmetry at $n=\pm M$, effective DFT period (including two zero samples at n=0 and $n=-2M$) $4M$;
- DST type IV: Odd symmetries at $n=-½$, even symmetry at $n=M-½$, no zero samples, effective DFT period $4M$;

Again, DST-II and DST-III are inverses of each other, whereas DST-I and DST-IV are identical with their inverse transforms. Relevant in terms of data compression are DST-I and DST-IV, which match the properties of boundary prediction problems, where a set of M subsequent samples is predicted from the same boundary sample, such that the prediction error increases with larger distance from the boundary (see section 5.2.4) – in case of an AR(1) process, the DST in an optimum way removes correlation from the prediction error, specifically

- The DST-I is best suitable in case of two-sided prediction (block of M samples predicted from boundary samples at both ends) [JAIN 1976]. The first basis function from the following set of length M is a half sine wave with maximum in the center of the block; this matches the properties of the prediction error which can be expected to be maximum around position M/2 (farthest from the boundary samples used for prediction)

$$t_k^{\text{DST-I}}(n)=\sqrt{\frac{2}{M+1}}\cdot\sin\left[\frac{\pi}{M+1}(k+1)(n+1)\right] \quad ; \quad 0\le\{n,k\}<M\ ; \quad (2.265)$$

- The DST-IV is best suitable for one-sided prediction (block of M samples predicted from boundary samples at the beginning); The first basis function from the following set of length M is a quarter sine wave with maximum by the end of the block,

$$t_k^{\text{DST-IV}}(n)=\sqrt{\frac{2}{M}}\cdot\sin\left[\frac{\pi}{M}\left(k+\frac{1}{2}\right)\left(n+\frac{1}{2}\right)\right] \quad ; \quad 0\le\{n,k\}<M\,. \qquad (2.266)$$

The two-dimensional DCT is widely used in image and video compression (e.g. in standards like MPEG, JPEG, H.261/2/3). Mathematically precise formulations of the 2D DCT-II and its inverse 2D DCT-III over a rectangular signal block of size M_1M_2 are (with factors C_0 defined separately for the two dimensions, following (2.261))

$$c_{k_1k_2} = \frac{2}{\sqrt{M_1M_2}} C_0^{(k_1)} C_0^{(k_2)} \sum_{n_1=0}^{M_1-1} \sum_{n_2=0}^{M_2-1} s(n_1,n_2) \cos\left[k_1\left(n_1+\frac{1}{2}\right)\frac{\pi}{M_1}\right] \cos\left[k_2\left(n_2+\frac{1}{2}\right)\frac{\pi}{M_2}\right]; \quad (2.267)$$

$$s(n_1,n_2) = \frac{2}{\sqrt{M_1M_2}} \sum_{k_1=0}^{M_1-1} \sum_{k_2=0}^{M_2-1} C_0^{(k_1)} C_0^{(k_2)} c_{k_1k_2} \cos\left[k_1\left(n_1+\frac{1}{2}\right)\frac{\pi}{M_1}\right] \cos\left[k_2\left(n_2+\frac{1}{2}\right)\frac{\pi}{M_2}\right]. \quad (2.268)$$

Fig. 2.41 shows *basis images* of different separable 2D transforms, as defined by (2.247).

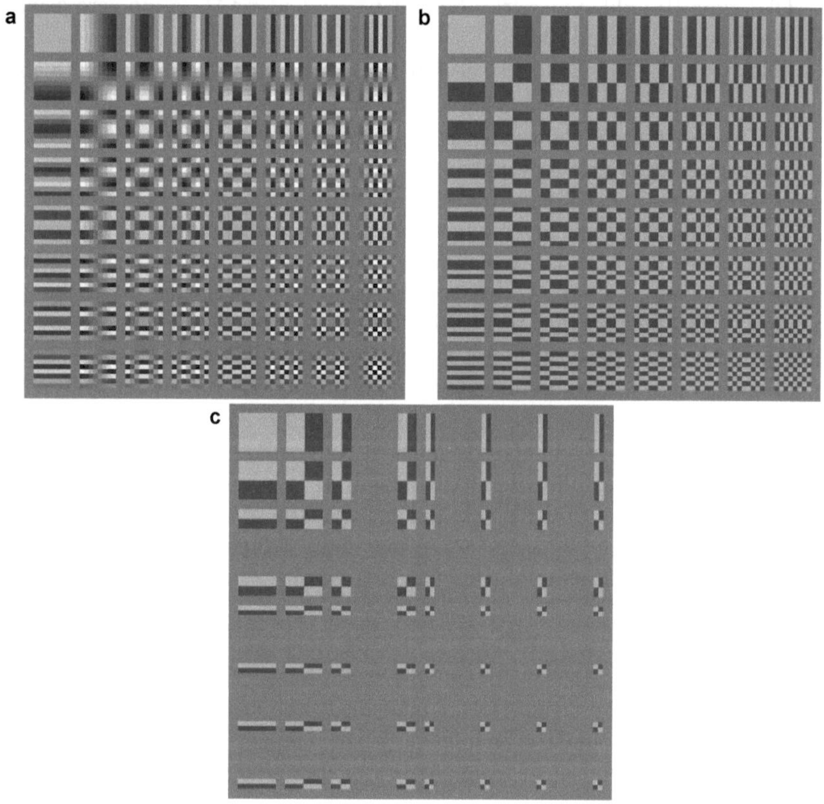

Fig. 2.41. 2D basis images of transforms. **a** DCT **b** Walsh **c** Haar

Integer transforms. Rectangular (binary) basis transforms allow to be computed without multiplications. Rectangular-basis transforms also guarantee perfect reconstruction of signals from a transform coefficient representation of finite bit precision[65]. Furthermore, the basis functions of rectangular transforms allow

[65] However, it should be observed that the necessary bit precision for lossless representation increases by $\log_2 M$ bits compared to the original signal representation, even in the case of Haar and Walsh transforms

efficient representation of discontinuities such as sharp edges, whereas they give only a poor representation of smoothly increasing amplitudes or smooth periodic structures. Sinusoidal transforms are better capable to approximate the latter types of signals by minimum error, but the trigonometric functions cannot be implemented up to full mathematical precision; rounding errors may occur, which could even affect the property of orthogonality. As a compromise, (non-binary) integer transform bases can be designed, which capture smoothly varying signal behavior better than rectangular basis functions, and retain orthogonality properties even with low word length integer arithmetic. One example for this class is the following transform of length $M=4$, which is used in the Advanced Video Coding standard (cf. Sec. 7.8) [MALVAR ET AL. 2003],

$$\mathbf{T}^{\text{int}}(4)=\begin{bmatrix}1 & 1 & 1 & 1\\ 2 & 1 & -1 & -2\\ 1 & -1 & -1 & 1\\ 1 & -2 & 2 & -1\end{bmatrix}. \tag{2.269}$$

Different normalization factors $1/2, 1/\sqrt{10}, 1/2, 1/\sqrt{10}$ have to be applied for orthonormality of the respective basis vectors of (2.269), but the necessary scaling can be combined with quantization (if used for compression). A truly orthonormal/identical-norm integer transform (the square root scaling factor could be transferred to the inverse transform), approximating a length-4 DCT is defined by the following matrix

$$\mathbf{T}^{\text{int}}(4)=\frac{1}{\sqrt{676}}\begin{bmatrix}13 & 13 & 13 & 13\\ 17 & 7 & -7 & -17\\ 13 & -13 & -13 & 13\\ 7 & -17 & 17 & -7\end{bmatrix}. \tag{2.270}$$

$$\mathbf{T}^{\text{int}}(8)=\frac{1}{\sqrt{1352}}\begin{bmatrix}13 & 13 & 13 & 13 & 13 & 13 & 13 & 13\\ 19 & 15 & 9 & 3 & -3 & -9 & -15 & -19\\ 17 & 7 & -7 & -17 & -17 & -7 & 7 & 17\\ 9 & 3 & -19 & -15 & 15 & 19 & -3 & -9\\ 13 & -13 & -13 & 13 & 13 & -13 & -13 & 13\\ 15 & -19 & -3 & 9 & -9 & 3 & 19 & -15\\ 7 & -17 & 17 & -7 & -7 & 17 & -17 & 7\\ 3 & -9 & 15 & -19 & 19 & -15 & 9 & -3\end{bmatrix}. \tag{2.271}$$

A truly orthonormal integer transform of block length $M=8$, with a similar construction as (2.270) proposed in [WIEN 2003], is defined by the transform matrix in (2.271). Herein, some of the basis functions can be constructed as mirror-symmetric extensions of the transform (2.270) with $M=4$, or the latter can be generated by using first halves of each second basis function in (2.271), as indicated by the boxes.

Similar constructions of 'nested' sets of transform basis functions for different transform block sizes have not been found yet beyond M=8, if the property of strict orthogonality shall be retained, However, if the orthogonality constraint is slightly released, such that $0 \neq \mathbf{t}_i^{\mathrm{T}}\mathbf{t}_j << \mathbf{t}_i^{\mathrm{T}}\mathbf{t}_i$ for $i \neq j$, similar constructions are possible. The HEVC standard contains the definition of integer approximations of the DCT for block sizes $M=4,8,16,32$, where the different-length basis functions are nested exactly the same way as above; also the norms of the basis vectors are almost equal and consistent over the various transform block sizes, such that no specific quantization needs to be employed. As an example, the transform matrix for $M=16$ is shown here, and the nested DCT functions for $M=8$ are highlighted in the leftmost eight columns [BUDAGAVI ET AL. 2013]:

$$\mathbf{T}^{\text{int}}(16)=\left[\begin{array}{rrrrrrrrrrrrrrrr}
64 & 64 & 64 & 64 & 64 & 64 & 64 & 64 & 64 & 64 & 64 & 64 & 64 & 64 & 64 & 64 \\
90 & 87 & 80 & 70 & 57 & 43 & 25 & 9 & -9 & -25 & -43 & -57 & -70 & -80 & -87 & -90 \\
89 & 75 & 50 & 18 & -18 & -50 & -75 & -89 & -89 & -75 & -50 & -18 & 18 & 50 & 75 & 89 \\
87 & 57 & 9 & -43 & -80 & -90 & -70 & -25 & 25 & 70 & 90 & 80 & 43 & -9 & -57 & -87 \\
83 & 36 & -36 & -83 & -83 & -36 & 36 & 83 & 83 & 36 & -36 & -83 & -83 & -36 & 36 & 83 \\
80 & 9 & -70 & -87 & -25 & 57 & 90 & 43 & -43 & -90 & -57 & 25 & 87 & 70 & -9 & -80 \\
75 & -18 & -89 & -50 & 50 & 89 & 18 & -75 & -75 & 18 & 89 & 50 & -50 & -89 & -18 & 75 \\
70 & -43 & -87 & 9 & 90 & 25 & -80 & -57 & 57 & 80 & -25 & -90 & -9 & 87 & 43 & -70 \\
64 & -64 & -64 & 64 & 64 & -64 & -64 & 64 & 64 & -64 & -64 & 64 & 64 & -64 & -64 & 64 \\
57 & -80 & -25 & 90 & -9 & -87 & 43 & 70 & -70 & -43 & 87 & 9 & -90 & 25 & 80 & -57 \\
50 & -89 & 18 & 75 & -75 & -18 & 89 & -50 & -50 & 89 & -18 & -75 & 75 & 18 & -89 & 50 \\
43 & -90 & 57 & 25 & -87 & 70 & 9 & -80 & 80 & -9 & -70 & 87 & -25 & -57 & 90 & -43 \\
36 & -83 & 83 & -36 & -36 & 83 & -83 & 36 & 36 & -83 & 83 & -36 & -36 & 83 & -83 & 36 \\
25 & -70 & 90 & -80 & 43 & 9 & -57 & 87 & -87 & 57 & -9 & -43 & 80 & -90 & 70 & -25 \\
18 & -50 & 75 & -89 & 89 & -75 & 50 & -18 & -18 & 50 & -75 & 89 & -89 & 75 & -50 & 18 \\
9 & -25 & 43 & -57 & 70 & -80 & 87 & -90 & 90 & -87 & 80 & -70 & 57 & -43 & 25 & -9
\end{array}\right] \tag{2.272}$$

From the matrix (2.272) and the construction of the nested shorter transforms, it can be recognized that the basis functions of the DCT alternate as even and odd symmetric functions with amplitude-identical coefficients at both sides (as can be expected from the type-II construction, see above). Only the even functions are useful for the construction of the shorter DCT transforms, as they again give an alternating even/odd set of functions. Likewise, the even functions of the next larger transform can be constructed by symmetric extension of all functions.

Another example for an almost orthogonal transform based on integer coefficients of basis functions is the approximation of the type-IV DST (2.266) which is used as alternative transform for this purpose with block length M=4 in the HEVC standard,

$$\mathbf{T}^{\text{int}}(4) = \begin{bmatrix} 29 & 55 & 74 & 84 \\ 74 & 74 & 0 & -74 \\ 84 & -29 & -74 & 55 \\ 55 & -84 & 74 & -29 \end{bmatrix}. \tag{2.273}$$

An optimum transform – KLT. Linear transforms for multimedia signal compression should approximate a signal as accurate as possible, using lowest possible number of transform coefficients. Often, the energy of the reconstruction error is used as criterion for optimality. Assuming that only T out of U coefficients shall be retained to represent the signal, the reconstruction error of a 1D transform is

$$e(n) = s(n) - \frac{1}{A}\sum_{k=0}^{T-1} c_k t_k^*(n). \tag{2.274}$$

This gives an energy of the error over the complete block

$$\begin{aligned} \|\mathbf{e}\|^2 &= \sum_{n=0}^{M-1} e^2(n) \\ &= \sum_{n=0}^{M-1} \left[s^2(n) - 2s(n)\frac{1}{A}\sum_{k=0}^{T-1} c_k t_k^*(m) + \left| \frac{1}{A}\sum_{k=0}^{T-1} c_k t_k^*(n) \right|^2 \right]. \end{aligned} \tag{2.275}$$

Substituting in (2.275) under the condition that $s(n)$ is real-valued

$$s(n) = \frac{1}{A}\sum_{l=0}^{U-1} c_l t_l^*(m) = \left[\frac{1}{A}\sum_{l=0}^{U-1} c_l^* t_l(n) \right]^* = \frac{1}{A}\sum_{l=0}^{U-1} c_l^* t_l(n) \tag{2.276}$$

gives

$$\|\mathbf{e}\|^2 = \sum_{n=0}^{M-1} \left[s^2(n) - \frac{2}{A^2}\sum_{l=0}^{U-1}\sum_{k=0}^{T-1} c_k t_k^*(n) c_l^* t_l(n) + \left| \frac{1}{A}\sum_{k=0}^{T-1} c_k^* t_k(n) \right|^2 \right], \tag{2.277}$$

by which, using the orthogonality condition (2.240),

$$\|\mathbf{e}\|^2 = \sum_{n=0}^{M-1} s^2(n) - \frac{1}{A}\sum_{k=0}^{T-1} |c_k|^2. \tag{2.278}$$

(2.278) shows that for $T = U = M$, $\|\mathbf{e}\|^2 = 0$, such that again the energy of the signal over M values can be determined from the coefficients of the linear orthogonal transform. This corresponds to the condition (2.244), however it is more general now as it shows that by omitting transform coefficients, the error energy in the reconstruction is exactly the energy of these coefficients, scaled by the normalization factor A.

Consequently, the optimum transform can be derived under the condition that the energy of the error shall be *minimized* if reconstruction from a finite number of coefficients is performed. This is equivalent to *maximizing* the energy over the first T coefficients. Now assume that samples from a stationary random process shall be transformed, expected squared values are taken of the first T coefficients, such that the following condition can be formulated (for simplicity, the orthonormal case is considered):

$$\begin{aligned}\sum_{k=0}^{T-1}\mathcal{E}\left\{\left|c_k\right|^2\right\}&=\sum_{k=0}^{T-1}\mathcal{E}\left\{c_k c_k^*\right\}=\sum_{k=0}^{T-1}\mathcal{E}\left\{\sum_{n=0}^{M-1}s(n)t_k(n)\sum_{m=0}^{M-1}s(m)t_k^*(m)\right\}\\&=\sum_{k=0}^{T-1}\mathcal{E}\left\{\sum_{m=0}^{M-1}\left(\sum_{n=0}^{M-1}s(n)s(m)t_k(n)\right)t_k^*(m)\right\}\\&=\sum_{k=0}^{T-1}\left[\sum_{m=0}^{M-1}\left(\sum_{n=0}^{M-1}\mu_{ss}\left(\left|m-n\right|\right)t_k(n)\right)t_k^*(m)\right].\end{aligned}\tag{2.279}$$

The maximum of this expression is approached if the innermost parenthesis fulfills the condition

$$\sum_{n=0}^{M-1}\mu_{ss}\left(\left|m-n\right|\right)t_k(n)=\lambda_k t_k(m),\tag{2.280}$$

which is achieved by establishing the transform basis as the set of eigenvectors of the discrete autocovariance sequence. Substituting (2.280) into (2.279) further gives

$$\sum_{k=0}^{T-1}\mathcal{E}\left\{\left|c_k\right|^2\right\}=\sum_{k=0}^{T-1}\lambda_k\underbrace{\left[\sum_{m=0}^{M-1}t_k(m)t_k^*(m)\right]}_{=1},\tag{2.281}$$

which shows that the related eigenvalue λ_k represents the power of coefficient c_k. The basis functions of this *Karhunen-Loève transform* (KLT) need specific adaptation by the autocovariance statistics of a given signal, and would be globally optimum for Gaussian (e.g. autoregressive) stationary processes. Formulating (2.280) for the entire set of basis functions, these can be computed as eigenvectors $\boldsymbol{\phi}_k$ of the autocovariance matrix (2.157)[66], which must then be constructed as an MxM matrix containing the covariance function samples $\mu_{ss}(0)$ ... $\mu_{ss}(M-1)$:

$$\begin{aligned}&\mathbf{C}_{ss}\boldsymbol{\phi}_k=\lambda_k\boldsymbol{\phi}_k\\&\quad\text{with}\quad\boldsymbol{\phi}_k=\left[\phi_k(0)\quad\phi_k(1)\quad\cdots\quad\phi_k(M-1)\right]^{\mathrm{T}}\quad;\quad 0\le k<U.\end{aligned}\tag{2.282}$$

Alternatively, the conjugates[67] of the eigenvectors $\boldsymbol{\phi}_k^*$ can be defined to establish the rows of transform matrix $\mathbf{T}^{\mathrm{KLT}}$ [68]

---

[66] We have assumed zero-mean property here and in the subsequent equations; however, the power component related to the mean would typically concentrate in coefficient c_0, such that the basic proof for the optimality of the KLT does not change.

[67] Definition by conjugates of eigenvectors makes the KLT consistent with the DFT definition and with correlation analysis; actually, the DFT is the optimum transform for perfect

$$\left[\boldsymbol{\phi}_k^*\right]^{\mathrm{T}} \mathbf{C}_{ss}{}^{\mathrm{T}} = \lambda_k \left[\boldsymbol{\phi}_k^*\right]^{\mathrm{T}}. \tag{2.283}$$

(2.283) can be written in matrix notation

$$\begin{bmatrix} \phi_0^*(0) & \phi_0^*(1) & \cdots & \phi_0^*(M-1) \\ \phi_1^*(0) & \ddots & \ddots & \\ \vdots & \ddots & & \vdots \\ \phi_{U-1}^*(0) & & \cdots & \phi_{U-1}^*(M-1) \end{bmatrix} \begin{bmatrix} \mu_{ss}(0) & \mu_{ss}(1 & \cdots & & \mu_{ss}(M-1) \\ \mu_{ss}(1) & \mu_{ss}(0) & \mu_{ss}(1) & \ddots & \\ \vdots & \mu_{ss}(1) & \ddots & \ddots & \vdots \\ & & \ddots & & \\ \mu_{ss}(M-1) & & \cdots & & \mu_{ss}(0) \end{bmatrix}$$

$$= \begin{bmatrix} \lambda_0 & 0 & 0 & \cdots & 0 \\ 0 & \lambda_1 & 0 & & \\ 0 & 0 & \lambda_2 & \ddots & \vdots \\ \vdots & & \ddots & \ddots & 0 \\ 0 & & \cdots & 0 & \lambda_{U-1} \end{bmatrix} \begin{bmatrix} \phi_0^*(0) & \phi_0^*(1) & \cdots & \phi_0^*(M-1) \\ \phi_1^*(0) & \ddots & \ddots & \\ \vdots & \ddots & & \vdots \\ \phi_{U-1}^*(0) & & \cdots & \phi_{U-1}^*(M-1) \end{bmatrix}$$

$$\Leftrightarrow \mathbf{T}^{\mathrm{KLT}} \mathbf{C}_{ss} = \boldsymbol{\Lambda} \mathbf{T}^{\mathrm{KLT}}. \tag{2.284}$$

Multiplying both sides of (2.284) by the inverse transform matrix retains the diagonal matrix $\boldsymbol{\Lambda}$ on the right side, which is populated by the eigenvalues λ_k:

$$\mathbf{T}^{\mathrm{KLT}} \mathbf{C}_{ss} \left[\mathbf{T}^{\mathrm{KLT}}\right]^{\mathrm{H}} = \boldsymbol{\Lambda} = \mathrm{Diag}\left\{\lambda_k\right\} \;\Rightarrow \sum_{k=0}^{U-1} \mathcal{E}\left\{\left|c_k\right|^2\right\} = \mathrm{tr}\left\{\boldsymbol{\Lambda}\right\} = \mathcal{E}\left\{\sum_{n=0}^{M-1} s^2(n)\right\}. \tag{2.285}$$

$\boldsymbol{\Lambda}$ in (2.285) can also be regarded as the 'optimum transform' of the autocovariance matrix, whereby a statistical representation of the discrete spectral samples c_k is generated. While the correlation inherent in the signal is indicated by the fact that $\mathbf{C}_{ss}$ is not a diagonal matrix, the diagonal shape of $\boldsymbol{\Lambda}$ indicates that no correlation is present anymore between the spectral samples.

2.7.3 Efficiency of transforms

An important criterion for judging the efficiency of a transform is concentration of as much signal energy as possible in as few transform coefficients as possible [CLARKE 1985]. The related *energy packing efficiency* η_e is the normalized ratio of energy, contained within the first T out of U transform coefficients:

---

cyclic signals, such as signals composed from sinusoids, each with a period being an exact fraction of the analysis block length.

68 This provides consistency with transform basis definitions used so far, which could be interpreted as a correlation test between the signal samples and the respective basis function. In case of a complex basis, it is necessary to test against the complex conjugate.

$$\eta_e(T) = \frac{\sum_{l=0}^{T-1} \mathcal{E}\left\{c_l^{\,2}\right\}}{\sum_{k=0}^{U-1} \mathcal{E}\left\{c_k^{\,2}\right\}}. \tag{2.286}$$

The KLT maximizes the energy packing efficiency, as it is optimized using this criterion, see (2.279). Another aspect is the *decorrelation efficiency* η_c, determined from the autocovariance matrix $\mathbf{C}_{ss}$ and its transform[69],

$$\mathbf{C}_{cc} = \mathcal{E}\left\{\mathbf{c}\mathbf{c}^{\mathrm{H}}\right\} = \mathcal{E}\left\{(\mathbf{T}\mathbf{s})(\mathbf{T}\mathbf{s})^{\mathrm{H}}\right\} = \mathbf{T}\mathcal{E}\left\{\mathbf{s}\mathbf{s}^{\mathrm{T}}\right\}\mathbf{T}^{\mathrm{H}} = \mathbf{T}\mathbf{C}_{ss}\mathbf{T}^{\mathrm{H}}. \tag{2.287}$$

The decorrelation efficiency is then defined as[70]

$$\eta_c = 1 - \frac{\sum_{\substack{k=0\\(k\neq l)}}^{U-1} \sum_{l=0}^{U-1} \left|\mu_{cc}(k,l)\right|}{\sum_{\substack{k=0\\(k\neq l)}}^{M-1} \sum_{l=0}^{M-1} \left|\mu_{ss}(k,l)\right|}. \tag{2.288}$$

For the case of the KLT, $\mathbf{C}_{cc}$ is the eigenvalue matrix $\mathbf{\Lambda}$ in (2.285) where all entries with $k\neq l$ are zero, hence η_c has a maximum of 1. This means that the KLT achieves optimum decorrelation when optimized for a signal that posesses certain autocovariance statistics. For other transforms than the KLT, also linear statistical dependencies (non-zero correlation) between coefficients of the discrete transform representation may be present.

2.7.4 Transforms with block overlap

A linear transform can be interpreted as a convolution of the signal, using impulse responses which are the time reversed and complex conjugate basis functions. Unlike conventional convolution, the computation of transform coefficients needs only to be performed at each Mth position in the case of transforms without block overlap, which can also be interpreted as subsampling of the convolution output. In the spectral domain, the generation of the transform coefficient can therefore be interpreted as multiplication of the signal spectrum by the Fourier transfer functions of the respective basis vectors. The transform coefficients are carrying information related to *all frequencies* which are passing through their respective Fourier-domain transfer functions. Fig. 2.42 shows the Amplitude transfer func-

---

[69] When the number of basis functions equals the number of samples, the square matrices $\mathbf{C}_{ss}$ and $\mathbf{C}_{cc}$ are both of same size.

[70] An exception is the case of zero-correlation (white noise), where a 0/0 division would happen. Formally, the decorrelation efficiency would be 0 in that case.

tions computed from several basis vectors of the DCT, block length $M = 8$. Beneath a pass-band, each of the functions has significant side lobes, indicating that the frequency separation property of the DCT in this case is rather poor.

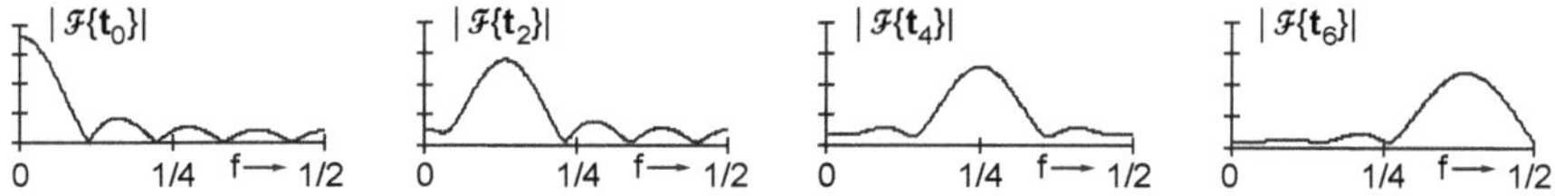

Fig. 2.42. Fourier amplitude spectra of DCT basis vectors $\mathbf{t}_0$, $\mathbf{t}_2$, $\mathbf{t}_4$, $\mathbf{t}_6$

Longer impulse responses (or basis functions) can improve the spectral cut-off and stop-band suppression. Windowing enforces the basis functions to roll off smoothly towards the tails and provides spectra with less energy in the side lobes compared to hard truncated functions. To prevent loss of information which is close to the tails of the window, basis functions of neighbored blocks need to overlap, such that synthesis can be performed by an overlap-add approach. The hop size between the start positions of two subsequent windows must not be larger than the number of transform coefficients U, such that the number of samples in the signal is not larger than the number of transform coefficients. With a hop size smaller than U, the transform would be over-complete; therefore, typically hop size U is used. This principle of *block-overlapping transforms* can still establish an orthogonal system of basis functions. Real-valued cosine-modulated functions combined with an appropriate weighting window are e.g. used in the *lapped orthogonal transform* (LOT) [MALVAR, STAELIN 1989] and in the TDAC transform (*time domain aliasing cancellation*) [PRINCEN, BRADLEY 1986]; more generally, this family of transforms is denoted as *cosine modulated filter banks* or *modified DCT*; a prominent application domain is audio signal compression (cf. Sec. 8.2.1).

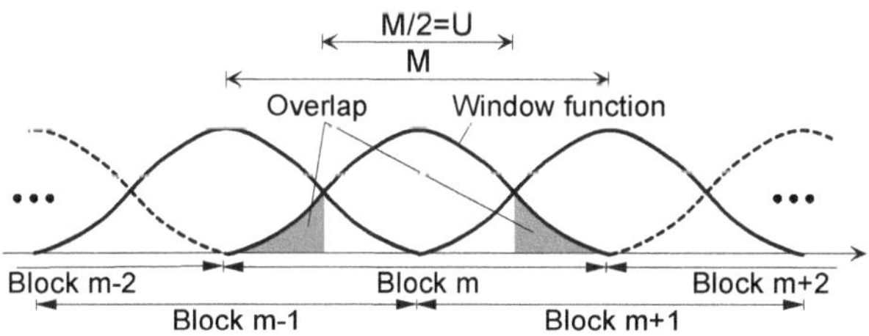

Fig. 2.43. Positions of analysis blocks with their overlapping window functions in a block-overlapping transform, M=2U

Example: TDAC transform [PRINCEN, BRADLEY 1986]. Here, decomposition is performed into U=M/2 frequency bands, the basis functions are based on an orthonormal version of the type-IV DCT (2.264), have a length M=2U and are defined as[71]

[71] Other overlap factors are possible, as long as the condition (2.290) holds.

$$t_k(n) = \sqrt{\frac{4}{M}} w(n) \cos\left[\frac{2\pi}{M}(k+0.5)(n+0.5-U/2)\right]. \qquad (2.289)$$

The orthonormality of the underlying DCT is not changed when multiplying the even or odd symmetric basis functions with an even-symmetric window function $w(n)$, and therefore the synthesis again uses the same set of basis functions. By this, the corresponding value of the window function is multiplied twice to a sample position in the overall signal flow. Therefore, when the entire sequence of window functions from all blocks is superimposed, their squared values must sum up to unity to achieve perfect reconstruction (see example from Fig. 2.44). Assuming a window function which is nonzero for $0 \le n < M-1$, this is fulfilled when the following conditions hold true in cases $M \le 2U$[72] and transition width $M-U$:

$$w^2(n) + w^2(n+U) = 1 \text{ for } 0 \le n < M-U \text{ and } w(n) = 1 \text{ for } M-U \le n < U. \qquad (2.290)$$

This also guarantees that the cosine basis functions are still orthonormal when considered jointly across all blocks,

$$\sum_{m=-M/U}^{M/U} \sum_{n=0}^{M-1} t_k(n+mU) t_l(n+mU) = \begin{cases} 1, & k = l \\ 0, & k \ne l. \end{cases} \qquad (2.291)$$

Typically, symmetric windows $w(n)$ are used, where $w(n)=w(M-1+n)$[73]. An example fulfilling (2.290) for the case $U=M/2$ (even M) is the sine window

$$w(m) = \sin\left(\frac{\pi}{M} \cdot (m+0,5)\right). \qquad (2.292)$$

Fig. 2.43 shows the Fourier-domain amplitudes of different TDAC basis functions using the sine window in case $U=8$, $M=16$. The side lobes of the spectra are largely reduced as compared to the DCT case in Fig 2.42, whereas the main lobes have become broader, which effects spectral overlap to occur mainly between directly neighbored frequency bands.

In the time domain, the block overlap causes smooth transitions between adjacent blocks instead of discontinuities in case of non-overlapping transforms, when coarse quantization or discarding of coefficients occurs during coding. This is particularly beneficial in case of periodic signals extending over block boundaries

---

[72] For $M>2U$, the hop size would be so small that more than two blocks overlap; in that case, the sum of squares from all window functions has to be constant.

[73] This symmetry is reasonable when the same window function is used over all transform blocks. This is however not necessary; moreover, orthonormality is still achieved whenever the entire time sequence over all squared window functions sums up to one. Furthermore, it is also possible to apply the squared window function *only during analysis* or *only during synthesis*, and apply a flat weighting (rectangular window with same overlap) at the other end. These properties allow adaptive switching of window lengths depending on signal properties, as often used in audio compression, or adaptive switching between overlapping and non-overlapping transforms, as used for image compression in the JPEG-XR standard.

(where phase discontinuities are avoided) and for signals with constant or smoothly increasing amplitude (avoiding unnatural amplitude discontinuities). On the other hand, the block overlap can be disadvantageous when the signal has discontinuities of the amplitude.

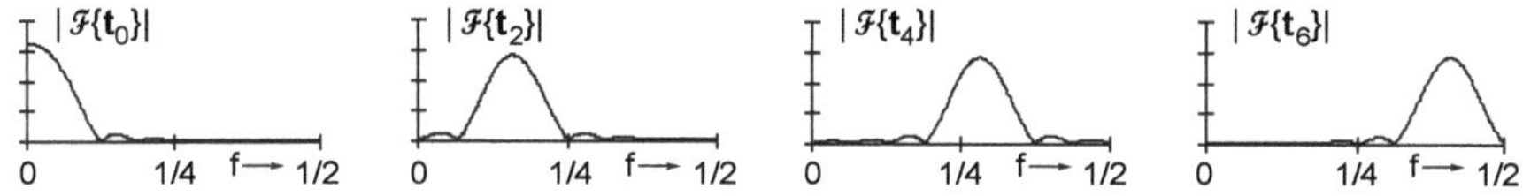

Fig. 2.44. Fourier amplitude spectra of TDAC basis vectors $\mathbf{t}_0$, $\mathbf{t}_2$, $\mathbf{t}_4$, $\mathbf{t}_6$

In principle, the transition shape of the window at the left and right block boundary can also be different, still achieving perfect reconstruction under the condition that the complementary shape (squares summing up to unity) is used in the corresponding adjacent block. This enables switching between windows/transforms of different length M or overlapping and non-overlapping transform basis functions locally, still retaining perfect reconstruction.

2.8 Filterbank transforms

The general principle of a *filterbank transform*, with linear block and overlapping transforms as special cases, is shown in Fig. 2.45. Interpretation of a linear transform analysis as parallel convolution operation with sub-sampling was given in Sec. 2.7.4; the inverse transform (synthesis) can be interpreted in a similar way. Generalizing this principle without explicit consideration of block segmentation and analysis hop sizes allows to formulate properties of the basis functions (filter impulse responses) with even more flexibility.

The frequency analysis is performed using U parallel filters. Direct usage of the filter output samples would give a representation which is *over-complete* by a factor of U. Therefore, the output signals of the different frequency bands are subsampled (decimated) and those retained are used as transform coefficients. The maximum (critical) decimation factor providing a complete representation of an arbitrary signal and thus enabling perfect reconstruction is equal to the number of subbands (U:1), such that the total number of coefficients equals the number of samples in $s(n)$. During synthesis, the signal is reconstructed by *interpolation* of the different subband signals and subsequent superposition of all components.

When comparing the DCT and its overlapping variants in Sec. 2.7.4, the aspect of spectral separation properties was discussed. With a set of ideal equal-bandwidth filters, U non-overlapping frequency bands cover a bandwidth of $f_\Delta = 1/(2U)$ each, such that alias-free critical sub-sampling and reconstruction could be applied according to sampling theory. This is however not possible if causal filters or filters with finite impulse response shall be used. With non-ideal filters and critical sub-sampling, overlaps of frequency bands occur, as schematically shown in Fig. 2.46. Fig. 2.46a shows the amplitude transfer function of a lowpass filter[74], which is shifted in frequency to provide the transfer functions of modulated bandpass and highpass filters. The corresponding layout of the spectrum (up to half sampling rate of the original signal) is shown in Fig. 2.46b.

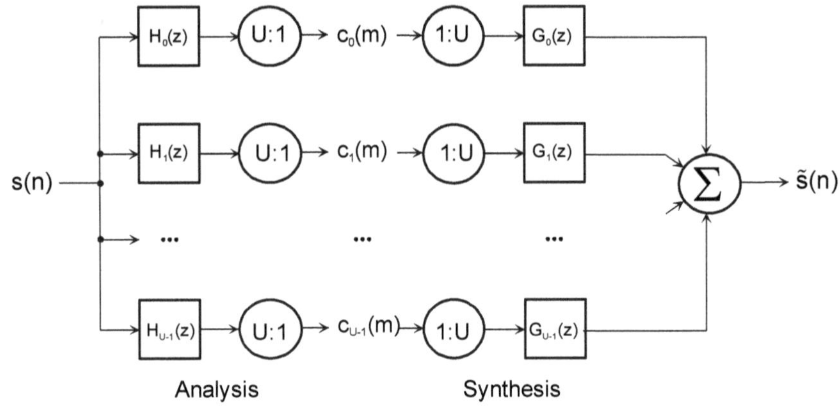

Fig. 2.45. Subband analysis and synthesis system, U frequency bands

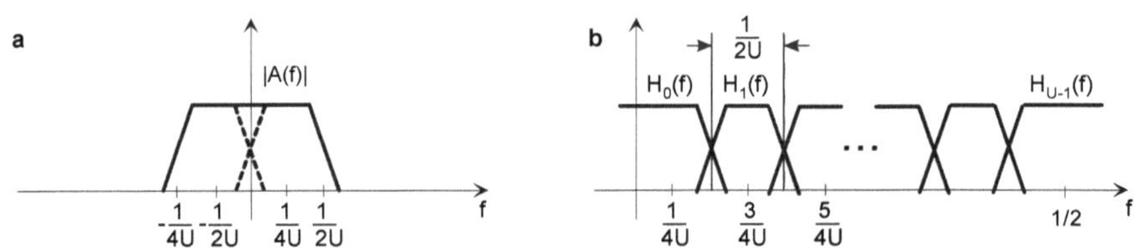

Fig. 2.46. a Lowpass filter **b** overlapping modulated bandpass filters

2.8.1 Decimation and interpolation

In case of discrete-time signals, scaling of the time axis has to be combined with downsampling (decimation) or upsampling (interpolation). The generation of a discrete signal $s_U(n)$, which is decimated by a factor U compared to the sampling of $s(n)$, is performed by discarding samples. The first step can be described as a multiplication by a train of Kronecker impulses

---

74 This lowpass filter can actually be interpreted as a superposition of two complex-conjugate bandpass filter transfer functions at centre frequencies $\pm 1/(4U)$. This allows to define bands of equal width in the range $0 \leq |f| < 1/2$. If the positive and negative parts of the spectra were regarded separately, the corresponding impulse responses would be complex.

$$s_{\delta_U}(n) = s(n) \sum_{m=-\infty}^{\infty} \delta(n - mU). \tag{2.293}$$

Subsequently, only each Uth value (i.e. one of the non-zero values) is retained without further information loss,

$$s_U(m) = s(mU) = s_{\delta_U}(mU). \tag{2.294}$$

The signals, $s(n)$, $s_{\delta_U}(n)$ and $s_U(m)$ are shown for the case U=2 in Fig. 2.47 left. The Fourier spectrum of the discrete Kronecker impulse sequence is a periodic sequence of Dirac impulses,

$$\sum_{m=-\infty}^{\infty} \delta(n - mU) \circ\!\!-\!\!\bullet \frac{1}{|U|} \sum_{k=-\infty}^{\infty} \delta\left(f - \frac{k}{U}\right). \tag{2.295}$$

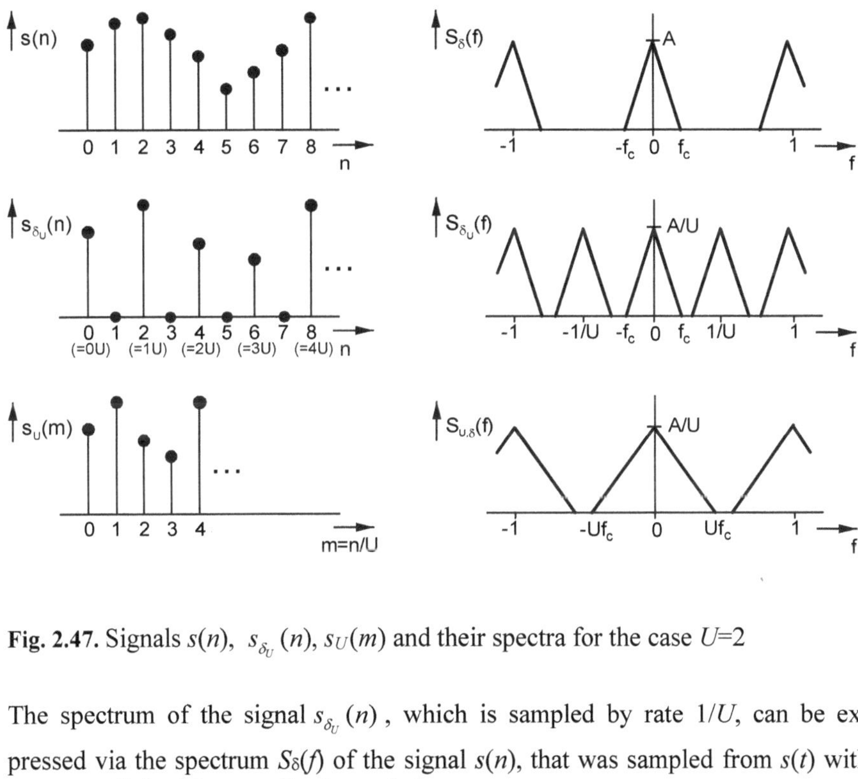

Fig. 2.47. Signals $s(n)$, $s_{\delta_U}(n)$, $s_U(m)$ and their spectra for the case U=2

The spectrum of the signal $s_{\delta_U}(n)$, which is sampled by rate $1/U$, can be expressed via the spectrum $S_\delta(f)$ of the signal $s(n)$, that was sampled from $s(t)$ with spectrum $S(f)$ with normalized rate f= 1, as

$$\begin{aligned} S_{\delta_U}(f) &= S_\delta(f) * \frac{1}{|U|} \sum_{k=0}^{U-1} \delta\left(f - \frac{k}{U}\right) \\ &= \frac{1}{|U|} \sum_{k=0}^{U-1} S_\delta\left(f - \frac{k}{U}\right) = \frac{1}{|U|} \sum_{k=-\infty}^{\infty} S\left(f - \frac{k}{U}\right). \end{aligned} \tag{2.296}$$

An identical spectrum $S_{\delta_U}(f)$ would show when the signal had originally been sampled by a rate $1/U$ (relative to f=1). If the signal was band limited to a maximum frequency $f_c = 1/(2U)$ before subsampling, no alias occurs. Computation of the spectrum is also possible directly from the subsampled signal,

$$
\begin{aligned}
S_{U,\delta}(f) &= \sum_{m=-\infty}^{\infty} s_U(m)\mathrm{e}^{-\mathrm{j}2\pi mf} = \sum_{m=-\infty}^{\infty} s_{\delta_U}(mU)\mathrm{e}^{-\mathrm{j}2\pi mf} \\
&= \sum_{n=-\infty}^{\infty} s_{\delta_U}(n)\mathrm{e}^{-\mathrm{j}2\pi n\frac{f}{U}} = S_{\delta_U}\left(\frac{f}{U}\right),
\end{aligned}
\tag{2.297}
$$

and therefore

$$
S_{U,\delta}(f) = \frac{1}{|U|}\sum_{k=-\infty}^{\infty} S\left(\frac{f-k}{U}\right). \tag{2.298}
$$

In (2.297) and (2.298), the frequency is re-normalized by the new sampling rate $1/U$ which means that the frequency axis of $S_{U,\delta}(f)$ is scaled by a factor U compared to the frequency axis of $S_{\delta_U}(f)$. Fig. 2.47 shows the respective spectra next to the corresponding signals.

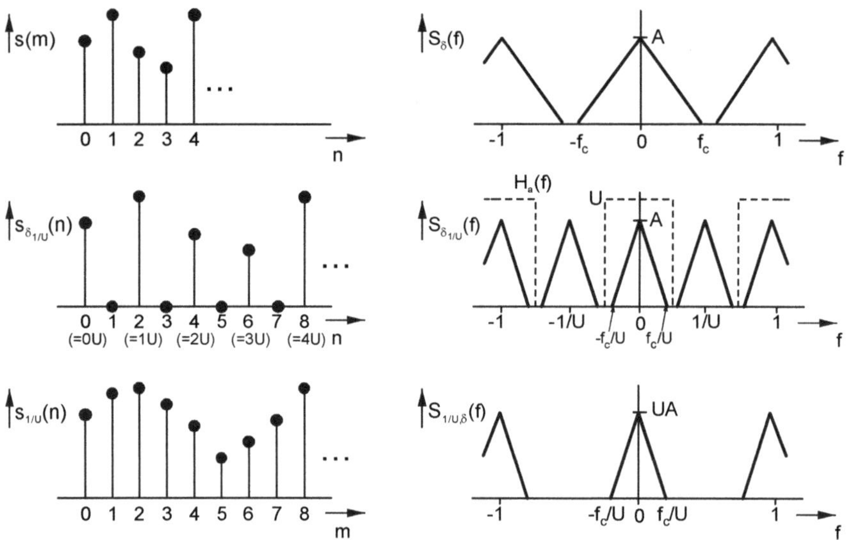

Fig. 2.48. Signals $s(m)$, $s_{\delta_{1/U}}(n)$, $s_{1/U}(n)$ and their spectra, example of upsampling by U=2

In interpolation, the increase of sampling rate by a factor U is achieved by inserting U–1 zero values between the available samples (Fig. 2.48):

$$
s_{\delta_{1/U}}(n) = \begin{cases} s\left(\dfrac{n}{U}\right) & \text{for} \quad m = \dfrac{n}{U} \in \mathbb{Z} \\ 0 & \text{else.} \end{cases} \tag{2.299}
$$

The related spectrum is scaled by a factor $1/U$ compared to the original spectrum $S_\delta(f)$,

$$S_{\delta_{1/U}}(f)=\sum_{n=-\infty}^{\infty}s_{\delta_{1/U}}(n)\mathrm{e}^{-\mathrm{j}2\pi nf}\,, \tag{2.300}$$

or alternatively

$$\begin{aligned}S_{\delta_{1/U}}(f)&=\sum_{m=-\infty}^{\infty}s_{\delta_{1/U}}(mU)\mathrm{e}^{-\mathrm{j}2\pi mUf}=\sum_{m=-\infty}^{\infty}s(m)\mathrm{e}^{-\mathrm{j}2\pi mUf}\\&=S_{\delta}(Uf)=\sum_{k=-\infty}^{\infty}S(Uf-k)=\sum_{k=-\infty}^{\infty}S\left[U\left(f-\frac{k}{U}\right)\right].\end{aligned} \tag{2.301}$$

When the sampling rate is re-normalized to f=1, U spectral copies (including the original baseband) appear in the range $-1/2\le f<1/2$. Lowpass filtering with cut-off frequency f_c=1/(2U) has to be applied to eliminate the U–1 alias copies and to generate the interpolated signal $s_{1/U}(n)$. Amplitude scaling by a factor of U is further necessary,

$$S_{1/U,\delta}(f)=S_{\delta_{1/U}}(f)H_{\mathrm{a}}(f)\ \text{ with }\ H_{\mathrm{a}}(f)=U\mathrm{rect}(Uf)*\sum_{k=-\infty}^{\infty}\delta(f-k)\,. \tag{2.302}$$

In the time domain, the impulse response of the lowpass filter (in ideal case a discrete-time sinc function) interpolates the missing values, leaving the originally available sampling positions m from (2.299) unchanged:

$$h(n)=\mathrm{si}\left(\frac{\pi n}{U}\right). \tag{2.303}$$

The spectrum of the interpolated signal $s_{1/U}(n)$ is

$$S_{1/U,\delta}(f)=|U|\sum_{k=-\infty}^{\infty}S\big[(f-k)U\big]=|U|S(Uf)*\sum_{k=-\infty}^{\infty}\delta(f-k)\,, \tag{2.304}$$

being identical to the spectrum of a signal that would have been originally sampled with a rate which is higher by a factor U,

$$s_{\delta_{1/U}}(t)=s(t)\sum_{n=-\infty}^{\infty}\delta\left(t-\frac{n}{U}\right)=\sum_{n=-\infty}^{\infty}s_{1/U}(n)\delta\left(t-\frac{n}{c}\right)\ \text{ with }\ s_{1/U}(n)=s\left(\frac{n}{U}\right). \tag{2.305}$$

The operations of decimation and interpolation described so far are only applicable with integer factors U. By combinations it is however possible to implement down- and upsampling by any rational factors, e.g. sampling rate conversion by a factor U_1/U_2 can be achieved by performing interpolation by a factor of U_1 followed by decimation by a factor U_2.

Interpolation in discrete time can be interpreted similar to continuous-time interpolation (2.51), it is however only performed at pre-defined positions. Before decimation, it is usually necessary to perform lowpass filtering to avoid alias, unless the signal is already appropriately band limited for the new sampling rate.

In case of re-sampling with rational factors, it is only necessary to compute those samples which would be retained after the second (decimation) step. This can be achieved by defining a set of interpolation filters typically having a target cut-off frequency

$$f_c = \min\left\{\frac{1}{2}, \frac{U_1}{2U_2}\right\}, \tag{2.306}$$

where the filters in the set have to be designed to support all phase shifts that can occur between the existing sampling positions. The number of filters to be defined for the non-existing re-sampling phase positions is $N_{\mathrm{Ph}}=\max\{U_1-1,U_2-1\}$. However, due to the fact that in case of rational re-sampling factors always two of the phase positions are mirror symmetric relative to the original sampling grid (e.g. 1/4 and 3/4=1−1/4), it is usually only necessary to design $\lfloor N_{\mathrm{Ph}}/2\rfloor+1$ different filters and re-use them with mirrored impulse response for the corresponding other position (an example with the interpolation filters of HEVC can be found in Tab. 7.1).

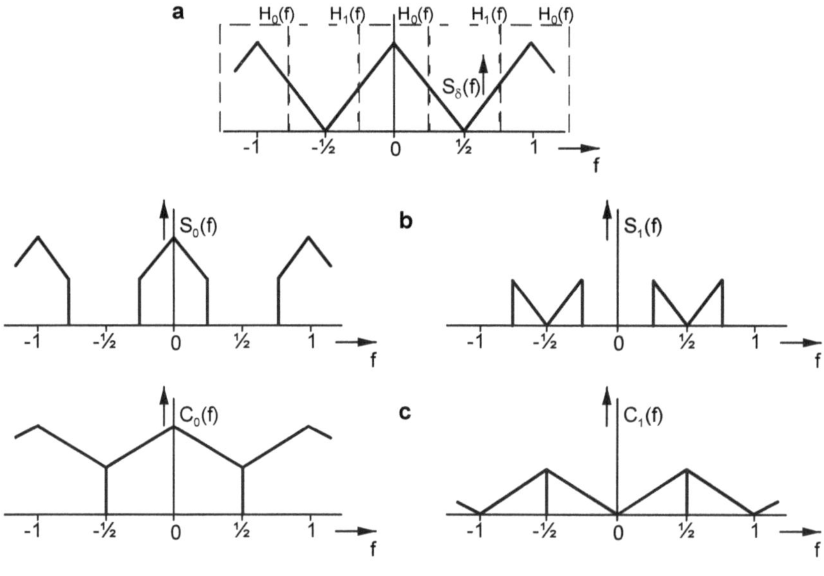

Fig. 2.49. Decomposition of a signal into decimated lowpass and highpass components **a** Signal spectrum **b** Spectra after lowpass/highpass filtering and multiplication by Kronecker impulse train **c** Spectra after sub-sampling

Further, the processes of decimation and interpolation are not restricted to lowpass signals as discussed so far, but can be applied to any appropriately band limited signals (e.g. bandpass outputs from the filter bank), such that no spectral overlaps occur. Fig. 2.49 illustrates decimation with U=2 applied in parallel to the *low* and *high frequency bands*, separated under assumption of ideal filters here. Fig. 2.49a shows the spectrum of the original signal, Fig. 2.49b the results after filtering and multiplication by the discrete sampling function (2.293), denoted as $S_k(f)=S_\delta(f)H_k(f)$, k=0 and k=1 for low and high components, respectively. By discarding the zero samples (Fig. 2.49c), the spectra $C_k(f)$ are expanded by the factor U=2. Observe that after sub-sampling the highpass spectra $C_1(f)$ appear over an *inverted frequency axis* around frequency zero, i.e. spectral components which originally were close to $f=1/2$ now appear around $f=0$, whereas components which were originally around $f=1/4$ are mapped into proximity of $f=1/2$ after sub-sampling. This phenomenon of frequency inversion likewise occurs in case of multiple-band filter banks within each *odd-indexed* band[75]

In the following sub-sections, it will be shown that perfect reconstruction can indeed be achieved even if the sub-sampling of the particular bands cannot be performed alias-free, i.e. different from the concept of Fig. 2.49, non-ideal filters are used for the separation. It is however then necessary to design the filter banks for the analysis and synthesis stages jointly, such that alias components are eliminated when the interpolated signals are superimposed; standalone alias-free interpolation of the different frequency band signals is no longer possible.

2.8.2 Properties of subband filters

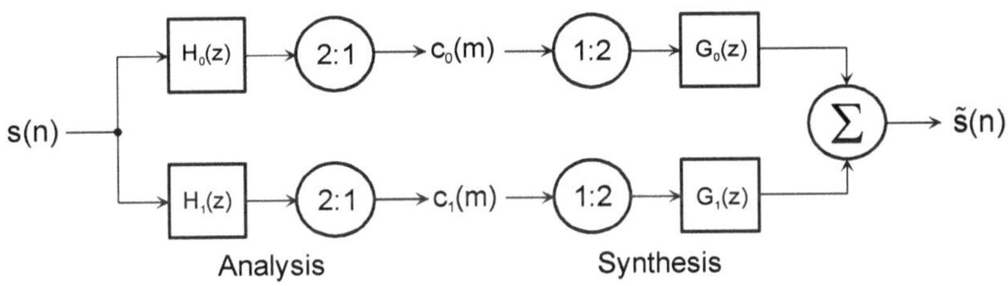

Fig. 2.50. Subband analysis system with U=2 frequency bands

Frequency transfer functions of the analysis filters from a critically-sampled filterbank overlap in case of non-ideal filters with finite impulse responses, such that alias can occurs by sub-sampling. Let $H_k(z)$ express the z-domain transfer functions of the analysis filters, $G_k(z)$ those of synthesis (interpolation) filters. For the case of $U=2$, which applies to the subband system in Fig. 2.50, only one lowpass band (k=0) and one highpass band (k=1) are generated. Hence, from (2.296), only one additional spectrum appears at $f=1/2$ or $z=-1$ due to subsampling after the filter operations $H_k(z)S(z)$. Over the complete system chain, the following spectrum appears after synthesis:

$$\tilde{S}(z)=\underbrace{\frac{1}{2}\left[H_0(z)G_0(z)+H_1(z)G_1(z)\right]S(z)}_{\text{baseband components}}+\underbrace{\frac{1}{2}\left[H_0(-z)G_0(z)+H_1(-z)G_1(z)\right]S(-z)}_{\text{alias components}}. \tag{2.307}$$

75 This gives ground for yet another interpretation about the correlations between pairs of even-indexed or odd-indexed coefficients that can often be observed in block transforms. The bands overlap in frequency, which is one of the causes for correlation between formally orthogonal components after sub-sampling. On the other hand, as even and odd bands appear by original and reversed frequency order, linear relations are lost, such that the correlation is cancelled out again.

Quadrature mirror filters (QMF). In (2.307), the upper part expresses the components from the baseband spectrum, while the lower term contains alias components, which shall be eliminated. This can be achieved if the lower term has a value of zero. In the QMF construction, the highpass analysis filter $H_1(f)$ is derived from the lowpass filter $H_0(f)$ by time reverting, modulating by a discrete cosine with $f = 1/2$ and shifting the impulse response. Due to the symmetry in the lowpass transfer function around $f = 0$, lowpass and highpass functions are symmetric around the point $f = 1/4$ after the modulation. Modulation, time reversal and shift establish them as a system of *orthogonal functions*. Typical QMF relationships of the different filters in the signal domain and the spectral domains of f- and z-transfer functions are listed in Table 2.1. The relevant mapping relationships for the impulse responses and the Fourier and z transfer functions are also given in the lower part of the table.

Table 2.1. Definition of quadrature mirror filters (QMF): Relationships of orthogonal lowpass and highpass analysis and synthesis filters, expressed by impulse responses, z- and Fourier spectra

	$a(n)$	$A(f)$	$A(z)$
H_0	$h_0(n)=a(n)$	$H_0(f)=A(f)$	$H_0(z)=A(z)$
H_1	$h_1(n)=(-1)^{-1-n}\cdot a(-1-n)$	$H_1(f)=e^{-j2\pi f}\cdot A(1/2-f)$	$H_1(z)=z^{-1}\cdot A(-z^{-1})$
G_0	$g_0(n)=a(-n)$	$G_0(f)=A(-f)$	$G_0(z)=A(z^{-1})$
G_1	$g_1(n)=(-1)^{n+1}\cdot a(n+1)$	$G_1(f)=e^{j2\pi f}\cdot A(f-1(2)$	$G_1(z)=z\cdot A(-z)$
Inversion	$h(-n)$	$H(-f)=H^*(f)$	$H(z^{-1})$
Modulation	$(-1)^n h(n)$	$H(f-1/2)=H^*(1/2-f)$	$H(-z)=H(z\cdot e^{-j\pi})$

Substituting $z=\exp(j2\pi f)$ in (2.307) gives

$$\begin{aligned}\tilde{S}(f) &= \frac{1}{2}\left[H_0(f)G_0(f)+H_1(f)G_1(f)\right]S(f) \\ &+\frac{1}{2}\left[H_0(f-1/2)G_0(f)+H_1(f-1/2)G_1(f)\right]S(f-1/2).\end{aligned} \tag{2.308}$$

If the common model filter $A(f)$ as defined in Table 2.1 is used,

$$\begin{aligned}\tilde{S}(f) &= \frac{1}{2}\left[A(f)A(-f)+A(1/2-f)A(f-1/2)\right]S(f) \\ &+\frac{1}{2}\left[A(f-1/2)A(-f)+e^{j\pi}A(-f)A(f-1/2)\right]S(f-1/2).\end{aligned} \tag{2.309}$$

The alias component at $f=1/2$ is eliminated, and the condition

$$A(f)A^*(f)+A(1/2-f)A^*(1/2-f)=|A(f)|^2+|A(1/2-f)|^2=2 \qquad (2.310)$$

gives perfect reconstruction at the output. (2.310) is generalized to the case of an arbitrary number of U subbands by

$$\sum_{k=0}^{U-1}|H_k(f)|^2=U. \qquad (2.311)$$

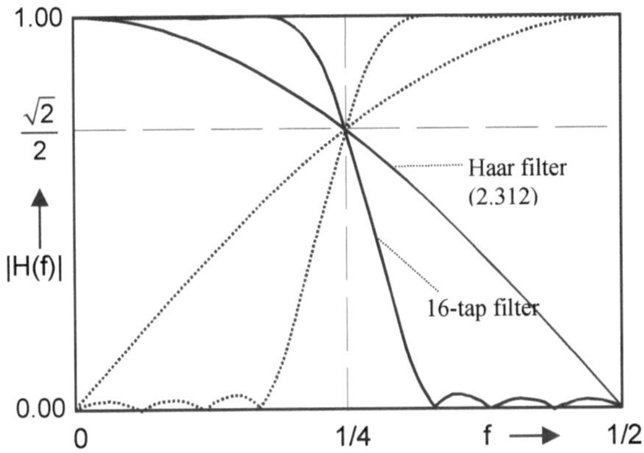

Fig. 2.51. Fourier magnitude transfer functions of filters from (2.312) and a 16-tap filter[76] [— lowpass ··· highpass]

Whereas the alias components are eliminated perfectly, the condition (2.311) for mathematically perfect reconstruction of the signal can only be fulfilled for two specific cases of QMF:

- Impulse response lengths are identical to the number of subbands U (which is the special case of block transforms, e.g. for U=2 the Haar filter basis);
- Ideal pass/stop band filters, which would require infinitely extended impulse responses (i.e. sinc function or modulated versions thereof).

Example: Haar filter basis. For the case $U=M=2$, the Haar filter defines the basis functions of almost any orthonormal block transforms, including DCT, Walsh, Hadamard, Haar transforms and the KLT as optimized for an AR(1) process. The z transfer functions according to Table 2.1 are[77]

---

[76] The 16-tap lowpass FIR filter has the z transfer function

$$H_0(z)=0.007\cdot z^7-0.02\cdot z^6+0.002\cdot z^5+0.046\cdot z^4-0.026\cdot z^3-0.099\cdot z^2+0.118\cdot z+0.472$$
$$+0.472\cdot z^{-1}+0.118\cdot z^{-2}-0.099\cdot z^{-3}-0.026\cdot z^{-4}+0.046\cdot z^{-5}+0.002\cdot z^{-6}-0.02\cdot z^{-7}+0.007\cdot z^{-8}.$$

[77] Another definition $H_1(z)=\frac{\sqrt{2}}{2}-\frac{\sqrt{2}}{2}\cdot z^{-1}$ (sign permutation) may alternatively be used.

$$H_0(z) = F(z) = \frac{\sqrt{2}}{2} + \frac{\sqrt{2}}{2} \cdot z^{-1}$$
$$H_1(z) = z^{-1} \cdot F(-z^{-1}) = -\frac{\sqrt{2}}{2} + \frac{\sqrt{2}}{2} \cdot z^{-1}. \tag{2.312}$$

The Fourier spectrum gives with (2.311)

$$\left|H_0(f)\right|^2 + \left|H_1(f)\right|^2 = \left(\frac{\sqrt{2}}{2}\right)^2 \cdot (2\cos \pi f)^2 + \left(\frac{\sqrt{2}}{2}\right)^2 \cdot (2\sin \pi f)^2 = 2. \tag{2.313}$$

The disadvantage however is the flat decay of the amplitude transfer function, due to the short length of the filters, causing poor frequency separation property and eventually strong alias in the sub-sampled signals. Other finite-length filters constructed by the conditions of Table 2.1 will not fulfill (2.311) perfectly. The design of such filters is made as a compromise between frequency separation properties for alias suppression in the subbands, and a reconstruction error which should be kept as low as possible, such that

$$U - \sum_{k=0}^{U-1} \left|H_k(f)\right|^2 \overset{!}{=} \min . \tag{2.314}$$

Fig. 2.51 shows the amplitude transfer functions of the filters (2.312) and a pair of 16 tap QMF filters originally suggested in [Johnston 1980], where for the latter the value that can be computed from (2.314) is in the range of 10^{-4}.

To define more general conditions for alias-free *and* lossless reconstruction, the constraint of QMF, where H_0 and H_1 are mirror-symmetric, can be released. From (2.307), the elimination of the alias component is also achieved if the following conditions are met[78]:

$$G_0(z) = \pm z^m \cdot H_1(-z),$$
$$G_1(z) = \mp z^m \cdot H_0(-z). \tag{2.315}$$

Substituting (2.315) into (2.307) gives

$$\tilde{S}(z) = \frac{1}{2}\left[H_0(z)H_1(-z) - H_1(z)H_0(-z)\right]S(z)z^m , \tag{2.316}$$

which gives as condition for perfect reconstruction

$$K(z) - K(-z) = 2z^{-m} \quad \text{with} \quad K(z) = H_0(z)H_1(-z). \tag{2.317}$$

Two types of filters, which are determined from (2.315)-(2.317), are introduced in

---

[78] Both combinations ($\pm / \mp$) are possible. The following equations in the explanation of PRF use the first option ($\pm$), whereas e.g. (2.325) uses ($\mp$).

the following sub-sections. The term z^{-m} expresses an arbitrary shift which may occur anywhere in the analysis/synthesis chain. In image processing, filtering is often performed such that the current sample of the signal is weighted by the center sample of the impulse response (in case of odd length) or by one of the two center samples (in case of even length). As all filters introduced here are of FIR type[79], this has the effect that the reconstructed pictures are not spatially shifted.

Perfect reconstruction filters (PRF). For this type of filter, the basis functions of lowpass and highpass can be orthogonal, but the highpass impulse response may no longer be a modulated version of the lowpass response, no mirror symmetry exists. Typically, the resulting frequency bands have unequal widths. Besides the property of guaranteed perfect reconstruction, the filters have linear phase property, and with appropriate selection of the filter coefficients can be implemented using integer computations. (2.317) can be expressed by the following condition,

$$\det\left(\mathbf{K}(z)\right) = 2z^{-m} \quad \text{with} \quad \mathbf{K}(z) = \begin{bmatrix} H_0(z) & H_0(-z) \\ H_1(z) & H_1(-z) \end{bmatrix}. \tag{2.318}$$

The factorization of $P(z)$ into $H_0(z)$ and $H_1(-z)$ is now reduced into a problem to factorize the matrix $K(z)$, which shall have a determinant expressing a shift by m samples and multiplication by a factor of 2. The factorization is simplified, if the z polynomials are decomposed into polyphase components, where sub-responses of subscripts A and B contain only the even and odd samples of the impulse response, respectively:

$$H_k(z) = H_{k,\mathrm{A}}(z^2) + z^{-1}H_{k,\mathrm{B}}(z^2)\,. \tag{2.319}$$

Writing the polyphase components of the filter pair into the following polyphase matrix[80],

$$\mathbf{H}(z) = \begin{bmatrix} H_{0,\mathrm{A}}(z) & H_{0,\mathrm{B}}(z) \\ H_{1,\mathrm{A}}(z) & H_{1,\mathrm{B}}(z) \end{bmatrix}, \tag{2.320}$$

(2.318) will be fulfilled if (2.320) has $\det(\mathbf{H}(z^2))=z^{1-m}$. The following construction of polyphase matrices was suggested in [VETTERLI, LEGALL 1989]; observe that the leftmost matrix is the polyphase matrix of the Haar filter (2.312) which is further extended by the z polynomials expressed in the matrix product,

$$\mathbf{H}(z) = \begin{bmatrix} 1 & 1 \\ 1 & -1 \end{bmatrix} \cdot \prod_{p=1}^{P-1} \begin{bmatrix} 1 & 0 \\ 0 & z^{-1} \end{bmatrix} \cdot \begin{bmatrix} 1 & \alpha_p \\ \alpha_p & 1 \end{bmatrix}. \tag{2.321}$$

[79] For IIR subband filters, see e.g. [SMITH 1991].

[80] For a deeper discussion of polyphase systems, see Sec. 2.8.3.

$\mathbf{H}(z)$ is complemented by its inverse, which represents the polyphase components of the synthesis filters,

$$\mathbf{G}(z) = \begin{bmatrix} G_{0,A}(z) & G_{1,A}(z) \\ G_{0,B}(z) & G_{1,B}(z) \end{bmatrix} = \frac{1}{2}\begin{bmatrix} 1 & 1 \\ 1 & -1 \end{bmatrix} \cdot \prod_{p=1}^{P-1} \left[\begin{bmatrix} z^{-1} & 0 \\ 0 & 1 \end{bmatrix} \cdot \begin{bmatrix} 1 & -\alpha_p \\ -\alpha_p & 1 \end{bmatrix} \cdot \frac{1}{1-\alpha_p{}^2} \right]. \tag{2.322}$$

The impulse response length of the filters $H_k(z)$ and $G_k(z)$ will then be $2P$.

Examples. For $P = 1$, the result from (2.321) and (2.322) is the Haar filter pair (2.323), using the alternative form of $H_1(z)$ as defined in the footnote on p. 123. For $P = 2$, the following set of filters is computed [LEGALL, TABATABAI 1988):

$$\begin{aligned} H_0(z) &= \frac{1}{\sqrt{2(\alpha^2-1)}}(1+\alpha z^{-1}+\alpha z^{-2}+z^{-3}), \\ H_1(z) &= \frac{1}{\sqrt{2(\alpha^2-1)}}(1+\alpha z^{-1}-\alpha z^{-2}-z^{-3}), \\ G_0(z) &= -H_1(-z) = \frac{1}{\sqrt{2(\alpha^2-1)}}(-1+\alpha z^{-1}+\alpha z^{-2}-z^{-3}), \\ G_1(z) &= H_0(-z) = \frac{1}{\sqrt{2(\alpha^2-1)}}(1-\alpha z^{-1}+\alpha z^{-2}-z^{-3}). \end{aligned} \tag{2.324}$$

As an example, the normalization factor is 1/4 for $\alpha = 3$, which enables a division-free integer implementation. With (2.317), $K(z)-K(-z)=2z^{-3}$.

Biorthogonal filters. In the PRF construction described above, lowpass and highpass filter kernels are always of same length, and orthogonality still applies due to the Haar polyphase matrix in combination with the other symmetric matrix entries in (2.321). Even this relationship between the bases H_0 and H_1 can be waived[81]; (2.316) only requires the analysis highpass H_1 to be a '$-z$'-modulated version of the synthesis lowpass G_0, and the synthesis highpass G_1 shall be a '$-z$'-modulated version of the analysis lowpass H_0. Hence, a *bi-orthogonal* relationship shall exist between the pairs of analysis highpass / synthesis lowpass filters and the analysis lowpass / synthesis highpass filters. In this case, as for the following example sets of filters, lowpass and highpass impulse responses can also have

---

[81] Orthogonality is however an important property regarding encoding of frequency coefficients, see (5.47).

different lengths. Linear-phase properties are retained when the filters themselves have symmetric impulse responses, but are not required in general[82],

$$H_0^{(5/3)}(z)=\frac{1}{8}\left(-z^2+2z+6+2z^{-1}-z^{-2}\right);\quad H_1^{(5/3)}(z)=\frac{1}{2}\left(-1+2z^{-1}-z^{-2}\right),$$
$$G_0^{(5/3)}(z)=\frac{1}{2}\left(z+2+z^{-1}\right);\quad G_1^{(5/3)}(z)=\frac{1}{8}\left(-z^3-2z^2+6z-2-z^{-1}\right);\qquad (2.325)$$

$$\begin{aligned}H_0^{(9/7)}(z)&=0.027z^4-0.016z^3-0.078z^2+0.267z+0.603\\&\quad+0.267z^{-1}-0.078z^{-2}-0.016z^{-3}+0.027z^{-4},\\H_1^{(9/7)}(z)&=0.091z^2-0.057z-0.591+1.115z^{-1}\\&\quad-0.591z^{-2}-0.057z^{-3}+0.091z^{-4}.\end{aligned}\qquad (2.326)$$

Biorthogonal filters are often employed in the *Discrete Wavelet transform* (see Sec. 4.4.4). Certain constraints should be observed in the design, in particular that an iterative application of the lowpass filter on scaled (sub-sampled) signals shall still have the effect of a (stronger) lowpass filter.

2.8.3 Implementation of filterbank structures

If filterbanks are implemented according to the direct structures introduced so far, the complexity of realization for subband analysis and synthesis is considerably higher than with block transforms using fast transform algorithms. Methods which reduce the computational complexity are introduced here.

Cascaded two-band systems. If two-band systems from Fig. 2.50 are configured in a cascaded tree consisting of T subsequent stages, each output signal from a preceding stage of the cascade is again decomposed into two more narrow subbands, and a complete decomposition into $U=2^T$ subbands can be realized as shown in Fig. 2.52. Intermediate results are used as input to several filters at the subsequent stage, and the later stages use increasingly sub-sampled signals, which significantly reduces operations compared to a system with parallel filters. Due to the frequency inversion occurring in highpass band sub-sampling (see Fig. 2.49), any frequency band that stems from an odd number of highpass filter / decimation steps will be frequency inverted. For the subsequent level, it is therefore necessary to exchange the sequence of filters H_0 and H_1 if an arrangement of subbands by increasing frequency order is desirable[83].

---

82 Both filters are sometimes modified, multiplying H_0 by $\sqrt{2}$ and dividing H_1 by $\sqrt{2}$, which almost approaches orthonormality at least for the case of the 9/7 filter. A shift m=1 is used w.r.t. (2.315). By the lengths of their lowpass/highpass analysis filter kernels, these two filter pairs are denoted as 5/3 and 9/7, respectively.

83 This has an analogy with the distinction between Walsh and Hadamard transforms, where the iterative Hadamard development ignores the frequency reversion.

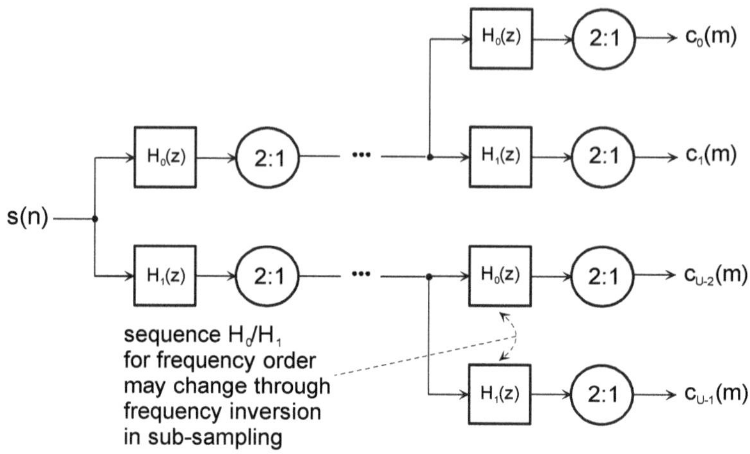

Fig. 2.52. Realization of subband analysis filter in a cascade from 2-band systems

Exploitation of filter symmetries. If symmetric (linear phase) filters are used, duplicate multiplications can be avoided, where samples have to be multiplied several times by identical factors. If the highpass basis function is a modulated version of the lowpass or uses the same multiplication factors (as in the cases of QMF and PRF types), yet another degree of freedom exists to reduce multiplications further by a factor of up to 2 by using results jointly (see also Problem 2.18).

Polyphase systems. Only each Uth sample will be retained by subband analysis after the filtering and sub-sampling steps. Hence, the convolution does not need to be performed at positions which are discarded anyway. This leads to a reduction of operations by a factor of U. The structure of a *polyphase system* is shown in Fig. 2.53 using an example of $U=2$. Sub-sampling is performed *prior to filtering*, whereby the signal is decomposed into U *polyphase components*, which establish a set of sample sequences each sub-sampled at a different phase position. Further, it is necessary to decompose the filter impulse responses into polyphase components, such that instead of a length-P filter, U partial filters of lengths either $\lfloor P/U \rfloor$ or $\lfloor P/U+1 \rfloor$ are obtained. If the subband filter impulse response h_k is decomposed into U polyphase components $h_{k,\mathrm{A}}(m)$, $h_{k,\mathrm{B}}(m)$, ... , U partial filters of transfer functions $H_{k,\mathrm{A}}(z)$, $H_{k,\mathrm{B}}(z)$, ... are given (see Fig. 2.53a for the case $U=2$). Similarly, it is not necessary to apply multiplications on zero values inserted for interpolation filtering at the synthesis stage. This can be realized by performing the interpolation filtering step within the polyphase components, and compose the different phase positions into the reconstructed signal only in a last step. In fact, the expansion of the signal is performed *after filtering*, but no zero values are actually inserted, as the polyphase components from all partial filters fill the corresponding gaps. The same reduction of the number of multiply/add operations by a factor of U is also achieved in synthesis (Fig. 2.53b).

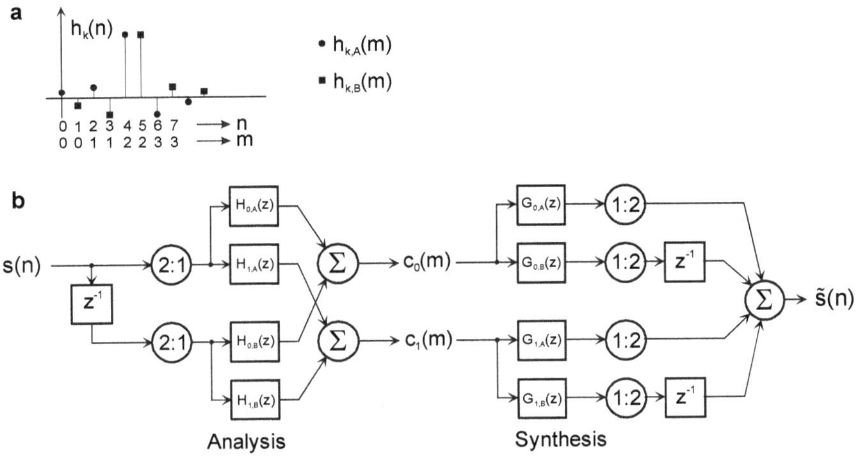

Fig. 2.53. Realization of subband analysis and synthesis by polyphase systems, U=2
a Separation of impulse response into partial terms $h_A(n)$ und $h_B(n)$
b Structure of the overall polyphase system

For a system with $U=2$, the polyphase components of a signal $s(n)$ are sequences of even samples $s(2m)$ and odd samples $s(2m+1)$. In the z transform domain, the following relationships apply:

$$\begin{aligned} s(2m) = s_A(m) &\overset{z}{\circ\!\!-\!\!\bullet} S_A(z) = \frac{1}{2}\left[S\left(z^{1/2}\right) + S\left(-z^{1/2}\right)\right], \\ s(2m+1) = s_B(m) &\overset{z}{\circ\!\!-\!\!\bullet} S_B(z) = \frac{1}{2}\left[z^{1/2}S\left(z^{1/2}\right) - z^{1/2}S\left(-z^{1/2}\right)\right], \\ s(n) &\overset{z}{\circ\!\!-\!\!\bullet} S(z) = S_A(z^2) + z^{-1}S_B(z^2) \quad ; \quad m = \left\lfloor \frac{n}{2} \right\rfloor. \end{aligned} \tag{2.327}$$

Here, the subscripts A and B relate to the even and odd polyphase components, respectively. Formally, the components of the z transform can be combined in the following vector notation,

$$\mathbf{S}(z) = \begin{bmatrix} S_A(z) \\ z^{-1}S_B(z) \end{bmatrix}. \tag{2.328}$$

The same procedure can be applied to the z polynomials of the filter impulse responses. As the convolution in the signal domain corresponds to a multiplication in the z domain, the filtering of the even/odd signal spectra by the respective filter transfer functions can be expressed as

$$\underbrace{\begin{bmatrix} C_0(z) \\ C_1(z) \end{bmatrix}}_{\mathbf{C}(z)} = \underbrace{\begin{bmatrix} H_{0,\mathrm{A}}(z) & H_{0,\mathrm{B}}(z) \\ H_{1,\mathrm{A}}(z) & H_{1,\mathrm{B}}(z) \end{bmatrix}}_{\mathbf{H}(z)} \underbrace{\begin{bmatrix} S_{\mathrm{A}}(z) \\ z^{-1}S_{\mathrm{B}}(z) \end{bmatrix}}_{\mathbf{S}(z)}. \tag{2.329}$$

For the synthesis part, a similar principle applies. Writing the reconstructed signal by

$$\tilde{\mathbf{S}}(z) = \begin{bmatrix} \tilde{S}_{\mathrm{A}}(z) \\ z^{-1}\tilde{S}_{\mathrm{B}}(z) \end{bmatrix} \quad ; \quad \tilde{S}(z) = 2\left[\tilde{S}_{\mathrm{A}}(z^2) + z^{-1}\tilde{S}_{\mathrm{B}}(z^2)\right], \tag{2.330}$$

the synthesis filter step can be expressed as

$$\underbrace{\begin{bmatrix} \tilde{S}_{\mathrm{A}}(z) \\ z^{-1}\tilde{S}_{\mathrm{B}}(z) \end{bmatrix}}_{\tilde{\mathbf{S}}(z)} = \underbrace{\begin{bmatrix} G_{0,\mathrm{A}}(z) & G_{1,\mathrm{A}}(z) \\ G_{0,\mathrm{B}}(z) & G_{1,\mathrm{B}}(z) \end{bmatrix}}_{\mathbf{G}(z)} \underbrace{\begin{bmatrix} C_0(z) \\ C_1(z) \end{bmatrix}}_{\mathbf{C}(z)}. \tag{2.331}$$

Combining (2.329) and (2.331), the condition for perfect reconstruction is

$$\begin{bmatrix} G_{0,\mathrm{A}}(z) & G_{1,\mathrm{A}}(z) \\ G_{0,\mathrm{B}}(z) & G_{1,\mathrm{B}}(z) \end{bmatrix} \begin{bmatrix} H_{0,\mathrm{A}}(z) & H_{0,\mathrm{B}}(z) \\ H_{1,\mathrm{A}}(z) & H_{1,\mathrm{B}}(z) \end{bmatrix} = \mathbf{G}(z)\mathbf{H}(z) = \mathbf{I}, \tag{2.332}$$

from which the following relationships are determined:

$$\begin{aligned} G_{0,\mathrm{A}}(z)H_{0,\mathrm{A}}(z) + G_{1,\mathrm{A}}(z)H_{1,\mathrm{A}}(z) &= 1 \\ G_{0,\mathrm{A}}(z)H_{0,\mathrm{B}}(z) + G_{1,\mathrm{A}}(z)H_{1,\mathrm{B}}(z) &= 0 \\ G_{0,\mathrm{B}}(z)H_{0,\mathrm{A}}(z) + G_{1,\mathrm{B}}(z)H_{1,\mathrm{A}}(z) &= 0 \\ G_{0,B}(z)H_{0,\mathrm{B}}(z) + G_{1,\mathrm{B}}(z)H_{1,\mathrm{B}}(z) &= 1. \end{aligned} \tag{2.333}$$

These are fulfilled by the following conditions,

$$\begin{aligned} H_{0,\mathrm{A}}(z) = G_{1,\mathrm{B}}(z) \quad &; \quad H_{0,\mathrm{B}}(z) = -G_{1,\mathrm{A}}(z); \\ H_{1,\mathrm{A}}(z) = -G_{0,\mathrm{B}}(z) \quad &; \quad H_{1,\mathrm{B}}(z) = G_{0,\mathrm{A}}(z), \end{aligned} \tag{2.334}$$

which by substitution into (2.333) gives the additional condition

$$H_{0,\mathrm{A}}(z)H_{1,\mathrm{B}}(z) - H_{0,\mathrm{B}}(z)H_{1,\mathrm{A}}(z) = \det\left(\mathbf{H}(z)\right) = 1. \tag{2.335}$$

Using (2.328) to express the polyphase filters in the (not downsampled) z domain, (2.333) is equivalent to (2.315), while (2.335) is equivalent to (2.317). A special case of the polyphase transform is observed for $\mathbf{H}(z) = \mathbf{G}(z) = \mathbf{I}$, which is the so-called *lazy transform* where the 'subband' signals $c_0(m)$ and $c_1(m)$ would simply be the polyphase components generated without any lowpass or highpass filtering.

Lifting implementation. Subband filters described by polyphase components can be implemented in a *lifting structure* [DAUBECHIES, SWELDENS 1998] as shown in Fig.

2.54. The first step of the lifting filter is a decomposition of the signal into its even- and odd-indexed polyphase components by the lazy transform. Then, the two basic operations are *prediction steps* $P(z)$ and *update steps* $U(z)$. The prediction and update filters have simple impulse responses typically of length 2 or 3; the number of steps necessary and the values of coefficients in each step are determined by a factorization of biorthogonal filter pairs. Finally, normalization by factors a_{Low} and a_{High} is applied.

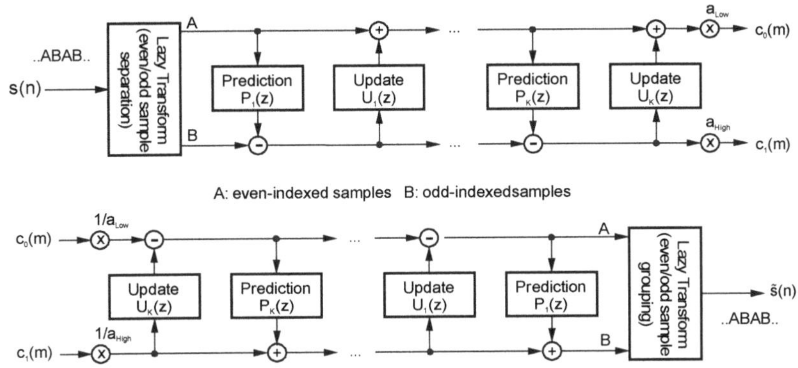

Fig. 2.54. Lifting structure of a subband filter (top: analysis; bottom: synthesis).

The construction of the prediction and update filter kernels can best be started from the polyphase representation. Assume that decomposition of a signal has been performed by a polyphase filter matrix $\mathbf{H}^0(z)$ (which could be the identity matrix $\mathbf{I}$ for the lazy transform in the beginning). If a *prediction step* is performed using the filter transfer function $P(z)$, the result is identical to a filter expressed by the polyphase matrix

$$\mathbf{H}^{\text{pr}}(z) = \underbrace{\begin{bmatrix} 1 & 0 \\ -P(z) & 1 \end{bmatrix}}_{\mathbf{P}(z)} \cdot \mathbf{H}^0(z)$$
$$= \begin{bmatrix} H_{0,\text{A}}(z) & H_{0,\text{B}}(z) \\ H_{1,\text{A}}(z) - P(z)H_{0,\text{A}}(z) & H_{1,\text{B}}(z) - P(z)H_{0,\text{B}}(z) \end{bmatrix}. \quad (2.336)$$

The complementary synthesis filter guarantees perfect reconstruction, such that $\mathbf{G}^{\text{pr}}(z)\mathbf{H}^{\text{pr}}(z) = \mathbf{I}$ when $\mathbf{G}^0(z)\mathbf{H}^0(z) = \mathbf{I}$,

$$\mathbf{G}^{\text{pr}}(z) = \mathbf{G}^0(z) \cdot \begin{bmatrix} 1 & 0 \\ P(z) & 1 \end{bmatrix} = \begin{bmatrix} G_{0,\text{A}}(z) + P(z)G_{1,\text{A}}(z) & G_{1,\text{A}}(z) \\ G_{0,\text{B}}(z) + P(z)G_{1,\text{B}}(z) & G_{1,\text{B}}(z) \end{bmatrix}. \quad (2.337)$$

Similarly, a single *update step* can be formulated as

$$\mathbf{H}^{\text{up}}(z)=\underbrace{\begin{bmatrix}1 & U(z)\\ 0 & 1\end{bmatrix}}_{\mathbf{U}(z)}\cdot\mathbf{H}^{0}(z)=\begin{bmatrix}H_{0,\text{A}}(z)+U(z)H_{1,\text{A}}(z) & H_{0,\text{B}}(z)+U(z)H_{1,\text{B}}(z)\\ H_{1,\text{A}}(z) & H_{1,\text{B}}(z)\end{bmatrix}, \tag{2.338}$$

where the complementary synthesis filter is

$$\mathbf{G}^{\text{up}}(z)=\mathbf{G}^{0}(z)\cdot\begin{bmatrix}1 & -U(z)\\ 0 & 1\end{bmatrix}=\begin{bmatrix}G_{0,\text{A}}(z) & G_{1,\text{A}}(z)-U(z)G_{0,\text{A}}(z)\\ G_{0,\text{B}}(z) & G_{1,\text{B}}(z)-U(z)G_{0,\text{B}}(z)\end{bmatrix}. \tag{2.339}$$

Using (2.336)-(2.339) iteratively starting by a lazy transform, the equivalent polyphase matrix after a number of subsequent prediction and update steps is the concatenated product of all matrices, e.g. for a number of L subsequent prediction and update steps

$$\mathbf{H}(z)=\begin{bmatrix}a_{\text{Low}} & 0\\ 0 & a_{\text{High}}\end{bmatrix}\prod_{l=1}^{L}\begin{bmatrix}1 & U_l(z)\\ 0 & 1\end{bmatrix}\begin{bmatrix}1 & 0\\ -P_l(z) & 1\end{bmatrix}. \tag{2.340}$$

Vice versa, it is possible to factorize a given polyphase matrix containing higher-order z polynomials into a series of prediction/update matrices with only simple, low-order polynomials. Separation of single prediction and update steps from a given (complete) polyphase matrix $\mathbf{H}(z)$ will result in the following expression according to (2.336) and (2.338):

$$\mathbf{H}(z)=\begin{bmatrix}1 & 0\\ -P(z) & 1\end{bmatrix}\cdot\mathbf{H}^{-\text{pr}}(z)\quad;\quad\mathbf{H}(z)=\begin{bmatrix}1 & U(z)\\ 0 & 1\end{bmatrix}\cdot\mathbf{H}^{-\text{up}}(z). \tag{2.341}$$

The factorization is always possible, as the determinant of any of the single prediction and update matrices is one, and hence inversion is possible. By polynomial division, the result can be computed step by step, and the factorization typically terminates when only a diagonal matrix with normalization factors a_{Low} and a_{High} is left.

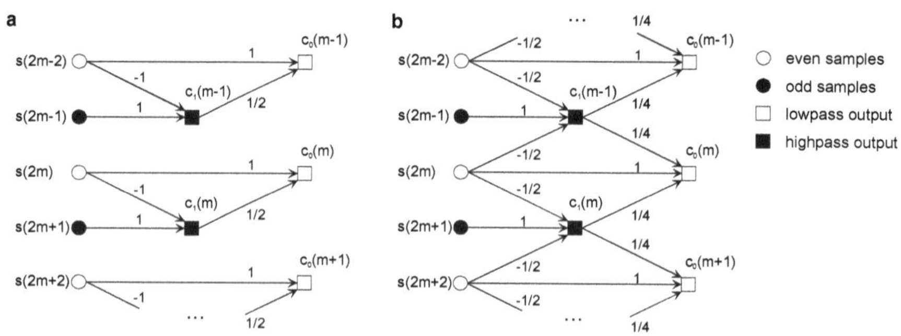

Fig. 2.55. Lifting flows of **a** Haar basis (2.312) **b** biorthogonal 5/3 filter (2.325)

Examples. The biorthogonal 5/3 filter from (2.325) can be expressed by the following polyphase matrix, which is further factorized into one normalization, one prediction and one update matrix

$$\mathbf{H}(z)=\begin{bmatrix}-\frac{1}{8}z+\frac{3}{4}-\frac{1}{8}z^{-1} & \frac{1}{4}z+\frac{1}{4}\\ -\frac{1}{2}-\frac{1}{2}z^{-1} & 1\end{bmatrix}=\underbrace{\begin{bmatrix}1&0\\0&1\end{bmatrix}}_{\mathbf{A}}\cdot\underbrace{\begin{bmatrix}1&\frac{1}{4}z+\frac{1}{4}\\0&1\end{bmatrix}}_{\mathbf{U}(z)}\cdot\underbrace{\begin{bmatrix}1&0\\-\frac{1}{2}-\frac{1}{2}z^{-1}&1\end{bmatrix}}_{\mathbf{P}(z)}. \tag{2.342}$$

Here, $a_{\text{High}}=a_{\text{Low}}=1$, $P(z)=½(z^{-1}+1)$ and $U(z)=¼(1+z)$. Another example is for the Haar filter[84], where $a_{\text{Low}}=\sqrt{2}$, $a_{\text{High}}=\sqrt{2}/2$, $P(z)=-1$ and $U(z)=1/2$:

$$\mathbf{H}(z)=\frac{\sqrt{2}}{2}\begin{bmatrix}1&1\\-1&1\end{bmatrix}=\underbrace{\begin{bmatrix}\sqrt{2}&0\\0&\sqrt{2}/2\end{bmatrix}}_{\mathbf{A}}\underbrace{\begin{bmatrix}1&\frac{1}{2}\\0&1\end{bmatrix}}_{\mathbf{U}(z)}\underbrace{\begin{bmatrix}1&0\\-1&1\end{bmatrix}}_{\mathbf{P}(z)}. \tag{2.343}$$

The lifting structure can also be interpreted by a signal flow diagram, which is shown in Fig. 2.55 for the examples of a Haar filter (2.343) (without considering the normalization factors) and the biorthogonal 5/3 filter (2.342).

The lifting structure further allows definition of *nonlinear subband filters*. A simple example is usage of rank-order filters like median or weighted median filters in prediction and update steps [CLAYPOOLE ET AL. 1997]

2.8.4 Wavelet transform

The continuous-time *wavelet transform* (WT) is defined by the convolution equation

$$\mathcal{W}_s(t,f)=\int_{-\infty}^{\infty}s(\tau)\psi_f(t-\tau)\mathrm{d}\tau\,, \tag{2.344}$$

being based on bandpass filter kernels

$$\psi_f(t)=\frac{1}{\sqrt{\alpha}}\cdot\psi\left(\frac{t}{\alpha}\right)\quad\text{with}\quad\alpha=\frac{f_0}{f}. \tag{2.345}$$

The function $\psi(\cdot)$ is the *mother wavelet*, which is a bandpass filter of center fre-

[84] In the case of the Haar filter, the usage of the lifting approach does not to give an advantage in terms of complexity for signal decomposition, which is due to the fact that the polyphase polynomials already are of order zero before the factorization. This method is however relevant in motion-compensated temporal-axis wavelet filtering, cf. Sec. 7.3.2, and can also be used to avoid bit-depth extension of the transformed representation.

quency f_0, which is time-scaled by the factor α when intended to operate at a different frequency [RIOUL, VETTERLI 1991].

The continuous WT in (2.344) is not useful for practical signal analysis. It is highly overcomplete, being defined for an infinite number of instances both of time and frequency positions. In the *discrete wavelet transform* (DWT), the analysis shall only be performed for discrete (sampled) signal positions, and only for a discrete set of frequencies. The commonly used method is defining a set of basis functions by a *dyadic frequency sampling scheme*, where the upper band limits f_k and the distances of sampling positions t_k used for the respective frequency bands are defined over power-of-two relationships, such that the frequency partitioning has octave-band style. Assume that U frequency bands are defined[85] by

$$\alpha_k = 2^{U-k}, \quad f_k = \frac{1}{\alpha_k T}, \quad t_k(n) = \alpha_k nT \quad \text{with } 0 \le k < U. \tag{2.346}$$

The distances between discrete center frequencies of the analysis are no longer constant, and the effective bandwidth[86] $\Delta f_k = [\, f_k - f_{k-1} \,]$ of the frequency bands is increased by a factor of 2 when incrementing k. Simultaneously, the distance between analysis positions $\Delta t_k = [\, t_k(n) - t_k(n-1) \,]$ decreases by a factor of two. This means that for higher frequency bands (higher k), the temporal resolution becomes more precise, while less precision in the resolution of the frequency axis is achieved. This is illustrated in Fig. 2.56 for both cases of an idealized DWT and a discrete short time Fourier transform (STFT), which is typically implemented via windowed DFT or DCT analysis. Using the definitions in (2.346), the DWT coefficient of discrete frequency k and position n is defined as

$$c_k(n) = \frac{1}{\sqrt{\alpha_k}} \int_{-\infty}^{\infty} s(\tau)\psi\left(\frac{\tau - nT}{\alpha_k}\right) d\tau \,. \tag{2.347}$$

Remark that the basis functions defined here to compute the DWT are time-continuous and have the purpose to perform filtering for band limitation, whereas the convolution is only defined at discrete positions, such that sampling is implicitly included. As in (2.346), T is the sampling distance corresponding to the reso-

---

[85] In principle, the number U could become arbitrarily high, however for discrete signals of finite length N, at least one sampling position $t_u(m)$ should be retained in the last step. The condition $t_1(1) - t_1(0) \le NT$ gives e.g. $U_{\max} = \log_2 N$ for cases where N is a power of 2, or $\lfloor \log_2 N \rfloor + 1$ otherwise. For practical applications, a much lower (pre-defined) number of bands is used for discrete wavelet decomposition. To be consistent with previous notation, we use the variable k as an index that increases with the frequency (starting with $k=0$ for the lowest frequency, whereas $k=U$ would be the original signal (without wavelet decomposition).

[86] The term 'effective bandwidth' is not precisely defined, except for the case of ideal filters. One possible way of interpretation is the width of a rectangular function with identical maximum amplitude and total integration area as the filter's Fourier transfer function has.

lution accuracy when all frequency bands are used (i.e. original sampling before DWT decomposition is applied).

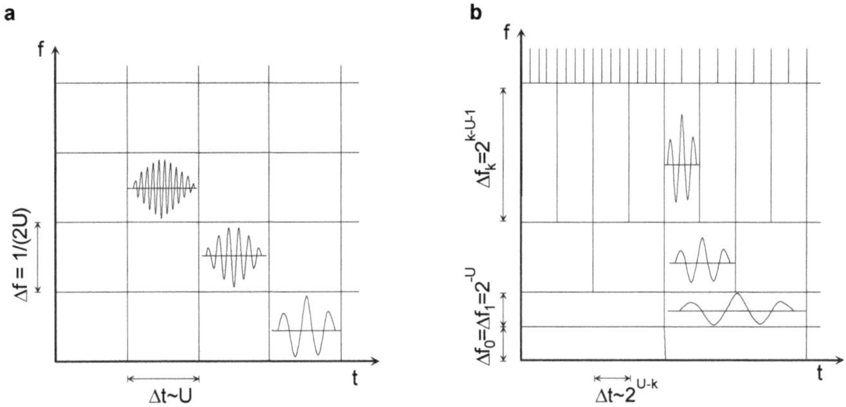

Fig. 2.56. Resolution accuracy in signal and frequency domains **a** for STFT **b** for DWT

The DWT allows reconstructing the signal by different resolution levels (scales). In a more abstract sense, the frequency domain representation up to half sampling rate can be constructed from a set of *scale spaces* and a set of *wavelet spaces*, each of which is related to one of the dyadic resolution levels (see Fig. 2.57). When the scale space $\mathcal{V}_k$ represents a certain bandwidth resolution of a (sampled) signal $s_k(n)$, the next-lower scale space $\mathcal{V}_{k-1}$ represents a signal $s_{k-1}(n)$ with half number of samples and half bandwidth. The scale space $\mathcal{V}_U$ represents the signal $s(n)=s_U(n)$ with maximum possible resolution, relating to a sampling distance $T=1$, which then corresponds to the frequency cut-off $|f|=1/2$. To achieve the perfect approximation, the wavelet space $\mathcal{W}_k$ must be an orthogonal complement which contains the residual between two adjacent scale spaces:

$$\mathcal{V}_{k+1} = \mathcal{V}_k \oplus \mathcal{W}_k \quad \text{and} \quad \mathcal{V}_k \perp \mathcal{W}_k \tag{2.348}$$

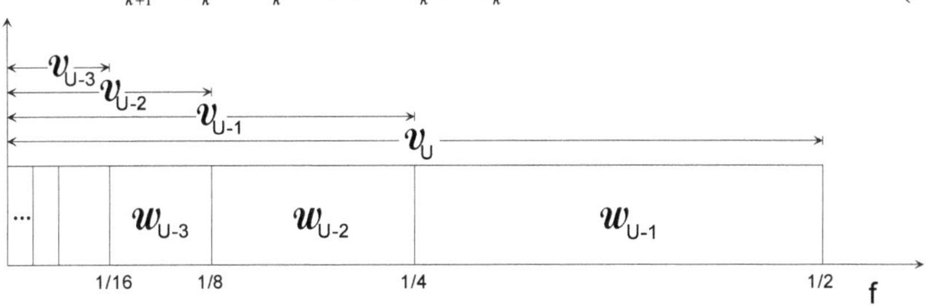

Fig. 2.57. Layout of dyadic scale and wavelet spaces by partitioning of the frequency axis

If the conditions in (2.348) hold true, all lower-frequency wavelet spaces must be orthogonal as well. All details which are lost when reducing the resolution from $\mathcal{V}_k$ to $\mathcal{V}_{k-1}$ are found in $\mathcal{W}_{k-1}$. Iteratively, an arbitrary scale space can be expressed as a direct sum of all lower-indexed wavelet spaces, where the summation is terminated by the lowest-resolution scale space[87]:

$$\mathcal{V}_k = \mathcal{W}_{k-1} \oplus \mathcal{W}_{k-2} \oplus \ldots \oplus \mathcal{W}_1 \oplus \mathcal{V}_1 \,. \tag{2.349}$$

The analysis of the signal, i.e. the decomposition into components which relate to the respective scale and wavelet spaces, is performed by *scaling functions* $\varphi(\tau)$ and *wavelet functions* $\psi(\tau)$. The scaling function is in principle a lowpass filter which is used to generate a lower-resolution representation, e.g. to construct $\mathcal{V}_{k-1}$ out of $\mathcal{V}_k$.

As $\mathcal{V}_{k-1} \subset \mathcal{V}_k$, any function in $\mathcal{V}_{k-1}$ can be expressed as a linear combination of basis functions $\varphi_k(\tau)$ related to the scale space $\mathcal{V}_k$. Therefore, also the scaling function in $\mathcal{V}_{k-1}$ can be described by the *refinement equation* expressing a superposition of scaling functions in $\mathcal{V}_k$:

$$\varphi_{k-1}(\tau) = \sum_m h_0(m)\varphi_k(\tau - m\alpha_k T) \,. \tag{2.350}$$

As for the wavelet space $\mathcal{W}_{k-1} \subset \mathcal{V}_k$ is also valid, an associated wavelet function can be generated similarly by the *wavelet equation*

$$\psi_{k-1}(\tau) = \sum_m h_1(m)\varphi_k(\tau - m\alpha_k T) \,. \tag{2.351}$$

Likewise, the operations (2.350)/(2.351) can be reversed, such that the next-higher scaling function (representing a signal of higher resolution) shall be reconstructed from a current level's scaling and wavelet functions as

$$\begin{aligned}
\varphi_k(\tau) &= \sum_m g_{0,\mathrm{A}}(m) \cdot \varphi_{k-1}(\tau - m\alpha_{k-1}T) + \sum_m g_{1,\mathrm{A}}(m) \cdot \psi_{k-1}(\tau - m\alpha_{k-1}T) \\
\varphi_k(\tau - \alpha_k T) &= \sum_m g_{0,\mathrm{B}}(m) \cdot \varphi_{k-1}(\tau - m\alpha_{k-1}T) + \sum_m g_{1,\mathrm{B}}(m) \cdot \psi_{k-1}(\tau - m\alpha_{k-1}T),
\end{aligned} \tag{2.352}$$

where A and B denote the even and odd polyphase components of the discrete filter functions.

The iterative development of scaling and wavelet functions shall now be illustrated for the simplest possible orthogonal wavelet basis, which is the Haar basis. The refinement and wavelet equations to perform the mapping from $\mathcal{V}_{k+1}$ into $\mathcal{V}_k$ and $\mathcal{W}_k$, using the discrete filter coefficients (2.312), give

---

[87] The termination by a scale space is necessary if the analyzed signal is finite, or if the delay occurring by the analysis shall be finite, as is typically always the case in multimedia signal processing and analysis. Theoretically, a scale space could also be established from an infinite series of sub-ordinate wavelet spaces. To make the notation consistent with the previous frequency representations that are introduced, the signal in the lowest-resolution scale space $\mathcal{V}_1$ is either denoted as $s_1(n)$ or $c_0(n)$.

$$\begin{aligned}\varphi_{k-1}(\tau) &= \underbrace{\frac{\sqrt{2}}{2}}_{h_0(0)}\varphi_k(\tau) + \underbrace{\frac{\sqrt{2}}{2}}_{h_0(1)}\varphi_k(\tau-\alpha_k T) \\ \psi_{k-1}(\tau) &= \underbrace{\frac{\sqrt{2}}{2}}_{h_1(0)}\varphi_k(\tau) \underbrace{-\frac{\sqrt{2}}{2}}_{h_1(1)}\varphi_k(\tau-\alpha_k T)\end{aligned} \quad . \tag{2.353}$$

The scaling function $\varphi_U(\tau)$ in $\boldsymbol{\mathcal{V}}_U$ is a rectangle ('hold element' in sampling) of length T and amplitude 1. Fig. 2.58 shows the weighted superposition of two copies of this scaling function, resulting in the scaling and wavelet functions in $\boldsymbol{\mathcal{V}}_{U-1}$ and $\boldsymbol{\mathcal{W}}_{U-1}$, respectively. If this is performed iteratively, both functions are scaled to double width and are amplitude-scaled by another factor of $\sqrt{2}$ with each subsequent iteration step. For this case, the convergence into the final shape of scaling and wavelet functions is already achieved after one iteration (which is due to the fact that the shape of the scaling function will always remain the rectangle, regardless how wide it may become).

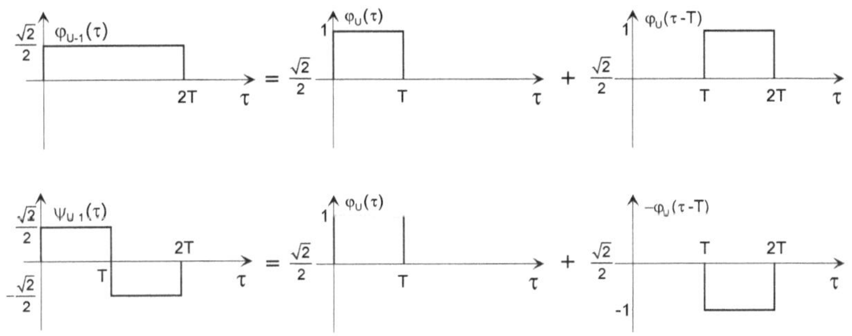

Fig. 2.58. Development of next-higher level scaling and wavelet functions for the Haar basis

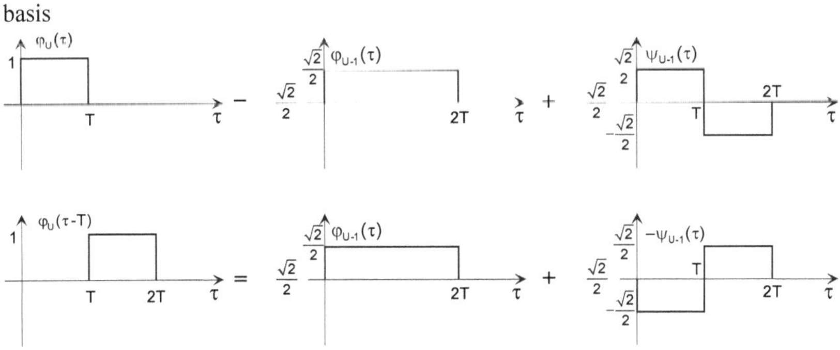

Fig. 2.59. Reconstruction of next-lower level scaling functions for the Haar basis

Now, reconstruction of the different copies of the scaling function in $\boldsymbol{\mathcal{V}}_k$ shall be performed from the scaling and wavelet functions in $\boldsymbol{\mathcal{V}}_{k-1}$ and $\boldsymbol{\mathcal{W}}_{k-1}$. The related equations are

$$\varphi_k(\tau) = \underbrace{\frac{\sqrt{2}}{2}}_{g_0(0)} \varphi_{k-1}(\tau) + \underbrace{\frac{\sqrt{2}}{2}}_{g_1(0)} \psi_k(\tau) \,;\, \varphi_{k+1}(\tau - \alpha_k T) = \underbrace{\frac{\sqrt{2}}{2}}_{g_0(1)} \varphi_{k-1}(\tau) - \underbrace{\frac{\sqrt{2}}{2}}_{g_1(1)} \psi_{k-1}(\tau) \,. \quad (2.354)$$

This process of reconstruction is shown in Fig. 2.59.

(2.350) and (2.351) are the key equations of the DWT. They can be used to determine discrete lowpass and highpass analysis filter coefficients $h_0(k)$ and $h_1(k)$ of a filter bank system. Assume that continuous-time scaling and wavelet functions shall be orthogonal. If such functions can be constructed iteratively using discrete filter coefficients $h_0(k)$ and $h_1(k)$, any signal decomposition performed by these coefficients in a filterbank system would be orthogonal as well in case of large number of iterations, even if the impulse responses $h_0(k)$ and $h_1(k)$ may not be orthogonal. Furthermore, the scaling functions, even though they play conceptually a similar role as a band-limiting lowpass filter in conventional sampling, do not necessarily need to provide perfect band separation. When only the full set of spaces is relevant in a wavelet representation, orthogonality needs to be observed just between the underlying continuous scaling and wavelet functions, which is a much weaker condition than non-overlapping frequency bands. The orthogonality of the decomposition is guaranteed if the following condition holds true[88]:

$$\int_{-\infty}^{\infty} \varphi(\tau)\psi(\tau)\,d\tau = 0 \text{ where } \varphi(\tau) = \lim_{U\to\infty} \varphi_1(\tau) \text{ and } \psi(\tau) = \lim_{U\to\infty} \psi_1(\tau) \,. \quad (2.355)$$

This leads to conditions which can be used to design biorthogonal filter pairs. The continuous scaling and wavelet functions can be used to reconstruct (interpolate) continuous signals from the samples in the DWT domain. The discrete coefficients according to (2.350) and (2.351) can also directly be used to perform all underlying operations directly in the sampled signal domain (i.e. compute DWT from a sampled signal). Assume that a discrete approximation of the signal is available in some resolution scale k as $s_k(n)$. The scaling coefficients of the next-coarser approximation (representing a signal of half resolution or half number of samples) are then computed as

$$s_{k-1}(n) = \sum_m h_0(m) s_k(2n - m) \,, \quad (2.356)$$

and the complementary wavelet coefficients are

---

[88] The limit transitions in the following equation assume that the iterative construction of the scaling and wavelet functions could be continued ad infinitum (not stopping at k=0, which would be the case with a finite number of U bands, but rather continue with negative values). In contrast to the previous definitions in (2.350)-(2.354), time-axis scaling by a factor of 2 may be performed during each iteration of the continuous scaling and wavelet functions (corresponding to the subsampling in the discrete filterbank) to prevent infinite extension. This also implies, that even starting from an initial rectangular scaling function, final functions $\varphi(\tau)$ and $\psi(\tau)$ are becoming smooth with appropriate choice of the coefficients h and g, provided that the initial function has lowpass characteristics.

$$c_{k-1}(n) = \sum_m h_1(m) s_k(2n-m) . \tag{2.357}$$

This decomposition can be computed iteratively, starting by $s_U(n) \equiv s(n)$. Actually, each level of this decomposition is identical to the decomposition of a signal into low- and high-frequency subbands as introduced in Sec. 4.4.2. However, in contrast to the cascaded system from Fig. 2.52, only the low frequency output (the next lower scale signal s_{k-1}) is subject to further decomposition. Using the corresponding synthesis functions, it is possible to compute the reconstruction of the signal by inverting the sequence of recursion defining the *inverse DWT* (IDWT):

$$s_k(n) = \sum_m g_0(n-2m) s_{k-1}(m) + \sum_m g_1(n-2m) c_{k-1}(m) . \tag{2.358}$$

Note that (2.356)-(2.358) implicitly include polyphase operations in the expression of the discrete convolutions. As shown earlier, perfect reconstruction is possible if the synthesis coefficients $g_0(k)$ and $g_1(k)$ are related to $h_0(k)$ and $h_1(k)$ by bi-orthogonality (4.170). However, if $h_0(k)$ and $h_1(k)$ are chosen such that the continuous scaling and wavelet functions are orthogonal, it can be concluded that the sequences of discrete scaling and wavelet coefficients will also be orthogonal, even if the filter basis may only be biorthogonal (the latter being sufficient to achieve perfect reconstruction).

Fig. 2.60 shows the block diagram of a DWT analysis/synthesis filter bank, and a schematic layout of the resulting frequency decomposition, which can be described as an *octave-band structure*. For consistency with the notation used in case of other transforms, the signal relating to the scale space $\mathcal{V}_1$ is denoted as c_0 (instead s_1), while the designations of the wavelet coefficients relating to wavelet spaces $\mathcal{W}_k$, $k = 1, \ldots, U-1$, are retained as c_k as above.

If the Haar basis (2.312) is used, the resulting decomposition is exactly the same as for the Haar transform, cf. (2.249). However, longer filter impulse responses can provide a better frequency separation and also better alias suppression in the scaled signal versions.

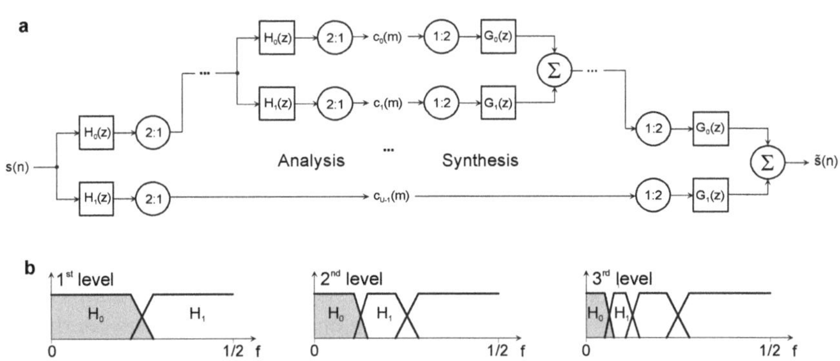

Fig. 2.60. **a** Octave-band filter bank system for DWT and IDWT **b** Octave-band frequency layout (3 levels of analysis)

For many classes of signals, in particular for natural image signals, the higher accuracy of frequency resolution for the lower-frequency bands provides a good fit with signal models. According to the AR(1) model with $\rho\rightarrow 1$, significantly more low-frequency than high-frequency components can be expected. For the high-frequency components, accurate frequency analysis is less important than an *accurate localization of detail*, in particular if a signal potentially exhibits discontinuities, as it is the case in edge areas (which are not adequately captured by the AR model). The fact that discontinuities in the signal appear at various resolution levels, and therefore also across wavelet bands at the same location, is denoted as *scaling property*.

2.8.5 Two- and multi-dimensional filter banks

The simplest realization of a two- or multi-dimensional filter bank is the *separable* method, where the analysis and synthesis filters are a product of horizontal and vertical filters. For the 2D case, the basis functions for the frequency band of index k_1 in horizontal and k_2 in vertical direction can be described as

$$h_{k_1,k_2}(n_1,n_2) = h_{k_1}(n_1)h_{k_2}(n_2) \text{ and } g_{k_1,k_2}(n_1,n_2) = g_{k_1}(n_1)g_{k_2}(n_2). \qquad (2.359)$$

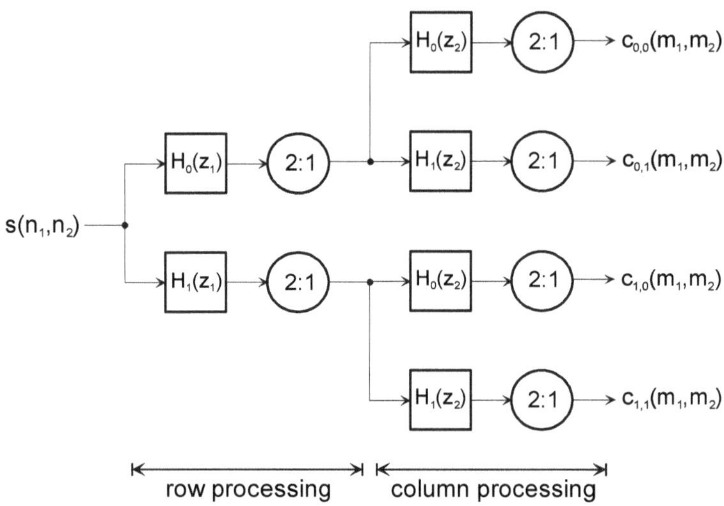

Fig 2.61. 2D system for decomposition into four frequency bands

With U_1U_2 bands, the total sub-sampling factor is $|\mathbf{U}| = U_1U_2$ in case of critical sampling. Separable 2D systems with 2-band decomposition structures per dimension can be realized sequentially, such that filtering and sub-sampling is at first performed over one dimension. Only a reduced number of samples then needs to be fed into the second directional decomposition stage. A block diagram with the case $U_1=2$, $U_2=2$, $|U|=4$, is shown in Fig. 2.61.

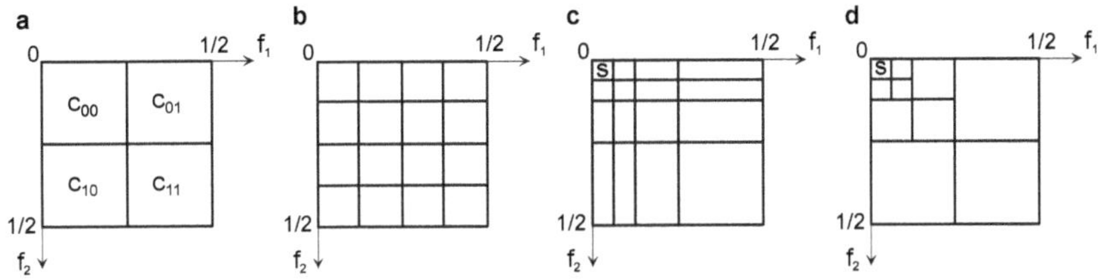

Fig. 2.62. Layout of 2D frequency bands. **a** 4 band elementary decomposition **b** 16 bands of equal bandwidth **c** Separable octave-band, 16 bands **d** 2D DWT, 10 bands (**S** = scaling band)

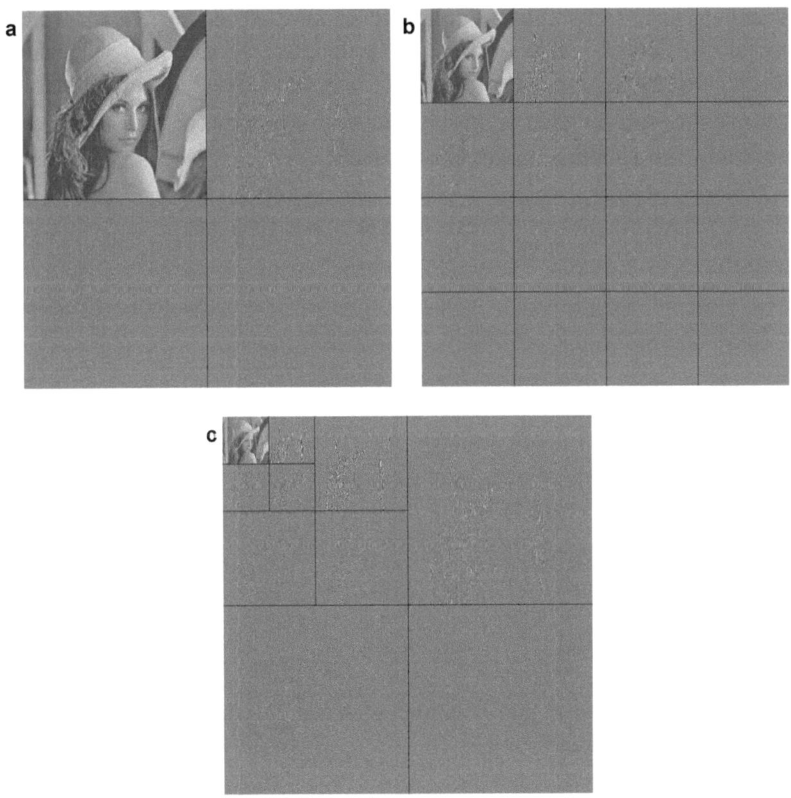

Fig. 2.63. Decomposition of an image into subband pictures (amplified by factor 4, except c_{00}) **a** relating to Fig. 2.62a **b** relating to Fig. 2.62b **c** relating to Fig. 2.62d

Fig. 2.62a depicts the related layout of subbands in the 2D frequency domain. This basic 4-band decomposition structure of Fig. 2.61 can then again be applied iteratively to respective (sub-sampled) outputs of the previous stage. For a case where all 4 subbands are equally decomposed in the next level, Fig. 2.62b shows a layout example with 16 bands. Fig. 2.62c is an example where a wavelet-style octave-band decomposition is applied fully separable over both dimensions, which is equivalent with the Haar transform scheme in Fig. 2.41c. Fig 2.62d shows the layout which is commonly denoted as *2D DWT*, where only the low-pass output c_{00} of the 4-band system is subject to further 4-band decomposition etc. In Fig. 2.62c/d, 'S' denotes the scaling band of lowest resolution, which represents a sub-sampled version of the picture.

Fig. 2.63 shows results of subband and wavelet decomposition applied to an image signal, where the different sub-sampled subband pictures are shown in the positions of their corresponding frequency partitions in Fig. 2.62.

It is also possible to realize non-separable 2D filter banks. Fig. 2.64 shows an example of a 2D decimation by a factor of 2, where a subband system decomposes a rectangular-grid (separable) sampled signal into two components of quincunx sampling. To describe such systems, the principles introduced in the context of multi-dimensional sampling can be used. If $s_k(\tilde{n}_1,\tilde{n}_2)$ is the original signal and $s_{k-1}(n_1,n_2)$ the sub-sampled signal, the relationship between the indices can be expressed by the sampling matrix **U** such that[89]

$$\begin{bmatrix}\tilde{n}_1\\ \tilde{n}_2\end{bmatrix}=\begin{bmatrix}u_{11} & u_{12}\\ u_{21} & u_{22}\end{bmatrix}\begin{bmatrix}n_1\\ n_2\end{bmatrix} \quad ; \quad \tilde{\mathbf{n}}=\mathbf{U}\mathbf{n}\,; \quad \mathbf{n}=\mathbf{U}^{-1}\tilde{\mathbf{n}}\in\mathbb{Z}\,. \tag{2.360}$$

The factor of sub-sampling, and hence the number of spectral copies (original plus alias spectra) is equal to the absolute determinant

$$|\mathbf{U}|=|u_{11}u_{22}-u_{21}u_{12}|\,. \tag{2.361}$$

The related frequency sampling matrix $\mathbf{F}=[\mathbf{U}^{-1}]^{\mathrm{T}}$ points to the positions of periodic spectral copies, where alias may occur. In analogy with (2.297), the z transform of the decimated signal is

$$S_{k-1}(z_1,z_2)=\frac{1}{|\mathbf{U}|}\sum_{k_1=0}^{U_1-1}\sum_{k_2=0}^{U_2-1}S\left(\mathrm{W}^{-(f_{11}k_1+f_{12}k_2)}\,z_1^{\,f_{11}}z_2^{\,f_{12}},\mathrm{W}^{-(f_{21}k_1+f_{22}k_2)}\,z_1^{\,f_{21}}z_2^{\,f_{22}}\right)$$
$$\text{with}\quad \mathrm{W}=\mathrm{e}^{\mathrm{j}2\pi}\ \text{ and }\ U_i=\max_{j=1,2}\{|u_{ij}|\}. \tag{2.362}$$

The reverse operation is an interpolation by factor U^*, which is a generalization of (2.299) using the parameters in **U**

$$S_k(z_1,z_2)=S_{k-1}(z_1^{\,u_{11}}z_2^{\,u_{21}},z_1^{\,u_{12}}z_2^{\,u_{22}})\,. \tag{2.363}$$

---

[89] The following considerations are strictly valid for integer subsampling factors u_{ij}, as otherwise an additional sub-sample phase shift would be necessary which would require an additional interpolation step.

Example. The sampling matrix $\mathbf{T}_\mathrm{q}$ for the case of quincunx decimation and the related frequency sampling matrix $\mathbf{F}_\mathrm{q}$ in analogy with (2.63) and Fig. 2.64 are expressed as

$$\mathbf{U}_\mathrm{q} = \begin{bmatrix} 2 & 1 \\ 0 & 1 \end{bmatrix} \quad ; \quad \mathbf{F}_\mathrm{q} = \left[\mathbf{U}_\mathrm{q}^{-1}\right]^\mathrm{T} = \begin{bmatrix} \frac{1}{2} & 0 \\ -\frac{1}{2} & 1 \end{bmatrix}. \tag{2.364}$$

The z transform of the decimated signal is

$$S_{k-1}(z_1, z_2) = \frac{1}{2}\sum_{k=0}^{1} S_k\left(\mathrm{W}^{-\frac{1}{2}k} z_1^{\frac{1}{2}}, \mathrm{W}^{\frac{1}{2}k} z_1^{-\frac{1}{2}} z_2\right) \text{ with } \quad \mathrm{W} = \mathrm{e}^{\mathrm{j}2\pi}. \tag{2.365}$$

To realize a non-separable decimation, it is typically necessary to use non-separable filters. The quincunx decimation can be performed using the following biorthogonal pair of 2D filter matrices [KOVACEVIC, VETTERLI 1992], where according to the conditions of biorthogonal filters, the highpass $\mathbf{H}_1$ is operated with a one-sample delay either horizontally or vertically relative to the lowpass $\mathbf{H}_0$.

$$\mathbf{H}_0 = \frac{1}{32}\begin{bmatrix} 0 & 0 & -1 & 0 & 0 \\ 0 & -2 & 4 & -2 & 0 \\ -1 & 4 & 28 & 4 & -1 \\ 0 & -2 & 4 & -2 & 0 \\ 0 & 0 & -1 & 0 & 0 \end{bmatrix} \quad ; \quad \mathbf{H}_1 = \frac{1}{4}\begin{bmatrix} 0 & -1 & 0 \\ -1 & 4 & -1 \\ 0 & -1 & 0 \end{bmatrix}. \tag{2.366}$$

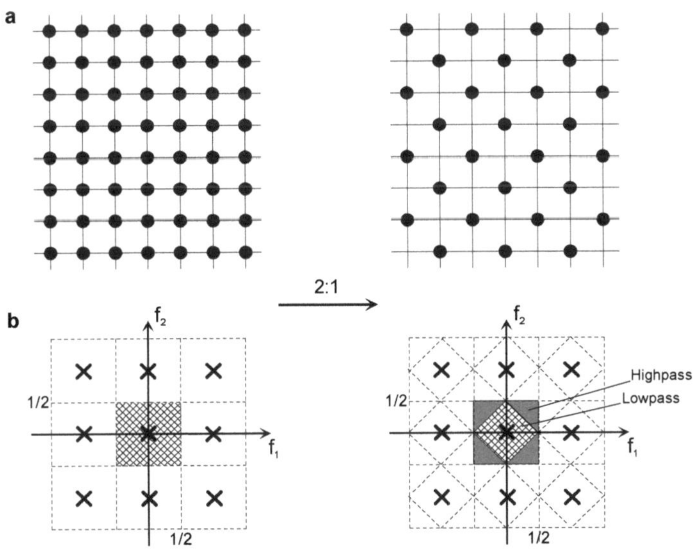

Fig. 2.64. Non-separable 2D system with 2:1 quincunx decimation
a Sub-sampling schema in the spatial domain **b** Layout of frequency bands

As in (2.325), the kernels of lowpass and highpass filters are of different size. Applying the relationships $G_0(\mathbf{z})=H_1(-\mathbf{z})$ and $G_1(\mathbf{z})=-H_0(-\mathbf{z})$ from (2.316), the synthesis filters are determined by multiplication (modulation) with alternating signs. For symmetric 2D filters, this is realized such that impulse response values with an odd sum of indices are multiplied by −1, i.e.

$$\begin{aligned} g_0(n_1,n_2) &= (-1)^{n_1+n_2}\, h_1(n_1,n_2), \\ g_1(n_1,n_2) &= (-1)^{n_1+n_2+1}\, h_0(n_1,n_2). \end{aligned} \tag{2.367}$$

The resulting synthesis filter matrices are

$$\mathbf{G}_0 = \frac{1}{4}\begin{bmatrix} 0 & 1 & 0 \\ 1 & 4 & 1 \\ 0 & 1 & 0 \end{bmatrix} \quad ; \quad \mathbf{G}_1 = \frac{1}{32}\begin{bmatrix} 0 & 0 & 1 & 0 & 0 \\ 0 & 2 & 4 & 2 & 0 \\ 1 & 4 & -28 & 4 & 1 \\ 0 & 2 & 4 & 2 & 0 \\ 0 & 0 & 1 & 0 & 0 \end{bmatrix}. \tag{2.368}$$

2.8.6 Pyramid decomposition

The DWT is a *multi-resolution scheme* for signal representation. This means that by using more higher-frequency wavelet bands, the resolution of the reconstructed signal is increased; in a critically sampled (typically dyadic) wavelet representation, the total number of coefficient samples equals the number of samples in the full-resolution signal, regardless of the depth of the wavelet tree. An alternative type of multi-resolution methods are the *pyramid schemes. U* signal representations with different sampling resolutions are generated by filtering and downsampling, *in addition* to the original (full) resolution. In principle, arbitrary downsampling factors are possible, even though out of complexity reasons and to avoid excessive over-completeness, dyadic factors are often chosen when pyramid schemes are used in compression, unless different up/downsampling is needed e.g. to support multiple spatial resolutions.

Whereas the scheme of *Gaussian pyramid* generates the different resolution representations as independent entities (in case of dyadic resolutions this would correspond to the scale spaces in Fig. 2.57), the *Laplacian pyramid* establishes a differential representation, which can be interpreted as a set of bandpass channels (this would roughly correspond to the wavelet spaces in Fig. 2.57). However, in contrast to the DWT approach, no downsampling is applied to the bandpass components, which means that the representation (in terms of number of samples) is over-complete, as the spectrum below the respective pass band should be approximately void. On the other hand, alias and frequency reversion are avoided by omitting the downsampling, which can be beneficial in terms of coding, e.g. when shift invariance is required as in motion compensated prediction.

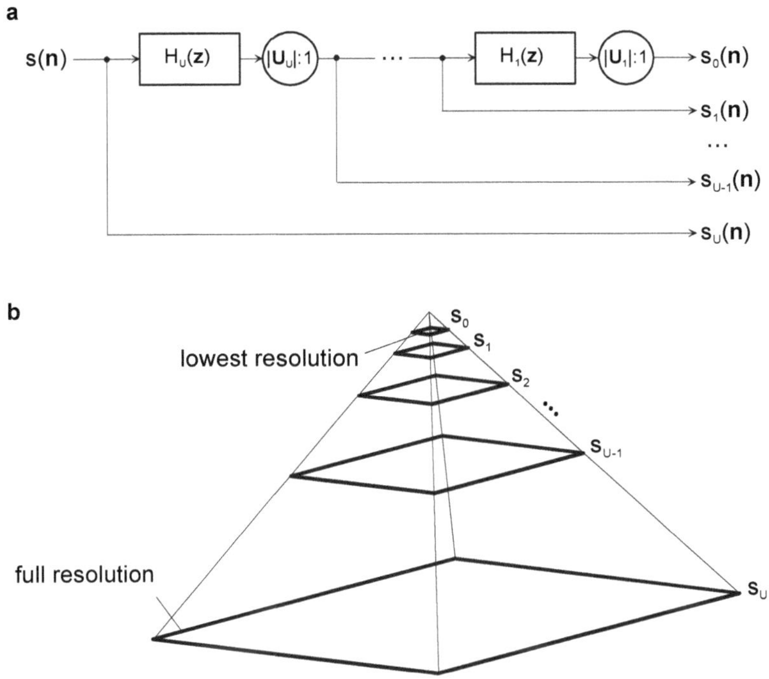

Fig. 2.65. a Generation of the Gaussian pyramid representation (typically identical filters and identical subsampling schemes described by **U** are used throughout the levels). **b** Illustration of images sizes (dyadic scheme) as levels of a pyramid

Gaussian pyramid. All resolution levels can be used independently, i.e. no lower resolution level is needed if a finer resolution level shall be used. The generation of the Gaussian pyramid representation is performed by elementary building blocks consisting of lowpass filtering followed by decimation described by a sampling matrix **U** (see Fig. 2.65a; as an example, for a 2D signal with horizontal/vertical subsampling factors $U_1=U_2=2$, the total subsampling ratio is 4:1 with $|\mathbf{U}|=4$). This is performed in an iterative cascade through all levels of the pyramid, starting from the base and terminating at the top (see Fig. 2.65b). By cascading U elementary building blocks, a total of U+1 resolution levels (including the original resolution $s(\mathbf{n})=s_U(\mathbf{n})$) are generated. The signal $s_{k-1}(\mathbf{U}^{-1}\mathbf{n})$ is obtained by lowpass filtering and subsampling, $s_{k-1}(\mathbf{U}^{-1}\mathbf{n})=s_k(\mathbf{n})*h(\mathbf{n})$[90]. The concept is similar to the processing of scale-space components in the wavelet transform (Fig. 2.57), but exhibits redundancy due to the fact that the coarser resolutions establish subspaces

[90] This convolution is to be performed with reference to **n** coordinates, but only at positions where $\mathbf{U}^{-1}\mathbf{n}$ consists of integer numbers (i.e. the positions still existing after subsampling). In case of non-dyadic subsampling, **U** itself could contain non-integer numbers. In that case, it would be necessary to include a position-dependent sub-sample phase shift (interpolation) in the lowpass filter impulse response.

of the finer resolutions instead of being orthogonal complements. The resulting representation is therefore significantly redundant and over-complete, and not as such very suitable for the purpose of compression.

The method is denoted as *Gaussian* pyramid, because filters approximating a Gaussian-shaped impulse response are often used as lowpass filters prior to decimation in this context. The convolution of two Gaussian functions results in a Gaussian of extended length. Hence, the effect of the cascaded system at a later stage is approximately equivalent to the usage of *one* Gaussian filter with a longer width of the impulse response (lowpass with lower cut-off frequency)[91]. The implementation complexity in the cascaded pyramid system is however much lower due to the intermediate sub-sampling operations.

An example for simple approximation of a non-separable 2D Gaussian with a short kernel is given by the filter matrix[92]

$$\mathbf{H}_G = \frac{1}{8}\begin{bmatrix} 0 & 1 & 0 \\ 1 & 4 & 1 \\ 0 & 1 & 0 \end{bmatrix}. \tag{2.369}$$

Laplacian pyramid. Each resolution level (except for the smallest scale image) is represented by a difference signal. The principle as applied for generation of the difference signals is shown in Fig. 2.66a. Firstly, the lower-resolution signal $s_{k-1}(\mathbf{n})$ is generated as in the case of the Gaussian pyramid. Then, it is upsampled and filtered by a lowpass interpolation filter to generate a prediction and compute the difference (prediction error) signal[93],

$$\hat{s}_k(\mathbf{n}) = s_{k-1}(\mathbf{U}^{-1}\mathbf{n} \uparrow \mathbf{n}) * g(\mathbf{n}), \quad e_k(\mathbf{n}) = s_k(\mathbf{n}) - \hat{s}_k(\mathbf{n}). \tag{2.370}$$

For reconstruction, the difference is added to the prediction from the next-coarser signal (Fig. 2.66b). If U elementary building blocks are arranged in a cascaded structure, a total of U+1 resolution levels is represented by U difference signals $e_1(\mathbf{n})\ldots e_U(\mathbf{n})$ and one strongly-scaled signal $s_0(\mathbf{n})$. The reconstructed signals

---

91 This statement ignores the alias which can occur due to the subsampling, depending on the spectrum of the signal. On the other hand, the Gaussian function is non-negative, which gives a penalty in terms of the sharpness of the frequency cut-off, but prevents from ringing at signal discontinuities (e.g. edges in images).

92 A typical primitive 1D approximation of a Gaussian filter function is the binomial filter with the coefficient vector $\mathbf{h}=[\frac{1}{4}\ \frac{1}{2}\ \frac{1}{4}]^T$. (2.369) represents a superposition of a horizontal and a vertical binomial filter. Iterated convolution gives longer binomial functions, which by tendency give an approximation of a sampled Gaussian according to the central limit theorem. For 2D, typically separable filters are used.

93 This convolution is performed at all positions where $\mathbf{n}$ is integer, where $\mathbf{U}^{-1}\mathbf{n}\uparrow\mathbf{n}$ expresses that in the upsampled s_{k-1} zero values are inserted where $\mathbf{U}^{-1}\mathbf{n}$ is not an integer number. In case of non-dyadic subsampling, $\mathbf{U}$ may itself contain non-integer numbers and it may be necessary to additionally include a position-dependent sub-sample phase shift in the filter impulse response.

$s_1(\mathbf{n}) \ldots s_U(\mathbf{n})$ at the different pyramid levels are equivalent to the output of the Gaussian pyramid. Reconstruction always must start at the lowest resolution level and requires U sequential operations.

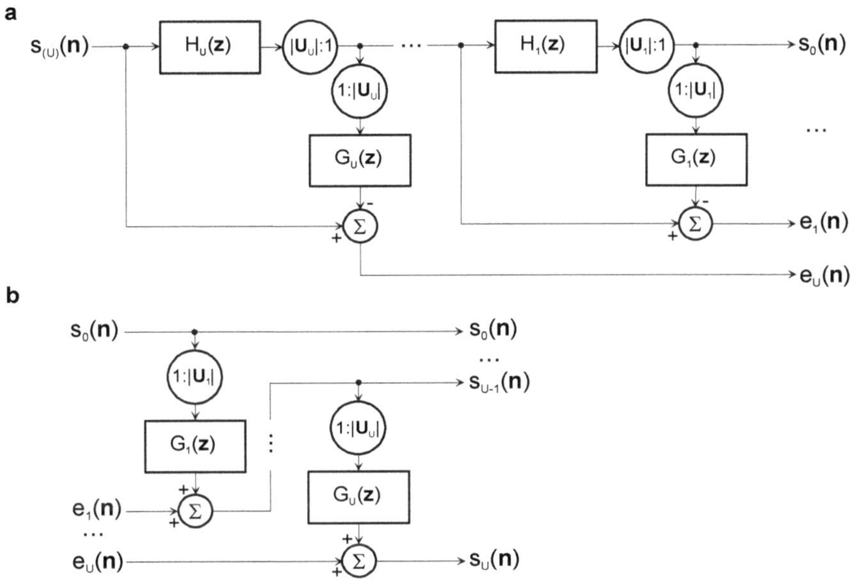

Fig. 2.66. Laplacian pyramid representation: **a** analysis **b** synthesis

Assuming (almost) alias-free subsampling and high-quality interpolation, the difference between the signal $s_k(\mathbf{n})$ and the output of the filter (2.369) would be close to the prediction $\hat{s}_k(\mathbf{n})$ from (2.370), which could then be generated directly by the filter

$$\mathbf{H}_L = \begin{bmatrix} 0 & 0 & 0 \\ 0 & 1 & 0 \\ 0 & 0 & 0 \end{bmatrix} - \mathbf{H}_G = \frac{1}{8}\begin{bmatrix} 0 & -1 & 0 \\ -1 & 4 & -1 \\ 0 & -1 & 0 \end{bmatrix}. \tag{2.371}$$

This filter kernel provides an approximation of the local second derivative of the signal and is denoted as *Laplacian filter operator*. From this, the differential pyramid is also called *Laplacian* pyramid [BURT, ADELSON 1983]. In principle, this pyramid represents second derivatives of the signal within different scale spaces, which could also be interpreted as bandpass-filtered (or highpass-filtered for $e_U(\mathbf{n})$) versions of $s(\mathbf{n})$[94].

---

94 When subsampling is omitted and filtering is done by concatenating Gaussian impulses responses, the signals $e_k(n)$ are also entitled as *differences of Gaussians* (DoG), which is approximately equal to filtering by 2nd derivatives of Gaussians (Laplacian of Gaussian, LoG). Such representations are largely over-complete in terms of number of samples and

Unlike the wavelet transform, the differential signals $e_k(\mathbf{n})$ of the Laplacian pyramid are not orthogonal complements. Firstly, $e_k(\mathbf{n})$ and $s_k(\mathbf{n})$ must be correlated, as the prediction error contains all detail information that is not predictable from $s_{k-1}(\mathbf{n})$. Second, due to the usage of non-ideal filters the prediction errors $e_k(\mathbf{n})$ and $e_{k-1}(\mathbf{n})$ over the different levels may be correlated as well; furthermore, structures with wide spectra such as edges and pulses would also appear in the prediction errors over a variety of k values. In general however, this redundancy will be significantly lower than in the case of the Gaussian pyramid. Furthermore, an over-completeness is inherent to the pyramid schemes in terms of number of samples. For example, if U pyramid levels are used for a 2D (image) signal, the total number of samples to be represented grows by a factor of

$$\sum_{u=0}^{U}\left(\frac{1}{4}\right)^{u}<\frac{4}{3}, \tag{2.372}$$

as compared to the number of samples in the original signal. In contrast to that, block and block-overlapping transforms, filterbank and DWT transforms can use critical sampling, such that the overall number of frequency coefficients is identical to the number of signal samples. However, block and wavelet transform, though not over-complete, need to make trade-offs between lowpass and highpass filters in order to achieve perfect reconstruction. This can invoke other effects (in particular aliasing in bands and additionally frequency reversion in highpass bands), which may even be more severe than the disadvantage of increased number of samples. When using a pyramid representation in the context of encoding, the over-completeness seems to be a disadvantage in first place, but the aforementioned alias-bearing effects of critically sampled representations can be avoided. Furthermore, the redundancy between various components in the pyramid can be utilized and removed by coding. Therefore, the differential pyramid has turned out to be efficient as a compression method, particularly in the context of scalable (multi-resolution) representations of image and video signals, where additional methods such as prediction over another dimension can remove the redundancy. Nevertheless, it should be noted that the over-completeness causes a penalty in terms of larger complexity, as more samples need to be processed.

---

are not usually used in coding, but rather when using multi-resolution representations for feature analysis (see [MCA, SEC. 4.4])

2.9 Problems

Problem 2.1.

a) Determine a condition for alias-free hexagonal sampling (Fig. 2.11c).
b) What is the ratio of the area of the base band, as compared to rectangular sampling with horizontal and vertical sampling distances equal to the vertical distance in the hexagonal case?
c) What is the ratio of the horizontal sampling distance in hexagonal sampling, as compared to the rectangular sampling case of b) ?
d) Compute the determinant of the sampling matrix $\mathbf{T}_{\text{hex}}$, normalized by the vertical sampling distance. Discuss the relationship of this value with the results from parts b) and c).

Problem 2.2.

a) Show that the quincunx grid (Fig. 2.11d) can be constructed by superposition of two rectangular grids which are offset by $T_1|T_2$, and each having sampling distances $2T_1|2T_2$ horizontally | vertically.
b) Compute the periodic spectrum from this construction, and show that it is identical to the spectrum found via the sampling matrix (2.63).

Problem 2.3.

A two-dimensional cosine (2.1) of horizontal frequency $F_1=1/(3T)$ is sampled by a quincunx grid.
a) Compute the 2D Fourier spectrum.
b) Determine the upper limit for vertical frequency $|F_2|$ guaranteeing alias free sampling.
c) Which horizontal frequency becomes visible after ideal lowpass filter reconstruction from the sampled signal, if the vertical frequency is $F_2=1/(3T)$?

Problem 2.4.

For the generalized Gaussian PDF (2.126),
a) Show that $\gamma=2$ gives the Gaussian normal PDF (2.127).
b) Show that $\gamma=1$ gives the Laplacian PDF (2.128).
c) With $\Gamma(c)=\Gamma(c+1)/c$, which PDF can asymptotically be expected for $\gamma\rightarrow\infty$?
[use values $\Gamma(3)=2$; $\Gamma(1)=1$; $\Gamma(1.5)=\sqrt{\pi}/2$; $\Gamma(0.5)=\sqrt{\pi}$].

Problem 2.5.

A one-dimensional, stationary zero-mean process $s(n)$ with Gaussian PDF has an autocovariance function $\mu_{ss}(k)=\sigma_s^2\rho^{|k|}$.
a) Construct the autocovariance matrix $\mathbf{C}_{ss}$ of size 3x3 .
b) Show that for $\rho=0$: $p_3(\mathbf{x})=p_{\mathcal{N}}(x_1)p_{\mathcal{N}}(x_2)p_{\mathcal{N}}(x_3)$. Here, $p_3(\mathbf{x})$ is a vector Gaussian PDF (2.156) for vector random variables $\mathbf{x}=[x_1\ x_2\ x_3]^{\mathrm{T}}$, and $p_{\mathcal{N}}(x_i)$ shall be Gaussian normal distributions (2.127).

Problem 2.6.

a) Combined random instantiations from two event sets S_1 and S_2 shall be statistically independent, i.e. $\Pr(S_1,S_2)=\Pr(S_1)\Pr(S_2)$. Show the following relationships for this case: $H(S_1|S_2)=H(S_1)$; $H(S_2|S_1)=H(S_2)$; $I(S_1;S_2)=0$.
b) Instantiations drawn from the two event sets S_1 and S_2 shall now always be identical. Show that $H(S_1|S_2)=0$; $H(S_2|S_1)=0$; $I(S_1;S_2)=H(S_1)=H(S_2)$.

Problem 2.7.

The joint PDF of two Gaussian processes $s_1(n)$ and $s_2(n)$ shall be defined by (2.153). Further, $\sigma_{s_1}=\sqrt{2}\sigma_{s_2}$.

a) Determine the joint PDF for the cases of uncorrelated signals ($\rho_{s_1s_2}(\mathbf{0})=0$) and fully dependent signals ($\rho_{s_1s_2}(\mathbf{0})=1$).
b) Determine the conditional PDF $p_{s_2s_1}(x_2|x_1;\mathbf{0})$ for the general case first, then specifically for the two cases of a).

Problem 2.8.

The eigenvalues of the matrix $\mathbf{C}=\begin{bmatrix}1 & \rho\\ \rho & 1\end{bmatrix}$ are $\lambda_1=1+\rho$ and $\lambda_2=1-\rho$.

a) Following (A.20), determine $\mathbf{\Phi}_1$ und $\mathbf{\Phi}_2$ of $\mathbf{C}$ such that they establish an orthonormal base $\mathbf{\Psi}=[\,\mathbf{\Phi}_1\ \mathbf{\Phi}_2\,]$ according to (A.24).
b) Sketch the eigenvectors within a coordinate system of axes x_1, x_2.
c) Determine the inverse $\mathbf{\Psi}^{-1}$.
d) Compute the determinant of $\mathbf{C}$, and compare the result against the product of the eigenvalues.

Problem 2.9.

For the AR(1) model from (2.189) and (2.190), prove the validity of the autocorrelation and variance properties (2.191) and of the spectral properties (2.192).

Problem 2.10.

For statistical modeling of a 1D signal, an AR(1) model of variance σ_s^2 and correlation coefficient $\rho_{ss}(1)=0.95$ is used. By using the autocorrelation function of the model, a linear predictor shall be optimized. Determine the coefficients $a(1)$ and $a(2)$ for a predictor filter of order $P=2$ by solving the Wiener-Hopf equation (2.207).

Problem 2.11.

For linear 2D prediction of a separable AR(1) process with $\rho_1=\rho_2=0.95$, a non-separable predictor filter is used which implements $\hat{s}(n_1,n_2)=0.5\,s(n_1-1,n_2)+0.5\,s(n_1,n_2-1)$ as prediction equation. Determine the variance of the prediction error signal, and compare against the variance of the innovation signal $v(n_1,n_2)$.

Problem 2.12.

A video sequence consists of pictures which are unchanged except for global translation motion. The picture information is an output from a 2D separable AR(1) model generator. Parameters are $\rho_1=\rho_2=0.95$. From one picture to the next, translation shift by $k_1=7$ horizontally and $k_2=3$ is observed. Different methods of linear prediction shall be compared using the criterion of prediction error variance:

a) Spatial prediction, separable predictor according to (2.226);
b) Temporal prediction $\hat{s}(n_1,n_2,n_3)=s(n_1,n_2,n_3-2)$;
c) Motion-compensated temporal prediction $\hat{s}(n_1,n_2,n_3)=s(n_1-k_1,n_2-k_2,n_3-1)$.

Problem 2.13.

When defined as a block transform, the Haar transform has $U^*=\log_2 M+1$ 'basis types', which are subsequently described by an index $u^*=0,1,...,\log_2 M$. From each basis type, M^* basis functions are developed, each of which has only one internal flip of the sign for the cases $u^*>0$. M^* is 1 for $u^*=0,1$ and 2^{u^*-1} for the other basis types. Basis functions determined from the same basis type are indexed by $i=0,1,..., M^*-1$ subsequently, they are non-overlapping. The orthonormal transform basis set is

$$t_k^{\text{Haar}}(n)=\begin{cases}\text{ha}(n-i\frac{M}{M^*}) \text{ for } i\frac{M}{M^*}\le n<(i+1)\frac{M}{M^*}\\ 0, \text{ else}\end{cases}$$

with

$$k=\begin{cases}k^* & \text{for} \quad k^*=0,1\\ M^*+l & \text{for} \quad k^*>1\end{cases} \quad \text{and} \quad \text{ha}(n)=\sqrt{\frac{M^*}{M}}\cdot(-1)^{\left\lfloor\frac{2^{k^*}n}{M}\right\rfloor}.$$

The *Walsh basis* consists of $K=M$ basis functions with constant length M. The kth function has k flips between positive and negative values. The function for $k=0$ consists of M positive constant values. The development of remaining Walsh functions is performed recursively, starting from $t_1(n)$. The number of recursions necessary to generate all basis functions is $\log_2 M-1$. During one recursion step, all basis functions developed in the previous step are scaled (which is done by eliminating each second sample), and then combined into new basis functions, once periodically and once 'antiperiodic', i.e. mirrored. The number of new basis functions is doubled by each iteration step, and with ν flips in the scaled function, two new functions can be generated, one of which has $2\nu-1$ sign flips, the other 2ν flips. The process of recursion can be described as follows, where the periodic/antiperiodic combinations are implemented by multiplying the scaled functions by ± 1:

Let

$$t_k^{\text{Wal}}(n)=t_k^{\text{Rad}}(n) \quad \text{for} \quad 0\le k<2, \quad 0\le n<M$$
$$k^*=1, M^*=2, K^*=\log_2 M, P(0)=-1.$$

While $k^*<K^*$

$$
\begin{array}{l}
\{ \\
\quad \text{For } 0 \le i < \log_2 M^* : \\
\qquad t_{\text{scal}}(n,i) = t^{\text{Wal}}_{(M^*+2i)/2}(2n,i) \\
\qquad t^{\text{Wal}}_{M^*+2i+j}(n) = \begin{cases} t_{\text{scal}}(n,i) & \text{for} \quad 0 \le n < M/2 \\ P(i)^{j+1} \cdot t_{\text{scal}}(n - M/2, i) & \text{for} \quad n \ge M/2 \end{cases} \qquad \text{for } j = 0,1 \\
\quad \text{For the next step, set } P(2i+j) = -P(i)^{j+1},\ M^* = 2M^*,\ k^* = k^* + 1. \\
\}
\end{array} \tag{2.373}
$$

The factor $P(i)$ has the effect that from a periodic basis function (having even number of flips), an antiperiodic function is generated at first in the next step and vice versa. This guarantees an ever increasing number of flips, equal to the index of the function. The recursion process is illustrated for the case $M = 8$ in Fig. 2.67

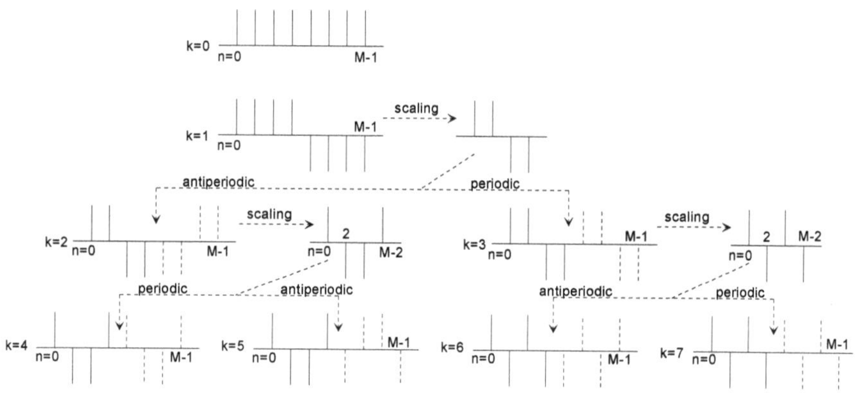

Fig. 2.67. Development of the Walsh transform basis

a) Construct the transform matrices of 1D Haar and Walsh transforms for M=4.
b) Transform the following image matrix by the related separable 2D transforms using (2.246):

$$
\mathbf{S} = \begin{bmatrix} 18 & 4 & 2 & 4 \\ 18 & 4 & 2 & 4 \\ 2 & 4 & 2 & 4 \\ 2 & 4 & 2 & 4 \end{bmatrix}.
$$

c) Interpret the results. Discuss in particular which transform better compacts the given image.

Problem 2.14.

a) Determine the transform matrix of a 1D DCT for M=3, and show that the transform is orthonormal.
b) Set up the autocorrelation matrix (2.157) of size 3x3 for an AR(1) model. The matrix shall then be transformed by the 1D DCT of a), using (2.287).

c) For a model of variance σ_s^2, two different cases ρ=0.9 and ρ=0.5 shall be considered to fill the matrix $\mathbf{C}_{cc}$. Give an interpretation of the differences you observe in the matrix entries.
d) For both cases from c), compute the trace (A.21) of the autocorrelation matrices and their transformed counterparts. Give an interpretation of the result.

Problem 2.15.

a) Determine the transform matrix of a 1D Haar transform for M=4.
b) Set up the autocorrelation matrix (2.157) of size 4x4 for an AR(1) model, and apply its transformation by the Haar transform of a), using (2.287).
c) Give an interpretation about the remaining correlations between transform coefficients.

Problem 2.16.

a) Determine the Fourier transfer functions $\mathcal{F}\{\mathbf{t}_k\}$ for the basis vectors of a transform

$$\mathbf{t}_0 = \left[\frac{\sqrt{2}}{2} \quad \frac{\sqrt{2}}{2}\right]^T, \qquad \mathbf{t}_1 = \left[-\frac{\sqrt{2}}{2} \quad \frac{\sqrt{2}}{2}\right]^T.$$

b) Prove the orthogonality of this basis system.
c) Show that the functions $|\,\mathcal{F}\{\mathbf{t}_k\}\,|$ of both basis vectors have a mirror symmetry around the frequency f= 1/4.
d) Show that $|\,\mathcal{F}\{\mathbf{t}_0\}\,|^2 + |\,\mathcal{F}\{\mathbf{t}_1\}\,|^2$ = const.

Problem 2.17.

a) Determine the basis vectors of the block-overlapping transform according to (2.289)-(2.292) with settings U=2, M=4. [Hint: To simplify expressions of the trigonometric functions, use constants cos(3π/8)=sin(π/8)=A and cos(π/8)=sin(3π/8)=B; consider for which other cases identical values ±A or ±B would appear.]
b) Show the orthogonality of the basis system.
c) Determine the Fourier transfer functions $\mathcal{F}\{\mathbf{t}_k\}$. Do the basis functions have a linear-phase property?
d) Would a realization of this transform by a fast algorithm be possible?

Problem 2.18.

a) Show the validity of the orthogonality property for linear-phase QMF systems for the following cases of filters:

$$i)\ H_0(z) = A\cdot z^2 + B\cdot z + C + C\cdot z^{-1} + B\cdot z^{-2} + A\cdot z^{-3}$$

$$ii)\ H_0(z) = A\cdot z^2 + B\cdot z + C + B\cdot z^{-1} + A\cdot z^{-2}$$

b) Determine the z transform representations of polyphase filters $H_{0,A}$, $H_{0,B}$, $H_{1,A}$ and $H_{1,B}$ according to Fig. 2.53 for both filter configurations from a). Which number of multiplications per sample would be necessary at minimum?

Problem 2.19.

A zero-mean random signal $s(n)$ shall be modeled by an AR(1) process. To describe the parameters of the process, the value of spectral power density $\Phi_{ss,\delta}(f=1/2)=\sigma_s^2/9$ is given.

a) Determine the correlation parameter ρ, and the variance σ_v^2 of the white-noise innovation in dependency of σ_s^2.

The signal shall be decomposed into two polyphase components, where the sequences of even- and odd-indexed samples shall be processed independently by predictor filters of first order, $H(z)=az^{-1}$, as shown in Fig. 2.68.

b) Determine the optimum predictor coefficient a.
c) Determine the variance of the prediction error signal $e_e(n)$ of the even-indexed samples when the optimum a is used. From this, compute the coding gain $G=\sigma_s^2/\sigma_{e_e}^2$. By which factor will this gain be smaller, as compared to the optimum case of prediction for the AR(1) model (without polyphase decomposition) ?
d) Compute the covariance between the signals $e_e(n)$ und $e_o(n)$ when the optimum a is used.

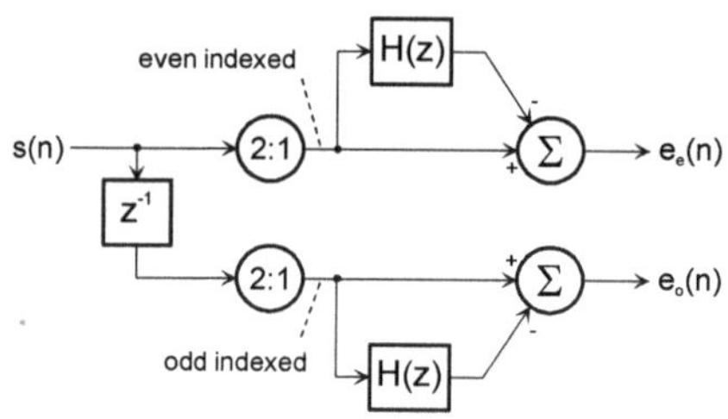

Fig. 2.68. Prediction within polyphase components

Problem 2.20.

An AR(1) process $s_{AR}(n)$ is characterized by the correlation parameter ρ=0.75 and the variance of the Gaussian innovation signal, σ_v^2=7.

a) Determine the variance of the AR process.

For linear prediction of $s_{AR}(n)$, a falsely adapted prediction error filter with transfer function $A(z) = 1 - z^{-1}$ is used (see Fig. 2.69).

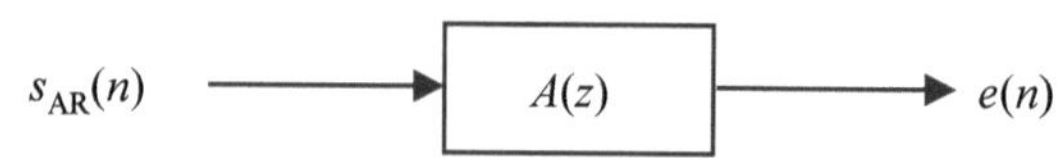

Fig. 2.69. Prediction of an AR(1) process

b) Compute the variance of the prediction error signal $e(n)$ and the coding gain.
c) Which would be the coding gain in case of optimum prediction? By which factor is the coding gain worse in the case of the falsely-adapted prediction from b)?
d) Determine the power density $\Phi_{ee,\delta}(f)$. Can $e(n)$ be a white noise process?
e) Determine a system (z transfer function and block diagram), which generates the optimum prediction error signal of lowest possible variance from $e(n)$.

Problem 2.21.

A binary signal shall be synthesized according to the block diagram given in Fig. 2.70. Herein, $v(n)$ is an uncorrelated zero-mean Gaussian process of variance σ_v^2. **T** is a threshold decision circuit with following characteristics:

$$b(n) = \begin{cases} 0 \text{ if } s(n) \le C \\ 1 \text{ if } s(n) > C \end{cases}$$

It is now to be assumed that $b(n)$ behaves as first order Markov process.

a) Determine the probabilities Pr(0) and Pr(1) depending on C.

b) For C=0, determine the probabilities Pr(1|0) and Pr(0|1) depending on ρ.

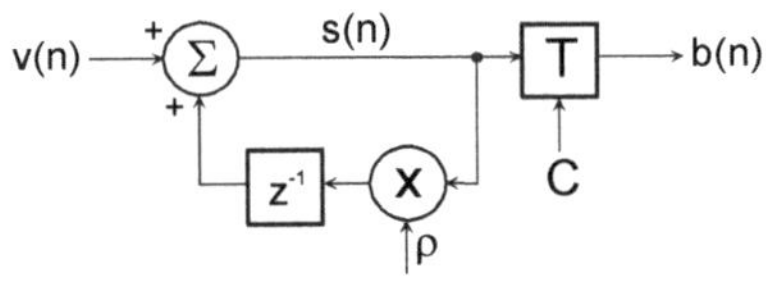

Fig. 2.70. Circuit for generating a binary signal

3 Perception and Quality

Optimization of multimedia signal coding algorithms for best subjective quality requires considering perceptual properties of vision and hearing. This allows to categorize the information contained in signals as being relevant or irrelevant components. Physiological properties of the human perception system, e.g. characteristics of the human visual and auditory system have a direct impact here. In particular masking effects, which are well known to be in effect both in vision and hearing, will be discussed in this chapter. Certain elements of low-level human vision and hearing have a high similarity with technical systems for signal capture and recognition, and can well be interpreted as methods of signal and information analysis and processing. However, variations between individuals can be large even at low level. Metrics for measuring the objective and subjective quality of multimedia signals (including some which intend to match with human perception) are introduced in the last part of this chapter.

3.1 Properties of vision

3.1.1 Physiology of the eye

Light rays intruding the eye (Fig. 3.1) project a natural scene from the outside world. These are forwarded to the retina via a lens which is protected by the cornea. In the retina, the picture from the outside is projected to light-sensitive receptors which perform a kind of sampling. Two types of receptors exist, which are the *rod cells* (around 120 million units), and the *cone cells* (around 6 million units). Both types of receptors contain light-sensitive pigments, where the rods are specialized in acquisition of the brightness, while the cones are sensitive to both brightness *and* color; three different types of cones are tuned for highest sensitivity in the red, green and blue primary ranges of the color spectrum, respectively. The distribution of the receptors is not uniform over the retina. In the area with highest density of receptors, the *fovea centralis*, mostly cones are present. At the

© Springer-Verlag Berlin Heidelberg 2015
J.-R. Ohm, *Multimedia Signal Coding and Transmission*,
Signals and Communication Technology, DOI 10.1007/978-3-662-46691-9_3

periphery of the retina, a mixture of cones and rods can be found. Fig. 3.2 shows typical densities of rods and cones on the human retina.

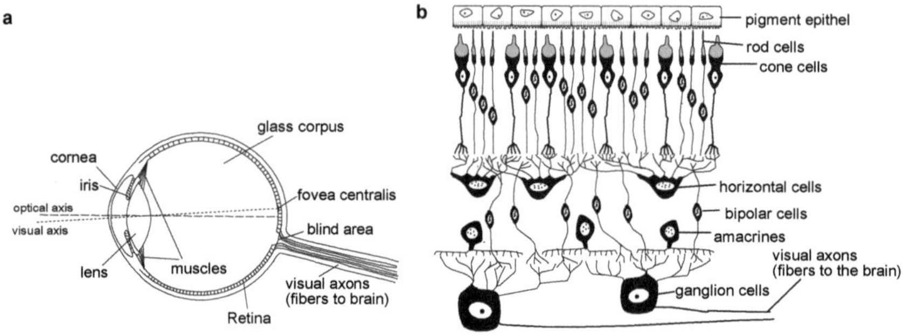

Fig. 3.1. Physiology of the human eye **a** Overall structure **b** Structure of the retina

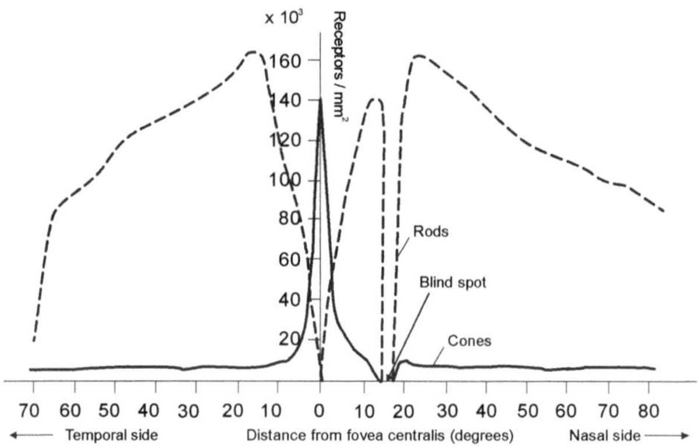

Fig. 3.2. Density of rods and cones on the retina (after [PURVES ET AL. 2012])

The information from the receptors has to be forwarded to the brain via nerves. For this purpose, the intensity of light is transformed into electrical information. An electro-chemical reaction modulates the frequency of short electrical pulses depending on the light flux received by the receptor. Only when differences between adjacent receptor cells appear, information is conveyed to the next level. A hierarchy of four different types of neurons exists in the retina to perform the analysis of local (dis-)similarity in a hierarchical fashion: *horizontal cells*, *bipolar cells*, *amacrine cells* and *ganglion cells*[1]. The first layer of neurons only propagates an excitation signal to the next layer when the difference between the neighbored receptor inputs is *above a threshold.* Further layers proceed similarly by evaluating inputs from neurons of the previous layer. Approximately 800,000 nerve fibers finally perform the transport of visual information from the eyes to

[1] The schematic drawing in Fig. 3.1b is physiologically not fully correct, as the horizontal and bipolar cells are rather found within the outer layers of the retina.

the brain, which indicates that a kind of data compression is performed already within the retina, as the number of receptors is higher by a factor of ≈150. The area where the nerve fibers are attached to the retina is not populated by receptors; this is called the *blind spot*.

The visual nerves propagate the information towards the visual center of the brain, the *cortex praestriata*. It is assumed that certain precincts of the visual cortex are responsible to process and analyze the information originating from related areas of the retina. Depending on the information which is input, a theory of visual perception assumes that specialized structures in each precinct are responsible to detect edges, lines, slits etc. of different orientations and resolutions [HUBEL, WIESEL 1962] [MARR 1982]. This theory of vision supposes that the early visual processing stages of the brain perform similar operations as technical computer vision or image processing systems, e.g. oriented edge analysis and detail (frequency) analysis to discover the width of structures. To prove this theory, experiments applied on primates and cats have shown that specialized areas of the brain (denoted as *hypercolumns*) are primarily active if horizontal or vertical stripes are shown to the animal subject. It is assumed that similar properties apply in case of human brains, where only EEG acquisition can be applied that is less precise than electrode-based acquisition of brain activity.

It can also be found that specialized areas of the visual cortex are responsible to process information and detect basic structures separately from the inputs of the left and right eye. Presumably, further processing of the visual information is performed in the brain areas below the cortex. This includes detection and recognition of more complex shapes and objects, stereoscopic vision etc. Finally, as *differences of excitations* are processed by priority, objects moving over time draw higher attention than objects which are static within a scene.

3.1.2 Sensitivity functions

The amplitude sensitivity of the *Human Visual System* (HVS) follows *Weber's law* [WEBER 1834], which states that a perceivable difference in brightness ΔL has a constant ratio compared to absolute brightness:

$$\frac{\Delta L}{L} = \text{const.} \approx 0.02\,. \tag{3.1}$$

This has traditionally been utilized in technical systems for processing and display of visual information. An example is the nonlinear amplitude transfer characteristic of video cameras, cathode ray tubes and film material, which in bright areas tolerates additive noise on the medium that carries the visual information. In addition to the global ratio $\Delta L / L$, noise and other local fluctuations are masked in areas of high detail, while they are becoming more visible in flat (constant) amplitude areas. This property is however somewhat in congruency to sensitivity functions defined over spatial frequency, which will be discussed below.

Subjective perception of contrast is dependent on angular orientation and on spatial frequency (or detail) of a structure. Regarding angular orientation, experiments have shown that humans are less capable to recognize amplitude differences of diagonal stripe structures, as compared to horizontal and vertical structures with the same stripe period. Due to this effect, more preference should be given to horizontal and vertical structure orientations.

The visual sensitivity as depending on spatial frequency and directionality is typically measured by exposing test subjects to periodic signals (e.g. sinusoids or stripes). Just noticeable amplitude differences for a given stripe wavelength and orientation are then mapped into the value of a frequency-dependent sensitivity function. The spatial frequency measurement unit used in the observation of a regular stripe or sinusoid pattern should be normalized, such that it becomes independent of the viewing distance. An appropriate unit to measure this is *cycles per degree* of the viewing angle (*cy/deg*), as illustrated in Fig. 3.3a. Assume a stripe structure is viewed from a distance D, such that the distance from one stripe to the next, the wave length λ, maps into an angle φ. For half viewing distance $D/2$, the width of the same stripes as projected on the retina would be doubled, or the same width would be perceived if a stripe structure had only half wavelength $\lambda/2$. The spatial frequency f_{cy} is defined as the reciprocal value of φ,

$$\varphi = 2\arctan\frac{\lambda}{2D} \quad \Rightarrow f_{cy} = \frac{1}{\varphi}\left[\frac{\text{cy}}{\text{deg}}\right]. \tag{3.2}$$

An example about the dependency between display size, sample density and visible spatial frequency measured in *cy/deg* is presented in Problem 3.1.

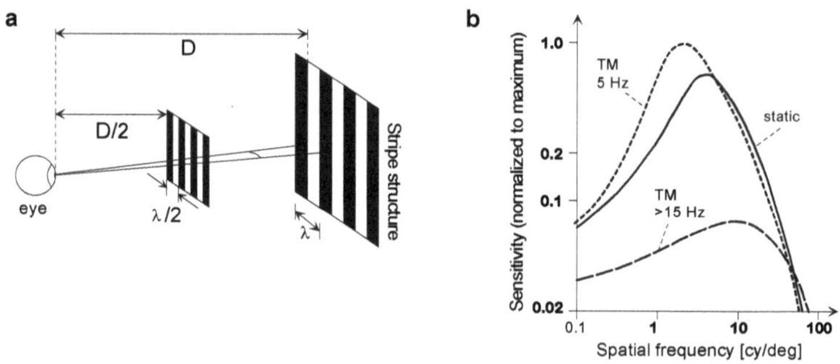

Fig. 3.3. **a** Definition of spatial frequency by *cy/deg* **b** Contrast sensitivity depending on spatial frequency for static case and with two different temporal modulations (after [ROBSON 1966])

In Fig. 3.3b, typical sensitivity curves are shown over the spatial frequency. Slightly different results on this are reported in the literature, however the qualitative properties as shown in Fig. 3.3b can be found consistently. In particular, the unit of measurement is often vague, such that it should be normalized for a maximum sensitivity level. For static stripe patterns (no change over time), the maximum sensitivity is generally reported to be around 3–5 *cy/deg*, and significant decay of the sensitivity function is observed towards lower and higher frequencies; the lower sensitivity in higher frequencies can directly be interpreted by the resolution limitation of the HVS, due to the fact that the number of receptor cells on the retina is finite. The decay towards lower frequencies may also be explained by the physiology of the retina, where low variations of amplitude fall below the difference perception threshold in the early neuronal processing stages. As a consequence, higher coding errors can be allowed for higher-frequency coefficients in transform coding, as these can be expected to be masked by the HVS. This property is often exploited both in still picture coding and video coding. However, it should be observed that the actual spatial frequency that is perceived depends on the observer's distance from the display (cf. Fig. 3.3a and Problem 3.1), which is not known beforehand by the time when the signal is encoded. It is also changing when the observer is interactively zooming into details at certain regions of the picture. Therefore, it is advisable to assume some worst case, e.g. for a digital display find the minimum viewing distance where a typical observer would not recognize pixel differences any more, and then align the corresponding number of pixels per degree *NP* with the point *NP*/2 on the sensitivity curve (since this is the maximum frequency that is possible according to the sampling theorem).

Spatial sensitivity is not independent of temporal sensitivity. For example, the effect of *line flicker* is observed in interlaced video: A slight decay of amplitudes from one field period to the next is recognized as spatial variation between adjacent lines, which becomes visible in large uniform areas. This effect is however quite specific for interlaced display devices, and is indeed rather an aliasing problem. If lines of one field are black and lines of the other field are white, the impression is even more annoying to the eye. To measure spatio-temporal transfer functions of the human visual system, a spatial pattern of certain frequency (e.g. a sinusoid or stripe) is modulated over time. Fig. 3.3b illustrates this effect, showing in addition a spatial frequency sensitivity function for a value close to the maximum of temporal sensitivity, which reportedly is somewhere in the range between 5 and 10 Hz for typical human subjects; the qualitative behavior is the same as for the static case, but the maximum is shifted towards slightly lower spatial frequency. The temporal sensitivity function decays drastically for temporal frequencies above 10 Hz[2], however in this case, even though globally at a lower level, the maximum of spatial sensitivity is again shifted towards a higher spatial frequency [ROBSON 1966].

Effects of spatio-temporal sensitivity functions have hardly been utilized in video data compression so far, which is partially due to their strong dependency on viewing conditions. [GIROD 1993] points out that the sensitivity for temporal frequency changes dramatically, if the observer is allowed to *track the motion.* The eye then performs a motion compensation, and is hence able to detect informally much higher spatio-temporal frequency components; in principle, the temporal spectral component is 'de-skewed', the effects discussed in Sec. 2.1.2 are reversed. The structure of a fast moving object in a movie or video can indeed

[2] Without this decay, it would not be possible to perceive fluent motion in movies.

only be perceived when it is tracked; when the subject is enforced to fixate one position on the screen, strange perceptual effects (caused by alias) become visible. In particular, it can be concluded that for parts of the scene which can be well tracked, such as smooth-moving objects, coding errors will have a higher visibility than in case of 'chaotic' motion, e.g. objects with fast changes in illumination and shape, deformations etc.

Another aspect that is hardly considered so far in image and video compression is the perception of *visually relevant differences*. Probably, the HVS has a good ability to perceive *regular structures* of images, such as edges, lines and periodic patterns, where it is annoying when unexpected irregularities appear due to coding distortions. The same is true when regular structures such as block grids become visible where they are not expected. On the other hand, variations in more irregular texture structures and noise are somewhat irrelevant, as long as the structure is perceived as visually identical. Properties that could indicate the presence of irregular types of structures are local stationarity and Gaussianity [BALLÉ 2012], whereas regular structures are either sparse in time (such as edges, lines) or sparse in frequency (such as periodic patterns), and usually also possess a high regularity of phase measurements. For video, the problem can be extended over the time axis, where it is often annoying that the movement of content is somewhat irregular, (local variations or jerkiness), whereas the observer would expect smooth motion. In this context, variation of artifacts over time is also annoying and usually more seriously perceived than artifacts in a single picture.

3.1.3 Color vision

The retina holds three different types of cones which are specifically tuned to the acquisition of the red, green and blue primaries. The entire color spectrum however consists of visible light components of many different tones (frequencies). The visible spectrum covers wavelengths between approximately 400 nm (boundary of ultraviolet range) and 700 nm (boundary of infrared range). Mixtures of spectral components appear as unique (but non-primary) color tones. Perception of such mixtures by the visual system is possible, as the receptors are not performing narrowband analysis of the primaries; the wavelengths of the primaries can be interpreted as approximate center frequencies of bandpass filters with spectral overlap. Fig. 3.4 shows qualitative receptor potentials, which relate to bandpass filter functions of red, green and blue receptors, where green has largest sensitivity. In general, due to the overlaps, color tones located at one of the three primaries are perceived as very saturated, but mixtures between only two of the primaries can also reach high saturation. However, special cases exist where color tones cannot uniquely be identified. An example are spectra which consist of multiple narrow peaks with sharp cut-off. This limits the perception of certain colors, e.g. golden and brown tones, which are subject to rather subjective judgement about their naturalness.

It can be assumed that a preliminary interpretation of color is performed by an early stage of vision, in the neurons of the retina. As the number of cones is lower than the number of rods, the spatial resolution of color perception is lower than it is the case for brightness reception. As a consequence, encoding of chrominance components in images can usually be done by lower resolution than for the luminance (cf. Sec. 1.3.1). In the multiplication factors used to transform *RGB* into luminance and chrominance components, additional properties of the HVS are taken into account. It can be observed that one dominant component may suppress the influence of the other two. This is specifically true for the green component, which falls into the spectral range where the sensitivity of the HVS is highest, while for the blue component it is lowest. In particular, highly saturated greenish tones of different hue can much better be differentiated than it is the case for bluish tones.

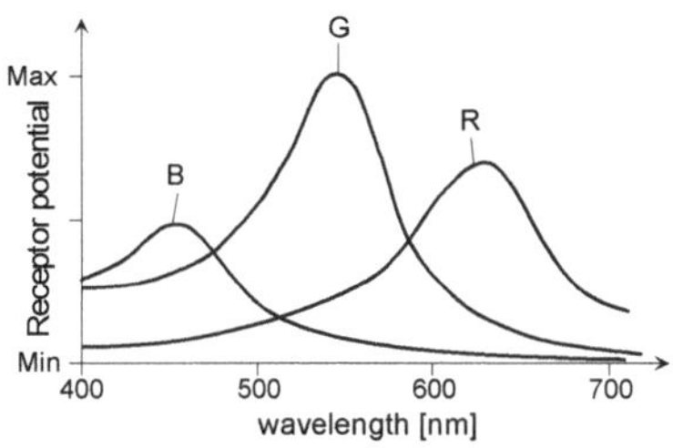

Fig. 3.4. Sensitivity transfer functions of the three types of cone receptors (qualitative)

Color perception is one of the most subjectively biased parts of the HVS, and can easily be jeopardized. Even if relatively false colors are presented to an observer, the visual sense may adapt after a short period of time and perceive them as 'natural'. The human observer is also able to adapt on illuminant conditions, such that objects for which the color is known are perceived subjectively like *having* this color, which is probably mainly due to associative processes in the brain. Further, the color sensitivity curves largely vary among different individuals. In the interest of a 'physically objective' representation of colors, it would be necessary to use more than 3 color channels for representation of the visible color spectrum, and also use a physical reference relating to illumination conditions (see [MCA, SEC. 4.1]).

3.1.4 Binocular vision and reproduction

The HVS is capable to perceive the depth distance of objects in a scene by binocular vision. The basic principle is shown in Fig. 3.5a. If an object is closer to the eyes (A), the projection to the retina has a larger distance from the optical axis than for the case where the object is farther away (B). The difference in positions of the same objects' projection to the left and right eye is the binocular *parallax*, which is reciprocally proportional with the depth distance. This would however only apply for the case where both optical axes are in parallel, converging at infinity, i.e. if the observer would fixate a point at infinite distance. Usually, an observer will fixate an object of interest, the optical axes would converge at this finite distance. The projection of the object will be close to the fovea centralis of the retina in both eyes, and the parallax will be close to zero here; it will however become positive or negative for depth distances which are closer or farther away from the convergence distance. Furthermore, the eyes' lenses will also accommodate by the distance of this object, other parts of the scene might be perceived out of focus. Other points for which the parallax is also zero are positioned on an elliptic surface called *horopter*, and obviously, points on the horopter would also be in focus. The actual position and convexity of the horopter depends on the depth distance of the convergence point.

The process by which the brain actually estimates the depth distance is probably learned from experience, evaluating the parallax, as well as the convergence and lens accommodation of the eyes. Technical systems can perform a similar action by depth analysis from a stereoscopic camera system (cf. MCA, SEC. 4.8). It should however be noted that the criterion of parallax is becoming more insignificant with increased depth distance. Other criteria such as common size of known objects, perspective or lighting modified by haze probably also play an important role when a human subject assesses large depth distances. Some of these secondary aspects are probably already becoming relevant for depth distances beyond 5m.

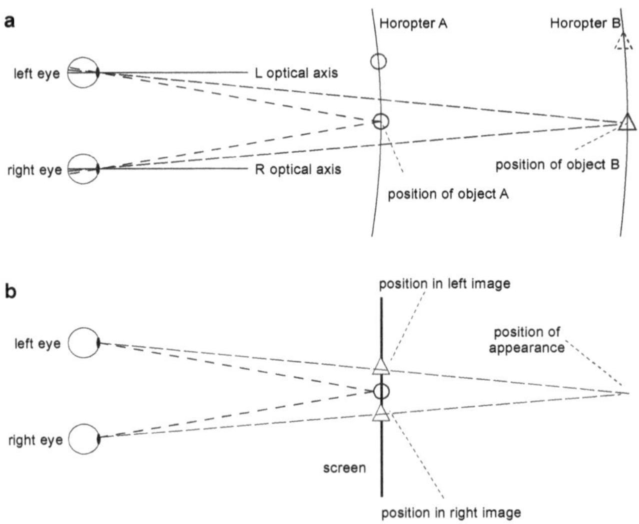

Fig. 3.5. Principles of **a** binocular vision and **b** display

The principle of binocular display is inverting the vision principle. Assume that a display exists that is capable to independently emit light to the left and right eye[3].

---

[3] This can be achieved by special glasses which in principle separate a multiplex light signal coming from the display. However, autostereoscopic and holographic displays which

Then, it depends on how large the distance between the positions of corresponding information displayed for the left and right eye is, whether the observer interprets the content to be at the depth plane of the display (the former object A with parallax shift zero in this case), or whether it is interpreted to be behind or in front of the display. However, if the parallax range within the same scene is too large, this may be in contradiction with the experience of the subject (focusing to the depth position of the screen, while eye convergence is behind or in front). In case of extreme divergence, this can cause fatigue and eye strain; as such, it should be avoided in stereo and 3D movie productions.

Another property of binocular vision that can be exploited for coding is the disability to perceive distortions differently in both eyes. Generally, this can be utilized by coding only one of the views to full quality or resolution, whereas the stereoscopic illusion is fully retained. This may however not be applicable if one of the eyes of a human subject has reduced ability anyway; furthermore, the primary view should be changed from time to time to avoid fatigue.

3.2 Properties of hearing

3.2.1 Physiology of the ear

A global schematic diagram of the human ear is shown in Fig. 3.6. It can roughly be described by the parts of *outer ear*, *middle ear* and *inner ear*. The task of the outer ear is to receive acoustic waves from the outside world and forward them via the *auditory canal* towards the *eardrum*. Length and shape influences the resonance frequency of the canal, taking influence on the frequency sensitivity peak of the auditory sense which is between 1 KHz and 4 KHz. The main task of the middle ear, starting at the eardrum, is a transformation of the air sound pressure into the hydrodynamic pressure modality which is processed by the inner ear. This is achieved by the three *ossicles* (*hammer*, *anvil*, *stirrup*), which transfer the pressure from the eardrum towards the *oval window*, the entrance of the inner ear. The inner ear is entirely filled by fluid, and all operations are effected by hydromechanic oscillations. The central part of the inner ear is the *cochlea* (which looks like a snail's house of about 2½ windings). It consists of three parallel channels, also denoted as *scales*. These are separated by membranes, where the *scala media* (the middle one) is filled by another fluid than the *scala vestibuli* and the *scala tympani*.

The *basilar membrane*, which plays a central role in audible perception, is connected to the scala tympani. It is further connected to *Corti's organ*, which contains the receptor cells. By the construction of the scales, which have different

---

are usually based on a dense spatial multiplex of multiple views are also appearing (see MCA, SEC. 7.5) widths from the input towards the inner parts of the cochlea, the basilar membrane performs a *frequency-space-transformation*. This means that different tone frequencies will effect maximum resonance at different locations of the basilar membrane, giving an effect that can be interpreted like a filter bank of bandpass systems. The oscillations of the basilar membrane are received by Corti's organ through *hair cells*, where the attached synapses perform a transformation into nervous pulses by a chemical reaction. The transmission towards the brain is performed by around 30,000 nerve fibers. It is interesting to note that a significant amount of audible information processing is already done before the information is directed towards the brain. In particular the relatively raw frequency analysis which is performed by the basilar membrane should have a significant effect on psycho-acoustic sensitivity functions discussed in the following sections.

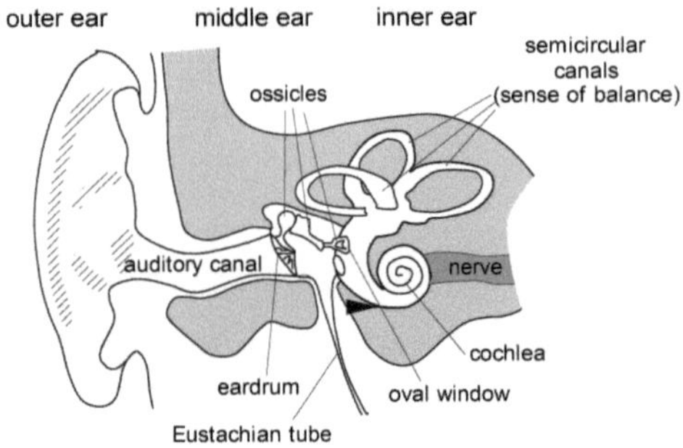

Fig. 3.6. Schematic diagram of the human ear

3.2.2 Sensitivity functions

In hearing, spectral properties (related to analysis of the 1D temporally-dependent signal) and amplitude (sound level or loudness) are the most important parameters for subjective impression. While in spectral color mixtures, only single color tones are perceived, the sense of hearing is able to finely separate single audible tones or sounds according to spectral and temporal properties. This is in particular true for *harmonic sounds*, where spectra are composed by a pitch frequency and its harmonics, a case that applies to many types of musical instruments and also to voiced segments of speech. An important criterion for the physical intensity of audio signals is the *sound pressure level*

$$L = 20 \cdot \log \frac{p}{p_0} \; dB \,, \tag{3.3}$$

where the normalization factor $p_0 = 20\mu$Pa relates to a sound intensity of 10^{-12} W/m^2, which references to the *threshold of hearing* at a frequency of 1 KHz of average human subjects. In general, the threshold of hearing is frequency dependent; in particular, very low and very high frequencies can only be perceived when the sound pressure level is considerably higher. Fig. 3.7 shows the frequency-dependent threshold of hearing, the upper *threshold of pain*, and typical ranges of frequency and sound levels of music and speech. The spectral range of sounds audible by humans approximately covers the frequency range between 16 Hz and 16 KHz.

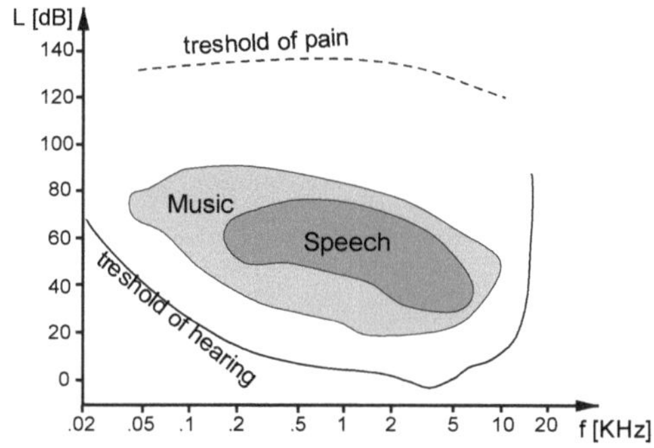

Fig. 3.7. Parameters of human hearing depending on sound pressure level and frequency (adapted after [ZWICKER 1982])

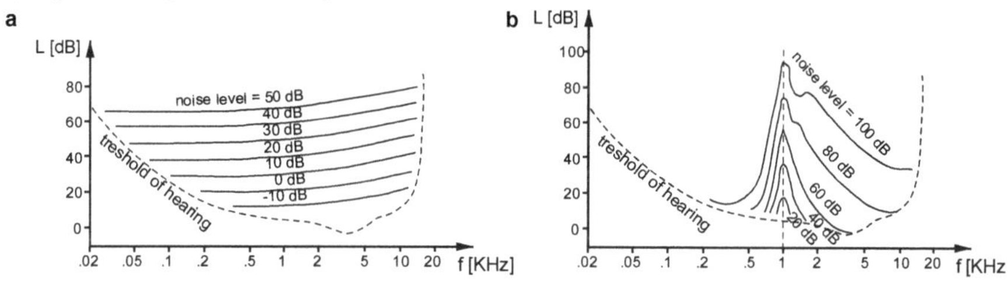

Fig. 3.8. Masked threshold of hearing with interfering signals of different sound pressure levels: **a** white noise **b** narrow-bandwidth noise of mid frequency 1 KHz (adapted after [ZWICKER 1982])

Most psycho-acoustic effects that can be exploited in audio signal compression are based on differentiation and masking properties of the human auditory sense. Distinct sounds are differentiated by their spectral properties. This is however only true if no other sound source is present which acts as interference or *masking signal*. Further, the strength of the masking effect depends on the sound pressure level of the masking signal. Fig. 3.8a shows the frequency-dependent *masked threshold of hearing*, where the masking signal is broadband white noise of sound pressure levels between –10 dB and +50 dB. It is evident that the threshold for audibility of other tones is increased substantially above the original threshold of hearing. White noise as masking signal, however, is not very significant for encoding of musical signals. To determine effects of mutual masking for natural signals, it is more useful to investigate threshold(s) when the masking signal is narrow-bandwidth noise. This is shown in Fig. 3.8b for the example of an interfering narrow-bandwidth signal of mid frequency 1 KHz and different sound pressure levels. Here, the masking has higher influence on frequencies which are above the frequency of the interfering signal.

The masking effect is also noticeable when the interfering signal is not of narrow-bandwidth nature, but a harmonic signal. Fig. 3.9 shows that masking is in effect over the entire range of the harmonic spectrum, starting from the pitch frequency. The effect of higher disturbance towards higher frequencies can be observed again.

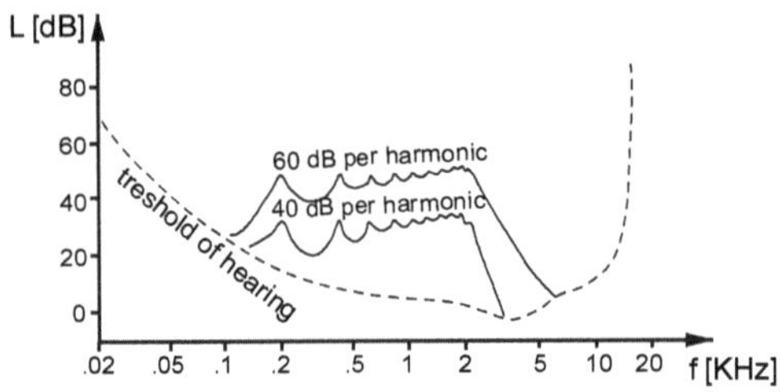

Fig. 3.9. Frequency-dependent masking in case of harmonic signal as interference, pitch frequency 200 Hz (adapted after [ZWICKER 1982])

It has further been shown in psycho-acoustic experiments that a group of pure sinusoids within a narrow frequency band is perceived like a narrow-bandwidth noise signal in this frequency band. It was also found that two pure tones have to be sufficiently different in frequency to be perceived as separate. This leads to the conclusion that the frequency resolution of the audible sense is finite. For criteria which better reflect the psycho-acoustic properties, the definition of *frequency groups* was introduced. It is assumed in this context that the human hearing system does not analyze sounds of discrete frequency, but is rather operating like a filter bank of bandpass systems. Each filter is tuned to a specific mid frequency f_0, having bandwidth Δf_G. The bandwidths shall be defined such that discrete frequencies (sinusoids) within the same frequency group are not perceived as disparate sounds. It turns out that neither the mid frequencies f_0 establish a linear scale on the frequency axis, nor are the bandwidths Δf_G equal over all groups. Fig. 3.10a shows results from hearing experiments, where the width of the distinguishable frequency groups is approximately constant up to f= 500 Hz ($\Delta f_G \approx 100$ Hz), and is roughly linear increasing with frequency for the range above this frequency. By this, it can be concluded that the range of the audible spectrum should be separated into approximately 24 distinguishable frequency groups, assuming that the human auditory sense would judge the spectrum of a broad-band sound (e.g. a complex music signal) according to the spectral hull of signal power of a set of discrete frequency groups. Based on this assumption, the *Bark scale* was defined, where the artificial unit Bark[4] expresses the *tone-ness* of a sound, being a percep-

[4] named after the acoustic scientist BARKHAUSEN.

tual equivalent of frequency. An approximation of the bold line in Fig. 3.10a is expressed by the following formula (after [ZWICKER 1982]; alternative approximations can be found in [TRAUNMÜLLER 1990]):

$$\frac{f}{Bark} = 13\arctan\left(0.76\frac{f}{KHZ}\right) + 3.5\arctan\left(\frac{1}{7.5}\frac{f}{KHZ}\right). \tag{3.4}$$

The result is then quantized into an integer value. Fig. 3.10b shows the mapping of the audible range of frequencies onto the discrete values of the Bark scale, ranging from 0–24, which define the lower and upper bounds of the 24 frequency groups. Comparing two sounds with regard to their Bark-scale related spectral energy distribution leads to a similarity measure which much better approximates the properties of the human hearing system.

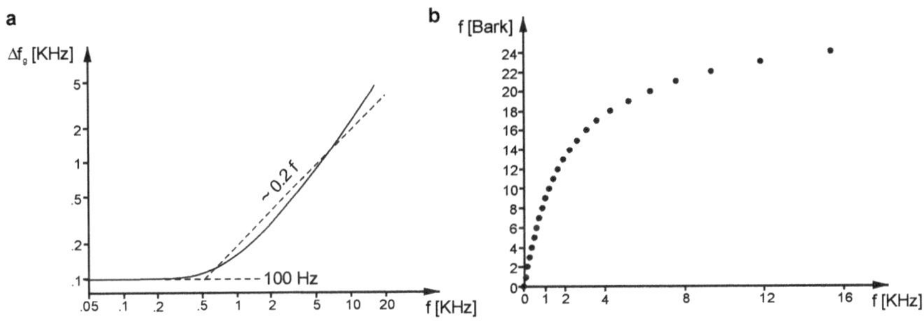

Fig. 3.10. **a** Relationship between frequency and frequency group width within the audible spectral range **b** Definition of discrete frequencies on the *Bark scale* (adapted after [ZWICKER 1982])

A similar approach is used by the *Mel scale* (first described in [STEVENS, VOLKMAN 1940]), which models the perception of harmonic signals (instead of narrowband noise signals as used for the measurements that were used in defining the Bark scale), using a linear frequency mapping up to approximately 1 kHz, and logarithmic mapping above that point.

The *static* properties of audibility effects can primarily be characterized in the frequency domain as done so far. In addition, a *dynamic* behavior of masking can be found, which is in particular effective if masking sounds are switched on or off. This has to be analyzed in the time domain. Fig. 3.11 shows the temporal *pre and post masking effects* (which are dynamically changing) and the *simultaneous masking effect* (which is more static). Here, an interfering sound (noise signal) was switched on for a duration of 200 ms. The effect of post masking is quite plausible, as the hearing system has adapted itself to the masking sound, such that it takes a short period until another tone (which was masked before) is detected. The pre-masking effect is likely due to processing delays within the chain from ear to brain, where loud sound events get higher priority and may interrupt the processing of previous events.

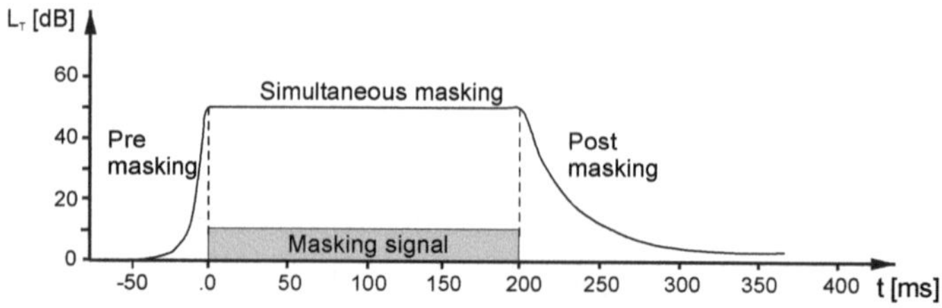

Fig. 3.11. Pre and post masking effects when an interfering signal (white noise) is switched on and off (adapted after [ZWICKER 1982])

3.3 Quality metrics

In optimization of multimedia codecs, quality measurements can quantify the distortion that is introduced by a compression algorithm. Therefore, availability of quality metrics is important for judging the benefits or comparing compression algorithms, and for making encoder decisions, i.e. generate a coded representation which provides minimum distortion for a given amount of bits (or minimum amount of bits for a given distortion). When the purpose of the media compression is human consumption, it is appropriate to use quality metrics which as good as possible match human perception[5]. According to the basic method of computation, quality metrics can be classified as follows:

- Full-reference quality metrics compare the processed (decoded) signal against a reference (original or ideally expected) signal. They are typically straightforward to compute by similarity measurements, which can be quantified to provide an accurate prediction of the perceived quality. However, the quality can only be given relative to the reference signal, which may not be available in some application scenarios, e.g. when standalone testing at the receiver end has to be performed.
- Reduced-reference quality metrics solve the aforementioned problem by transmitting additional parameters (metadata) which were extracted from the reference signal, and are intended to be compared against corresponding parameters extracted from the decoded signal; the deviation between the reference and the decoded signals is then concluded from the difference of the parameters. Typical parameters which are useful in reduced-reference metrics are low-level feature descriptors such as local mean or variance, frequency hull, harmonicity, color distribution, edge sharpness etc. Another option is additional lossless transmission of selected parts of the signal as side infor-

[5] If the purpose is usage by a machine (e.g. in a computer vision system), other criteria may apply, such as conservation of relevant visual features.

mation, such that a partial full-reference comparison can be made against the decoder output; this of course requires that the parts are representative enough to conclude about the overall quality.

– No-reference quality metrics require only the processed (decoded) signal as input for the evaluating processes. They are typically based on some prior knowledge about expected compression artifacts (such as: blocking, blurring, noise, etc.) which are detected and quantitatively measured in the decoded signal to calculate an overall quality index. Therefore, no-reference metrics are usually designed for a set of specific distortion types that may not be generalized for evaluating any type of distortions.

Basically, these principles are applicable both for objective (quality measured by signal processing) as well as subjective (quality measured by judgement of human subjects) approaches[6]. In testing of compression algorithms, full-reference metrics are most relevant. This applies to both automatic measurements and subjective assessment, where in the latter case often a reference with known quality is presented to the test subjects. Reduced-reference and no-reference metrics are often used to conclude about the overall quality in case of lossy transmission.

3.3.1 Objective signal quality metrics

Minimization of *squared error* or *Euclidean distance* between amplitudes of the reference $s(\mathbf{n})$ and the reconstruction $\tilde{s}(\mathbf{n})$ is widely applicable and simple to compute for various types of multimedia signals. If the difference between reference and reconstructed samples

$$\mathcal{E}\left\{\left(s(\mathbf{n})-\tilde{s}(\mathbf{n})\right)^2\right\} \qquad (3.5)$$

is interpreted as a zero-mean noise process[7], the related quality measurement can be expressed as *Signal-to-Noise Ratio* (SNR), which is usually mapped into a logarithmic (deciBel = dB) scale:

$$\mathrm{SNR}\left[\mathrm{dB}\right]=10\lg\frac{\mathcal{E}\left\{\left(s(\mathbf{n})-m_s\right)^2\right\}}{\mathcal{E}\left\{\left(s(\mathbf{n})-\tilde{s}(\mathbf{n})\right)^2\right\}}\approx 10\lg\frac{1}{|\mathbf{N}|}\sum_{\mathbf{n}}\frac{\left(s(\mathbf{n})-m_s\right)^2}{\left(s(\mathbf{n})-\tilde{s}(\mathbf{n})\right)^2}\,. \qquad (3.6)$$

(3.6) takes reference to the variance of the signal, and $|\mathbf{N}|$ is the number of samples in an empirical measurement. SNR measurements relating to the signal power or energy (for finite signals) are less usual, because the result could be highly biased by the signal mean value. For audio and speech signals, the zero-mean

---

[6] In the latter case, the reduced reference approach could be implemented by asking test subjects about specific annoying artifacts they perceived as unnatural in the decoded signal, without knowing an artifact-free reference.

[7] The zero-mean property of $q(\mathbf{n})$ is a reasonable assumption, unless $\tilde{s}(\mathbf{n})$ is systematically modified in amplitude, such that the difference is biased.

property applies anyway. Image and video signals are strictly positive-valued[8] and bounded within an amplitude range A_{max}. The mean of a given picture can by itself be regarded to be a random variable with local deviations. A metric which is not biased by the mean or variance of individual pictures is the *Peak SNR* (PSNR)

$$\text{PSNR}\left[\text{dB}\right] = 10\lg\frac{1}{|\mathbf{N}|}\sum_{\mathbf{n}}\frac{A_{max}^2}{\left(s(\mathbf{n})-\tilde{s}(\mathbf{n})\right)^2}\,. \tag{3.7}$$

PSNR has been established as a widely used metric for image and video distortion comparison. A more general form of SNR which additionally allows individual position-dependent weighting of errors is the *weighted SNR*

$$\text{WSNR}\left[\text{dB}\right] = 10\lg\frac{1}{|\mathbf{N}|}\sum_{\mathbf{n}}\frac{r(\mathbf{n})w(\mathbf{n})}{w(\mathbf{n})\left(s(\mathbf{n})-\tilde{s}(\mathbf{n})\right)^2}\,, \tag{3.8}$$

where the reference $r(\mathbf{n})=A^2$ in case of PSNR or $r(\mathbf{n})=[s(\mathbf{n})-m_s]^2$ in case of (3.6). The weight $w(\mathbf{n})$ could be used dependent on local structure (e.g. different weights for edge samples or textured areas in pictures). It can also be used to perform the SNR computation only for selected sample positions by choosing a value of one for relevant and zero for irrelevant positions. This would e.g. allow comparison of two pictures over an arbitrary-shaped area, rather than a rectangular picture of size $N_1 \mathrm{x} N_2$.

In PSNR, A_{max} is the maximum amplitude, or could also be interpreted as the maximum possible difference that can occur between two distinct amplitude values from signals s and $\tilde{s}$. Therefore, the PSNR can never be less than zero, even when completely different pictures are compared. Even though SNR measurements roughly match subjective quality assessment, the following observations can be made:

- Sample-wise difference measurements are highly sensitive against phase shifts, whereas delays (in case of images: spatial shift), i.e. linear phase shifts are uncritical for good perceptual quality, though SNR criteria might suggest poor quality.
- PSNR does not take into account masking of noise and other artifacts in cases of high amplitudes in $s(\mathbf{n})$ (or, more specifically, by highly-detailed structures in case of pictures). From this point of view, the measure as given in (3.6) is favorable, but even too simplistic to reflect more complex noise masking relationships which can depend on frequency properties, or spatio-temporal properties, or a combination thereof. For example, coding noise that varies over time is observed as very annoying, even though it may hardly be noticeable within a single video picture.

---

[8] In fact, this is not true for chrominance components such as C_b and C_r. In practice however, digital chrominance components are often represented as positive integer values by adding a bias compensating the maximum negative amplitude; e.g. in case of an 8-bit representation, a value of 128 is added. Chrominance components of a given image are typically not zero-mean signals, as they are biased by the individual color characteristics: In case of a 'reddish' image, the component C_r of (1.6) would clearly have a positive mean value.

Segmental SNR computation. SNR is based on average measurements, which do not take into account *fluctuations* of quality. A signal with only one highly distorted segment or region may be judged as being of bad quality, which may not at all be unveiled by global SNR computation (averaged over the entire signal). As possible workaround solutions, additional criteria such as SNR measurements over local segments, computation of minima, maxima, variance or other criteria which express fluctuations have been proposed. The *segmental SNR*, which is the mean or expected value of SNR results from equal-size segments [JAYANT, NOLL 1984], was originally proposed for speech and audio distortion analysis, where the ratio is computed based on the variances of the signal versus the distortion over the limited number of samples **n** within each segment n_s, where xSNR expresses one of the definitions e.g. of (3.6)-(3.8), and N_s is the total number of segments,

$$\mathrm{SNRSEG}=\frac{1}{N_s}\sum_{n_s=0}^{N_s-1}\mathrm{xSNR}=10\lg\frac{1}{|\mathbf{N}|}\sum_{\mathbf{n}}\frac{r(\mathbf{n},n_s)}{\left(s(\mathbf{n},n_s)-\tilde{s}(\mathbf{n},n_s)\right)^2}$$

$$=10\lg\left[\prod_{n_s=0}^{N_s-1}\frac{1}{|\mathbf{N}|}\sum_{\mathbf{n}}\frac{r(\mathbf{n},n_s)}{\left(s(\mathbf{n},n_s)-\tilde{s}(\mathbf{n},n_s)\right)^2}\right]^{\frac{1}{N_s}}=-10\lg\left[\prod_{n_s=0}^{N_s-1}\frac{1}{|\mathbf{N}|}\sum_{\mathbf{n}}\frac{\left(s(\mathbf{n},n_s)-\tilde{s}(\mathbf{n},n_s)\right)^2}{r(\mathbf{n},n_s)}\right]^{\frac{1}{N_s}}. \tag{3.9}$$

Herein, $r(\mathbf{n},n_s)$ is the reference for computation, which is a constant in case of PSNR or the instantaneous variance in case of ordinary SNR. From the lower line in (3.9), SNRSEG is computed from the geometric average of the non-logarithmic signal to noise ratios of the individual segments; comparing cases with same average of logarithmic (dB) values, SNRSEG increases when larger fluctuation across segments occurs. In case of using the computation via (3.6), this can even be desirable, as it may take into account that distortion could be less perceivable in segments with high signal variance (due to masking effects).

SNR averages over sets. The formal principle of segmental SNR is also often used for quality measurement of video sequences, where each of N_3 pictures could be interpreted as a segment, and logarithmic (dB) values are computed separately for each picture and averaged afterwards, out of simplicity reasons. As long as PSNR is almost constant over all pictures, this is not critical. However, if by purpose the PSNR is made lower for some pictures and higher for other pictures by re-allocating bit rate, this can be misleading. Therefore, when comparing different video compression algorithms, it is also advisable to compare the amount of PSNR fluctuations along with average values.

In comparing algorithms, it can be desirable to refer to single-value criteria, e.g. achieving an assessment of an algorithm's performance over various bit rates. A method to achieve this for codecs was proposed in [BJØNTEGAARD 2001] and further refined in [BJØNTEGAARD 2008]. It can best be interpreted graphically, by plotting measurements of PSNR values over a logarithmic rate axis. The graph is interpolated by a polynomial function (e.g. third order function when four measurements are available). When two codecs are compared, the average rate difference, or alternatively the average PSNR difference can be computed by analysing the area between the two interpolated graphs. The values, often denoted as 'Bjøntegaard difference rate' (BD rate) or 'Bjøntegaard difference PSNR' (BD PSNR) have to be considered carefully in cases where the interpolated graphs are crossing, or when outliers are present in the measurements (e.g. untypical behaviour at certain rate points, deviation from convexity), because in such cases the polynomial interpolation can produce unreasonable results.

Averaging of SNR values over different items of a test data set is also often used, as it simplifies comparison of algorithms by investigating one single number. The same considerations as given under (3.9) apply, i.e. the average of logarithmic values could be misleading depending on the amount of fluctuations of the single values.

Another possible need of averaging occurs when a signal consists of multiple components, such as luminance and chrominance in pictures, multiple views or multiple channels in audio. Again, this is uncritical when the components have identical meaning and similar reconstruction quality. However, specifically chrominance (B-Y and R-Y difference) components are typically simpler in structure than the luminance, can be easier compressed by using a significantly lower amount of bits for achieving identical or even higher PSNR values. Comparing compression algorithms which decode luminance and chrominance at different levels of quality can be misleading. No ideal solution exists for this case, even more when considering that the perceivable effect of luminance and chrominance artifacts is quite different. A possible workaround could be made by computing the average mean squared error over all samples from all components (which inherently would give a lower weight to the chrominance due to the subsampling), and perform the transformation into logarithmic (dB) domain jointly for all components. However, even when the PSNR is measured as average in the RGB domain with 4:4:4 sampling, it may not fully reflect the subjective visual quality, since the sensitivity of the human visual sense is higher in the green part of the spectrum.

Perceptually adapted metrics. Out of reasons given above, and particularly due to the fact that sample-based differences do not fully match with human perception, SNR measurements do not directly coincide with subjective quality impression, though usually indicating a reasonable trend. Development of better perceptually-weighted quality measurement criteria is a topic of ongoing research, where however, at least for still pictures and video, no single method has established as a commonly recognized standard; even though several methods have been described in recommendations of standardization bodies as further discussed below, those are mostly meant as informative, often specifying several methods which can be combined or alternatively used, and often not even precisely specifying the exact method of how the metrics are computed. A typical assessment of the methods is made by running subjective tests which give Mean Opinion Score (MOS) values (see Sec. 3.3.2), and comparing the results for coincidence with the quality metric. Criteria for such comparison are normalized correlation coefficients computed after appropriate curve fitting by regression, and/or Spearman rank correlation coefficients [SPEARMAN 1904][9]. The different metrics usually perform better than PSNR, however a definite judgement is difficult, as the results highly depend on the data sets that were used for the subjective tests (in terms of quality ranges, variety of items and types of distortions). For a reasonable review about methodology for quantitative assessment of various methods, the interested reader is referred to ITU-R Rec. BT.1676 entitled '*Methodological framework for specifying accuracy and cross-calibration of video quality metrics*'. For the purpose of this book, the main intention is to give an overview of possible approaches that are likely capable of improving the coherence between a metric's value and human perception. A more systematic investigation of different methods can be found in

- ITU-T Rec. J.144 '*Objective perceptual video quality measurement techniques for digital cable television in the presence of a full reference*' which includes a comparison of PSNR and 8 different perceptually tuned full-reference metrics;
- ITU-R Rec. BT.1683 '*Objective perceptual video quality measurement techniques for standard definition digital broadcast television in the presence of a full reference*' [ref] specifies and compares four different approaches.
- ITU-R Rec. BT.1907 '*Objective perceptual video quality measurement techniques for broadcasting applications using HDTV in the presence of a full reference signal*', extending the latter for the case of full HD video.

Without trying to describe the different methods in detail, some of the specific elements which improve the coherence of measurable criteria with subjective quality evaluation, but require careful combination and appropriate weighting, are:

- Matching of reference, i.e. (if unknown) identify the matching picture in the time sequence, and perform search for an area of the reference picture which best matches the area of the distorted picture; compute PSNR and other criteria based on this match;
- Performing global or local gain (contrast), level (brightness) and color (hue, saturation) adjustment between reference and distorted picture, or compare those parameters to quantify the degradation;
- Excluding areas with zero error from the PSNR or other criteria computation;
- Investigate the variation of PSNR or other criteria over the whole sequence;
- Performing a frequency domain transform (often 2D pyramid or wavelet-like) and compute PSNR based on frequency weighting, investigate the amount of blurring that has occurred, or use the local properties for frequency-dependent masking functions;
- Application of edge analysis and generation of an edge map, which is then compared between reference picture and distorted picture, or used to generate criteria for perceptibility (or masking) of distortions around edges; as a sim-

---

[9] In the latter, the population of subjective measurements and the results from computing the quality metric are ordered in lists by increasing values, and the normalized covariance coefficient is computed comparing the ranks in the lists rather than the values themselves.

ple variant, take only those samples into account in PSNR computation, where relevant edges are detected ('Edge PSNR');
- Identification of codec-specific distortions, such as blockiness by identifying larger deviations at regular block edge positions;
- Segmentation of pictures into planar regions, edge and texture regions, applying different masking/visibility criteria;
- Performing motion analysis, in order to apply temporal masking models (e.g. less visibility of errors in case of fast motion), or identify the presence of jerkiness (i.e. local motion that changes inconsistent with the frame rate);
- Application of spatio-temporal filters to consider spatio-temporal masking effects;
- Identification of regions of interest (e.g. human faces) which would attract human observers, where errors are more strongly weighted.

Another family of metrics, which could also be interpreted as a variant or subset of some of the methods listed above, and being relatively simple to compute, is the *Structure Similarity Index* (SSIM) (see overview in [WANG, BOVIK 2009]). At its core, it is based on measuring mean m, variance σ^2 and covariance μ over areas of the (original) reference signal s and the distorted version $\tilde{s}$. The following criteria $\alpha_1 \dots \alpha_3$ explore the changes of 'luminance', 'contrast' and 'structure', respectively[10]:

$$\alpha_1(s,\tilde{s}) = \frac{2m_s m_{\tilde{s}} + \beta_1}{m_s^2 + m_{\tilde{s}}^2 + \beta_1};\ \alpha_2(s,\tilde{s}) = \frac{2\sigma_s \sigma_{\tilde{s}} + \beta_2}{\sigma_s^2 + \sigma_{\tilde{s}}^2 + \beta_2};\ \alpha_3(s,\tilde{s}) = \frac{\mu_{s\tilde{s}} + \beta_3}{\sigma_s \sigma_{\tilde{s}} + \beta_3} \quad (3.10)$$

The purpose of the parameters β_i is for providing a non-linear threshold such that low activity areas get lower weight; typical setting are $\beta_1=\beta_2=1$ and $\beta_3=\beta_2/2$. The final SSIM between s and $\tilde{s}$ is then computed as combination

$$\mathrm{SSIM}(s,\tilde{s}) = [\alpha_1(s,\tilde{s})]^{\gamma_1}[\alpha_2(s,\tilde{s})]^{\gamma_2}[\alpha_3(s,\tilde{s})]^{\gamma_3}, \quad (3.11)$$

where the exponential weights can be used to give the three criteria different preferences; however, often $\gamma_i=1$ is used.

SSIM quality is measured in two levels: Region level and picture level. At the region level, it is first applied independently to the Y, C_b and C_r components, whose contributions are combined into a local quality criterion for region j by using a weighted summation

$$\overline{\mathrm{SSIM}}_j(s,\tilde{s}) = w_{\mathrm{Y}}\mathrm{SSIM}(s_{\mathrm{Y}},\tilde{s}_{\mathrm{Y}})_j + w_{\mathrm{C_b}}\mathrm{SSIM}(s_{\mathrm{C_b}},\tilde{s}_{\mathrm{C_b}})_j + w_{\mathrm{C_r}}\mathrm{SSIM}(s_{\mathrm{C_r}},\tilde{s}_{\mathrm{C_r}})_j \quad (3.12)$$

At the picture level, a weighted combination of values from all N_R regions is computed[11],

[10] Herein, α_1 and α_2 are based on the idea of measuring the 'flatness' of mean and variance, which becomes one when the underlying values of s and $\tilde{s}$ are identical; α_3 is the normalized covariance.

[11] Typically $w_j=1$, but adaptation of parameters can be used if it is assumed that certain regions should be weighted differently, e.g. in the center part of the picture

$$Q_{\mathrm{P}} = \frac{\sum_{j=1}^{N_{\mathrm{R}}} w_j \overline{\mathrm{SSIM}}_j(s,\tilde{s})}{\sum_{j=1}^{N_{\mathrm{R}}} w_j}. \tag{3.13}$$

Optionally, in case of video sequences, the values Q_{P} of the pictures can be further averaged.

Since the HVS's sensitivity is non-uniform over spatial frequency, it is straightforward to extend SSIM for measuring the criteria in bandpass channels. The most relevant approach is the complex wavelet SSIM (CW-SSIM) [WANG, SIMONCELLI 2005]. Since wavelet coefficients have typically zero mean, only the criteria α_2 and α_3 from (3.10) are useful, where the first compares magnitude and the second compares phase of complex wavelet coefficients in the original and reconstructed pictures. The latter can further be exploited by different weighting, considering the fact that phase shifts should stay consistent in case of relevant structures such as edges, whereas they may change more randomly for noise-like textures.

Reduced-reference models which are usable for assessing the quality of a video only at the receiver end are described in ITU-T Rec. J.246 '*Perceptual visual quality measurement techniques for multimedia services over digital cable television networks in the presence of a reduced bandwidth reference.* The recommended method is based on the observation that the 'Edge PSNR' investigated in ITU-T Rec. J.144 (see above) gives a reasonable performance w.r.t. matching the subjective quality, while computing PSNR values only for selected edge samples. In the case of a reduced-reference metric, this is used by transmitting original amplitudes of a few edge samples as side information, such that the quality can be approximately estimated by comparing them against the corresponding decoded samples. ITU-R Rec. BT.1908 '*Objective video quality measurement techniques for broadcasting applications using HDTV in the presence of a reduced reference signal*' extends the method for HDTV resolution sources.

If specific codecs are tested, the presence and strength of typical degradations they produce (e.g. blocking, ringing artifacts or jerkiness) can also be used as criteria in implementing metrics without reference. This however requires application of models for motion trajectories and edges, as well as prior knowledge about properties of a codec, e.g. the position of a block grid.

In the area of audio, due to the fact that properties of the human auditory perception have more systematically been investigated and described (particularly in terms of temporal and frequency-domain masking effects), a more widely accepted and reportedly reliable metric[12] has been defined in ITU-R Rec. BS.1387 '*Method for objective measurements of perceived audio quality*', also denoted as PEAQ. The recommendation specifies a method of extraction and also provides a software implementation of the basic version; however, it should be noted that

---

[12] at least for higher quality ranges, which are the relevant operation target of audio codecs

commercial implementations of the whole PEAQ framework exist which even better match with human perception. The basic version is implementing a listening model in the Fourier (FFT) domain, which in terms of frequency masking (described via a noise-to-mask ratio) also takes into account modulation, loudness and harmonic structure of the error, and a general detection probability of distortions (relevant for the case where the compression shall be perceptually lossless, i.e. no perceivable distortions). Combination of different criteria is performed by an artificial neural network (ANN). The advanced version, besides the FFT method, also implements a filterbank-based frequency analysis, which in particular allows better perceptual modeling of the coding noise loudness, taking into account the additional disturbance that could occur due to entirely missing frequency components.

ITU-T Rec. P.862 '*Perceptual evaluation of speech quality (PESQ): An objective method for end-to-end speech quality assessment of narrow-band telephone networks and speech codecs*' defines a related approach, again assuming presence of the full reference. The method is partially based on the Intermediate Reference System (IRS) which had previously defined in Rec. P.830, in order to simulate typical distortions that could occur in the context of subjective tests, which can be used here for investigating typical distortions when comparing the reference versus the signal under test. Time alignment is made as one of the first steps. Silence periods are excluded from the comparison. Further criteria include comparison of the pitch power density, comparison of the loudness density and calculation of the disturbance density, i.e. superimposed noise (all computed in a perceptually adapted manner via the Bark frequency scale).

3.3.2 Subjective assessment

When a reliable evaluation about the perceived signal quality shall be achieved, it is inevitable to perform formal subjective assessment tests involving a representative ensemble of test subjects. For some categories of signal compression algorithms, such as audio compression attempting to approach transparent (perceptually lossless) quality, this is the only viable assessment method, objective measures like SNR would not be useful at all.

In subjective assessment, an ensemble of test subjects are asked to give their opinions about the perceived quality. The answers are statistically evaluated, where criteria as mean value and variance of opinions are computed to determine the average results and their significance. Rejection of clear outliers is usually part of the process. The qualification of the test subjects (expert or non-expert) is also of high importance, but their proper selection could partially depends on the goal of the test.

Further, any methodology for quality assessment needs to define a quality scale. For the case where the goal of transparent quality is investigated, a binary decision is sufficient. When the material under test is expected to exhibit quality losses, the five-grade quality/impairment attribute scale as defined in Table 3.1 is widely used for all kinds of uni-modal media signals (speech, audio, image, video), as well as their multimedia combination.

By evaluating the answers gained from the ensemble of test subjects, it is possible to obtain results of *Mean Opinion Score* (MOS) assessments. Typical subjective tests consist of a training phase and a test phase. Before the test actually starts, subjects are screened for defects e.g. in listening capabilities, visual acuity or color blindness. Test results of each subject are statistically evaluated, and a subject can be withdrawn from the ensemble if turning out to be a systematic outlier, or if answers are fundamentally different when the same item is assessed a second time.

The score categories are generic enough to allow usage for universal purposes, e.g. general quality, perceptibility of speech, readability of text, but it is also possible to ask for very specific cases such as naturalness of spatial illusion in case of multi-channel audio or stereoscopic video, color artifacts etc. Of course, this requires to give specific hints to the test subjects during training. Also in terms of algorithms tested, the methodologies are quite generic and can be used to assess compression methods as well as channel error correction, de-noising etc.

Table 3.1. Score and related subjective assessment criteria used in MOS tests

Score	Quality	Impairment
5	Excellent	Imperceptible
4	Good	Perceptible, but not annoying
3	Fair	Slightly annoying
2	Poor	Annoying
1	Bad	Very annoying

Subjective tests should be reproducible, otherwise their value would be limited. Definition of assessment conditions for different types of multimedia signals has a long history. As an example, methodologies defined by ITU-R Rec. BT.500-13 *'Methods for the Assessment of the Quality of Television Pictures'*[13] are described here with more detail. Basically, the following aspects are specified:

- Viewing conditions for assessment are defined for home viewing and lab environments; this includes ambient lighting conditions, preferred viewing distances and angles, properties of displays such as luminosity, resolution and aspect ratio;
- Appropriate selection of test material, and length of sequences;

[13] ITU-R BT.500 originally covers the case of SD resolution quality assessment. For HD resolution, the corresponding recommendation is ITU-R BT.710, which however defines exactly the same testing methodologies except for the setup of viewing and ambient conditions. Further, ITU-R BT.2022 describes modifications of the viewing setup in subjective tests for the case of flat panel displays, whereas the original BT.500 was designed in 1974 when only CRT display technology was available.

- Selection of an ensemble of test subjects, which could be expert or non-expert viewers, depending on the purpose of a given test. A group of 15 subjects is typically the minimum number to perform the assessment and obtain statistically relevant results; due to the possibility that subjects may be rejected due to their behavior during the test, a larger number is advisable;
- Quality and impairment scales as defined in Table 3.1;
- As the interpretation of scales can be application specific, and to adjust the scale, so-called 'anchor' examples can be included in the test (either identified or hidden), which e.g. would mark the upper or lower scale of expected quality, and subjects are instructed to utilize the full range of scales appropriately during a training phase[14];
- Test methods (main part of BT.500, see further discussion below), and mapping of application scenarios where these methods would be appropriate; arrangement and duration of test sessions;
- Procedures to evaluate the results of the assessment in terms of computing mean scores, confidence intervals and criteria to reject individual subjects.

Depending on the quality of the material to be evaluated, on the application environment (e.g. offline or live), and on the availability of one or several references, different test methods are defined as follows:

- *Double Stimulus Impairment Scale* (DSIS): Viewers are shown multiple sequence pairs consisting of a *reference* (R) and a *test* (T) sequence, each of 10 sec duration. Subjects are asked to score the amount of impairment in the test sequence as compared to the reference by using the *impairment scale* of Table 3.1. The sequence of display R-*break*-T-(*break*-R-*break*-T-)*vote*. During the breaks of 3 sec, a mid-gray level is shown, the voting break is longer (5-11 sec). The second iteration is optional. This kind of test is often used e.g. to evaluate coding algorithms in the lower range of data rates, where it would be clear that the impairment as compared to the original source would be severe. Alternatively, the reference could be a standard coding algorithm with a reasonable level of quality (not lower than any method under test).
- *Double Stimulus Continuous Quality Scale* (DSCQS): Unlike DSIS, this is intended for reliably testing higher ranges of quality, potentially including investigation for perceptually lossless behavior. Test sequences A and B of duration 10 sec are presented in a schema A-*break*-B-*break*-A-*break*-B-*vote*. Voting can start when the second round begins. One of A and B is the original source, the other is processed material, but the observer is unaware which is which. Attributes for scores are given by the *quality scale* of Table 3.1.
- *Single Stimulus* (SS): No reference is available, and mapping to a quality or impairment scale is achieved either directly or by showing with multiple repetition and verifying the consistency of votes. Alternative numerical scales can also be used to provide a finer resolution.

---

[14] During the training, different material than in the later test shall be used to avoid biased votes. Generally, the test sessions have to be time limited to avoid fatigue of subjects, which could also cause unreasonable voting.

- *Stimulus comparison* (SC): When comparison of only two algorithms A vs. B is made, this can be done directly, either showing pairs of items simultaneously on two different screens, in a split-screen mode, or in A-B/B-A sequence[15] . Instead of using quality or impairment scales, a 7-grade direct comparison scale −3...+3 is used with 0 meaning equal quality, +/− meaning better/worse, and grades 1/2/3 indicating slightly/clearly/much.
- *Continuous Quality Evaluation* (CQE): This test methodology is usable on live material or can also be used to judge the quality of video streaming e.g. over the Internet, where rapid fluctuations of quality may occur. A common reference sequence is not explicitly available. A series of video sequences is presented once to the viewer, who evaluates the picture quality instantaneously (in real time) by moving a slider. Possible methods are with single stimulus (SSCQE) with slider position corresponding to a score on the quality scale, or with simultaneous double stimulus, which can also be used to measure impairment compared to a reference. Samples of the slider position are typically taken twice each second. When multiple algorithms shall be compared, the slider positions of corresponding time points, where the same material was shown, have to be matched. The overall quality can be assessed by averaging (integrating) a subject's votes over time, where additionally the analysis of variation (e.g. standard deviation) can be useful to assess quality fluctuations.

ITU-T P.910 '*Subjective Video Quality Assessment Methods for Multimedia Applications*' (2008) is similar, defining the following test methods, and allowing a more continuous scaling of quality values instead of the classical 5-scale MOS):

- 'Absolute Quality *Rating'* (ACR), similar to SS from BT.500, and ACR-HR (hidden reference) which allows to determine differential MOS (DMOS) relative to a reference which is shown among the other material under test and unknown to the viewers
- 'Degradation Quality *Rating'* (DCR), similar to DSIS from BT.500, i.e. quality is compared against the original reference which is known to the subjects.
- 'Pairwise comparison' (PC), where the same sequence is presented as processed by two different systems either sequentially, or (in case of small picture formats) simultaneously on one screen. In case of N>2 algorithms to be compared, a total of $N(N-1)$ comparisons (each against each other) needs to be run. Ratings like in the SC method of BT.500 can be used as well.

The general methods of BT.500, though originally defined for (analog) SD TV, are applicable to a large variety of video, if necessary applying small adaptations of the procedure. For the case of stereo and 3D video, more considerations are necessary e.g. in order to separate effects caused by an algorithm from display effects, and the necessity to investigate effects of stereo vision in the presentation, or when screening the test subjects. Recommendations ITU-R BT.1438 '*Subjective assessment of stereoscopic television pictures*' and BT.2021 '*Subjective*

---

[15] In any of these cases, care should be taken that the position of A and B is randomly changed, in order to compensate for possible influences of the viewing situation.

methods for the assessment of stereoscopic 3DTV systems' take these issues into account.

For speech compression and processing, ITU-T Rec. P.800 'Methods for subjective determination of transmission quality' plays a similar role. Specification of the listening environment, as well as requirements for the test sentences are included. Intelligibility of speech is also an important criterion in this case.

For generic audio signals, the following specifications are relevant:

- ITU-R BS.1116: '*Methods for the subjective assessment of small impairments in audio systems including multichannel sound systems*';
- ITU-R BS. 1284: '*General methods for the subjective assessment of sound quality*'.

Specific criteria for quality judgement in audio include absence of noise as well as criteria that affect the signal itself, such as preservation of timbre, transparency, sound balance, stereo impression etc.

Other specific recommendations relate to testing the quality of combinations of different media, e.g.

- ITU-R BT.1788 '*Methodology for the subjective assessment of video quality in multimedia applications*';
- ITU-R BS.1286 '*Methods for the subjective assessment of audio systems with accompanying picture*'.

These methods are again based on the aforementioned specifications, and there is no need to make a detailed discussion here.

3.4 Problems

Problem 3.1.

A vertical sinusoidal test pattern is shown on an HD display (N_2=1080 lines) with height H. A viewer is observing from a distance D=3H.

a) Determine the wavelength of a sinusoid on the screen, when the spatial frequency is 5 *cy/deg* according to (3.2).
b) What is the period of the sinusoid, counted in sampling units?
c) Are the results of a) and b) changing when H is doubled?
d) What should be the block length of a type-I DCT (2.258), such that the basis vector $\mathbf{t}_1$ approximately represents a spatial frequency of 5 *cy/deg*?
e) Assume that the HVS of an observer has an ideal cut-off frequency of 60 *cy/deg*. At which viewing distance D=xH would a pattern of alternating (black/white) lines be perceived as gray?

4 Quantization and Coding

By sampling, multimedia signals are discrete in space and/or time. Transformed equivalents of these signals, e.g. frequency coefficients or prediction error signals, and feature values or parameters related to these signals are formally represented by continuous (or large number of discrete) amplitude values. For compact coding, further quantization has to be employed, which causes distortion. The quantization step size and the range of amplitudes determine the necessary number of levels and the rate necessary for binary representation. The relationship between rate and distortion is expressed by the rate distortion function, for which a lower bound is expressed by the source coding theorem. An analytic derivation of the R-D bound is straightforward for stationary processes, which is useful in understanding basic limitations of source coding for real world signals. Statistical dependencies within sources also need to be considered in the rate-distortion dependency. Basic methods suitable to approach the rate-distortion bound for minimum-rate representation of continuous and discrete sources are also discussed in this chapter, including various approaches of entropy coding, vector quantization, tree and trellis coding.

4.1 Scalar quantization and pulse code modulation

Conversion of continuous-value sampled signals $s(\mathbf{n})$ into discrete values $\tilde{s}(\mathbf{n})$ requires quantization, which performs a mapping into a discrete alphabet (cf. (2.122)-(2.125)). For digital representation, each state of the quantizer output y_j (letter of the alphabet) needs to be represented by a codeword (a uniquely distinguishable sequence of bits). If the process of quantization and encoding is applied independently to samples, the representation is denoted as *Pulse Code Modulation* (PCM).

As the quantization characteristic (Fig. 2.21) is an irreversible mapping function, it is not possible to uniquely reconstruct the original value in general, causing a quantization error

© Springer-Verlag Berlin Heidelberg 2015

J.-R. Ohm, *Multimedia Signal Coding and Transmission*,
Signals and Communication Technology, DOI 10.1007/978-3-662-46691-9_4

$$s_D(\mathbf{n}) = s(\mathbf{n}) - \tilde{s}(\mathbf{n}). \tag{4.1}$$

For coding by a finite number of bits it is necessary to define a finite set of quantized values. Therefore, only a finite amplitude range of width A_{max} can be supported. When values outside of this range are present in the source, overload occurs, where no upper bound of the quantization error may exist, unless the source itself is restricted to a maximum amplitude. To abstain from this effect, it is assumed in a first consideration that the signal to be quantized shall be uniformly distributed within the given range A_{max}, and this range is uniformly partitioned into quantization intervals of width Δ, such that in total A_{max}/Δ intervals need to be represented. For the direct coded representation, a binary number of length

$$B = \left\lceil \log_2\left(\frac{A_{max}}{\Delta}\right)\right\rceil \text{ bit} \tag{4.2}$$

is then necessary. The quantization error signal $s_D(\mathbf{n})$ can be interpreted as additive noise which is superimposed (with negative sign according to the definition (4.1)) to the original signal. This noise will also pass through the lowpass filter if reconstruction is performed from the quantized samples, and will appear like additive noise with flat spectral characteristics within the bandwidth of the reconstruction (assuming that the quantization of the signal samples is performed independently and corresponding quantization error samples would therefore be statistically independent[1]).

Assuming that the signal $s(\mathbf{n})$ stems from a stationary process with uniform PDF, and the reconstruction values y_j are at the centers of the quantization intervals (distance $\Delta/2$ from each of the boundaries), the random values $s_D(\mathbf{n})$ have a PDF

$$p_{s_D}(q) = \frac{1}{\Delta}\operatorname{rect}\left(\frac{q}{\Delta}\right). \tag{4.3}$$

According to (2.121), the quantization error has a variance

$$\sigma_{s_D}^2 = D = \mathcal{E}\left\{s_D^2(nT)\right\} = \frac{1}{\Delta}\int_{-\Delta/2}^{\Delta/2} q^2\,\mathrm{d}q = \frac{\Delta^2}{12}, \tag{4.4}$$

which grows quadratic with the quantization step size. Likewise, the power of the signal $s(\mathbf{n})$ is

$$\sigma_s^2 = A_{max}^2/12. \tag{4.5}$$

With (4.2), $A_{max}=\Delta 2^B$, the signal to noise ratio is

$$\frac{\sigma_s^2}{D} = \frac{(\Delta 2^B)^2/12}{\Delta^2/12} = 2^{2B}. \tag{4.6}$$

or expressed logarithmically (deciBel)

$$10\lg(\sigma_s^2/D) \approx B\cdot 6.02\text{dB}. \tag{4.7}$$

Doubling the number of quantization steps requires one additional bit for direct

---

[1] This may no longer be the case with sources where the samples themselves are highly correlated, and subsequent samples have similar amplitudes and are also represented by identical quantized values.

coding and improves the SNR by approximately 6 dB. It is also possible to express the minimum necessary number of bits as (potentially non-integer number)

$$B \geq \frac{1}{2}\log_2(\sigma_s^2 / D) . \tag{4.8}$$

All considerations so far are strictly valid for quantization of uniformly-distributed sources, where no overload occurs and the quantization error has uniform distribution as well. In case of sources without amplitude limitations (e.g. Gaussian distributed), the range $A_{\max}$ has to be defined such that the probability of overload is reasonably small.

For the case of an arbitrary quantizer with J reconstruction values y_j and quantization interval boundaries x_j the following mapping rule holds,

$$\tilde{s}(\mathbf{n}) = y_j \quad \text{for} \quad x_j \leq s(\mathbf{n}) < x_{j+1} \quad \text{with} \quad x_0 = -\infty \, , x_J = +\infty \, . \tag{4.9}$$

The PDF of the quantized signal is given in (2.123), whereas the quantization error PDF can be characterized through superposition of the piecewise PDFs of all quantization intervals

$$p_{s_{\mathrm{D}}}(q) = \sum_{j=0}^{J-1} p_{s_{\mathrm{D}}}^{(j)}(q) \tag{4.10}$$

which can be mapped from a shifted segment of the original PDF, using (2.12)

$$p_{s_{\mathrm{D}}}^{(j)}(q) = p_s(q - y_j)\left[\varepsilon(y_j - x_j - q) - \varepsilon(y_j - x_{j+1} - q)\right] . \tag{4.11}$$

Especially for the case of uniform quantization of a uniform source with $x_{j+1}-x_j=\Delta=\text{const.}$, and $y_j=x_j+\Delta/2=x_{j+1}-\Delta/2$, the resulting PDF is (4.3).

For the case where the PDF of the original signal is non-uniform, the amplitude range $A_{\max}$ can be designed such that the occurrence of overload is negligible (e.g. it is common to select the width $A_{\max} = 2\cdot(3\ldots4)\sigma_s$ in case of Gaussian PDF). If, furthermore, Δ is sufficiently small such that $p_s(x)$ is approximately constant within any interval, (4.3) and (4.6) are still reasonably approximating the properties of the quantization error when uniform quantization is used.

The index $0 \leq j < J$ uniquely identifies the value from the discrete alphabet. The procedure of quantization is a mapping $s(\mathbf{n}) \rightarrow j(\mathbf{n})$, the reconstruction[2] is the reverse mapping $j(\mathbf{n}) \rightarrow s_{\mathrm{Q}}(\mathbf{n}) = y_{j(\mathbf{n})}$. If high-precision discrete samples are re-quantized into a lossy representation, the basic effects are not significantly different from the quantization of continuous signals; therefore, it is not treated as a separate case here.

Uniform quantization. Fig. 4.1 shows different examples of uniform quantization characteristics, and the corresponding quantization error characteristics which give the mapping from the input signal amplitude $s(\mathbf{n})$ to the quantization error amplitude $s_{\mathrm{D}}(\mathbf{n})$, i.e. the difference between the quantization characteristic and a zero-distortion function $y=x$. The different cases include one quantization range

---

[2] Sometimes denoted as 'de-quantization' which is somewhat misleading as the quantization process is a non-revertible operation.

for only positive amplitudes (as e.g. used for PCM representation of monochrome image signals), and two different versions of zero-point symmetry, one of which supports a reconstruction value of zero.

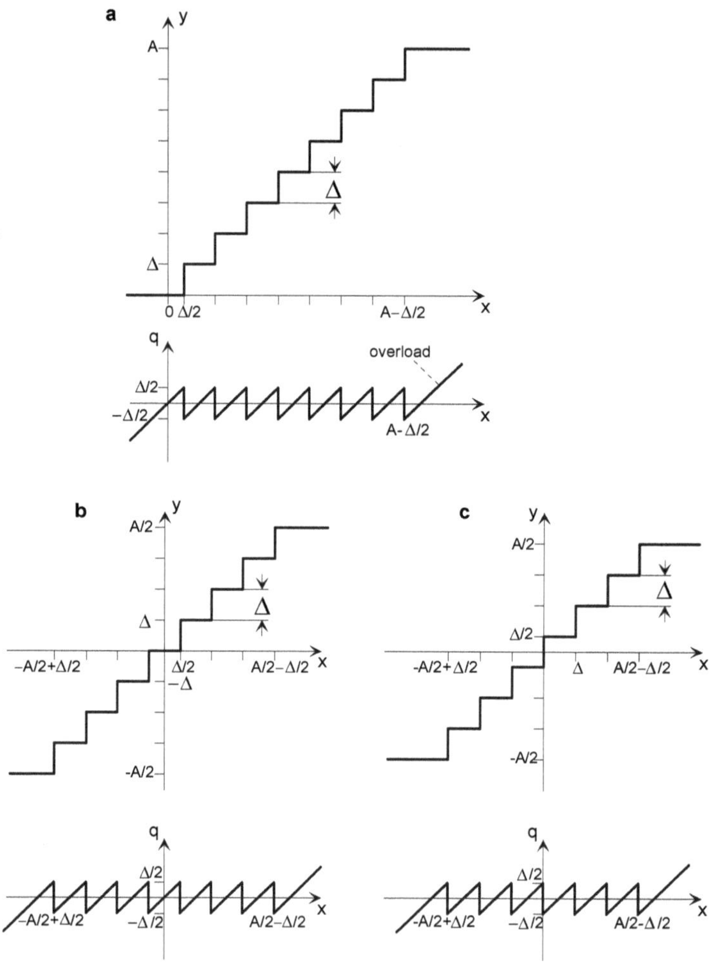

Fig. 4.1. Uniform scalar quantization characteristics, and quantization error characteristics in dependency of the signal amplitude variable: **a** quantizer for positive-only signal, and **b**/**c** quantizers which are symmetric around value zero (with/without zero reconstruction)

If uniform quantization is used, the processes of quantization and reconstruction are extremely simple due to the linear relationship between the list of reconstruction values and the indices. For computation of the index $j(\mathbf{n})$ from the signal value $s(\mathbf{n})$, it is sufficient to divide the value by Δ and perform integer (floor) rounding. Optionally, an offset shift OFFQ can be included in the quantization step, and another offset shift OFFR in the reconstruction step[3]. To compute the reconstruction value $\tilde{s}(\mathbf{n})$, scaling of the index by Δ and reverse offset shift must be performed. To achieve symmetry within positive and negative value ranges, it is advantageous to separate the sign[4]

$$j(\mathbf{n}) = \min\left(\left\lfloor \frac{|s(\mathbf{n})| - \text{OFFQ}}{\Delta} \right\rfloor, J_{\max}\right)$$
$$\tilde{s}(\mathbf{n}) = \left(j(\mathbf{n})\Delta + \text{OFFR}\right)\operatorname{sgn}(s(\mathbf{n})). \tag{4.12}$$

For the examples of Fig. 4.1, $\Delta = A_{\max}/J$ in all cases, OFFQ$=-\Delta/2$ and OFFR$=0$ for Figs. 4.1a/b, OFFQ$=0$ and OFFR$=\Delta/2$ for Fig. 4.1c. In case of Fig. 4.1a, $J_{\max}=J-1$ can be an even or odd number, while due to symmetry reasons J must be odd for Fig. 4.1b ($J_{\max}=(J-1)/2$) and even for Fig. 4.1c ($J_{\max}=J/2-1$).

Non-uniform quantization. The application of a quantizer with non-uniform step size (Fig. 4.2a) can be useful if the signal $s(\mathbf{n})$ has a non-uniform PDF. Further, quantization by non-uniform step sizes can be used to implement nonlinear amplitude-mapping functions in combination with quantization. This can e.g. be useful if quantization errors are perceived as more severe at low amplitude ranges. For example, non-uniform quantization with a logarithmic mapping characteristic is typically used in PCM of speech signals, implementing larger quantization step sizes for higher amplitudes, where the noise sensitivity of the human ear is known to be lower (masking effect). The measurable quantization error variance may however be increased by non-uniform quantization, such that it is mainly useful under perceptual quality aspects.

As an example, the so-called *a-law* companding (formally applicable to continuous amplitudes before quantization) is defined by[5]

$$k(x) = \begin{cases} x/A_{\max} \cdot a/(1+\ln a), & 0 \le |x/A_{\max}| \le 1/a \\ \left(1+\ln\left(a|x/A_{\max}|\right)\right)/(1+\ln a)\operatorname{sgn}(x), & 1/a \le |x/A_{\max}| \le 1 \end{cases} \tag{4.13}$$

In speech quantization, it is used with a value $a=87.56$. Definition of an expansion characteristic, which reconstructs the original samples, is straightforward.

Optimum non-uniform quantization (in terms of minimizing the distortion) can be designed by using the *Lloyd-Max Algorithm* [LLOYD 1957, MAX 1957], which is

---

[3] OFFQ modifies the interval boundary, whereas OFFR also allows defining the 'dead zone' quantization of Fig. 4.2b.

[4] No sign needs to be encoded (can be set to 1) for the case of positive-only quantization in Fig. 4.1a, and for the reconstruction value zero in Fig. 4.1b.

[5] According to standard ITU-T G.711 (see Sec. 8.1.3). The a-law compression / expansion characteristic is used throughout European countries, whereas other countries (e.g. US) use the μ-law characteristic which is formally different but following a similar logarithmic companding concept.

based on a criterion for minimization of the squared error specifically for a given signal PDF. If a signal value x is mapped by an arbitrary quantizer into the nearest reconstruction value y_j from a codebook $\mathcal{C}=\{y_j; j=0, \dots, J-1\}$, the index is

$$j(\mathbf{n}) = \underset{y_j \in \mathcal{C}}{\arg\min} (s(\mathbf{n}) - y_j)^2 . \tag{4.14}$$

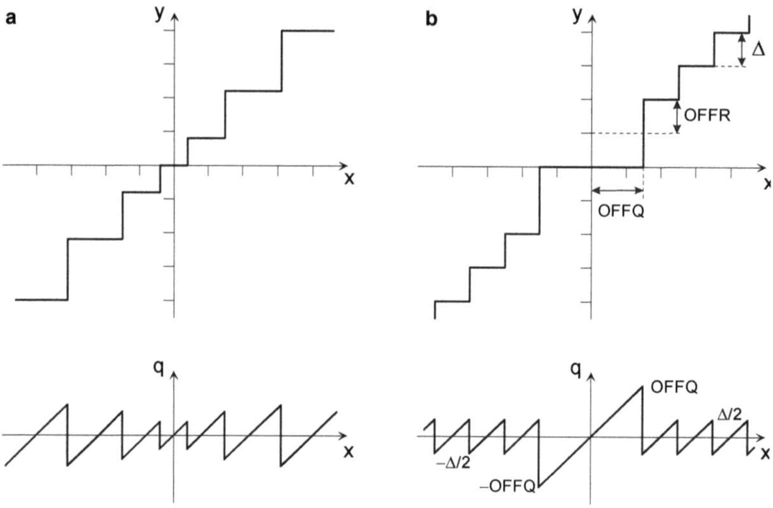

Fig. 4.2. Quantization characteristics **a** of non-uniform quantization **b** of dead zone quantization

Using

$$\text{Prob}[x < s(\mathbf{n}) \le x + \mathrm{d}x] \approx p_s(x)\mathrm{d}x , \tag{4.15}$$

the contribution to the squared quantization error for the case of $s(\mathbf{n})$ within an infinitesimal interval $[x; x+\mathrm{d}x]$ around y_j is

$$(y_j - x)^2 p_s(x)\mathrm{d}x . \tag{4.16}$$

The variance of the quantization error is given by integration within the boundaries of each interval and summation over all intervals,

$$\sigma_{s_D}^2 = \sum_{j=0}^{J-1} \underbrace{\int_{x_j}^{x_{j+1}} (y_j - x)^2 p_s(x)\mathrm{d}x}_{\Pr(S_j) D_{\mathcal{C}}(S_j)}. \tag{4.17}$$

With the condition for minimization

$$\frac{\mathrm{d}\sigma_{s_D}^2}{\mathrm{d}y_j} \overset{!}{=} 0 \text{ for } j=0,1,\ldots,J-1, \tag{4.18}$$

differentiation under the integral (4.17) gives

$$\int_{x_j}^{x_{j+1}} 2(y_j - x)p_s(x)\mathrm{d}x = 0 \Rightarrow \; y_{j,\mathrm{opt}} = \int_{x_j}^{x_{j+1}} xp_s(x)\mathrm{d}x \Big/ \int_{x_j}^{x_{j+1}} p_s(x)\mathrm{d}x\,. \tag{4.19}$$

The reconstruction values y_j of the optimum quantizer are identical to the linear mean expectations that can be determined from the PDF within the respective quantization intervals. The condition of minimum quantization error w.r.t. the interval boundaries gives

$$\begin{aligned} &\frac{\mathrm{d}\sigma_{s_D}^2}{\mathrm{d}x_j} \overset{!}{=} 0 \text{ for } j=0,1,\ldots,J-1 \\ &\Rightarrow \frac{\mathrm{d}}{\mathrm{d}x_j}\left[\int_{x_{j-1}}^{x_j} (y_{j-1}-x)^2 p_s(x)\mathrm{d}x + \int_{x_j}^{x_{j+1}} (y_j - x)^2 p_s(x)\mathrm{d}x\right] = 0 \end{aligned} \tag{4.20}$$

With

$$\frac{\mathrm{d}}{\mathrm{d}u}\int_a^u f(x)\mathrm{d}x = f(u) \Rightarrow \;(y_{j-1}-x_j)^2 p_s(x_j) - (y_j - x_j)^2 p_s(x_j) = 0 \tag{4.21}$$

and finally

$$x_j = \frac{1}{2}(y_{j-1} + y_j)\,. \tag{4.22}$$

The optimum decision thresholds x_j are points with equal distance from the two adjacent reconstruction values.

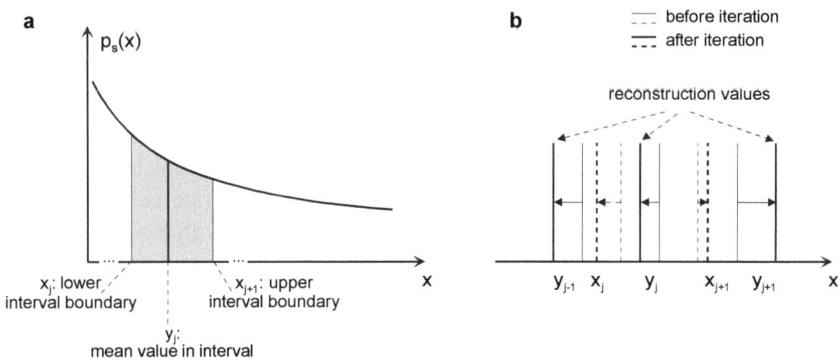

Fig. 4.3. **a** Optimum position of a reconstruction value, depending on the shape of the PDF within a quantization interval **b** update of reconstruction values and interval boundaries by one iteration

Unlike uniform quantization, the optimized reconstruction value is no longer at the center position of the interval, but it is the mean of signal values allocated to this interval, or the mass centroid of the respective PDF partition (Fig. 4.3a). When a set of new values $y_{j,\text{opt}}$ is computed following (4.19), the optimum boundaries will also move according to (4.22) (Fig. 4.3b); both steps contribute to minimization of distortion. As a consequence, the optimum solution can be approached iteratively, where by each iteration step the new sets of y_j and x_j values can never give worse performance. If the overall distortion D is not changing significantly any more, the optimum is reached and the procedure is terminated. The expected values within the intervals can either be determined using a known signal PDF or a training set of representative samples.

Non-uniform quantization has the disadvantage that the quantization decision is more complex than with uniform quantization where a simple scaling of values can be used to determine the codeword index, see (4.12). Alternatively, reconstruction values can be arranged off-centre by a constant offset shift, trying to define a common value over all intervals, such that the minimization of distortion from (4.19) still partially applies. This can be implemented by proper adaptation of the OFFQ and OFFR values in (4.12).

Such an approach provides a *dead zone quantization characteristic* as shown in Fig. 4.2b, which increases the probability of zero-quantized values. This can be beneficial when the zero values are encoded by a very low rate as compared to non-zero values (e.g. the case in transform coefficient coding of pictures). An optimum decision on such an approach can however not only be based on distortion criteria, but must likewise take into account rate criteria (cf. Sec. 4.3).

4.2 Coding theory

4.2.1 Source coding theorem and rate-distortion function

The source coding theorem [SHANNON 1959] can be summarized as follows:

- In encoding a discrete, memory-less source, a distortion not larger than D shall be allowed. Then, coding is not possible with a rate below the rate-distortion bound $R(D)$, and a code of rate $R=R(D)+\varepsilon$ with $\varepsilon>0$ can be designed;
- $R(D)$ is a convex function which can be uniquely inverted as the distortion-rate function $D(R)$. This means that the lower bound can also be formulated w.r.t. the distortion, i.e. when a rate R is available, it is not possible to encode with distortion lower than $D(R)$
- One method to approach $R(D)$ is by using a code which applies joint coding of K samples, where the block length K shall be sufficiently large.

The determination of $R(D)$, the *rate-distortion function* (RDF) for generic sources is only possible by numerical approximations[6]. An analytic description can be derived for certain sources, e.g. for the case of an uncorrelated, stationary Gaussian process $v(n)$, i.e. Gaussian white noise of variance σ_v^2. For a squared-error distortion metric D, its encoding requires a minimum rate per sample [BERGER 1971]

$$R_v(D) = \frac{1}{2}\log_2 \frac{\sigma_v^2}{D} . \qquad (4.23)$$

Occasionally, when the distortion D equals the quantization error variance according to (4.4) with uniform scalar quantization of a source with uniform PDF, the same rate was found in (4.8). However, in the case of a Gaussian source, the rate-distortion bound cannot be achieved without additional effort in quantization and entropy coding. The uncorrelated Gaussian source is important, as in terms of the rate-distortion behavior and the complexity of a coding system that is able to reach this bound, it establishes a *worst case* among all memory-less sampled sources of same variance σ^2 [BERGER 1971]. Further, (4.23) only applies to cases $D \leq \sigma_v^2$, as otherwise the logarithmic function would give a negative rate. A more general expression can hence be formulated as[7]

$$R_v(D) = \max\left(0, \frac{1}{2}\log_2 \frac{\sigma_v^2}{D}\right) \quad \Rightarrow \quad D_v(R) = \sigma_v^2 \cdot 2^{-2R} \ , \ R \geq 0 . \qquad (4.24)$$

4.2.2 Rate-distortion function for correlated signals

For a one-dimensional correlated (AR type) Gaussian process $s(n)$, the RDF can be determined by spectral averaging [BERGER 1971]

$$R_s(\Theta) = \int_{-1/2}^{1/2} \max\left(0, \frac{1}{2}\log_2 \frac{\phi_{ss}(f)}{\Theta}\right) \mathrm{d}f = \int_{0}^{1/2} \max\left(0, \log_2 \frac{\phi_{ss}(f)}{\Theta}\right) \mathrm{d}f . \qquad (4.25)$$

As the power spectrum allows interpreting the correlated process by an infinite number of independent spectral components, which would also be Gaussian distributed, the argument function (4.24) can be used for spectral integration[8]. If the spectral power within a given frequency range Δf is above the threshold Θ, a rate contribution according to (4.23) is assigned, otherwise the area under the frequency-dependent rate function is zero. In the latter case, the distortion contribution for a given frequency f is equal to the corresponding area under the power spectrum;

---

[6] An algorithm to compute the approximation of the RDF from a set of data samples was proposed in [BLAHUT 1987]

[7] For R=0, no information about the signal is conveyed; the squared distortion metric is upper bounded by σ_v^2, the maximum possible uncertainty about the signal.

[8] provided that the power spectrum does not contain Dirac impulses, for which no logarithm is defined.

in the former case, the distortion equals $\Theta\cdot\Delta f$. When the values of the power spectrum within the range $0 \le |f| < 1/2$ are only partially above the threshold, the integration in (4.25) has to be computed piece-wise. If all values are above the threshold, the overall distortion D equals Θ. For the more general case, the distortion $D_s \le \Theta$ needs to be determined by piece-wise integration:

$$D_s(\Theta) = \int_{-1/2}^{1/2} \min\left[\Theta, \phi_{ss}(f)\right] \mathrm{d}\, f\,. \tag{4.26}$$

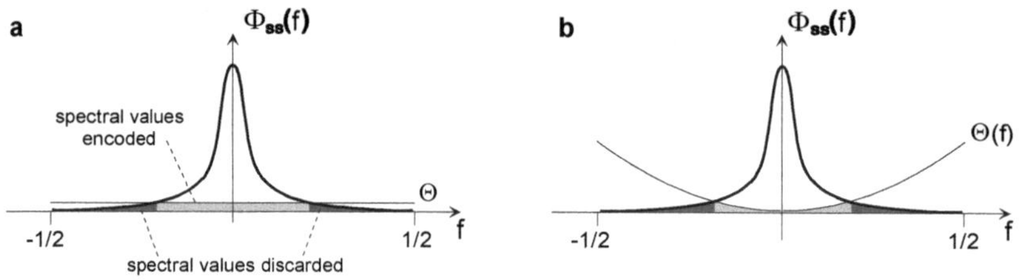

Fig. 4.4. Interpretation of the rate distortion function for an AR(1) process: **a** Equal distortion over all spectral components **b** Spectrally weighted distortion threshold function

Fig. 4.4a illustrates the total distortion $D_s(\Theta)$ by shaded areas below the argument function of (4.26). The distortion shall not become larger than the signal's power density at any frequency. The spectral components of the dark-shaded areas are discarded: In the given example, the signal loses high-frequency components, which is a suitable strategy for low-rate encoding of sources with lowpass characteristics, even better combined with downsampling. Fig. 4.4b additionally shows the example of a *frequency-weighted threshold function*[9].

Example: R(D) of an AR(1) process. The AR(1) process (Sec. 2.6.1) is described by the correlation coefficient ρ and variance of the innovation signal σ_v^2. From (2.159), if $\phi_{ss}(f) > \Theta$ throughout the range $0 \le |f| < 1/2$, the rate distortion function is

$$\begin{aligned} R_s(D) &= \frac{1}{2}\int_{-1/2}^{1/2} \log_2 \frac{\sigma_s^2(1-\rho^2)}{D\big(1-2\rho\cos(2\pi f)+\rho^2\big)}\mathrm{d}f \\ &= \frac{1}{2}\int_{-1/2}^{1/2} \log_2 \frac{\sigma_s^2(1-\rho^2)}{D(1+\rho^2)}\mathrm{d}f - \frac{1}{2}\int_{-1/2}^{1/2} \log_2\left(1-\frac{2\rho\cos(2\pi f)}{1+\rho^2}\right)\mathrm{d}f \\ &= \frac{1}{2}\log_2 \frac{\sigma_s^2\left(1-\rho^2\right)}{D} = \frac{1}{2}\log_2 \frac{\sigma_v^2}{D} = R_v(D). \end{aligned} \tag{4.27}$$

---

[9] Frequency-weighted functions are often used in encoding of multimedia signals. Due to spectral masking inherent in human senses of vision and hearing, higher distortion can be allowed for certain components. In principle, this also allows to define distortion metrics and RDF dependencies which are better adapted to human perception.

Therefore, $R_s(D)$ for the correlated AR(1) process with variance σ_s^2 can directly be expressed as the RDF $R_v(D)$ of its white-noise innovation signal[10]. This is however only valid for the case where all spectral components are above Θ. As the power spectrum (2.192) of the AR(1) process steadily decreases towards the minimum value at $|f| = 1/2$, the validity of (4.27) is constrained to the range of low distortion

$$D \leq \frac{1-\rho}{1+\rho}\sigma_s^{\,2} = D_{\max}(\rho)\sigma_s^{\,2} . \tag{4.28}$$

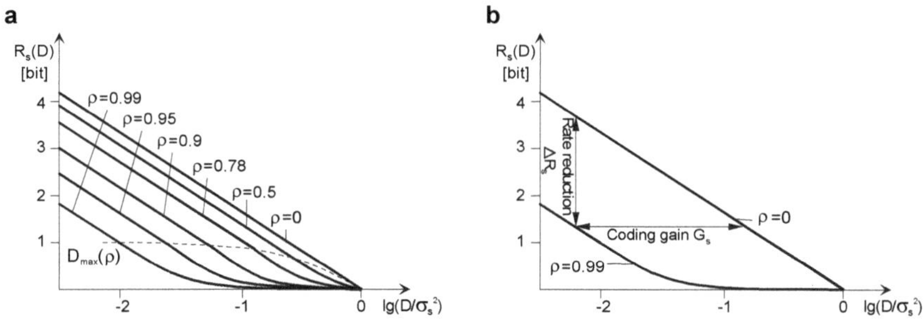

Fig. 4.5. **a** $R(D)$ of an AR(1) process over different parameter values ρ
b gain in rate and gain in distortion by exploitation of correlation properties
[after CLARKE 1985]

If this condition holds, also $D=\Theta$. For larger D, the parametric dependency on Θ in (4.25) and (4.26) can be solved as follows for the AR(1) process for which a value f_Θ can be found such that $\phi_{ss}(f) < \Theta$ for $|f| > f_\Theta$, and therefore

$$R_s(\Theta) = \int_0^{f_\Theta} \log_2 \frac{\phi_{ss}(f)}{\Theta} \mathrm{d}f \quad ; \quad D_s(\Theta) = 2\left[\int_0^{f_\Theta} \Theta \,\mathrm{d}f + \int_{f_\Theta}^{1/2} \phi_{ss}(f)\,\mathrm{d}f\right]$$
$$\text{with} \quad \Theta = \frac{\sigma_v^{\,2}}{1-2\rho\cos(2\pi f_\Theta)+\rho^2} \Rightarrow f_\Theta = \frac{1}{2\pi}\arccos\frac{1}{2\rho}\left(1+\rho^2-\frac{\sigma_v^{\,2}}{\Theta}\right). \tag{4.29}$$

Fig. 4.5a illustrates $R_s(D)$ graphs for the AR(1) process with different parameters ρ. The D axis is normalized by the variance of the process and logarithmically scaled, such that the $R_v(D)$ relationship (4.23) maps to a line. The dotted line interconnects the values $D_{\max}(\rho)$ of (4.28); above $D_{\max}$, from (4.27), the $R_s(D)$ graphs of the correlated process are lines in parallel to the $R_s(D)$ of the uncorrelated process ($\rho = 0$). The corresponding value of minimum rate is

$$R_{\min}(\rho) = R\big(D_{\max}(\rho)\big) = \log_2(1+\rho) . \tag{4.30}$$

The distance between the $R(D)$ plots of $\rho=0$ and $\rho>0$ can be interpreted as either a gain in terms of rate (vertically) or in terms of distortion (horizontally), see Fig.

---

[10] This applies for any AR(P) process.

4.5b. This gain however decreases when the distortion grows larger than $D_{max}(\rho)$. Such a breakdown of rate-distortion performance can indeed be observed in encoding of multimedia signals at low rates. By using a coding scheme which (unlike PCM) utilizes the correlation between samples of an AR(1) process while observing the constraint (4.28), a maximum reduction of rate

$$\Delta R_s = \frac{1}{2}\log_2\left(\sigma_s^2 / D\right) - \frac{1}{2}\log_2\left(\sigma_v^2 / D\right) = -\frac{1}{2}\log_2\left(1-\rho^2\right) \tag{4.31}$$

can be achieved, keeping the same distortion as for the case of independent encoding of samples. (2.208) can be used to determine the variance of the innovation signal. An alternative expression is $\Delta R_s = -1/2 \log_2(|\mathbf{C}_{ss}|/\sigma_s^2)$, valid for any AR process, where the absolute determinant of the autocovariance matrix (2.209) gives the variance of the innovation. The same reduction in rate respectively entropy can be found in (2.188), where for the AR(1) process with K=2

$$H_s = \frac{1}{2}\log\left((2\pi e)^2 \left| \begin{bmatrix} \sigma_g^2 & \rho\sigma_g^2 \\ \rho\sigma_g^2 & \sigma_g^2 \end{bmatrix} \right| \right) = \log\left(2\pi e \sigma_g^2\right) + \frac{1}{2}\log\left(1-\rho^2\right). \tag{4.32}$$

Comparison to (2.187) indicates that the entropy of a Gaussian AR(1) vector process with K=2 is by $\frac{1}{2}\log\left(1-\rho^2\right)$ lower than for a Gaussian white noise vector process of same variance[11].

Alternatively, the *coding gain*

$$G_s = \frac{1}{\gamma_s^{\;2}} = \frac{\sigma_s^{\;2}}{\sigma_v^{\;2}} = \frac{1}{\sigma_s^{\;2} - \sum_{p=1}^{P} a(p)\mu_{ss}(p)} \tag{4.33}$$

expresses the factor by which the distortion decreases when a coding scheme utilizes the correlation[12], as compared to independent (PCM) encoding of samples at the same rate. By substituting (4.28) into (4.27), it is found that this gain can be realized when $R \geq R_{min}$. The coding gain is also the reciprocal value of the *spectral flatness measure* (SFM) [JAYANT, NOLL 1984]

$$\gamma_s^{\;2} = \frac{2^{\left(\int_{-1/2}^{1/2} \log_2 \phi_{ss}(f)\,\mathrm{d}f\right)}}{\sigma_s^{\;2}} = \frac{2^{\left(\int_{-1/2}^{1/2} \log_2 \phi_{ss}(f)\,\mathrm{d}f\right)}}{\int_{-1/2}^{1/2} \phi_{ss}(f)\,\mathrm{d}f}, \tag{4.34}$$

---

[11] The rate reduction by $-\log_2(1-\rho^2)$ only applies to one of the two samples in the case of encoding vectors with K=2, as the other sample still requires encoding of its entire information. If more samples are jointly encoded, the rate reduction converges into (4.31).

[12] In case of Gaussian (AR) processes, all statistical dependencies are linear and can be expressed as correlations. See also (5.45) for a direct relation of the log spectrum integral and the covariance matrix for that case.

Numerator and denominator are equal and the coding gain becomes unity if all spectral components contribute equally to the overall power, which is the case for white (usually Gaussian) noise, the case of maximum flatness. In *any other case* (non-flat spectral distribution), the geometric mean is lower than the arithmetic mean, such that a coding gain >1 can be realized. Observe that the SFM in the given form can only be determined for processes where all spectral power components are larger than zero.

(4.25) can be reformulated as

$$R_s(\Theta) = \frac{1}{2}\int_{-1/2}^{1/2} \log_2\left[\max\left(\Theta, \phi_{ss}(f)\right)\right]\mathrm{d}f - \frac{1}{2}\log_2\Theta . \tag{4.35}$$

Obviously, the integral is similar to that in the numerator part in (4.34), however it is considered here that spectral components that fall below Θ do not contribute to the rate, nor do they contribute to the coding gain, as they can e.g. not be used for prediction. Therefore, a modification of (4.34) that also considers the case where spectral components are discarded can be formulated as

$$\gamma_s^2(\Theta) = \frac{2^{\left(\int_{-1/2}^{1/2}\log_2 \max[\phi_{ss}(f),\Theta]\mathrm{d}f\right)}}{\int_{-1/2}^{1/2}\max\left[\phi_{ss}(f),\Theta\right]\mathrm{d}f} = \frac{1}{G_s(\Theta)} . \tag{4.36}$$

4.2.3 Rate-distortion function for multi-dimensional signals

The RDF for a κ-dimensional AR model is found by generalization of (4.25)

$$R_s(\Theta) = \int_{-1/2}^{1/2}..\int_{-1/2}^{1/2} \max\left(0, \frac{1}{2}\log_2\frac{\phi_{ss}(\mathbf{f})}{\Theta}\right)\mathrm{d}^\kappa\,\mathbf{f} , \tag{4.37}$$

and (4.26)

$$D_s(\Theta) = \int_{-1/2}^{1/2}\cdots\int_{-1/2}^{1/2} \min\left[\Theta, \phi_{ss}(\mathbf{f})\right]\mathrm{d}^\kappa\,\mathbf{f} . \tag{4.38}$$

Even if the AR model and its spectrum are separable, $R(D)$ may not be implicit except within the range of low distortion, where no spectral components are discarded by encoding. As an example, a 2D separable AR(1) process is specified by horizontal and vertical correlation coefficients ρ_1 and ρ_2, where according to the relationship between the variances of the process $s(n_1,n_2)$ and the innovation signal $v(n_1,n_2)$

$$R_s(D) = \frac{1}{2}\log_2 \frac{\sigma_s^{\ 2}\left(1-\rho_1^{\ 2}\right)\left(1-\rho_2^{\ 2}\right)}{D} = \frac{1}{2}\log_2 \frac{\sigma_v^{\ 2}}{D}. \tag{4.39}$$

The validity of (4.39) is now constrained to cases of low distortion[13]

$$D \leq D_{\max}(\rho_1,\rho_2) = \frac{\left(1-\rho_1\right)\left(1-\rho_2\right)}{\left(1+\rho_1\right)\left(1+\rho_2\right)}\sigma_s^{\ 2}, \tag{4.40}$$

which is a lower value than for the 1D case (4.28). The maximum coding gain by exploiting correlation along both horizontal and vertical directions is

$$G_s = \frac{1}{\left(1-\rho_1^{\ 2}\right)\left(1-\rho_2^{\ 2}\right)}, \tag{4.41}$$

but can only be realized for rates $R \geq R(D_{\max}) = \log_2(1+\rho_1) + \log_2(1+\rho_2)$. The reduction in rate then is $\Delta R_s = -\frac{1}{2}\log_2(1-\rho_1^{\ 2}) - \frac{1}{2}\log_2(1-\rho_2^{\ 2})$. Within the low-distortion range, the correlations in both dimensions contribute independently to the overall gain in the case of the separable model.

Example. For a 2D separable AR(1) source with $\rho_1 = \rho_2 = .95$, an additional reduction of distortion by 10.1 dB or reduction of rate by 1.68 bit/sample can be achieved over the 1D case. As compared to PCM (not exploiting any correlation), the gain is 20.2 dB by distortion, or 3.36 bit/sample by rate[14].

For real-world multimedia sources, the stationarity assumption usually does not apply. These could rather be interpreted by *switched* or *composite source models*, where statistics vary locally. For example, disparate regions of high detail and regions of low detail co-exist in images. For such cases, it can additionally be necessary to encode and transmit *side information*, which must likewise be considered when determining rate-distortion behaviour.

The 'local' rate-distortion functions can give different contributions to the overall rate/distortion point. An appropriate strategy (in terms of optimizing the global rate-distortion behaviour) is encoding all local components such that the benefit by spending additional bits is equal, which is achieved by selecting an operation point where all 'local' rate-distortion functions have identical slope.

---

[13] cf. Problem 4.1

[14] This model analysis matches fairly well the gain which can e.g. be achieved in lossless coding of still images; typically, an 8 bit/sample PCM image representation can be lossless encoded (without any modification of samples) into 3-5 bit/sample by methods of linear predictive coding.

4.3 Rate-distortion optimization of quantizers

A continuous-amplitude signal is mapped by quantization into a signal of discrete amplitude levels. Another case is a signal having discrete amplitudes originally, which is *re-quantized*, i.e. mapped into a signal of less amplitude levels. To represent the indices i which express the reconstruction values in case of digital transmission, the rate of *entropy* (2.171) is necessary at minimum, where the set of indices establishes an alphabet with finite number of symbols. As practical coding methods, different entropy coding schemes will be described in Sec. 4.4. Treating quantization and entropy coding as separate entities is by tendency only useful (even though still suboptimum) in the range of high rates, where the number of symbols in the alphabet is large. Particularly at low rates, the optimization of *quantization and encoding* should better be regarded as a *combined problem*. For example, assume a value that is close to the boundary between two quantization intervals. Then, a reconstruction causing slightly higher distortion could be preferable, if a reduced rate gives benefit. In the process which optimizes the quantization (e.g. (4.16)-(4.22)) or during the quantization itself it is then necessary to consider an additional rate constraint. Let $R_{\mathcal{C}}(S_j)$ be the number of bits that is spent by code $\mathcal{C}$ when quantization index j is selected, i.e. the discrete source is in state S_j. Then, the mean rate is

$$R_{\mathcal{C}} = \mathcal{E}\{R_{\mathcal{C}}(S_j)\} = \sum_{j=0}^{J-1} \Pr(S_j) R_{\mathcal{C}}(S_j) . \tag{4.42}$$

Modifying (4.17) by formulation as a constrained problem that can be solved by *Lagrangian optimization*, with distortion $D_{\mathcal{Q}}$ caused by a quantizer and the rate $R_{\mathcal{C}}$ consumed by a code,

$$D^*_{\mathcal{Q},\mathcal{C}} = D_{\mathcal{Q}} + \lambda R_{\mathcal{C}} = \sum_{j=1}^{J} \Pr(S_j)\left[D_{\mathcal{Q}}(S_j) + \lambda R_{\mathcal{C}}(S_j)\right]. \tag{4.43}$$

Depending on the factor λ, the influence of $D_{\mathcal{Q}}$ and $R_{\mathcal{C}}$ can be varied, where $\lambda=0$ gives (4.17), i.e. no rate constraint. The optimum factor λ can be found by derivation of (4.43) over the rate,

$$\frac{\partial D^*_{\mathcal{Q},\mathcal{C}}}{\partial R_{\mathcal{C}}} = \frac{\partial D_{\mathcal{Q}}}{\partial R_{\mathcal{C}}} + \lambda_{\text{opt}} \overset{!}{=} 0 \Rightarrow \lambda_{\text{opt}} = -\frac{\partial D_{\mathcal{Q}}}{\partial R_{\mathcal{C}}} . \tag{4.44}$$

The optimum λ for a given rate is therefore given by the slope of the distortion-rate function $D_{\mathcal{Q}}(R_{\mathcal{C}})$[15]. In contrast to the rate-distortion bound which can analytically be determined only for certain classes of stationary processes, the local

---

[15] As $R(D)=D^{-1}(R)$, the optimum λ an also be interpreted as the reciprocal negative slope of the RDF for a given distortion point D.

slope of $D_Q(R_C)$ has to be estimated. This can be achieved by introducing an *operational rate distortion function* (ORDF), which is searched during the iterative optimization of a quantizer or during the quantization decisions. For simplicity, it is reasonable to assume that a code exists which produces a rate that is close to the entropy after quantization, i.e. setting $R_C(S_j)=I(S_j)$. Then, for a given quantizer and source probability statistics, a point of D_Q (4.17) and R_C (4.42) can be determined which has a certain position in the *R-D* plane. Testing different configurations of quantizers, all R_C/D_Q points which contradict the assumption of convexity (that must hold for any RDF) can be rejected as being sub-optimum. This can be determined when all *R*/*D* points from a set of investigated quantizers are pairwise interconnected by linear interpolation, and only points which are members of the lowest interconnection graph are retained as ORDF (see Fig. 4.6). Further refinement (possibly finding an even lower bound) is possible by adjusting λ according to the current slope of the ORDF.

This Lagrangian multiplier method can also be used, particularly during the process of optimizing a quantizer e.g. by the Lloyd algorithm (4.18)-(4.22) [CHOU, LOOKABAUGH, GRAY 1989]. This is generally denoted as *entropy constrained quantization* (ECQ). When rate constraints are employed during encoding by a given quantizer/encoder configuration, the actual resulting rate allocation can also be used instead of the entropy-based estimate.

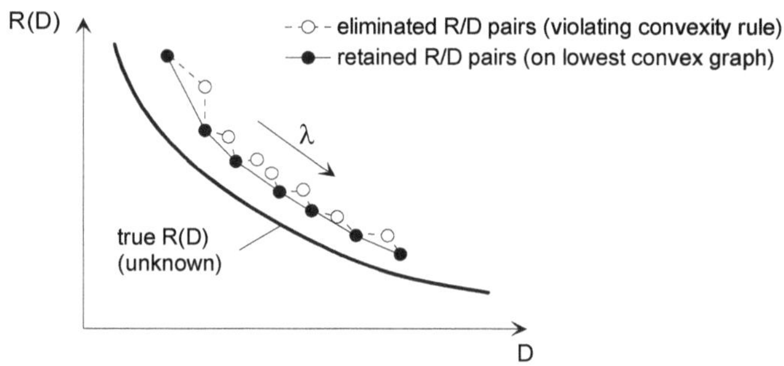

Fig. 4.6. Development of the ORDF based on convexity constraint

If sources with locally variable statistics are encoded and a certain rate budget can be spent, it should first be invested for components which give the highest value in terms of distortion minimization, i.e. with largest slope of the ORDF at the given target point. This is typically guaranteed by choosing a fixed value of λ in (4.43) for rate-constrained quantization of all components, regardless of their local statistics.

A specific case, where a close to optimum performance in terms of rate-distortion criteria can be found systematically, occurs in case of uncorrelated sources with exponential, specifically Laplacian PDF. In [SULLIVAN 1996] it is shown that by appropriate choice of a deadzone quantizer (OFFR and OFFQ values in (4.12)] an optimum in rate-distortion sense can be found in a non-iterative fashion.

Embedded quantization. The distortion due to quantization has an approximate squared relation with the quantization step size Δ, see (4.4)[16]. Assume now that the residual error shall be *re-quantized* by another quantizer of half step size $\Delta/2$. The number of reconstruction values J is doubled when the characteristic of the second quantizer divides each of the previous quantization intervals into two halves of equal width. If the reconstruction values are assumed to be positioned at the centers of quantization intervals, each reconstruction value y_j for the quantization of step size Δ branches into positions $y_j \pm \Delta/4$ for the quantization with step size $\Delta/2$. The embedding of two quantization characteristics under these assumptions is shown in Fig. 4.7; dotted lines represent the quantization with half step size. Fig. 4.7a shows a configuration for quantization of only positive values, while Fig. 4.7b gives the example of dead zone quantization (cf. Fig. 4.2b); the dead zone width is 2Δ here, and the zero-quantization threshold is Δ. In the next finer re-quantization, the dead zone is divided into three intervals, one of which is a new dead-zone with half width Δ, the other two are 'normal' quantization intervals of width $\Delta/2$.

Fully embedded quantization can perform the method of re-quantization over multiple levels, starting with only two quantization intervals for the case of positive-valued signals, and three intervals for the case of dead-zone quantization. This is shown in Fig. 4.8; the reconstruction values are always positioned at the centers of the quantization intervals. The final result of the quantization is an integer-value index which has a word length equal to the number of levels; the first level relates to the *most significant bit* (MSB), the last level to the *least significant bit* (LSB) of the integer word. This index can then be truncated starting from the LSB, which gives a coarser quantization with lower number of levels. The entire quantization could indeed be performed by the binary tree decision. A much simpler way is employing an ordinary uniform quantization using the interval width (quantization step size) of the last level (LSB), where the total number of quantization intervals has to be a power of two, 2^B. Then, implicitly, each level of the embedded quantization corresponds to the related bit position in a positive integer of word length B. This corresponds exactly to the scheme in Fig. 4.8a which shows the quantization of positive-amplitude signals. The quantized representation of the amplitude range between MIN and MAX is becoming more accurate with each additional bit becoming available. This is denoted as a *bit-plane representation*, where a bit-plane is constructed by all bits of a certain significance level from all available samples of a signal. One single bit-plane can be interpreted as a signal consisting of binary samples, and by combination of all bit-planes, the full multi-level representation of the samples is established.

---

[16] This may also depend on signal statistics, e.g. if many values are close to zero and quantized into zero reconstruction, the average distortion is significantly lower than e.g. $\Delta^2/12$.

The dead-zone scheme in Fig. 4.8b allows zero, positive and negative reconstruction values, where the zero value does not require a sign bit. In a magnitude/sign representation, the sign needs only to be conveyed once, together with the non-zero bit of highest significance, which can be at any bit-plane position. The relevance of the sign bit is therefore also depending on the bit-plane position where the significance of the sample is found.

It should be observed that the embedded quantizer designs introduced here are assuming the placement of reconstruction values at the centers of the intervals (see Fig. 4.8b). Otherwise, the quantization partitions might not be consistently embedded in a way that the quantization error is decreased with each additional bit, and that no better reconstruction value might exist in a neighboring interval. Another possibility achieving a similar effect as embedded quantization is *re-quantization* (with low number of intervals usually) of the quantization error produced by a previous quantizer. This may lead to overlapping quantization intervals, except for the case of uniform step sizes and centered reconstruction values. It however highly depends on the design of the subsequent entropy coding whether this redundancy impacts the rate.

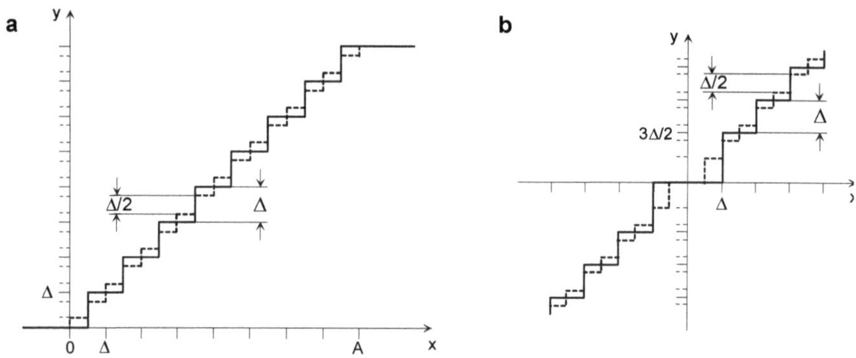

Fig. 4.7. Quantization characteristics for the case of re-quantization by step size Δ/2
a Asymmetric positive-only range of quantization **b** Symmetric dead-zone quantization

Embedded quantization and bit-plane coding are used in *scalable representation* of quantized signals, where the more significant bits are a representation which allows reconstruction by a higher distortion at a lower rate. It is however not possible to move freely on the rate-distortion curve by adding or omitting bits from the bit-plane representation. More specifically, even when efficient entropy coding is applied to the bit-plane information, the rate-distortion curve can *exactly* be mapped only at rate points of complete bit planes and may lose performance otherwise. This phenomenon is known as the problem of *fractional bit-planes* [TAUBMAN 2000].

By rate-distortion optimized bit-plane truncation, re-ordering of information in the bit planes can be performed, such that samples or components which contribute highest reduction of distortion at minimum cost of bits are conveyed first. Such strategies can partially compensate the fractional bit-plane phenomenon.

This is however only achievable if the method of bit-plane coding allows clustering the information into components of more or less significance. Such strategies are e.g. applied by the EBCOT algorithm used in the JPEG 2000 standard (cf. Sec. 6.4.4).

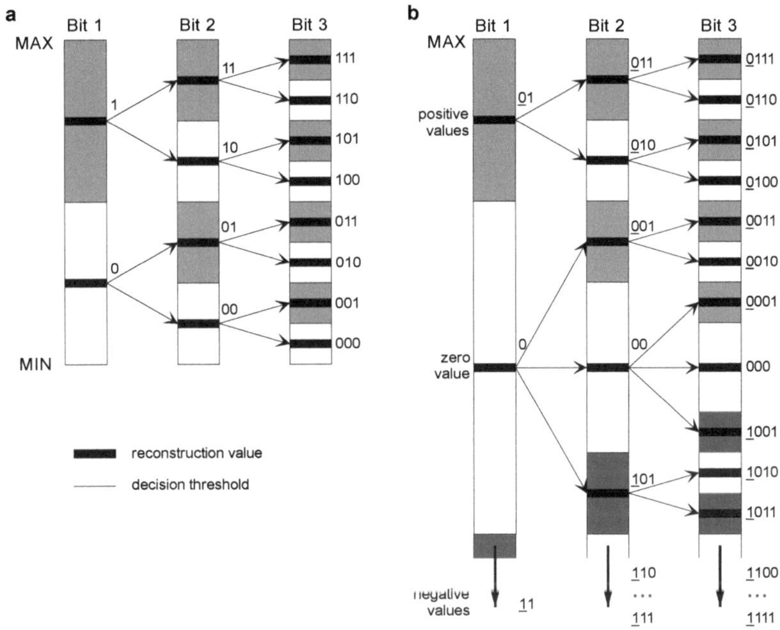

Fig. 4.8. Embedded quantization structures and associated bit-plane code **a** for positive-valued signals **b** with dead zone for sign/magnitude signals; sign bit underlined

4.4 Entropy coding

4.4.1 Properties of variable-length codes

The *entropy* $H(S)$ of (2.171) is the minimum rate, by which a discrete source with source alphabet $S=\{S_0,S_1,..,S_{J-1}\}$ can be lossless encoded. The goal of *entropy coding* is to define a code $\mathcal{C}=\{C_0(S_0),C_1(S_1),\ldots\}$, which allows approximately approaching the rate of entropy. In principle, this is possible by using *Variable-Length Codes* (VLC). Encoding by the code $\mathcal{C}$ means that each element S_j of the source alphabet is uniquely mapped to a codeword C_j (binary string) consisting of $Re(S_j)$ bits[17], which as good as possible should match the self information of S_j,

[17] Restriction is made to binary code alphabets here, however in general coding can be implemented with multiple-letter alphabets. An example is the *Morse code* used in telegra-

i.e. $Re(S_j) \approx I(S_j) = -\log_2 \Pr(S_j)$. Codes of variable length must be decodable not only codeword by codeword, but also if a concatenated *stream of codewords* is received by the decoder. This imposes the following constraints:

- *Non-singularity*, which means that no duplicate codewords shall exist;
- *Unique decoding*, which means that by receiving a string of bits it shall be possible to identify start and end positions of each codeword within the stream.

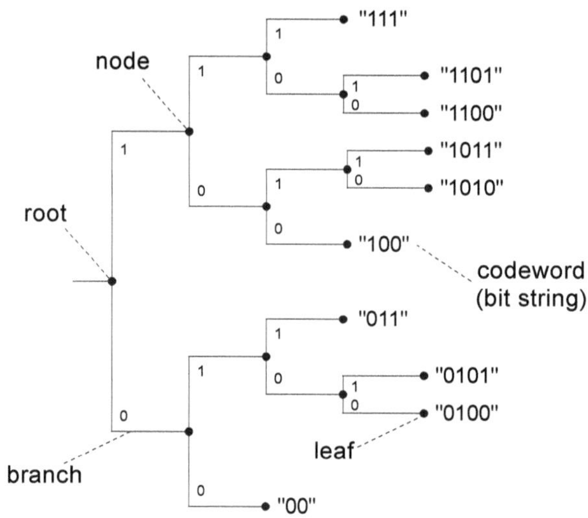

Fig. 4.9. Code tree representing a prefix-free code of J=10 codewords

An additional simplifying property is *instantaneous decoding*, which means that a codeword can be identified as being unique as soon as it is received entirely, without reference to subsequent elements of the bit stream. An important class of codes that are unique and instantaneously decodable are *prefix-free codes*. These are constructed such that never a valid bit string representing a full codeword shall be the prefix of another codeword's bit string. As an example two codes are given, $\mathcal{C}=\{0,10,110,111\}$ which is uniquely decodable, and $\mathcal{C}=\{0,\underline{10},\underline{10}0,111\}$ which is not, as the string '10' could either be the entire second codeword, or the prefix of the third. The decoding process can be interpreted as parsing of a *code tree* (Fig. 4.9). In case of binary coding, the tree typically consists of two-way *branches*, each of which represents one bit received. In principle, a branch could come without a sibling, but such branches can be eliminated without affecting the decodability of the code. The branches connect *nodes*; the *root* and the *leaves* repre-

---

phy consisting of four code 'letters' (short and long beep, short and long break). Each of these relates to a certain duration in time, such that the code alphabet itself consist of symbols of variable length, but additionally a variable number of code symbols is used for different source letters based on their probability of occurrence.

sent the beginning and termination of valid codewords[18]. The entire *path* of branches between root and leaf signifies the bit string associated with one codeword.

For a given set of J binary codewords (bit strings S_j) having lengths $\mathcal{R}_e(S_j)$ (corresponding to the number of branches in the tree from the root to the respective leaf), it is always possible to construct a valid prefix-free code if *Kraft's inequality* applies [COVER, THOMAS 1991][19]

$$\sum_{j=0}^{J-1} 2^{-\mathcal{R}_e(S_j)} = C \leq 1 . \tag{4.45}$$

The value C=1 indicates that a corresponding code tree would be fully populated, i.e. branches emanating from any node always have siblings. A code with C<1 has a certain redundancy, which may be desirable to avoid errors in case of erroneous bit reception), whereas for $C>1$ the condition that every codeword has to be assigned a unique sequence of branches from the root to its leave cannot be met.

The expected mean rate by using a variable length code to encode a source alphabet $\mathbf{S}=\{S_j; j=0, \dots , J-1\}$ is

$$R = \mathcal{E}\{\mathcal{R}_e\} = \sum_{j=0}^{J-1} \Pr(S_j)\mathcal{R}_e(S_j) \geq H(\mathbf{S}) = -\sum_{j=0}^{J-1} \Pr(S_j)\log_2 \Pr(S_j) . \tag{4.46}$$

The self information $I(S_j)=-\log_2 \Pr(S_j)$ is not generally integer, whereas bits formally do not exist by fractional units, i.e. $\mathcal{R}_e(S_j)$ has to be an integer number when mapping shall be performed by a fixed set of codewords. This means that only for the specific case where all $\Pr(S_j)=2^{-m}$ (fractional powers of two), the rate of entropy (2.171) can exactly be approached by a simple code mapping. An example design for a code of variable length is the *Shannon Code*, where

$$\mathcal{R}_e(S_j) = \begin{cases} I(S_j) & \text{if } \Pr(S_j) = 2^{-m} \\ \lfloor I(S_j) \rfloor + 1 & \text{else.} \end{cases} \tag{4.47}$$

A prefix-free code having this bit allocation by guarantee exists, however it may come with a value $C<1$ in Kraft's inequality, which means that some of the codewords can be longer than would actually be needed for unique decoding. Nevertheless, the design can by guarantee come close to the entropy by

---

[18] If parsing of an entire *stream of codewords* is considered, the leaf terminating one codeword usually establishes the root of the next codeword. In the case of non-instantaneous codes, it may happen that a node is identified as being a leaf only by subsequent bits. An example is the code $\mathcal{C}$={00,10,11,110}, which is not prefix-free but unique.

[19] In (4.45), condition $C=1$ holds true, if the code tree is fully occupied, i.e. no 'empty branches' remain in the code tree. If $C<1$, strings exist that are not used to represent letters from the alphabet; codes are indeed sometimes constructed such that additional strings may be reserved for re-synchronization or other purposes.

$$H(S) \leq R_e < H(S) + 1 . \tag{4.48}$$

A straightforward method to approach the entropy closer is using a combination of K source symbols (letters) into vectors $\mathbf{s}=[s(1)\ s(2)\ \ldots\ s(K)]^T$. Then, the excessive bit penalty from (4.48) is dispersed over K source symbols, and a Shannon code should exist which requires a rate per source symbol

$$\frac{1}{K} \cdot H_K(S) \leq R_e < \frac{1}{K} \cdot H_K(S) + \frac{1}{K} \leq H(S) + \frac{1}{K} , \tag{4.49}$$

where $H_K(S)$ denotes the entropy per vector, which is defined as a K-dimensional version of the joint entropy (2.172), where according to (2.173) the upper bound $H_K(S) = K \cdot H(S)$ would apply when the source symbols $s(k)$ within $\mathbf{s}$ are statistically independent. In any other case, the entropy of the vector will be *lower than* the K-fold entropy of single source symbols, and joint statistical properties are implicitly exploited in the design of the code. Hence, in any case, the entropy of the source can be approached by a margin $\varepsilon = 1/K$, in case $K \to \infty$ it is theoretically possible to come arbitrarily close. However, due to the fact that all possible combinations need to be supported, the vector code requires a total of J^K distinct codewords. Therefore, the complexity of such codes (in terms of size of code tables, decisions to be made etc.) grows *exponentially* by the length of the vectors and can hardly be managed.

Methods of entropy coding should not only approach the entropy rate closely, but also keep the complexity in terms of computation and memory usage reasonably low. It is further desirable to exploit statistical dependencies between source samples, and being capable of adaptation for varying source statistics. (4.46) can be re-written as

$$R_e = -\sum_{j=0}^{J-1} \Pr(S_j) \log_2 2^{-R_e(S_j)} = -\sum_{j=0}^{J-1} \Pr(S_j) \log_2 \left[C \cdot \Pr^*(S_j) \right] \quad \text{with} \quad \Pr^*(S_j) = \frac{2^{-R_e(S_j)}}{C} , \tag{4.50}$$

where the constant C is defined as in (4.45). The difference between the rate produced by a variable-length code and the entropy is

$$\begin{aligned} R_e - H(S) &= -\sum_{j=0}^{J-1} \Pr(S_j) \log_2 \left[C \cdot \Pr^*(S_j) \right] + \sum_{j=0}^{J-1} \Pr(S_j) \log_2 \Pr(S_j) \\ &= \sum_{j=0}^{J-1} \Pr(S_j) \log_2 \frac{\Pr(S_j)}{\Pr^*(S_j)} + \log_2 \frac{1}{C} . \end{aligned} \tag{4.51}$$

For codes which operate at the upper bound ($C=1$) of Kraft's inequality (4.45), the rightmost term is zero, and the left term is the *relative entropy*. In fact, this given code would be optimum for a source where probability $\Pr(S_j)=\Pr^*(S_j)$, and the relative entropy expresses the rate penalty which occurs if the actual probability distribution deviates from the assumption under which the code was designed.

4.4.2 Huffman code design

The Huffman code design [HUFFMAN 1952] is more efficient than the Shannon approach, since it allows using codeword lengths $\mathcal{R}e(S_j) < I(S_j)$, while the property of an instantaneous decodable prefix-free code is retained. In the following procedure, it is only necessary to know the probabilities of the letters S_j in the source alphabet S:

1. Construct a list $\mathcal{L}$ of probabilities $\Pr(S_0)$, $\Pr(S_1)$,..., $\Pr(S_{J-1})$. Initially, each list field relates to exactly one source symbol. The indices j of the source letters are associated with bit strings $\{C_j; j=0,\ldots J-1\}$ of the code $\mathcal{C}$. Initially, the bit strings consist of zero bits each.
2. In $\mathcal{L}$, determine the two list entries of smallest probabilities. The bit string(s) associated with those two entries are prefixed by ‘0’- and ‘1’-bits, respectively.
3. The two entries processed in step 2 are removed from $\mathcal{L}$. The sum of their probabilities is associated with a new list entry. This new entry is further associated with all source letter indices and bit strings which were previously represented by the two deleted entries.
4. If $\mathcal{L}$ contains only one entry, this will be the root node, such that the code design is finalized. Otherwise, the procedure iterates over steps 2-4.

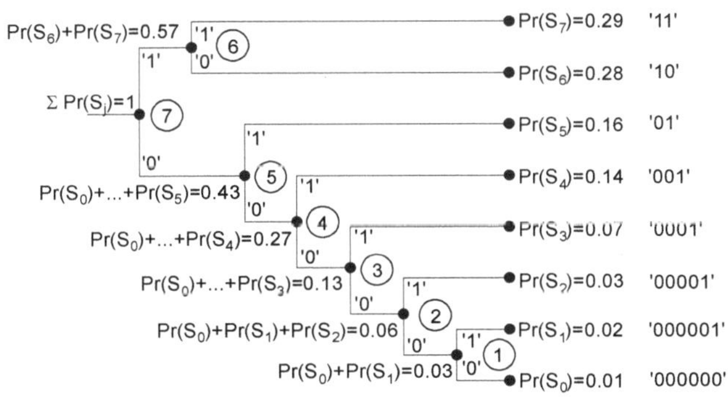

Fig. 4.10. Example construction of a Huffman code tree (7 steps)

Note that the code grows in reversed order; expressed in terms of a code tree, this means that the code design starts at the leaves and propagates towards the root[20].

[20] Another design which grows the tree from the root is the Shannon-Fano method (proposed independently by C. E. SHANNON and R. M. FANO in 1949). Herein, an ordered list of probabilities is iteratively partitioned into two halves with approximately equal sum probability. When only one entry is left in a split partition, this relates to a leaf. The Huffman design does not pre-determine sub-tree allocations by initial probability ordering and there-

For an ordered list of probabilities of source symbols with $\Pr(S_0) \leq \Pr(S_1) \leq \ldots \leq \Pr(S_{J-1})$, an optimum code shall give rate allocations $R(S_0) \geq R(S_1) \geq \ldots \geq R(S_{J-1})$. On the other hand, for a code tree where each branch has a sibling ($C=1$ in Kraft's inequality (4.45)), the two longest codewords must always have even lengths. As both conditions are observed in each iteration of the Huffman design, the code is close to optimum within its class of instantaneously decodable prefix-free codes.

Fig. 4.10 shows the design of a Huffman code for an example with 8 source symbols of different probability, also showing the iterated steps 1-7 of the code tree construction. For each step in the procedure, the probability value currently stored in the list is given as well.

In Huffman coding, the upper bounds (4.48) and (4.49) basically still apply, however it can be shown that the upper bound is never met. The following approximation gives an estimate of the maximum penalty over the entropy rate [COVER, THOMAS 1991]:

$$\begin{aligned} R_e &< \frac{1}{K}\left(H_K(S)+\Pr_{\max}\right) && \text{if } \Pr_{\max} < 0.5 \\ R_e &< \frac{1}{K}\left(H_K(S)+\Pr_{\max}+0.086\right) && \text{if } \Pr_{\max} \geq 0.5 \end{aligned} \quad ; \Pr_{\max} = \max_{j=1\ldots J^K}\left[\Pr(S_j)\right] \tag{4.52}$$

For the case of combining source samples into vectors of length K, $\Pr_{\max}$ is typically low, such that the entropy rate is approached closely. However, the number of code symbols in $\mathcal{C}$ and hence the complexity of design, encoding and tree parsing/decoding grows exponentially with K[21], which is often not rewarding when comparing the benefit.

4.4.3 Systematic variable-length codes

The Huffman code design can generate extremely long code strings for *untypical* source symbols (or source symbol vectors) which occur with low probability. Firstly, if the sum of probabilities Pr* of all untypical cases is below a certain margin, their occurrence hardly will have any significant effect on the rate, even if the length of the codewords assigned to them deviates from the original self information (since $-x\log x \rightarrow 0$ for $x \rightarrow 0$). Second, from the complexity point of view, it may be preferable to not consider untypical cases at all in the design of VLCs. Some methods following this paradigm are described here, by which reasonable performance of the code and moderate complexity can be compromised. Besides a reduction of the code table size, the decoding complexity can also be reduced by introducing more *regularity* in the code design. For example, subsets of symbols can be identified having common prefixes to allow simplified parsing

fore is more flexible, in many cases providing a code rate which comes closer to the entropy.

[21] The code table including all combinations shall have the size J^K.

of the code string. Another aspect is the *universality* of the code design, which means that the same code syntax can be adapted to sources of different probability distributions, or be used similarly for different components in a coded representation. This will in particular be important when adaptive entropy coding shall be used; it is however also favorable in terms of complexity, when the same decoding unit (hardware or software) is able to decode different source components, and it is not 'hard-wired' for one specific VLC.

Table 4.1. VLC table for encoding of 'AC' coefficients and separate encoding of exceptional RUN/LEVEL combinations not contained in the VLC table (excerpt from the MPEG-1 standard)

1. VLCs for run length and quantization index, 's'=sign

*) differently defined for first and last coefficient
) largest LEVEL value in the table *) largest RUN value in the table

RUN	LEVEL	code	RUN	LEVEL	code
EOB	--	10	9	1	0000101s
0	1	1s / 11s *)	0	5	00100110s
1	1	011s	...	...	...
0	2	0100s	13	1	00100000s
2	1	0101s	0	7	0000001010s
0	3	00101s	...	...	...
3	1	01111s	21	1	000000010110s
4	1	00110s	0	12	0000000011010s
1	2	000110s	...	...	...
5	1	000111s	0	40**)	000000000010000s
...	...	...	...	...	...
ESCAPE	--	000001	31***)	1	0000000000011011s

2. Fixed-length code, separate coding of run-length and quantizer level index. Invoked by and following the ESCAPE codeword: 6 bit for run length, 8 bit for $|\text{index}|<128$, 16 bit for $128\leq|\text{index}|\leq256$.

RUN	code	LEVEL	code
0	000000	-255	1000000000000001
1	000001	...	...
2	000010	-128	1000000010000000
...	...	-127	10000001
...	...	...	...
...	...	127	01111111
...	...	128	0000000010000000
...	...	...	...
63	111111	255	0000000011111111

Partial fixed-length coding. Fixed-length coding only for untypical source symbols can be realised by using a special VLC codeword for an ESCAPE symbol, as prefix which is followed by a *fixed-length suffix*. The fixed length code must have

$\log_2 J^*$ bit if J^* is the number of untypical symbols to be supported. For simplicity, it can also be constructed in a way that all source symbols (including the typical ones for which a direct VLC expression exists) can be covered by the fixed length code. In this case, either some codes are void or it would be up to the encoder to decide whether to use the VLC or the ESCAPE/fixed-length code, the latter being an unreasonable choice.

A code of this type is instantaneously decodable and can be expressed by a code tree, where a binary tree with bit-depth of the fixed-length code unfolds beyond the node related to the ESCAPE symbol. The VLC length for the ESCAPE codeword should be set to approximately $-\log_2(\Pr^*_{\mathrm{untyp}})$ depending on the overall probability $\Pr^*_{\mathrm{untyp}}$ of untypical symbols. As an example, an excerpt of a VLC table from the MPEG-1 standard, designed for coding DCT coefficients in an 8x8 transform block, is shown in Table 4.1. This combines source symbols RUN and LEVEL into a vector (cf. Sec. 5.3.1). Combinations of short RUN and small LEVEL are expected to occur more frequently due to the statistical properties of transform coefficients and are represented by the shortest codewords. In case of ESCAPE (6 bit code), the RUN and LEVEL symbols are encoded independently, with a fixed-length code of 6 bit for RUN, and LEVEL by either 8 or 16 bit, depending on amplitude value. The total length of codewords in the untypical cases is therefore either 20 or 28 bit.

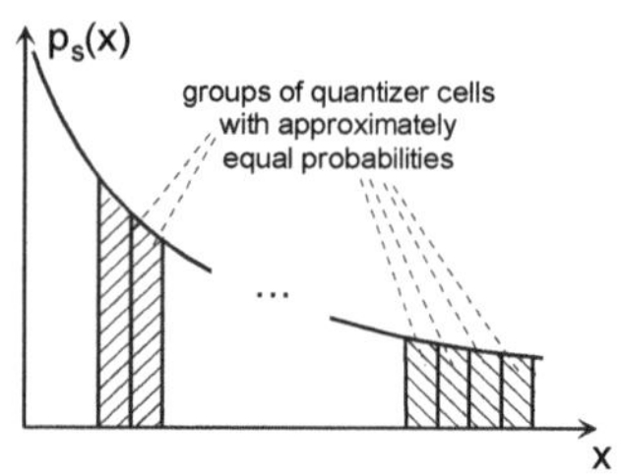

Fig. 4.11. Groups of quantizer cells which will be coded with equal numbers of bits in a VLC design for the case of exponentially decaying source distribution and uniform quantization

Universal variable-length codes. Source models of Gaussian PDF and its generalized extension (2.126) show an exponential decay of probabilities towards higher amplitude levels. Markov chains are useful as models for correlated binary sources, which also have an exponential decay of probabilities for increasing run-lengths according to (2.164). Optimum VLCs for such sources therefore need to provide an exponentially increasing VLC length towards source symbols of higher values. According to (2.122), the probability $\Pr(S_j)$ of a quantized source symbol is equivalent to the area under the continuous PDF within the quantization interval, and the optimum bit assignment is $-\log_2\Pr(S_j)$ (equal to self information). In exponential distributions, the tangential slope of the PDF decreases towards higher amplitudes. Hence, it is straightforward that the number of adjacent quantized source symbols which have to be encoded by using the same number of bits in the

VLC codewords also increases exponentially towards higher amplitude levels when uniform quantization is used (see Fig. 4.11).

A simple example of UVLC are the *unary codes*, which in their simplest construction assign a sequence of j '0' bits and one terminating '1' bit to the source symbol S_j ('code A' construction), where it is assumed that the source symbols are ordered by decreasing probability with increasing j. An alternative is to start and stop with a '1' bit and insert $j-1$ '0' bits between; the special case of $j=0$ just uses a single '0' bit ('code B' construction). In both cases, $j+1$ bits are assigned to S_j, such that the unary code would reach the entropy for a dyadic-exponentially decreasing probability distribution $\Pr(S_j)=2^{-j-1}$ (see Tab. 4.2). In principle, a unary code can extend to arbitrary length; if the number of symbols J is fixed, the last bit can be skipped for the last symbol, e.g. the codes shown in Tab. 4.2 can still be decoded when $J=4$, and '000' for code A or '100' for code B would be used for S_3; this is a *truncated unary code.*

Table 4.2. Construction of a unary code

Source symbol	Code A	Code B
0	1	0
1	01	11
2	001	101
3	0001	1001
...	...	...

Other types of codes which provide systematic increase of length and/or number of same-length codewords are the *Golomb-Rice codes* (GR) and the *Exponential Golomb codes* (EG), which are constructed by combining a unary string with a fixed-length string. While the numbers of same-length codewords is constant in case of GR codes, it is exponentially increasing with the codeword length for EG codes. Therefore, to support a source behaviour as illustrated in Fig. 4.11, EG codes are the better choice. An example of a construction on the basis of an EG code is shown in Fig. 4.12, where as an additional option the UVLC string (pattern 1, 01, 001 etc.) and the fixed-length string (with varying number of bits b_i depending on the unary code) are interleaved. This has the advantage that each codeword is terminated by a '1', and two subsequent '1' bits only appear at the end of a codeword (i.e., a string '11|0' uniquely marks the boundary between two codewords); furthermore, all codeword consist of an odd number of bits. This way, the code provides an additional capability of detection and recovery from transmission errors. However, no codewords of even length exist, such that a deviation of the codeword length assignment from the self information could become as large as 1 bit in worst case. Another variant of GR code can be constructed from a variable-length unary prefix and a fixed-length suffix, which however would no longer have the inherent error detection and resynchronisation capability.

a)

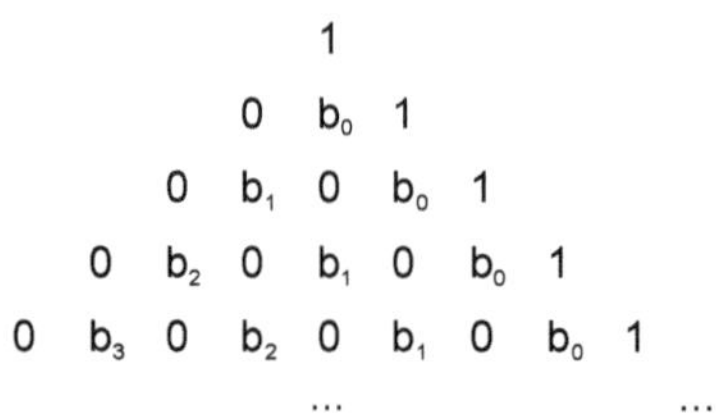

b)

Source symbol	Codeword
0	1
1	001
2	011
3	00001
4	00011
5	01001
6	01011
7	0000001
8	0000011
9	0001001
10	0001011
...	...

Fig. 4.12. Universal VLC construction based on an interleaved EG code design: **a** construction schema **b** code table

Table 4.3. Construction of RVLC from GR and EG codes

Source symbol	Reversible GR code		Reversible EG code	
	Prefix	Suffix	Prefix	Suffix
0	0	00	0	0
1	0	01	0	1
2	0	10	101	0
3	0	11	101	1
4	11	00	111	0
5	11	01	111	1
6	11	10	10001	0
7	11	11	10001	1
8	101	00	10011	0
9	101	01	10011	1
10	101	10	11001	0
11	101	11	11001	1
12	1001	00	11011	0
13	...	...	11011	1
...	...	...	...	...

Reversible variable-length codes. Variable-length codes introduced so far are prefix-free codes. The decoder needs to parse the bit string sequentially, corresponding to the paths in the code tree which originate from the root. When a leaf of the tree is reached, the current codeword terminates. If a bit error occurs, a wrong path having different length could be selected, such that the decoder may interpret an earlier or later termination of the codeword, and the next codeword(s) will be misinterpreted as well. Conventional prefix-free codes cannot be parsed in reverse direction. *Reversible Variable-Length Codes* (RVLC) allow this [TAKASHIMA, WADA, MURAKAMI 1995] [WEN, VILLASENOR 1997] [GIROD 1999]. In addition to the prefix condition, a *suffix condition* is formulated, which requires that no codeword's suffix (starting from the last bit) shall coincide with any longer codeword's suffix, such that the code can instantaneously be decoded in reverse order. RVLC design methodologies are known for Huffman, GR and EG codes, where the latter two have the most efficient solutions. A systematic construction can be made for codewords of length $Re(S_j)$ which are partitioned into a prefix of length $Re(S_j)-L$ and an L-bit fixed-length extension. For GR codes the suffix condition is fulfilled, if the prefix starts and ends by a '1', whereas all other bits of the prefix are '0'. A prefix of length one is '0'[22]. For EG codes, this condition can be modified such that the prefix also starts and ends by a '1', but the zero constraint may only be applied to prefix bits at odd positions (corresponding to the unchanged bit positions in Fig. 4.12a). Bit strings can be found which use GR or EG codewords as prefixes fulfilling these conditions. Examples of a GR code with L=2 and an EG code with L=1 are shown in Table 4.3[23].

Application of RVLC (as other systematic code designs) imposes a penalty on efficiency due to the constraints imposed on the codeword length allocations, which can be evaluated by (4.51). An RVLC design which can almost close this gap (however being less systematic) was proposed in [GIROD 1999].

4.4.4 Arithmetic coding

Arithmetic coding is widely used in recent standards for multimedia signal compression. Unlike Huffman codes or systematic VLCs, where a direct mapping of separate codewords from the bit stream with source symbols is made via a table or tabulated construction, arithmetic decoding can rather be interpreted as a *sliding window* method, where encoder and decoder are in a unique state in each step, but instantaneous decoding may not be possible by the time when an encoder releases bits based on the state of the source symbol. The decoder interprets the information carried by newly arriving bits along with the previous state, and outputs decoded source symbols whenever the interval boundary rules as subsequently described are met. The term *arithmetic coding* is related to the underlying computation which has be performed with a sufficiently high arithmetic precision, in order to achieve performance close to the rate of the entropy.

The simple type of the *Elias code* [ELIAS 1975] is presented here as example case of arithmetic compression for a non-uniformly distributed binary source producing symbols 'A' and 'B' with probabilities Pr('A') and Pr('B')=1−Pr('A'). Assuming statistical independence between subsequent source outputs, the probabilities are Pr('A')^2, (1−Pr('A'))^2, (1−Pr('A'))Pr('A') and Pr('A')(1−Pr('A')) for sequences 'AA', 'BB', 'BA' and 'AB', respectively. To perform the code mapping, *probability intervals* having widths equal to the probabilities of the associated source symbols or their sequences are constructed. Initially (at the 'root' of the code), only one interval i_0=[0,1] of width one exists. This interval is then sub-divided into subsequently narrower intervals, each expressing by its width the probability of a given source symbol sequence. For ex-

---

[22] The suffix condition is also fulfilled, if the role of '0' and '1' bits is interchanged.

[23] The prefixes of the RVLC are complete EG codewords. To fulfill the suffix condition, the code pyramid of Fig. 4.12 must be modified as '0', '1$b_0$1', '1$b_1$0$b_0$1', '1$b_2$0$b_1$0$b_0$1' etc.

ample, if the first source symbol is $s(1)$='A', the associated interval is $ip_1=[0;\Pr(\text{'A'})]$, while for the source symbol $s(1)$='B', the interval is $ip_1=[1-\Pr(\text{'B'});1]=[\Pr(\text{'A'});1]$. If e.g. the first source symbol is $s(1)$='B' and the second is $s(2)$='A', the next step of sub-division would be $ip_2=[\Pr(\text{'A'});\Pr(\text{'A'})+\Pr(\text{'B'})\Pr(\text{'A'})]=[\Pr(\text{'A'});\Pr(\text{'A'})+(1-\Pr(\text{'A'}))\Pr(\text{'A'})]$.

Arithmetic coding in principle represents the position of the probability interval, which is uniquely associated with the related source symbol within a given sequence. With non-uniform source probability distribution, the grid of probability intervals is also non-uniform. Efficient entropy coding should provide a bit stream where no further redundancy exists, i.e. it should have the properties of a first order Bernoulli process (2.131). A unique mapping between the non-uniformly distributed source symbol sequence (represented by the probability intervals) into a uniformly distributed and independent binary code sequence is achieved by comparison of the *non-uniform* probability intervals ip_n against the *uniform grid* of *code intervals* ic_m.

One code interval associated with an instantiated sequence $\mathbf{b}_m=[b_1,b_2,b_3, ..., b_m]$ of m bits is defined as

$$\begin{aligned} ic_m &= \left[0.5\cdot b_1 + 0.25\cdot b_2 + ... + 2^{-m} b_m \; ; \; 0.5\cdot b_1 + 0.25\cdot b_2 + ... + 2^{-m} b_m + 2^{-m}\right] \\ &= \left[\sum_{k=1}^{m} 2^{-k} b_k , \sum_{k=1}^{m} 2^{-k} b_k + 2^{-m}\right]. \end{aligned} \tag{4.53}$$

The rule of encoding is to release bits $b_1...b_m$, when a probability interval ip_n is completely contained in a code interval ic_m, i.e. if the following conditions hold true, expressed via lower and upper interval boundaries:

$$ip_{n,\text{lo}} \geq ic_{m,\text{lo}} \;\text{ AND }\; ip_{n,\text{up}} \leq ic_{m,\text{up}} . \tag{4.54}$$

As furthermore the width of interval ip_n equals the probability of the source symbol sequence of length n, the associated self information can be expressed as

$$I(ip_n) = -\log_2\left(ip_{n,\text{up}} - ip_{n,\text{lo}}\right) \geq -\log_2\left(ic_{m,\text{up}} - ic_{m,\text{lo}}\right) = -\log_2 2^{-m} = m . \tag{4.55}$$

From (4.54), the code interval ic_m is always larger or equal compared to the probability interval ip_n. From (4.55), the number of bits released up to this step is then always less or equal to the self information of the source sequence. Hence,

- The number of bits sent is smaller or equal to the accumulated self information over the whole sequence of source symbols;
- The sequence of source symbols can eventually not yet fully be decoded from the bits released up to here, unless the probability intervals perfectly match with the code intervals, in which case the number of released code bits equals the self information.

Decoding uses the reverse principle of encoding. It is tested whether the interval ic_m, associated with the received sequence of code bits fully fits into an interval ip_n associated with a source sequence, i.e. whether the following condition holds:

$$ip_{n,\mathrm{lo}} \le ic_{m,\mathrm{lo}} \text{ AND } ip_{n,\mathrm{up}} \ge ic_{m,\mathrm{up}} . \tag{4.56}$$

If the sequence of source symbols grows sufficiently long, the intervals will get arbitrarily small, such that the entropy can also be approached arbitrarily close. This however is the basic constraint of arithmetic coding: As the computation of the interval boundaries can only be performed by finite arithmetic precision, it is not possible to approach the entropy perfectly. However, even for case of finite arithmetic precision the margin over the entropy rate can be very small (see below).

Fig. 4.13 graphically shows the grids of probability intervals and code intervals for the case Pr('A')=1/4. The following two example sequences of source symbols are actually lowest and highest probability cases at step n=3:

- Sequence 'AAA': For $s(1)$='A', the probability interval ip_1=[0;1/4] falls exactly into the code interval ic_2=[0;1/4]. Bits b_1=0 and b_2=0 are encoded. Likewise, ip_2=[0;1/16] falls exactly into ic_4=[0;1/16] and ip_3=[0;1/64] falls exactly into ic_6=[0;1/64], which means that in total the code bit sequence '000000' is released up to this point. This is the special case where both encoding and decoding can be performed instantaneously, which is however caused by the specific fact that probability and code intervals exactly match. Here, the probability Pr('A')=1/4 is an integer power of 2, for which also the instantaneously decodable Huffman code gives optimum performance.
- Sequence 'BBB': Neither ip_1=[1/4;1] nor ip_2=[7/16;1] entirely fall into any code interval, both are expressing self information of less than one bit. The next sub-division ip_3=[37/64;1] is contained in ic_1=[1/2;1]. With $s(3)$='B', b_1='1' is released. This example shows that a sequence of higher probability causes significantly less bits to be sent; however when receiving b_1='1', it will only be possible to decode the first two source symbols $s(1)$='B' and $s(2)$='B': ic_1=[1/2;1] falls entirely into probability intervals ip_1=[1/4,1] and ip_2=[7/16;1], whereas the third source symbol cannot be decoded yet. Partial information about $s(3)$ which was carried by b_1 is retained in the state of the decoder, such that it can uniquely be decoded by some future step.

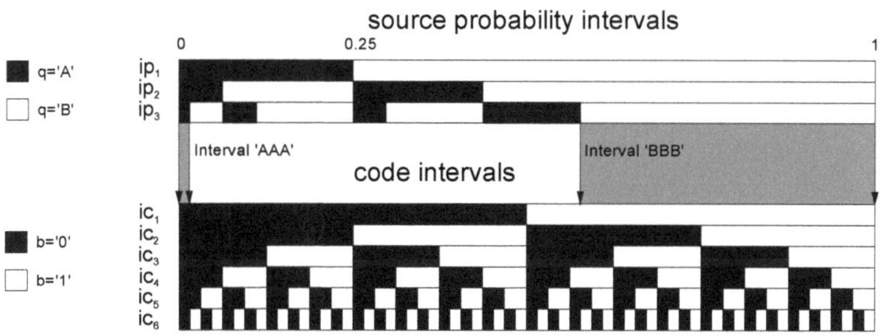

Fig. 4.13. Code intervals and probability intervals in the example discussed above

This example illustrates that in arithmetic coding a *variable number of source symbols* is mapped into a *variable number of code bits*. It can be observed that untypical sequences of source symbols, such as long chains of the less probable letter 'A', are encoded almost instantaneously without imposing a critical burden on the complexity. It is not necessary to parse code trees of large depth.

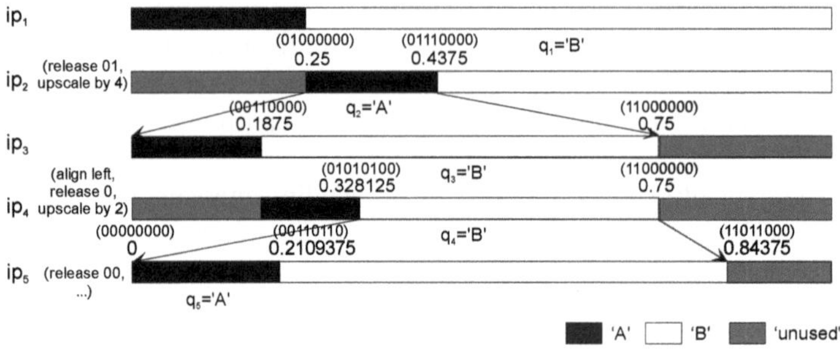

Fig. 4.14. Integer scaling, release of bits and alignment of probability intervals into the larger [0,1] range; the source sequence is 'BABBA'

As however the intervals would become increasingly narrow, the precise representation by a finite number of bits is not possible (and cannot even be derived with arbitrary arithmetic precision), such that discrete positions have to be used instead. In the sequel, a method is described which can be related with a uniform quantization of the interval boundary positions and integer arithmetic.

Condition (4.54) can easily be checked, when the probability interval boundaries are represented as integer numbers (covering the amplitude range 0...1) with bit depth $B \geq m$, where (4.54) is equivalent to the test that the difference between the m leading bits (starting from MSB) of the integer numbers representing $ip_{n,\mathrm{lo}}$ and $ip_{n,\mathrm{up}}$ is zero. Then, those m leading bits representing $ip_{n,\mathrm{lo}}$ can be instantaneously released. For decoding, it is sufficient to construct the probability interval boundaries and decode the source sequence up to the point of the corresponding interval and any interval above.

In practical realizations, neither the encoder nor the decoder need to construct the complete set of intervals; it is only necessary to track the path corresponding to the current source symbol sequence arriving at the encoder and of the associated bit sequence received by the decoder. One interpretation is that after each successful encoding or decoding step (i.e. when intervals are found to match), the recently-active probability interval is expanded to full scale [0,1] (see Fig. 4.14).The interval computations are to be performed by integer precision again. The step of interval expansion can implicitly be combined with the integer rounding of probability intervals[24]. After a probability interval corresponding to a source symbol has been determined by comparing the integer representation of its lower and upper boundaries, all leading bits can be released as described above. An example with an 8 bit unsigned integer accuracy is shown in Fig. 4.14, where

[24] For more detail on efficient implementation of arithmetic encoders and decoders, refer to [WITTEN, MCNEAL, CLEARY 1987] and [PENNEBAKER ET AL. 1988].

the relevant bit representations of probability intervals are also shown. With ip_2=[01000000, 01110000] the leading two bits match and can be released as representing the corresponding code interval. Subsequently, the remaining interval can be magnified by a factor of 4, which is performed by adding two zero bits to the integer representation. This results however in an unused range at values above 0.75. In principle, magnification and releasing of bits can be performed whenever the remaining probability interval range becomes less or equal 0.5. However, when an unused range is left of the remaining interval, it can be useful to left-align it, which is done with ip_4=[01010100,11000000] by subtracting the lower interval boundary, such that the new interval becomes [00000000,01101100], the leading zero bit matches and can be released, followed by magnification by a factor of two by inserting a zero bit. In the given example, Pr(A)=0.25 and Pr(B)=0.75, both of which have an error-free integer representation, such that interval quantization errors never occur. In general, it may not be possible to represent any probability value by their integer precision. If computation is performed using integer arithmetic of B bit precision, the rounding causes a deviation from entropy, which is typically implementation dependent (e.g. by the rounding strategy, for which other methods than the one presented here may be used.), but in worst case guarantees a margin of (see [PENNEBAKER ET AL. 1988])

$$R_e \geq H + 2^{2-B} \text{ bit.} \tag{4.57}$$

The 'loss' of two bits as compared to the smallest integer step size 2^{-B} can be explained due to the fact that rounding errors can occur at both lower and upper interval boundaries with maximum total deviation of 2^{-B} in interval width, and that the unused ranges after alignment also can be up to a width of 0.5, which effectively corresponds to losing precision of one more bit. Nevertheless, using sufficient arithmetic precision, the entropy is approached to a much smaller margin than usually achievable by Huffman or systematic code designs, while the arithmetic encoding and decoding processes are quite regular in implementation.

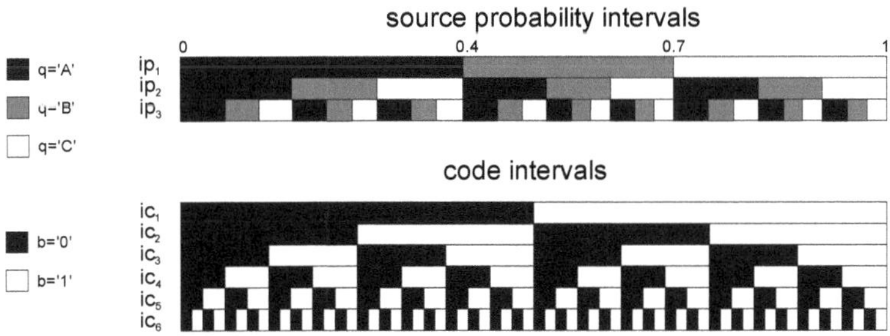

Fig. 4.15. Example of probability and code interval grids for a source with 3 different symbols

The method of arithmetic coding is not restricted to binary sources as discussed so far. When a source alphabet of an arbitrary finite number J of source symbols shall be encoded, the binary code intervals remain identical, whereas for a source alphabet of J symbols each step of sub-division will generate J new probability intervals. Fig. 4.15 shows an example of a three-letter alphabet with probabilities Pr('A')=0.4, Pr('B')=Pr('C')=0.3.

Unlike Huffman coding or other types of prefix-based variable length codes, it is not necessary to use or store a code table. The probability values are *directly used* to run the procedures of interval subdivision in encoding and decoding. This simplifies usage of *dynamically changing* probability values, which means that arithmetic encoding and decoding devices are *universally applicable* to any source statistics and alphabets, and *easily adaptable* for changing source statistics.

4.4.5 Adaptive and context-dependent entropy coding

In the previously described methods of entropy coding, the probability distribution of the source was assumed to be known a priori and should not be subject to change during encoding. For the case of stationary sources, this allows the design of an efficient code approaching the entropy rate. As multimedia signals include non-stationary components, this can lead to sub-optimum solutions. In worst case, it may happen that the rate is considerably increased by a badly designed variable-length code which does not fit the statistics of the source. This problem can be solved by *adaptive entropy coding*.

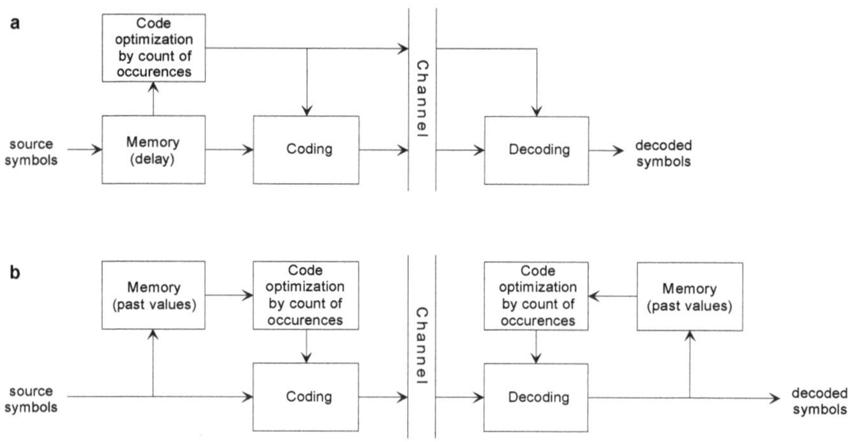

Fig. 4.16. Adaptive entropy coding: **a** forward adaptation **b** backward adaptation

The core approach for compression is not different from the methods described so far, but the code tables or probability parameters have to be adapted using the actual source characteristics. This can be done by one of the following principles:

- *Forward adaptation* (Fig. 4.16a): For a given segment of the source, occurrences are measured and used for code adaptation. Either the probability maps resulting from the measurements, or the related code parameters must be conveyed to the decoder as side information, which means they are part of the encoded representation. The disadvantages of this method are two-fold: The source signal has to be stored temporarily to determine the occurrence values, which causes a delay; the rate increase for code adaptation should be justified by the better compression.

- *Backward adaptation* (Fig. 4.16b): As entropy coding is lossless, both the encoder and the decoder can measure occurrences of source symbols over identical windows from *already decoded data.* If the signal does not show significant statistic variation, it can be used to determine the current statistics in adaptation of entropy coding. A disadvantage occurs in case of transmission errors, where encoder and decoder could run out of synchronization, using different parameters for adaptation, which can cause uncontrollable decoding errors. On the other hand, no transmission of side information is necessary.

In all algorithms for entropy coding introduced so far, first-order probabilities, joint or conditional probabilities can be used to design the codes. If conditional probabilities are used, it is possible to achieve a more efficient coding of sources in which subsequent source symbols or vectors thereof are *statistically dependent.* This is even applicable if they are not encoded as joint units. Context-dependent entropy coding is extensively used in the arithmetic coding parts of various coding standards such as JPEG 2000, AVC and HEVC.

Assume that a *context* $\mathcal{C}$ is established from previously decoded source symbols. In simplest case, this context could be the state of the directly preceding source symbol, which is then used to pre-determine the probabilities of values in the current state. The goal is then to approach the *conditional entropy* of the source alphabet S relating to the known context $\mathcal{C}$, which gives according to (2.178)

$$H(S\,|\,\mathcal{C}) = H(S) - I(S;\mathcal{C}) \le H(S)\,. \tag{4.58}$$

When the encoder output represents K different information components, multiple contexts $\mathcal{C}_k$, k=1...K, can be required, each requiring M_k different conditional probability values. At the time of encoding and decoding, it is then necessary to switch to the context and its associated probability values, depending on the next information component to be decoded. Furthermore, in *context adaptive entropy coding*, adaptation of the context definitions or context selection can be done depending on the source properties, or on previously decoded source symbols. Even though the concepts and implementations of coding and decoding procedures are not changed in principle, this imposes additional effort in managing the code tables (as in Huffman coding) or probability tables (as in arithmetic coding), and in retaining synchronization between encoder and decoder.

An alternative way of encoding multiple-letter alphabets by entropy coding is a two-step procedure, where in a first step of *binarization* a unique mapping into an intermediate alphabet of binary codewords is made, and the binary letters ('bins') are compressed close to the entropy in a second step of context-dependent and adaptive binary coding. As only the probability of the next bin is derived from its binary context, the analysis of statistical dependencies becomes simpler than with multi-letter alphabets, and identical procedures can be applied to various components of different meaning. The step of *binarization* can include the following approaches:

- The integer representation of a signal $s(\mathbf{n})$ is separated into sign and magnitude as $|s(\mathbf{n})|=b_0(\mathbf{n})\cdot 2^0+b_1(\mathbf{n})\cdot 2^1+\ldots+b_{B-1}(\mathbf{n})\cdot 2^{B-1}$, $s(\mathbf{n})=|s(\mathbf{n})|\cdot \mathrm{sgn}(\mathbf{n})$. The B bit-plane values $b_i(\mathbf{n})$ establish a binary representation, where the probabilities of an element being '0' or '1' can be related to contexts of values from the next-higher bit plane or previously decoded values in the same bit plane. This concept is e.g. applied in the EBCOT algorithm of JPEG 2000 (cf. Sec. 6.4.4).
- Representation by a simple VLC which systematically leaves statistical dependencies after the mapping of source symbols into binary codewords. Examples for this are the universal and systematic VLC methods described in Sec. 4.4.3. By defining contexts either within or across binary codewords, a context-dependent binary arithmetic coder is able to further compress such representations significantly. An example for this is the CABAC algorithm which is subsequently described.
- Furthermore, the binarization itself can be switched, depending on the amplitude level. For example, the number of suffix bits used in a Golomb-Rice code could be varied depending on the amplitude level or even in a signal dependent fashion, when it is found that a larger number of different source symbols have similar probability.

Context-adaptive binary arithmetic coding (CABAC) [MARPE ET AL. 2003] is a core element of the AVC and the HEVC video compression standards. The scheme can universally be applied to different components and syntax elements representing the source signal (coding modes, motion vectors, transform coefficients etc.). As shown in Fig. 4.17, CABAC is applied after binarization, using context modeling of the resulting 'bins', and binary arithmetic coding at the core. Binarization maps non-binary syntax elements to sequences of bits referred to as bin strings. The bits of a bin string can then either be processed in an arithmetic coding mode or in a *bypass* mode, for those where the benefit of arithmetic coding would be marginal (e.g. sign information, or bins for which the binarization itself already provides efficient entropy coding). Context modeling can be characterized such that for a given state of information available at the decoder (e.g. values a/b from neighbored blocks, which have to be determined from previously decoded bin), the probability of the next bin to be '0' or '1' is determined. In arithmetic coding mode, the arithmetic coding engine simply uses the probability as determined from the context model when encoding the bin. In many cases, only the most probable values of a syntax element (e.g. low amplitudes which are expressed by one or few leading bits of a UVLC binarization) are actually encoded using external context modeling. The compression performance of the arithmetic-coded bins is continuously improved by updating of the corresponding probability estimate with each occurrence of a context model, based on the bin value (see below). The bypass coding engine inserts the bypass coded bins in the bit stream, maintaining the probability interval ranges of the arithmetic coding engine in a simplified way by inserting intervals of width 0.5, and still doing up-scaling and left alignment of interval ranges when bits are released (similar to Fig. 4.14). By this, it is avoided to terminate and re-start the arithmetic coding engine each time when bins are coded in bypass mode.

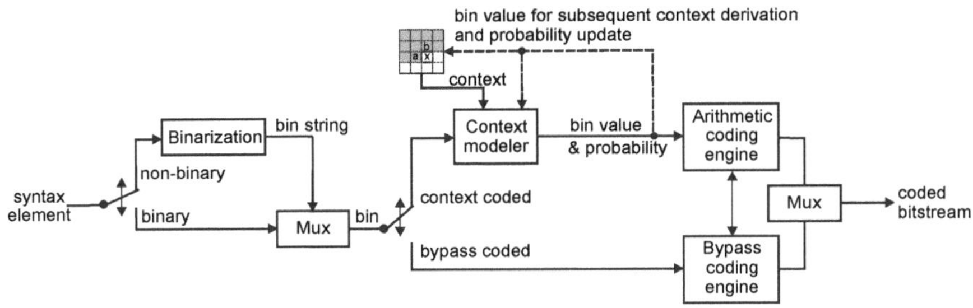

Fig. 4.17. CABAC block diagram

Another simple approach for adaptive entropy coding without redesign of code tables (and therefore also easily applicable with Huffman coding and other approaches relying on code table mapping) is *codeword swapping*. Here, re-allocation of syntax elements to codewords is performed depending on probability counts, such that the most frequent source alphabet letters are assigned with shortest codewords. This can also be combined with context-based coding. As however the code is pre-designed, there is no guarantee that the codeword lengths match the self information of all source letters; therefore such approaches are usually inferior compared to the more flexibly changeable arithmetic coding approaches.

Methods for backward-adaptive probability estimation. For adaptive entropy coding, it is necessary to determine the probabilities of the different source symbols, potentially in their context dependency. This is discussed here for the example of binary coding (two-letter alphabet) by counting the frequency of occurrence. Assume that $\mathrm{Ct}(0|\mathbf{b};\mathcal{C}_k)$ and $\mathrm{Ct}(1|\mathbf{b};\mathcal{C}_k)$ are the zero- and one-occurrence counts found with the context $\mathcal{C}_k$ having a value of **b**, accumulated from past values in case of backward estimation. The *scaled-count estimate* of the probability is

$$\hat{\Pr}(0\,|\,\mathcal{C}_k=\mathbf{b}) = 1-\hat{\Pr}(1\,|\,\mathcal{C}_k=\mathbf{b}) = \frac{\mathrm{Ct}(0\,|\,\mathcal{C}_k=\mathbf{b})+\Delta\cdot\Pr_{\mathrm{init}}(0\,|\,\mathcal{C}_k=\mathbf{b})}{\sum_{j=0}^{1}\left[\mathrm{Ct}(j\,|\,\mathcal{C}_k=\mathbf{b})+\Delta\cdot\Pr_{\mathrm{init}}(j\,|\,\mathcal{C}_k=\mathbf{b})\right]} \quad (\Delta>0)\,. \tag{4.59}$$

The values $\Pr_{\mathrm{init}}$ effect an initialization of the probability estimate, which will give a reasonable result even when no single instance was found previously. These are reflecting an initial assumption about the probabilities of zero- and one-values in case of binary coding[25]. In fact, it is only necessary to estimate probabilities of either zero- or one-values. The factor Δ controls the speed of the adaptation. For higher Δ, the probability estimate will roughly retain the initial value unless a significant number of counts is found.

The concept straightforwardly extends to non-binary sources with J different source symbols and context state $\mathcal{C}_k=\mathbf{x}$:

$$\hat{\Pr}\left(S_j \mid \mathcal{C}_k=\mathbf{x}\right)=\frac{\operatorname{Ct}\left(S_j \mid \mathcal{C}_k=\mathbf{x}\right)+\Delta \cdot \Pr_{\text{init}}\left(S_j \mid \mathcal{C}_k=\mathbf{x}\right)}{\sum_{i=1}^{J}\left[\operatorname{Ct}\left(S_i \mid \mathcal{C}_k=\mathbf{x}\right)+\Delta \cdot \Pr_{\text{init}}\left(S_i \mid \mathcal{C}_k=\mathbf{x}\right)\right]} . \tag{4.60}$$

The probability estimation methods of (4.59) and (4.60) require to maintain a counter for each possible state of each context and re-normalize the result for an update of the probability after each occurrence of that state.

Even though in case of probabilities for binary sources (4.59) it is sufficient to maintain the counters only for one of the two states ('0' or '1') of the source, this can still be memory-consuming when the number of contexts is large. The probability estimate can also be implemented via a table lookup, where only a finite number of discrete 'probability states' are updated with each occurrence of the less probable or more probable source symbol (bin) for a given context. An approach with 64 discrete probability states is used in the implementation of AVC's and HEVC's CABAC [MARPE ET AL. 2003] This way, the memory for storing the current probability estimate is limited to 6 bit per context model, and the computation becomes much simpler by replacing multiplications and divisions by predefined next-state transition updates, depending on each occurrence of the less probable bin (LPB) or the more probable bin (MPB).

4.4.6 Entropy coding and transmission errors

Variable length codes are rather sensitive in case of transmission errors. By deletion or flipping of one single bit, a wrong interpretation can occur such that a different prefix is recognized, which may belong to a codeword of different length. The problem likewise affects Huffman-type VLCs and arithmetic codes[26]. In principle, the wrong interpretation can propagate from one code symbol to the next, until finally an invalid code symbol is identified, or until it is recognized that the number of decoded source symbols is larger than expected. In both cases, it is

---

[25] When the probability initialization state is explicitly signaled as side information, the scheme would in principle be a combination between forward and backward adaptation from the decoder's perspective; however, such an initialization can also be used as part of a re-synchronization mechanism, where in order to re-start synchronous encoder and decoder operation after data losses, the encoder explicitly conveys the probability state (as learned from a backward adaptation) from time to time.

[26] Systematic VLCs (cf. Sec. 4.4.3) can be designed such that effects of error propagation are minimized, or that partial recovery can be performed at the decoder.

almost impossible to identify the position of the original erasure, such that often the whole decoded sequence needs to be considered as invalid. The following example explains different possible cases in the context of a prefix-free code.

Example: Propagation of errors in case of a variable-length code of prefix-free (truncated unary) type. The code string representation of four source symbols shall be 'A'→'0', 'B'→'10', 'C'→'110' und 'D'→'111'. An example source sequence may be 'ABCD', such that the bit string '0|10|110|111' is sent. In the following cases, the respective distorted bit is underlined.

Case A: Received '011110111', interpretation of the decoder: '0|111|10|111' ⇔ 'ADBD'. 2 letters are wrong, however the error is occasionally compensated, i.e. the last letter 'D' is again decoded correctly. No error is detected by the decoder, as the code stream can be separated into valid codewords. The number of decoded source symbols is correct as well.

Case B: Received '110110111', interpretation of the decoder: '110|110|111' ⇔ 'CCD'. All decoded symbols are wrong, the error is not detected as the code stream can be separated into valid codewords, unless the decoder would know to expect 4 symbols. In fact, the last letter 'D' is decoded correctly, but it may be wrongly interpreted as it is appearing at the wrong position.

Case C: Received '010010111', interpretation of the decoder: '0|10|0|10|111' ⇔ 'ABABD'. Similar to case *B*, but the number of letters is too high.

Case D: Received '010110101', interpretation of the decoder: '0|10|110|10|1' ⇔ 'ABCB'. The last letter is wrong, however the error is recognized, as the bit stream cannot entirely be associated with valid codewords, the last bit remains isolated.

In some cases, the decoding process could synchronize again with a correct starting bit of a valid codeword string, as in cases *A*-*C*. Before that occurs, it is however quite probable that a wrong number of source symbols are decoded (cases *B* and *C*). To detect and at least partially recover from errors, one of the most effective counteractions is the introduction of *resynchronization* mechanisms. A resynchronization code consists of a bit string which has to be unique, i.e. will not be allowed to be part of any other string of the variable length code, nor be reproducible by concatenation of different code strings. For systematic construction, it is often defined as a long string of only '0' or '1' bits, which can easily be forbidden within other codewords by systematic construction of the code[27]. By using resynchronization, it is possible to detect the occurrence of transmission errors with a high reliability, and re-start the decoding process from that point.

---

[27] A simple approach avoiding start code emulation with a start code of Nx'0' is inserting a stuffing bit '1', when the next codeword could lead to a sequence with more than N '0' bits. The stuffing bit can then be identified and ignored in normal decoder operation.

4.4.7 Lempel-Ziv coding

Lempel-Ziv (LZ) codes are used for universal lossless compression. Due to flexible adaptation mechanisms, these are in principle applicable for compression of any kind of digital data. For generic compression of multimedia signals, LZ codes are of limited value however, as mechanisms that are directly using models of signal statistics can typically achieve better results; further, for *lossy* compression, a tight integration of quantization and entropy coding would be required.

LZ codes are often applied to compress text files, the variant LZ78 is also used in the modem standard V.42 and in the graphics/image representation format GIF. Technology-wise, LZ coding takes a different approach than the entropy coding methods introduced so far, as *fixed-length codewords* are used to encode source symbols combined into vectors of *variable length*. This is achieved by installing a *dictionary* which contains source symbol vectors of high expected probability. In particular the adaptive modifications of LZ are relevant: LZ77 (having fixed length of the dictionary, but automatic adaptation of its content) [ZIV, LEMPEL 1977], LZ78 [ZIV, LEMPEL 1978] and LZW [WELCH 1984], supporting dynamic adaptation and growth of dictionary size. Here, a sliding window over past values is used to update the dictionary from occurrence counts of recently decoded source symbols (see Fig. 4.18). Source symbol vectors (words) which cannot be found in the dictionary are encoded as single letters by a fixed-length code that is following an 'ESCAPE' symbol.

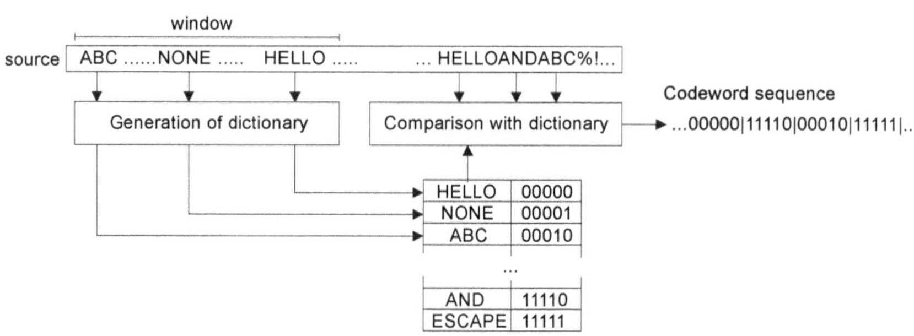

Fig. 4.18. Principle of dictionary adaptation in Lempel-Ziv-Welch coding

Due to the fixed length of codewords, error propagation in case of transmission errors cannot occur by wrong identification of codeword starting points, but the number of decoded source symbols may be wrong, and a wrong adaptation of the dictionary can be caused at the decoder. As source symbols are combined in variable-length vectors, the reconstructed source sequence may indeed have less or more elements than the original, which can be identified by the decoder, when the expected number of source symbols does not match the number of decoded symbols.

The different entropy coding methods can be categorized by the way *how* they obtain variable codeword lengths which are as close as possible to the self-information of the source symbols. From a high-level view, it is possible to use *fixed or variable length of source-symbol vectors* and/or *fixed or variable length of codewords*. The overview given in Table 4.4 shows that the methods of run-length coding and arithmetic coding exploit most flexibility, as both the input and output of the code mapping are based on variable-length methods. This may be one reason for the competitive performance of these approaches. One fundamental strategy in obtaining low complexity is to concentrate the effort on source states with high probability and try to reach the self information of these as close as possible, which usually is only achievable if they are combined as vectors; on the other hand, extremely rare 'untypical' source states are not causing significant penalty, even when they are not coded close to the rate of their self information.

Table 4.4. Properties of different entropy-coding methods

	Source vectors	
Codewords	fixed-length	variable-length
fixed-length	Fixed-length coding (FLC), systematic VLC with 'escape' condition	Run-length coding with FLC, LZ coding
variable-length	Shannon & Huffman coding, systematic VLC with 'typical' condition	Run-length coding with VLC, arithmetic coding

4.5 Vector quantization

Vector quantization (VQ) combines samples for joint encoding already during the quantization step. Fig. 4.19 illustrates how vectors can be formed from 1D or 2D signals. If groups of K samples are combined into vectors, a sequence of vectors $\mathbf{s}(\mathbf{m})$ is available for quantization and encoding. In case of 1D signals, the vectors are often composed by natural sample order, $\mathbf{s}(\mathbf{m})=[s(mK),s(mK+1),\ldots, s(K(m+1)-1)]^{\mathrm{T}}$, where $m=\lfloor n/K \rfloor$ as shown in Fig. 4.19a. Other methods of vector formation are possible, e.g. alternating or interleaved packing of samples into vectors[28]. In VQ of 2D image signals, vectors are mostly defined as rectangular or quadratic blocks of dimension $K=M_1 \mathrm{x} M_2$ (see Fig. 4.19b). VQ can not only be applied directly to the signal, but can also be used for representation of transform

[28] This can be advantageous for purposes of scalability, robust transmission etc., but will reduce the compression efficiency in case of signals where statistical dependency between direct neighbors is high.

coefficients, prediction error signals, filter coefficients, motion vectors or other parameters which have to be encoded.

4.5.1 Basic principles of VQ

A typical block diagram of vector quantization is shown in Fig. 4.20. A *codebook* $\mathcal{C}=\{\ \mathbf{y}_j\ ;j=1,..,J\ \}$ contains a set of J different K-dimensional *codewords* (reconstruction vectors) $\mathbf{y}_j=[y_j(0),y_j(1),...,y_j(K-1)]^{\mathrm{T}}$. The encoder has to find the codeword $\mathbf{y}_i$ which is most similar to an incoming vector $\mathbf{s}\equiv\mathbf{s}(\mathbf{m})$. The output of the encoder is the index $i\equiv i(\mathbf{m})$ associated to the codeword; if necessary, the discrete index values can be subject to further entropy coding[29]. If the Euclidean distance is used as a distortion metric,

$$i=\underset{\{j|\mathbf{y}_j\in\mathcal{C}\}}{\arg\min}\left[\mathrm{d}_2^2(\mathbf{s},\mathbf{y}_j)\right]\ \text{ with }\ \mathrm{d}_2^2(\mathbf{s},\mathbf{y}_j)=\sum_{k=0}^{K-1}\left[s(k)-y_j(k)\right]^2=\left[\mathbf{s}-\mathbf{y}_j\right]^{\mathrm{T}}\left[\mathbf{s}-\mathbf{y}_j\right]. \quad (4.61)$$

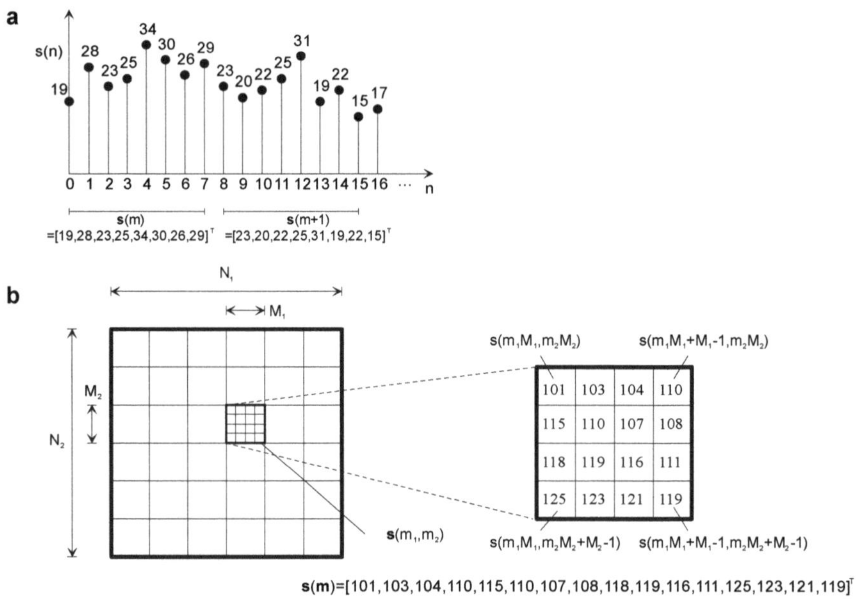

Fig. 4.19. Examples for combinations of samples into vectors **a** in 1D signals **b** in 2D signals

[29] Additional constraints can be imposed during codebook design, such as equal probability of codeword usage over the whole set, or optimization based on entropy constraints (cf. Sec. 4.3). If fixed-length coding of indices is used, high robustness is achieved in particular when transmission is made over error-prone channels. Transmission errors will then not take influence on subsequent source symbol reconstructions, as it would usually be the case when variable-length codes are used.

The decoder uses the index i to pick the vector $\mathbf{y}_i$ from an identical codebook; the output $\tilde{\mathbf{s}}(\mathbf{m}) = \mathbf{y}_{i(\mathbf{m})}$ is then placed in the reconstructed signal at position $\mathbf{m}$. The vectors $\mathbf{s}$ and $\mathbf{y}$ can be interpreted as points within a K-dimensional Euclidean signal space $\mathcal{R}^K$. The norms $\|\mathbf{s}\|_2$ and $\|\mathbf{y}\|_2$ express distances from the origin of $\mathcal{R}^K$. The square root of (4.61) is the distance between $\mathbf{s}$ and $\mathbf{y}$. For the case of two-dimensional vectors this is illustrated in Fig. 4.21.
The most computational demanding task of the VQ encoder is the search for the optimum reconstruction $\mathbf{y}$. The complexity of the decoder is very low (table lookup using the index), whereas the memory necessary to store the codebook depends on its size.

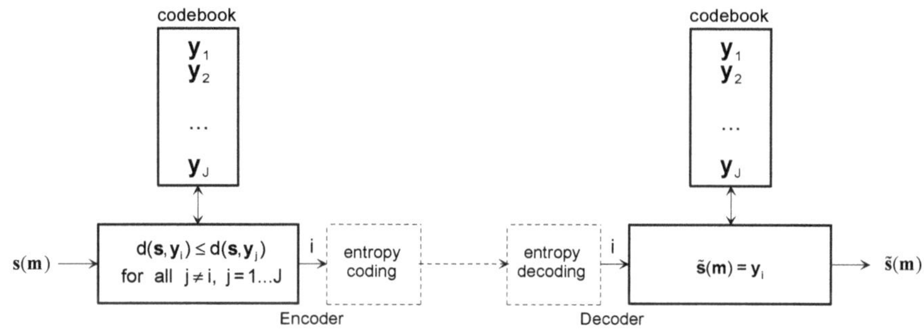

Fig. 4.20. Structure of a vector quantizer (encoder and decoder)

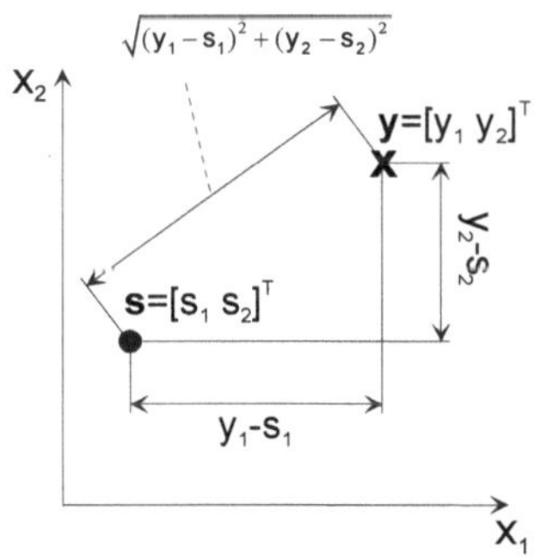

Fig. 4.21. Interpretation of the Euclidean distance between vectors **s** and **y** in $\mathcal{R}^2$.

Fig. 4.22 shows an example layout of a codebook in $\mathcal{R}^2$ by the positions of the reconstruction vectors $\mathbf{y}_j$. From (4.61) it is possible to determine the boundaries of J distinct partitions in $\mathcal{R}^2$ which are associated with the codewords $\mathbf{y}_j$. For the case of a Euclidean distance criterion, these partitions are denoted as the *Voronoi regions*. The net of boundaries is the *Voronoi net*. It is a complementary representation of the *Delaunay net*, which is constructed from the connecting lines between direct neighbors $\mathbf{y}_j$. The lines of the Voronoi net are perpendicular with the lines of the Delaunay net, and divide the latter into equal halves. The Euclidean distance from any $\mathbf{s}$ positioned on the Voronoi net towards any of the adjacent $\mathbf{y}_j$ is exactly equal.

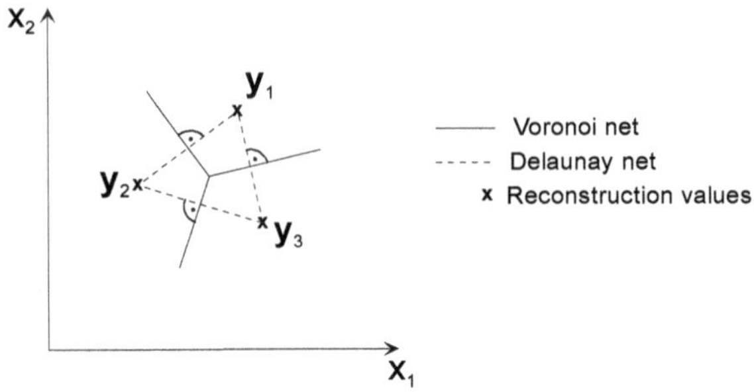

Fig. 4.22. Voronoi net and Delaunay net in $\mathcal{R}^2$, example codebook of J=3 codewords $\mathbf{y}_j$

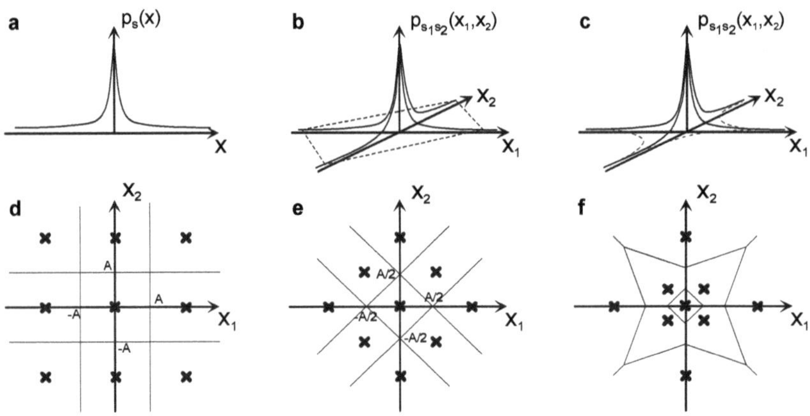

Fig. 4.23. Interpretation about the efficiency of VQ: **a** Example PDF of a signal **b-c** joint PDF of two statistically independent samples (GGD with $\gamma=1$ (**b**) and $\gamma<1$ (**c**)) **d** Voronoi regions for a scalar quantizer **e-f** Voronoi regions of codebooks with same number J=9, uniformly (**e**) and non-uniformly distributed (**f**) reconstruction values.

One advantage of VQ, as compared to a scalar quantization using the same number of reconstruction values, is illustrated in Fig. 4.23. Fig. 4.23a shows the non-uniform PDF $p_s(x)$ of a signal, which could e.g. be a generalized Gaussian distribution (GGD) (2.126). Assuming that samples of a stationary process are statistically independent, and two samples over random variables x_1 and x_2 are combined into a random vector variable $\mathbf{x}$, the resulting joint (vector) PDF is $p_\mathbf{s}(\mathbf{x})=p_s(x_1)p_s(x_2)$. This is shown in Fig. 4.23b/c for the case $\gamma=1$ (Laplacian) and another case with $\gamma<1$. If a scalar quantizer with $J=3$ is applied to the samples separately, possible reconstruction value positions are as shown in Fig. 4.23d, some of which are placed at positions where both samples' amplitudes would be large, which is less probable from the joint PDF. A vector quantizer has more degrees of freedom to place reconstruction values at positions with higher probability, by establishing interdependencies between the values. E.g., if $p_s(x)$ is Laplacian, $p_s(\mathbf{x})$ would be equal at positions $|x_1|+|x_2|$=const., for which the codebook layout in Fig. 4.23e is a better choice. This is still uniform in terms of having equal distances between neighbored reconstruction values, but can no longer be described by independent scalar quantization[30]. Such constellations with equally-shaped Voronoi cells are classified as *uniform* or *Lattice VQ* codebooks. If it is even less likely that two high amplitudes occur jointly, as for the GGD with $\gamma<1$, the configuration in Fig. 4.23e would be even better suitable. Its reconstruction values are however *non-uniformly* distributed. The fact that VQ can achieve better performance by concentrating reconstruction values in subspaces of higher probability is denoted as *codebook shape advantage*. For example, in case of correlated Gaussian distributed sources the outer limitation should be a K-dimensional hyper ellipsoid, with special case of a hyper sphere for uncorrelated samples. For uncorrelated Laplacian distributed sources, equal probabilities can be found on planes where $\Sigma|x_k|$=const., the corresponding codebook scheme (in case of uniform codebooks) has been denoted as *pyramid VQ* in [FISCHER 1986].

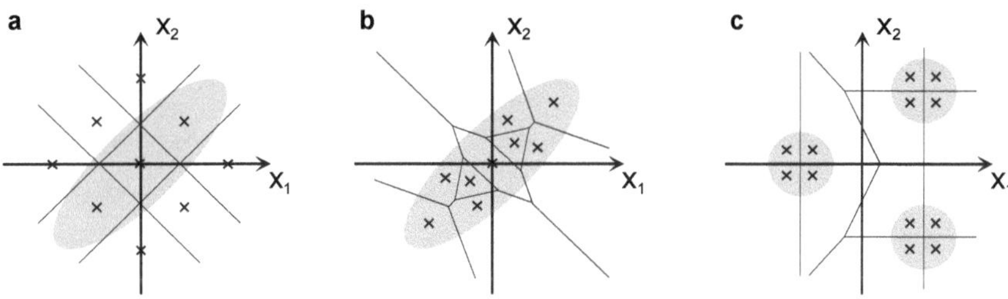

Fig. 4.24. Capability of VQ to exploit statistical dependencies: **a** Layout of a uniform codebook (Voronoi regions of highest probability in case of correlated signals are shaded) **b** Layout of a non-uniform codebook for correlated signals **c** Layout of a non-uniform codebook for an example of nonlinear statistical dependencies in a mixture distribution

Assume that values within a vector are instantiations of the same stationary process (e.g. characterized by a Gaussian vector PDF (2.156)) and that they are correlated (linearly statistically dependent). Then, clusters of random vectors around the principal axis $x_1=x_2=\ldots$ within $\mathcal{R}^K$ can be expected. This can be considered in the design of entropy coding for the *uniform* codebook VQ in Fig. 4.23d, where shorter VLC codewords can be used to encode the indices associated with the shaded Voronoi regions in Fig. 4.24a. A *non-uniform* codebook can also directly be optimized such that the correlation within the signal or the shape of the joint/vector PDF is taken into account by placing reconstruction values only in highly probable subspaces (Fig. 4.24b). This includes the capability to exploit

---

30 In fact, the same set of reconstruction values would be possible by application of a Haar transform with U=2, which rotates the coordinate axis of $\mathcal{R}^2$ by 45^0, and performing subsequent scalar quantization of transform coefficients; see Problem 4.7

nonlinear statistical dependencies, where global covariance may be zero, but $p_s(x_1,x_2) \neq p_s(x_1)p_s(x_2)$; an example of a non-uniform codebook optimized for a mixture distribution is shown in Fig. 4.24c.

If a codebook of J reconstruction values is used, it is possible to represent the information by a rate of $\log_2 J$ bit/vector or $(\log_2 J)/K$ bit/sample without entropy coding. In the subsequent sections, the principles of designing vector quantization codebooks with uniform and non-uniform characteristics are described.

4.5.2 VQ with uniform codebooks

If the codebook describes a regular grid of reconstruction values, the VQ scheme is denoted as *uniform-codebook* or *lattice VQ*[31]. For the following illustrations the two-dimensional space $\mathcal{R}^2$ is used. Fig. 4.25a shows a rectangular grid of reconstruction values, the lattice $\mathbf{Z}_2$. Reconstruction values represent a 2-dimensional grid of integers. This expresses the vector counterpart of a uniform scalar quantizer, where the axes would be scaled by the quantizer step size Δ. Fig. 4.25b shows the lattice $\mathbf{A}_2$, which for K=2 describes the optimum uniform VQ, showing Voronoi regions of hexagonal structure[32].

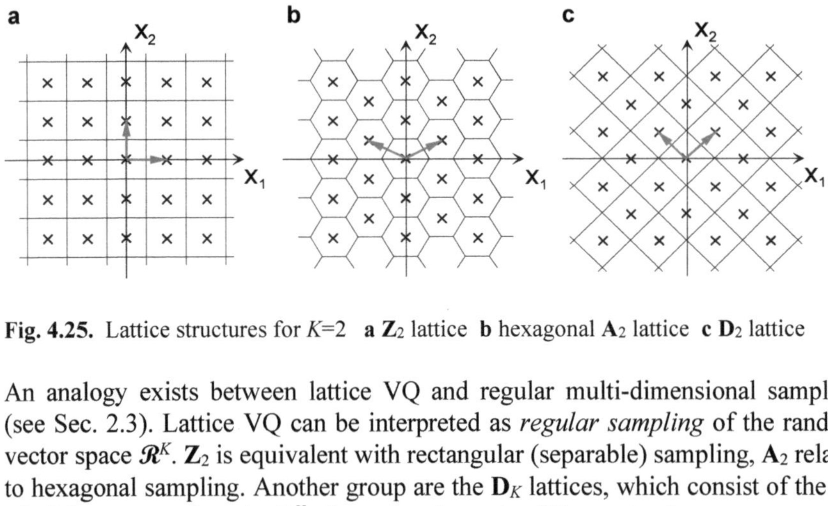

Fig. 4.25. Lattice structures for K=2 **a** $\mathbf{Z}_2$ lattice **b** hexagonal $\mathbf{A}_2$ lattice **c** $\mathbf{D}_2$ lattice

An analogy exists between lattice VQ and regular multi-dimensional sampling (see Sec. 2.3). Lattice VQ can be interpreted as *regular sampling* of the random vector space $\mathcal{R}^K$. $\mathbf{Z}_2$ is equivalent with rectangular (separable) sampling, $\mathbf{A}_2$ relates to hexagonal sampling. Another group are the $\mathbf{D}_K$ lattices, which consist of the set of all integer numbers in $\mathcal{R}^K$ where the elements of the vector have an even sum. $\mathbf{D}_2$ (Fig. 4.25c) is equivalent with quincunx sampling in $\mathcal{R}^2$.

---

[31] An excellent reference on lattice VQ is [CONWAY, SLOANE 1988].

[32] This lattice, for a given number of reconstruction values (or given area of the elementary Voronoi regions) describes the densest packing within 2D, such that the average distance between the centroid of the Voronoi region and points falling into this region is smaller than in the case of $\mathbf{Z}_2$.

For any vector size K, an optimum lattice should provide the densest packing of identically-shaped Voronoi cells. The shape of the Voronoi cell would then be the best possible approximation of a K-dimensional hyper sphere (where cells truly shaped as hyper spheres would *not* allow seamless packing, empty spaces would occur between the spherical surfaces). Generally, the approximation error as compared to a hyper sphere decreases by increasing the vector dimension K. Optimum lattices can often systematically be determined from error-correcting block codes, which have the property of equal Hamming distances between pairs of nearest valid codewords. For example, the lattice $\mathbf{\Lambda}_{16}$ which is optimum for K=16 is constructed from a Reed-Muller code, and the lattice $\mathbf{\Lambda}_{24}$ is derived from an extended Golay code [CONWAY, SLOANE 1988].

The regular lattice structure of reconstruction values can be described by a *generator matrix* $\mathbf{G}=[\mathbf{g}_0\ \mathbf{g}_1]$, which for the examples of lattices $\mathbf{Z}_2$, $\mathbf{A}_2$ and $\mathbf{D}_2$ can be defined as[33]

$$\mathbf{G}_{\mathbf{Z}_2}=\begin{bmatrix}1 & 0\\ 0 & 1\end{bmatrix}=\mathbf{I}\ ;\ \mathbf{G}_{\mathbf{A}_2}=\frac{2}{\sqrt{3}}\begin{bmatrix}\frac{\sqrt{3}}{2} & -\frac{\sqrt{3}}{2}\\ \frac{1}{2} & \frac{1}{2}\end{bmatrix}\ ;\ \mathbf{G}_{\mathbf{D}_2}=\frac{\sqrt{2}}{2}\begin{bmatrix}1 & 1\\ -1 & 1\end{bmatrix}. \quad (4.62)$$

As in the sampling matrices $\mathbf{T}$ of (2.60), the columns of $\mathbf{G}$ are the *basis vectors* $\mathbf{g}_k$ of the lattice, for which the set of reconstruction values can be obtained by a K-dimensional vector of integer index values i_k, k=1...K by

$$\mathbf{\Lambda}=\{\mathbf{y}:\mathbf{y}=i_1\mathbf{g}_1+i_2\mathbf{g}_2+...+i_K\mathbf{g}_K\}\Rightarrow\mathbf{y}=\mathbf{G}\mathbf{i},\ \mathbf{i}=\begin{bmatrix}i_1 & i_2 & \cdots & i_K\end{bmatrix}^{\mathrm{T}}. \quad (4.63)$$

As the number of discrete points defined by $\mathbf{\Lambda}$ could in principle be infinite, it is necessary to limit the range of values i_k for encoding. Then, a codebook with finite number of reconstruction values $\mathbf{y}_i$ is obtained. The encoding step of lattice VQ is rather simple, as the generator matrix can also be interpreted to define a transformation of the coordinates in $\mathfrak{R}^K$. For quantization, the following strategies can be applied (also in combination):

- The inverse transformation $\mathbf{G}^{-1}$ is applied to the signal values, which could then be quantized in a $\mathbf{Z}_K$ lattice (i.e. scalar) and give the index values. This however would only be suitable if $\mathbf{G}$ establishes an orthogonal basis, as otherwise the Euclidean distance is not preserved and the shape of the Voronoi regions is modified by the mapping.

---

33 Several definitions of *dual lattices* often exist, which are addressing the same lattice grids in principle, but would lead to another range of quantization when addressed by a finite set of index values $\mathbf{i}$ as in (4.63). Observe that $|\mathbf{G}|$ determines the areas or volumes of the Voronoi cells, which have been normalized to unity in all cases of (4.62). Fast algorithms for quantization however do not need to perform the scaling by the irrational numbers.

- The grid of reconstruction values is decomposed into I rectangular-sampled 'cosets'. In a first step, the optimum value for each coset is found by scalar quantization, in a second step, only I optimum values have to be compared by explicit computation of the Euclidean distance relative to the vector **s** that is to be quantized.

For most common types of lattices (in particular for the optimum lattices in K=2,3,4,8,12,16 und 24), efficient algorithms are known for fast quantization. Decoding can be performed directly by (4.63).

A lattice VQ with densest packing of Voronoi cells for a given K is optimum in the sense that for a given volume of the cell, the average Euclidean distance between the centroid and points in the cell is minimized when the shape of the cell is as close as possible to a sphere. This is called the *space filling advantage* of vector quantization, which along with the previously mentioned *codebook shape advantage* provides a second reason why vector quantization is generally better than scalar quantization with subsequent entropy coding (even when the latter is done by combining separately quantized scalars into vectors).

To still better adapt a lattice quantizer for encoding of signals with non-uniform PDF, *companding* of signal values can be performed, One approach is to apply a nonlinear amplitude-mapping function to the signal samples prior to quantization, and perform inverse mapping after reconstruction from the points of the uniform lattice [SIMON 1998]. The equivalent vector quantizer with non-uniform distribution of the reconstruction values is shown in Fig. 4.26b. Another method is encoding the amplitude gain a of the vector separately by a non-uniform scalar quantizer, and encode positions on the K–1 dimensional spherical hull of radius a by mapping them into the reconstruction grid of a K–1 dimensional lattice VQ, as shown in Fig. 4.26c [KRÜGER ET AL. 2011]. Both solutions do not provide all degrees of freedom that are given by generic non-uniform VQ codebooks, but have the advantage of being less complex.

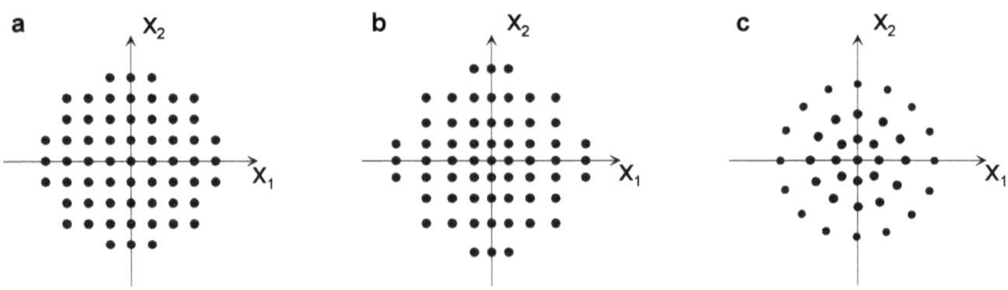

Fig. 4.26. Structures of lattice VQ codebooks with spherical shapes.
a uniform lattice **b** non-uniform lattice by companding **c** non-uniform lattice with separate gain factor for sphere radii

4.5.3 VQ with non-uniform codebooks

To design codebooks for non-uniform VQ, the *generalized Lloyd algorithm* (GLA), also known as *Linde-Buzo-Gray algorithm* [LINDE, BUZO, GRAY 1980] or one of its numerous variants are usually applied. The principle is a vector extension of the Lloyd algorithm for designing optimum scalar quantizers as described in (4.16)-(4.22). For VQ design, a *training set* is often used instead of a model distribution, which provides the advantage of implicit modeling the empirical statistics of the source (including dependencies between samples in a vector). Unlike the scalar case, the GLA does not by guarantee converge to a global optimum, since the Euclidean distortion function which is minimized can have multiple local minima due to the multi-dimensional dependency.

The training set has to be selected carefully, showing the typical occurrence and variety of signals which later shall be encoded by the quantizer. Otherwise, the codebook may be over-trained, and the quality could suffer when data outside of the training set are encoded.

As in the original Lloyd algorithm, the GLA invokes an iterative procedure. Within one iteration, let $S_j=\{{}^i\mathbf{s}_j\}$ denote the countable set of M_j vectors, which fall into the Voronoi region of $\mathbf{y}_j$. Instead of using statistical expectation as in (4.16)-(4.22), the counts of values are used in our example as empirical probability estimates, such that

$$D_j=\frac{1}{M_j}\sum_{i=1}^{M_j}\sum_{k=1}^{K}({}^is_j(k)-y_j(k))^2 \quad ; \quad \hat{\Pr}\left(S_j\right)=\frac{M_j}{\sum_j M_j}. \tag{4.64}$$

Deriving D_j over $\mathbf{y}_j$ gives the reconstruction vector which minimizes the distortion[34],

$$y_{j,\mathrm{opt}}(k)=\frac{\sum_{i=1}^{M_j}{}^is_j(k)}{M_j} \quad \Rightarrow \quad \mathbf{y}_{j,\mathrm{opt}}=\frac{\sum_{i=1}^{M_j}{}^i\mathbf{s}_j}{M_j}. \tag{4.65}$$

$\mathbf{y}_{j,\mathrm{opt}}$ is the arithmetic mean over the vectors in S_j or asymptotically, $\mathbf{y}_{j,\mathrm{opt}}=\mathcal{E}\{\mathbf{s}_j\}$. An example how the algorithm works with a small training set is illustrated in Fig. 4.27 for a vector quantizer of K=2 and J=4. Training data samples ${}^i\mathbf{s}_j$ are marked as '•', reconstruction values $\mathbf{y}_j$ as '**x**'. For initialization, a symmetric uniform codebook is used here (Fig. 4.27a); in this case, the boundaries of the Voronoi regions are the coordinate axes of $\mathcal{R}^2$. Replacing the $\mathbf{y}_j$ by the mean values (centroids) of the samples in the previous Voronoi regions causes modification of the Voronoi net (Fig. 4.27b). By this, some training vectors may be re-allocated to a different set $\{\mathbf{s}_{j'}\}$. Hence, in the next iteration (Fig. 4.27c) it is again possible to reduce the overall distortion D by a further optimization of the reconstruction vectors and subsequently of the boundaries (Voronoi net). By both steps, D can never increase from one iteration to the next.

---

34 The same principle can be applied with a model such as the Gaussian vector PDF, where the optimum vector would become, in analogy with (4.19), $\mathbf{y}_{j,\mathrm{opt}}=\mathcal{E}\{\mathbf{x}^{(j)}\}$, the expected value over the current Voronoi region.

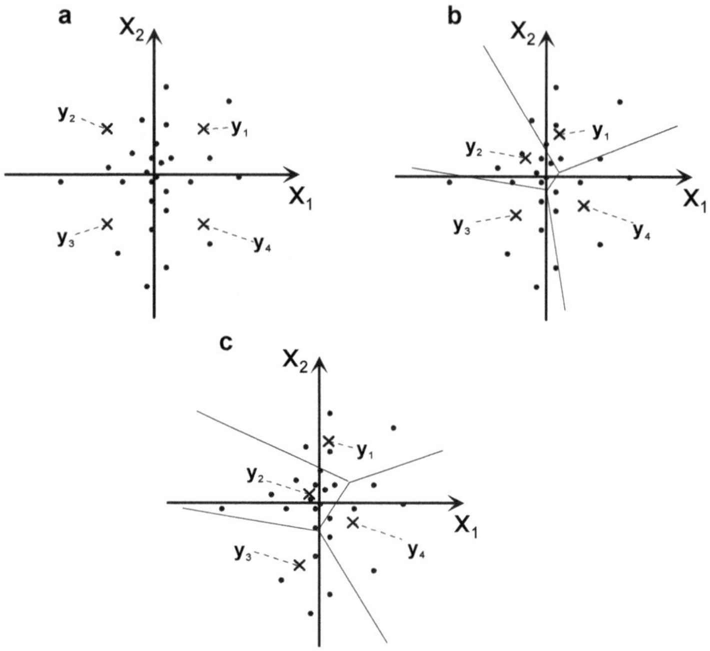

Fig. 4.27. Graphical illustration of two iterations of the generalized Lloyd algorithm.

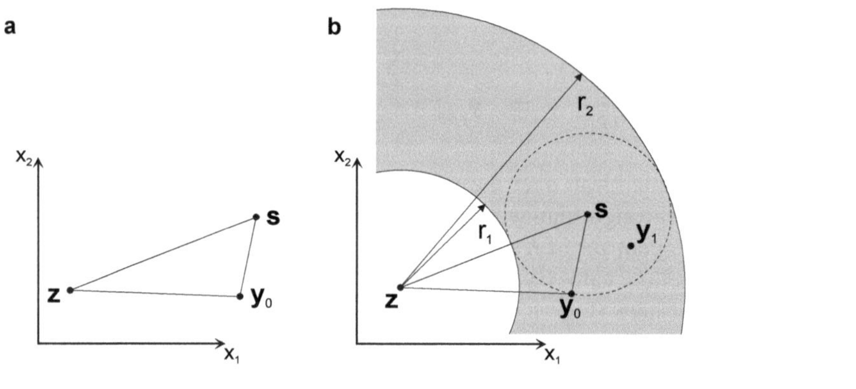

Fig. 4.28. Graphical interpretation **a** of the triangular inequality **b** of a method to limit the search to a specific range of the codebook, according to (4.69)

Formally, it is necessary to compare each source vector $\mathbf{s}$ against *each* $\mathbf{y}_j$ e.g. by the criterion (4.61). The resulting complexity in terms of multiplications and additions then is J operations per sample, or KJ operations per vector. With $R=(\log_2 J)/K \Rightarrow J=2^{RK}$, the codebook size and search complexity grows exponentially with the rate and the vector size. Hence, it is not practical to apply VQ algorithms with unstructured non-uniform codebooks, when a large rate is required. A possible solution would be the reduction of K, which however may reduce the performance[35]. Strategies to resolve this problem exclude those $\mathbf{y}_j$ from the search which most probably are no good candidates to encode a given $\mathbf{s}$, i.e. it needs to be identified that they are not in close proximity in $\mathfrak{R}^K$. This can be achieved by designing *structured codebooks* (see section 4.5.4), which however may have disadvantages in terms of compression performance. Another method of identifying relevant subspace areas without losing compression performance is presented in the following paragraph, which however does typically provide search complexity reduction only by a factor of roughly 2.

Let a *fix point* $\mathbf{z}$ be defined, and distances $\|\mathbf{z}-\mathbf{y}_j\|$ between the fix point and all codebook entries $\mathbf{y}_j$ be computed in advance and stored along with the codebook. Further, an arbitrary *initial vector* shall be selected from the codebook, e.g. the vector $\mathbf{y}_1$. Then, according to the *triangular inequality* (see graphical interpretation in Fig. 4.28a) the following conditions apply regarding the vector $\mathbf{s}$:

$$\begin{aligned}\|\mathbf{z}-\mathbf{y}_1\| &\geq \|\mathbf{z}-\mathbf{s}\| - \|\mathbf{s}-\mathbf{y}_1\| =: r_1 \\ \|\mathbf{z}-\mathbf{y}_1\| &\leq \|\mathbf{z}-\mathbf{s}\| + \|\mathbf{s}-\mathbf{y}_1\| =: r_2.\end{aligned} \tag{4.66}$$

Now, for the best match $\mathbf{y}_i$ available in the codebook,

$$\|\mathbf{s}-\mathbf{y}_i\| \leq \|\mathbf{s}-\mathbf{y}_1\| \tag{4.67}$$

giving by combination with (4.66)

$$\begin{aligned}\|\mathbf{z}-\mathbf{y}_i\| &\geq \|\mathbf{z}-\mathbf{s}\| - \|\mathbf{s}-\mathbf{y}_i\| \geq r_1 \\ \|\mathbf{z}-\mathbf{y}_i\| &\leq \|\mathbf{z}-\mathbf{s}\| + \|\mathbf{s}-\mathbf{y}_i\| \leq r_2.\end{aligned} \tag{4.68}$$

Consequently, it is sufficient to perform the search only for those $\mathbf{y}_j$ where the following condition holds:

$$r_1 \leq \|\mathbf{z}-\mathbf{y}_j\| \leq r_2 . \tag{4.69}$$

This can easily be checked by once computing r_1 and r_2, and compare the range they express against the distances relative to the fix point, which are pre-computed. However, the efficiency of such an approach, i.e. the percentage of $\mathbf{y}_j$ which can actually be excluded, depends on the selection of $\mathbf{z}$ and is to some extent random by the choice of the initial vector $\mathbf{y}_1$. A graphical interpretation is given in Fig. 4.28b. It may be useful to test several candidate vectors instead of just one $\mathbf{y}_1$, and choose one for which the margin between r_1 and r_2 becomes as small as possible. Further, several fix points $\mathbf{z}$ can be used in case of large codebooks, but the basic principle remains similar.

---

[35] A small vector length K is more disadvantageous if the signal samples contained in the vector are statistically dependent. Further, for non-stationary sources, it may be advantageous to use different codebooks or also different rates and vector lengths for different characteristics of data to be encoded.

4.5.4 Structured codebooks

Complexity reduction can be achieved by breaking down the decision process in a sequence of partial decisions, which can be expressed as a decision tree, related to a pre-defined structure in the codebook, such that not all branches are investigated. Some of these methods also implicitly provide a *scalable representation*, where the initial level(s) can be used for a raw approximation, the remaining level(s) as refinement (enhancement) information achieving lower distortion.

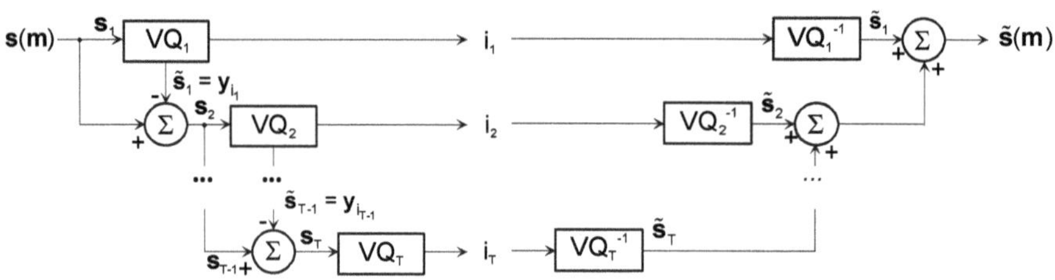

Fig. 4.29. Multi-stage vector quantization (T levels)

Multi-stage VQ [Ho, Gersho 1988] as shown in Fig. 4.29 concatenates T sequential stages (levels) of vector quantizers. Levels 2 and higher encode residuals (coding errors) of previous levels. With initial settings $\tilde{\mathbf{s}}_0 = 0$ and $\mathbf{s}_0 = \mathbf{s}$, the input signal of the encoder in level t is $\mathbf{s}_t = \mathbf{s}_{t-1} - \tilde{\mathbf{s}}_{t-1}$. For each level, a specific codebook $\mathcal{C}_t$ containing J_t reconstruction vectors is provided. For $t \geq 2$, these have to be designed for encoding vectors of residual values. The full representation is the set of indices i_t of the partial codebooks. The overall data rate (without entropy coding of indices) is

$$R = \frac{1}{K} \sum_{t=1}^{T} \log_2 J_t \quad [b/\text{sample}]. \tag{4.70}$$

The reconstruction $\tilde{\mathbf{s}}$ is the superposition of vectors from all levels,

$$\tilde{\mathbf{s}} = \sum_{t=1}^{T} \mathbf{y}_{i_t} . \tag{4.71}$$

Decoding could be terminated at any level, the index information is inherently scalable. It is however not guaranteed that the outputs of the different levels are statistically independent in case of non-uniform codebooks. The shape of a Voronoi cell of the reconstruction selected by the first level determines the maximum coding error, and therefore codebook entries of the next level which are designed for larger errors would typically not be considered. This could be exploited by context-dependent or joint coding of indices, which is however difficult to realize due to the large number of possible combinations. Sub-optimality is further caused by the fact that the codebook sizes J_t have to be chosen *before* the codebook training, and codebooks have to be optimized successively starting from level 1. It can hardly be guaranteed that the set of residual codebook vectors provides the capability to correct any errors occurring at the previous level, due to the irregular shape of Voronoi regions. Most of these effects could be avoided by additional constraints in the codebook design, e.g. sizes of Voronoi regions being approximately equal, which however again would impose sub-optimality as compared to the original GLA design (in particular, in the first level). In summary, typically multi-level VQ methods are inferior in terms of coding performance as compared to a VQ codebook generated by a single-level design, however the latter requires much larger storage when using the same rate for the representation.

The scheme of a *tree-structured codebook* is shown in Fig. 4.30a. Similar to in multi-level VQ, the search for the best reconstruction vector is performed by progressing over several refinement steps, but in a more general structure of a *code tree*, where each parent node has individual siblings (in contrast to multi-level VQ, where an identical refinement codebook is used independent of previous level reconstructions). If the tree consists of T levels, and each node at level t–1 can branch into I_t siblings for the subsequent level t, the total number of vector comparison operations is

$$M = \sum_{t=1}^{T} I_t \,, \tag{4.72}$$

whereas the total number of vectors to be stored in the codebook at level t is[36]

$$J_t = \prod_{u=1}^{t} I_u \Rightarrow J_{\text{total}} = \sum_{t=1}^{T} J_t \,, \tag{4.73}$$

where J_{total} expresses the overall number of vectors to be stored in the tree-structured codebook.

The total number of possible choices up to the last level is J_T, which requires

$$R = \log_2 J_T = \sum_{t=1}^{T} \log_2 I_t \,, \tag{4.74}$$

for representation (without entropy coding). In case of a *binary tree*, I_t=2 over all levels, and only M=2log$_2 J_T$ comparisons (instead of J=J_T in case of full search in a single-level codebook with the same number of J vectors) are necessary. With tree-structured codebooks, the complexity of VQ only grows *linearly* with the rate, in contrast to the case of exhaustive (full) search, where an exponential growth was observed. As a disadvantage, the memory requirements for storing the tree-structured codebook (4.73) is larger than for a single-level codebook, since reconstruction vectors for all intermediate levels have to be stored, i.e. growing more than exponentially with the rate[37].

---

[36] It is assumed here that the number of splits I_t would be fixed in each level which leads to a uniform tree structure. It may however also be variable in case of non-uniform trees (see Sec. 4.5.6.

[37] For multi-level VQ, the storage requirement only grows linearly with the rate through the number of levels, i.e. the necessary memory is smaller than for a single-level codebook. This is the penalty for allowing individual refinements at each node of the tree, which

Tree-structured codebooks can be designed by *splitting* (see Fig. 4.30b), where the number of siblings I_t branching from a node at level $t-1$ has to be defined in advance. In Fig. 4.30b, the example of a regular binary tree structure with I_t=2 over all levels is shown. The initial reconstruction vector $\mathbf{y}_{0,1}$ at the root of the tree can be computed as the global mean from a training set. The associated Voronoi region is the entire space $\mathcal{R}^K$, which is divided into two sub-spaces (see Fig. 4.30b)[38]. Within each hemisphere, the sibling vectors $\mathbf{y}_{1,1}$ and $\mathbf{y}_{1,2}$ are then optimized according to (4.65), see Fig. 4.30c. Each of the two new Voronoi regions is again split, such that from the second level of the codebook tree, 4 reconstruction vectors are available (Fig. 4.30d). As a splitting procedure, *additive splitting*

$$\mathbf{y}_{\text{split},1} = \mathbf{y} - \varepsilon \cdot \mathbf{1} \quad ; \quad \mathbf{y}_{\text{split},2} = \mathbf{y} + \varepsilon \cdot \mathbf{1} \quad ; \quad 0 < \varepsilon \ll 1 \tag{4.75}$$

was used in the given example; another approach is *multiplicative splitting*

$$\mathbf{y}_{\text{split},1} = \mathbf{y} \cdot (1+\varepsilon)^{-1} \quad ; \quad \mathbf{y}_{\text{split},2} = \mathbf{y} \cdot (1+\varepsilon) \quad ; \quad 0 < \varepsilon \ll 1. \tag{4.76}$$

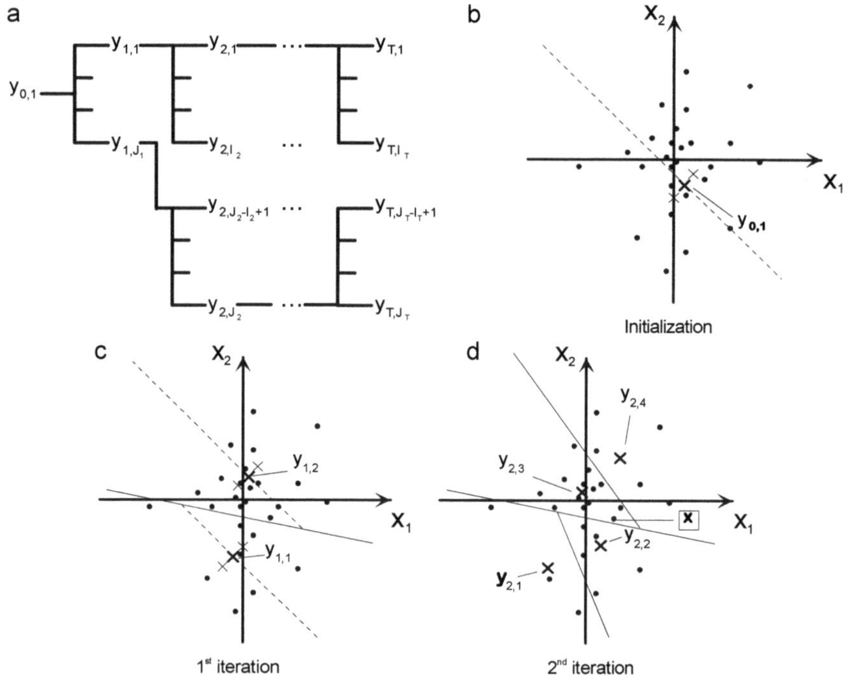

Fig. 4.30. a Tree-structured codebook **b-d** First iteration steps illustrating the 'splitting' of quantizer partitions for the case of a binary tree structure

avoids the major causes of the sub-optimality of multi-level VQ. This will become more clear through the design procedure described below, where each quantizer cell is individually split into finer partitions by the next step.

[38] In case of correlated samples within the vector, it is best to perform the sub-division orthogonal to the principal axis of the mass distribution of the training set. This is in principle done both by the additive and multiplicative splitting procedures described below.

If it is necessary to split into more than two vectors, a set of orthogonal basis functions can be employed in (4.75)/(4.76) instead of the unity vector, e.g. using the Hadamard transform (2.251) (functions to be weighted by $\pm\varepsilon$). The split procedure described here is also useful in generating *initial codebooks* for a single-level VQ design, where the vectors at tree levels $t<T$ would be discarded afterwards. Also in this case, running several iterations of codebook optimization after each splitting step can provide better optimized codebooks.

It can nevertheless occur that at a subsequent level a vector $\mathbf{y}_{t,j}$ exists, which is closer (by Euclidean distance) to the signal vector $\mathbf{s}$ than the vector $\mathbf{y}_{t,i}$ reached by the tree-based decision. This is caused by the fact that layouts of quantizer partitions at previous decision levels restrict the subspaces of subsequent searches. As an example, in Fig. 4.30d the signal vector '☒' is originally closest to the reconstruction vector $\mathbf{y}_{2,2}$; indeed $\mathbf{y}_{2,3}$ is selected, because at the first level of the tree, $\mathbf{y}_{1,2}$ was chosen. The decision boundary between the quantizer partitions of $\mathbf{y}_{2,2}$ and $\mathbf{y}_{2,3}$ which are siblings from different nodes is *not a Voronoi line*. Therefore, tree-structured VQ may be suboptimum in achieving distortion minimization, when compared to a full search performed over all end nodes of the tree.

Gain/Shape VQ. This scheme [SABIN, GRAY 1984] uses standardized reconstruction vectors $\breve{\mathbf{y}}_j$ with norm $\left\|\breve{\mathbf{y}}_j\right\|=1$, which are multiplied by a gain factor σ_i. The task is to find the pair $(\sigma_i,\breve{\mathbf{y}}_j)$ which minimizes the Euclidean distance compared to an input vector $\mathbf{s}$. This metric can be expressed by

$$d_2(\mathbf{s};\breve{\mathbf{y}}_j,\sigma_i)=\left\|\mathbf{s}-\sigma_i\cdot\breve{\mathbf{y}}_j\right\|_2^2=\mathbf{s}^{\mathrm{T}}\mathbf{s}-2\sigma_i\mathbf{s}^{\mathrm{T}}\breve{\mathbf{y}}_j+\sigma_i^2\breve{\mathbf{y}}_j^{\mathrm{T}}\breve{\mathbf{y}}_j\,. \tag{4.77}$$

This expression can first be minimized independently of the choice for the gain factor, if $\mathbf{s}^{\mathrm{T}}\breve{\mathbf{y}}_j$ becomes maximum; this is the *cross correlation* between the input vector and the respective codebook vector. Finally, by deriving (4.77) over σ_i and due to $\breve{\mathbf{y}}_j^{\mathrm{T}}\breve{\mathbf{y}}_j=1$, the optimum σ_i is chosen as the closest quantized representation of $\mathbf{s}^{\mathrm{T}}\breve{\mathbf{y}}_j$.

Mean separating VQ. In case of signals where a locally varying mean value is present (e.g. images), it can be removed prior to VQ encoding, such that the residual codebook vectors only represent detail changes. In *mean separating VQ* (Fig. 4.31a), the block mean value over the vector $\mathbf{s}(\mathbf{m})$ is computed, subtracted from the elements of the vector and encoded separately. After decoding, the reconstructed mean value is added to the samples of the zero-mean reconstruction vector. If R_{M} bit are used to encode the mean values, the total rate per vector (without entropy coding) becomes $R_{\mathrm{M}}+\log_2 J$. Further, predictive coding of mean values can be used.

4.5.5 Adaptive VQ

Codebooks can also be adapted; the effect is similar as with adaptive entropy coding (i.e. better compression for signals with varying statistics). However, the adaptation is typically forward driven (codebook to be transmitted as side information), since VQ is usually lossy and no implicit reconstruction of the original signal is available at the decoder side which would be usable for codebook training. As the amount of side information can be extremely large in case of large codebooks, it is also possible to adapt only a subset of the codebook as necessary, and/or to apply a decorrelating compression scheme (such as predictive or transform coding) to code the codebook entries, in case when the samples in the vectors are correlated.

In case of signals with locally varying statistics, it is furthermore beneficial to allocate variable rate (i.e. varying codebook size), depending e.g. on the local variance. Variable-rate coding can be achieved implicitly by entropy coding or explicitly by codebook switching. In particular, the latter case is beneficial, as it additionally may allow to use smaller sub-codebooks and thereby to reduce the search effort at the encoder (provided that the decision for the codebook switching is done in advance of VQ coding.

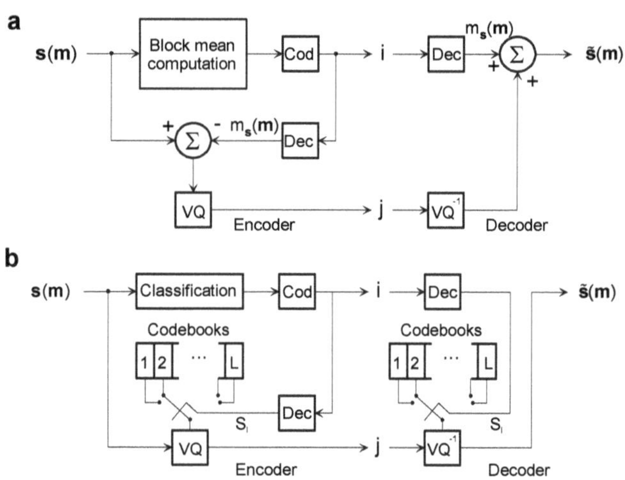

Fig. 4.31. Specific types of VQ **a** Mean-separating VQ **b** Classifying VQ

Classifying VQ. In classifying VQ [Ramamurthy, Gersho 1986], a *feature-based classification* of the signal vectors is performed prior to encoding (Fig. 4.31b). This can be based on a composite-source model, assuming that the source consists of different types of sub-sources with specific feature characteristics; optimization of codebooks is then also done separately for each sub-source type. Example classification criteria are *block variances*, dominant *edge direction* within the blocks (in case of images), or voiced/unvoiced classification (in case of speech). Based on the classification result, a class label S_l, indicates one out of L classes. For each class, a specific *sub-codebook* $\mathcal{C}_l=\{\mathbf{y}_{j,l};\ j=1,2,...,J_l\}$ is provided. The *vector index* j and the *class index* l must be conveyed to the decoder. The reconstruction vector is then addressed from a super-codebook which is the unification of all sub-codebooks

$$\mathcal{C}=\bigcup_{l=1}^{L}\mathcal{C}_l=\left\{\mathbf{y}_{j,l};j=1,2,...,J_l;l=1,2,...,L\right\}. \tag{4.78}$$

The minimum rate depends on the probabilities of the different classes $\Pr(S_l)$ and the probabilities of related code symbols $\Pr(j|S_l)$ by the respective entropy,

$$R\leq-\sum_{l=1}^{L}\frac{\Pr(S_l)}{K_l}\left[\underbrace{\log_2\Pr(S_l)}_{\text{class index entropy}}+\underbrace{\sum_{j=1}^{J_l}\Pr(j\,|\,S_l)\log_2\Pr(j\,|\,S_l)}_{\text{vector index entropy}}\right]\leq\sum_{l=1}^{L}\frac{\Pr(S_l)}{K_l}\log_2\frac{J_l}{\Pr(S_l)}. \tag{4.79}$$

Here, also the case is considered where different block sizes (number of samples per vector K_l) are used in the classes. If the distortion D_l occurs in class S_l, the total distortion over all classes is

$$D=\sum_{l=1}^{L}\Pr(S_l)\cdot D_{S_l}\ . \tag{4.80}$$

Classifying VQ allows to use a higher rate for classes which are more difficult to code. Due to the prior classification, better adaptation of the codebooks to locally changing (non-stationary) source statistics is achieved, and the encoding complexity is reduced, as only the subset $\mathcal{C}_{S_l}$ of the super-codebook (4.78) needs to be searched.

Finite State VQ (FSVQ). When the vector $\mathbf{s}(\mathbf{m})$ is processed, coder and decoder are in a certain state $S_{l(\mathbf{m})}$, which can e.g. be determined by the constellation of reconstruction vectors decoded previously in the neighborhood, or by feature information that can be deduced from the reconstruction. If the number of possible values l instantiated in $S_{l(\mathbf{m})}$ is limited to L, encoder and decoder can be interpreted as *finite-state machines*. Assume that in any state S_l, only a subset of possible codebook vectors, organized within a *state codebook* $\mathcal{C}_{S_l}$, are accessible. The unification of all state codebooks forms the super codebook

$$\mathcal{C}=\bigcup_{l=1}^{L}\mathcal{C}_{S_l}\quad;\quad\mathcal{C}_{S_l}=\left\{\mathbf{y}_{j,S_l};j=1,2,...,J_{S_l}\right\}. \tag{4.81}$$

By the formulation in (4.81), it is not yet determined whether reconstruction vectors can be members of different state codebooks; it is also possible that the state codebooks vary in size. If the $\mathcal{C}_{S_l}$ are overlapping sets, the total number of reconstruction vectors will be lower than the sum over all J_{S_l} from the single states, and the memory size can be reduced by using a mapping table. The computation of the code symbols $j(\mathbf{m})$ and the follow-up state $S_{i(\mathbf{m}+1)}$ are performed according to the relations[39]

$$j(\mathbf{m}) = \arg\min_{\mathbf{y}_j \in \mathcal{C}_{S_i(\mathbf{m})}} d(\mathbf{s}(\mathbf{m}), \mathbf{y}_j) \quad ; \quad i(\mathbf{m}+1) = f(j(\mathbf{m}), i(\mathbf{m})) \,. \tag{4.82}$$

In state S_l, J_{S_l} different code symbols can be selected. This requires a significantly lower rate as compared to the address space necessary for the entire super codebook $\mathcal{C}$. The performance highly depends on the definition of the *next-state function* $f(j, S_l)$. By this function – dependent on the previous state and the previous transmitted code symbol – the state codebooks $\mathcal{C}_{S_l}$ are defined. These should consist of subsets of reconstruction vectors, which are most likely to be selected within a certain state configuration. General methods for optimization of FSVQ codebooks are described in [DUNHAM, GRAY 1985], [FOSTER, GRAY, DUNHAM 1985].

The block diagram of an FSVQ encoder and decoder is shown in Fig. 4.32a. Observe the structural similarity with classifying VQ (Fig. 4.31b) – FSVQ could indeed be interpreted as a backward-adaptive version of classifying VQ, not requiring encoding and transmission of the information about the selected subcodebook.

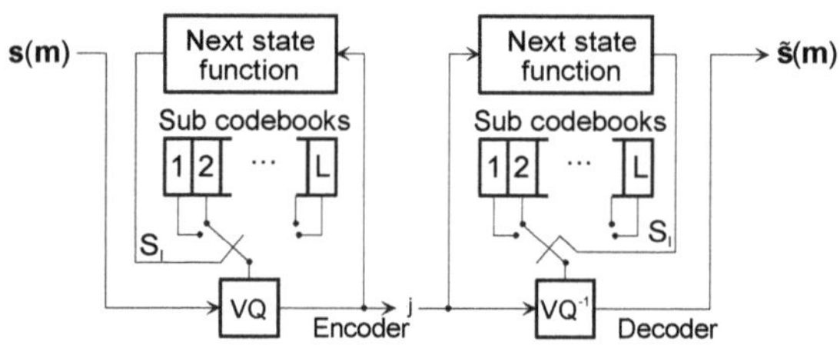

Fig. 4.32. FSVQ encoder and decoder

4.5.6 Rate-constrained VQ

Methods which are similar to schemes for optimization of scalar quantization considering rate constraints (Sec. 4.3) can also be applied in the context of VQ. In the GLA (Sec. 4.5.3), *minimization of distortion* during each iteration step is guaranteed; it might however happen that the *rate increases* (as e.g. measured by entropy of vector indices) during a distortion-minimizing iteration step of the codebook generation. Considering the rate-distortion performance, no guarantee

[39] For $\mathbf{m}=(m_1,m_2)$, $\mathbf{m+1}$ means e.g. $(m_1+1,\ m_2)$ in the case of row-wise sequential processing.

for improvement exists per iteration. Therefore, a modified distortion metric can be used similar to (4.43), taking into account the rate $R_e(S_j)$[40]:

$$D^{*}_{\mathrm{e},e}(\mathbf{s},\mathbf{y}_j)=\left[\mathbf{s}-\mathbf{y}_j\right]^{\mathrm{T}}\left[\mathbf{s}-\mathbf{y}_j\right]+\lambda R_e(S_j)\,. \tag{4.83}$$

The Lagrangian multiplier λ expresses the weight by which the rate shall influence the decision, the optimum is given by (4.44). At the decision boundary between two quantizer partitions, this distance metric must be equally balanced regarding the two reconstruction vectors. Hence, depending on λ, the boundary between the quantizer cells related to the reconstruction vectors with indices j_1 and j_2 is now placed according to the following condition, where $\mathbf{s}_{\mathrm{b}}$ denotes any vectors on the decision boundary[41]

$$\left[\mathbf{s}_{\mathrm{b}}-\mathbf{y}_j\right]^{\mathrm{T}}\left[\mathbf{s}_{\mathrm{b}}-\mathbf{y}_j\right]+\lambda R_e(S_{j_1})=\left[\mathbf{s}_{\mathrm{b}}-\mathbf{y}_j\right]^{\mathrm{T}}\left[\mathbf{s}_{\mathrm{b}}-\mathbf{y}_j\right]+\lambda R_e(S_{j_2})\,, \tag{4.84}$$

which can be re-written as

$$\begin{aligned}\mathbf{y}_{j_1}{}^{\mathrm{T}}\mathbf{y}_{j_1}-\mathbf{y}_{j_2}{}^{\mathrm{T}}\mathbf{y}_{j_2}+2\mathbf{s}_{\mathrm{b}}{}^{\mathrm{T}}\left(\mathbf{y}_{j_2}-\mathbf{y}_{j_1}\right)&=2\left(\mathbf{s}_{\mathrm{b}}-\frac{\mathbf{y}_{j_1}+\mathbf{y}_{j_2}}{2}\right)^{\mathrm{T}}\left(\mathbf{y}_{j_2}-\mathbf{y}_{j_1}\right)\\&=\lambda\left(R_e(S_{j_2})-R_e(S_{j_1})\right).\end{aligned} \tag{4.85}$$

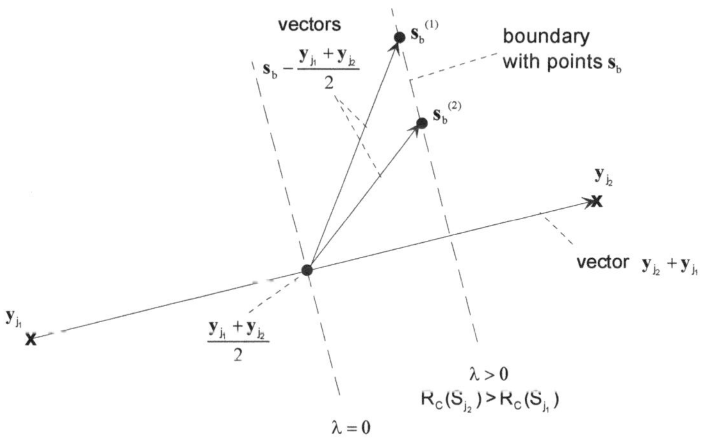

Fig. 4.33. Positions of two neighbored reconstruction vectors, and boundary between the quantizer regions in rate-constrained VQ

---

[40] During codebook design, the rate of codeword S_j can best be estimated by the empirical self information (2.170), computed from normalized occurrence counts; if rate constraints are applied during encoding, the actual rate of a VLC scheme used to encode the codebook indices can be used.

[41] This expresses a line in $\mathcal{R}^2$, a plane or hyper plane in higher-dimensional $\mathcal{R}^K$

An illustration for the case of $K=2$ is given in Fig. 4.33. For $\lambda=0$ ('ordinary' VQ without rate-constraint criterion), the boundary is a Voronoi line which includes the point $(\mathbf{y}_{j_1}+\mathbf{y}_{j_2})/2$ and is perpendicular to the Delaunay vector $\mathbf{y}_{j_2}-\mathbf{y}_{j_1}$. For $\lambda>0$, the boundary is shifted along the Delaunay line towards the reconstruction vector requiring larger rate, but remains perpendicular. For the case of equal rates, the boundary is a Voronoi line, regardless of λ. Otherwise, in cases where $\mathbf{s}$ has approximately equal Euclidean distances from both reconstruction vectors, the vector requiring lower rate is preferred.

An optimization of the VQ for a specific entropy coding rate can also be achieved by tuning the parameter λ[42].

Rate-distortion optimization for tree-structured codebooks. Tree-structured VQ as introduced in Sec. 4.5.4 uses symmetric splitting with equal depth of the code tree for all paths. A binary representation of the tree paths corresponds to a fixed-length code. If asymmetric code trees of different path length as in entropy coding (Sec. 4.4.1) are used, a direct combination with variable-length coding can potentially provide better results under rate-distortion criteria. In *tree pruning algorithms* [CHOU, LOOKABAUGH, GRAY 1989] [KIANG ET AL. 1992], branches are removed from a full regular tree depending on their effectiveness. Branches which have least impact on reducing distortion compared to the rate cost are removed first; they would only offer a low slope on the operational rate-distortion function (cf. Sec. 4.3). *Growing* a tree [RISKIN, GRAY 1991] step by step is the opposite design approach with similar goals. In growing, the quantizer partitions are not split at once at a certain depth level of the tree, but rather one at a time, by which the largest decrease of distortion versus rate is achieved. Let D_j be distortions (4.64), and $\Pr(S_j)$ the probabilities of tree nodes (leaves) accepted so far by the growing algorithm. Both values can be determined empirically from a training set. Further, let $t(j)$ express the depth of the tree at the given position. If partition S_j would be split into I sub-partitions S_{j_1} ... S_{j_I} (typically I=2), the decrease of distortion is

$$\Delta D(S_j) = D_j - \sum_{i=1}^{I} \frac{\Pr\left(S_{j_i}\right)}{\Pr\left(S_j\right)} D_{j_i} \tag{4.86}$$

New index codewords j_i can be generated by appending $\log_2 I$ bit to the parent node's index codeword j. Splitting is always performed for the cell with maximum of $\Delta D(S_j)$ which could be at any position/depth of the tree. The procedure is continued until some pre-set termination criterion (e.g. limit of distortion) is reached. Observe that this typically gives both an asymmetric tree structure and a variable-length code representation of the vectors residing on the leaves of the tree. The final index codewords can readily be used as variable-length code strings and represent a unique prefix-free code.

---

[42] For growing λ, the rate decreases, and for $\lambda\to\infty$, $R\to 0$.

4.6 Trellis-coded quantization

In VQ, the encoder performs an instantaneous mapping from a block of samples to a codeword, which is constructed from elementary code symbols (bits or groups of bits). The sequence of codewords is independent, which is the typical case of a block code. An alternative is *sliding block coding*, where code symbols are shared by several codewords or may influence the reconstruction of several subsequent samples. Therefore, the encoder choice for a code symbol cannot be made instantaneously, but rather with a decision delay[43]. A typical principle of decoding in a sliding-block method is shown in Fig. 4.34. Herein, a code symbol $i(m)$ is combined with its L predecessors $i(m-1)...i(m-L)$ to generate an output $\tilde{s}(m)$, which can by itself be a scalar or a vector value. Linear or nonlinear mapping functions can be used. The parameter L is the *constraint length* of the sliding block code.

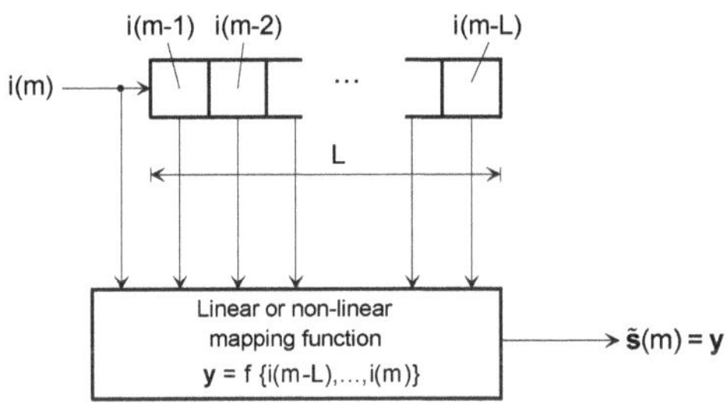

Fig. 4.34. Decoder in sliding block coding

An important sub-class of sliding block codes used in source coding are *trellis codes*. In *trellis coded quantization* [MARCELLIN, FISCHER 1990], the mapping function is a codebook of scalar or vector reconstruction values, and a sequence of $L+1$ code symbols establishes a codeword (codebook index). If each code symbol can be instantiated by J different values, the codebook $\mathcal{C}=\{\mathbf{y}_j; j=1,\ldots,J^{L+1}\}$ will map the sequence of length $L+1$ into one of its J^{L+1} entries for reconstruction. The configuration of possible decoder states is shown in Fig. 4.35 by an example of $J=2$, $L=2$. The state transition graph as shown in Fig. 4.35a gives a *trellis diagram*. In each decoding step, one out of $2^3=8$ different reconstruction vectors

---

43 'Delayed decision coding' could be beneficially applied in any cases as well where recursive (state) dependencies exist at the decoder, e.g. in predictive coding [MODESTINO, BHASKARAN 1981], context-based coding etc. In the current section, we restrict the usage of the term 'sliding block coding' to cases where the code design itself is tied to its usage.

$\tilde{s}(m)$ is generated, which are related to the 8 *branches* of the trellis that point towards the possible J^L=4 next states of the diagram. The mapping table is shown in Fig. 4.35b. Due to the finite-state property with L=2, code symbols before $i(m–2)$ will no longer have any direct influence on $\tilde{s}(m)$. A *prologue* of L code symbols initializes until the first state that is relevant for encoding is reached.

The operation of the trellis encoder can best be interpreted if the possible sequences of code symbols $i(m)$ are mapped to *paths* in the trellis diagram (Fig. 4.35a). It would in principle be necessary to compare all possible paths in the trellis to find the optimum code symbol sequence. The decision is made by accumulating a cost criterion over the different paths, typically minimizing the path distortion. However, before the optimum code symbol $i(m)$ can be determined by the encoder, at least L further decoding steps have to be evaluated. By the *Viterbi algorithm* described in the following sub-section, the number of paths to be compared in trellis coding is always *upper bounded* by J^{L+1}, which is identical to the number of codewords in the codebook.

a

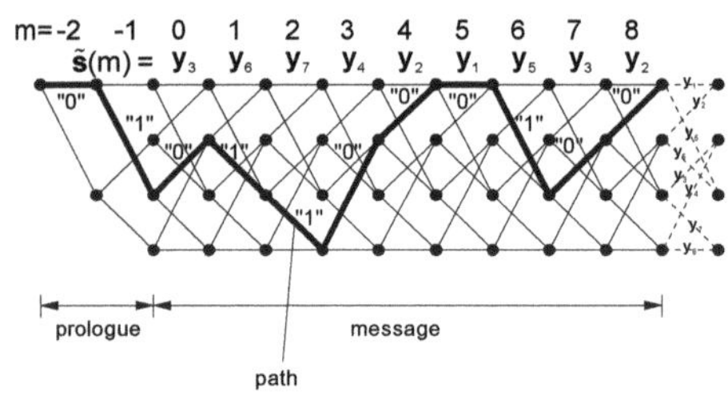

b

( )	i(m)	i(m-1)	i(m-2)
y_1	0	0	0
y_2	0	0	1
y_3	0	1	0
y_4	0	1	1
y_5	1	0	0
y_6	1	0	1
y_7	1	1	0
y_8	1	1	1

Fig. 4.35. **a** Trellis diagram for J=2, L=2 and **b** related mapping table, combining code symbols to codewords and associated reconstruction.

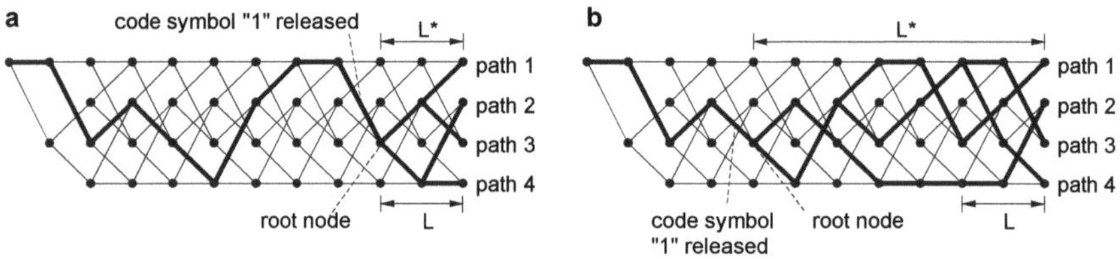

Fig. 4.36. Viterbi algorithm. Retained paths from the trellis diagram and position of the root node **a** in case of minimum possible $L^*=L=2$ **b** in case $L^*>L$

Viterbi algorithm. In the trellis diagram (Figs. 4.35 and 4.36), each node can be reached from J different paths, and each path has J possible extensions towards nodes of the following state. As however an extension of a node adds the same amount of distortion to any of the incoming paths of that node, the path with minimum distortion occurring up to the position of the node will be the optimum among all incoming paths of this node. The *globally optimum path* must be one of the optimum paths leading towards the J^L different nodes of the current level of the trellis. As presented in [VITERBI 1967], during each encoding step it is therefore only necessary to compare J incoming paths and select the best among them at each node. If the $M=J^L$ candidate paths of all nodes at one level originate from *one identical root node*, any code symbols which are allocated to paths ahead of this root node must belong to the globally optimum path and can finally be released from this encoding step. The root node resides L^* steps in the past, wherein the number L^* can be variable (see the examples in Fig. 4.36a/b), but the search complexity is nevertheless limited. In total, per encoding step (and hence on average per code symbol released) no more than $JM=J^{L+1}$ comparison operations (w.r.t. the distortion metric) will be necessary.

***M*-algorithm.** This algorithm retains a pre-selected fixed number of M optimum paths in each search step of a multi-path search. Its complexity compared to a single-path search is therefore linearly increased by the number M, which makes it very regular, parallelizable and suitable for hardware implementations [MOHAN, SOOD 1986]. The algorithm can flexibly be adapted to the requirements of specific multi-path, tree- or trellis-based encoding approaches. In the latter case, with $M<J^L$, it can even be less complex than the Viterbi algorithm, where it is no longer guaranteed to approach the global optimum. Even though increasing the value M usually provides better performance, this may not be justified beyond a certain margin when considering the additional complexity.

Codebook optimization. Optimization of codebooks for trellis-coded quantization (TCQ) can be performed by a variant of the GLA described in Sec. 4.5.3. Here, the Viterbi algorithm is used during the encoding step of each iteration, while the optimization of codewords is still similar [STEWART, GRAY, LINDE 1982], [AYANOGLU, GRAY 1986]. Without significantly modifying the methods described in Sec. 4.5.5, it is also possible to run codebook optimization for TCQ using *entropy* or *rate constraints* [FISCHER, WANG 1992].

4.7 Problems

Problem 4.1.

a) Show for the AR(1) process the relationship of the spectral flatness measure $\gamma_s^2=\sigma_v^2/\sigma_s^2$ under assumption of 'low distortion' according to (4.28).
b) Show the validity of (4.31). Which gain in rate can be achieved when coding a 2D separable AR(1) process, if the 2D correlation properties are exploited ? Compute the values of the gains in the 1D and 2D cases with $\rho_1=\rho_2=0.95$.

c) Which is the rate at the boundary of the range of 'low distortion' according to (4.40) for the separable 2D AR(1) process? Compute the value of this rate for $\rho_1=\rho_2=0.95$.
d) The distortion parameter Θ in (4.25) shall be chosen such that in coding of a 1D AR(1) process, a band limitation to *i)* $f_{max}=1/8$, *ii)* $f_{max}=1/4$ and *iii)* $f_{max}=1/2$ is achieved. Determine the Θ values for these frequencies f_{max}, where the AR(1) process is defined by *I)* $\rho=0.5$ and *II)* $\rho=0.95$.

Problem 4.2.

The probabilities of the 4 symbols of a discrete source alphabet are Pr(A)=0.4; Pr(B)=Pr(C)=Pr(D). Determine the code symbol lengths of a Shannon code design, and sketch the code tree. Further, design a Huffman code. Compute the entropy and the rates that are necessary for both codes. Why is the Huffman code design more efficient?

Problem 4.3.

A memoryless binary process [0,1] is characterized by the first order probabilities Pr(0)=0.25 and Pr(1)=0.75.
a) Compute the entropy.
b) Determine the joint probabilities for pairs of samples Pr(0,0), Pr(0,1), Pr(1,0) and Pr(1,1) and for triplets of samples Pr(0,0,0), Pr(0,0,1), ... , Pr(1,1,1).
c) Determine the rates using an entropy coding design by the Huffman method, using first order probabilities, then for vectors of samples the second order and third order joint probabilities as determined in b). Compute the differences compared to the entropy.

The samples shall now be statistically dependent (same first order probabilities), which is characterized by a binary Markov chain (Fig. 2.24). One of the two transition probabilities is Pr(1|0)=0.5.

d) Determine the probabilities Pr(0|1), Pr(1|1) and Pr(0|0).
e) Find the probability values of sample pairs Pr(0,0), Pr(0,1), Pr(1,0) and Pr(1,1).Then, determine the code symbol lengths for a Huffman code, compute the rate and give an interpretation in comparison to the result of c).
f) Compute the entropy according to (5.3), and give an interpretation in comparison with the results of c) and e).

Problem 4.4.

A discrete source is arithmetic encoded. The source outputs two different letters 'A' and 'B' with probabilities Pr(A)=0.8, Pr(B)=0.2.
a) Compute the entropy.
b) Sketch the layout of probability intervals within the range [0,1] for all combinations of 3 subsequent source symbols, assuming statistical independency. For each interval, compute the lower and upper boundaries.

Encoding shall be performed by an arithmetic unit which has a precision of 3 bit.

c) Sketch the code interval raster for three subsequent code bits. Perform rounding of the probability interval boundaries from b) according to the 3 bit precision. In this case, do not further subdivide probability intervals which exactly match a code interval, as the encoding process can start independently afterwards. Sketch a code tree relating to all intervals that are retained. For each path within the tree, identify the string of code bits that would be transmitted along the path, and the sequence of source symbols that is expressed.

d) What is the resulting rate (per source symbol)? Compare against the entropy, and check whether the value is within the guaranteed bound (4.57) of arithmetic encoding, regarding the given precision of arithmetic operations.

Problem 4.5.

A binary source with memory shall be encoded by a run-length code. An analysis of the source has given the parameters of a 2-state Markov chain model: Pr(0|1)=0.8 ; Pr(1|0)=0.4.

a) The source outputs the sequence '011000010101100 0'. Determine the run-length codes for both methods illustrated in Fig. 5.1a/b.
b) Compute the probabilities for '0' and '1' sequences of lengths 1,2 and 3.
c) Determine the probabilities for run lengths 0,1,2 and 3 for the method of Fig. 5.1b. The run lengths shall be encoded by a Huffman code, where an ESCAPE symbol expresses run lengths larger than 3. Determine the code string lengths for the run lengths 0-3 and for ESCAPE.
d) What is the average rate per run length symbol, when the remaining run lengths up to 63 are coded by a fixed length code?

Problem 4.6.

In a VQ constellation using vector length K=2 and J=2 reconstruction values, the codebook shall be optimized using the Generalized Lloyd algorithm (see Sec. 4.5.3) The training set is $\mathcal{X}=\{[-1\ 0]^T;[-3\ -2]^T;[\ 1\ 1]^T;[-5\ -4]^T;[0\ 1]^T;[1\ 0]^T;[2\ 2]^T\}$. Initially, a codebook with vectors $\mathbf{y}_0=[-1\ -1]^T$ and $\mathbf{y}_1=[1\ \ 1]^T$ shall be used.

a) Determine the optimized $\mathbf{y}_0$ and $\mathbf{y}_1$ in two iteration steps.
b) For the second iteration, determine the optimized vectors for an entropy constrained VQ method (see Sec. 4.5.5) with value λ=65. The self information $I(S_j)$ shall be determined from the occurrence counts of the $\mathbf{y}_j$ selections in the first iteration. From the first iteration, determine the boundary line between the two quantizer cells for cases λ=0 and λ=65. Sketch the boundary lines and reconstruction vectors in $\mathcal{R}^2$.

Problem 4.7.

The PDF $p_\mathbf{s}(x_1,x_2)$ of two-dimensional random vectors $\mathbf{s}(n)$ shall be uniform as sketched in Fig. 4.37.

a) Express the PDF in dependency of x_1 and x_2.
b) Determine the probability of signal values with $|x_1| \geq A/2$.

A vector quantizer with the following set of reconstruction values is used for quantization: $\mathbf{y}_1=[-a,0]^T$; $\mathbf{y}_2=[a,0]^T$; $\mathbf{y}_3=[0,-a]^T$; $\mathbf{y}_4=[0,a]^T$.

c) Sketch the Voronoi regions and determine the optimum value of a.
d) Before quantization, a linear block transform shall be performed, such that the vector quantizer can be replaced by a scalar quantizer applied to transform coefficients. Which orthonormal block transform is suitable for this purpose? Sketch the Voronoi regions in the transformed signal space.

e) Determine the variance of the quantization error for the optimum value of a.

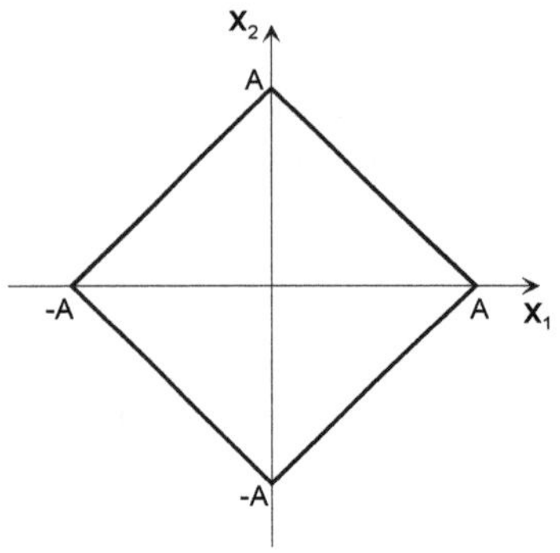

Fig. 4.37. Boundary of a 2D uniform vector PDF

5 Methods of Signal Compression

This chapter introduces fundamental concepts of signal coding, as they are commonly applied for various types of multimedia signals (image, video, graphics, audio, speech) and associated components (e.g. depth maps of images/video). A common property is correlation between samples, which can be removed by signal processing, performing a mapping into an equivalent representation which exposes less statistical dependencies between samples and is sparse, i.e. concentrating information in few relevant samples. The main approaches for this are predictive coding, where the original signal is mapped to a prediction error signal, and transform coding, where a mapping into transform coefficients is made. Sparseness is important, as it allows representing many values by zero, which can be encoded quite efficiently. On the other hand, multimedia signals can hardly be classified as stationary random processes, they typically change local statistical properties. Therefore, adaptation of the compression methods is important, both in predictive and in transform coding. Though the methods of adaptation are often specific for a certain type of multimedia signals and are with more detail discussed in upcoming chapters, common principles are introduced here.

5.1 Run-length coding

Most types of multimedia signals and related components require representation by multiple amplitude levels. Examples requiring direct binary representation include two-tone images and any kind of on-off information e.g. for representation of shapes and contours, or for signalling of on/off switching flags. Methods of entropy coding can directly be applied to binary signals, where in simple cases of sequential dependencies, the conditional probabilities can be modeled by discrete Markov chains (see Sec. 2.5.4).

It is also possible to encode a binary signal into a multi-level signal with less samples using run-length coding (RLC), which expresses positions of *changes* between the two levels in a binary sample sequence. This is a lossless, reversible transformation of a two-level signal into a multi-level signal, the latter typically

© Springer-Verlag Berlin Heidelberg 2015
J.-R. Ohm, *Multimedia Signal Coding and Transmission*,
Signals and Communication Technology, DOI 10.1007/978-3-662-46691-9_5

having a lower number of samples. If combined with variable-length coding, RLC can also be interpreted as a method of conditional entropy coding for the two-level signal. An advantage of the run-length transformation is the implicit joint encoding of binary samples, where the complexity of the code only grows linearly by the maximum run-length[1]. The run length indicates the number of subsequent samples having an equal level in the binary signal. Two different methods are commonly used, which are illustrated in Fig. 5.1; it depends on the properties of the binary signal which of these methods would be the better choice:

- Method A: The run-length code signifies the number of subsequent samples of same value, i.e. in principle the positions of transitions ('0'→'1', '1'→'0') are signaled (Fig. 5.1a). It is further necessary to convey the starting level (here: '1') to the decoder. The smallest possible run length is one. This method is preferable if both levels have approximately identical probability, or if the probability to remain in the states is significantly higher than the probability of a state change.
- Method B: The run-length code describes the number of subsequent samples only for a default level (level 0 in Fig. 5.1b), where a sample of the non-default level implicitly follows after that. A run-length value of zero is required to express that another sample of the non-default level follows immediately. It is not necessary to convey the starting level: This is assumed to be the default level; otherwise, the first run-length has to be zero, as in the example shown. This method is efficient if the default level has significantly higher probability than the non-default level, and/or if longer runs in the less probable level are not typical.

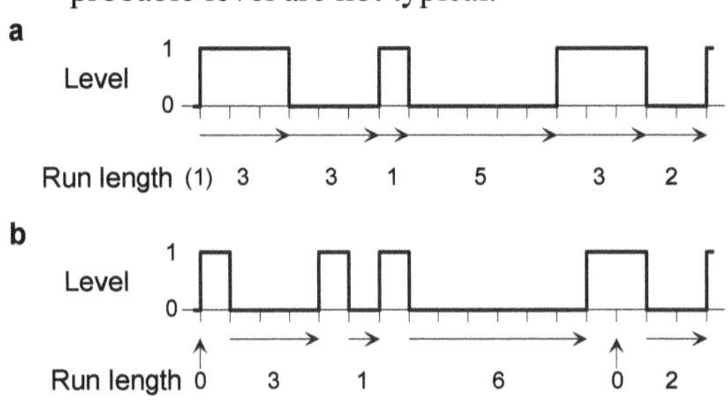

Fig. 5.1. Run-length coding **a** Method A **b** Method B [with default level zero, Pr(0)>Pr(1)]

Run-length coding of 1D binary sequences can be modeled by the 2-state Markov chain (Sec. 2.5.4), for which the entropy is determined by (2.182). Accordingly, the entropy of the two-level signal, separate for the two states, is

[1] In contrast, if binary samples are combined into vectors for the purpose of entropy coding, the complexity would grow exponentially with the vector length. The design of a run-length code, even in tandem combination with VLC of the run length, could have a much lower complexity than direct VLC encoding of the binary sequence.

$$H(S_1) = -\Pr(0\,|\,1)\log_2 \Pr(0\,|\,1) - \Pr(1\,|\,1)\log_2 \Pr(1\,|\,1) \tag{5.1}$$

$$H(S_0) = -\Pr(1\,|\,0)\log_2 \Pr(1\,|\,0) - \Pr(0\,|\,0)\log_2 \Pr(0\,|\,0)\,, \tag{5.2}$$

and the total entropy is the average,

$$H(S) = \Pr(0)H(S_0) + \Pr(1)H(S_1)\,. \tag{5.3}$$

The transition probabilities of the Markov chain model can be adapted from statistics of a given signal by counting the occurrences of run-lengths from a training set, and then using (2.164) to determine best-fitting parameters. Extension to more efficient coding of run-length values for 2D or multi-dimensional signals can be made by defining a binary context $\mathcal{C}=\mathbf{b}$ as in (2.167), which includes more values for determining transition probabilities, e.g. samples from the line above in a binary image (cf. Sec. 6.1).

A coding scheme approaching the entropy rate according to (5.3) can be designed using the expected run length probabilities as determined by (2.164). Observe that (2.164) defines an exponential distribution of probabilities, such that longer run lengths have higher self information and will require longer codewords. Usually, this will lead to code constructions which can be matched by systematic codes such as Golomb-Rice or Exp-Golomb codes (cf. Sec.4.4.3).

The probabilities $\Pr_0(l)$ and $\Pr_1(l)$ of '0'- and '1'-sequences of length l are described in (2.164). They can be used to determine the probabilities of the run-length symbols for the two methods in Fig. 5.1. For the scheme of Fig. 5.1a, though the count of '0' and '1' runs is equal, the probabilities of the lengths in the two states can be different. For runs with lengths $l>0$ in states '0' and '1', entropy coding should be designed according to the probabilities

$$\Pr_0(l) = \Pr(1\,|\,0)\big[1-\Pr(1\,|\,0)\big]^{l-1} \quad ; \quad \Pr_1(l) = \Pr(0\,|\,1)\big[1-\Pr(0\,|\,1)\big]^{l-1}\,. \tag{5.4}$$

For the scheme of Fig. 5.1b, the probability of two subsequent '1' symbols (to be represented by run length $l=0$) is

$$\Pr(l=0) = \Pr(1\,|\,1). \tag{5.5}$$

The probabilities of run lengths $l>0$ are determined assuming that a change from '1' into '0' occurs, and a run of length l in state '0' follows, which gives with (2.164)

$$\Pr(l) = \Pr(0\,|\,1)\Pr_0(l) = \Pr(0\,|\,1)\Pr(1\,|\,0)\underbrace{\big[1-\Pr(1\,|\,0)\big]}_{\Pr(0|0)}{}^{l-1}. \tag{5.6}$$

Run-length information is highly vulnerable by transmission losses. This can even become more severe, when combined with variable-length coding. Hence, the usage of resynchronization mechanisms, as discussed in Sec. 4.4.6, is important.

5.2 Predictive coding

Predictive coding is applied in various speech, image and video compression schemes, either to the signal itself, or to side information parameters such as motion vectors, quantization step sizes etc. Also transform coefficients are often predictively encoded. Further, predictive coding is often applied for redundancy removal in lossless compression of signals and parameters.

5.2.1 Open-loop and closed-loop prediction systems

In Fig. 2.30, the general principle of linear prediction has been shown. In general, even when the input amplitude values $s(\mathbf{n})$ are discrete (PCM representation), prediction error samples $e(\mathbf{n})$ may be of continuous amplitude, such that entropy coding cannot directly be applied to the prediction error signal. By enforcing quantized estimates $\hat{s}(\mathbf{n})$ from the same finite set of discrete values as used for the PCM representation of $s(\mathbf{n})$ both at the encoder and decoder, $e(\mathbf{n})$ will be from a finite discrete set as well, can be directly entropy coded, and the decoded $e(\mathbf{n})$ can be used for perfect reconstruction of $s(\mathbf{n})$ (lossless coding case). Compared to independent sample-wise encoding (PCM), a coding gain can be realized which should be in the order of magnitude as derived in (4.33), (4.31) and (4.41) for the case of AR processes.

A different situation occurs, when a more aggressive quantization needs to be applied to $e(\mathbf{n})$ for the purpose of lossy coding. The prediction error samples are mapped into reconstruction values $e_Q(\mathbf{n})$, as shown in Fig. 5.2a[2], such that a quantization error $e_D(\mathbf{n})=e(\mathbf{n})-e_Q(\mathbf{n})$ occurs. The resulting effect can be expressed in the spectral domain for any linear prediction system, where frequency spectra are defined for the original signal as $S(\mathbf{f})$, the reconstructed signal as $\tilde{S}(\mathbf{f})$, the prediction error signal as $E(\mathbf{f})$, the quantized prediction error signal as $E_Q(\mathbf{f})$, the predictor filter as $H(\mathbf{f})$ and the quantization error as $E_D(\mathbf{f})$. Then, the difference between $S(\mathbf{f})$ and $\tilde{S}(\mathbf{f})$ can be determined as

$$\tilde{S}(\mathbf{f}) = \frac{E_Q(\mathbf{f})}{1-H(\mathbf{f})} = \frac{E(\mathbf{f})}{1-H(\mathbf{f})} - \frac{E_D(\mathbf{f})}{1-H(\mathbf{f})} = S(\mathbf{f}) - \frac{E_D(\mathbf{f})}{1-H(\mathbf{f})}, \tag{5.7}$$

by which

$$S(\mathbf{f}) - \tilde{S}(\mathbf{f}) = \frac{E_D(\mathbf{f})}{1-H(\mathbf{f})} \quad \Rightarrow \quad s(\mathbf{n}) - \tilde{s}(\mathbf{n}) = e_D(\mathbf{n}) * b(\mathbf{n}). \tag{5.8}$$

In (5.8), $b(\mathbf{n})$ is the impulse response of the recursive synthesis filter with transfer function (2.219), $B(\mathbf{f})=1/[1-H(\mathbf{f})]$. The effect of convolving the quantization

[2] The system shown in Fig. 5.2a is also denoted as D*PCM [JAYANT, NOLL 1984].

error by the impulse response of the synthesis filter is denoted as *drift*. If the source $s(\mathbf{n})$ is highly correlated, as it is the case with image signals, the filter typically has a long impulse response such that the drift could become severe, leading to a significant increase of the reconstruction error.

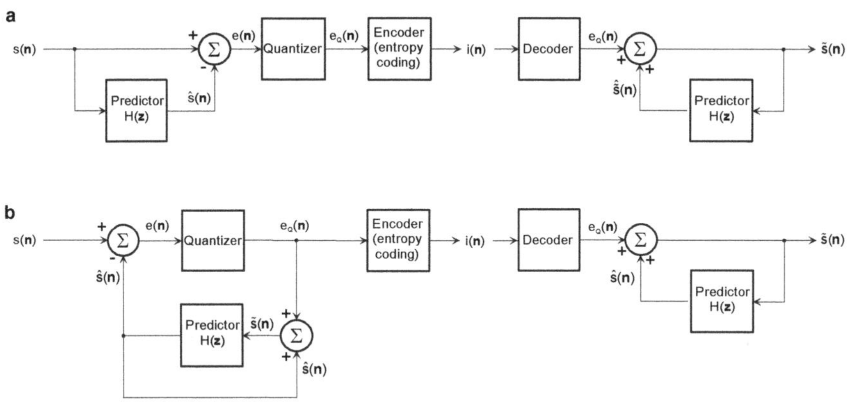

Fig. 5.2. Encoders and decoders of predictive systems with quantization of the prediction error signal: **a** open loop **b** closed loop (DPCM)

The system from Fig. 5.2a has an *open loop* structure, the recursion at the decoder is not under control from the encoder side. As a consequence, the estimates $\hat{s}(\mathbf{n})$ and $\hat{\hat{s}}(\mathbf{n})$, the latter derived from the reconstructed values, diverge. A *Differential Pulse Code Modulation* (DPCM) system, which is a closed-loop approach and compensates this deficiency, is shown in the schematic diagram of Fig. 5.2b. The predictor uses estimates $\hat{s}(\mathbf{n}) = \hat{\hat{s}}(\mathbf{n})$ for the sample $s(\mathbf{n})$, which are computed from the available reconstruction $\tilde{s}(\mathbf{n})$ identically at the encoder and decoder. The *prediction error*

$$\tilde{e}(\mathbf{n}) = s(\mathbf{n}) - \hat{\hat{s}}(\mathbf{n}) \tag{5.9}$$

is quantized into the value $\tilde{e}_Q(\mathbf{n})$, which is then encoded and transmitted. The decoder performs the following operation to generate the reconstructed value

$$\tilde{s}(\mathbf{n}) = \tilde{e}_Q(\mathbf{n}) + \hat{\hat{s}}(\mathbf{n})\,. \tag{5.10}$$

For generating an identical estimate $\hat{\hat{s}}(\mathbf{n})$, the decoding step (5.10) has also to be executed at the encoder side. As a result, the quantization error between $\tilde{e}(\mathbf{n})$ and $\tilde{e}_Q(\mathbf{n})$ becomes identical with the reconstruction error (coding error) between $s(\mathbf{n})$ and $\tilde{s}(\mathbf{n})$,

$$\tilde{e}(\mathbf{n}) - \tilde{e}_\mathrm{Q}(\mathbf{n}) = \left[s(\mathbf{n}) - \hat{\tilde{s}}(\mathbf{n}) \right] - \left[\tilde{s}(\mathbf{n}) - \hat{\tilde{s}}(\mathbf{n}) \right] = s(\mathbf{n}) - \tilde{s}(\mathbf{n}) = s_\mathrm{D}(\mathbf{n}). \quad (5.11)$$

Only for the case $\tilde{e}_\mathrm{Q}(\mathbf{n}) = \tilde{e}(\mathbf{n})$, i.e. when lossless encoding of $\tilde{e}(\mathbf{n})$ is performed, closed- and open-loop methods give identical results.

The transfer function of the predictor filter is still following (2.219), however the input-output relationship of a DPCM encoder loop can no longer be described as linear system, since the mapping $s(\mathbf{n}) \rightarrow e_\mathrm{Q}(\mathbf{n})$ includes the nonlinear operation of quantization. In analogy with (4.33), the coding gain of a DPCM system is defined as the ratio of variances between original signal and prediction error signal,

$$G = \frac{\sigma_s^{\,2}}{\sigma_{\tilde{e}}^{\,2}}. \quad (5.12)$$

The DPCM solution of the drift problem does not come without penalty. The related effect is known as *quantization error feedback*. Formulating (5.9)-(5.11) in the frequency domain as

$$\tilde{E}_\mathrm{D}(\mathbf{f}) = \tilde{E}(\mathbf{f}) - \tilde{E}_\mathrm{Q}(\mathbf{f}) = S(\mathbf{f}) - \tilde{S}(\mathbf{f}) \;\text{ and }\; \tilde{E}_\mathrm{Q}(\mathbf{f}) = \tilde{S}(\mathbf{f})\big(1 - H(\mathbf{f})\big), \quad (5.13)$$

gives

$$\begin{aligned} \tilde{E}_\mathrm{Q}(\mathbf{f}) &= \left[S(\mathbf{f}) - \tilde{E}_\mathrm{D}(\mathbf{f}) \right]\left[1 - H(\mathbf{f})\right] \\ \tilde{E}(\mathbf{f}) &= \tilde{E}_\mathrm{Q}(\mathbf{f}) + \tilde{E}_\mathrm{D}(\mathbf{f}) = \underbrace{S(\mathbf{f})\left[1 - H(\mathbf{f})\right]}_{E(\mathbf{f})} + \tilde{E}_\mathrm{D}(\mathbf{f}) H(\mathbf{f}). \end{aligned} \quad (5.14)$$

Here, $E(\mathbf{f})$ would be the prediction error spectrum with the open-loop case using the same predictor. The prediction error in closed-loop DPCM

$$\tilde{e}(\mathbf{n}) = e(\mathbf{n}) + \tilde{e}_\mathrm{D}(\mathbf{n}) * h(\mathbf{n}), \quad (5.15)$$

indicates that the DPCM prediction error signal $\tilde{e}(\mathbf{n})$ (compared to the open-loop case) additionally contains the convolution of the quantization error $\tilde{e}_\mathrm{D}(\mathbf{n})$ by the impulse response of the predictor filter. As this is statistically independent of $e(\mathbf{n})$, it is unavoidable that the variance increases and the prediction gain of DPCM (5.12) is lower than the maximum possible gain (4.33) in case of lossy encoding. The convolution also has the effect that the prediction error becomes correlated, unlike the innovation signal of an AR process which would be white noise. This means that further compression of the prediction error could be made possible by exploiting the feedback phenomenon, which is however not easy to achieve due to the potentially large number of states that would need to be traced in look-ahead decisions. Further, this effect increases for lower rates. For the extreme case of quantizing all samples into zero, the entire information is lost such that $\tilde{E}_\mathrm{Q}(\mathbf{f}) = 0$, $\tilde{E}_\mathrm{D}(\mathbf{f}) = S(\mathbf{f})$ and finally $\tilde{S}(\mathbf{f}) = 0$. With $D = \sigma_s^2$, $R = 0$ this is the point where all rate-distortion graphs in Fig. 4.5 converge; also prediction becomes useless, as the decoder does not have any information about previous samples available.

5.2.2 Non-linear and shift-variant prediction

Though for the case of using LSI systems in prediction it is helpful to analyze the effects in the frequency domain, prediction can be performed by non-linear and shift-variant systems as well. This can be particularly useful when the signal has locally varying statistical properties, and/or higher-order statistical dependencies which cannot be modeled by covariance analysis. Examples for non-linear and/or shift-variant predictors are[3]

- Median and weighted median predictors
- Predictors imposing rules that control the influence of certain samples in the prediction, e.g rejection of samples that are regarded as 'outliers', or give preference to samples that are expected to provide best extrapolation according to direction, motion, pitch frequency etc.
- Predictors using extrapolation based on non-linear function fitting
- Higher-order polynomial (e.g. Volterra filter) predictors
- Linear predictors with local adaptation of coefficients
- Predictors using signal-dependent shift or scaling of sampling positions
- Predictors using neural networks.

Schemes using an appropriate *local adaptation* can be expected to give a performance gain when signals are non-stationary, regardless of the prediction method being linear or non-linear. Components of predictor adaptation are included in the block diagram of Fig. 5.3. Based on local analysis and estimation of signal statistics, predictor parameters are determined typically with the target of minimizing the prediction error variance. Examples are the optimization of linear predictors based on autocovariance statistics (see Sec. 2.6.2) and motion estimation. The basic methods applied in adaptive prediction are:

- Forward adaptation: The optimization of parameters is performed using samples from the segment/region to be encoded, and it is necessary to convey the prediction filter parameters as side information to the decoder.
- Backward adaptation: Adjustment of predictor coefficients is performed from signal values and prediction error values which are already known to the decoder (previously received and reconstructed). It is not necessary to encode and transmit predictor parameters, since the operation can identically be performed at the encoder and decoder sides; no overhead information increases the bit rate, but the method may be more sensitive to local change of signal statistics and transmission errors.

---

[3] Quantization of the prediction $\hat{s}(\mathbf{n})$ would also introduce a nonlinearity, which can however be neglected if the precision is high.

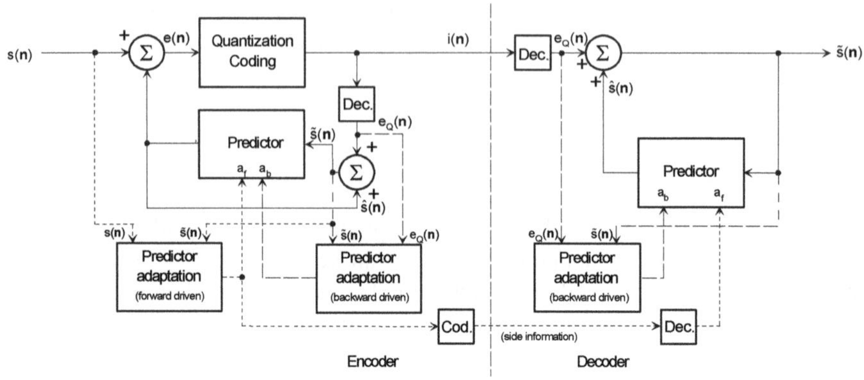

Fig. 5.3. DPCM encoder and decoder, integrating components for predictor adaptation

Usage of forward-adaptive and backward-adaptive methods is not exclusive, and it can be advantageous to combine elements from both. For example, it may be useful to explicitly invoke a certain backward adaptive element at the decoder only when signaled by a flag, to be determined by the encoder, which can evaluate its benefit directly. Likewise, as methods of backward adaptation by tendency save the transmission of side information, parameters determined by knowledge available at the decoder could also be used for efficient encoding ('prediction of the side information'). In such schemes, the encoder should precisely replicate the decoder actions, as deviation between adaptations performed at the encoder and decoder could cause additional drift effects.

5.2.3 Effects of transmission losses

As discussed above, drift generally occurs when different predictions are used at the encoder and decoder sides in predictive coding. Even when operating in a closed loop, the encoder could make wrong assumptions about the information which is available for prediction at the decoder, when *transmission losses* occur in error-prone channels. To avoid interrelationships between quantization error feedback and transmission error propagation, lossless encoding is assumed in the subsequent analysis, such that $E_Q(\mathbf{f})=E(\mathbf{f})$ and $\tilde{S}(\mathbf{f})=S(\mathbf{f})$ if no errors occur on the channel. Assume that by data loss a component $\hat{E}(\mathbf{f})$ is superimposed to the reconstructed prediction error before being fed into the synthesis filter. The reconstruction signal is then influenced as follows:

$$\tilde{E}(\mathbf{f})=E(\mathbf{f})+\hat{E}(\mathbf{f})$$
$$\tilde{S}(\mathbf{f})=\frac{E(\mathbf{f})}{1-H(\mathbf{f})}+\frac{\hat{E}(\mathbf{f})}{1-H(\mathbf{f})}=S(\mathbf{f})+\frac{\hat{E}(\mathbf{f})}{1-H(\mathbf{f})}. \quad (5.16)$$

The additional interference component in the reconstructed signal is the convolution of the channel error component $\widehat{E}(\mathbf{f})$ by the impulse response of the (recursive) synthesis filter. If $\widehat{e}(\mathbf{n})$ has characteristics of impulse-like noise (e.g. random bit errors), the reconstruction contains superimposed impulse responses of the synthesis filter, originating from positions where errors occur. In case of total loss, $\widehat{E}(\mathbf{f}) = -\tilde{E}_Q(\mathbf{f})$, which causes a similar (increasing over time) drift effect as in open-loop coding (5.8),

$$S(\mathbf{f}) - \tilde{S}(\mathbf{f}) = \frac{-\widehat{E}(\mathbf{f})}{1-H(\mathbf{f})} = \frac{\tilde{E}_Q(\mathbf{f})}{1-H(\mathbf{f})} \quad \Rightarrow \quad s(\mathbf{n}) - \tilde{s}(\mathbf{n}) = \tilde{e}_Q(\mathbf{n}) * b(\mathbf{n}). \quad (5.17)$$

5.2.4 Vector prediction

Predictive coding can be applied on groups of samples such that a vector prediction $\hat{\mathbf{s}}(\mathbf{m})$ is computed from past (decoded) samples $\tilde{\mathbf{s}}(\mathbf{m}')$, generating a residual vector,

$$\mathbf{r}(\mathbf{m}) = \mathbf{s}(\mathbf{m}) - \hat{\mathbf{s}}(\mathbf{m}) \text{ with } \hat{\mathbf{s}}(\mathbf{m}) = f\left\{\tilde{\mathbf{s}}(\mathbf{m}') \mid \mathbf{m}' < \mathbf{m}\right\}. \quad (5.18)$$

A block diagram of this vector extension of DPCM is shown in Fig. 5.4. The benefit of such a method could be similar as for the case of VQ overcoming scalar quantization; furthermore, residual vector codebooks could be designed in a way that statistical properties of the prediction error signal (including properties of quantization error feedback which occurs at lower rates) could be better exploited.

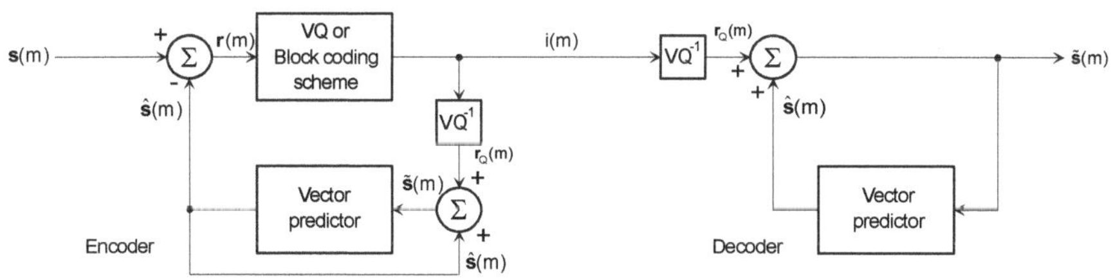

Fig. 5.4. Vector predictive coding (closed loop scheme, encoder and decoder)

The fact that samples cannot be predicted from other samples within the same vector implies that the prediction error samples are no longer obtained by *one-step* prediction as assumed so far in DPCM. A 1D vector prediction of samples from an AR(1) process shall be considered here, where a current K-dimensional vector $\mathbf{s}=[s(n+1)\ldots s(n+K)]^{\mathrm{T}}$ is predicted and mapped into a residual error vector $\mathbf{r}=[r(1)\ldots r(K)]^{\mathrm{T}}$. Due to the AR(1) property, all samples of the current vector are best predicted only from the closest (last) sample $s(n)$ of the previous vector[4].

[4] Actually, the prediction should be performed from the reconstructed sample $\tilde{s}(n)$ in a closed-loop method. We assume lossless compression here to study the effect of vector prediction without further quantization error feedback.

Taking into account the covariance between samples, the optimum prediction for the samples of the current vector is

$$\hat{s}(n+k) = \rho^k s(n) \Rightarrow r(k) = s(n+k) - \rho^k s(n) \; ; \; k > 0 . \tag{5.19}$$

The variance increases, i.e. the prediction becomes worse for samples that are farther away from $s(n)$. The variance of the kth sample in the residual is[5]

$$\mathcal{E}\{r^2(k)\} = \mathcal{E}\{s^2(n+k)\} + \rho^{2k}\mathcal{E}\{s^2(n)\} - 2\rho^k \underbrace{\mathcal{E}\{s(n)s(n+k)\}}_{\sigma_s^2\rho^k} = \sigma_s^2\left(1-\rho^{2k}\right), \tag{5.20}$$

which increases over k. Autocovariance values between the samples in $\mathbf{r}$ are

$$\begin{aligned}\mathcal{E}\{r(l)r(k)\} &= \mathcal{E}\{\left(s(n+l)-\rho^l s(n)\right)\left(s(n+k)-\rho^k s(n)\right)\}\\ &= \underbrace{\mathcal{E}\{s(n+l)s(n+k)\}}_{\sigma_s^2\cdot\rho^{|k-l|}} - \rho^l \underbrace{\mathcal{E}\{s(n)s(n+k)\}}_{\sigma_s^2\cdot\rho^k}\\ &\quad - \rho^k \underbrace{\mathcal{E}\{s(n)s(n+l)\}}_{\sigma_s^2\cdot\rho^l} + \rho^l\rho^k \underbrace{\mathcal{E}\{s^2(n)\}}_{\sigma_s^2}\\ &= \sigma_s^2\left(\rho^{|k-l|} - \rho^{|k+l|}\right).\end{aligned} \tag{5.21}$$

This, the samples within the prediction error vector are correlated, though the correlation is weaker than for samples within an original AR(1) random vector[6]. Therefore, additional compression gain is achieved when decorrelation is applied to $\mathbf{r}$, e.g. by employing a linear transform, or joint quantization (VQ). In the latter case, the VQ codebook design should inherently take into account the statistical dependencies within the residual vectors $\mathbf{r}$.

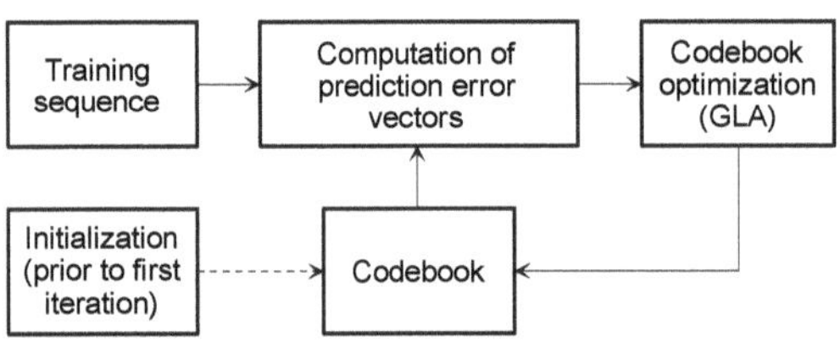

Fig. 5.5. Codebook design in vector DPCM

For best adaptation of the codebook to the prediction error signal statistics (where in particular the properties of within-vector correlations and the rate-dependent quantization error feedback are important), the *closed-loop codebook design* shown in Fig. 5.5 is used during the GLA iterations. The mapping from original

---

[5] Furthermore, vector prediction is shift variant, i.e. the residual depends on the start position of the vector.

[6] This effect similarly applies to the case of 2D vector prediction (cf. Sec. 6.3.2).

image signal values into prediction error vectors is renewed by each iteration in this case [CUPERMAN, GERSHO 1985].

If lossless coding is performed, a stable prediction can be generated from original samples (as they are perfectly reconstructed by the decoder). In this case, preceding samples *within the vector* $\mathbf{s}(\mathbf{m})$ can be used to predict subsequent samples, and the residual vector $\mathbf{r}$ can be constructed from a series of prediction error samples $e(\mathbf{n}) = s(\mathbf{n}) - \hat{s}(\mathbf{n})$, with $\hat{s}(\mathbf{n}) = f\{s(\mathbf{n}') \mid \mathbf{n}' < \mathbf{n}\}$, as in conventional sample-wise prediction, equivalent to the system shown in Fig. 2.30b. If for the 1D case $\mathbf{s}(m)$ was a random vector from an AR(1) process, in contrast to (5.21) samples in the residual vector $\mathbf{r}(m)=\mathbf{e}(m)$ will be uncorrelated, the variance $\sigma_e{}^2=\sigma_s{}^2(1-\rho^2)$ would be independent of the position within in the vector and its starting position (shift invariance applies).

The approach of using one-step (sample wise) prediction while arranging the prediction error samples in vectors $\mathbf{e}(\mathbf{m})$ shall now be extended to the case of lossy coding. A synthesis filter as in Fig. 2.30c can easily be excited by samples drawn from a sequence of vectors. However, in case of lossy coding, due to the convolution of the quantization error (5.8) that also occurs between samples of the current vector, a direct mapping between $\mathbf{e}(\mathbf{m})$ and a coded version $\tilde{\mathbf{e}}$ (e.g. vector drawn from a codebook) does no longer coincide with the reconstruction error $\mathbf{s}(\mathbf{m}) - \tilde{\mathbf{s}}(\mathbf{m})$. A possible solution would be an encoder performing reconstruction of vectors $\tilde{\mathbf{s}}(\mathbf{m})$ from each possible coded version $\tilde{\mathbf{e}}$ of the excitation, which is denoted as *analysis by synthesis* (A/S) coding. Such an approach is however fairly complex, as in addition to the distortion comparison, a convolution (recursive filter) operation is necessary for each $\tilde{\mathbf{e}}$.

A possible simplification can be achieved when the prediction is decomposed into a 'deterministic' part $\hat{\mathbf{s}}$ which only uses the samples from previously decoded vectors $\tilde{\mathbf{s}}(\mathbf{m}')$ as in (5.18), and another part which complements the influence of preceding prediction error samples in the current vector $\mathbf{e}(\mathbf{m})$. This decomposition is possible in case of linear prediction, where the superposition principle holds. For the lossless case, the mapping of prediction error (innovation) samples $\mathbf{e}$ into a vector of signal samples $\mathbf{s}$ can be expressed by a linear matrix operation $\mathbf{s} = \mathbf{Be} + \hat{\mathbf{s}}$, where $\hat{\mathbf{s}}$ signifies the initial prediction from past decoded samples, and $\mathbf{Be}$ expresses the influence of sample-to-sample prediction in $\mathbf{s}$, i.e. it is determined by the synthesis filter. Note that $\hat{\mathbf{s}}$ is identical for any choice of $\mathbf{e}$, and the reconstruction is identical to the original $\mathbf{s}$. Therefore, it is possible to define the following mapping to the residual signal in (5.18), which would be generated only from the 'deterministic' part,

$$\mathbf{r} = \mathbf{s} - \hat{\mathbf{s}} = \mathbf{Be} \Rightarrow \mathbf{e} = \mathbf{B}^{-1}\left[\mathbf{s} - \hat{\mathbf{s}}\right]. \tag{5.22}$$

The advantage is that in the domain of $\mathbf{r}$ the coding error is identical to the reconstruction error $\mathbf{s} - \tilde{\mathbf{s}}$ even in the lossy coding case. The linear matrix mapping further implies, if a quantized representation $\tilde{\mathbf{e}}$ is used for $\mathbf{e}$, such that the reconstruction error is

$$\mathbf{s}-\tilde{\mathbf{s}}=\mathbf{r}-\tilde{\mathbf{r}}=\mathbf{B}\left[\mathbf{e}-\tilde{\mathbf{e}}\right]. \tag{5.23}$$

In terms of squared error (Euclidean distance) this gives

$$\left[\mathbf{s}-\tilde{\mathbf{s}}\right]^{\mathrm{T}}\left[\mathbf{s}-\tilde{\mathbf{s}}\right]=\left[\mathbf{e}-\tilde{\mathbf{e}}\right]^{\mathrm{T}}\mathbf{B}^{\mathrm{T}}\mathbf{B}\left[\mathbf{e}-\tilde{\mathbf{e}}\right]. \tag{5.24}$$

$\mathbf{B}^{\mathrm{T}}\mathbf{B}$ can therefore be used as a weighting matrix when determining the squared error in the prediction error (**e**) domain[7]. Compared to the A/S approach, this can be less complex if $\mathbf{B}^{\mathrm{T}}\mathbf{B}$ can be pre-computed (which is not the case for adaptive prediction) or if it would be relatively sparse (which cannot be expected when it expresses a recursive filter with an impulse response longer than the current vector). Alternatively, the weighting can be performed in the frequency domain (entitled as 'frequency domain noise shaping'), where a weighting of the squared coding error by the squared Fourier transfer function of the synthesis filter is applied. Since in adaptive linear prediction the impulse response of the synthesis filter represents an approximation of the spectral shape of the encoded signal, such a weighting can further be used to exploit masking properties (quantization error less perceivable). This is one of the reasons why in particular for speech signals open loop prediction can well be applied by partially ignoring the drift effect.

5.2.5 Delayed-decision predictive coding

In case of VQ or other codebook-type presentation of the innovation vectors **e**, a search for the best match in (5.24) is applied over a discrete set. However, in general the number of choices could tend towards infinity, which could be the case when the vector **e** is input to transform coding with large number of quantization choices, or if a similar scheme is used to evaluate the effect of a current quantization decision to any subsequent samples as a method of 'delayed decision coding'. As an example for this, [SCHUMITSCH ET AL. 2003] proposed a video encoder which maps the effect of motion compensated prediction into a linear matrix (conceptually similar to **B**, but relatively sparse in that case, since every sample only takes influence on few samples in subsequent pictures), and solve the problem of finding $\tilde{\mathbf{e}}$ s.t. $\left\|\mathbf{r}-\mathbf{B}\tilde{\mathbf{e}}\right\|^2=\min$ by linear regression. Rate constraints can be included in such an optimization as well. Nevertheless, such schemes also turn out to be extremely complex in terms of computational effort.

At the decoder side, A/S coding has only moderate complexity, regardless if it is used in the context of vector prediction, or for delayed-decision operation at the encoder[8]. If the degrees of freedom in the encoding of recursively dependent samples are large, as it is usually the case for multi-dimensional sources, A/S encoding with full search is not manageable any more in terms of complexity. As an alternative, the comparison can be restricted to the most promising cases, and

---

[7] (5.23) is equivalent to the convolution (5.8), expressed as matrix operation over one vector.

[8] This is often implemented as multi-pass coding, investigating the propagation of different encoder options.

multiple encoding passes are run to select the best. If the dependency of decisions can be mapped into a graph structure such as a tree or trellis (cf. Sec. 4.6), multi-path search algorithms can also be used for that purpose. An example for this is *tree coding of correlated sources*, where an AR synthesis filter is used as code generator (see Fig. 2.27), which is fed by an innovation signal from a codebook; this could be represented by scalar or vector values. If J different code symbols are used at each encoding/decoding step, the possible states of the decoder can be interpreted as a Jary code tree (Fig. 5.6).

Determining the globally optimum path through this tree would be highly complex, but sub-optimum algorithms, which retain a certain number M of best choices of nodes at each level have a complexity independent of the decision constraint length L and are efficient for this purpose [ANDERSON, MOHAN 1984] [OHM, NOLL 1990].

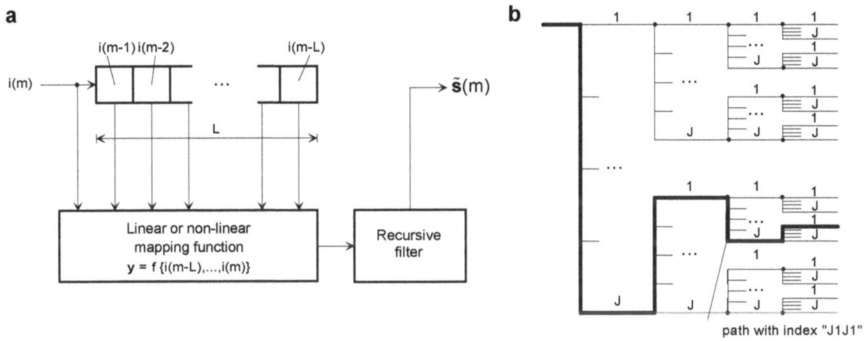

Fig. 5.6. **a** Decoder and **b** code tree in tree coding

5.2.6 Prediction in multi-resolution pyramids

If a lower-resolution version of the signal is available, it can be used to obtain an estimate $\hat{s}(\mathbf{n})$ for the current position in a higher-resolution version. In simplest case, the lower-resolution version can be generated by skipping samples which shall be predicted from other, retained samples. An example with two levels of hierarchy (subsampling 4:1 used to predict missing samples in subsampling 2:1, subsampling 2:1 to predict missing samples in full resolution) applied to a 1D signal is shown in Fig. 5.7[9].

The principle is equivalent to the Laplacian pyramid scheme in Fig. 2.66a, but omitting the anti-alias downsampling filter $H(\mathbf{z})$. The interpolation filter $G(\mathbf{z})$ still needs to be applied for the prediction of the missing samples in the next higher layer[10]. In contrast to the original Laplacian pyramid scheme, due to the fact that

---

[9] Only the case with dyadic (factor 2) subsampling will be analyzed in detail in this section, but other cases can be operated similarly.

[10] In the example of Fig. 5.7, two adjacent samples are averaged to predict the sample in between, whereas the retained samples are unchanged. Then, the interpolation filter is a

the retained samples are unchanged, no differential information is needed to reconstruct them, such that this is a critically-sampled multi-resolution representation (same number of samples accumulated over all pyramid levels as the original signal has).

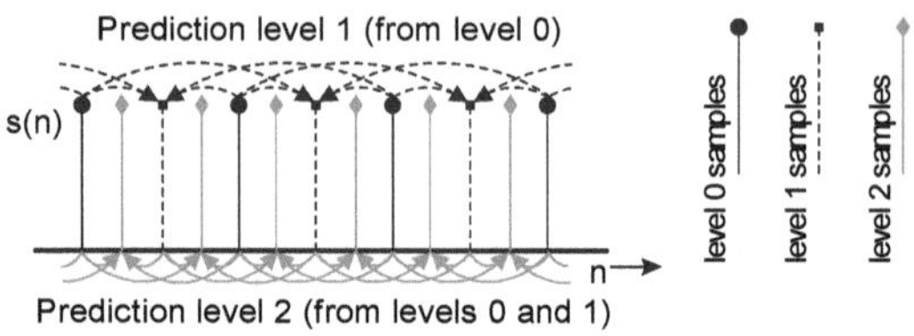

Fig. 5.7. Principle of prediction in multi-resolution pyramid (3 levels, two-sided prediction from 4 samples) in a 1D signal

Another interpretation would be that the downsampling signal establishes one polyphase component, which is used to predict the other polyphase component(s). However, alias would typically occur by unfiltered downsampling, which typically decreases the performance of the prediction when the signal contains high-frequency components. This behaviour can best be analyzed in the frequency domain. Due to the downsampling without filtering, alias occurs such that in the case of a 1D signal

$$\hat{S}(f)=\frac{1}{2}\left[S(f)+S(f-1/2)\right]G(f), \tag{5.25}$$

where $G(f)$ is the Fourier transfer function of the interpolation filter. With $G(f)=2\cos^2(\pi f)$ as in the example of Fig. 5.7,

$$E(f)=S(f)-\hat{S}(f)=S(f)\left[1-\cos^2(\pi f)\right]-S(f-1/2)\cos^2(\pi f). \tag{5.26}$$

If $S(f)$ has extreme lowpass characteristics, the alias spectrum $S(f-1/2)$ would be largely suppressed by the cosine-square filter which is a comb filter approaching zero for $f\to 1/2$. Otherwise, the alias component deteriorates the quality of the prediction[11].

If a lowpass filter is applied before downsampling to suppress the alias component, the downsampled signal is no longer a polyphase component of the higher resolution, the lower resolution samples cannot directly be inserted for reconstruction, and prediction needs to be performed for all positions in the higher resolution. As a consequence, the representation is over-complete (Laplacian pyramid, cf. Sec. 2.8.6). The prediction is

linear interpolator with $\mathbf{g}=[1/2\ 1\ 1/2]^{\mathrm{T}}$, such that $|G(f)|=[1+\cos(2\pi f)]=2\cos^2(\pi f)$. An even lower quality producing more alias would come by hold interpolation (sample copy) with $\mathbf{g}=[1\ 1]^{\mathrm{T}}$, $|G(f)|=2\cos(\pi f)$. Higher-quality interpolators could be used as well.

[11] The scheme discussed here is applied in the context of video compression in hierarchical B picture prediction (see Fig. 7.8b). In that case, by using motion compensation, the spectrum over the time axis usually fulfills the desirable property mentioned here.

$$\hat{S}(f) = \frac{1}{2}\left[S(f)H(f) + S(f-1/2)H(f-1/2\right]G(f), \tag{5.27}$$

such that[12]

$$\begin{aligned} E(f) &= S(f) - \hat{S}(f) \\ &= S(f)\left[1 - H(f)G(f)\right] - S(f-1/2)H(f-1/2)G(f). \end{aligned} \tag{5.28}$$

An optimum prediction for this case requires usage of ideal lowpass filters $H(f)$=$G(f)$=rect($2f$) to suppress the alias components and retain all signal components below $|f|=1/4$ in the prediction. In that case, the residual signal would only contain the high-frequency components[13]. The residual signal has the same resolution as the original signal at the respective pyramid level, furthermore it will be correlated, where in case of ideal filtering

$$\phi_{ee}(f) = \phi_{ss}(f)\,\mathrm{rect}\left(2f - 1/2\right) \bullet\!\!-\!\!\circ\, \varphi_{ee}(m) = \varphi_{ss}(m) * \left[\frac{1}{2}\mathrm{si}(\pi m/2)\left(-1\right)^m\right]. \tag{5.29}$$

Both open- and closed-loop prediction concepts can be applied in pyramid prediction structures. In a closed-loop system, the encoder has to perform local decoding of the lower resolution before using it for the prediction of the higher resolution. To counteract the disadvantage of an over-complete pyramid and avoid overly increased bitrate, it is appropriate to exploit redundancy between the residual samples[14]. With multi-resolution methods, the sequence of dependent predictions is typically shorter than in sample-recursive prediction. As a consequence, the decoded result is less affected by possible channel errors, provided that the samples which take influence on many samples (i.e. the lower resolution layers) are appropriately protected e.g. by channel coding mechanisms.

In *open-loop prediction*, quantization errors are propagated into the subsequent dependent decoding steps similar to (5.8). In the case of a pyramid, the prediction synthesis filter is the interpolation filter, where the variance of the quantization error propagation can be determined from the interpolator's impulse response. In case of an ideal interpolator (sinc function) the amplification of the quantization error variance per sample depends on the upsampling ratio, but it will be band limited in the samples of the next-higher resolution level(s), and therefore the total error tends to become larger in the low-frequency range of the spectrum

---

[12] Depending on signal properties, it may be useful to employ a downsampling filter with a phase shift of half a sample, to obtain symmetry in the prediction of the (even/odd) polyphase positions of the full-resolution signal.

[13] This means that in the ideal filter case also the prediction error signal could in principle be subsampled, similar as in subband systems, see Fig. 2.49c. This would however introduce shift variance which may be undesirable depending on the context where it is used (e.g. in combination with motion compensation).

[14] This could be done by transform coding or by context-dependent entropy coding. Note that no prediction recursion is applied between the residual samples within a given level of the pyramid. As neighbored samples in the higher resolution are predicted from the same lower-resolution samples, the associated residuals can be redundant.

when the full resolution is reconstructed (see Fig. 5.8 for the case of 4:1 downsampling with K=3, i.e. two differential pyramid levels).

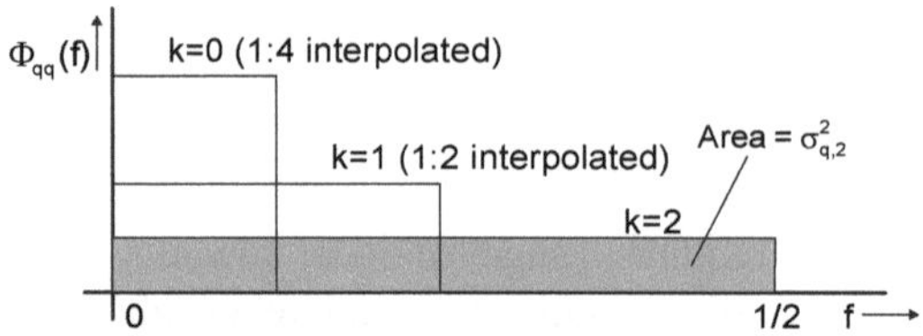

Fig. 5.8. Effect of open-loop quantization error propagation in a multi-resolution pyramid for case of identical quantization in all levels and assuming ideal interpolation filters in a 1D signal (case of K=3 levels shown). The total error increases by a factor of K, but is not uniformly distributed over frequencies.

Since in a differential pyramid the errors of all levels accumulate and can be regarded as statistically independent, the total reconstruction error variance for the highest resolution adds from quantization error variances over all levels (where k=0 denotes the downscaled signal, $k=1\ldots K-1$ are the differential levels of the prediction),

$$\sigma_{\mathrm{q,tot}}^2 = \sum_{k=0}^{K-1} \sigma_{\mathrm{q},k}^2 \,. \tag{5.30}$$

However, due to the fact that the quantization error propagated from the lower layers has lowpass characteristics, the overall reconstruction error will not be spectrally flat (even when the quantization error spectrum was flat in the individual levels, as shown in Fig. 5.8). Therefore, it is appropriate to quantize the lower-resolution levels for less distortion, which will be discussed next.

Assume that a process with Gaussian PDF (where also the residuals are Gaussian with rate-distortion function (4.23)) shall be encoded. The variances of the components in the different levels can be computed from the power spectra and the down- and upsampling filter transfer functions, e.g. for the case of a dyadic decomposition into one subsampled signal (k=0) and one residual (k=1),

$$\begin{aligned}\sigma_{e,0}^2 &= \int_{-1/4}^{1/4} \left[H^2(f)\,\Phi_{ss}(f) + H^2(f-1/2)\,\Phi_{ss}(f-1/2) \right] \mathrm{d}f \\ \sigma_{e,1}^2 &= \int_{-1/2}^{1/2} \Phi_{ss}(f) - \left[H^2(f)\,\Phi_{ss}(f) + H^2(f-1/2)\,\Phi_{ss}(f-1/2) \right] G^2(f)\,\mathrm{d}f.\end{aligned} \tag{5.31}$$

Now, consider the 'low distortion case', where a positive rate is allocated to each level of the pyramid, up to the full-resolution residual. Then, due to the fact that the number of samples doubles with each additional resolution level,

$$R \sim \frac{1}{2}\sum_{k=0}^{K-1} 2^k \log_2 \frac{\sigma_{e,k}^2}{\sigma_{\mathrm{q},k}^2} = \frac{1}{2}\sum_{k=0}^{K-1} 2^k \log_2 \sigma_{e,k}^2 - \frac{1}{2}\log_2 \prod_{k=0}^{K-1} \left(\sigma_{\mathrm{q},k}^2\right)^{2^{-k}} \,. \tag{5.32}$$

The total rate becomes minimum when the right expression is maximized, which is the case when the lower-resolution levels are quantized for less distortion,

$$\left(\sigma_{\mathrm{q},k}^2\right)^{2^{-k}} = \text{const.} \Rightarrow 2^{-k}\log_2 \sigma_{\mathrm{q},k}^2 = \text{const.} \Rightarrow \sigma_{\mathrm{q},k} \sim 2^{k-1}. \tag{5.33}$$

From this result, the quantization step size which linearly influences the standard deviation of the quantization error, should be doubled with each pyramid level. The total quantization error occurring at the highest resolution can then be computed from (5.30),

$$\sigma_{\mathrm{q,tot}}^2 = \sum_{k=0}^{K-1} 2^{k-K+1}\sigma_{\mathrm{q},K-1}^2 = 2^{-K+1}\left(2^K-1\right)\sigma_{\mathrm{q},K-1}^2 = \left(2-2^{1-K}\right)\sigma_{\mathrm{q},K-1}^2, \tag{5.34}$$

which is upper bounded by twice the quantization error variance occurring when encoding only the highest-resolution level signal. This disadvantage would however only occur when the signal to be encoded was spectrally flat (white noise). In the more typical case of correlated signals, the disadvantage occurring due to this increase will pay off when the rate saving due to the lower prediction error variances $\sigma_{e,k}^2$ in (5.32) compensates for it. Note, however, the additional disadvantage of the dyadic over-complete pyramid structure where up to the double number of samples (compared to the full resolution) needs to be encoded. In an extreme case, the signal has low-frequent structure (either globally as stationary process or locally as real-world signal), where the residual variances in the higher levels could be close to zero. In that case, no rate needs to be spent for the higher levels, and their contribution to the overall error could be far below the $\sigma_{q,k}^2$ assumed above. Unless alias is present from down- and upsampling, the quantization noise propagated from the lower levels is low frequent, but it is still beneficial to reduce it as much as possible, which means that the decision whether the residual at a given level is totally discarded or quantized into non-zero values should use thresholds that are increased with the level index k.

With closed-loop prediction over pyramid levels, quantization errors do not accumulate, such that the reconstruction error equals the quantization error of the highest resolution level $\sigma_{\mathrm{q},K}^2$. However, with (5.15), the quality of the estimate $\hat{s}_k(\mathbf{n})$ is decreased due to the quantization error $\sigma_{\mathrm{q},k-1}^2$ from the next lower resolution level, which increases the variance of the prediction error samples of the higher resolution level $\sigma_{e,k}^2$ by $\sigma_{\mathrm{q},k-1}^2$. The resulting total rate is

$$R_{\mathrm{total}} = \frac{1}{2}\sum_{k=0}^{K-1} 2^k \log_2 \frac{\sigma_{e,k}^2+\sigma_{\mathrm{q},k-1}^2}{\sigma_{\mathrm{q},k}^2} = \frac{1}{2}\sum_{k=0}^{K-1} 2^k \log_2\left(\sigma_{e,k}^2+\sigma_{\mathrm{q},k-1}^2\right) - \frac{1}{2}\log_2 \prod_{k=0}^{K-1}\left(\sigma_{\mathrm{q},k}^2\right)^{2^{-k}}. \tag{5.35}$$

Again, the last term in the right-hand equation is maximum under condition (5.33). However, there is no guarantee that R_{total} is minimized for this setting, as the other term introduces a dependency with the variance distribution of the residual signals over the various pyramid levels, and also depends on the value of the quantization error itself, i.e. it becomes rate dependent. However, using the value of (5.33) for the quantization error variances of the levels shows that the rate may only marginally be increased since

$$R_{\text{total}} = \frac{1}{2}\sum_{k=0}^{K-1} 2^k \log_2 \frac{\sigma_{e,k}^2 + \sigma_{q,k-1}^2}{\sigma_{q,k}^2} = \frac{1}{2}\sum_{k=0}^{K-1} 2^k \log_2 \left(\frac{\sigma_{e,k}^2}{\sigma_{q,k}^2} + \frac{1}{4} \right). \tag{5.36}$$

For signals that can be well predicted (low-detailed content) it is even more appropriate to spend the whole rate budget for finer quantization of the lowest resolution component, enforcing a good prediction and achieving a low reconstruction error in the high-resolution reconstruction.

In the considerations so far, it was however neglected that due to the down- and upsampling filters,

- the residual signal at level k may in ideal case not contain any low-frequency components, but such components now appear due to the quantization of the downsampled signal that is used for prediction[15];
- The power spectrum of the quantization error from the downsampled signal is amplified by a factor of two in the low-frequency range, but does not contain any high-frequency components (cf. Fig. 5.7).

From this point of view, it would be most appropriate to quantize the next lower level k–1 by a factor of $\sqrt{2}$ finer, such that the variance of the propagated quantization error at low frequencies equals the variance of the quantization error in the high frequencies that would occur in the level k itself. However, it should be noted that the actual optimum settings are dependent on the down- and upsampling filters, and on the power spectrum of the signal to be encoded. Therefore, the optimum for a given signal can only be determined by appropriate rate-distortion optimization (using a $D+\lambda R$ criterion) for a given operational point of the RD function.

5.3 Transform coding

Transform coding (TC) comprises encoding methods based either on orthogonal block transforms, or filterbank / wavelet transforms. The basic principles of these transforms were introduced in Sec. 2.7 and 2.8; this section now concentrates on the aspects of gain expected from discrete transform coding, i.e. *quantization* and *encoding* of transform coefficients, in a way that is not specific for the different types of multimedia signals.

In image and video transform coding, the area of the signal is usually first partitioned into relatively small segments. In image coding, these are often square or rectangular blocks of fixed or variable sizes. In audio coding, the signal is partitioned into segments with fixed or variable lengths, overlapping segments are used as well.

---

[15] The quantization error feedback from (5.15) is still relevant here.

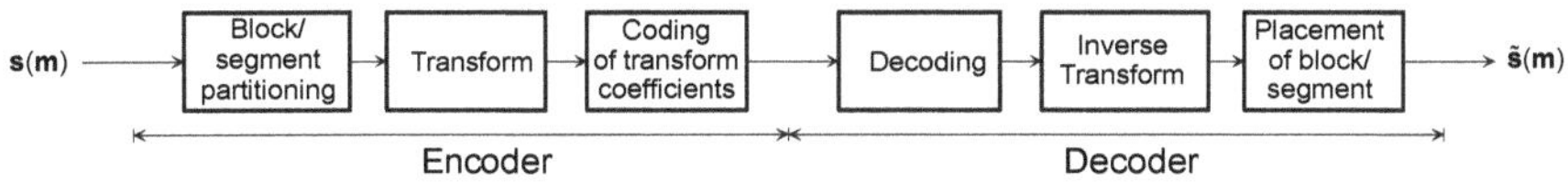

Fig. 5.9. Overall scheme of block-based transform coding and decoding

Fig. 5.9 shows a generic block diagram of transform coding schemes. The signal needs to be stored in a buffer until the number of samples necessary to apply the transform is available. After transformation, coefficients are quantized and encoded (usually by entropy coding). At the decoder, the decoding and reconstruction of the quantized coefficients is followed by the inverse transform, finally the decoded segments are placed at the respective position and combined as output.

5.3.1 Gain in discrete transform coding

The decorrelating effect of discrete linear transforms was analyzed in detail in Sec. 2.7.3. In the following, mostly orthonormal transforms are assumed. For other orthogonal transforms, normalization / scaling may be necessary to achieve correspondence between distortion in the signal and frequency domains. Assuming Gaussianity, to encode the real-valued coefficient $c_{\mathbf{k}}$ by an expected distortion D_{TC}, the rate

$$R_{\mathbf{k}} = \max\left[0, \frac{1}{2}\log_2 \frac{\mathcal{E}\{c_{\mathbf{k}}^2\}}{D_{\mathrm{TC}}}\right] \tag{5.37}$$

has to be provided according to (4.24). If the max(·) function is neglected[16] and constant distortion is assumed over all coefficients, the mean rate is

$$R_{\mathrm{TC}} = \frac{1}{|\mathbf{U}|}\sum_{\mathbf{k}} \frac{1}{2}\log_2 \frac{\mathcal{E}\{c_{\mathbf{k}}^2\}}{D_{\mathrm{TC}}} = \frac{1}{2|\mathbf{U}|}\log_2\left[\prod_{\mathbf{k}} \frac{\mathcal{E}\{c_{\mathbf{k}}^2\}}{D_{\mathrm{TC}}}\right]; \quad |\mathbf{U}| = U_1 U_2 \cdots U_K . \tag{5.38}$$

If the image is encoded without exploiting the correlation properties (i.e. by PCM, and for strict definition, also assuming a Gaussian PDF), the following relationship exists between rate and distortion D_{PCM}, which can be determined by (4.23):

$$R_{\mathrm{PCM}} = \frac{1}{2}\log_2 \frac{\sigma_s^2}{D_{\mathrm{PCM}}} \Rightarrow D_{\mathrm{PCM}} = \frac{\sigma_s^2}{2^{2R_{\mathrm{PCM}}}} . \tag{5.39}$$

Assuming identical rates $R_{\mathrm{TC}}=R_{\mathrm{PCM}}$, and substituting (5.38) into (5.39) gives

[16] This is valid in the case of high rates where $\mathcal{E}\{c_{\mathbf{k}}^2\} \geq D_{\mathrm{TC}}$ over all **k**.

$$D_{\mathrm{PCM}} = \frac{\sigma_s^2}{2^{2\frac{1}{2|\mathbf{U}|}\log_2 \prod_{\mathbf{k}} \frac{\mathcal{E}\{c_{\mathbf{k}}^2\}}{D_{\mathrm{TC}}}}}. \tag{5.40}$$

The ratio of D_{PCM} over D_{TC} can be interpreted as the *coding gain* achieved by the discrete transform, which turns out to be the ratio of *arithmetic* and *geometric* mean values of the squared expected values over the discrete set of coefficients[17],

$$G_{\mathrm{TC}} = \frac{D_{\mathrm{PCM}}}{D_{\mathrm{TC}}} = \frac{\frac{1}{|\mathbf{U}|}\sum_{\mathbf{k}} \mathcal{E}\{c_{\mathbf{k}}^2\}}{\left[\prod_{\mathbf{k}} \mathcal{E}\{c_{\mathbf{k}}^2\}\right]^{\frac{1}{|\mathbf{U}|}}}. \tag{5.41}$$

The coding gain of block transform coding (5.41) can also be interpreted as a discrete counterpart of (4.33) and the related spectral flatness measure (4.34). The coding gain becomes larger, when the spectral power is better concentrated in a low number of coefficients. From the definition of the spectral flatness measure, it can be shown for an AR process that

$$\log_2 \sigma_v^2 = \int_{-1/2}^{1/2} \cdots \int_{-1/2}^{1/2} \log_2 \phi_{ss}(\mathbf{f}) \mathrm{d}^\kappa \mathbf{f}. \tag{5.42}$$

In (5.41), the numerator equals σ_s^2 for the case of an orthonormal transform, as can be concluded from (2.159) and (2.160). The denominator can be re-written as[18]

$$\begin{aligned} &\log_2\left[\prod_{\mathbf{k}} \mathcal{E}\{c_{\mathbf{k}}^2\}\right]^{\frac{1}{|\mathbf{U}|}} = \frac{1}{|\mathbf{U}|}\sum_{\mathbf{k}} \log_2 \mathcal{E}\{c_{\mathbf{k}}^2\} \geq \frac{1}{|\mathbf{U}'|}\sum_{\mathbf{k}'} \log_2 \mathcal{E}\{c_{\mathbf{k}'}^2\} \text{ when } |\mathbf{U}'| \geq |\mathbf{U}| \\ &\Rightarrow \lim_{|\mathbf{U}|\to\infty} \frac{1}{|\mathbf{U}|}\sum_{\mathbf{k}} \log_2 \mathcal{E}\{c_{\mathbf{k}}^2\} = \int_{-1/2}^{1/2} \cdots \int_{-1/2}^{1/2} \log_2 \phi_{ss}(\mathbf{f}) \mathrm{d}^\kappa \mathbf{f} \end{aligned} \tag{5.43}$$

In the case of an optimal decorrelating transform applied to a correlated Gaussian process, the values $\mathcal{E}\{c_{\mathbf{k}}^2\}$ are the eigenvectors of the covariance matrix $\mathbf{C}_{ss}$. As however from rules of linear algebra the product of eigenvectors is equal to the determinant of the covariance matrix, it can be further concluded that

---

[17] In an orthonormal transform, the arithmetic mean over the expected coefficient energies within a block is identical with the power (variance) of the (zero-mean) signal. The following condition holds for any orthogonal transform, as the factor A from (2.237) would appear likewise in numerator and denominator, and hence can be neglected in the computation of the coding gain.

[18] Under the assumption that the sequence of coefficients $c_{\mathbf{k}}$ provides an approximation of the power density spectrum.

$$\sum_{\mathbf{k}} \log_2 \mathcal{E}\left\{c_{\mathbf{k}}^{\ 2}\right\} = \log_2 \left|\mathbf{C}_{\mathbf{ss}}\right| \tag{5.44}$$

with definition of the covariance matrix as in (2.209), or asymptotically (i.e. for a covariance matrix of large extension[19]

$$\lim_{|K|\to\infty} \frac{1}{|K|} \log\left|\mathbf{C}_{\mathbf{ss}}\right| = \int_{-1/2}^{1/2} \log \phi_{ss}\left(f\right) \mathrm{d}f \tag{5.45}$$

For an orthonormal transform the total energy as well as the number of coefficients grows linearly with block size $|\mathbf{U}|$. Therefore, the upper expression in (5.43) will only stay constant for different transform block sizes applied to the same signal, when the spectral energies are uniformly distributed. In cases of non-uniformly distributed spectral energy, the logarithmic scaling provides lower values on average when the block length is increased[20]. However, the 'spectral entropy' expressed by (5.43) is (for the case of an AR process) lower-bounded by the continuous-frequency expression in (4.34). Another interpretation is that even for the optimum transform (2.284), the discrete values $\mathcal{E}\{c_{\mathbf{k}}^2\}$ are transformed from a *finite* autocovariance sequence, which becomes periodically or symmetrically extended, depending on the properties of the transform basis. In contrary, $\phi_{ss}(\mathbf{f})$ is the Fourier transform of an autocovariance sequence of *infinite* extension, which decays towards zero. This implies that short-range block transform coding cannot take into account those statistical signal dependencies which span beyond the block / segment size, i.e. between adjacent blocks. It can be concluded that (for an AR process)

$$\lim_{|\mathbf{U}|\to\infty} \left[\prod_{\mathbf{k}} \mathcal{E}\left\{ c_{\mathbf{k}}^{\ 2}\right\}\right]^{\frac{1}{|\mathbf{U}|}} = \sigma_v^{\ 2} \Rightarrow \left[\prod_{\mathbf{k}} \mathcal{E}\left\{ c_{\mathbf{k}}^{\ 2}\right\}\right]^{\frac{1}{|\mathbf{U}|}} \geq \sigma_v^{\ 2} \quad \Rightarrow \quad G_{\mathrm{TC}} \leq G_{\mathrm{AR}}, \tag{5.46}$$

where G_{AR} is the maximum possible coding gain (4.33) of an AR model. This conclusion can however not be generalized unconditionally to the case of any signals, as it assumes stationarity of the random process; for example, images show changes of the local amount of detail, change of mean, edges etc. Audio signals have transients, variation of loudness etc. Such changes justify usage of smaller transform block sizes, however the penalty (5.46) due to the lack of exploitation of longer-term statistical dependencies between adjacent blocks, should be considered as well when deciding to switch.

---

[19] This is formulated here for the 1D signal case. For a discussion on this identity, see R. M. GRAY, *Toeplitz and Circulant Matrices* (http://ee.stanford.edu/~gray/toeplitz.pdf), or refer to (2.188) which provides a similar relation w.r.t. rate reduction.

[20] see as an example Problem 5.2.

5.3.2 Quantization of transform coefficients

After application of a one- or multidimensional transform, the coefficients have to be quantized and encoded. To reconstruct the signal, the inverse transform is applied on quantized coefficients (see Fig. 5.10). The error energy between the original signal $\mathbf{s}$ and the reconstructed signal $\tilde{\mathbf{s}}$ depends on the energy of the quantization error $\mathbf{q}$, which is the difference between the coefficient vector $\mathbf{c}$ and its quantized version $\tilde{\mathbf{c}}$. Using the concepts of (2.244), this gives[21]

$$\begin{aligned}\sigma_e^2 &= \mathcal{E}\left\{(\mathbf{s}-\tilde{\mathbf{s}})^{\mathrm{T}}(\mathbf{s}-\tilde{\mathbf{s}})\right\} = \frac{1}{A}\mathcal{E}\left\{(\mathbf{s}-\tilde{\mathbf{s}})^{\mathrm{T}}\mathbf{T}^{\mathrm{H}}\mathbf{T}(\mathbf{s}-\tilde{\mathbf{s}})\right\} \\ &= \frac{1}{A}\mathcal{E}\left\{(\mathbf{c}-\tilde{\mathbf{c}})^{\mathrm{H}}(\mathbf{c}-\tilde{\mathbf{c}})\right\} = \frac{1}{A}\mathcal{E}\left\{\mathbf{q}^{\mathrm{H}}\mathbf{q}\right\} = \frac{1}{A}\sigma_q^2 .\end{aligned} \tag{5.47}$$

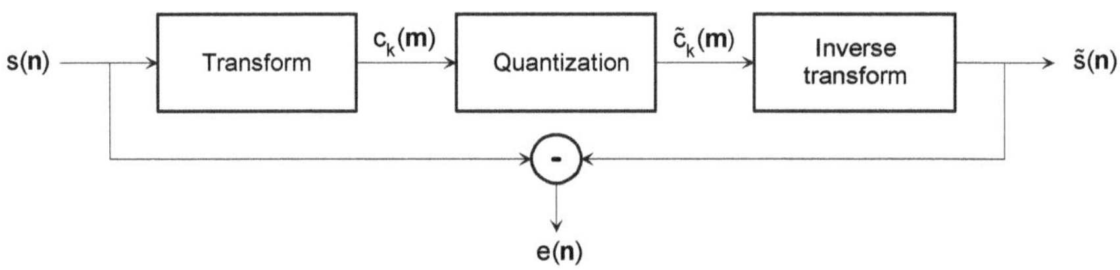

Fig. 5.10. Principle of data compression by transform coding

For non-orthogonal transforms, (5.47) does not apply, which means that it is not straightforward to map the distortion introduced by encoding of the coefficients directly to the distortion appearing in the reconstructed signal. Instead,

$$\sigma_e^2 = \mathcal{E}\left\{(\mathbf{s}-\tilde{\mathbf{s}})^{\mathrm{T}}(\mathbf{s}-\tilde{\mathbf{s}})\right\} = \mathcal{E}\left\{(\mathbf{c}-\tilde{\mathbf{c}})^{\mathrm{H}}\mathbf{T}^{-\mathrm{H}}\mathbf{T}^{-1}(\mathbf{c}-\tilde{\mathbf{c}})\right\} = \frac{1}{A}\mathcal{E}\left\{\mathbf{q}^{\mathrm{H}}\mathbf{W}\mathbf{q}\right\}, \mathbf{W} = \mathbf{T}^{-\mathrm{H}}\mathbf{T}^{-1}, \tag{5.48}$$

which indicates that the vectors of squared errors from individual transform coefficients need to be weighted by the matrix $\mathbf{W}$ to obtain the optimum distortion-minimizing decision. This is quite straightforward for the case where the synthesis basis vectors in $\mathbf{T}^{-1}$ are orthogonal but have different norms. In that case, $\mathbf{W}$ is still a diagonal matrix, the quantizer decisions for the different transform coefficients do not depend on each other, and the weighting can simply be implemented by applying different quantization step sizes to coefficients c_k such that $\mathbf{W} = \mathrm{diag}\{\Delta_k^{-2}\}$. If this is not the case, as e.g. for bi-orthogonal transforms, the matrix $\mathbf{W}$ causes dependencies between the elements of the transform-domain quantization error vector $\mathbf{q}$ regarding their effect on the reconstruction error $\mathbf{e}$.

21 Note that the following proof is based on the condition $\mathbf{T}^{\mathrm{H}}\mathbf{T}\sim\mathbf{I}$, such that the linear dependency of the quantization error in the transformed domain and the reconstruction error is strictly valid only for the case of orthogonal transforms with basis functions of identical norms.

Ultimately, it might then be necessary to compare every possible combination of individual transform coefficient quantization to find the best overall decision, which would bring scalar quantization into the range of vector quantization in terms of complexity. Therefore, it is advantageous to design transforms where **W** is close to a diagonal matrix such that remaining dependencies can be ignored.

Scalar quantization can easily be adapted (if necessary) to the PDF of individual transform coefficients. Firstly, the amount of bits necessary for the quantized representation of a given transform coefficient can be approached by entropy coding or adaptive quantization. Second, the weighting matrix **W** can also be used to implement *frequency dependent* quantization e.g. based on perceptual models. It should however be observed that squared error criteria are typically increased by this.

Transform-domain vector quantization can be used as well. This can be expected to preserve the advantages of VQ (see Sec. 4.5.1) in transform coding. In principle, it is possible to build vectors from spectral coefficients of only *one frequency band* **k** at different positions,

$$\mathbf{c}_{\mathbf{k}} = \left[c_{\mathbf{k}}(\mathbf{m}), c_{\mathbf{k}}(\mathbf{m}+1), \ldots\right]^{\mathrm{T}}, \tag{5.49}$$

or from *different frequency bands* at the same position **n**

$$\mathbf{c}(\mathbf{n}) = \left[c_{\mathbf{k}}(\mathbf{n}), c_{\mathbf{k}+1}(\mathbf{n}), \ldots\right]^{\mathrm{T}}. \tag{5.50}$$

The method (5.49) can be denoted as *intra-band VQ*, and the method (5.50) as *inter-band VQ*:

- In intra-band VQ, vectors should be arranged by samples where most statistical dependencies are expected, e.g. for horizontal low frequency coefficients it is appropriate to use vectors of larger horizontal extension.
- Inter-band VQ combines coefficients from different spatial bands of the same spatial location into vectors. Hence, if inter-band dependencies are still existent after the transform, these will inherently be exploited when an appropriate codebook design is made.

Adaptive quantization of transform coefficients – 'energy picking'. When PCM-like uniform quantization is applied in the transform domain, the number of bits to be allocated to a certain transform coefficient can be determined from the variance of the underlying random process, see e.g. (4.24). Such an approach would however only be reasonable in case of stationarity, where it can e.g. be guaranteed that the percentage of overload cases is controllable. As multimedia signals often have locally varying statistics, it is rather useful to adapt the rate allocation e.g. segment-wise or block-wise. When a fixed distortion is chosen, direct application of (4.24), using the local variances of transform coefficients, establishes a typical case of variable bit rate coding. Beyond the varying number of bits for coding the coefficients, the bit allocation needs to be conveyed as side information, which can e.g. be achieved by representing the spectral hull (or other properties such as harmonic multiples of a pitch frequency) in a compact way. The way of doing this efficiently is specific for the type of signal and therefore does not need to be discussed here. When variable rate coding is not an option, a typical algorithm to distribute a fixed budget of bits over a set of transform coefficients is 'energy picking' [JAYANT, NOLL 1984]. The rationale behind is the logarithmic nature of the RD function, which for same distortion requires one additional bit when the variance of a transform coefficient is quadrupled. This can be utilized in the following procedure:

1. In the given set, search for the coefficient with highest value c_k^2 and allocate one more bit from the bit budget. Then, replace $c_k^2 \to c_k^2/4$.
2. If the entire bit budget is spent, the bit allocation is finished. Otherwise, loop over 1.

Note that this can also lead to zero bit allocation, in which case the corresponding transform coefficient is quantized as zero. The method can either be applied block-wise or on ensembles of coefficients by using $\mathcal{E}\{c_k^2\}$.

Quantization adjustment in wavelet coding. The considerations about the coding gain, which were introduced in the context of block transform coding in Sec. 5.3.1, are not directly transferable to the case of wavelet coding due to the following reasons[22]:

- The direct mapping of the coding error in the frequency coefficients to the reconstruction error in the image domain as introduced in (5.47), is strictly valid only for the case of orthogonal transforms, however could be approximately valid for biorthogonal filter banks when the depth of the wavelet transform tree is large.
- The formulation of the coding gain in (5.41) assumes frequency representations of equal bandwidth and equal number of samples.

To study the effect of quantization or coding errors which are added to the signal in the wavelet domain, mapping of the quantization errors $Q_0(f)$ and $Q_1(f)$ into the reconstruction error $E(f)$, as occurring in the lowpass and highpass bands of a 2-band system, can be formulated as

$$E(f) = G_0(f)Q_0(2f) + G_1(f)Q_1(2f)\,. \tag{5.51}$$

Taking the squared absolute values and assuming statistical independency of the two transform quantization errors gives

$$\begin{aligned}|E(f)|^2 &= |G_0(f)|^2\,|Q_0(2f)|^2 + |G_1(f)|^2\,|Q_1(2f)|^2 \\ &+2\,\mathrm{Re}\{G_0{}^*(f)Q_0{}^*(2f)G_1(f)Q_1(2f)\}\,.\end{aligned} \tag{5.52}$$

From this, it can be concluded that

---

[22] For the case of the Haar transform (2.249) which could be interpreted either as a block transform or as a wavelet transform, the coding gain formulation (5.41) is valid; however, the coefficients of higher frequency bands within a block can be correlated, e.g. in case of a ramp-shape signal (see also Problem 2.15).

$$\sigma_e^{\,2} = \int_{-1/2}^{1/2} \mathcal{E}\left|E(f)\right|^2 \mathrm{d}f = \sigma_{q_0}^2 \underbrace{\sum_n g_0^{\,2}(n)}_{w_0} + \sigma_{q_1}^2 \underbrace{\sum_n g_1^{\,2}(n)}_{w_1} + 2\mu_{q_0q_1}(0) \underbrace{\sum_n g_0(n) g_1(n)}_{w_{12}}, \tag{5.53}$$

which under the simplified assumption of flat spectra in $Q_k(f)$ gives

$$\sigma_e^{\,2} = \sigma_{q_0}^2 w_0 + \sigma_{q_1}^2 w_1 + 2\mu_{q_0q_1}(0) w_{12} \;. \tag{5.54}$$

Firstly, this means that the quantization errors are weighted by the squared norms of the reconstruction filter impulse responses. If further the lowpass and highpass interpolation filters are orthogonal, w_{12}=0, such that the additional covariance component is zero. The same applies if the quantization errors are statistically independent and spectrally white, which usually can be assumed at high rate points. If however subbands are partially discarded, quantization errors become correlated with the signal and can also become mutually correlated particularly within the overlapping frequency ranges of the analysis filters. This effect which is dependent both on signal properties and the distortion operational point, is ignored for simplicity in the following paragraphs, and the covariance of the quantization errors is assumed to have no influence to the reconstruction. This assumption becomes even more justified when the scaling band '0' is not directly coded, but further decomposed within the wavelet tree, which should however be assumed to be sufficiently deep in this case.

Since the reconstruction is run over the entire wavelet tree, the number of samples is varying over the various wavelet bands in a dyadic fashion, and the quantization errors residing at the lower-frequency levels of the tree will be weighted by multiple filter operations. For a 1D wavelet transform with T levels of the wavelet tree, scaling coefficient c_0 and wavelet coefficients $c_1 \ldots c_T$, the resulting reconstruction error is (see simplified signal-flow scheme in Fig. 5.11)

$$\sigma_e^{\,2} = \sigma_{q_0}^2 w_0^{\,T} + \sum_{t=1}^{T} \sigma_{q_t}^2 w_0^{\,T-t} w_1 \;. \tag{5.55}$$

An extension to multi-dimensional wavelet schemes is straightforward.

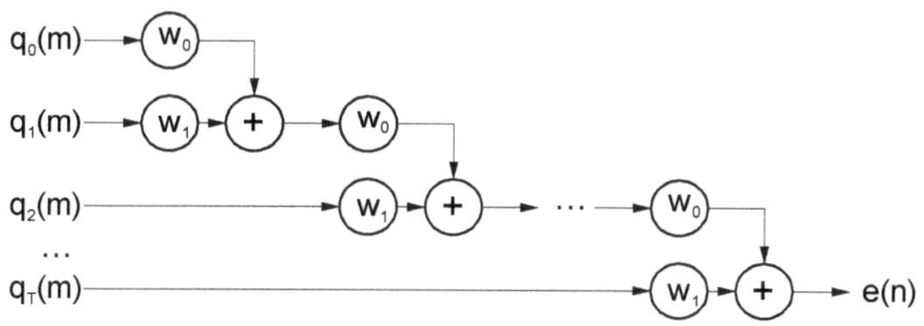

Fig. 5.11. Signal flow and weighting of quantization errors from wavelet bands into the reconstruction for a 1D wavelet tree

As the squared quantization step size is proportional with the quantization error variance, quantization step sizes need to be adapted by the reciprocal square root of the respective subband's weight in (5.55), which gives the optimum quantizer step size[23] for each wavelet band, normalized by a global quantizer scaling factor Δ,

$$\Delta_0 = \frac{\Delta}{\sqrt{w_0^{\,T}}} \quad ; \quad \Delta_t = \frac{\Delta}{\sqrt{w_0^{\,T-t} w_1}} = \Delta_0 \cdot \sqrt{\frac{w_0^{\,t}}{w_1}} \text{ for } t > 0 . \tag{5.56}$$

For the case $w_0=w_1=1$, all quantization step sizes would be equal. This would occur in cases of orthonormality, or with unity gain of lowpass and highpass filters. Some bi-orthogonal filters, e.g. the 9/7 filter (2.326), are rather close to orthonormality, such that uniform weighting of quantization errors is approximately correct.

5.3.3 Coding of transform coefficients

Quantization of transform coefficients can be done similar as in PCM (however applied to discrete coefficient samples in the transform domain). Beyond the step size Δ, the amplitude range of the quantization has to be selected according to the statistics of the different coefficients. Encoding by such a method can be interpreted in terms of the coding gain (5.41), the expected concentration of information into few coefficients causes unequal bit allocation across the coefficients that e.g. represent different frequency bands. This can be achieved as follows:

- Method 1: For signals with known PDF (e.g. Laplacian, Gaussian) and variance, the optimum range of quantization (in terms of minimizing the distortion for a given number of quantization levels) can be determined. If however a certain frequency band has low variance and the range is set to a narrow margin, coefficients with high amplitude (when occurring in rare cases) would cause overload of the quantizer, leading to increased error after the inverse transform (which is hardly noticeable in terms of average coding error).
- Method 2: If the same full amplitude range and uniform quantization is used for all coefficients, entropy coding can naturally provide unequal bit allocation across the different frequency bands, according to their statistics. For frequency bands with less variance, the coefficient values will more frequently fall into the low levels or the zero-reconstruction zone of the quantizer, which then can be encoded using fewer bits.

In case of stationary processes (such as AR models), the design can be based on global statistics, and both methods would produce approximately equal results in terms of average rate-distortion performance. However, multimedia signals typically deviate from stationarity in a way that groups of samples which fit a common statistical model are finite, and model parameters change between such groups. For example, the instantaneous spectral power density of speech and au-

---

[23] under condition of squared error criteria and without frequency weighting of coding errors.

dio signals varies depending on spoken words or musical tones played, and more rapid changes occur at dedicated time positions as transients (temporal sparseness). In images, amplitude transitions occur at well localized edge positions, whereas flat or textured areas are wider and have a spectral appearance that is more similar to that of stationary processes. Except for transients, measured spectra are often extremely sparse, i.e. the relevant information concentrates in only few transform coefficients, but it needs to be identified which these are. In case of transients, if their position is identified, usage of fewer transform coefficients with short basis functions can also keep the total number of non-zero values low. Generally, the number of bits necessary to provide a certain minimum distortion (i.e. the $R(D)$ function) varies locally, e.g. depending on amount of detail, loudness, frequency of occurrence of changes etc.[24].

When it is expected that a large percentage of transform coefficients shall be quantized to the zero value, special care should be taken to fit the zero-value representation properly into the whole entropy coding concept. Specific methods for signalling positions of zero-value coefficients are often applied in advance to the coding of the remaining (non-zero) coefficients. Examples for such methods are:

- Signalling of 'significance' e.g. by run-length coding or an explicit flag. In run-length coding, after each non-zero coefficient an information about the number of subsequent zero coefficients is sent (for example, the method depicted in Fig. 5.1b is often used in image transform coding); in explicit signalling, the entropy of such a flag would be low provided that many coefficients are not significant.
- If it can be expected that only few relevant high frequency components appear (e.g. in the case of local low detail in image signals), a cut-off frequency beyond which all coefficients are set to zero can be specified by signalling the 'last' non-zero coefficient within a local area.
- Coefficients can be combined into 'frequency groups' where special signalling mechanisms are designed to express that the whole group is zero. This approach in principle uses a rough approximation of the spectral hull to classify ranges of the coefficient spectrum into 'zero' or 'non-zero'.

It should be noted that any of the aforementioned methods can be combined with rate-distortion criteria, e.g. setting an entire coefficient group to zero when only one or few isolated coefficients would formally fall into the non-zero range, but expressing this would be too costly in terms of rate.

Typical linear (block-based or subband/wavelet) transforms are only able to utilize linear statistical dependencies (correlation) from the transformed representation. In terms of the methods described above for coding of local properties, non-linear dependencies can also be exploited for coding. As an example, the

---

24 A case that is not covered here is a class of signals which are neither sparse in the time/spatial domain (such as transients) nor in the frequency domain. This is the case for noise-like components, which are however often irrelevant for detailed perception, and could either be suppressed or synthesized without changing the subjective appearance of the reconstruction.

probability of (non-)zero coefficients occurring can be deduced from the presence of other (non-)zero coefficients either in the local neighborhood or in neighbored frequency bands, since in cases of low local detail, no coefficients would be present at all. Such dependencies can be exploited in conditional (context dependent) entropy coding e.g. for the significance signalling of transform coefficients.

When the PDF of transform coefficients is symmetric around the zero value (as it is often the case in multimedia signals), a sign has to be encoded, but in terms of probability statistics it has a self information of one bit whenever a non-zero amplitude value occurs. Therefore, the sign of non-zero quantized transform coefficients is often encoded separately from quantized magnitude information.

If the discrete transform is applied locally (e.g. separately on segments or blocks of the signal) statistical dependencies (including linear dependencies i.e. correlation) between the coefficients in adjacent blocks often still exist. This can be exploited by application of predictive coding (e.g. DPCM) across block boundaries. However, it depends on the properties of the transform whether prediction can be successfully applied in the transform domain.

Embedded quantization and bit-plane coding as described in Sec. 4.3 can also be applied to transform coefficients. For coefficients with zero-symmetry PDF, it is useful to encode amplitude and sign separately. Simple methods combine context-dependent entropy coding or run-length coding with bit-plane coding; this starts from the MSB, proceeding down to the LSB. No explicit quantization step is necessary, when the whole procedure is applied to an integer rounded representation of orthonormal transform coefficient magnitudes, where the full representation over all bit planes relates to the lowest distortion level that can be achieved. The relevance of the sign depends on the highest bit-plane where a significant non-zero amplitude value is found. For typical PDFs (Gaussian, Laplacian) with concentration around zero, it can be expected that less significant values are found towards the more significant bit planes, and even less when the variance of a coefficient's frequency position is low.

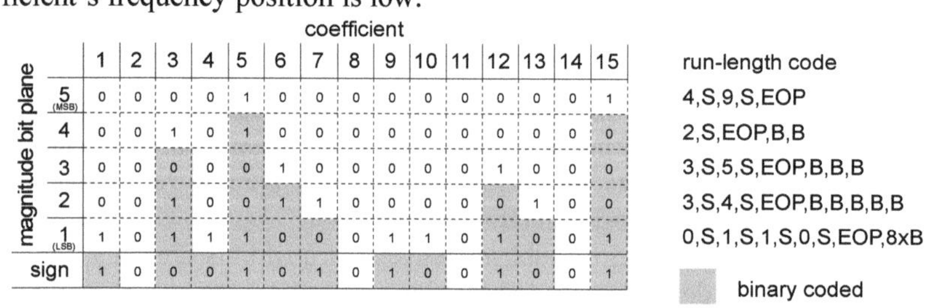

Fig. 5.12. Bit-plane coding of transform coefficients using run-length coding (significance scan), binary sign coding and binary amplitude refinement coding

A typical method of run-length/bit-plane coding for a group of 15 coefficients is shown in Fig. 5.12. The end of each bit-plane scan is indicated by an *End of Plane* (EOP) symbol, which fulfills a similar functionality as the 'last' coefficient coding described above. After expressing the presence of a significant coefficient in a given bit plane via the run-length value, 'S' indicates that a sign bit has to be encoded whenever the highest significant non-zero bit related to a coefficient's amplitude is found. In the lower bit planes, all coefficients that were already found to be significant in one of the higher bit planes are no longer included in the significance scan of the run-length. All less significant bits of coefficients that are already known as significant are encoded in a binary representation 'B' at the end of the respective bit plane's information ('refinement scan'). The refinement resolves the shape of the PDF more and more locally, such that the occurrence of '0' and '1' bits in this part of the information becomes more and more uniform towards the lower bit planes, and application of entropy coding also is less effective and omitted here.

Note that the significance scan can also be implemented by context-dependent entropy coding. Further, in particular for the higher bit planes, usage of (context-dependent) entropy coding can give additional compression for the refinement part of the information. In general, contexts known at the decoder through previous decoding steps may include any information from higher bit planes or preceding information from the same bit plane, from coefficients of other frequency bands or from coefficients of the same frequency band at neighboring positions.

The method of bit-plane coding is particularly beneficial for *scalable* coding of transform coefficients. It however has the disadvantage of being more complex than a single-stage entropy coding method, as multiple scans are necessary both at the encoder and decoder.

5.3.4 Transform coding under transmission losses

As the synthesis of transforms used in coding is equivalent to feeding the coefficients into non-recursive filters (with impulse responses a.k.a. basis functions), the effect of errors (coding or transmission errors) spreads into all samples that are covered by the extension of the synthesis basis functions. Errors appear as patterns of synthesis basis functions which are superimposed to the decoded signal. This effect occurs for any types, irrespective of block, overlapping-block or wavelet/subband transform types. Consequently, bits which contribute most to the reduction of distortion in the rate-distortion sense will also have the highest impact on distortion when disturbed by channel errors. If no ordered relevance of bits is available, spectral models can be used to determine which coefficients have highest energy and are therefore most vulnerable, producing high errors when lost. A reasonable strategy is allocating stronger (unequal) error protection to such relevant components during transmission.

Splitting of the bit stream into sub-streams which represent components of different relevance can be used systematically. This is generally denoted as *data partitioning*, or, in the context of separating different frequency components in transform coding, as *frequency partitioning*.

5.4 Bit streams with multiple decoding capability

In the encoding methods discussed so far, it was usually assumed that the resolution of a signal is given at the encoder input and shall be decoded at the decoder output with the same target resolution, and with the best quality that is achievable by the given rate. Transmission services with uncertainties regarding channel conditions (throughput, errors) or client device types (e.g. different spatio-temporal resolution supported by decoder and display) benefit from coded representations where bit streams can easily be adapted, e.g. if necessary protected anywhere in the transmission chain, or giving support for partial and flexible decoding as necessary to feed replay devices of different capability or resolution. In such cases, it is desirable to give the provided bit stream a structure that allows to decompose it into separate or nested parts, each of which is capable for self-contained decoding of the same signal with a given rate and associated resolution.

5.4.1 Simulcast and transcoding

In simulcast (for *simul*taneous broad*cast*) bit streams are provided separately for each rate and/or resolution. Each stream is self-decodable without dependency on the other streams, and the total rate to be transmitted or stored is the sum of the single stream rates. Simulcast streams are usually providing rather coarse granularity of bit rates and resolutions, to avoid too much information duplication and overhead. If more fine-granular adaptation of instantaneous rate is needed, it is necessary to apply *transcoding*, which means decoding the signal and encoding it again by the target rate. In order to keep the complexity low, the following approaches are often applied in this context:

- Re-using side information parameters (such as motion vectors and motion information in video, LPC parameters in speech, spectral hull / frequency dependent quantization information in audio)
- In case of transform coding: Transcoding in the transform domain, by applying re-quantization of transform coefficients. This avoids applying an inverse transform in decoding, and forward transform in subsequent encoding.

Due to the recursive prediction synthesis process, a similar approach is not possible in case of predictive coding, where re-quantization of prediction error values would incur drift. The same applies when predictive coding methods are combined for better compression with transform coding. Generally, when different information components need to be considered, transcoding will usually be compromising in terms of rate-distortion optimization.

5.4.2 Scalable coding

Unlike simulcast, *embedding of bit streams* is used for scalability, which means that all parts of the stream are typically necessary to decode at full resolution or highest quality, whereas nested sub-streams are used for reduced resolution(s). Even though scalability typically causes an overhead bit rate compared to the case where such a multiple-decoding functionality is not enabled (i.e. one single-layer stream which can only be fully decoded to one resolution), it consumes usually lower rate than simulcast due to the fact that information from lower-resolution streams is also utilized in decoding higher resolutions and therefore less redundancy exists in the overall representation.

The scalable information consists of a lowest-resolution and lowest-quality base layer, and one or several enhancement layers, the highest of which provides the full resolution and quality. In principle, the number of layers could be rather large, i.e. scalable methods exist where the sub-stream partitioning is made with fine granularity (at the expense of additional signalling overhead at least). Scalable coding should allow flexible post-encoding modification of bit stream sizes and decoder output, being more efficient than simulcast and less complex than transcoding.

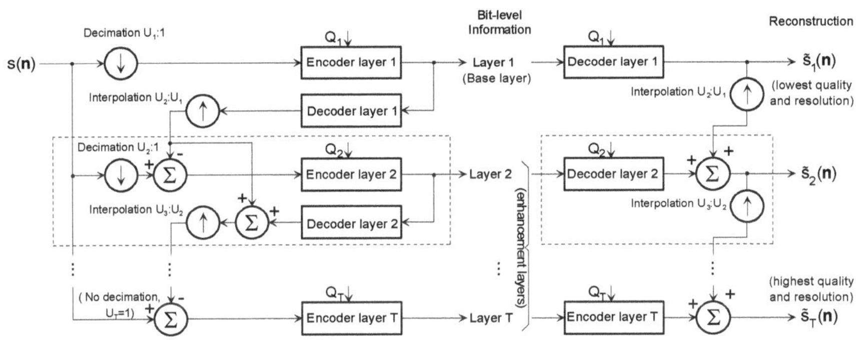

Fig. 5.13. Principle of layered coding with *T* layers[25]

Two important principles to generate a scalable representation are:

- Layered coding (Fig. 5.13), where a low-to-high layer recursion is applied, such that a lower layer decoding is used to predict and encode the signal in one or several higher layer(s). Typical approaches for layered coding are multi-resolution pyramids to achieve resolution variation (Sec. 5.2.6) and cascaded quantization (i.e. re-quantization of quantization error).
- Embedded coding, which from the beginning generates the compressed information in a way that the lower resolution or lower quality can implicitly be obtained by only using a part of the stream. Examples for this are wavelet decomposition for resolution variation (Sec. 2.8.4) and bit-plane coding for quantization with variable distortion (Sec. 4.3).

In layered coding, the downsampling to feed the different sub-sampled resolutions into the different encoding stages can be done either in parallel (as shown in Fig.

[25] Decimation and interpolation elements are optional. Appropriate filters should be included in these steps.

5.13), or in a cascaded sequence starting from the highest resolution, as it was introduced in the context of the Laplacian pyramid (Fig. 2.66). If two subsequent layers have the same resolution, the decimation/interpolation elements can be omitted. Generally, by supplementing more building blocks of the intermediate-level type, an arbitrary number of scalable layers can be realized. Of course, the particular action to be taken in decimation and interpolation depends on the mode of scalability, e.g. motion-compensated processing can be implemented for frame rate upsampling.

5.4.3 Multiple-description coding

Multiple-description coding (MDC) represents a signal by multiple self-contained streams, loosely depending on each other such that the quality improves whenever more streams are available for decoding. A certain level of quality is guaranteed by any of the streams received standalone, i.e. MDC does not follow the hierarchical principle (with base layer and sub-ordinate enhancement layers) of scalable coding. The three basic methods are *MDC quantization, MDC transforms* and *MDC prediction* methods. Only examples of 2-stream MDC methods are presented here, but extensions to higher numbers of partial descriptions are possible.

MDC quantizers. The idea behind MDC quantization is usage of two quantizers with coarser quantization step sizes but different offsets [VAISHAMPAYAN 1993]. If the information from both is combined, more exact conclusions can be drawn about *where* the original position of the value was in the quantizer cells (e.g. upper half or lower half). This is equivalent to the information conveyed by a single quantizer of finer step sizes. Systematic design of such quantizers can be made, an example is given by the matrix in Fig. 5.14. The equivalent finer quantizer has 22 steps, which can be uniquely identified when both code symbols i_1 and i_2 are received. Each standalone MDC stream conveys only information about 8 coarse steps. For example, if only i_1=2 is received, the actual finer quantizer value could have been either 5, 7, or 8. The best reconstruction choice then depends on the source statistics; for example, in case of uniform probability, the average of the reconstruction values associated with those three indices should be the optimum output. The efficiency of the MDC quantizer in terms of source compression (as compared to a single quantizer) relates to the number of cells that are populated in the matrix. In the given example, 2x3=6 bit/sample have to be transmitted in MDC, while only $\log_2 22 \approx 4.45$ bit/sample would be necessary if a monolithic quantizer was used, i.e. the MDC overhead is roughly 33%. Extreme cases would be population of the main diagonal only (which means duplicate transmission of information over the two paths) and full population (which does not cost anything in compression efficiency, but neither provides much evidence about the value if only one code index is received).

index i_2 received from second stream \ index i_1 received from first stream	0	1	2	3	4	5	6	7
0	**1**	**3**	X	X	X	X	X	X
1	**2**	**4**	**5**	X	X	X	X	X
2	X	**6**	**7**	**9**	X	X	X	X
3	X	X	**8**	**10**	**11**	X	X	X
4	X	X	X	**12**	**13**	**15**	X	X
5	X	X	X	X	**14**	**16**	**17**	X
6	X	X	X	X	X	**18**	**19**	**21**
7	X	X	X	X	X	X	**20**	**22**

Fig. 5.14. Schema of an MDC quantizer designed for two separate streams; numbers in bold are the quantization indices of the corresponding single-description quantizer

MDC transforms. Here, an uncorrelated set of transform coefficients shall be generated within each description, but pair-wise correlation of coefficients may be given across different descriptions [Goyal et al. 1998]. An extreme case fulfilling this paradigm is made by sending identical descriptions over both channels, which obviously doubles the total rate, but guarantees full reconstruction if only one stream is received. Another simple example would be to generate polyphase (even/odd indexed) sequences of samples from a signal, and apply a transform coder to both of them separately for transmission over two channels. This means that each second sample will be missing if only one stream is received. Further, the two transforms will be less efficient than one, as the correlation is lower within the separate polyphase sequences[26]. In [Wang et al. 2001], optimum *pair-wise decorrelating transforms* are developed from the background of a KLT design. Nevertheless, due to the inter-stream correlation which by purpose exist. MDC transforms are inferior in compression efficiency, compared to a transform optimized for a single stream transmission.

MDC prediction. If a transmission error occurs at any sample in a prediction chain, all subsequent samples are affected, unless the chain is terminated by a systematic refresh of the decoder state. Similarly to the transform principle discussed above, separate prediction sequences can be used for the even- and odd-indexed samples, in order to provide two separate MDC streams. In case of losses in one stream, at least half of the samples can be decoded. Further, it can be tried to achieve partial recovery of the other samples by interpolation, which usually gives better quality than erroneous prediction. Fig. 5.15 shows an example as similarly suggested in [Apostolopoulos 2000] for the case of error-prone video using motion-compensated prediction. The same principle is applicable to more-dimensional series of samples, e.g. arranged as two quincunx grids in 2D, or to alternating sample groups (1D vectors or 2D blocks).

[26] For an analysis of correlation statistics in even/odd polyphase sequences, refer to Problem 2.19.

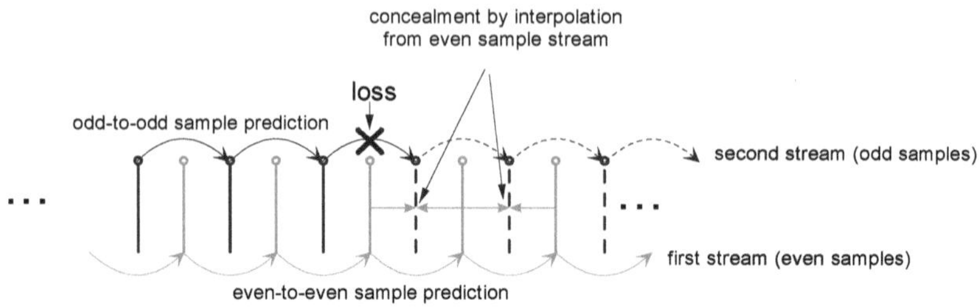

Fig. 5.15. MDC prediction with separate even/odd picture prediction chains, interpolation for partial error recovery and systematic refresh for resynchronisation

5.5 Distributed source coding

In most source coding concepts, optimization decisions made by the encoder are expecting a 'deterministic' behaviour of the decoding algorithm. This applies to the different variants of prediction, including inter-layer prediction in scalable coding. This further applies to context based entropy coding of various information components, which could be re-interpreted as a concept where information that the decoder already has available can be exploited to reduce the data rate. An abstract view on this is shown in Fig. 5.16a, where according to the basic concepts of information theory the rate should be larger or equal to the conditional entropy $H(\mathbf{S}\,|\,\hat{\mathbf{S}})$, with $\hat{\mathbf{S}}$ subsuming all prior knowledge that the decoder could have about **S**, i.e. interpreting the decoder as a state machine with next-state transition behavior controlled by the information sent from the encoder.

As a simple example, assume two pictures $\mathbf{S}_1$ and $\mathbf{S}_2$ of a video sequence which are statistically dependent. According to the basic principles of information theory, a meaningful rate range for encoding of $\mathbf{S}_2$ should be within the margin $H(\mathbf{S}_2)\geq R_2\geq H(\mathbf{S}_2|\mathbf{S}_1)$, the rate for encoding $\mathbf{S}_1$ should be within $H(\mathbf{S}_1)\geq R_1\geq H(\mathbf{S}_1|\mathbf{S}_2)$, but the total rate should not become lower than the joint entropy $R_1+R_2\geq H(\mathbf{S}_1,\mathbf{S}_2)$. According to the theory of *distributed source coding* (DSC), two dependent sources can be encoded by *independent encoders* observing the given single-source rate bounds, when only a decoder exists which receives a total rate which is at minimum the joint entropy, and the decoder should still be able to decode both sources perfectly, regardless of the distribution of rate into the two streams (Fig. 5.16b), but under the assumption that the codes are properly designed for that purpose. These are the basic bounds derived through the *Slepian-Wolf theorem* [SLEPIAN, WOLF 1973] for lossless decoding and extended by the *Wyner-Ziv theorem* [WYNER, ZIV 1976] for lossy decoding. Both theorems apply for two as well as an arbitrary number of dependent sources, however the proof is only possible for sources where the rate distortion function is well defined, e.g. for the case of Gaussianity and stationarity.

The basic idea of DSC is similar to channel coding, where it is assumed that the true state of a distorted symbol has to be estimated by the decoder, which is possible if redundancy is available. Therefore in DSC, like in channel coding, the decoder has to solve an estimation problem from the information received, whereas the encoder behaves deterministically and can be implemented in a low complexity fashion (e.g. without fully analyzing and systematically removing the redundancy between samples, and independently encode somewhat incomplete information instead, expecting that the decoder would know the modality of incompleteness and tries to recover the complete information by utilizing the redundancy itself). This could e.g. be interesting in cases where encoders have to operate with low battery power, such as wireless sensor networks for surveillance applications.

Besides the example given here of independent encoding and dependent decoding of samples, independent encoding in different resolutions as done in scalable coding is another example where the DSC theorems in principle show that it should be possible to encode a multi-resolution representation by a total rate which is not higher than the rate of the highest resolution alone (as this would subsume all other resolutions).

The real problem is how to achieve this practically. It has been shown that for two dependent Gaussian sources partitioning a quantization space into cosets and only transmitting the cosets information for the second source (expecting no large amplitude deviations) is a viable strategy. Approaches similar to turbo decoding, where the sources are used for alternating estimation from each other, have been investigated as well. Unfortunately, most of the methods which were shown to be close to the optimum expected by the DSC theory, are no longer applicable to non-stationary sources. Practical DSC schemes, which have for example been proposed for video compression, seem to be performing well only for cases where the redundancy between pictures is extremely high. For a review on DSC in the area of video coding, see [GIROD ET AL. 2005].

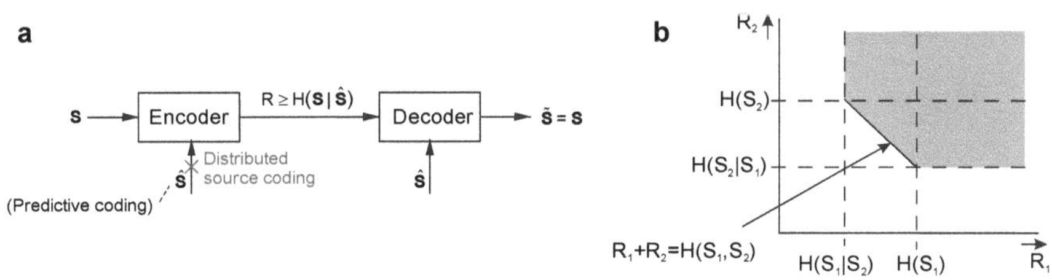

Fig. 5.16. **a** Distributed source coding (lossless case): In contrast to predictive coding, only the decoder has prior information about the source **b** lower rate bounds in case of independent coding of two correlated sources

Beyond the lack of practical applicability to multimedia signal coding in general and video signals specifically (where it would be most useful due to the high redundancy between adjacent pictures), the theory of DSC unveils that decoders performing tasks for source estimation could work close to optimum even under difficult conditions such as data losses or reconstruction errors, e.g. performing similar operations as the estimation of the source for error concealment.

5.6 Problems

Problem 5.1.

An AR(1) process described by correlation coefficient $\rho=\sqrt{3}/2$ and variance $\sigma_S^2=4$ is coded by DPCM. Fig. 5.17 shows the case of one-dimensional linear prediction; in the two-dimensional case, a separable AR(1) process with $\rho_1=\rho_2$ and appropriately optimized prediction shall be assumed.

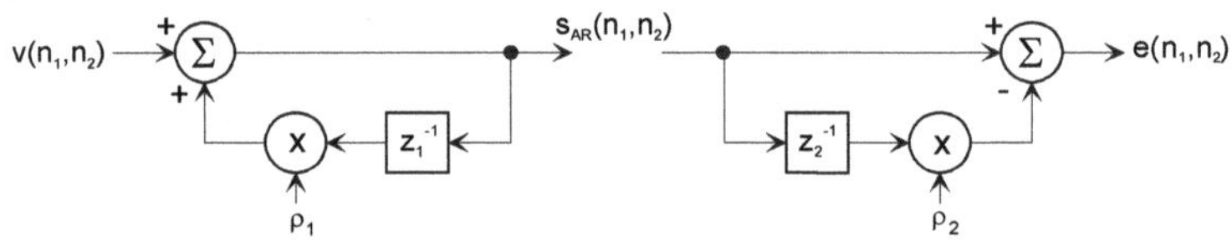

Fig. 5.17. Prediction and synthesis of an AR(1) process

a) Compute the variance of the prediction error $e(n_1,n_2)$ and the coding gain for the 1D and the 2D cases.
b) By how many bit/sample can the rate be reduced in the low distortion range fulfilling (4.28) or (4.40), when *i*) one-dimensional *ii*) two-dimensional DPCM is applied?

c) Instead of the conventional DPCM encoder structure (Fig. 5.2b), a modified structure in Fig. 5.18 shall be used. Supplement the '+' and '–' signs at the summations, such that structures in Figs. 5.2b and 5.18 produce exactly the same bit stream $i(\mathbf{n})$.

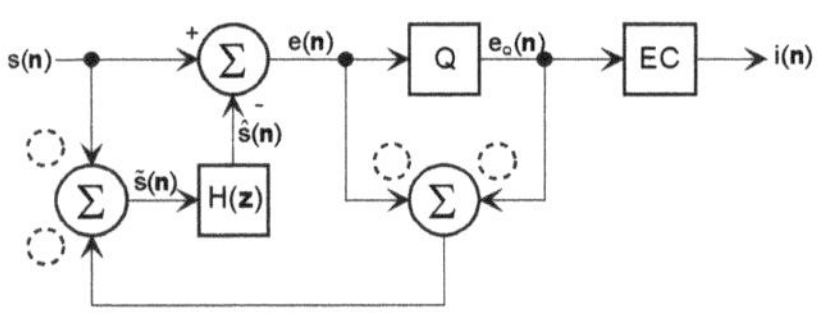

Fig. 5.18. Alternative DPCM encoder structure

Problem 5.2.

Compute the coding gain for DCT transform coding of a 1D or separable 2D AR(1) process with correlation parameters $\rho_i=0.5$ and $\rho_i=0.95$. Four cases of 1D and 2D-separable transforms shall be considered: $U=2$, $U=3$, $U_1 \times U_2=2\times2$, $U_1 \times U_2=3\times3$. Compare the results against the maximum achievable coding gain for an AR(1) model for the 1D and 2D cases. [Hint: Use the result from Problem 2.14 to determine the variances of coefficients for the DCT block lengths $U_i=3$.]

Problem 5.3.

A linear block transform of block size $M=U=2$ is defined by the following transform matrix:

$$\mathbf{T}(2)=\begin{bmatrix}1/2 & 1/2\\ 1 & -1\end{bmatrix}.$$

a) Determine the inverse transform matrix $\mathbf{T}^{-1}$.
b) Are the basis vectors of the transform orthogonal? Describe a transform matrix of an orthonormal transform $\mathbf{R}=[\mathbf{r}_0{}^{\mathrm{T}}\,\mathbf{r}_1{}^{\mathrm{T}}]^{\mathrm{T}}$, such that $\mathbf{r}_0\sim\mathbf{t}_0$ and $\mathbf{r}_1\sim\mathbf{t}_1$.
c) An AR(1) process is characterized by σ_s^2 and ρ. It shall be transformed block-wise by $\mathbf{T}$. Determine the variances of the transform coefficients $\mathcal{E}\{c_0{}^2\}$ and $\mathcal{E}\, c_1{}^2\}$.
d) A quantization error vector $\mathbf{q}=[\,q_0\ q_1\,]^{\mathrm{T}}$ is superimposed to the transform coefficients. Compute the resulting error $\tilde{\mathbf{s}}-\mathbf{s}$ in the reconstructed signal (expressed as two-dimensional vectors). How shall the quantization step sizes of uniform quantizers be chosen, such that the contributions of q_0 and q_1 to the energy of the reconstruction error would become equal?
e) The transform $\mathbf{T}$ is used as basis for a wavelet transform using 2 decomposition steps, i.e. depth 2 of the wavelet tree. This can then be interpreted as a block transform with $M=4$. Sketch a signal flow diagram related to the analysis of a single block. Also, express the related transform matrix $\mathbf{T}(4)$. Is this transform orthogonal or orthonormal?

6 Intra-picture Coding

Still (or intra-picture) image coding includes compression of binary images and images with multiple amplitude levels, the latter case either monochrome or with multiple color components, as typically captured by cameras or image scanners. Beyond that, compression schemes are also applied to other sensor data such as magnetic resonance intensity in medical imaging, infrared, x-ray, depth captured by range sensors, or to synthetic content, specifically generated by computer graphics or captured from screens. Various methods such as run-length coding and conditional entropy coding for binary images; vector quantization, predictive coding and transform coding for multi-amplitude images are presented in more detail. Further sections discuss lossless image compression, synthesis based coding, fractal coding and 3D image coding. As examples for the overall combination of building blocks, still image compression standards like JPEG, JPEG 2000 and the intra-picture compression of AVC and HEVC video coding standards are discussed in detail.

6.1 Compression of binary pictures

A binary image has only two amplitude levels, e.g. being a pure black and white image; typical examples of binary images are scanned text pages or two-tone prints. Further, shape masks expressing geometric properties of content are also often represented as binary images. Transmission of binary images has a long history in telecommunication. Long-distance transmission of binary pictures was already performed using specific coding in telegraphy during the 19th century, and fax services were the first wide-spread application for transmission of binary picture information over conventional (analog) telephone lines.

6.1.1 Compression of bi-level images

By a threshold operation, a binary image can be generated from a multi-level image as

© Springer-Verlag Berlin Heidelberg 2015

J.-R. Ohm, *Multimedia Signal Coding and Transmission*,
Signals and Communication Technology, DOI 10.1007/978-3-662-46691-9_6

$$b(\mathbf{n}) = \begin{cases} 0 \text{ if } s(\mathbf{n}) < \Theta \\ 1 \text{ if } s(\mathbf{n}) \geq \Theta. \end{cases} \tag{6.1}$$

By determining optimum threshold values Θ, it can be avoided that relevant picture structures disappear. Concerning the statistical behavior, it can be expected that the binary image inherits properties of statistical dependency between adjacent samples (e.g. correlation) from an original gray-value image (cf. Problem 2.21). The coherence between adjacent samples in the binary signal is best modeled by Markov chains (cf. Sec. 2.5.4).

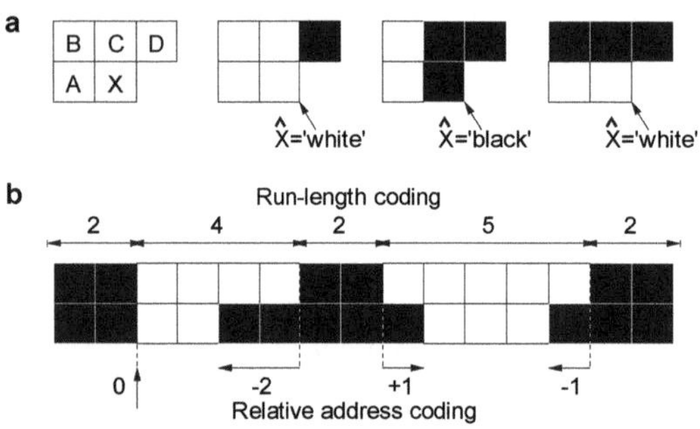

Fig. 6.1. a Prediction of binary samples, prediction topology (left) and examples of three different prediction rules **b** Relative address coding

Run-length coding. The method of run-length coding as introduced in Sec. 5.1 can be applied to 2D signals by performing sequential row-wise processing. However, additional mechanisms are required to utilize the redundancy (correlation) in vertical direction, i.e. between adjacent lines. The following methods can be applied:

– *Prediction*: Previously-encoded samples either from the current row, or from the row above can be used to predict the current sample marked as **X**. It is necessary to use only samples already known to the decoder, which has to perform the same prediction. In Fig. 6.1a, it is assumed that four samples **A-D** are used, which can have a total of 16 different configurations. For each configuration, a prediction rule has to be defined; examples of three different rules are also illustrated in Fig. 6.1a. Another approach would be explicit signalling of the value when no reasonable prediction is possible from the neighbor values. The prediction error could be coded as '0', when the actual value fits with the prediction and '1' otherwise. A high occurrence of zero-values would be expected if the prediction is successful, such that it is advantageous to apply the run-length method of Fig. 5.1b subsequently[1].

[1] Application of such prediction schemes is not limited to combination with run-length coding; any other entropy-coding method can be used as well to encode the prediction error signal, which comes with a certain probability of '0'=prediction successful and '1'=prediction failure. In the combination with context-based arithmetic coding, it is even

- *Relative Address Coding* (RAC) and *Relative Element Address Designate* (READ): The addresses of transitions (white→black, black→white) are encoded relative to the address of a transition in the previous row (Fig. 6.1b); only for the first row, direct run-length coding is necessary.
- *Line skipping*: If a row is completely identical with the previous one, this is indicated by a specific flag or code.

In the context of transform coding, also *modified sequential scans* are used with 1D run-length coding to exploit similarities in close neighborhoods both horizontally and vertically. Often, scanning is performed within small 2D blocks and then over a sequence of these blocks (see Figs. 6.19 and 6.22). Peano-Hilbert scanning (Fig. 6.19c) is an approach which establishes groups of samples within square patterns in a hierarchical fashion to visit neighbors in both dimensions as early as possible. Another approach making a compromise between visiting both horizontal and vertical neighbors similarly early is stripe-wise column scanning as used in JPEG2000 (Fig. 6.27).

Template prediction. Fine-structured drawings and graphics produced by using halftoning or pattern-based shading show frequent changes between white and black samples, which may be systematic but are poorly handled by RLC, where the run durations are expected to be exponentially decreasing (implicit in first-order Markov models). In such cases, the method of *template prediction* can be used instead of rule-based prediction (see Fig. 6.2a): A template is a pattern from the neighborhood of the current sample position. Within a certain surrounding range of samples previously transmitted, the best match for this template (which excludes the current sample itself) is searched. An identical search is necessary both at the encoder and the decoder. The prediction for the current sample is the value which is found at the related reference position of the best match[2]. If the actual value of the current sample deviates from the prediction, a '1' is encoded as prediction error, otherwise a '0'. The method is more complex (due to the search both at encoder and decoder), but the prediction result will typically be better than the rule-based nearest-neighbor prediction shown in Fig. 6.1a. Encoding of the symbols '0' and '1' is based on probabilities of the cases 'prediction successful' and 'prediction failure' when the goal is lossless coding of the binary picture.

It is not always necessary to perform lossless coding of the binary picture; e.g. a sample that falls in the category 'prediction failure' can be inverted if the error is tolerable and lower amount of coded bits can be achieved. Furthermore, template

more suitable to define the expected probabilities for 'X=0' and 'X=1' individually for each of the 16 neighbor-sample contexts. Such a 'statistical prediction' can even be adapted by specific image properties to provide further rate reduction, and would then be very similar to context-based arithmetic encoding (cf. Fig. 6.2b).

[2] As a matching criterion, the Hamming distance (count of samples that have a different value than the template) is typically used. If several matches with same Hamming distance exist, the frequency of occurrence counts for '0' and '1' at the reference position can also be used as a probability estimate in the subsequent entropy coding.

prediction can likewise be applied to a local group (vector) of neighbored samples.

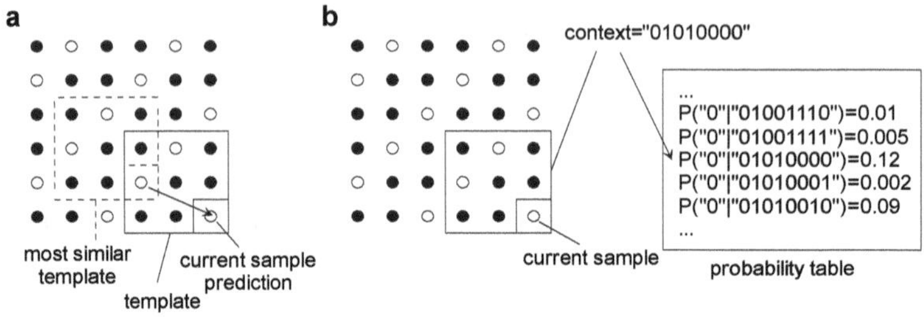

Fig. 6.2. a Template prediction **b** context-based arithmetic encoding

Context-based arithmetic encoding (CAE). CAE exploits the conditional probabilities, taking into account the state from a local area of already decoded neighbored samples (cf. Sec. 4.4.5). The effect is similar with prediction methods, particularly with template prediction, when the same neighborhood is used. However,

- firstly, a direct combination is made between the (nonlinear, i.e. conditional probability based) prediction step and the step of entropy coding;
- second, the prediction is not 'hard', but rather made in a statistical sense, taking into account the probability by which for a given state of the template a state '0' or '1' is to be expected for the current sample, whereas the prediction schemes usually take global probabilities of the cases 'prediction successful' and 'prediction failure';
- third, in template prediction, the model is trained empirically in an adhoc fashion from the area of the search range, where many of the possible template states may not even be present; in contrast, CAE uses a table which contains the probabilities of the current sample being '0' or '1' for each possible state of the template; this can be updated with conventional methods as described in Sec. 4.4.5.

An example is shown in Fig. 6.2b. The context $\mathcal{C}$ is usually a pre-defined group of neighbor samples, but could also be adapted by certain rules deduced from decoded values, e.g. directional characteristics. If the neighborhood is defined over P samples, 2^P different configurations can occur from their respective binary states; for Fig. 6.2b, $P=8$. For each of the possible configurations, the conditional probability has to be available by which the current sample would have a value '0' or '1'. A fixed probability table can be used, where only one of the two probabilities needs to be stored in the table due to the relationship $\Pr(1|\mathcal{C})=1-\Pr(0|\mathcal{C})$. Arithmetic coding is directly employed using this probability table; additionally, adaptation of probability values can be used, as discussed in Sec. 4.4.5. CAE is likewise applicable to binary signals or to bit-planes (e.g. the latter is used in the JPEG2000 standard).

As a general remark, there is no need to run CAE in a lossless coding mode. For example, the current sample may be encoded as its opposite value if this is beneficial to reduce the amount of coded bits. This is however only useful when the effect of the sample's value in encoding of subsequent samples (where it will become member of the context) is taken into account. For example, removal of isolated opposite-level samples or straightening of edges may effect a simplification of the structure and therefore lowers the expected number of bits. The corresponding distortion criterion is again the Hamming distance.

Stacks of binary planes for multi-level coding. Fig. 6.3b shows an example of the first three bit planes of a gray-level image, with the mapping illustrated in Fig. 6.3a. The structural detail increases towards the LSB, which is explainable due to the larger local variations in small amplitude ranges, and influence of noise (see problem 6.6). As a consequence, binary coding methods which exploit the coherence within the signal are requiring higher rates, when applied to bit planes of lower significance. On the other hand, coherences between bit planes could additionally be exploited, e.g. larger local variations can be expected in regions where already the next-higher bit plane shows a large amount of changes.

Fig. 6.3. a Contributions of single bits to overall gray levels of a 3 bit discrete amplitude representation. **b** Left to right: Bit planes 1 (MSB), 2 and 3 (LSB) of a 3-bit PCM signal.

Halftoning. These methods are often used in printing for the purpose of mapping a multiple-amplitude signal into a binary image signal with higher sample density to obtain a gray-level illusion. For this purpose, the density of black samples is increased in dark areas and decreased in bright areas. If the dots are packed sufficiently dense, establishing local high-frequency patterns, the observer's visual system by its lowpass characteristics performs a smoothing such that the ratio of white versus black dots is perceived as gray[3]. Due to the systematic usage of patterns, which are constructed by elementary groups of sample constellations, application of neighborhood-based context models or template prediction can be used for systematic data compression. In terms of modeling, this could be interpreted as a Markov Random Field (cf. Sec. 6.6.2) where only few state transitions are likely to appear.

---

[3] Halftoning methods are known, which are optimum for subjective impression, see e.g. [MEŞE, VAIDYANATHAN 2000].

6.1.2 Binary shape coding

The rough position of a region is typically described by a bounding rectangle given by position, width and height. The exact information about the shape of an arbitrary region can be defined globally, or within a coordinate system of a *bounding rectangle* for which the absolute location in the picture and the width/height (in samples) need to be signaled additionally. For each sample, a binary flag is then set with the following meaning:

$$b(\mathbf{n})=\begin{cases}1: \text{sample belongs to the region}\\ 0: \text{sample does not belong to the region.}\end{cases} \tag{6.2}$$

Methods for binary image coding as presented in Sec. 6.1 can directly be used for shape coding. For example, binary *context-based arithmetic coding* (CAE) is employed by the MPEG-4 standard to describe the shape of arbitrary regions, either in single images or in an image sequence, where in the latter case the context is extended to the third (temporal) dimension.

Quad tree representation. This method [SAMET 1984] provides a block-based approximation of a region, but can also reach the granularity of a sample-accurate representation to represent a binary shape. In the original quad-tree approach, the code expresses whether a block of size M_1xM_2 is sub-divided into 4 smaller blocks of size $M_1/2$x$M_2/2$ each, or if it shall be kept without split. Any of the smaller blocks can again be split into four siblings etc., such that a hierarchy of quad-block sub-partitions is mapped by a hierarchical tree structure. An example of a region shape is shown in Fig. 6.4a. The associated quad tree is outlined in Fig. 6.4b. Each node of the tree has 4 branches, if the associated block is further sub-divided. If no sub-division is made, the tree terminates, i.e. a leaf is signaled.

The information about the sub-division is represented by the *quad tree code*, where in the example of Fig. 6.4 nodes (quadrangles) are represented by '0' bits, while leaves (circles) are represented by '1' bits. This code is uniquely decodable like a prefix-free code, as each '0' must be followed by at least four additional bits. In the example shown, the decoder would interpret the tree as follows, where the parentheses express the hierarchy levels of parsing: 0(10(1111)10(1110(**1111**))). If the maximum depth of the quad-tree levels is predefined (e.g. three levels in the given example), some bits for signalling (printed bold in the given example) could be omitted, as it is implicit that no further split can occur. If e.g. an image shall be sub-divided into blocks of maximum size 64x64 samples, minimum size 8x8 samples, the smallest possible code consists of 1 bit (a single '1' would indicate that the maximum unit is not sub-divided), while the maximum number of bits would be 21 for overall splitting into 8x8 (1x '0' for splitting at level 64x64, 4x '0' at level 32x32, 16x '0' at level 16x16; then, it is uniquely indicated that *all* blocks shall be sub-divided into smallest size of 4x4). Any other variant which stops splitting at higher levels would consume less bits.

Variants of the quad-tree method also include non-square partitions, which for simplicity may only be invoked at the last level of splitting. As an example, refer to the splitting modes for prediction units in the HEVC standard, as shown in Fig. 7.31[4].

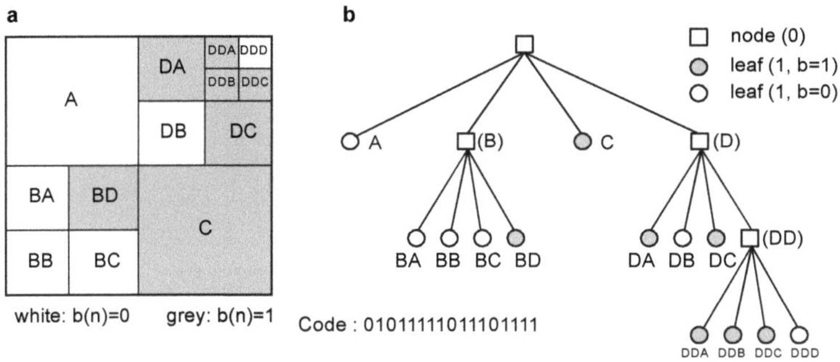

Fig. 6.4. a Quad-tree sub-division of an image **b** associated quad tree with code

If the quad tree shall be used for binary shape representation, it is necessary to assign the information b from (6.2) to the leaves, which means that all samples within the corresponding quadrangle are associated with $b(\mathbf{n})=0$ or $b(\mathbf{n})=1$. This is illustrated by the gray-shaded areas and nodes in Fig. 6.4. If multiple regions (e.g. several object shapes) shall be characterized, usage of multiple labels is straightforward.

Quad-tree-like signalling of segmentation is also widely used for purposes of variable block-size coding, e.g. variable block size transform coding, variable block size prediction, etc. In such cases, encoded transform coefficients or encoded prediction parameters such as motion vectors would be associated with the leaves of the quad-tree.

All information components (tree splitting information as well as the attached region property information) can be further compressed by entropy coding, depending on the probability of splitting/not splitting. In terms of region membership, it is also beneficial to apply context-based entropy coding, with probabilities depending on already decoded neighbors' memberships. Furthermore, *merge* signalling can be used for efficient coding of the attached information in cases where it is identical with another leaf representing a neighbored partition (even if that resides at another level or in another branch of the quad-tree).

Beyond block-based splitting and merging, partitions can be established at even finer granularity for better alignment with sample-accurate region boundaries. An example of *geometric partitioning* is shown in Fig. 6.5, where the basic idea is to express the boundary by a straight line. 2 parameters are necessary to express the separation line either by its start and end points on the block boundary, or by one position and one angular orientation parameter. When sub-partitioning of a block

[4] the two original quad-tree splitting options (no split, regular quad-split) are included as 'MxM' and 'M/2xM/2'

is restricted to separation lines that are strictly parallel with block boundaries, signalling only requires one parameter and one flag for horizontal/vertical orientation (i.e. 0 or 90 degree angles). Geometric partitioning could also be combined with quad-tree coding, when it would only be invoked as an alternative at the last level of quad-tree splitting.

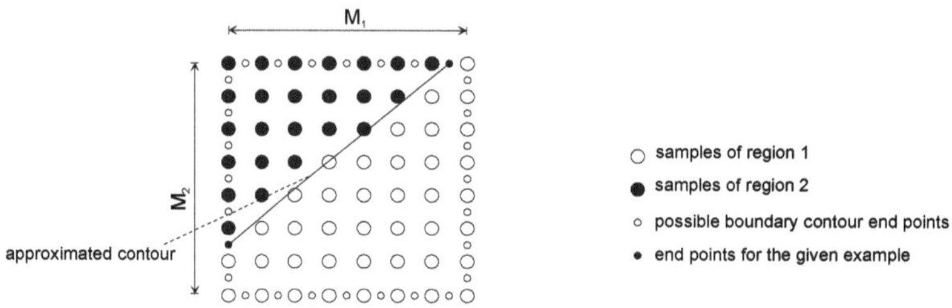

Fig. 6.5. Geometric partitioning based on line separation of a rectangular block into two sub-regions

The method of sub-dividing a 2D picture into regular rectangular block structures is not limited to usage in image shape or structure representation; an extension into the third (temporal or volumetric) dimension are *'octree'* codes [MEAGHER 1980].

6.1.3 Contour coding

Chain codes. Discrete contours (see example in Fig. 6.6a) can be encoded by *chain codes* [FREEMAN 1970] from a start point up to an end point. A chain code connects directly neighbored contour positions, which in the case of rectangular sampling is restricted to be one of the eight nearest neighbors (Fig. 6.6b)[5]. If start and end points of the contour are direct neighbors, it is *closed.* The coordinate reference of at least one point (typically the starting point) has to be encoded explicitly when the contour is to be associated with a position in a picture.

For direct chain coding, application of *entropy coding* will only be useful if certain contour directions have higher probability, which could be the case for regular geometric forms, consisting of smooth contours with preference orientations, for which differential coding described below becomes even more efficient.

---

[5] This corresponds to the system $\mathcal{N}_2^{(2)}$ of (2.107). Similar approaches using only four neighbors by employing the $\mathcal{N}_1^{(1)}$ system are also possible, but would require two links in case of diagonal contour orientations. Assuming that all contour directions are equally probable, the expected mean rate will be the same for both types of neighborhood systems.

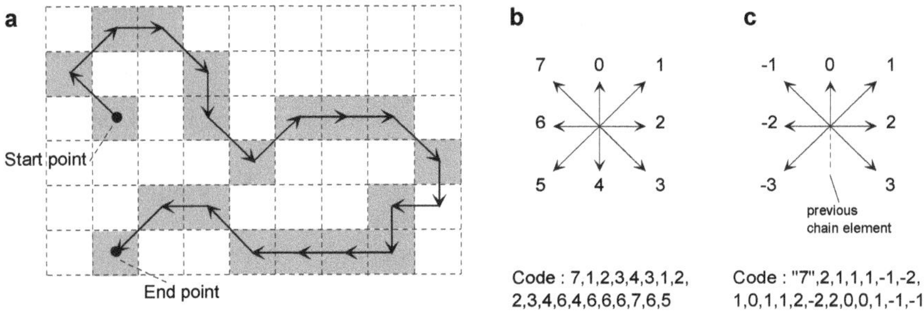

Fig. 6.6. **a** Discrete contour and its chain description; interconnection directions for an 8-neighborhood **b** in direct chain coding **c** in differential chain coding

Differential chain codes. Those describe changes of the contour direction from one chain element to the next [KANEKO, OKUDAIRA 1985]. If no 180° (back) turns are allowed, the differential code only needs to support 3 different continuation directions in the case of the $\mathcal{N}_1^{(1)}$ system, and 7 different directions in the case of the $\mathcal{N}_2^{(2)}$ system (Fig. 6.6c). In particular for smooth contours, the combination of differential chain coding and entropy coding is highly efficient, where the 0° direction (straight continuation) is expected by highest probability, while directions '–3' and '+3' would only be selected at extreme corner locations which can be expected to occur rarely. Typically, in the case of the $\mathcal{N}_2^{(2)}$ system, discrete contours can be represented by differential chain/entropy coding using around 1.5 bit per contour position.

To further reduce this rate, chain codes can be extended to encode combinations of subsequent elements. Also in this case, more smooth contours will require less bits for representation, as e.g. chains of double or triple continuations in the same direction would occur by higher probabilities. Systematic smoothing of contours could be employed for lossy coding at lower rates.

Other methods which can be applied for lossy contour encoding are based on *interpolation* of contours from sparsely located control points, which need no longer to be direct neighbors, nor do they need to coincide with discrete coordinate positions. By these methods, also hierarchical representations with variable accuracy of the contour representation can be realized[6]. In this context, it is possible to represent the contour as a sequence of point coordinates, which are then processed as finite or (in case of closed contours) cyclic signals; these are typically smooth, as neighbored point coordinates are by nature similar. Therefore, transformation into the frequency domain and representation by transform coefficients is also possible and gives a compact representation (cf. MCA, SEC. 4.5).

---

[6] One criterion to judge the similarity of two contours or shapes is the size of the *area between*, which can be determined from an exclusive-or combination of the corresponding binary shape images. Another possibility which interprets the points of a contour as a set and determines the distance between two sets by identifying closest pairwise matching is the *Hausdorff distance* (cf. [MCA, SEC. 5.3]).

6.2 Vector quantization of pictures

VQ can be applied directly to image signals, as was shown in Fig. 4.19b. If the values within the square blocks defined as vector input to the VQ encoder are statistically dependent, this can be implicitly exploited by designing suitable codebooks using the generalized Lloyd algorithm (Sec. 4.5.3).

If the overhead needed to transmit the codebook is not too high, the coding performance can be improved by using adaptive codebooks (see Sec. 4.5.5), specifically designed for one picture. This also applies for image sequences, where images within one shot can be expected to be quite similar, such that the rate needed for transmission of the entire codebook or some adapted vectors can be held reasonably small as compared to the overall rate.

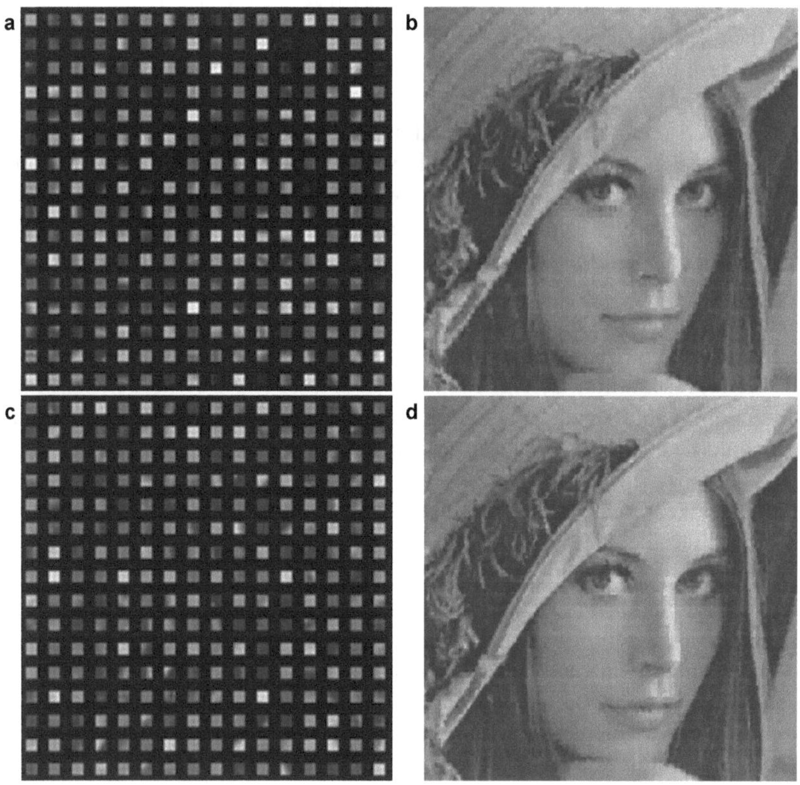

Fig. 6.7. VQ in image domain, codebook size J=256 **a** Codebook generated by GLA procedure from a training set of 15 pictures **b** VQ reconstruction of a picture not included in training set **c** codebook generated from a single picture **d** VQ reconstruction of this image

The influence of the codebook design on the quality of the reconstruction is illustrated in Fig. 6.7. Codebooks designed for an image-domain VQ with J=256 vectors and $M_1 \times M_2$=4x4 samples vector size are shown in Figs. 6.7a/c; each of the small squares is one codebook entry. The example in Fig. 6.7a was generated using a training set of 15 pictures. Fig. 6.7b shows the reconstruction of a picture that was not contained in the training set. In contrast, Fig. 6.7c shows a codebook which was specifically designed for this one picture, the corresponding reconstruction is shown in Fig. 6.7d. Some vectors have structures that are specific to that picture, e.g. optimized for diagonal edges at the brim of the hat. Therefore, the reconstruction shows much less blocking artifacts, however the codebook needs to be signaled as side information.

The block-separate VQ method described has the following disadvantages here:

- *Blocking artifacts* could become visible in the reconstructed signal, most typically in flat areas and around edges;
- Due to relatively small block sizes which have to be used to keep codebook size and algorithmic complexity low, significant redundancy can exist between neighboring blocks.

The following approaches can be followed to reduce these effects (see also Sec. 4.5.4, 4.5.5 and 6.9):

- *Boundary filtering* (post processing) or usage of block overlaps in the construction of the vectors;
- Separate coding of the mean values by mean-separating vector quantization (MSVQ), utilizing redundancy by predicting mean values;
- Usage of pre-classification, in particular for edge areas, employing specific codebooks that are specifically designed to better reconstruct edges of similar orientation;
- Usage of finite-state vector quantization [ARAVIND, GERSHO 1986], where due to the two-dimensional structure of images even more information is available for definition of the next state function than it is the case for 1D signals (see *side match* method described in the subsequent paragraph).

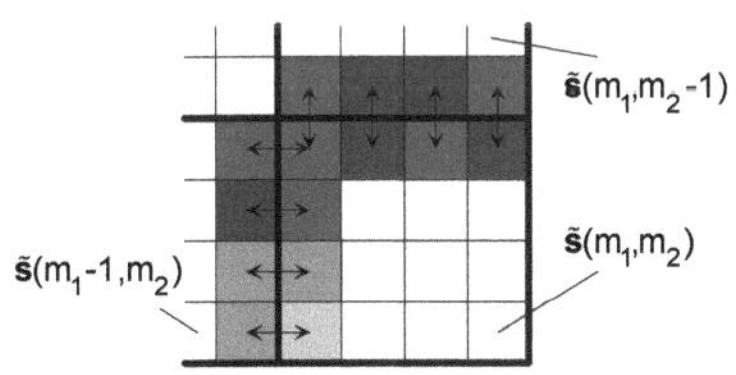

Fig. 6.8. Side-match FSVQ

Example: Side-match FSVQ. The method originally introduced in [KIM 1992] uses a super codebook of J reconstruction vectors which is designed as for conventional VQ. The number J_s=J^* of code symbols is set constant in all states. To determine the next-state function in dependency on the reconstruction vectors which were used in the neighbor blocks left and above, those J^* vectors are selected as candidates for the current block, for which the *lowest differences* in amplitude occurs at the boundaries between the actual vector and previously decoded neighbors at left and top (see arrows and shaded areas in Fig. 6.8).

6.3 Predictive coding of pictures

In predictive coding of images, open-loop coding (Sec. 5.2.1) is usually avoided (except for the lossless coding case where it is identical to the closed-loop approach). According to the high correlation of adjacent image samples, optimum 1D and 2D synthesis filters would have long impulse responses, such that the amplification of the quantization error by (5.8) is significant in the open-loop case, and closed-loop coding (DPCM) is the better choice. The following paragraphs introduce methods of predictive coding that are specific for 2D image signals.

6.3.1 2D prediction

The computation of estimates shall only use values already known to the decoder due to causality considerations. For example, assuming that the sequence of processing starts in the top left corner and proceeds row-wise over an image, the following prediction equation applies in DPCM when using a quarter-plane filter of order P_1P_2-1 [cf. (2.224)],

$$\hat{s}(n_1,n_2)=\sum_{\substack{p_1=0 \\ (p_1,p_2)\neq(0,0)}}^{P_1-1}\sum_{p_2=0}^{P_2-1} a(p_1,p_2)\tilde{s}(n_1-p_1,n_2-p_2)\,. \tag{6.3}$$

The predictor filters may be separable or non-separable, local adaptation of filter coefficients can be done for minimization of prediction error variance, either by using a forward-adaptive (2.207) or backward-adaptive (2.223) approach. Beyond that, asymmetric half-plane filters (as in Fig. 6.9) still allow a causal processing sequence; they are providing additional information from the top-right neighbor sample(s).

An example for a very simple prediction of image samples uses a (weighted) average from the nearest horizontal and vertical neighbors, (typically $a_1+a_2=1$),

$$\hat{s}(n_1,n_2)=a_1 s(n_1-1,n_2)+a_2 s(n_1,n_2-1)\,. \tag{6.4}$$

Another example is median prediction, where the median value from the decoded samples of some causal neighborhood is used,

$$\hat{s}(n_1,n_2)=\text{median}\left\{\tilde{s}(m_1,m_2)\,|\,(m_1,m_2)\in\mathcal{N}_{\text{causal}}(n_1,n_2)\right\} \tag{6.5}$$

For median prediction, a relatively small number of three or five neighbored samples is mostly used; weighted median filters or hybrid median/FIR filters can be applied as well. For natural images, predictors with fixed settings of predictor coefficients will not generally be optimum. Beyond the Wiener-Hopf equation (2.206) and backward adaptation using the LMS algorithm (2.223), simple ways of adaptation use switching of prediction modes based on a small set of choices. In backward-adaptive switching, a causal neighborhood can be analyzed e.g. regarding edge presence and orientations to adapt the predictor geometry by pre-set rules [RICHARD, BENVENISTE, KRETZ 1984]. An example is given in Fig. 6.9. The sample **X** shall be predicted by a combination of the neighbored decoded samples **A-D**. Depending on local edge orientation analysis (which may be done exclusively based on the samples A-D, but could also involve a larger area of previously decoded samples), the estimate $\hat{X}$ is determined in a way such that prediction across the edge is likely to be avoided.

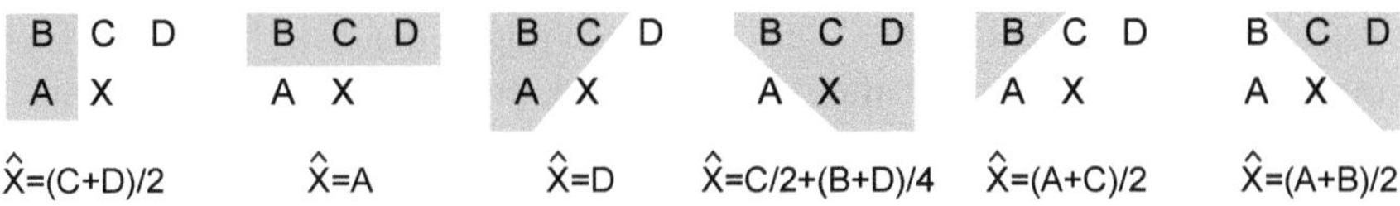

Fig. 6.9. Examples of edge positions and orientations and example prediction rules

In forward-adaptive switching, the predictor choice has to be encoded as side information. This is usually done for groups of samples. For example, in the lossless mode of the JPEG standard, the predictor signalling is valid for a group of subsequent samples, until a new setting is made. Typically, forward-adaptive switching should also include a case where *no prediction* is performed. A backward adaptive switched prediction scheme is implemented in the lossless and near-lossless coding standard JPEG-LS. Combination between backward and forward adaptation would also be possible, where explicit signalling is only done when the mode as determined by implicit backward analysis is suboptimum. In general, none of the methods described so far is exclusive, e.g. combination of switched prediction at edge positions and median prediction elsewhere is possible as well.

Another method of non-linear prediction is *template prediction*. The assumption is that similar structures appear systematically in a wider neighborhood, e.g. due to texture similarity. The neighborhood is to be defined from samples that are already available at the decoder ('search region' **Π**). The template **Θ** is a causal neighborhood of the current sample (excluding the sample itself). The prediction is taken as copy from the position which gives best matches by the corresponding template (Fig. 6.10), with $\Delta(\cdot)$ being a distance or similarity function,

$$\hat{s}(n_1,n_2)=\left\{\tilde{s}(m_1,m_2)\,|\min_{(m_1,m_2)\in\Pi}\Delta\big(\Theta(n_1,n_2),\Theta(m_1,m_2)\big)\right\}. \tag{6.6}$$

The method can be extended to vector prediction, where a group of samples is predicted through an adjacent template.

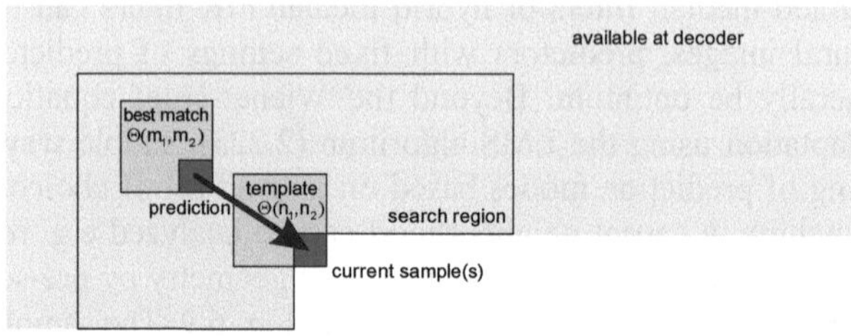

Fig. 6.10. Principle of template matching prediction

6.3.2 2D block prediction

Joint prediction of local groups of samples is particularly useful when it shall be possible to achieve lower data rates by exploiting redundancy between such groups. As the residual signal after vector or block prediction is typically correlated (see Sec. 5.2.4), VQ [BHASKARAN 1987] or transform coding can be employed for subsequent compression.

Typically, samples are combined from rectangular or square blocks for joint prediction (Fig. 6.11a), but 1D arrangements can be used as well, e.g. taking samples from the same line or column [CAO ET AL. 2013], or diagonal neighbors [OHM, NOLL 1990][7].

It depends on the sequence of processing at the decoder, which neighbored blocks are already available at the decoder. Mostly, only boundary samples from the blocks on top and at left, i.e. closest neighbors of the current block's samples are used. Prediction for the samples of an entire local block can be performed by

- a dedicated set of linear predictors designed individually for each sample of the current block (see example in Fig. 6.11b), which can be interpreted as weighted averaging from available boundary samples; this can also be supported by adaptation with or without side information signalling;
- A 2D surface function that is jointly extrapolated from the decoded boundary samples (Fig. 6.11b); in simplest case, the function f(A,B,...,H,I) can be the mean value of the boundary samples (denoted as *DC prediction*), but higher-order functions such as linear amplitude planes or ramps can also be used (denoted as *planar prediction*);
- Copy prediction, where another block that is already available at the decoder is used for predicting the current block (Fig. 6.11c); in this case, the displacement offset is either explicitly signaled as side information (denoted as *displacement prediction* [KONDO ET AL. 2004]) or *intra block copy* (IBC). Alternatively, the displacement can be derived by template matching both at the decoder and encoder [TAN ET AL. 2006].

---

[7] The advantage of such 1D arrangements is that direct neighbors can be available for prediction for all samples in the vector, still retaining a causal processing sequence over vectors.

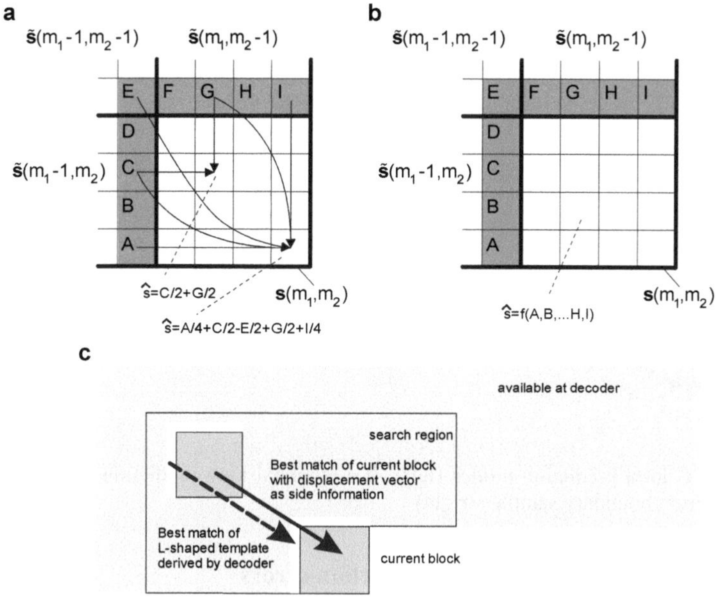

Fig. 6.11. Methods of block-wise vector prediction: **a** using dedicated linear predictors for each sample **b** using an amplitude (polynomial) surface function determined from boundary samples **c** copying another block available at the decoder by template matching or explicit signalling of displacement

A specific method of the first case with adaptation signaled as side information is *directional prediction* which is used for intra picture compression in the AVC and HEVC standards. Here, prediction is performed by extrapolating available boundary samples along a specific direction. In HEVC, intra prediction allows 33 directional modes (compared to only 8 in AVC), plus DC and planar surface prediction from decoded boundary samples of adjacent prediction blocks. The different directional modes of HEVC are shown in Fig. 6.12. If available, samples from the blocks which are positioned at top right and bottom left of the current block can also be used. If not available, these samples are in a first step extrapolated from samples of other available blocks, the result being used for prediction. When the directional extrapolation does not necessarily coincide with the actual sample positions of the current block, sub-sample interpolation needs to be performed. Furthermore, a lowpass smoothing filter can be applied to the boundary samples before using them for extrapolation.

For efficient coding, the side information for intra prediction modes (directional or DC/planar) is also predicted by deriving most probable modes (e.g. same or similar direction) from already decoded neighboring prediction blocks. Furthermore, due to the usage of same values for the prediction of all samples that are on one directional trajectory, the correlation properties of the 2D residual signal are similar as analyzed in (5.19)-(5.21). In HEVC, a type-IV 2D DST based on

(2.273) is used for further compression, which by its basis functions captures the property of the residual which grows in amplitude with increased distance from the prediction reference[8].

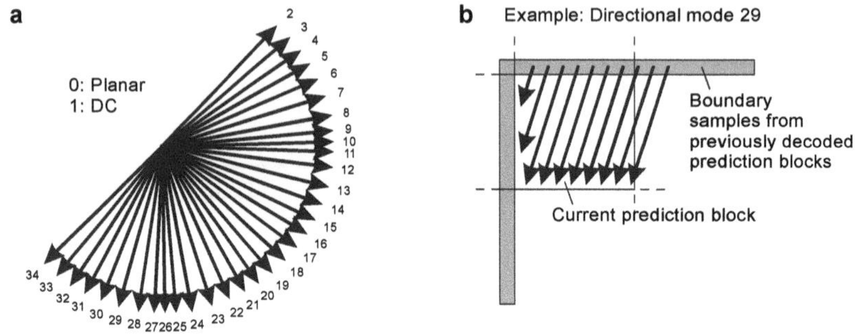

Fig. 6.12. **a** HEVC intra prediction modes (left) **b** directional intra prediction of current prediction block from boundary samples (right)

6.3.3 Quantization and encoding of prediction errors

Fig. 6.13 shows probability distributions measured from prediction error signals, which can roughly be modeled as Laplacian type (2.128). Assume that uniform scalar quantization with a set $\mathcal{V}=\{v_j;j=1,\ldots,J\}$ of reconstruction values and entropy coding is applied. Due to the PDF with high concentration around $e=0$, the entropy $H(\mathcal{V})$ could become significantly lower than $\log_2 J$. When the unquantized prediction error comes from a discrete set of amplitudes (by applying integer rounding and clipping to the prediction $\hat{s}(\mathbf{n})$), lossless encoding of a PCM source by DPCM is possible (in this case identical to an open-loop approach, since reconstruction is perfect). This approach is e.g. applied in the standards JPEG (lossless mode) and JPEG-LS.

Formally, the necessary bit-depth for representing the prediction error increases by one bit compared to the B-bit unsigned integer PCM representation with a value range $0 \ldots 2^B-1$. However, even though prediction error values $e(\mathbf{n})$ generally could take any value in the range $-2^B+1 \ldots 2^B-1$, this is no longer true when the actual prediction value $\hat{s}(\mathbf{n})$ is taken into account, only 2^B different values in the range $-\hat{s}(\mathbf{n}) \ldots 2^B-1-\hat{s}(\mathbf{n})$ can occur. Even more, in a 2's complement representation, there is an option to ignore the MSB (sign) bit, since the carry flag which would indicate the negative value range of $e(\mathbf{n})$ is anyway compensated

[8] When the prediction of the inner block can be performed from all four boundaries, a DST type-I (2.265) better matches the characteristics and can be used for efficient compression, see [MEIRI, YUDILEVICH 1981] and [FARRELLE, JAIN 1986].

during reconstruction when the same prediction $\hat{s}(\mathbf{n})$ is added. Therefore, it is not even necessary to increase the bit-depth of the computation [BOSTELMANN 1974].

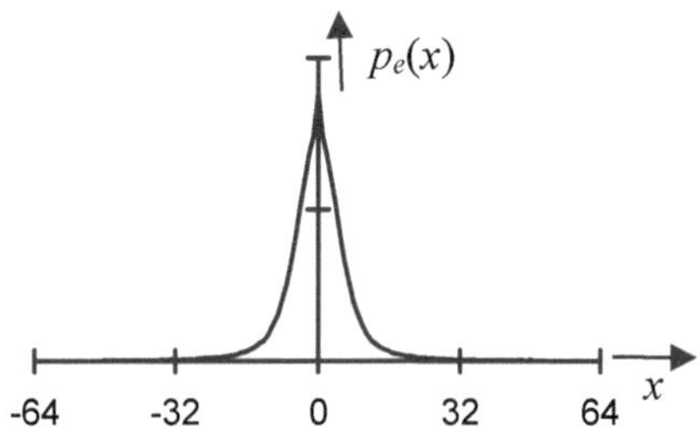

Fig. 6.13. Example PDF of a prediction error signal (separable 2D prediction)

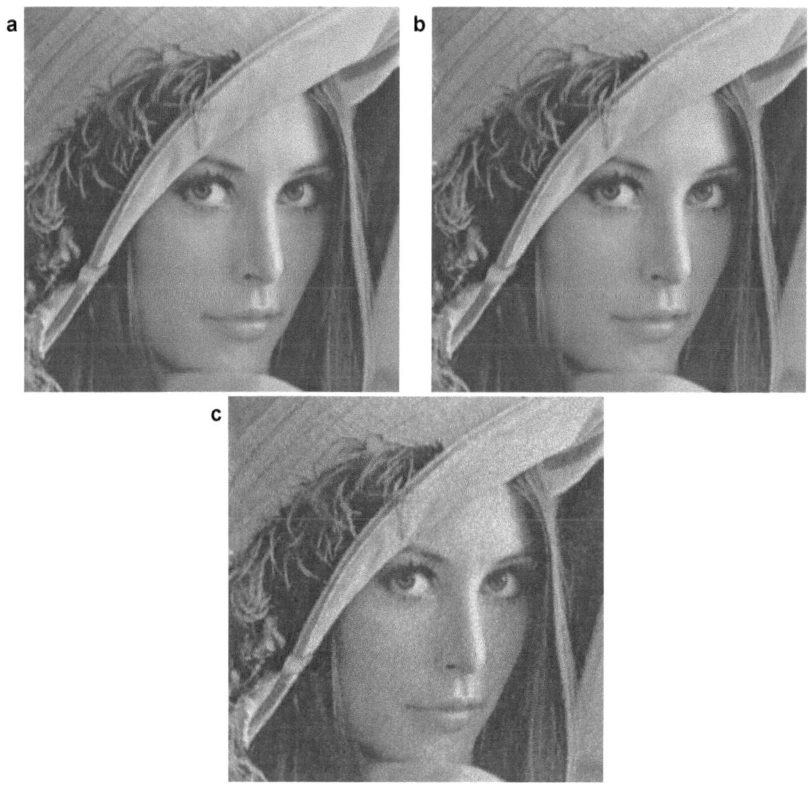

Fig. 6.14. DPCM using scalar quantization and entropy coding.
a J=511, $H(\mathcal{V})$=4.79 *b/s* **b** J=15, $H(\mathcal{V})$=1.98 *b/s* **c** J=3, $H(\mathcal{V})$=0.88 *b/s*

Fig. 6.14 shows reconstructed pictures obtained from DPCM with different values J, and corresponding entropy $H(\mathcal{V})$. Fig. 6.14a represents the case of lossless coding, whereas Fig. 6.14c uses only J=3 reconstruction values $\mathcal{V}=\{-V,0,+V\}$, where the zero reconstruction value becomes more probable with increasing V, and the entropy can go below 1 bit/sample.

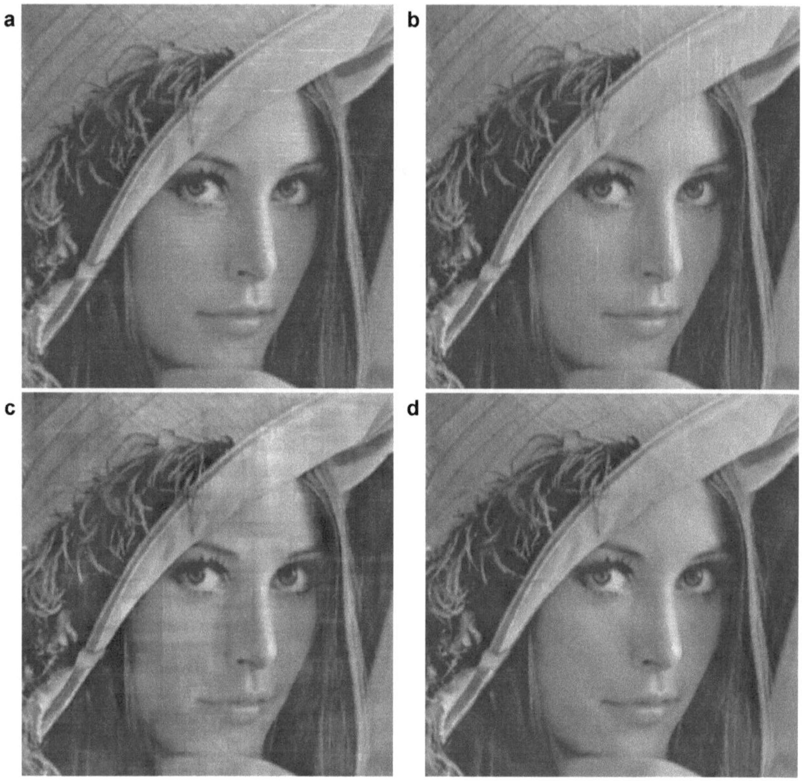

Fig. 6.15. Transmission error propagation in DPCM, random bit error probability $P_{\mathrm{err}}=10^{-3}$, no VLC coding: **a** 1D horizontal prediction **b** 1D vertical prediction **c** separable 2D prediction (3.15) **d** non-separable 2D prediction (3.17)

DPCM with look-ahead. Encoding decisions in closed-loop (DPCM) predictive coding schemes have influence on subsequent samples due to the usage of reconstructed values for the prediction. To exploit this, look-ahead or multi-pass coding can be used, investigating the influence of encoding decision to subsequent depending samples. However, the number of possible decisions is typically much too high to approach an optimum. Practical methods perform multiple encoding passes within a reasonably small set of best-performing candidates. Schemes for efficient tree-based candidate search in combination with vector prediction were introduced in [Hang, Woods 1985] and [Ohm, Noll 1990]. Generally, multi-pass methods only increase encoder complexity, whereas the decoder operation is not changed.

6.3.4 Error propagation in 2D DPCM

According to (5.17), a propagation with the impulse response of the synthesis filter occurs when channel errors affect the decoding, but are unknown at the encoder. Fig. 6.15 shows examples of error propagations in cases of 1D and 2D predictors. While in the first case, propagation can naturally only be observed along either the horizontal or vertical direction, in the latter case the interferences propagate over both dimensions.
As compared to the separable predictor, the non-separable predictor in Fig. 6.15d has less visible artifacts, because due to the choice of coefficients $a=a_1=a_2=0.5$, the influence decays exponentially by a^l over both dimensions (0.5; 0.25; 0.125; ...)[9]. For the separable filter using typically $\rho_i \rightarrow 1$, a much slower decay can be observed. On the other hand, the non-separable filter which has an advantage in terms of error resilience, is worse performing in compression[10]. In general, the problems of drift and error propagation are crucial in DPCM systems due to the recursive dependencies. Therefore, the state of the prediction loop should be re-initialized either periodically or in case of errors. The latter solution however would require a back channel to the encoder; both methods increase the rate and/or decrease the compression performance.

6.3.5 Coding of side information in 2D predictive coding

In case of forward-adaptive prediction, the encoder makes a decision about the predictor adaptation and has to convey the parameters to the decoder. Parameters that influence the decoding process, but are not by themselves part of the sample flow are denoted as *side information*. This includes, for example but not restricted to these categories, quantization step-size parameters, filter coefficients, predictor filter shape information (e.g. locations of past samples used for prediction, as in directional prediction), decomposition partitioning information e.g. in variable block size coding or wavelet tree depth, adaptation/pre-selection/switching of codes etc.

In state of the art compression schemes with large degree of adaptation by local picture content, 'side' information could make up a significant amount of the total bit rate. The usage of adaptive elements should therefore carefully be weighted w.r.t. the benefit in compression and the impact on complexity (encoder and decoder). This can in principle be achieved by *rate distortion optimization*, where the benefit in compression due to a specific adaptation made via the side information is weighted against the additional rate cost (cf. Sec. 4.3 and 7.2.7).

Further, due to the fact that side information can establish a significant amount of the total rate, and not all states of a parameter are equally probable, context-dependent entropy coding of side information also plays an important role.

---

[9] This can be investigated applying a similar approach as in Problem 6.2c.

[10] See Problem 2.11.

Finally, side information is usually associated with pre-defined local image regions, it may by itself be spatially redundant. Typical schemes that exploit such redundancy are spatial prediction of the parameters (linear or non-linear such as median prediction), explicit inheritance from a neighbor position. This can also be based on selection of a few likely candidates from the neighborhood with explicit signalling of the selected one. To explain and illustrate these principles, some methods of encoding side information parameters are discussed in the subsequent paragraphs.

Coding of filter parameters. Often, filter parameters e.g. in adaptive filtering or adaptive prediction are calculated to full mathematical precision, which is approximately implementable by floating point arithmetic on general purpose computers. In dedicated hardware implementation, usage of integer precision is more desirable for lower complexity. A straightforward way of implementing this is uniform (scalar) quantization of filter coefficients within the expected value range. However, the rounding applied to filter coefficients does not directly map to the deviation of the filtered signal. Moreover, performing quantization such that spectral (amplitude and phase transfer) characteristics of filters are retained is usually more important. In case of FIR filters, due to (2.113), a quantization error with uniform probability distribution over all filter coefficients would still give a flat distribution of spectral errors, whereas in (2.114), the filter coefficients influence the spectral shape via the denominator. Therefore, criteria to retain the spectral shape of the filter are particularly important in the case of prediction synthesis (IIR filters). As this is equivalent with keeping the variance of the prediction error as low as possible, a criterion which minimizes the first row of (2.209) subject to quantization of filter coefficients[11] can be applied. Another criterion, which retains the spectral shape as much as possible, is the Itakura distance [ITAKURA 1975], which can straightforwardly be extended into two dimensions [BALLÉ 2012].

When filter coefficients show a systematic behaviour, such as decreasing amplitudes from the center of a symmetric filter towards the outer end, it is also possible to use less bits for representing coefficients that are expected to be smaller. Furthermore, vector quantization can be used, where the encoder decision inherently allows using a distortion metric which is jointly applied to all coefficients, and codebooks can be optimized according to PDF statistics.

Coding of filter shapes in case of directional prediction. In the intra coding part of the HEVC standard, directional predictors are used with choice of orientations as shown in Fig. 6.12. Since some of the directions occur with higher probability depending on neighborhood constellations, context-dependent entropy coding allows efficient compression of the directional modes. Since there are 33 directional orientations, plus planar and DC (flat) prediction modes, three most probable modes (MPM) are determined in a first step for a given prediction block, depending on values available in already decoded neighbor blocks. Additional can-

[11] or the equivalent of (2.213)-(2.215) for the 2D case.

didates are added, which are either similar directions or (if not applicable), planar and DC modes as default. The three MPMs, which are likely selected in most cases under rate distortion optimization criteria, are coded via context dependent entropy coding, whereas the selection of one of the remaining 32 modes is signaled by a 5-bit fixed length code.

Coding of quantization parameters. In scalar quantization, the most important quantization parameter is the step size. Usually, a certain limited number of step sizes is pre-defined. The 'codebook' of step sizes can follow a uniform or non-uniform characteristic; for example, in AVC and HEVC, each increase of the quantization parameter (QP) value by 6 doubles the quantization step size with an approximately logarithmic scaling in between. Additional parameters that can be useful are

- offset of reconstruction values (particularly OFFR according to (4.12));
- additional offset of step size in chroma versus luma components;
- total number of steps (in particular to allow extension for higher bit-depth precision and up to lossless coding).

It is useful to enable mechanisms for adapting the quantization step size locally, e.g. allowing to encode a region of interest with higher quality. To save rate when using such feature, it is useful to perform differential coding of the step size relative to previous settings, or enable its change only when needed within a picture.

When arbitrary non-uniform quantizers shall be used, it is explicitly necessary to signal all reconstruction values, unless these represent samples of a systematic characteristic (e.g. logarithmic), in which case a more compact parametric representation is sufficient. In case of non-uniform vector quantizers, it is also necessary to encode the codebook, if it shall be adapted to the current picture's content. As however block vectors designed for picture-domain VQ usually show high redundancy within the vectors, compression schemes such as predictive or transform coding can again be used for a more compact representation (see Sec. 6.2).

Generally, *backward adaptation* of parameters is also possible, then exclusively using previously decoded information to deduce the parameters. To some extent, prediction, candidate-based and context-based methods of signalling parameters include elements from both forward and backward adaptation, utilizing the redundancy of local parameters that is inherent from the already decoded information, as well as explicit signalling whenever the derivation at the decoder end would fail. Such a combination is probably the most efficient solution for many types of adaptive coding, according to the current state of the art.

6.4 Transform coding of pictures

In 2D transform coding of images, the most important classes of transforms that are applied are (non-overlapping) block transforms, block-overlapping transforms and subband transforms, particularly wavelet transforms in the latter category.

6.4.1 Block transform coding

For block transform coding, the image signal is partitioned into blocks of size $M_1 \text{x} M_2$ (Fig. 6.16). If block sizes are constant as shown in the case of the Figure, the block with index (m_1,m_2) has the starting coordinate (top-left sample) at position $(n_1,n_2)=(m_1M_1,m_2M_2)$ of the image. For the case of a linear block transform, $U_1 \text{x} U_2$ transform coefficients are computed per block, where usually $U_i=M_i$. In block overlapping transforms, typically values $U_i<M_i\leq 2U_i$ are used. Meaningful block sizes in natural-image transform coding are between 4x4 and 32x32 samples, also depending on image size. The choice of the optimum block size depends mainly on the spatial correlation of the signal; for coding highly correlated signals, large blocks are advantageous, while in case of lower correlation smaller blocks are preferable:

- If the signal is highly correlated and the block size is too small, a high redundancy might still exist, which however can be difficult to identify and exploit in the transform domain due to the shift variance of the transform (actual values of coefficients vary with block start position);
- If the signal has low correlation, or if the transform block covers image subareas with large amount of changes (e.g. at edges), it is better to select smaller block sizes, which improves the concentration of information in a smaller number of coefficients.

Usage of variable block sizes, made switchable based on local properties, supports both cases.

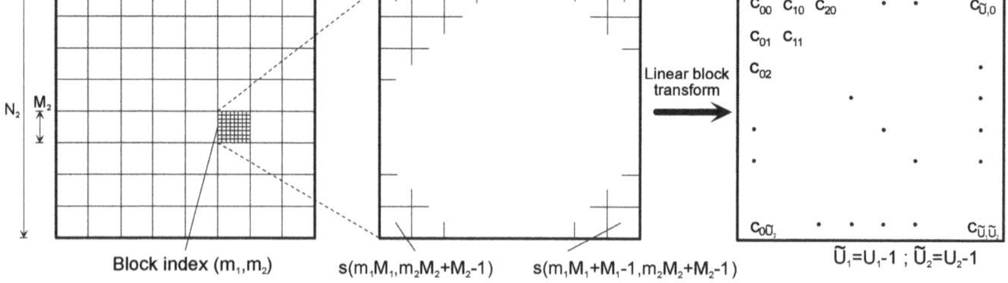

Fig. 6.16. 2D transform coding of equally-sized sub-image blocks

After the linear transform, though 'decorrelating', statistical dependencies remain as follows:

- Within the blocks: Local image structures have directional orientations such as edges, which lead to an expectation about high amplitudes of certain transform coefficients to occur jointly. However, in the case of most real-valued separable transforms this is occurring only for purely horizontal and vertical structures; furthermore, if a block area includes more random detail (such as unstructured or noisy texture areas), it is likely that more coefficients with higher amplitude can be expected to be present over wider frequency ranges; in contrary, for low-detailed areas, most coefficients will be close to zero, and it is rather unlikely that isolated non-zero coefficients are observed at frequency positions that are surrounded by many zero coefficients. This kind of local dependencies can be effectively exploited by an appropriate design of entropy coding with context dependencies across frequencies (transform coefficients) computed around the same local position.
- Across adjacent blocks: As image structures usually extend beyond (fixed) block boundaries, redundancy also exists between the blocks, which is however not easy to exploit in the transform domain (due to basis functions that are not precisely representing the signal beyond block boundaries). If simple prediction is not applicable, entropy coding can also take into account coefficient activity from adjacent blocks for context-dependent coding.

For simple inter-block dependencies, DPCM coding can be applied in combination with transform coding. The prediction can then be interpreted to be part of the encoding and decoding processes. Specific prediction techniques that can be applied in the DCT domain (and similar for other real-valued separable transforms) include:

- *Mean ('DC coefficient') prediction*: The coefficient $c_{0,0}$ carries information about the local mean value of the signal, which typically varies only gradually within a local neighborhood. It can be predicted using the respective DC coefficients of any previously decoded blocks, e.g. as shown in the configuration of Fig. 6.17a.
- *Horizontal/vertical detail ('AC coefficient') prediction*: In case of pure horizontal or vertical structures which extend beyond the block boundaries, transform coefficients in the next block into which the structure extends should be similar. For example, in case of horizontal edges, a vertical frequency coefficient c_{0u_2} can be expected to be similar to that of the left-hand block, whereas in case of vertical edges, a horizontal coefficient c_{u_10} could be predicted from the block above. As it can usually not be the case that both horizontally and vertically oriented structures appear simultaneously, switching can be performed (see Fig. 6.17b), where the dominant orientation can either be signaled or be derived from the energy distribution in the transform coefficients of the neighbored blocks.

Diagonal structures can hardly be predicted in the transform domain, in particular when real-valued separable transforms (such as the DCT) are used, which cannot have basis functions with a unique diagonal orientation. Therefore, directional structures can better be predicted in the image domain (see Sec. 6.3.2).

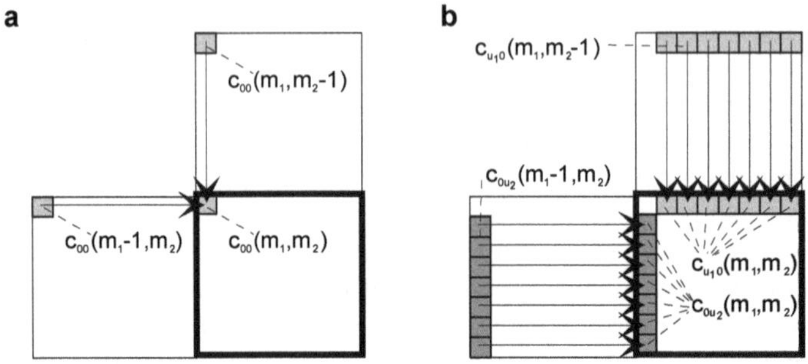

Fig. 6.17. Prediction of block transform coefficients **a** DC coefficient **b** AC coefficients

6.4.2 Quantization and coding of transform coefficients

Fig. 6.18 shows probability distributions of coefficients $c_{0,0}$, $c_{1,0}$ and $c_{2,0}$ from a 2D DCT of size 8x8. For sufficiently large sets of images, the DC coefficients $c_{0,0}$ show a wide range of amplitude distributions, with higher probability at the center of the range rather than the tails, similar to the PDF of the underlying images. For individual images, peaks at individual amplitudes often occur (dotted graph in Fig. 6.18a). On the other hand, DC coefficients can have high correlation between adjacent blocks and can efficiently be further compressed by predictive coding, where then the PDF of prediction errors can be expected similar to Fig. 6.13 (though with higher variance). The probability distributions of AC coefficients show a consistent concentration of values and symmetry around zero, when analyzed either over large sets or for single images. They are also assumed to be roughly Laplacian distributed by their global statistical behavior (see (2.128), with different variances for individual coefficient positions) [REININGER, GIBSON 1983]. When normalizing the AC coefficients by the instantaneous block variance, the PDF is close to Gaussian (normal) distribution, such that it can be concuded that the deviation from this is rather caused by the locally varying detail of image signals.

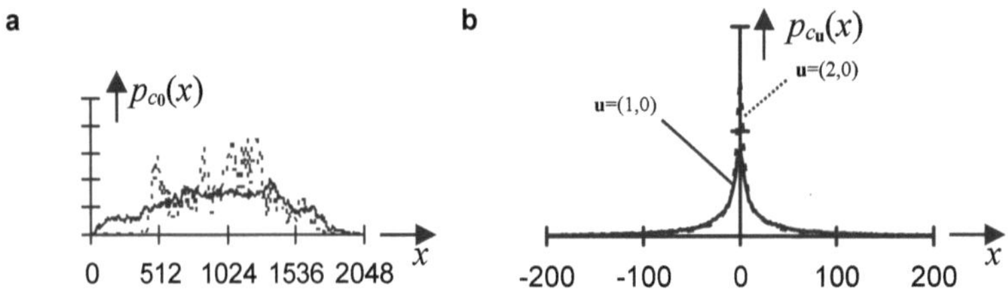

Fig. 6.18. PDF of 2D DCT coefficients **a** Coefficient $c_{0,0}$ of one single image (···) and averaged over 15 images (—) **b** Coefficients $c_{1,0}$ (—) and $c_{2,0}$ (···)

Quantization of transform coefficients. Transform coefficients typically have a larger range of discrete amplitudes as compared to the original (PCM) samples of the digital image signal, even if integer-type (e.g. (2.270), (2.271)) or rectangle-shape basis functions (such as Walsh, Hadamard) are used. Each coefficient is computed by a weighted superposition of samples, and all possible combinations of sample amplitudes have to be supported. Systematically, if M samples are combined and each of them has J possible discrete amplitudes, only J^M different combinations can occur, which applies likewise to the corresponding set of quantized transform coefficients. Nevertheless, even in the case of integer basis functions such as Haar or Hadamard, there is no systematic way to represent transform coefficients also with $\log_2 J$ bits each, since the superposition of samples increases the necessary integer word length; perfect reconstruction by the inverse transform would require full bit depth as to be determined by the bit-depth precision both of the signal and the basis functions by which it is multiplied[12]. Therefore, transform coding is less suitable for lossless compression than predictive coding.

In typical operation ranges (which could be close to lossless), coefficients are quantized after the transform. The spreading of the quantization error from a given transform coefficient $c_{\mathbf{u}}$ into the reconstructed signal is influenced by the inverse transform's synthesis basis $r_{\mathbf{u}}(\mathbf{n})$ (vector coordinates expressing 2D basis images here). As the reconstruction is a superposition of all transform coefficients, their quantization errors are superimposed, but can be assumed to be statistically independent in case of an orthogonal transform. Depending on the transform properties, this does not necessarily imply that reconstruction errors are uniformly distributed over all sample positions. For example, in case of the DCT, the sums of squared reconstruction basis functions tend to become larger towards the block boundaries. The maximum reconstruction error that could occur close to block boundary positions is also larger than in the center of the block. In case of block-separate transforms such as the DCT, skipping of higher-frequency basis functions often leads to so-called blocking artifacts, but also ringing is observed particularly in case of image edge structures.

Block overlapping transforms use basis functions decaying towards both ends, by which blocking artifacts are minimized, which could however cause additional edge smoothing when only low-frequency basis functions are used; ringing artifacts could also still be present with block overlapping transforms, particularly when edges are positioned close to the block center.

Any quantization method, such as uniform or non-uniform scalar quantizers, and vector quantizers can be applied to transform coefficients. Due to the expected concentration of amplitude values around zero, quantizers with zero reconstruction should be used, where the choice of the zero value can be interpreted as 'co-

---

[12] One approach to systematically limit the necessary bit-depth of the transform is by application of lifting algorithms and LSB truncation/rounding after each of the lifting steps which then still are perfectly reversible (cf. Sec. 2.8.3). However, if this is applied extensively, the orthogonality of the transform can be lost.

efficient not used for reconstruction'. For the case of uniform quantization, and assuming a Laplacian-like PDF of transform coefficients, optimum reconstruction values can be shifted to an off-centre position by using appropriate offset values in (4.12) to achieve optimum rate-distortion performance (see Sec. 4.3). Specifically, the case OFFQ<0 and OFFR=0 is interesting, which effectively introduces a deadzone during quantization, increases the number of zero-quantized values and gives slightly lower average reconstruction error for non-zero values.

Frequency-weighted quantization is justified under psycho-visual aspects (cf. Sec. 3.1.2). Based on perception properties, low-frequency coefficients should be quantized by using smaller step sizes than for higher-frequency coefficients, the threshold value by which it is decided whether a coefficient shall be non-zero is then also depending on the frequency. With global quantization scaling factor Δ and individual factors $w_{\mathbf{u}}$, the quantization step size of coefficient $c_{\mathbf{u}}$ is $\Delta_{\mathbf{u}}=w_{\mathbf{u}}\Delta$. However, frequency-weighted quantization does not necessarily improve the rate-distortion behaviour when measured by squared-error distortion metrics[13].

Table 6.1 shows an example of a frequency-dependent *quantizer weighting matrix* for the case of an 8x8 DCT. It should however be observed that optimum weighting highly depends on application properties, typical viewing distance etc., and can also be picture dependent.

Tab. 6.1. Example of a quantizer weighting matrix[14] for intra-picture coded DCT blocks of size 8x8

$G_{\mathbf{u}}$ for $c_{\mathbf{u}}$	u_1=							
u_2=	0	1	2	3	4	5	6	7
0	8	16	19	22	26	27	29	34
1	16	16	22	24	27	29	34	37
2	19	22	26	27	29	34	34	38
3	22	22	26	27	29	34	37	40
4	22	26	27	29	32	35	40	48
5	26	27	29	32	35	40	48	58
6	26	27	29	34	38	46	56	69
7	27	29	35	38	46	56	69	83

General remarks on transform coding for image compression. After quantization, bit-level coding has a significant effect on the encoded rate. The following

[13] See Problem 6.3; this comparison is however not balanced, as the SNR is a squared distortion metric which (for all cases with same absolute deviation error) becomes optimum when deviations are equally spread over all samples in the signal, or all frequency components in the transformed domain. This can also be proven due to the validity of Parseval's theorem (2.160).

[14] This example is the quantizer weighting matrix as recommended as default configuration in the standards MPEG-1 and MPEG-2 for the case of intra-picture coding of a transformed block. The values as shown in the table have to be divided by 16 and then multiplied by the quantizer scaling factor.

aspects require consideration in the design of a coding strategy for transform coefficients:

- Depending on the amount of local detail, the number of bits allocated to different areas of the image should be variable;
- Particularly at low bit rates, a high percentage of coefficients would typically be represented by the zero reconstruction value (see Fig. 6.18b). Hence, mechanisms have to be found that provide the 'significance on/off' signalling by a reasonably low amount of bits. Examples for this are signalling mechanisms for entire transform blocks or sub-blocks (coefficient groups) that are all-zero, and coding of significance maps using run-length coding or context-based entropy coding.
- An important distinction between natural images and statistical models like AR(1) is the lack of stationarity in the first; for natural images, variations between regions of lower and higher detail are typical.
- Statistical dependencies may exist either between the transform coefficients of the same block ('inter-band dependency') or between transform coefficients of adjacent blocks ('intra-band dependency'), or both in combination. Whereas spatial dependencies can to some extent be modeled by correlation (and be utilized e.g. by using prediction methods), dependencies across the different frequency bands are often non-linear, and can e.g. express the assumption that the occurrence of non-zero coefficients is more likely when already many other non-zero coefficients were observed. The same is true across blocks, i.e. it is more likely that several blocks of similar detail are close to each other. Context-based entropy coding is efficient in exploiting this.

'Zonal' coding. Early methods in transform coding of images were based on *zonal coding*, where non-zero coefficients are assumed to occur only in a specific zone of the transform domain, which may be statistically optimum under the assumption of stationary signal behaviour, but is hardly optimum when the amount of detail is locally varying. Adaptive methods with classification of block activity, e.g. [GIMLETT 1975] [CHEN, SMITH 1977] obviously perform better in this context. However, the decisions for quantization ranges of individual coefficients within each activity class are determined based on average energy distribution, there is no guarantee that distortion is within a certain limit for individual transform blocks, which can lead to locally visible artifacts.

Run-length coding. A scheme to describe positions of relevant (quantized as non-zero) coefficients more precisely for each block is based on *run-length coding*. Its application to 2D transform coding of images in combination with a zig-zag scan was first proposed in [CHEN, PRATT 1981/1984] As RLC is originally only suitable over 1D on/off sequences, it is necessary to define a *scan* within the 2D block of coefficients. If the run-length method of Fig. 5.1b is used, the number of runs equals the number of non-zero coefficients. In case of 2D DCT and similar 2D transforms, the *zig-zag scan* shown in Fig. 6.19a is advantageous, as it can be expected to detect relevant coefficients as early as possible due to the fact that it approximately scans the coefficients by increasing frequency order independent of orientation, and higher frequencies are more likely to be quantized into the zero value.

Scanning of 2D transform blocks. Unique scans over a 2D block of transform coefficients have to be defined with methods of signalling significance, as well as the sign and level information to be associated with the non-zero coefficients, in any case where sequential processing is applied. The proper definition of the scan sequence has implication on the possible exploitation of information that is already available at the decoder, e.g. in prediction or context-based arithmetic coding.

Switching of the scan mode for individual blocks can be useful. The row-wise scan sequence of Fig. 6.19b (respectively, if rotated by 90^0: column-wise) is advantageous if the block shows vertical (respectively: horizontal) edge structures, in which case high-energy coefficients can mainly be expected in the horizontal (respectively: vertical) orientation. The Peano-Hilbert scan (Fig. 6.19c) provides a dense space filling of 2D fields with minimum curve length [PEANO 1890] [HILBERT 1891]; it can be constructed by a nested hierarchy of sub-scans (indicated by the numbers 1-4 in the figure, where the first two nested levels are shown). The property of visiting close neighbors as early as possible can be advantageous when their dependencies shall be exploited.

Scanning sequences can also be useful at a higher level, e.g. for on/off signalling if *all* coefficients of a given block shall be zero. In case of large transform blocks, scanning can also be used at level of *sub-blocks.* Also here, it is necessary to define scan methods both between and within the sub-blocks. As an example, the HEVC standard uses variable DCT transform block sizes between 4x4 and 32x32, where scanning is performed over a sequence of 4x4 sub-blocks (e.g. 64 4x4 sub-blocks within a 32x32 transform block). Basically, the same scanning schemes (diagonal, horizontal or vertical) are applied both at the sub-block (within block) and coefficient (within sub-block) levels.

During the scan, it has to be checked whether coefficient amplitudes fall below the threshold of the zero quantizer cell (see Fig. 6.20). Run-length values in Fig. 6.20 indicate the numbers of zero-quantized coefficients between two coefficients quantized into non-zero values. Following this scheme, also the run-lengths for the different scan modes are given in the examples of Fig. 6.19. If no more non-zero coefficients can be found until the end of the block, this is usually indicated by an 'end of block' (EOB) or 'last coefficient' symbol[15].

The DC coefficient c_0 differs significantly by statistical properties and is therefore often treated separately from the AC coefficients, e.g. using DPCM coding.

---

[15] Alternatively, the 'last' position can be signaled explicitly, and the scan can start from there. Such an approach is used in transform coefficient coding of HEVC.

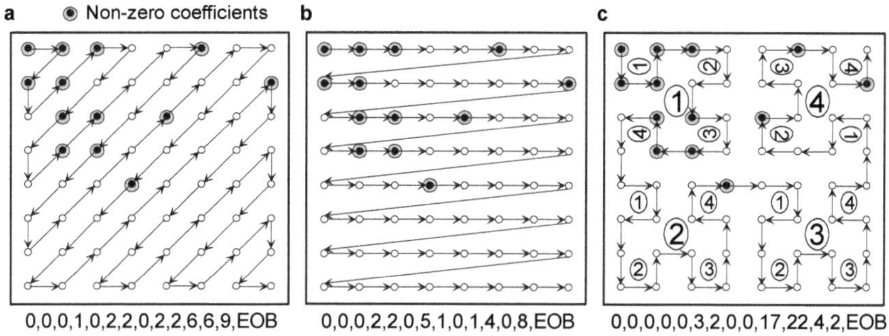

Fig. 6.19. Scanning schemes in run-length coding within a transform block: **a** zig-zag scan **b** row-wise scan **c** Peano-Hilbert scan

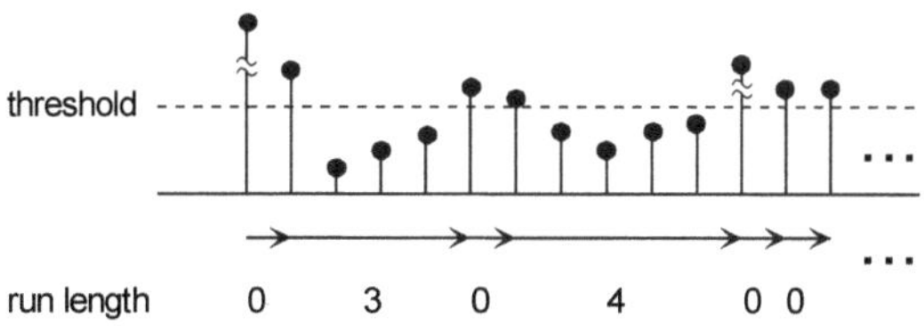

Fig. 6.20. Generation of run-length code sequence based on a zero-threshold decision

Signalling of all-zero blocks. In case of coarse quantization (lower data rates), it is likely that *all* coefficients of a transform block could be zero[16], which should be signaled with as few bits as possible. Methods to achieve this are:

- Usage of a 'coded block flag' (CBF) which in its 'on' state explicitly expresses that a block contains non-zero coefficients; this can also be subject to additional entropy coding, or combined coding of the CBFs of a group of several blocks as a 'coded block pattern' (CBP);
- Explicit coding of next non-zero block address;
- Run-length coding, i.e. signalling the number of zero blocks until the next non-zero block occurs within a specific scan sequence.

Note that run-length coding in this context could be interpreted either as a method of coding the CBF more efficiently, or for coding the next block address in a differential mode. Different scanning sequences can be defined, where at some global level, row-wise scanning of blocks is most practical. It is however also very efficient to group several transform blocks into a *macroblock* (e.g. in MPEG-1/-2/-4/AVC standards) or a *coding tree block* (in HEVC standard) structure, where the zero/non-zero signalling is performed sequentially over that structure, but jointly for all transform blocks within the current group.

[16] This case is even more likely when some kind of prediction (DC prediction, AC prediction, block prediction; motion-compensated prediction in case of video) is applied prior to transform coding, which is then designed to encode a prediction residual.

Significance map coding. Once a transform block is known to contain non-zero coefficients, it is necessary to identify their positions. Beyond the run-length coding method(s) described above, it is also possible to express the positions explicitly by a binary *significance map*. If it is sparsely populated, it can be expected that the entropy of the significance flags would be lower than 1 bit/coefficient. Unlike run-length coding, context-based entropy coding of a significance map can take into account statistical dependencies over groups of coefficients from the same block (or even from neighbored blocks), e.g. it can be expected that it is more likely to find a significant coefficient, when the presence of other significant coefficients in the vicinity was previously found and decoded.

Sign and level coding. Once a significant (non-zero) coefficient is found, it is necessary to code its amplitude and sign. If a symmetric PDF is assumed, positive and negative values of same amplitude are of equal probability, such that the entropy of the sign bit cannot be lower than 1 bit, and the sign can be coded in binary form. Further, less probability of higher amplitudes can be expected, as they are assumed to be approximately Laplacian distributed, and VLC coding adapted to such property is rather efficient. Typical approaches are described in the subsequent paragraphs.

Example 1: Combined coding of run-length, sign and level. In various image and video coding standards[17], entropy codes are used which combine the *run-length information* and the (absolute) *quantizer level information* of non-zero coefficients in a {RUN;LEVEL} VLC[18] of dimensionality K=2. Similar methods combining K=3 source symbols {RUN;LEVEL;EOB} are also used. The sign bit can simply be appended when assuming a zero-symmetric PDF. The entire information then consists of three components for *significance* (signaled by RUN and EOB), in the case of non-zero significance the *sign*, and the *amplitude* (LEVEL).

Statistical dependencies between transform coefficients can be exploited by defining appropriate entropy coding tables depending on the probability of joint occurrence of RUN/LEVEL combinations, or use adaptive entropy coding.

Example 2: Context-adaptive arithmetic coding of transform coefficients. The transform coefficient coding scheme used in HEVC encodes the position of the last non-zero transform coefficient in a transform block (TB, which could have a size of 4x4, 8x8, 16x16 or 32x32) as first syntax element. Regardless of the TB size, coding is done in a sequence of 4x4 sub-blocks, establishing *coefficient groups* in cases of TB sizes larger than 4x4. For each sub-block that is not entirely zero, a significance map, sign bits and levels for the non-zero transform coefficients have to be encoded.

The coordinate position of the last non-zero coefficient is coded once for the TB, and then, for each 4x4 sub-block (coefficient group) having at least one coef-

[17] E.g. JPEG, MPEG, H.26x, see Sec. 6.10 and 7.8

[18] For an example, see Table 4.1.

ficient preceding the last coefficient position, a significant group flag is transmitted, signalling that there is at least one non-zero coefficient in the group. Groups are scanned in diagonals (similar to the zig-zag scan, but not alternating), starting from the diagonal that contains the last coefficient. If a group is flagged as significant, a significance flag is encoded for each of its coefficients (provided that it is prior to the position of the last coefficient) further compressed by context-based entropy coding (CABAC, see Sec. 4.4.5). The context models for the significant coefficient flags are dependent on the coefficient position as well as the values of the two significant group flags at the right and the bottom. An example is shown in Fig. 6.21, where positions of several previously decoded transform coefficient samples in a diagonal up-right scan are evaluated to determine the probability of the current coefficient to be zero or non-zero.

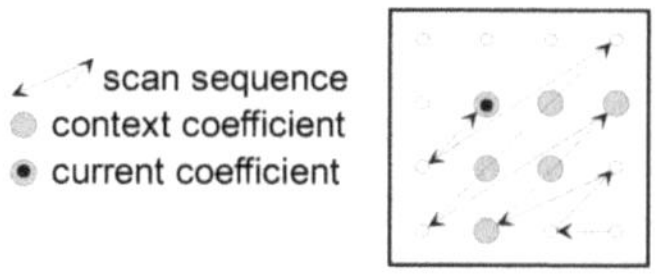

Fig. 6.21. Context formed by a group of previously coded coefficients in a diagonal up-right scan, proceeding from last (in frequency order) towards first coefficients

Coefficient scanning is performed in a diagonal up-right fashion per 4x4 sub-block, starting from bottom right, as shown in Fig. 6.22a for the case of a 16x16 TB. In cases of 4x4 and 8x8 TBs and intra prediction, horizontal and vertical scanning can also be performed, implicitly depending on the prediction direction. Fig. 6.22b shows an example of the respective scanning sequence in an 8x8 TB, which is again performed within 4x4 sub-blocks.

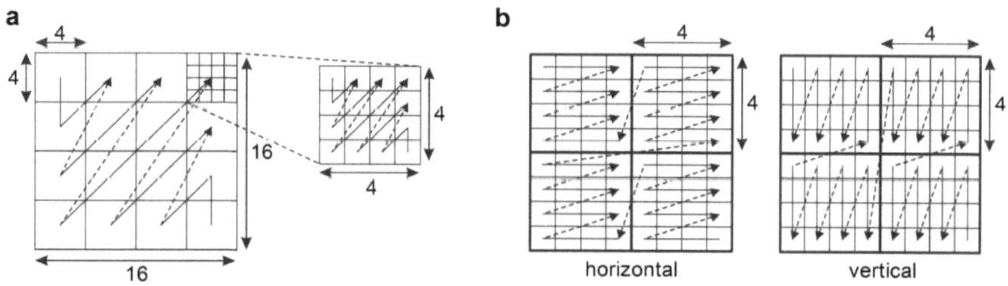

Fig. 6.22. a 4x4 diagonal sub-block scanning (example of a 16x16 TB) and coefficient scanning within the sub-blocks as implemented in HEVC **b** horizontal and vertical scanning as invoked for close-to-horizontal and close-to-vertical intra prediction directions, respectively (cf. Fig. 6.12; example of 8x8 TB)

For each position where the corresponding significant coefficient flag is equal to one, additionally a flag specifying whether the level value is greater than one is coded, and if this is found true, another flag signals whether the level is greater than two. Finally, if that latter condition is also true, the remaining level value is coded using a variant of a Golomb-Rice code, whereas the two ‘greater than’ flags are coded by CABAC, using context information from previously decoded coefficients.

In HEVC, sign bits are again coded as plain binary flags. However, a method denoted as *sign data hiding* is used for further compression improvement. The sign bits are coded conditionally based on the number and positions of coded coefficients. If there are at least two non-zero coefficients in a 4×4 sub-block and the difference between the scan positions of the first and the last non-zero coefficients is greater than 3, the sign bit of the first non-zero coefficient is inferred from the parity of the sum of the coefficient amplitudes, otherwise, the sign bit is coded normally. At the encoder side, this can be implemented by selecting one coefficient with an amplitude close to the boundary between two quantization intervals to be forced to use the other interval in cases where the parity would not indicate the correct sign of the first coefficient. This allows some of the sign bits to be saved – the encoder has to choose which transform coefficient amplitude can be altered such that the overall rate-distortion cost becomes minimum.

The JPEG standard. The JPEG standard, officially entitled ISO/IEC 10918 | ITU-T Rec. T.81 '*Digital compression and coding of continuous-tone still images*', has probably become the most widely used still image compression method, being based on a block transform (DCT) scheme (cf. Sec. 6.10).

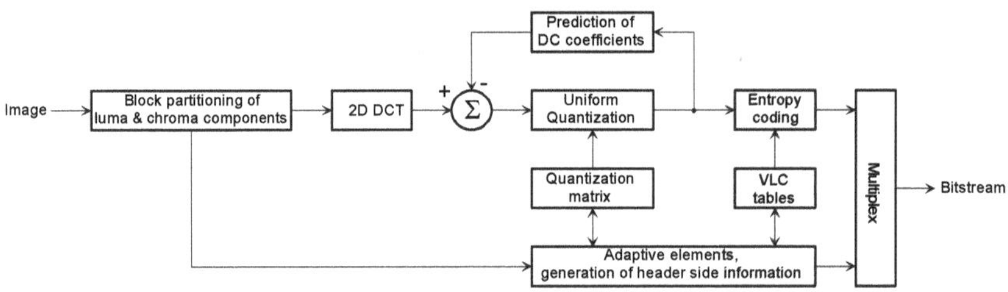

Fig. 6.23. Encoder of the DCT-based JPEG standard (sequential DCT mode)

The structure of a typical JPEG encoder is shown in Fig. 6.23. The core approach is adaptive block transform coding, using a 2D DCT of block size 8x8 samples. The DC coefficients can be predictively encoded by DPCM (using the DC coefficient from the left-hand and top-above blocks to compute the prediction). The AC coefficients are scanned in zig-zag order as shown in Fig. 6.19a, the non-zero values are quantized uniformly.

Two methods of entropy coding are used in the *baseline version* of JPEG, which defines fixed VLC tables separately for the luminance and the chrominance part of the information:

- *Coding of DC coefficients* (either predictive or non-predictive): A systematic VLC is used, where the prefix designates the number of bits necessary to represent the quantizer index, and the suffix is the index itself, encoded as an integer number of this given bit depth.

- *Coding of AC coefficients*: A combined run/level VLC similar to Table 4.1 is used, coefficients are scanned in zig-zag sequence. Only typical combinations of RUN and LEVEL are mapped into VLC symbols, otherwise an ESCAPE symbol is sent, in which case separate fixed-length representations of the RUN and LEVEL values are encoded. If no further non-zero coefficients are detected, an *end of block* (EOB) symbol is released.

In extended definitions contained in the JPEG standard[19], the encoder can determine optimized VLC tables for individual images; the parameters necessary for this adaptive entropy coding are then conveyed in the preamble of the bit stream. Besides table-based VLC, adaptive binary arithmetic coding and a method of bit-plane refinement (with up to 12 bits per coefficient) can be defined as options to further improve the compression performance and/or quality. The most important operating modes are as follows:

- *Sequential DCT mode*: This is the most frequently used mode of JPEG, where the transform blocks are scanned sequentially left-to-right, top-to-bottom. All encoded information about coefficients in each block is written to the bit-stream instantaneously in the given block sequences and using the zig-zag scan within the blocks. If a stream is received, the decoding process needs to follow the same sequence, and the reconstructed image is generated block-wise row by row as long as the transmission is not finished yet. Different coding processes are specified, the most common being ‘baseline sequential’. Another ‘extended sequential’ process uses adaptive Huffman and adaptive arithmetic coding, and up to 12 bit sample precision, but practically, its usage is quite uncommon.
- *Progressive DCT mode*: In principle, the compression algorithm is identical here with the sequential DCT mode, but the information bits are no longer released by ‘block order’. In *spectral-selection progression*, increasing frequency order is used, such that first the bits relate to DC coefficients of *all* blocks, then bits for all AC coefficients of first, second, third etc. diagonal of the zig-zag scan are written. If the progressive stream is successively transmitted, a first decoding can already be performed when only a low number of information bits have been received, giving a raw approximation of the *entire image*. This method of *progressive transmission* and decoding is particularly useful when low-rate channels are used, and an early display of the full image is desirable. The data rate as compared to the sequential DCT mode is not increased in principle (except for an overhead to guarantee the correct ordering of progression). It is however necessary to execute several decoder runs (one for each progression step executed by the display). The progressive DCT mode can also be used for error protection, where the need for protection of bits naturally results by the order in which they appear in the bit stream, and data partitioning is used in combination with unequal error protection (see

[19] These advanced methods, which are not part of the baseline (ISO/IEC 10918-1 | ITU-T Rec. T.81) but only defined in the standard's extensions (ISO/IEC 10918-3 | ITU-T Rec. T.84), are hardly used in practical implementations.

Sec. 9.3.2). Beside the spectral-selection progression, a *full progression* operation is also defined, which in addition allows refinement (by one or several bit-plane levels) of the quantization precision during subsequent progression steps. Different coding processes specified in the standard include Huffman or arithmetic coding, spectral selection or full progression, and 8 bit or 12 bit sample precision. However, the full progression modes typically produce considerably larger bitrate overhead.

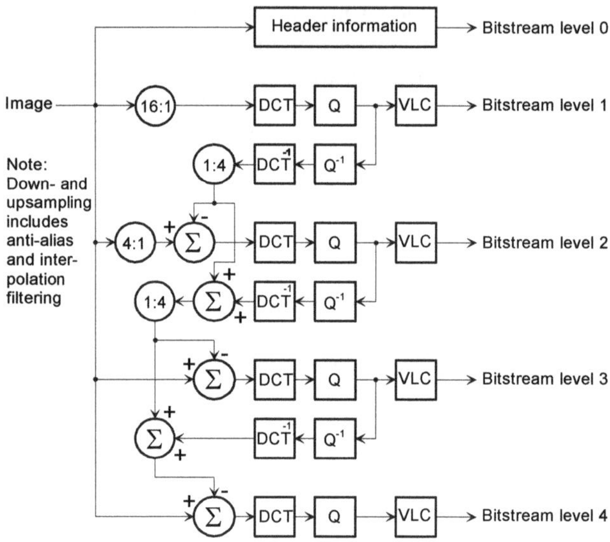

Fig. 6.24. Hierarchical-progressive mode of JPEG

Beyond the two DCT modes, the *hierarchical mode* uses a multi-resolution prediction (see Sec. 5.2.6), as shown in Fig. 6.24. The residual signals at the particular spatial resolution levels of the pyramid are separately encoded either by sequential or progressive DCT mode encoders. The process typically starts with a sub-sampled representation by a factor of four in both horizontal and vertical directions (factor 16 in total). Quantization step sizes can be selected independently; they influence the bit rates coming up with each of the levels. This mode increases the total bit rate for the full-resolution signal significantly compared to the sequential DCT mode, is more complex to implement, and has therefore not found wider usage so far.

Still picture profile of the HEVC standard. High Efficiency Video Coding (HEVC) (ISO/IEC 23008-2 | ITU-T Rec. H.265) is the most recent and most efficient compression standard for video, and also for still images when operated on single pictures. It is then based on a hybrid combination of 2D spatial block prediction and 2D transform coding, for which a general block diagram is shown in Fig. 6.25[20].

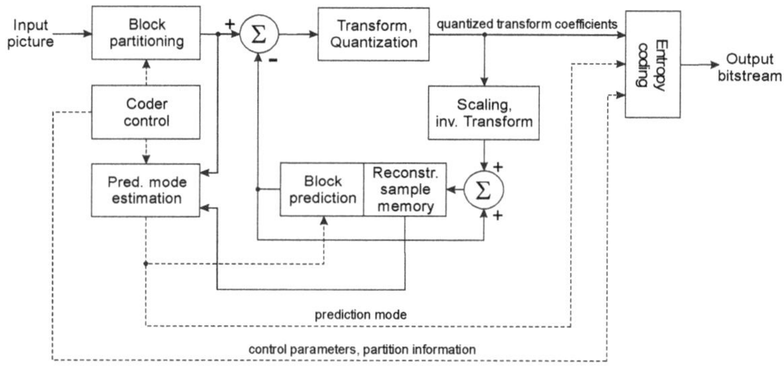

Fig. 6.25. Hybrid (2D block prediction / 2D transform) still picture coding

In the block partitioning stage, the input signal is processed within Coding Tree Blocks (CTB) of typical size 64x64 samples for the luma component. CTBs can be further split into square-shaped Coding Blocks (CB) of maximum size as large as the CTB size, and minimum size 8x8. The CB[21] splitting is signaled by a quad-tree code (see Sec. 6.1.2), and an intra-prediction mode according to Fig. 6.12 is signaled for prediction blocks which coincide with the CB, or in case of already using the smallest CB size, it can optionally be applied separately to four square-shaped sub-blocks. Below the CB level, further splitting into transform blocks (TB) can be performed. The actual prediction loop processing is run at the level of transform blocks (which here have to be smaller or equal in size than the prediction blocks for which the prediction mode selection is signaled). For TB sizes 8x8, 16x16 or 32x32, an integer approximation of the separable DCT (cf. Sec. 2.7.2) is applied, for TB size 4x4, a type-IV DST (2.273) is employed, which better matches the statistics of the prediction residual. Even though several adjacent TBs can use the same intra prediction mode, the prediction loop itself is run on a TB by TB basis to guarantee that nearby samples are available. The transformed residual is further quantized and entropy coded (as described in example 2 on p. 318).

Efficient hierarchical or progressive coding is also possible, when HEVC coding is used in spatial scalability mode, using a multi-resolution differential pyramid (sec 5.2.6), which means that for encoding a higher resolution, yet another

---

20 Note that the JPEG standard does not use spatial domain prediction (except in the lossless operation mode); however, it uses DPCM across blocks for the DC coefficients in the transform domain.

21 Formally, a CB only refers to one component (luminance or chrominance), whereas the term *coding unit* is used for all component's CBs at the same location. Similar terms are used for *transform unit* (TU) in relation to TB and *prediction unit* (PU) in relation to PB.

prediction is made available by upsampling the collocated reconstructed lower-resolution signal.

6.4.3 Overlapping-block transform coding

In terms of quantization and coding, coefficients computed from overlapping-block transforms (Sec. 2.7.4) can be treated similar to common block transforms. The transform representation is critically sampled. Depending on the basis functions, the coefficients could have a meaning in terms of frequency components in the signal. When block-overlapping transforms are used for 2D image compression [MALVAR, STAELIN 1989] [MALVAR 1991], one advantage compared to block transforms is that no discontinuous block edges exist or become visible as compression artifacts, and highest coding errors would rather occur at the centers (not at the edges) of basis functions, which may lead to less visibility. On the other hand, when the encoded signal itself has discontinuities (edges), ringing visibility can be increased due to the longer extension of the basis functions.

Overlapping block coding has been implemented in the standard *JPEG XR* (ITU-T Rec. T.832 | ISO/IEC 29199-2), where however the block overlap operation is implemented as pre- and post-filtering on top of a non-overlapping core transform, and can be invoked adaptively. The core transform is a 4x4 integer transform similar to the DCT (entitled as *photo core transform, PCT*). A secondary transform (entitled as *photo overlap transform, POT*) is applied in cases where overlap shall be used. For this, the block boundaries are offset by 2 samples horizontally and vertically. Then, averaging the coefficients of both transforms generates a similar result as a genuine overlapping transform. The DC coefficients of 16 blocks are processed by another PCT/optional POT step, such that the 'lowpass' component applies formally to a macroblock size of 16x16. Quantization and (adaptive Huffman based) entropy coding are used for representation of the transform coefficients.

A similar effect of reducing blocking artifacts could also be achieved by post filtering of the decoder output (see Sec. 6.9), however when included into the transform, it can directly target lower coding error compared to the original during the quantization decision at the encoder.

6.4.4 Subband and wavelet transform coding

In general, there is no fundamental difference in the quantization and encoded representation of coefficients regardless if they are generated by a filterbank, by a block overlapping or a block transform. Whereas the block (or overlapping) transforms in terms of their arrangement support a view where the unit of decomposition is a collection of transform coefficients representing various frequencies at the same location ('frequency over position'), the filterbank transform is often interpreted as providing sub-sampled signals of given frequency bands as the representative entity ('position over frequency'). Consequently, the first methods for 2D subband coding of images (where equal bandwidth was used over all frequencies) were designed such that the subbands were regarded as independent, where only intra-band dependencies were further taken into consideration (see e.g. [WOODS, O'NEILL 1986][WESTERINK ET AL. 1988][GHARAVI, TABATABAI 1988]). In particular, predictive encoding (DPCM) of the lowest-frequency coefficients can further improve the compression performance, as it is also the case for the 'DC' coefficient of a block transform. The lowest-frequency signal, though sub-sampled, still exhibits significant spatial correlation along both horizontal and vertical directions. Similarly, *directional correlation* can be observed within frequency bands where one spatial direction has lowpass characteristics. This can be exploited by horizontal/vertical directional prediction in the respective bands, which is conceptually comparable to the 'AC prediction' in block transform coding (Fig. 6.17). Locally joint appearance of non-zero coefficient values, both within bands and across bands, can further be used for efficient entropy coding of high frequency coefficients, where context-based schemes can also adapt to the directional orientation of the given band.

In summary, as for the case of block transforms, best compression performance can be obtained if both *inter-band* and *intra-band* statistical dependencies are exploited (which would correspond to *intra-block* and *inter-block*, respectively, in block transform coding). A barrier, which could inhibit efficient utilization of statistical dependencies between adjacent samples within one subband, is the alias from sub-sampling. Alias is also a problem in reconstruction, as it cannot be compensated (cf. (2.307)), when some frequency component being required for alias cancellation is suppressed during encoding. Beyond alias, the omission of high frequency components typically often causes *ringing artifacts* at amplitude discontinuities (edges) of the coded signal[22].

Wavelet transform coding [ANTONINI ET AL. 1992] has evolved as the most relevant subband coding approach for 2D images. The principal reasons for this are as follows:

- Very efficient methods exist to simultaneously utilize inter-band *and* intra-band dependencies between both scaling (lowpass) and wavelet (band-pass/highpass) coefficients;
- The Wavelet transform can use short basis functions with better localization property for the high-frequency components; therefore, it is advantageous in the vicinity of edges both in terms of compression performance and for minimization of the ringing problem; it effectively uses long basis functions for lower frequencies, which provide better exploitation of correlation e.g. in large flat areas;

---

22 'Low frequent ringing' could also appear in the wavelet transforms when several levels of wavelet coefficients are discarded. This may cause unnatural appearance in form of speckles. In general, ringing is more severe in subband transforms than it is in block transforms, specifically when longer impulse responses with sharp cut-off are used. Negative off-centre values in the impulse response (similar to the sinc function) are responsible for oscillations and overshoots at discontinuities.

- The Wavelet transform gives an implicit scalable representation of image signals when discarding higher frequency components, while still being a critically-sampled representation without overhead of samples (though it should be noticed, that due to constrained choices of filters, the sub-sampled pictures may bear alias).

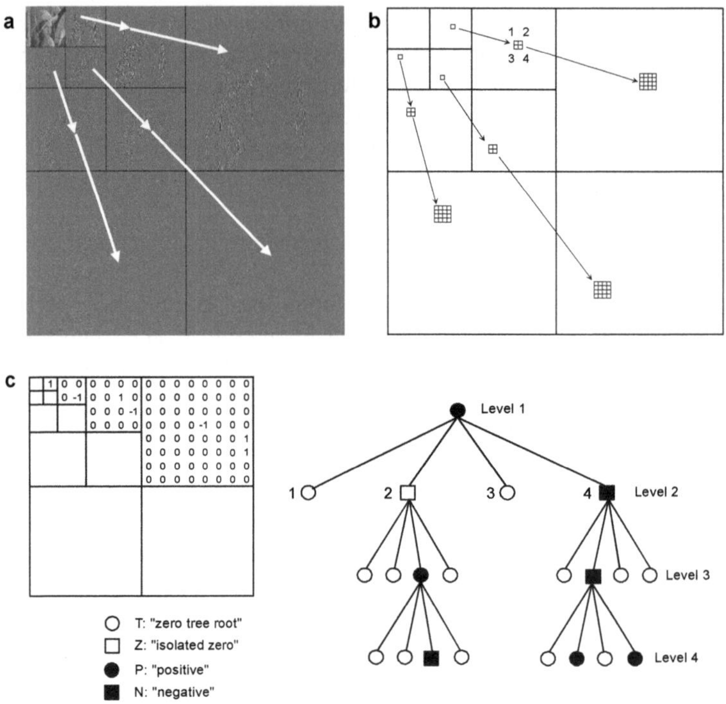

Fig. 6.26. a 2D wavelet decomposition **b** geometric referencing in zero tree coding **c** structure of the tree relationships within the MSB bit plane of the horizontal direction

Zero-tree coding. Fig. 6.26a illustrates the *localization property* of the wavelet transform which allows exploiting statistical dependencies *between* frequency bands. The wavelet coefficients represent frequency components at specific positions within the image which can be expected to be statistically dependent, subject to their directional orientation. If a high-amplitude sample is observed in a lower-frequency band at a specific location and with given orientation, it is likely that the next higher band of same or similar orientation also contains relevant information around that location, however with quadruple number of samples[23] (Fig. 6.26b). This effect can be observed in particular for vertical, diagonal and hori-

[23] Amplitudes are amplified by a factor of 4 in Fig. 6.26 to make structures in the wavelet components visible at all.

zontal structures (see Fig. 6.26a). Hence, conditional entropy coding can benefit on information whether directionally adjacent wavelet band of next-lower frequency contain significant energy at the respective position or not.

An early approach to use a data structure similar to a quad tree (cf. Sec. 6.1.2) for wavelet coding was proposed in [LEWIS, KNOWLES 1992]. The next relevant step was then made by the development of *embedded zero-tree wavelet* (EZW) coding [SHAPIRO 1993]. As in Fig. 6.26b, trees are defined separately over coefficient groups of horizontal, diagonal and vertical orientations. The method integrates the concept of embedded quantization (cf. Sec. 4.3), and bit-plane coding, within each bit-plane a scan of coefficients over the zero tree is performed. Each parent branches into four children at the next resolution level, with ordering as shown by numbers 1-4 in Figs. 6.26b/c. An example concerning the scan of the most significant bit plane over the horizontal orientation is given in Fig. 6.26c. Potential decisions of the quantizer (method of Fig. 4.8b) are for positive, negative or zero values, which can occur for the MSB at any coefficient position. The following four symbols are defined for coding the information of the given node in a given bit plane in the *dominant pass*:

- *Zero-tree root* ('T'): The value of the bit plane is zero at the given position, and at *all* subsequent branches down the entire tree.
- *Isolated zero* ('Z'): The value is zero at the given node position, but may be followed by a significant node at one or more of the subsequent branches anywhere in the tree below.
- *Positive* ('P'): The value has positive amplitude (but the same coefficient had been zero in more significant bit-planes).
- *Negative* ('N'): Same as 'P' with negative amplitude.

The 'P'/'N' coding jointly represents the sign and the information about coefficients that are newly found as significant (compared to previous passes). In this case, the tree is also further parsed towards higher resolutions. For coefficients that were already found significant (positive or negative) in higher bit planes, a *sub-ordinate* pass (refinement) is performed, which only encodes the branching into the upper or lower half of the next-finer embedded quantizer cell by binary '0' and '1' symbols (same as method of binary-coded information in Fig. 5.12c).

The information symbols 'T', 'Z', 'P', 'N', '0' and '1', as occurring at the respective coefficient positions within one bit plane, are re-ordered to obtain sequences which proceed from coarser to finer spatial resolution levels. The resulting sequence of symbols is then arithmetically encoded, taking into account the expected probabilities of the different symbols at the respective bit-plane levels[24]. Due to the re-ordering, it is possible to discard bits that relate to higher frequency bands, in cases where decoding shall be performed at less than full resolution.

EZW has been an ancestor for a whole lot of subsequent algorithm designs, which further improved its performance. *Set partitioning into hierarchical trees*

---

[24] In fact, arithmetic coding of the '0'/'1' refinement may not be necessary, and 'P' and 'N' could also be expected to have approximately equal probabilities, such that they might be replaced by a 'non-zero' symbol with binary coding of the sign bit.

(SPIHT) was presented in [SAID, PEARLMAN 1996]. The key concepts are partial ordering of the transformed coefficients by magnitude, using a set partitioning sorting algorithm, and ordered bit-plane transmission of refinement bits without arithmetic coding. The re-ordering of the information, which aims to send the most relevant bits first, introduces a certain amount of overhead information. Even though, SPIHT in most cases clearly outperforms EZW. In particular, by the 'relevant first' paradigm, the so-called problem of 'fractional bit-planes' (cf. Sec. 4.3) can be reduced to a minimum, when the algorithm is used to implement a fidelity-scalable representation of image signals. As a disadvantage, SPIHT did no longer provide an implicit mechanism to truncate bit streams for a reconstruction with reduced spatial resolution by discarding higher-frequency wavelet coefficients, because it is not possible to identify the transitions between zero-tree levels directly from the bit stream.

Context-adaptive binary arithmetic coding (CABAC), the arithmetic encoder implemented in the AVC and HEVC standards, was originally developed for adaptive entropy coding of wavelet coefficients over zero trees. The scheme proposed in [MARPE, CYCON 1999] allows flexible exploitation of intra- and inter-band dependencies in the bit planes of embedded-quantized wavelet coefficients, and bit-stream truncation for reduced spatial resolution as well.

Context-based codecs *not* using zero-tree correspondences over wavelet trees were also shown to provide excellent compression performance, in particular when proper context modeling and sophisticated rate-distortion optimization is performed, see e.g. [LI, LEI 1999]. An algorithm of this class (EBCOT, see description below) became the basis of the JPEG 2000 image compression standard, introducing a spatial block structure within subbands for the sake of region-based access. Complete omission of inter-band contexts as in EBCOT is attractive due to simpler realization of spatially scalable and locally accessible streams, but may lose performance for certain classes of images. A scheme which better compromises these aspects is *embedded zero block coding* (EZBC) proposed by [HSIANG, WOODS 2000]. EZBC performs a quad-tree-based set partitioning within subbands of a wavelet tree, signalling all-zero blocks in a hierarchical structure. The underlying context modeling allows flexible exploitation of statistical dependencies between coefficients *within* or *across subbands*. Bit-plane interleaving additionally improves the compression performance. Due to the block structure, EZBC provides a fully embedded solution supporting both spatial and bit-plane scalability, similar with EBCOT. However, EBCOT additionally supports localized access when only partially decoding the bit stream, such that it enables applications such as zooming into a picture on a screen without need to decode samples that are not to be displayed.

Embedded block coding with optimum truncation (EBCOT). EBCOT [TAUBMAN 2000] is an embedded entropy coding for wavelet coefficients, which has become the basis of the JPEG 2000 standard[25]. The embedded quantization with dead zone is identical to the methods described so far (see Fig. 4.8b), but in contrast to algorithms such as EZW, SPIHT and EZBC, coefficients over the different scales (frequency bands) are encoded basically independent of each other. The main motivation is for allowing random access to coefficients from different wavelet bands and also from local regions. To compensate for the disadvantage of not exploiting inter-band redundancy, EBCOT more extensively uses intra-band redundancy by context-based adaptive arithmetic coding. The method further applies a block structure for local grouping of wavelet coefficients, such that all information relating to certain local areas can be uniquely accessed and decoded independently[26]. *Precincts* are rectangular blocks which are typically constructed in a way that they allow to identify wavelet coefficients from the different bands to belong to the same spatial region. This means they contain less coefficients in the lower-frequency bands. *Code blocks* (CB) are independent and non-overlapping rectangular blocks, typically smaller than precincts[27], consisting of wavelet coefficients from only one subband.

The entropy coding within a CB is performed within bit planes of the embedded quantizer, and within each bit plane, different passes of binary coding are applied. The first pass is *significance propagation coding*, related to positions of coefficients which were insignificant (zero) in (previously decoded) higher bit planes, but which already had significant neighbors therein. The conditional probabilities of a coefficient to become significant in the current bit plane are determined from spatial neighborhood contexts in the next-higher bit plane. The contexts themselves are defined by horizontal, vertical and diagonal primitives from eight nearest neighbors, where it depends on the dominant orientation of the subband which neighbors are relevant. Fig. 6.27 illustrates the processing, where the following cases are excluded from the significance pass:

- Black – these *have* already been significant in the next higher bit plane and are processed in the subsequent *refinement pass*;
- Cross-marked – these don't have a significant previously-scanned neighbor within the context window, either from the same or higher bit plane; they are processed in the final *cleanup pass*.

The figure also shows a typical scan sequence, performed here along a stripe of the code block with a vertical height of 4 coefficient samples.

---

[25] It is possible here to give only a rough idea how EBCOT works, and how it is related to other methods discussed so far. For detailed aspects see [TAUBMAN, MARCELLIN 2001].

[26] Beyond codeblocks and precincts, JPEG2000 also allows to define *tiles*, which can be used to glue a larger picture together from smaller rectangular patches, e.g. a picture of size 8,000x8,000 could be defined via 16 tiles of size 2,000x2,000 each. Tiles are handled by the decoder as if they were independent pictures, the main purpose of tiles is to allow processing of pictures of extreme large size. For smaller resolutions, usually the entire picture is one tile.

[27] A typical CB size is 64x64 coefficient samples, constant over all subbands. At boundaries, CBs may be smaller if the picture or tile size is not a multiple of 64.

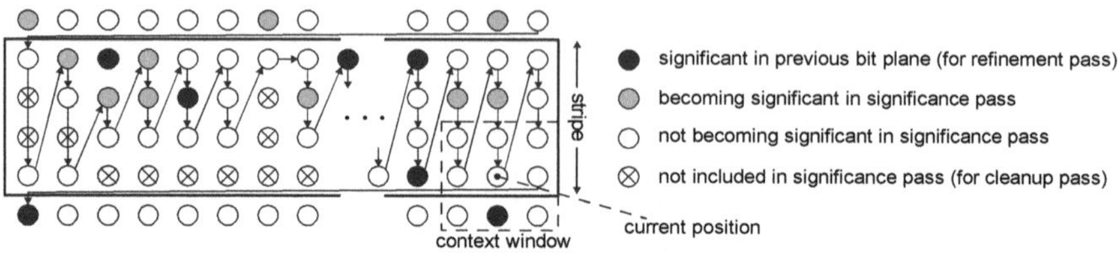

Fig. 6.27. Context window for the significance propagation pass in JPEG 2000 code blocks.

The significance scan can either use the *normal mode* of context-based arithmetic coding, or a *run mode*, where multiple adjacent coefficients staying insignificant are signaled by run-length coding. Once a new significant coefficient is found, its sign is coded as well. The second pass is *magnitude refinement coding* of coefficients which were already found to be significant in the higher bit planes. The third pass is the *cleanup pass* checking for significance of any coefficients that were not processed in the first two passes. Besides the context dependency, backward adaptation of the arithmetic coder's probability tables is continuously performed.

After processing entropy coding for each single CB, the EBCOT algorithm performs a re-ordering of the information from all CBs into layered sub-streams. This is done by using rate-distortion criteria, denoted as *post compression RD optimization*. Given a certain bit budget to be spent, it is analysed which benefit the next packet of bits from each code block gives (regardless if it covers one or several bit planes), to determine which choice provides the highest possible reduction of distortion. In this context, the bits resulting from the different scans can also be treated separately. At low rates, the magnitude refinement typically provides a more significant contribution to minimize the distortion with a given bit budget, whereas for higher rates, the significance propagation is often more important. Overhead information is necessary to describe the re-ordered stream structure, but this allows to achieve good performance at various operational rate-distortion points. In contrast to SPIHT, the problem of 'fractional bit-planes' is solved at least partially by this post-encoder optimization.

The *baseline mode* defined in part 1 of JPEG 2000 allows selection of two different bi-orthogonal wavelet filter pairs, both specified by their equivalent lifting structures (see Sec. 2.8.3):

- The 9/7 filter (2.326) is suitable for compression efficiency at low bit rates, but does not allow lossless reconstruction due to the necessary high precision of lifting steps;
- The 5/3 filter (2.317) has integer precision and allows synthesis with perfect lossless reconstruction from a finite set of quantization levels, whereas the compression performance at low rates is suffering, both in terms of subjective and objective performance, compared to the 9/7 filter.

The EBCOT algorithm which is used for coding (see above) re-orders coefficients in a rate-distortion optimized way. Beyond spatial and quality scalability, this allows flexible access to cropped regions of images. Various pointers are used to reference sub-stream entries. The sub-streams are structured into segments and packets, which can then uniquely be related e.g. to selective decoding of certain areas of the image, specific spatial resolution or quality levels. It can be signaled which portion of bits provides the next relevant information in the stream, but it is possible to retrieve the information either for a selected area (code block, precinct), or with priority of quality (bit plane). This increases the overall data rate of the representation, but is needed for the desired flexibility.

The highly-flexible bit stream structure of JPEG 2000 also allows providing efficient *mechanisms* for transmission error protection. Firstly, it is straightforward to protect the more relevant parts of the embedded stream selectively, e.g. by unequal error protection (cf. Sec. 9.3.2). Further, as the information within code blocks is self-contained (separately decodable and localizable), resynchronization is guaranteed even in the case of partial data losses.

Lossless coding using the Wavelet transform. The compression performance of wavelet coders highly depends on the transform basis. In principle lossless coding is possible if the output of the analysis filters provides a finite set of integer values, and synthesis filters give a perfect reconstruction. This can e.g. be realized if integer-precision biorthogonal filters (2.325) are used, or if rounding is applied in the prediction and update steps of a lifting structure implementation. On the other hand, integer filters provide a reasonable integer word length at the output only if the kernels are relatively short, which compromises for worse frequency separation between the bands and higher aliasing errors when subbands are discarded. Therefore, filters which are ideal for perfect reconstruction are less suitable for coding at lower rates, and scalable coding which would run from low rates up to lossless is only realizable by compromising the compression performance at one or the other end.

6.4.5 Local adaptation of transform bases by signal properties

The amount of local detail in image signals (influencing correlation and frequency characteristics) may abruptly change e.g. at edges or boundaries of objects. For large areas with regular structure (textured or uniform), the correlation properties of the signal can best be exploited by large neighborhoods, which could be realized by using larger blocks in block transform coding, or a larger number of decomposition levels in subband/wavelet coding. For areas with discontinuities such as edges, which have a wide spectrum, it is better to use small blocks, which also prevents edge ringing when coefficients are stronger quantized or discarded. For wavelet transform methods, the problem of many non-zero coefficients is partially solved by the *localization property* and the relatively short basis functions related to the highpass bands. Nevertheless, larger-area ringing can still be annoying since the effective length of lowpass basis functions increases exponentially with the number of levels in the wavelet tree. The main methods to cope with these problems are *variable block sizes* in block transform coding, and adaptation of the *transform decomposition depth* in wavelet transform coding. If frequency bands are enabled to be adapted by variable spectral resolution at any position in the image and for any frequency range, this is denoted as *wavelet packet decomposition.*

Variable block size transforms (VBST). In VBST, the main problem to be solved is the decision about the selection of local block sizes. A method based on the DCT was proposed in [VAISEY, GERSHO 1992]. Quadratic block sizes in the range between 4x4 and 32x32 samples are considered, where the block layout structure is encoded by a quad-tree (see Sec. 6.1.2, Fig. 6.4). Starting from the largest block size, this is done by analysis of mean value and variance, investigating whether these values show significant differences when measured for the larger and the next smaller sub-block size. Large variations indicate that a significant change of detail structure occurs, such that a division into smaller sub-blocks would be useful. It is generally necessary to track mean/variance deviations down to the smallest block size before making a decision; an example result obtained by a corresponding scheme is shown in Fig. 6.28.

More advanced approaches base the corresponding decisions on rate-distortion criteria, selecting the best sub-partition by taking into account the minimization of distortion versus the rate to be spent on transform coefficients as well as the side information for partition signalling and predictor adaptation.

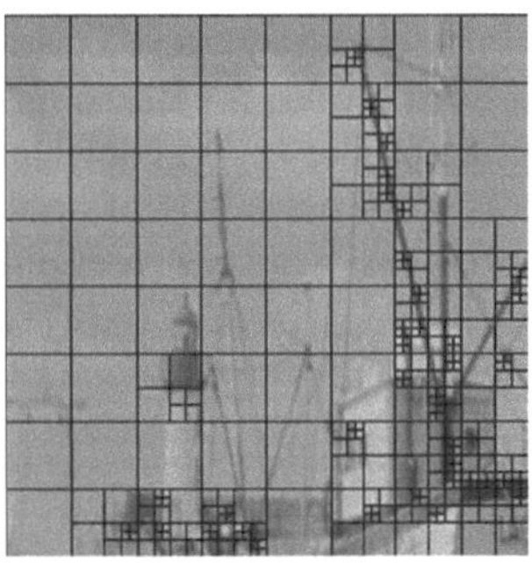

Fig. 6.28. Example of spatial block layout in variable block size transform coding, with DCT block sizes 4x4 ... 32x32 [courtesy by F. SPERLING]

In [WIEN 2003], a VBST method was proposed based on the integer-value transform basis (2.271)/(2.270), applied to the residual signal of a directional spatial prediction. Block partitions are either quadratic with sizes ranging between 4x4 and 16x16, or rectangular of sizes 4x8 and 8x16 with horizontal or vertical orientation. As the variable-block-size partitioning applies to both the directional prediction and the transform, the decision is made jointly based on rate-distortion criteria (comparison of several options by their RD performance).

The most advanced variable block size transform coding as of today is the method defined in the HEVC standard (see description on p. 318). Also here, the encoder has to check the split/no-split decision based on rate-distortion criteria such as $D + \lambda R$, with rate-dependent Lagrangian multiplier (see Sec. 4.3 and 7.2.7). An exhaustive search is hardly possible, but early termination conditions can be designed based on criteria such as homogeneity of regions (similar to the method described above). During the development of HEVC, even more degrees of freedom such as usage of non-square transform blocks of variable size had been investigated, but indeed no significant benefit had been demonstrated.

Wavelet Packet transform. Subband decomposition of equal-bandwidth frequency resolution over all bands and wavelet transform of octave-bandwidth frequency resolution can be interpreted as regular binary and dyadic tree structures. Other irregular decomposition structures are possible. Firstly, the depth of decomposition can be made signal adaptive. Second, unlike the case of the wavelet transform, higher-frequency bands could be further split based on an adaptation decision; this could for example be beneficial when narrow-band frequency components are present in a signal. The latter case is denoted as *wavelet packet transform*.

As criterion in the split/non-split decision for a given frequency band at a given location, rate-distortion optimization similar to (4.43) can be used, which also has to take into account the side information which is necessary to convey the decision to the decoder. It is however possible that no increased gain is observed by growing the tree by one level, but a significant gain could be realized if another split was performed at a higher level. Therefore, an alternative strategy could start from a uniform decomposition, and prune the tree by removing splits which do not give benefit. In [RAMCHANDRAN, VETTERLI 1993], tree-pruning methods are studied in combination with rate-distortion criteria. For a given operational point on the rate-distortion function, band-split decisions are preferred which contribute the steepest decrease of distortion versus rate.

Adaptation of basis functions. For block transform coding, DCT-type basis functions are often considered as being a reasonable trade off, providing good compaction for image signals without further adaptation. However, aggressive quantization or discarding of coefficients can lead to artifacts such as block edge visibility and ringing. These could be diminished by appropriate post processing (see Sec. 6.7), but an alternative is switching of basis functions depending on local image content. For example, JPEG-XR uses local switching between a block transform and a secondary block-overlapping transform with the goal to avoid blocking artifacts. Additional side information is necessary to convey the switching parameters to the decoder.

In case of wavelet filter kernels, bi-orthogonal filter designs offer a considerable degree of freedom when adapted under the constraint of a given signal's frequency characteristics, e.g. to maximize the concentration of information in the lowpass band. Another goal of adaptive filter designs could be alias suppression, using stronger lowpass filtering of the scaled versions of the signal, which indeed would shift additional energy into the higher-frequency wavelet bands [YANG, RAMCHANDRAN 2000][MAYER, HÖYNCK 2001].

Real-valued separable 2D transforms (regardless if DCT, subband, wavelet packet etc.) have the disadvantage of only being well suitable for pure horizontal/vertical structures, but lacking compression efficiency in case of diagonally orientated structures (and being even unable to differentiate between 45^0 and -45^0 orientations explicitly in the transform domain). Therefore, directional block transforms with bases specifically trained for the characteristics of residuals from directional intra prediction have been proposed denoted as *mode dependent directional transform* (MDDT). Another possibility is implementation of a secondary transform (after the DCT), which takes into account expected dependencies between coefficients, and re-organizing them for the case of specific directional orientations. An overview of various approaches can be found in [Xu et al. 2010]. In case of subband transforms, alternative basis functions exist which can be adapted specifically for detail characteristics of 2D signals like edges, corners etc. Some examples of this category have been proposed under names like *edgelets*, *contourlets*, *curvelets*, *ridgelets, shearlets* or *chirplets*. From results reported so far, usability for image compression may be limited since the necessary adaptation requires the encoding of side information, which does not trade off against the saving in bit rate achieved due to the improved basis function. Approaches which could avoid this by determining e.g. the directional orientation of basis functions from already decoded samples have also shown limited benefit so far, and also tend to be overly complex. For a review of various approaches, the interested reader is referred to [Ma, Plonka 2010].

Matching pursuit [Mallat, Zhang 1993] could also be interpreted as a combination of adaptive-basis transform coding and vector quantization. In this approach, a huge codebook of basis functions (which may be redundant, i.e. entries not orthogonal with each other) is searched for the best match (correlation) with a given signal. Then, the residual against the original is computed, and again the best match for the residual is searched, which can be iterated until a convergence in terms of coding error is reached. The information consists of the codebook indices. The approach has been shown to give good rate-distortion performance both for video coding [Neff, Zakhor 1997] and still picture coding [Ventura et al. 2006]. Its main drawback is the high complexity of the search. Furthermore, designing appropriate codebooks that are universally applicable and guarantee for approaching limited distortion is challenging.

6.5 Lossless and near-lossless picture coding

The decoder output of ‘mathematically’ lossless compression shall precisely reproduce the values of a source, e.g. samples originating from a sensor. Therefore, it is only possible to exploit the redundancy of the source picture in terms of statistical dependency between samples as well as their probability statistics. The limitations not allowing large compression ratios in the lossless case are as follows:

- Even though the dependency between neighbored samples in local image structures is typically high, noise is usually present (particularly in image captures from sensors), which produces random variations;
- Even though transform coding is a most effective method for lossy coding, it requires an increased bit depth range of the transform coefficients for lossless reconstruction, unless aggressive rounding is performed and specific structures (such as lifting) are used to revert the transform. Therefore, when operated in a lossless mode, a transform coding representation may require even more bits than a spatial domain PCM. Transforms with binary basis functions (Walsh, Haar) and small block sizes should be used to keep the problem to a minimum. However, spatial domain predictive coding schemes, using rounding of the prediction to the same bit depth as the signal to be predicted, are more suitable for lossless coding. These have the additional advantage that no quantization needs to be applied, open-loop schemes can be used (see Sec. 5.2.1).
- For efficient compression, entropy coding is necessary. Due to variable-length coding, the bit rate will vary locally. Therefore, the bit position cannot be mapped to the local position of the sample to be decoded, and selective decoding for local random access is not implicitly possible, unless additional pointers are inserted as side information. The same is basically true when using spatial prediction schemes, where it would be necessary to define a grid of positions where the prediction loop can be newly initialized for random access. In any case, increased number of bits will be the consequence.

If perfect lossless compression is not required, a near lossless (or visually lossless) mode can be introduced. Here, transform coding may still be inappropriate due to the effects mentioned above, however predictive coding also can no longer be operated in an open-loop fashion, or drift (5.8) is generated (which may however not be too severe, if the quantization error variance is low).

The JPEG standard operated in lossless mode uses a 2D DPCM with forward-adaptive switchable predictors and entropy coding of the residual signal. Coding processes are specified for adaptive Huffman or arithmetic coding of residuals.

Subsequently, the standard JPEG-LS was defined for better compression performance than the JPEG standard's lossless mode (see also Sec. 6.10). The method is mainly based on spatial prediction methods. JPEG-LS defines two coding modes, for which the selection is determined from a neighborhood context:

- *Run mode*: This assumes occurrence of a reasonable number of subsequent samples of identical amplitude values (or nearly identical in the case of near-lossless mode); run-length coding is performed, where the amplitude value has only to be encoded once per run.
- *Prediction mode*: This mode is rather preferred if significant fluctuations in amplitude occur in the neighborhood e.g. by noise influence. Backward-adaptive switchable predictor configurations are used. The residual (prediction error signal) is encoded by a VLC.

Generally, the run mode performs better for synthetic (computer graphics generated) content, whereas the prediction mode is better suitable for natural (camera captured or scanned) content. For entropy coding, a systematic VLC based on an exponential Golomb code design (cf. Sec. 4.4.3) is defined in the baseline version. The extended version further also specifies context-based arithmetic coding.

HEVC also defines a lossless coding mode, which for intra-only coding is based on bypassing the transform and the quantization. Still, the block-wise intra prediction (Sec. 6.3.2) is applied (including rounding), and the prediction residual is directly entropy coded, using the same scheme normally applied to transform coefficients; this implicitly includes the exploitation of neighborhood contexts from samples across diagonals, such that the probability of a prediction error sample being significant (non-zero) depends on neighbors' significance. Further improvement of the performance is possible when instead of the prediction from neighbor block boundary samples, also nearest neighbors within the current block are used in an open-loop operation. At the decoder, this does not necessarily require a recursive operation, since the impulse response of the IIR filter can be 'unrolled' as an FIR or linear matrix operation w.r.t. each sample of the current block. Such a scheme can also be applied as a secondary prediction to the residual ('residual prediction' or residual DPCM) of the common directional prediction approach.

In particular for content with uniform color areas, as often present in pictures which are generated synthetically by computer graphics approaches, an effective method for lossless and near lossless compression is using a codebook with few entries, which is constructed such that it contains the colors which appear locally. Instead of signalling the colors of samples directly, the index of the codebook entry is coded, which can again be done in combination with run-length coding, predictive coding and/or context-based entropy coding of palette indices. These schemes are denoted as *palette mode coding*. The construction of the codebook (palette) can be done on a picture by picture basis, or by local adaptation/updating the entries per region/block. In cases where a specific color entry is not contained in the codebook, it is necessary to send an ESCAPE syntax element and subsequently code the color values directly. In case of lossy coding, the most similar color from the palette can also be used.

6.6 Synthesis based picture coding

Most methods introduced so far target *sample fidelity* in encoding decisions. High sample fidelity as e.g. judged by SNR criteria also guarantees high *visual fidelity*, whereas low fidelity does not necessarily mean that the perceived visual quality is bad. For example, beyond relevant structures such as edges and corners which should not be changed by position and orientation, images may also contain more random elements where it is less important that exact amplitude values are reproduced at a given position, as long as the visual appearance is equivalent. In particular, this is the case for irregular texture areas showing e.g. water surfaces, clouds, smoke, skin, hair etc. The goal is to establish coded descriptions based on few parameters which would allow generating *visually similar* pictures[28] which do not carry obvious coding artifacts. Irregular texture is of particular interest in this context, since this type usually shows a broad, 'colored noise' like spectrum, and cannot be encoded efficiently by conventional methods such as transform or predictive coding, as the information does neither show sparseness in the spatial domain (as in case of edge-like structures) nor in the frequency domain (as in case of periodic structures).

In principle, also simplification of regular structures such as edges or periodic patterns could be applied, but these need to be represented by precise shape and position, and it might be difficult to perform such simplification correctly without violating the semantic meaning. Furthermore, saving of data rate compared to conventional schemes is likely less significant in these cases. On the other hand, regular structures are usually adjacent to or overlaid by noisy (irregular) components, and separating them could probably help in improving the compression. In some regard, synthesis-based coding could be seen to have some similarity with computer graphics, where simple, semantically relevant structures are generated in a deterministic fashion, and more random structures such as texture and noise are overlaid to make them appear photo-realistic. In a converse approach, noise or high-frequent texture can be removed from pictures to make them simpler to encode, retaining the semantically relevant structures (like a painted sketch). In particular, non-linear filters such as non-local means or anisotropic diffusion filters are suitable for this purpose (cf. MCA, SEC. 2.1.3).

The most important challenges which still need to be solved for implementing fully operational synthesis-based picture coding are

- Lack of reasonable distortion criteria which would allow to judge the visibility of errors when a part of an image is replaced by a synthetic reproduction;
- Automatic methods to separate semantically relevant structures which should be retained from irrelevant structures which could be removed or synthesized;
- Automatic finding of the correct model parameters, including the fitness of the model itself, e.g. selection of different models for structured or irregular, noise-like parts;
- Lack of reasonable description of rate-distortion criteria for a given operational point.

Fully synthetic coding is likely to be only applicable to a limited class of images with very clear structure (such as computer graphics generated content), but hybrid solutions (e.g. transform coder plus synthesis coder for specific regions) are a more realistic scenario. Note that similar methods are in principle applicable for transmission error concealment, where 'holes' (lost image areas) could be filled

[28] Basically, synthesis based on parametric descriptions is suitable to generate picture areas of arbitrary size. Therefore, the saving in terms of bit rate would be larger, the larger the synthesized region is.

by a synthetic replacement, provided that suitable model parameters are available at the decoder side (which means they can either be extracted by analyzing adjacent areas, or they are transmitted as additional side information for such purposes). Another application of hole-filling (by 'inpainting') is manipulation such as object removal and replacement by a suitable background.

In the following sub-sections, some methods that perform synthesis of structure or texture from either global or local parameters are described.

6.6.1 Region-based coding

The subjective relevance of different regions in an image varies; e.g. a typical human observer may first look at faces if they are present in a picture. Often the most relevant content is found in the center area anyway. A very simple adaptation of a compression scheme by the properties of regions of interest (ROI) could improve the quality e.g. by finer quantization or higher resolution of areas where relevant content is expected to be present.

Going beyond that, image compression can specifically be designed for better representation of region properties. This is particularly useful if regions of homogeneous color and texture are present, where the semantics of the picture are preserved by preserving the region shape, whereas samples within the regions themselves (provided that reasonable models are available) could be represented by few parameters and synthesized (see Sec. 6.6.2). Even though methods of this kind have a long history in picture coding, they somehow failed to meet the high expectations that were once formulated (see e.g. [KUNT ET AL. 1985]).

Geometric appearance of regions has to be encoded as *shape* (typically as binary mask expressing *which* samples belong to the region, see Sec. 6.1.2) or as *contour* (positions of boundary samples of the region, see Sec. 6.1.3). If needed by the application, the shape has to be represented in a sample accurate fashion. It should be noted that a region shape can also be expressed by a variable-size block structure, e.g. represented by a quad-tree code (see Fig. 6.4). Coding of shapes or contours can also be lossy, e.g. pre-processing may be performed that fills small holes or simplifies contour boundaries.

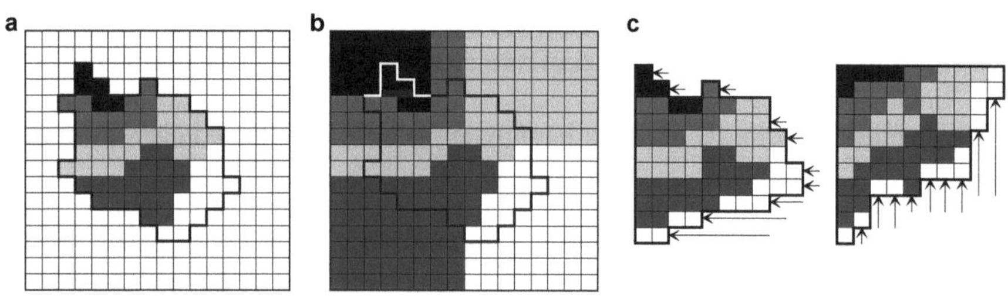

Fig. 6.29. **a** Arbitrary-shaped region, embedded within a larger rectangular block **b** Padding for extension of the signal to the size of a 16x16 block transform **c** Shift of rows and columns in the computation of a shape-adaptive transform

Compression schemes that target sample fidelity need to be modified to work with arbitrary shaped regions (instead of the commonly used rectangular blocks) Whereas sample-wise prediction can be used with minor modifications for reasonable prediction at region boundaries (and holes), transform coding schemes require more significant modifications, as basis functions are 2D images which need to follow the region's shape, and therefore are usually non-separable. One method to resolve this is *padding*, i.e. using an extension of the signal beyond the boundaries of the region, such that a square or rectangle matching the desired block size of the transform is generated. The transform then is over-complete, as the number of transform coefficients equals the number of samples in the original plus padded area, where samples in the latter are either redundant with original samples, or might even take arbitrary values since they are not needed as decoder output. However, the padding process itself can be designed such that as many coefficients as possible are coming close to zero[29]. Fig. 6.29b shows an example of a region padded up to a size of rectangular blocks by constant values, such that subsequent processing by a block transform of size 16x16 samples (or alternatively, four transforms over 8x8 samples) can be employed. Larger regions can consist of rectangular blocks in the inner part of the region and padded blocks at the boundary (see Fig. 6.30). A related scheme has been included for coding of pictures and video objects with arbitrary shape in the MPEG-4 Visual (ISO/IEC 14496-2) compression standard.

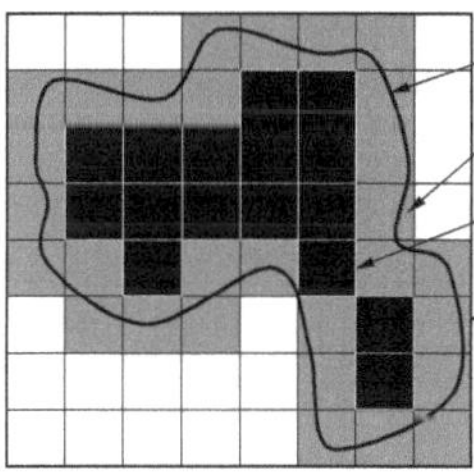

Fig. 6.30. Partitioning of a larger region into rectangular-shaped inner blocks and arbitrary-shaped boundary blocks

Orthogonal 2D transforms with non-rectangular, arbitrary-shaped basis images were proposed by [GILGE ET AL. 1989]. The basis images are computed similar to KLT eigenvectors based on covariance statistics, but individually for the given shape, and employing a correlation model for the texture. A simple variant which has good performance for small blocks is the *shape-adaptive DCT* [SIKORA, MAKAI 1995] illustrated in Fig. 6.29c. First, all rows of the region area (which is the same as in Fig. 6.29b) are left-aligned, and vertical transforms are computed over the length of each column. Subsequently, the columns are top-aligned followed by a

[29] Replication of samples from region boundaries (e.g. constant-value or mirror-symmetric extension, cf. Fig. 2.20) are useful methods to achieve this.

horizontal transform. This way, the transform is implemented as a separable sequence of 1D transforms of various different lengths[30].

6.6.2 Color and texture synthesis

When a region has constant amplitude and color, synthesis becomes trivial by flat filling (e.g. mean value of color). Smooth variations can be supported through linear interpolation or polynomial approximation of the color amplitude surface. However, flat or smoothly varying regions are not challenging for conventional coding methods anyway. Synthesis can be expected to be more advantageous for textured regions (including noisy components), as long as the texture itself exhibits randomness which is irrelevant for the visual appearance.

The random behaviour of texture can be formally described by a homogeneous Markov random field [CHELLAPPA, JAIN 1996], characterized by its conditional probability density

$$p_{s|\mathbf{s}}(x;\mathbf{n} \mid \mathbf{x};\mathcal{N}(\mathbf{n})), \tag{6.7}$$

where $s(\mathbf{n})$ is a sample with a probability of its random variable x influenced by a vector $\mathbf{s}$ constituted by samples from a neighborhood $\mathcal{N}(\mathbf{n})$ with random vector variable $\mathbf{x}$. Homogeneity means that samples are mutually members of their neighborhood, i.e. $\mathbf{m}\in\mathcal{N}(\mathbf{n}) \Leftrightarrow \mathbf{n}\in\mathcal{N}(\mathbf{m})$. The idea that the visual appearance of two sample fields drawn from the same MRF model is highly likely to be indistinguishable is the common basis for various texture synthesis methods (see review of various approaches in [NDJIKI ET AL. 2012])[31].

However, one problem is that an MRF can in principle not be described by a finite number of parameters (which are the conditional probabilities), unless an analytic PDF (e.g. Gaussian matching of an AR model) would be used, which may not be appropriate as a generic texture model. Another approach which appears suitable for textures holding structures of similar areas with characteristic amplitude changes (e.g. pebbles) is empirical sampling. The underlying assumption is that the conditional PDF (6.7) has only few relevant entries, which can be directly determined from an example picture.

Both approaches have their counterparts in prediction methods: AR modeling uses the same synthesis filter as linear prediction, whereas MRF sampling concurs

---

[30] Note that this version of shape-adaptive DCT is used in the MPEG-4 part 2 video compression standard, where however the transform block size is 8x8 samples, such that the region example shown here would need to be separated into 4 blocks. The design is not optimum due to the facts that the DCT coefficients of different transform block lengths do not express identical frequencies, and that additional irregular directional orientations can be introduced due to the shifts and amplitude variations resulting from the combination across various 1D transform lengths in the vertical transform step. Further, orthonormality is lost, and implementation of optimum quantization becomes shape dependent, which could significantly increase the complexity of encoding and decoding.

[31] Identity of visual appearance means that no side-by-side comparison (visual search for sample differences) should be allowed.

with template matching prediction (see Sec. 6.3.1). The main difference is that no prediction error is explicitly encoded (as would be necessary for sample fidelity), but a synthetic noise innovation with similar statistics as the prediction error is used instead.

For a certain class of texture (designated as 'micro texture', which includes sand, water surfaces, hair, grass etc.) a description as stationary Gaussian processes is likely to be suitable, such that Gauss-Markov random fields (GMRF) [RUE, HELD 2005] can be employed, where the conditional PDF is determined from a vector Gaussian PDF (2.156) via measurements of autocovariance.

AR synthesis. The stationary GMRF model is characterized by a Gaussian white noise field $v(\mathbf{n})$ with unity variance, amplified by factor σ and filtered by an all-pole filter $h(\mathbf{n})$,

$$s(\mathbf{n}) = \sigma v(\mathbf{n}) + s(\mathbf{n}) * h(\mathbf{n}), \tag{6.8}$$

where $h(\mathbf{0})$ shall be equal to zero. In the frequency domain, the synthesis is described as

$$S(\mathbf{f}) = \sigma V(\mathbf{f}) + S(\mathbf{f})H(\mathbf{f}) = \frac{\sigma}{1-H(\mathbf{f})}V(\mathbf{f}), \tag{6.9}$$

The power spectral density is

$$\phi_{ss}(\mathbf{f}) = \mathcal{E}\left\{\left|S(\mathbf{f})\right|^2\right\} = \frac{\sigma^2}{\left|1-H(\mathbf{f})\right|^2}\phi_{vv}(\mathbf{f}), \tag{6.10}$$

where typically $\phi_{vv}(\mathbf{f})=1$. The randomness of this unity-variance white noise innovation signal is only by the phase of its random instantiations. If the phase of the innovation is independent and identically distributed (i.i.d.) over all values $\mathbf{f}$, the phase of the output process will be i.i.d. as well. The conditional probability density of the GMRF model, with a neighborhood consisting of K samples is given by

$$p_{s|\mathbf{s}}(x \mid \mathbf{x};\mathbf{h};\sigma) = \left(\frac{1}{\sqrt{2\pi\sigma^2}}\right)^K e^{-\frac{\mathbf{h}^{\mathrm{T}}\mathbf{C}_{\mathbf{ss}}\mathbf{h}}{2K\sigma^2}} \quad \text{with } \mathbf{C}_{\mathbf{ss}} = \mathcal{E}\left\{\mathbf{s}\mathbf{s}^{\mathrm{T}}\right\}$$

$$\mathbf{s} = \left[s(\mathbf{n}), s(\mathbf{n}-\mathbf{m}_1),\ldots,s(\mathbf{n}-\mathbf{m}_K)\right]^{\mathrm{T}} \quad \text{and} \quad \mathbf{h} = \left[1,-h(\mathbf{m}_1),\ldots,-h(\mathbf{m}_K)\right]^{\mathrm{T}}. \tag{6.11}$$

The $\mathbf{m}_k$ values represent the relative distance from the current position in a causal neighborhood. Practically, the covariance matrix can be filled using measurements from a local texture region, it can then be used to determine the set of filter parameters $\mathbf{a}$ by solving the Wiener-Hopf equation (2.207). The gain parameter is estimated as

$$\sigma = \sqrt{\frac{1}{K}\mathbf{h}^{\mathrm{T}}\mathbf{C}_{\mathbf{ss}}\mathbf{h}}. \tag{6.12}$$

An advantage of this formulation (when compared to the conventional 'causal' AR model introduced in Sec. 2.6.1) is the fact that $h(\mathbf{n})$ can be any FIR filter that excludes the current sample position, such that in principle there is no restriction of being causal. The non-causal generalization is also denoted as *conditional AR*

(CAR) model [Rue, Held 2005]. In particular, symmetric 2D filter support leads to a Markovian homogeneity property of the generated field, which is then equivalent to a GMRF. Even though synthesis via the AR approach requires recursive filtering, which due to the mutual dependency of samples is not implementable for infinite fields, solutions for finite fields exist where appropriate boundary conditions need to be set. This can be implemented by factorizing a non-causal impulse response into causal parts starting from different directions [Chelappa, Kashyap 1985] [Ranganath, Jain 1985], or by solving a linear equation system which is equivalent to inverting a matrix formulation of the linear prediction process $\sigma v(\mathbf{n}) = s(\mathbf{n}) - s(\mathbf{n}) * h(\mathbf{n})$ of (6.8).

An operational approach which removes 'micro texture' of GMRF type on a block by block basis, and overlays a synthesis instead was presented in [Ballé, Stojanovic, Ohm 2011] and further detailed in [Ballé 2012]. The classification of this type of texture is performed by analysing the local coefficients of a complex Gabor wavelet filter bank, where it is assumed that the characteristics would be Gaussian if the kurtosis is below a threshold and the spectrum is non-sparse. Synthesis can be performed either directly by implementing the CAR model with additional boundary conditions, or by superimposing different scale components generated from a wavelet synthesis filter bank. Since the spectrum is non-sparse, DCT based transform coding is applied to represent the variances of the frequency bands in an efficient way (since the phase of the spectrum is irrelevant for the synthesis approach, the property can be utilized that the spectral amplitudes are highly redundant along the frequency axis). In the work, it is also shown that the Itakura distance [Itakura 1975], which enables quantifying the similarity of power spectra, approximately matches with visual perception for this class of texture.

MRF sampling. In MRF sampling, the output of the convolution $s(\mathbf{n})*b(\mathbf{n})$ in (6.8), which is a linear prediction operation from past samples, is replaced by a template matching prediction, i.e., a causal neighborhood from already synthesized samples is compared against templates of an example texture, and the sample that is at the corresponding position of the most similar template is used as prediction. Again, an innovation signal is added to generate the next synthesized sample. Variants of this method include averaging or weighted averaging of the predictions from several best matching templates. This method is also often used for 'inpainting' a similar texture in case of data losses, for picture manipulation where objects shall be removed and be replaced by a suitable background structure, or for hole filling in view synthesis.

Noise overlays. A specific case of 'texture' synthesis is superposition of artificial noise. As noise has a broad spectrum and particularly in transform coding low amplitude coefficients (where the noise may dominate over the signal) are quantized into zero value, compression can have a side effect of de-noising. This however may affect the photorealistic quality of decoded pictures, which can then be regenerated by superimposing synthetically generated noise. Furthermore, superimposed noise can also hide some other compression artifacts particularly in flat color areas, where the 'banding' effect (artificial visible amplitude discontinuities a block boundaries) might be annoying. This latter usage of noise overlays is also denoted as 'dithering'. A specific case where the suppression may be undesirable is compression of scanned film pictures. 'Film grain' noise which is coherent with the signal (strength depending on sample amplitudes) is considered to have a psycho-visual effect of sharpening images. On the other hand, for the purpose of a more efficient compression of the genuine picture content, it should better be removed. The AVC video compression standard includes signalling of supplemental information which allows parametric synthesis of a noise overlay (e.g. film grain noise) as post-decoder operation. The options include expressing the spectral characteristics of the noise via coefficients of an FIR (moving average) or IIR (autoregressive) filter model, as well as amplitude range (or standard deviation) of the synthetic innovation fed into these filters, and the types of additive (independent) or multiplicative (coherent) noise overlays to the signal.

6.6.3 Synthesis of illumination effects

In capturing natural scenes, characteristics of lighting play an important role. The color temperature of the illumination source can basically be adjusted through multi-channel amplitude transfer characteristics. As an approximation, a gain/offset adjustment

$$g(\mathbf{n}) = \alpha\, s(\mathbf{n}) + \beta \tag{6.13}$$

is often sufficient. For higher order functions, exponential characteristics similar to the gamma transfer in (1.5) or piecewise definitions are preferable rather than higher order polynomials due to their better stability. For color transfer, matrix formulations[32] such as

$$\begin{bmatrix} g_{\mathrm{R}}(\mathbf{n}) \\ g_{\mathrm{G}}(\mathbf{n}) \\ g_{\mathrm{B}}(\mathbf{n}) \end{bmatrix} = \begin{bmatrix} \alpha_{\mathrm{RR}} & \alpha_{\mathrm{GR}} & \alpha_{\mathrm{BR}} \\ \alpha_{\mathrm{RG}} & \alpha_{\mathrm{GG}} & \alpha_{\mathrm{BG}} \\ \alpha_{\mathrm{RB}} & \alpha_{\mathrm{GB}} & \alpha_{\mathrm{BB}} \end{bmatrix} \begin{bmatrix} s_{\mathrm{R}}(\mathbf{n}) \\ s_{\mathrm{G}}(\mathbf{n}) \\ s_{\mathrm{B}}(\mathbf{n}) \end{bmatrix} + \begin{bmatrix} \beta_{\mathrm{R}} \\ \beta_{\mathrm{G}} \\ \beta_{\mathrm{B}} \end{bmatrix} \tag{6.14}$$

can then be applied. Usually the illumination source itself does not appear in a picture, unless seen as specular reflection from an object surface. In particular when the illumination source is not diffuse and/or specular effects are present, significant modifications of the reflected light occur depending on the position and orientation of light source, object surface and camera (relative to each other). This could precisely only be modeled when a full 3D model of the scene and capturing/lighting situation would be available. However, in many cases, local

---

[32] Shown here for the case of an RGB representation; as however also the RGB-to-YC_bC_r transformations in (1.6)/(1.7) can be written as a matrix/vector multiplication, basically the same method can be applied. As however chrominance components are more invariant to illumination changes, a transfer could also be restricted to Y in this case (see also MCA, SEC. 4.1).

adaptation of one of the models above (e.g. position dependent gain and offset) is sufficient for such cases.

6.7 Fractal image coding

The theory of fractals assumes *self-similarity* (e.g. that the same content appears again with different scale in a picture, like zooming in with a 'magnifying glass'). To exploit this for image compression, it is necessary to define a *fractal transform* describing the mapping between the similar versions. Iterating the self-similarity principle, this would mean that a picture could be synthesized up to an infinite number of samples (or arbitrarily fine resolution) in a meaningful way. This however is not very useful when an arbitrary picture shall be represented by fractal transform parameters, which may only be valid for a finite region. An example for local application of fractal synthesis parameters determined from a real picture is the approach of fractal block coding described below.

The *collage theorem* establishes the foundation, by which a fractal transform is stable and a picture can be reconstructed with sufficient accuracy from the parameters.

Principle of fractal transforms. Assume that a certain area (e.g. a block) from an image shall be encoded. This target area shall be described as good as possible from a reference area contained in the same picture. In this process, the reference area may be modified by mapping *geometry* and *amplitude* (e.g. brightness and contrast).

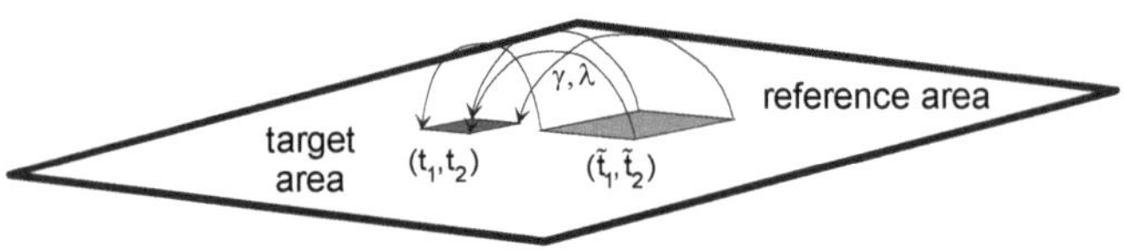

Fig. 6.31. Principle of a fractal block transform

The basic principle of a *fractal block transform* is shown in Fig. 6.31. In this case, a reference area with coordinates $(\tilde{t}_1,\tilde{t}_2)$ is mapped by a *geometric transform* $\gamma(\cdot)$ into a target block described by coordinates (t_1,t_2)[33]. Additionally, an *amplitude transform* $\lambda(\cdot)$ is applied, by which mean value and contrast of the reference area amplitudes are adapted to the actual values of the target block in the image to be

[33] The coordinate systems of reference and target can be exchanged, but the approach used here is more reasonable in a practical coding scheme where a unique mapping needs to be defined for each target sample. An example for a coordinate mapping is the affine transform (7.29), but a subset thereof could also be used.

decoded. The entire fractal transform for the given block area is then the functional combination of geometric and amplitude transform, $f(\cdot)=\gamma(\cdot)\circ\lambda(\cdot)$. The geometric transform is *contractive*, if no geometric distance between any two points in the target system (t_1,t_2) is larger than the distance between the corresponding points in the origin $(\tilde{t}_1,\tilde{t}_2)$. The amplitude transform in simplest case is a gain/offset scheme (6.13) with identical parameters applied to all samples of the block. It is contractive, if the variation of amplitudes is not increased during the mapping, which means $\alpha < 1$. The offset value β should be constrained such that every sample of the output image is in the amplitude range $0 \ldots A_{max}$, or clipping needs to be applied after the transform.

Collage Theorem. The *Collage Theorem* [BARNSLEY 1988] states that an iterative generation of a picture from an arbitrary source picture is possible by a fractal transform $f(\cdot)$ if

1. The description of the image from itself can be achieved by the transform with sufficiently small distortion;
2. The complete transform $f(\cdot)=\gamma(\cdot)\circ\lambda(\cdot)$ is contractive.

Proof of the Collage Theorem: Let **S** be an original picture, and $f(\mathbf{S})$ be the picture which is output when the fractal transform is applied *once* to the original. This can be seen as the optimum case of decoding from the fractal transform parameters $f(\mathbf{S})$, which have to be designed such that the picture can be described as good as possible from itself. Let now $\mathbf{G}=f^{(r)}(\mathbf{V})$ be the reconstructed picture after the rth iteration of the fractal transform applied to an arbitrary initial picture **V**. Further, $f^{(r)}(\mathbf{S})$ shall be the result after the rth application of the transform to the original picture **S**; observe that $f^{(r)}(\mathbf{S})$ probably will have a larger deviation than $f(\mathbf{S})$ from the original picture **S**, as in the iteration $2 \ldots r$ the original is not used. However, by the triangular inequality, the following relationship applies (where $\mathrm{d}(\cdot,\cdot)$ expresses the Euclidean distance):

$$\mathrm{d}(\mathbf{S},\mathbf{G}) \le \mathrm{d}\left(\mathbf{S}, f^{(r)}(\mathbf{S})\right) + \mathrm{d}\left(f^{(r)}(\mathbf{S}), f^{(r)}(\mathbf{V})\right). \tag{6.15}$$

Further,

$$\begin{aligned} \mathrm{d}\left(\mathbf{S}, f^{(r)}(\mathbf{S})\right) &\le \mathrm{d}\left(\mathbf{S}, f^{(1)}(\mathbf{S})\right) + \mathrm{d}\left(f^{(1)}(\mathbf{S}), f^{(2)}(\mathbf{S})\right) + \ldots + \mathrm{d}\left(f^{(r-1)}(\mathbf{S}), f^{(r)}(\mathbf{S})\right) \\ &\le (1+k+\ldots+k^{r-1})\,\mathrm{d}\left(\mathbf{S}, f^{(1)}(\mathbf{S})\right) \le (1-k)^{-1}\,\mathrm{d}\left(\mathbf{S}, f(\mathbf{S})\right) \end{aligned} \tag{6.16}$$

and

$$\mathrm{d}\left(f^{(r)}(\mathbf{S}), f^{(r)}(\mathbf{V})\right) \le k\,\mathrm{d}\left(f^{(r-1)}(\mathbf{S}), f^{(r-1)}(\mathbf{V})\right) \le \ldots \le k^r\,\mathrm{d}(\mathbf{S},\mathbf{V}). \tag{6.17}$$

Using (6.15), (6.16) and (6.17), the following upper bound condition holds for the remaining distortion:

$$d(\mathbf{S},\mathbf{G}) \le \underbrace{(1-k)^{-1}\, d(\mathbf{S}, f(\mathbf{S}))}_{\varepsilon} + k^r\, d(\mathbf{S},\mathbf{V})\,. \tag{6.18}$$

As according to the collage theorem k shall be smaller than 1, the distance between the original picture **S** and the output **G** of this iteration will converge towards ε after a sufficiently large number of iterations r, regardless from where the iterations start. This distance is lower bounded (for $k \to 0$) by the distance between **S** and $f(\mathbf{S})$. The convergence will be faster for smaller k, and for pictures **V** which are as close as possible to **S** (therefore, on average a mid-gray picture should be a good choice).

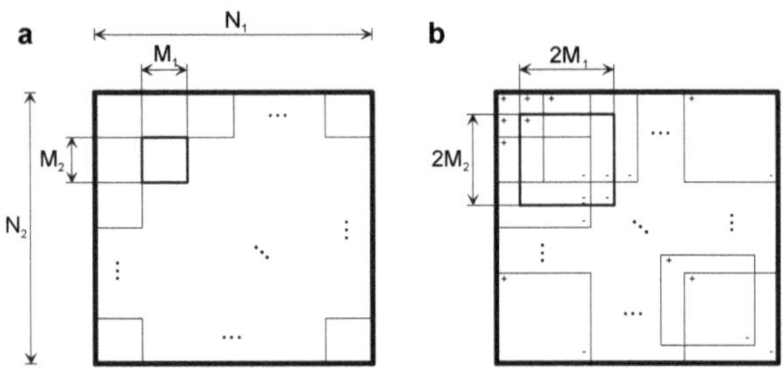

Fig. 6.32. Fractal block coding: **a** Positions of target blocks where parameters of the fractal transform must be defined (non-overlapping) **b** Example of possible selection of origin blocks (overlapping)

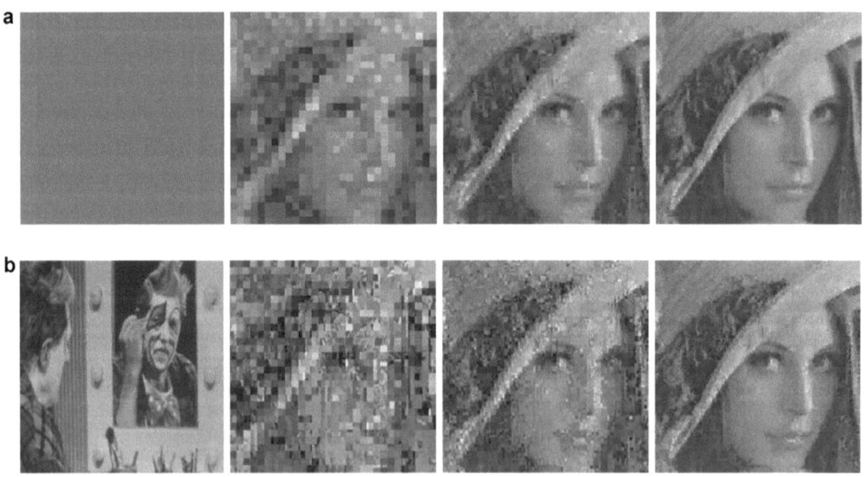

Fig. 6.33. Convergence of fractal decoding, first three iterations:
a Generated from uniform gray image **b** Generated from different image (courtesy by K. BARTHEL)

Fractal block coding was introduced in [JACQUIN 1992] implemented as shown in Fig. 6.32. The fractal code consists of the parameters for the geometric transform and the amplitude transform for each block. The encoder has to determine these parameters by analysis of the original image, which means to find the optimum fractal mapping of the image into itself, which is typically achieved by matching. By imposing certain limitations to the geometric transform, the search for the fractal parameters is speeded up and their representation made more compact:

- The target block areas are of equal size M_1xM_2 (denoted as *range blocks* in [JACQUIN 1992]);
- Contractivity of the geometric transform is fixed to a scaling factor 0.5 (factor 2 downsampling);
- The free parameters of the geometric transform are only translations, rotations of blocks by 0°, 90°, 180° and 270° and horizontal/vertical mirror mappings.

According to the proof of the collage theorem, the result of decoding cannot become better than the result by applying the fractal transform *once* to the original image. An example about convergence of the fractal decoding process is illustrated in Fig. 6.33. Identical transform parameters are used in both cases, however the initial picture **V** is once uniform gray, and once a picture with different structure than the picture to be decoded, i.e. the first case can be expected to be closer to **S**. It is obvious that the difference decreases significantly between both cases after only few iterations.

6.8 3D picture coding

The real world is three-dimensional, the image plane of a camera captures a 2D projection. Generally, 3D imaging uses multiple projections in order to generate a more complete vision of the 3D world. In principle, the plenoptic function (see Sec. 1.2) or a 3D model (like in computer graphics) could give a fully complete picture from any viewpoint. The following modalities are subsets thereof:

- Stereoscopic imaging, captured by 2 cameras, simulating the binocular vision from a fixed capturing perspective, and reproducible on a stereo display;
- Light-field imaging, captured by multiple cameras e.g. arranged as a 2D array, allows to generate various view points; displays allowing light-field rendering from a corresponding arrangement of multiple projectors also exist as prototypes;
- Volumetric imaging, where a full 3D coordinate space allows to indicate the presence of objects either as mass volume or by an outer hull, including various imaging features such as volumetric density, color, reflectance properties etc. Volumetric images are usually generated from multiple views by means of 3D reconstruction. A 3D volume can also be interpreted as a stack of multiple 2D images denoted as *slices*.

It is beyond the scope of this book to discuss all details of 3D imaging. To further compress the additional information that is due to one or more additional views, it is usually necessary to perform correspondence matching, such that prediction or transform coding can be applied between samples or groups of samples that have highest similarity. Conceptually, core approaches of multi-view compression are very similar to methods of video compression, where correspondence matching is used to perform *motion compensation* over the time axis in order to better exploit the coherence. For further reference, the reader is pointed to Sec. 7.6.

6.9 Reconstruction filtering

Coding algorithms often introduce unnatural, but to some extent systematic artifacts, and it can be targeted to reduce their visibility by reconstruction filtering[34]. When filter parameters are determined by the encoder which has both original and reconstructed picture available, objectively measurable quality improvement (e.g. PSNR) is possible by *adaptive post filtering*. In this case, it is necessary to encode filter parameters as side information (similar to approaches discussed in Sec. 6.3.5). The encoder determines filter parameters, which are best chosen when the effect of the decoding process is considered in combination with those parameters. Alternatively, clues for filter control can also be determined by analysis of the reconstructed signal at the decoder side, taking into account a model about the expected distortion. Practically, a combination of both (partially encoder and partially decoder controlled) may be the best approach, since the amount of side information should be kept low.

A classical de-noising approach with the target of optimizing SNR-based reconstruction quality is the *Wiener filter*. Beyond that, non-linear filtering methods may be better suitable to remove specific compression artifacts, which basically are a result of the non-linear quantization process. This can e.g. be achieved by controlling an amplitude mapping for reconstruction. Coding distortions can hardly be modeled as stationary additive noise process, since they are typically signal dependent, e.g. in presence of edge structures, or when locally high frequency components are removed, or codec dependent, e.g. when operations are executed over a fixed block grid. *De-blocking filters* and *de-banding filters* specifically target removal of artifacts introduced by block-based coding. *De-ringing filters* and *edge sharpening filters* reduce artifacts that are caused at amplitude discontinuities through frequency cut-off or failure of prediction. In the subsequent paragraphs, several post processing filter types that are commonly used are discussed[35].

---

[34] In case of video compression, similar approaches can also be used in the prediction loop, in order to improve prediction from other decoded pictures

[35] In hybrid video compression (cf. Sec. 7.2), these filters are often operated *within the prediction loop*. Even then, these are still reconstruction filters with the target of improving the quality of a picture that will be used for subsequent predictions. It is beyond the scope

Wiener filter. This is typically implemented as FIR filter with impulse response $h(\mathbf{n})$, designed by a variant of the Wiener-Hopf equation similar to the linear prediction case of (2.207), but with the decoded picture $\tilde{s}(\mathbf{n})$ as input and a reconstructed picture $\hat{s}(\mathbf{n}) = \tilde{s}(\mathbf{n}) * h(n)$ as output. The filter coefficients are determined by minimizing $\mathcal{E}[\, s(\mathbf{n}) - \hat{s}(\mathbf{n})\,]^2$, which leads to a formulation of the Wiener-Hopf equation in matrix notation as

$$\mathbf{c}_{\tilde{s}s} = \mathbf{C}_{\tilde{s}\tilde{s}}\mathbf{h} \Rightarrow \mathbf{h} = \mathbf{C}_{\tilde{s}\tilde{s}}^{-1}\mathbf{c}_{\tilde{s}s} \,, \tag{6.19}$$

which means that cross covariance between the original and the reconstructed pictures, and autocovariance of the reconstructed picture need to be determined to compute the coefficients. Unlike the linear prediction case, no inverse (recursive) filter is necessary in the reconstruction chain, such that symmetric filter supports using neighbor samples from all directions can be applied.

Due to the fact that side information is required to transmit the filter coefficients, and the benefit in terms of increasing quality (or achieving lower rate) should more than compensate for this, the degree of adaptation has to be chosen carefully. Global adaptation (once for the entire picture) gives some benefit [WITTMANN, WEDI 2007]. When the characteristics of coding distortions as well as the picture content are varying, local adaptation can be expected to do better, but would require significantly higher amount of side information. The rate of side information is influenced by

- the number of filter coefficients to be signaled, which highly depends on the size of the support region (filter geometry). Separable filters require fewer parameters than non-separable 2D filters, however they are less flexible in implementing directional filtering and therefore undesirable. By utilizing symmetries around the center sample, the number of coefficients can be reduced in a useful way, as shown in the examples of Fig. 6.34[36].
- the size of adaptation region: The number of parameters to be encoded scales reciprocally with the size of the adaptation region (e.g. number of samples per block in a block based adaptation scheme.
- the size of the discrete set of coefficients: To avoid explicit encoding of filter coefficients per region, a discrete set of filters can be optimized with coefficients encoded as side information at the picture level; for each region, it is then only necessary to signal which of the filters is selected. However, this procedure requires an a priori classification of blocks before the Wiener-Hopf equations can determine the different filters of the set. Criteria could be the

---

of this book to make an extensive discussion of non-linear reconstruction filters. Some of the types which could potentially also be used for the purpose of post-filtering, particularly edge-preserving and area-smoothing filters such as *median*, *bilateral*, *diffusion* and *non-local means* filter types are introduced in [MCA, SEC. 2.1.3].

[36] Most types of the example filter layouts shown in Fig. 6.34 were actually proposed and investigated for the purpose of adaptive loop filtering in video coding (cf. Sec. 7.2.2) during the development of the HEVC standard.

amount of local detail, directional characteristics etc. Alternatively, in a first step filters could be separately determined for each block, and then blocks with similar filters are clustered; this however requires at least two iteration steps at the encoder.

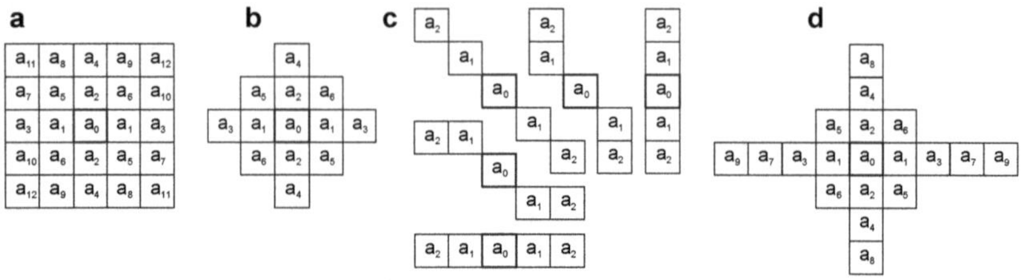

Fig. 6.34. Examples of non-separable 2D filter layouts for post filtering: **a** 5x5 square shape **b** 5x5 diamond shape **c** 5x1 directional (5 out of 16 possible directions shown) **d** 9x7 cross with 3x3 square center.

Finally, switching of filters (from a set) can also be performed without transmitting side information, by local analysis (or classification) of the signal $\tilde{s}(\mathbf{n})$ at the decoder side. This basically allows straightforward classification for solving the Wiener-Hopf equations separately for all the classes. However, it would still be useful to allow disabling the filtering entirely for certain regions of the picture, and those regions cannot be known beforehand. Therefore, an iterative optimization usually further improves the performance.

Sample adaptive offset (SAO). SAO was originally proposed as a process for loop filtering in the HEVC standard [Fu et al. 2012]. As it is basically doing post processing for lower reconstruction error and can be used for that purpose in still picture coding as well, it is discussed here.

SAO modifies reconstructed samples through a lookup table operation, where the lookup tables are designed by the encoder such that the difference between original samples $s(\mathbf{n})$ and filtered samples $\hat{s}(\mathbf{n})$ is minimized, when the decoded samples $\tilde{s}(\mathbf{n})$ are used as input. SAO parameters (lookup table and mode selection as explained below) are usually adapted on a region/block basis. If not disabled, SAO is operated either in one out of four edge modes, or in the banding mode. In the edge modes, depending on a local gradient criterion $\nabla\tilde{s}(\mathbf{n})$, a value β is retrieved from a lookup table to generate $\hat{s}(\mathbf{n}) = \tilde{s}(\mathbf{n}) + \beta\left(\nabla\tilde{s}(\mathbf{n})\right)$. The four edge modes are different in that the horizontal, the vertical or one of the two diagonal gradients is used to compute $\nabla\tilde{s}(\mathbf{n})$, two neighbors of the respective orientation are used. If no gradient is detected, β is set to zero; otherwise, the following four cases are distinguished and each mapped to its own value β, i.e. the following classification on $\nabla\tilde{s}(\mathbf{n})$ is executed by the decoder (see Fig. xx):

1. If $\tilde{s}(\mathbf{n})$ is a local minimum between the two neighbors, the corresponding value β from the lookup is added (positive sign);

2. If $\tilde{s}(\mathbf{n})$ is a local maximum between the two neighbors, the corresponding value β from the lookup is subtracted (negative sign);
3. If $\tilde{s}(\mathbf{n})$ is smaller than one of the neighbors and equal to the other neighbor, the corresponding value β from the lookup is added (positive sign for increasing edge);
4. If $\tilde{s}(\mathbf{n})$ is larger than one of the neighbors and equal to the other neighbor, the corresponding value β from the lookup is subtracted (negative sign for decreasing edge);

The first two cases could also be interpreted as a removal of outliers (and therefore have a de-ringing effect).

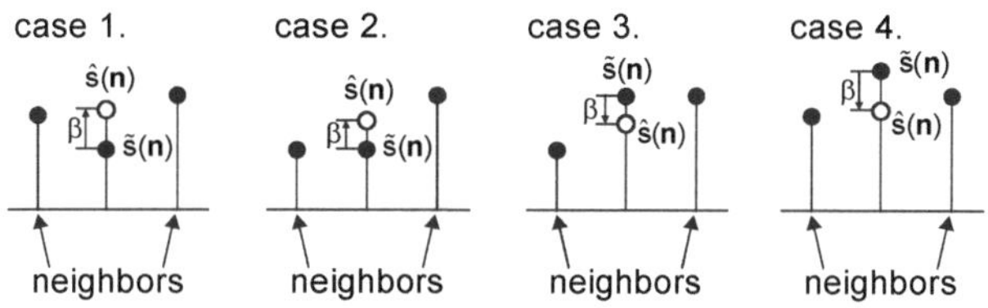

Fig. 6.35. Four cases of edge offset in SAO (corresponding to 1.-4. above).

In the banding mode, the selected offset value directly depends on the sample amplitude. The whole sample amplitude range is split into 32 uniform bands, i.e. the five MSBs of a sample are used to identify the band which it belongs to. Only those samples values which belong to one of four consecutive bands are modified by adding the values β from the offset table. The position of the four bands has to be selected by the encoder, and the index of the first band has also to be signaled. Four consecutive bands are usually sufficient due to the fact that most sample amplitudes in a local area tend to be concentrated in a relatively narrow amplitude range, particularly in flat areas where banding artifacts appear as most annoying.

The lookup table which has four entries in each of the cases, the mode and, in case of band offset, the starting position of the bands, have to be explicitly encoded. Alternatively, to reduce the amount of side information, they can be inherited from a neighbor region (to be signaled by a merge flag).

Deblocking filters. Due to coarse quantization at low bit rates, block-based coding often produces artificial discontinuities at block boundaries. Furthermore, due to the properties of DCT basis functions, coding errors tend to increase towards block edges. The principle of the deblocking filters defined in the AVC and HEVC standards is described here, which reflect the current state of the art and unveils the basic principles of detecting and removing unnatural amplitude transitions at block boundaries without destroying the signal.

Let the blocks on the two sides of a boundary be designated as P and Q (where these can be horizontal or vertical neighbors). Basically, the decision about deblocking is based on the difference between the samples p_0 and q_0 across the boundary, which are additionally compared against their respective neighbor in the same block, p_1 and q_1 (see Fig. 6.36). Filtering of the samples p_0 and q_0 is performed using a 1D filter across the boundary, when $| p_0 - q_0 |$ is smaller than a threshold α, and at the same time both $| p_1 - p_0 |$ and $| q_1 - q_0 |$ are smaller than another threshold β, which is usually significantly smaller than α. By these criteria, it is avoided that amplitude discontinuities (edges or textured detail) which are relevant part of the signal are destroyed. Both thresholds are increased when the quantization is coarser, such that larger differences would rather become explainable as compression artifacts. Additionally, differences $| p_2 - p_0 |$ and $| q_2 - q_0 |$ are computed, and if these are below the threshold β, filtering of positions p_1 and q_1 is performed additionally.

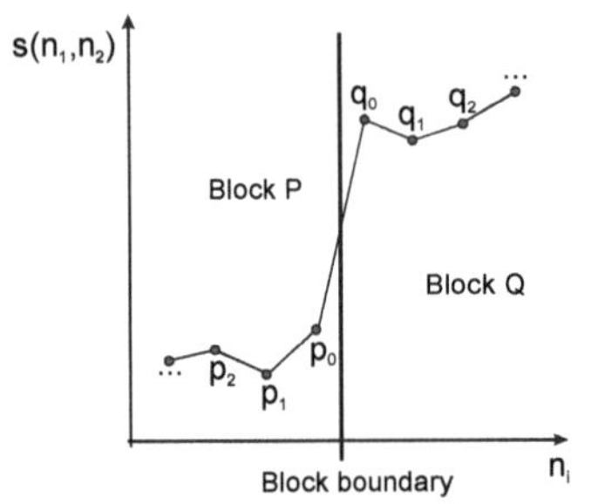

Fig. 6.36. Samples $p_0...p_2$ and $q_0...q_2$ at a block boundary, used to decide on the deblocking operations in AVC and HEVC.

Furthermore, the following criteria are taken into account to determine the strength (smoothing effect, length 3 or 5 taps) of the filter:

- Intra coding in at least one of the two blocks P and Q;
- Difference of motion vectors and reference picture indices (in case of inter-picture coding);
- Presence of non-zero coefficients.

The threshold values can be adjusted within a certain range, such that encoder designers can further optimize the behavior of the filter. A detailed description can be found in [LIST ET AL. 2003] for AVC and [NORKIN ET AL. 2012] for HEVC.

6.10 Still picture coding standards

The Joint Photographic Experts Group (JPEG) is a working group which was established jointly by the International Standardization Organization (ISO) and the International Telecommunication Union (ITU). In 1990, JPEG finalized the first version of the still image coding standard ISO/IEC 10918 (ITU-T T.81ff.) [PENNEBAKER, MITCHELL 1993]. It is sub-divided into the following parts[37]:

- ISO/IEC 10918-1 (twin text with ITU-T Rec. T.81) : *Requirements and guidelines*;
- ISO/IEC 10918-2 (twin text with ITU-T Rec. T.83): *Compliance testing*;
- ISO/IEC 10918-3 (twin text with ITU-T Rec. T.84): *Extensions*;
- ISO/IEC 10918-4 (twin text with ITU-T Rec. T.86): *Registration of JPEG profiles*;
- ISO/IEC 10918-5 (twin text with ITU-T Rec. T.871): *JPEG file interchange format*;
- ISO/IEC 10918-6 (twin text with ITU-T Rec. T.872): *Application to printing systems*.

The algorithm specified in part 1 has probably become the most widely used still image compression method, being based on a block transform (DCT) scheme since then. The modes 'sequential DCT', 'progressive DCT' and 'hierarchical progressive' were already described in Sec. 6.4.1, the lossless mode in Sec. 6.5.
In the extended coding process definitions of part 3, the encoder can determine optimized VLC tables for individual images; the parameters necessary for this adaptive entropy coding are then conveyed in the preamble of the bit stream. Further, adaptive binary arithmetic coding is defined which can further improve the compression performance. A mode of bit-plane coding, supporting up to 12 bit quantization precision for the transform coefficients, is also defined. However, none of these mechanisms has found any wide practical application.

Subsequently, JPEG defined another standard for better lossless compression performance than the previous standard's lossless mode. It is entitled '*Lossless and near-lossless compression of continuous-tone still images*' (nicknamed as JPEG-LS) and is sub-divided into the following parts:

- ISO/IEC 14495-1 (identical with ITU-T Rec. T.87): *Baseline*;
- ISO/IEC 14495-2 (identical with ITU-T Rec. T.870): *Extensions*.

The basic coding algorithms of JPEG-LS, *run mode* and *prediction mode* were described in Sec. 6.5. Beyond the exponential Golomb entropy coding of the baseline configuration, arithmetic coding is defined in the extensions.

Next in the series of JPEG standards is the '*JPEG 2000 image coding system*' of which is consisting of following parts:

- ISO/IEC 15444-1 (twin text with ITU-T Rec. T.800): *Core coding system*;
- ISO/IEC 15444-2 (twin text with ITU-T Rec. T.801): *Extensions*;
- ISO/IEC 15444-3 (twin text with ITU-T Rec. T.802): *Motion JPEG 2000*;
- ISO/IEC 15444-4 (twin text with ITU-T Rec. T.803): *Conformance testing*;
- ISO/IEC 15444-5 (twin text with ITU-T Rec. T.804): *Reference software*;

---

[37] In the description of this and subsequent standards, we mainly concentrate here on those parts which specify coding algorithms. With regard to file formats, the reader is referred to Sec. 9.4.3.

- ISO/IEC 15444-6 (twin text with ITU-T Rec. T.805): *Compound image file format*;
- ISO/IEC 15444-8 (twin text with ITU-T Rec. T.807): *Secure JPEG2000*;
- ISO/IEC 15444-9 (twin text with ITU-T Rec. T.808): *Interactivity tools, APIs and protocols*;
- ISO/IEC 15444-10 (twin text with ITU-T Rec. T.809): *Extensions for 3D data*;
- ISO/IEC 15444-11 (twin text with ITU-T Rec. T.810): *Wireless*;
- ISO/IEC 15444-12: *ISO base media file format*;
- ISO/IEC 15444-13 (twin text with ITU-T Rec. T.812): *Entry leve JPEG2000 encoder*;
- ISO/IEC 15444-14 (twin text with ITU-T Rec. T.813): *XML structured representation and reference.*

Part 1 defines a fully embedded still image codec based on wavelet transform technology, which is described in detail in Sec. 6.4.4. Part 2 supports the following features, which however are not commonly used:

- Usage of *arbitrary wavelet filters* by definition in the header;
- *More flexible quantization* of wavelet coefficients, including *psycho-visual weighting functions* and local adaptation of the embedded quantizer step size;
- Improved compression by *trellis-coded quantization* (cf. Sec. 4.6);
- Nonlinear pre-processing of images by *gamma correction* and *lookup tables*;
- Support of *additional color spaces*, including *multi-spectral* components (more than three);
- *Region of interest* (ROI) definition, which allows decoding of non-rectangular regions of arbitrary shape, also in combination with the scalability features; it is possible to increase the spatial resolution in decoding only within a specific ROI, or decode the ROI by higher quality of the embedded quantizer.

Part 3 defines *Motion JPEG 2000*, where a video sequence is encoded picture by picture independently by a JPEG 2000 coder, without utilizing inter-picture redundancy. This mainly includes mechanisms for accurate timing in video presentation and synchronization with other media types such as audio[38].

JPEG XR (eXtended Range), which originally had been a proprietary industry specification (Windows Media Photo), was approved as international standard in 2009 (ITU-T Recommendation T.832, ISO/IEC 29199, '*JPEG XR image coding system*'). One intention of the JPEG-XR design is better support for coding of graphics content [DUFAUX ET AL. 2009]. It consists of five parts:

- ISO/IEC 29199-1 (twin text with ITU-T Rec. T.Sup2): *System architecture*;

---

[38] Many implementations of so-called *Motion JPEG* based on ISO/IEC 10918 / ITU-T T.81 also exist in the market. These are usually proprietary solutions never officially defined by the JPEG committee, including mechanisms for time-synchronized decoding of a sequence of JPEG pictures to be displayed as video sequence.

- ISO/IEC 29199-2 (twin text with ITU-T Rec. T.832): *Image coding specification*;
- ISO/IEC 29199-3 (twin text with ITU-T Rec. T.833): *Motion JPEG XR*;
- ISO/IEC 29199-4 (twin text with ITU-T Rec. T.834): *Conformance testing*;
- ISO/IEC 29199-5 (twin text with ITU-T Rec. T.835): *Reference software*;

The algorithm defined in part 2 uses a small (4x4 size) block transform, optionally followed by a secondary transform with an offset of two samples in both directions, by this having the effect of an overlapping-block transform (see Sec. 6.4.3). JPEG XR uses a coding table based adaptive entropy coding and allows compression up to lossless quality.

The most recent standard (still under development) defined by JPEG is '*Scalable Compression and Coding of Continuous-Tone Still Images (JPEG XT)*', ISO/IEC 18477. It is less of a self contained specification rather than extending the first JPEG standard (ISO/IEC 10918/ITU-T T.81), to enable features arising by new market needs. These are in particular support for high dynamic range (HDR) image compression, lossy-to-lossless coding and privacy tools. Moreover, support for augmented reality applications is foreseen. The core coding part is unchanged, but enhanced by new building blocks such as specific amplitude mapping functions and extended bit depth as well as floating point representation necessary for HDR imaging. Currently, the following parts are foreseen[39]:

- ISO/IEC 18477-1: *Core coding system*;
- ISO/IEC 18477-2: *Coding of high dynamic range images*;
- ISO/IEC 18477-3: *Box file format*;
- ISO/IEC 18477-4: *Conformance testing*;
- ISO/IEC 18477-5: *Reference software*;
- ISO/IEC 18477-6: *IDR integer coding*;
- ISO/IEC 18477-7: *HDR floating point coding*;
- ISO/IEC 18477-8: *Lossless and near-lossless coding*;
- ISO/IEC 18477-9: *Alpha channel coding*.

The Joint Bilevel Images Group (JBIG), which has ceased to exist in the meantime, has defined the standard ISO/IEC 11544 (identical with ITU-T Rec. T.82), entitled '*Coded representation of picture and audio information – Progressive bi-level image compression*'. This standard allows lossy-to-lossless progressive transmission of bi-level images, and typically compresses for the lossless case by an approximately 40% rate reduction as compared to the commonly used G.3/G.4 fax standards [HAMPEL ET AL. 1992]. The most important algorithmic elements of JBIG are

- usage of 'line skipping', signaling that subsequent lines that are identical;
- arithmetic coding for all elements;

[39] Note: ITU-T Rec. numbers are not assigned yet for JPEG XT, but the twin text specifications can be expected to be in the T.8xx range.

- usage of different prediction methods, in particular template prediction (Sec. 6.1.1);
- progressive transmission capability, which means that a binary image can first be decoded and reconstructed by a lower resolution; by further bits arriving, the resolution can then gradually be increased.

For progressive transmission, a method is applied which is very similar to pyramid coding (see sec. 2.8.6), providing a lowpass filtered (hence alias reduced) representation at the lower resolution.

Further improvement was made in the definition of the successive standard JBIG–2 (ISO/IEC 14492, ITU-T Rec. T.88) entitled '*Lossy/lossless coding of bi-level images*', which in particular was designed for cooperative processing of documents in internet applications and document transmission over narrow-bandwidth wireless channels. The improvements are based on classification of areas from the image to be treated by specific encoding tools, such as text areas, halftone areas, line graphics areas etc.

Methods of still picture coding need to be also an inherent part of video compression, in particular for coding pictures where correlation over the time axis cannot be utilized, e.g. after a scene change or when dis-occluded areas appear. The collection of intra-picture coding tools of the High Efficiency Video Coding (HEVC) standard (ISO/IEC 23008-2, ITU-T Rec. H.265) has therefore also been defined as a standalone still picture coding profile. This establishes a high-performance still picture coding standard as well, and consequently, its tools for intra prediction and transform coding have extensively been presented in this chapter. An analysis of its compression benefit over other algorithms can be found in [NGUYEN, MARPE 2012].

6.11 Problems

Problem 6.1.

A common distortion measure in image and video coding is the *peak signal-to-noise ratio* (PSNR) (3.7). In case of an 8 bit digital signal as original source, value ranges are 0...255, such that the maximum amplitude is A=255. Let an original signal be given by an image matrix

$$\mathbf{S}=\begin{bmatrix}20 & 17 & 18\\ 15 & 14 & 15\\ 19 & 13 & 14\end{bmatrix}.$$

Reconstruction images are output by two different coding schemes as

$$\tilde{\mathbf{S}}_1 = \begin{bmatrix} 19 & 18 & 17 \\ 16 & 15 & 14 \\ 18 & 14 & 13 \end{bmatrix} \quad ; \quad \tilde{\mathbf{S}}_2 = \begin{bmatrix} 20 & 17 & 18 \\ 15 & 23 & 15 \\ 19 & 13 & 14 \end{bmatrix}.$$

For both reconstruction images, compute the mean absolute difference of samples $|s(\mathbf{n}) - \tilde{s}(\mathbf{n})|$ and the PSNR. Give an interpretation of the result. Which of the distortions would likely be more visible?

Problem 6.2.

Let an image matrix be given, $\mathbf{S} = \begin{bmatrix} 0 & 0 & 0 & 0 & 0 \\ 0 & 1 & 1 & 1 & 0 \\ 0 & 1 & 1 & 1 & 0 \\ 0 & 1 & 1 & 1 & 0 \\ 0 & 0 & 0 & 0 & 0 \end{bmatrix}$.

a) Compute the associated matrix $\hat{\mathbf{S}}$ of the prediction estimate within the area of the dotted rectangle for the cases
i) $\hat{s}(n_1,n_2) = s(n_1-1,n_2)$ *ii)* $\hat{s}(n_1,n_2) = 0.5\,s(n_1-1,n_2) + 0.5\,s(n_1,n_2-1)$
iii) $\hat{s}(n_1,n_2) = s(n_1-1,n_2) + s(n_1,n_2-1) - s(n_1-1,n_2-1)$.

b) Determine the values e in the prediction error matrix $\mathbf{E} = \mathbf{S} - \hat{\mathbf{S}}$ for the 3 cases from part a) of this problem:

c) Compute recursively the reconstruction matrices $\tilde{\mathbf{S}} = \mathbf{E}' + \hat{\mathbf{S}}'$ under the assumption that $\mathbf{E}'=\mathbf{E}$ at all positions except for the value $e'(1,1)=0$ (e.g. caused by transmission failure). In each step, the prediction $\hat{\mathbf{S}}'$ also has to be computed recursively from $\tilde{\mathbf{S}}$. Finally, compute the difference between $\mathbf{S}$ and $\tilde{\mathbf{S}}$. Give an interpretation of the result.

d) The prediction error signal is quantized into 3 reconstruction values $\mathcal{Q}=\{-1/3,0,1/3\}$. Again by recursive processing, compute the matrices of the related prediction $\hat{\mathbf{S}}'$, the prediction error $\mathbf{E} = \mathbf{S} - \hat{\mathbf{S}}$, the quantized prediction error $\tilde{\mathbf{E}}$ and the reconstruction $\tilde{\mathbf{S}}$. Finally, compute the differences $\mathbf{S} - \tilde{\mathbf{S}}$ and $\mathbf{E} - \tilde{\mathbf{E}}$. Interpret the results by comparing against the results from b).
[Hint: Before starting the recursive processing, the predictor memory shall be initialized by zero values at the boundary.]

Problem 6.3.

Computation of a 2D DCT over a block size of 4x4 gives the following matrix of transform coefficients, ordered as in Fig. 6.16a:

$$\mathbf{C} = \begin{bmatrix} 235 & 35 & 15 & 3 \\ -67 & 3 & 5 & -9 \\ -17 & 13 & -7 & 9 \\ 5 & 37 & 2 & 1 \end{bmatrix}$$

a) Perform a quantization by the weighting matrices $\mathbf{Q}_1$ und $\mathbf{Q}_2$ below, which express the quantization step sizes related to the particular coefficients. To compute the quantizer index $i_{\mathbf{u}}$ from coefficient $c_{\mathbf{u}}$ and for reconstruction of the quantized coefficients $\tilde{c}_{\mathbf{u}}$, use nearest-integer rounding,

$$i_{\mathbf{u}} = \left\lfloor \frac{c_{\mathbf{u}}}{q_{\mathbf{u}}} + 0.5 \right\rfloor \; ; \; \tilde{c}_{\mathbf{u}} = i_{\mathbf{u}} q_{\mathbf{u}}$$

$$\mathbf{Q}_1 = \begin{bmatrix} 8 & 8 & 8 & 8 \\ 8 & 8 & 8 & 8 \\ 8 & 8 & 8 & 8 \\ 8 & 8 & 8 & 8 \end{bmatrix} \; ; \; \mathbf{Q}_2 = \begin{bmatrix} 4 & 6 & 8 & 12 \\ 6 & 8 & 12 & 16 \\ 8 & 12 & 16 & 20 \\ 12 & 16 & 20 & 24 \end{bmatrix}$$

b) Run-length coding shall be applied to the zig-zag scanned coefficients (cf. Fig. 6.19a) for both cases of quantization. Determine the run-lengths if coding according to the method in Fig. 5.1b is used.

c) Entropy coding of quantizer indices (without run-length coding) shall be performed using the systematic VLC of Fig 4.12a and the following bit allocation table. Compute the resulting bit rates for both cases of quantization.

	0	±1	±(2..3)	±(4..7)	±(8..15)	±(16..31)	±(32..63)
Number of bits	1	3	5	7	9	11	13

d) Construct the table of a prefix-free code that has the given bit allocation.

e) Compute the distortion expressed by the Peak SNR (3.7), related to a peak amplitude A=255 for both cases of quantization. Which difference in rate could theoretically be expected for encoding of an AR(1) process, assuming that both cases of quantization are still in the linear range of the rate distortion function (4.40)?

Problem 6.4.

Black and white (binary) images shall be compressed. Herein, '*X*' is the current sample, and '*A*'/'*B*' are its left and top neighbors, respectively.

a) Determine the entropy per sample, when the following conditions hold:
i) $\Pr(X = A \mid A = B) = \Pr(X = A \mid A \neq B) = \Pr(X = B \mid A \neq B)$.
ii) $\Pr(X = A \mid A = B) = 3\Pr(X = A \mid A \neq B) = 3\Pr(X = B \mid A \neq B)$.

Now, the image patch of size 4x4 samples shown in Figure 6.37 shall be compressed.

b) Determine the empiric occurrences of black and white samples. Then, determine the entropy under the assumption that these empiric occurrences would match the actual probabilities.

c) Statistical independency of neighbored samples shall now be assumed. Design a Huffman code which combines each two samples into a vector before encoding. Which rate per sample is necessary now?

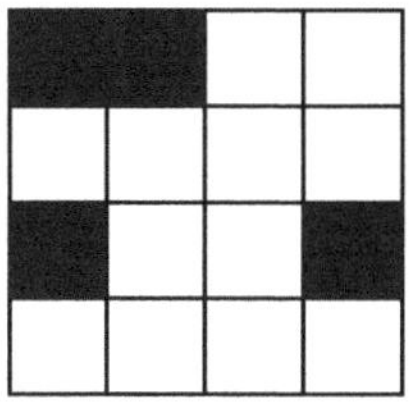

Fig. 6.37. Binary image patch to be compressed

Vectors can be configured either from horizontally-adjacent or from vertically-adjacent samples, as shown in Figure 6.38. In an adaptive entropy coding method, one additional bit is needed to signal which configuration is used. The code table from Figure 6.38 is used for encoding.

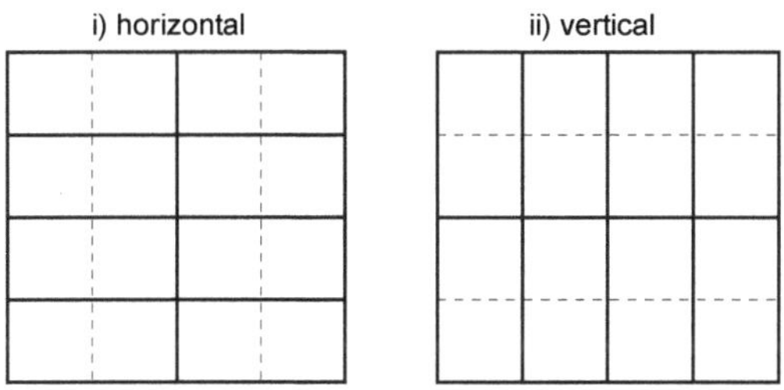

code table

'WW'	0
'BW'	10
'WB'	110
'BB'	111

Fig. 6.38. Combinations of two samples into vectors, and code table

d) Determine the occurrences of the different black/white combinations for both configurations *i)* and *ii)*, when the image from Figure 6.37 is encoded.
e) Determine for both configurations the number of bits needed to compress the image, using the code table. Discuss the advantage of adaptive coding.

Problem 6.5.

An AR(1) process is characterized by parameters $\rho=0.75$ and $\sigma_s^2=16$. For encoding, a rate $R=2$ bit/sample is available.

a) Determine the minimum possible distortion and the maximum coding gain, if correlation properties are exploited properly.
b) For encoding, a Haar transform kernel with $U=2$ is employed. Compute the variances of coefficients c_0 and c_1. Determine the coding gain. Which distortion occurs with a rate $R=2$?
c) The distortion D shall have a variance not higher than 1. Which is the necessary rate, *i)* according to the rate distortion function of an AR(1) process *ii)* if the Haar transform with $U=2$ is used?
d) A 2D separable AR(1) process of same variance with $\rho_1=\rho_2=\rho$ shall be coded. Determine the variances of the coefficients, using the results from b), and the coding gain of the separable transform for $U_1=U_2=2$.

Problem 6.6.

The correlation between two samples of a continuous-amplitude zero-mean signal is expressed by a uniform PDF within the amplitude range shown in Fig. 6.39. Similar to the property of Markov and AR(1) processes, the process shall have a one-sample memory, i.e. the dependency between pairs of subsequent samples completely expresses all dependencies. The amplitude shall be quantized into 8 intervals, expressed by sign bit, MSB and LSB of amplitude.

a) Determine the probability by which two subsequent samples have the same sign.
b) Determine the probability that the i) MSB and ii) LSB of two subsequent samples are identical.
c) Compute the entropy values for the three bits in terms of the entropy of a binary Markov process.

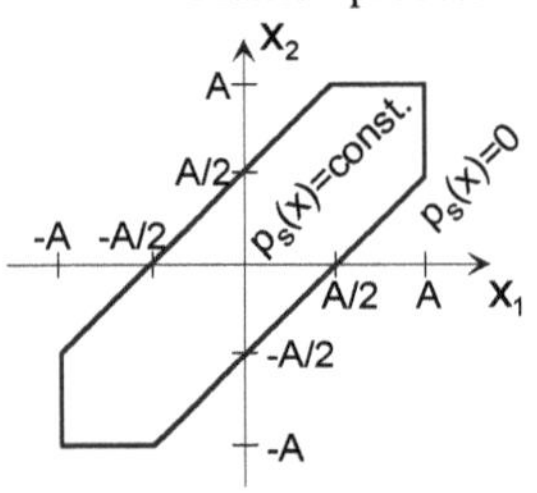

Fig. 6.39. Uniform joint PDF of 2 samples from a correlated signal

7 Inter-picture Coding

Inter-picture coding schemes are operating equivalent to intra-picture coding schemes presented in the previous chapter, as far as coding of 2D picture information is concerned. Similarity between (temporally or spatially) adjacent pictures can now be exploited by inter-picture compression, wherein techniques of displacement compensation play a key role. Firstly, high redundancy between subsequent pictures in a video sequence exists along the motion trajectory; furthermore, coding errors are less visible if they are consistent with the motion of the scene content. This chapter also introduces more simple schemes without motion compensation and shows their limitations. The most widely used video coding scheme combines motion-compensated prediction with spatial transform coding; this is denoted as hybrid video coding. The hybrid scheme is implemented in most of today's video coding standards, and its foundations are investigated in detail. Methods of advanced motion compensation are discussed, scalable and multi-view coding concepts are also introduced. As a possible alternative, motion-compensated subband/wavelet filterbanks are presented, extending transform-based compression into the dimension of the temporal axis of video sequences. The chapter also presents methods for mode and motion information encoding, which is of particular importance as the data rate to be spent for side information can reach a significant amount of the overall rate here.

7.1 Intra-picture and replenishment coding

Any of the intra picture coding methods described in chapter 6 can be used to encode the pictures of a video sequence (or a movie, also denoted as *motion picture sequence*) separately. Such methods are denoted as *intra-picture coding*[1], where no redundancies over time are exploited for compression. Unlike the case of still pictures intended for static display, decoding should be performed fast enough such that no pictures need to be skipped. Furthermore, tim-

[1] or short: Intra.

© Springer-Verlag Berlin Heidelberg 2015

J.-R. Ohm, *Multimedia Signal Coding and Transmission*,
Signals and Communication Technology, DOI 10.1007/978-3-662-46691-9_7

ing/synchronization information for sequential display output is necessary. Intra-only video coding could potentially produce larger visibility of coding errors due to temporal fluctuation (variation of artifacts over time, which is particularly visible if the compression scheme is shift variant and therefore giving different results when the content is moving, which is the case in most transform coding schemes)[2]. Usage of intra-only coding for moving pictures is therefore mostly useful in the range of high rates and high quality. Further, random variations between pictures (due to noise, granular motion) could inhibit straightforward exploitation of inter-picture redundancies, in which case intra-picture coding could be more beneficial. Finally, intra-only coding simplifies random access of arbitrary pictures, does not require additional picture memory, does not propagate transmission errors between pictures and is usually less complex than motion-compensated methods, such that it may be preferable under some application-dependent design choices.

If only small parts of a scene are moving (e.g. objects in front of a static background, which is often the case in videoconferencing, surveillance etc.), referencing previously decoded pictures for prediction can be significantly more efficient than intra-only schemes, even when no motion compensation is employed.

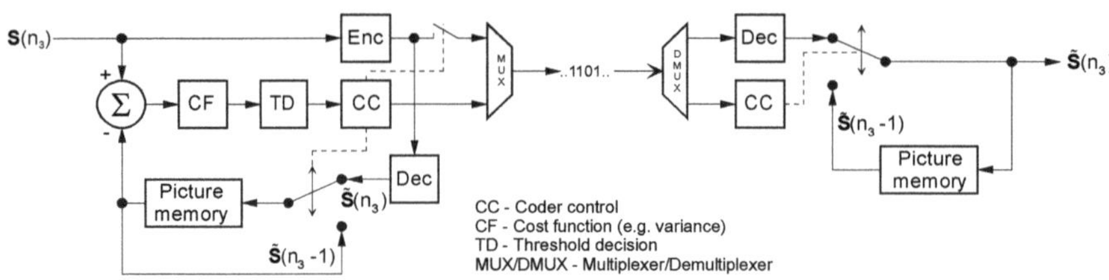

Fig. 7.1. Replenishment coder (left) and decoder (right); switch shown in a position where no copy is made, intra-coding mode is enabled

Replenishment is a scheme which can be combined with various still image coding techniques, provided they allow local on/off switching. The first picture of a sequence is entirely processed in intra coding mode, while for subsequent pictures all areas that show no or little changes compared to the preceding picture are copied from a decoded-picture memory. Areas with more changes (e.g. due to motion) are intra-mode coded without taking reference to a previous picture. The change decision is typically based on a threshold comparison, using cost functions such as absolute or squared difference between the original picture $\mathbf{S}(n_3)$ and a previous decoded picture $\tilde{\mathbf{S}}(n_3 - k_3)$[3]. The percentage of intra-coded areas, and

---

[2] Single pictures could be perceived to provide a good quality, though when displayed as moving video, coding errors become visible due to temporal fluctuation, e.g. when blocking artifacts are occurring at fixed grid positions whereas the picture is moving. In general this problem applies in any coding method that would not be shift invariant, e.g. filterbank transforms as well as block transforms.

[3] It is expressed here that several preceding pictures could be stored as reference.

thereby the major part of the rate can be controlled by varying the threshold value, as trade off versus the distortion which would occur in cases of larger deviation between the pictures. Side information needs to be conveyed to the decoder for signalling the change/no-change decision of each local region (where regions could e.g. be defined by a regular block grid). Schematic diagrams of encoder and decoder are shown in Fig. 7.1.

In case of scenes with larger amount of moving parts, the threshold would likely be surpassed in most areas, and the scheme by tendency acts like intra-picture coding. In the other extreme case of largely unchanged content, replenishment could be interpreted as a primitive inter-picture DPCM, with almost-zero prediction residual; not even a mechanism is foreseen to encode non-zero residual information.

Note: A replenishment scheme is defined in ITU-T rec. H.120 (1984ff.), which was originally designed for video transmission with 2 *Mb/s* compressed rate in video conferencing applications. 2D DPCM is used as intra-picture coding method; replenishment switching is performed within 16x16 block partitions.

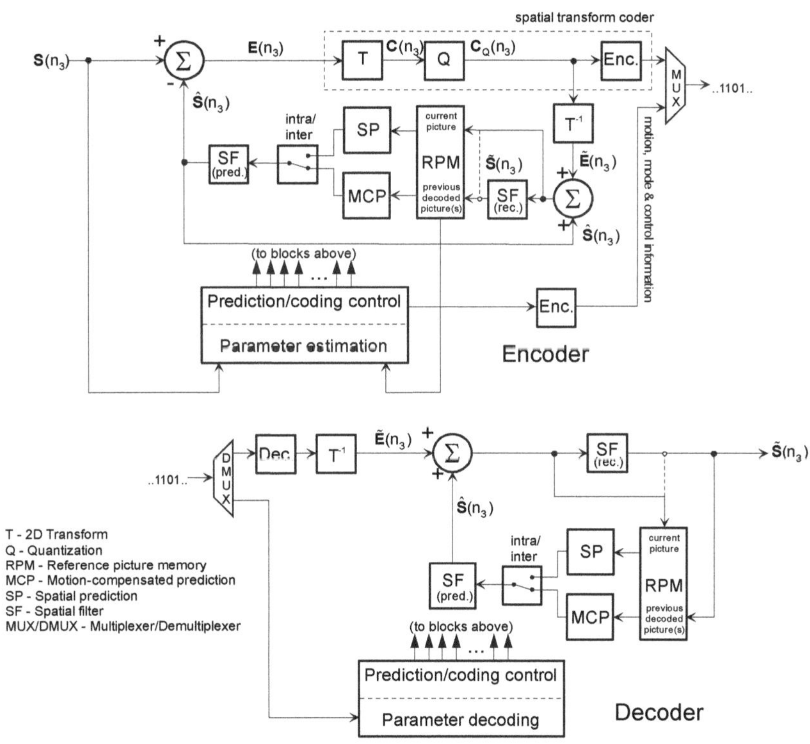

Fig. 7.2. Hybrid (predictive + 2D transform) video coding with motion compensation

7.2 Hybrid video coding

7.2.1 Motion-compensated hybrid coders

The term *hybrid coding* was introduced in multi-dimensional image and video coding to characterize the combination of different encoding principles applied in combination along different dimensions of the signal [HABIBI 1974]. The combination of directional spatial prediction with transform coding (see Sec. 6.3.2) is another 'hybrid' case where however both methods are run combined in *both* spatial dimensions. In video coding, a combination of DPCM (along the temporal axis) with 2D transform coding applied to the resulting residual signal in the spatial dimension is usually denoted as hybrid coding. This has become the basis for most video coding standards existing today. However, elements of predictive coding or DPCM can also be used to further exploit redundancy of the spatial dimensions, either in coding the motion compensated residual itself, or the side information such as motion parameters.

The structure of a hybrid video coder, combining motion-compensated prediction (cf. Sec. 2.6.2) or spatial intra-picture prediction with 2D spatial transform coding of the prediction residual signal, is shown in Fig. 7.2. For the prediction part, this is an adaptive DPCM structure. Usually, forward predictor adaptation is used, where the motion displacement and the spatial prediction parameters have to be estimated at the encoder and conveyed as side information[4]. The steps to be performed at the encoder are as follows:

1. For a given block or region, the best coding mode (e.g. intra or inter-picture coding) and the best parameters for that mode (e.g. displacement vectors, spatial prediction direction, reference picture index) need to be determined (estimated). In that context, the comparison between the current original region and a previously decoded reference region is mostly used. This is available in the decoded picture buffer anyway, would be identically used for reconstruction at the decoder and therefore allows to minimize the actual prediction error in the closed loop system. The associated functional block for parameter estimation and 'prediction/coding control' is typically the most complex part of the encoder.
2. The prediction estimate is computed for the current block or region, using the reference data from the decoded-picture buffer and parameters determined in the first step. Optionally, a spatial filter can be used to improve the prediction,

[4] Backward-adaptive motion-compensated prediction, where identical motion parameters need to be identically derived at the encoder and the decoder from previously decoded samples, has also been studied (initially suggested in [NETRAVALI, ROBBINS 1979/80]. It is usually reported to be worse performing than forward-adaptive schemes and to cause an additional burden in decoder complexity. More recently, it has been shown that encoder-controlled switching between forward and backward adaptation could give some advantage [KAMP, WIEN 2012], see p. 384.

interpolate samples etc. Finally, the prediction is subtracted from the original signal of the current block or region, to obtain the residual.
3. The residual is transformed, the transform coefficients are quantized and the result is (entropy) coded.
4. Reconstruction is performed by inverse mapping (reconstruction) of the transform coefficients, inverse transform and adding the decoded residual with the prediction. This reconstructed signal is needed as reference for subsequent predictions. Before writing it to the decoded-picture buffer, additional spatial filtering can be performed to reduce compression artifacts.

Steps 2 and 4 are likewise performed by the decoder, using the encoded parameters and transform coefficients as received. Adaptive elements typically allow switching of different options in the prediction/decoding loop. This includes[5]:

- Selecting the reference(s) for prediction: This could be either a picture at different time instance (previously decoded) for inter-picture prediction or previously decoded samples from the current picture for intra-picture prediction; in case where no reference is available, prediction should refer to default values which are either fixed or extrapolated from available data, or should be switched off entirely.
- Deciding whether one, two or more references should be combined (in the latter case, by weighted averaging in *multi-hypothesis* prediction, see Sec. 7.2.4).
- Choosing appropriate block or region partitioning, e.g. a variable-size block structure to be applied for prediction, transform, mode switching etc. (see Sec. 7.2.5).
- Defining the operational trade off in the local area w.r.t. quality versus rate (e.g. by selecting the quantizer step size, zeroing of residual blocks etc.). This is typically based on rate-distortion criteria (see Sec. 7.2.7)
- Optimizing parameters for filters in the prediction loop.

Additional spatial filters are shown at two different positions in the prediction loop of encoder and decoder (see also Sec. 6.7 and 7.2.2):

- The post filter has the main purpose to minimize the error between the reconstructed signal and the original. It could be applied outside the loop, but in terms of using the reconstructed picture as reference for subsequent pictures, it is rather beneficial to apply a post filter before storing the reconstructed picture in the (decoded) reference picture memory. As coding distortions are often caused by spatially variant (e.g. block structures) and non-linear (e.g. quantization) effects, non-LSI filters are better suitable for this purpose. Examples are de-blocking filters, de-ringing filters, edge-enhancement sharpening filters, de-banding filters, or noise-reduction filters.
- The prediction signal filter has the main purpose to improve the prediction signal after the core prediction process. For example, the motion compensation output may be inappropriate due to wrong motion estimation or occlu-

[5] Not all of these options are supported in typical realizations of hybrid coders as e.g. defined in video coding standards.

sions at object boundaries, or the result of intra prediction could suffer by irregular sharp edges. Therefore, this filter is often an additional spatial lowpass (smoothing) filter, but a Wiener filter approach (solving the Wiener-Hopf equation for a spatial filter that minimizes the prediction error or reconstruction error) is likewise appropriate.

Note that these filters do not necessarily need to be applied in every hybrid video coding scheme. Their benefit is further discussed in section 7.2.3. The prediction signal filter may be subsumed in the post filter if the motion-compensated prediction process itself does not produce artifacts that should be reduced. Some of the parameters listed above may also be conveyed at the picture level; for example, in case of scene changes, it is reasonable to apply intra-picture coding globally. Periodic insertion of intra-coded pictures can also be enforced for other purposes such as capability of direct access or error resilience.

If only past pictures (in capture/display order sequence) are used as prediction references, the hybrid coding scheme as shown in Fig. 7.2 has low delay capability, and pictures can be displayed instantaneously after decoding. Otherwise, pictures that succeed by display order could also be decoded first and be used as reference to improve the prediction of preceding pictures. This will however be sacrificed by an additional delay, both in coding and decoding (see Sec. 7.2.4). In any case, decoded picture buffer states at encoder and decoder should be identical, when used in the complementary prediction/reconstruction operations.

Due to the locally varying displacement shift and potentially local switching of the reference picture that is used for prediction, picture buffers with random memory access are necessary, and usage of cache memory is limited. Due to the picture dependencies, decoding has always to start with an intra-coded picture, or at least from intra-coded regions. In streaming applications, immediate random access may not be possible until such a case appears, but also in storage applications additional latency may be imposed due to the necessity of additional decoding steps.

7.2.2 Characteristics of inter-picture prediction error signals

In the subsequent analysis, the case of lossless coding is considered for simplicity, i.e. the original picture $\mathbf{S}(n_3)$ and the reconstructed picture $\tilde{\mathbf{S}}(n_3)$ (from which the prediction actually would be performed in closed-loop systems) are assumed to be identical. If a motion shift by k_1 samples horizontally and k_2 samples vertically occurs between picture n_3-k_3 and the current picture n_3, the optimum motion-compensated prediction generates a prediction error picture (ignoring picture boundary effects)[6]

$$r(n_1,n_2,n_3) = s(n_1,n_2,n_3) - s(\underbrace{n_1-k_1,n_2-k_2}_{\mathbf{n}-\mathbf{k}},n_3-k_3)\,. \tag{7.1}$$

[6] In this and following equations, vectors **n**, **k**, etc. are covering spatial coordinates only.

The residual picture $r(\mathbf{n},n_3)$ characterizes all components which cannot be predicted even by perfect motion compensation, such as new parts of a scene or uncorrelated noise. Without motion compensation, i.e. using identical coordinate positions of the other picture as prediction estimate, the prediction error signal would be

$$e(\mathbf{n},n_3)=s(\mathbf{n},n_3)-s(\mathbf{n},n_3-k_3)=r(\mathbf{n},n_3)+s(\mathbf{n}-\mathbf{k},n_3-k_3)-s(\mathbf{n},n_3-k_3). \quad (7.2)$$

To analyze the coding gain achievable by motion-compensated prediction, 2D power spectra of original pictures and of the prediction error pictures (both with and without motion compensation) shall now be compared. In hybrid coding, these are subject to subsequent spatial transform coding, such that their 2D spectral characteristics influence the overall effectiveness. For the case of 'ideal' motion compensation assumed in (7.1), the spatial prediction error power spectrum is the 2D spectrum of any unpredictable components for picture n_3[7]

$$\phi_{ee,\mathrm{mc}}(\mathbf{f},n_3)=\phi_{rr}(\mathbf{f},n_3). \quad (7.3)$$

If *no* motion compensation would be applied, the prediction error spectrum can be determined by Fourier transform of (7.2),

$$E(\mathbf{f},n_3)=S(\mathbf{f},n_3-k_3)\mathrm{e}^{-\mathrm{j}2\pi\mathbf{f}^{\mathrm{T}}\mathbf{k}}-S(\mathbf{f},n_3-k_3)+R(\mathbf{f},n_3)\,, \quad (7.4)$$

which can be re-formulated as

$$\begin{aligned}E(\mathbf{f},n_3)&=S(\mathbf{f},n_3-k_3)\left(\mathrm{e}^{-\mathrm{j}2\pi\mathbf{f}^{\mathrm{T}}\mathbf{k}}-1\right)+R(\mathbf{f},n_3)\\&=S(\mathbf{f},n_3-k_3)(-2\mathrm{j})\mathrm{e}^{-\mathrm{j}\pi\mathbf{f}^{\mathrm{T}}\mathbf{k}}\sin\left(\pi\mathbf{f}^{\mathrm{T}}\mathbf{k}\right)+R(\mathbf{f},n_3).\end{aligned} \quad (7.5)$$

The power density spectrum is invariant against the phase shift. With a constant (linear) motion shift, it can be regarded as approximately stationary over time, and asymptotically identical for all pictures, such that[8]

$$\begin{aligned}\phi_{ee,\text{no-mc}}(\mathbf{f})&=\mathcal{E}\left\{\left|E(\mathbf{f},n_3)\right|^2\right\}=4\,\phi_{ss}(\mathrm{f})\sin^2\left(\pi\mathbf{k}^{\mathrm{T}}\mathbf{f}\right)+\phi_{rr}(\mathbf{f})\\&=2\phi_{ss}(\mathbf{f})\left[1-\cos\left(2\pi\mathbf{k}^{\mathrm{T}}\mathbf{f}\right)\right]+\phi_{rr}(\mathbf{f}).\end{aligned} \quad (7.6)$$

---

[7] This statistical modeling analysis assumes a power density spectrum as asymptotically true expectation about the residual. In case of content which is unchanged except for motion shift, the residual would just contain new areas, which would still have a similar spectrum as original pictures (but would only appear locally). Practically, the motion compensated prediction error also includes noise components (which are temporally uncorrelated) and components where motion compensation fails. In principle, both correlated and uncorrelated components will be present in $e(\mathbf{n})$, but the spatial correlation is generally lower than in original images. For an investigation on these properties, refer to [SHISHIKUI 1992].

[8] Herein, the statistical independency of the residual at time n_3 from the picture at n_3-k_3 allows to abandon cross power spectra.

This prediction error spectrum depends on the spectrum of the image signal, such that the prediction error picture is correlated with the original picture in the case where no motion compensation is performed (except for the case of zero motion). The characteristics of this spectral component depend on the actual motion shift. $\phi_{rr}(\mathbf{f})$ still contains any components which cannot be predicted from the previous picture. It is observed that certain power spectrum values of the uncompensated prediction error can be larger by a factor of four, as compared to the original signal's power spectrum. This means that temporal prediction without motion compensation could perform *worse than intra-picture coding*, depending on the characteristics of the signal and the strength of motion[9].

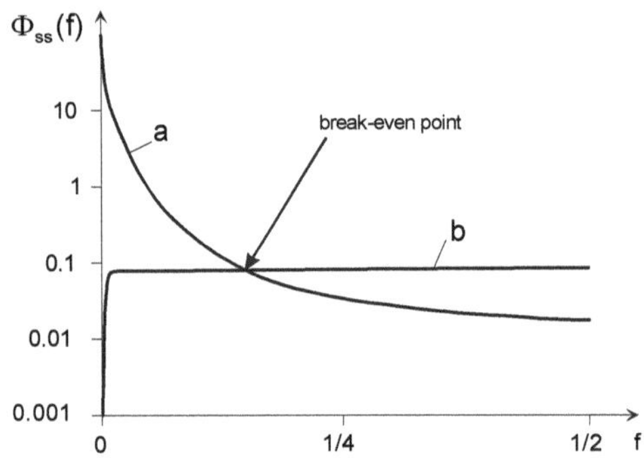

Fig. 7.3. 1D power density spectrum of an AR(1) model, ρ=0.96 (a) and related inter-picture prediction error power spectra (without motion compensation, without considering component ϕ_{rr}) for the case of motion shift k=1 (b)

As an example to illustrate this phenomenon, Fig. 7.3 shows a 1D power spectrum of an AR(1) model, and the prediction error spectrum according to (7.6) for shift k=1. If the deviation from the true motion is significant, the usage of inter-picture prediction is advantageous only for very low spatial frequencies. For the case k=1, and typical correlation with ρ=0.96 in Fig. 7.3, the break-even point is roughly at one third of the maximum spatial frequency. Inter-picture prediction of any higher-frequency components would deteriorate the coding performance. The same effect applies if motion compensation is using an erroneously estimated motion shift. More precise motion compensation increases the 'break-even point'.

In motion estimation and compensation, a deviation from the optimum often occurs. Subsequently, the displacement deviation error shall be denoted as $\mathbf{k}_\varepsilon=[k_{\varepsilon,1}, k_{\varepsilon,2}]^T$. The characteristics of the motion-compensated prediction error

[9] In worst case, a sinusoidal signal could e.g. move by half a period, such that the prediction error would double the amplitude, and have quadruple spectral power. If two subsequent pictures contain statistically independent components (e.g. camera noise or new content), the power spectra and variances would add up in the residual signal, such that the spectral power is doubled.

signal are then similar to the uncompensated case, but $\mathbf{k}_\varepsilon$ could be expected to be relatively small compared to the actual motion. In analogy with (7.6),

$$\phi_{ee,\Delta-\mathrm{mc}}(\mathbf{f}) = 2\phi_{ss}(\mathbf{f})\left[1-\cos\left(2\pi\mathbf{k}_\varepsilon^{\mathrm{T}}\mathbf{f}\right)\right]+\phi_{rr}(\mathbf{f}). \tag{7.7}$$

In practice, the deviation between estimation and actual motion would not be constant as in (7.7). Moreover, a superposition of different deviation related components occurs in the spectrum. The contributions can then be determined by the PDF of the motion deviation $p_{\mathbf{k}_\varepsilon}(\boldsymbol{\alpha})$ [GIROD 1987]. This can be expressed via the expected value with regard to $\mathbf{k}_\varepsilon$,

$$\begin{aligned}\phi_{ee,\Delta-\mathrm{mc}}(\mathbf{f}) &= 2\phi_{ss}(\mathbf{f})\left[1-\mathcal{E}_{\mathbf{k}_\varepsilon}\left\{\cos\left(2\pi\mathbf{k}_\varepsilon^{\mathrm{T}}\mathbf{f}\right)\right\}\right]+\phi_{rr}(\mathbf{f}) \\ &= 2\phi_{ss}(\mathbf{f})\left[1-\iint_{\boldsymbol{\alpha}} p_{\mathbf{k}_\varepsilon}(\boldsymbol{\alpha})\cos\left(2\pi\boldsymbol{\alpha}^{\mathrm{T}}\mathbf{f}\right)\mathrm{d}^2\boldsymbol{\alpha}\right]+\phi_{rr}(\mathbf{f}),\end{aligned} \tag{7.8}$$

or by interpreting the integral over α as real part of a 2D Fourier transform[10],

$$\phi_{ee,\Delta-\mathrm{mc}}(\mathbf{f}) = 2\phi_{ss}(\mathbf{f})\left[1-\mathrm{Re}\left\{\mathcal{F}\left\{p_{\mathbf{k}_\varepsilon}(\boldsymbol{\alpha})\right\}\right\}\right]+\phi_{rr}(\mathbf{f}). \tag{7.9}$$

7.2.3 Quantization error feedback and error propagation

The effects of quantization error feedback and propagation of transmission errors, as analyzed in the general context of DPCM (see Sec. 5.2.1 and 6.3.4), are likewise applying in case of motion-compensated inter-picture prediction. Quantization error feedback is expressed by the component $Q(\mathbf{f})H(\mathbf{f})$ from (5.14), where $H(\mathbf{f})$ is determined from (2.230) for the case of motion-compensated prediction. As the 2D power spectrum is invariant against motion shift of the signal at least in the case of translational motion, the 2D quantization error spectrum resulting by encoding of the previous picture can be modeled by the argument of the function in (4.38),

$$\phi_{qq}(\mathbf{f}) = \min\left[\Theta,\phi_{ss}(\mathbf{f})\right]. \tag{7.10}$$

This spectrum is superimposed to the spectrum of the motion-compensated prediction error (7.9). For the case of lossy coding with higher distortion, the quantization error spectrum becomes identical to the spectrum of the signal for any frequency components that fall below the threshold Θ (usually high frequencies with reference to the AR(1) model, Sec. 4.2.2). As a consequence, if those frequencies were eliminated in a reference picture due to lossy coding, they can never be used in prediction and will again fully appear in the motion-compensated prediction

---

[10] The PDF of deviations between estimated and actual motion is often assumed to follow a 2D independent Gaussian distribution [GIROD 1987/1993], its Fourier transform then has a real-valued 2D Gaussian shape as well.

error of the next picture[11]. Finally, the application of a spatial loop filter $A(z_1,z_2)$ from (2.230) has to be considered as part of the predictor filter. This includes noise reduction and/or interpolation filters[12]. Then, the prediction error signal for the case of constant deviation from the true motion is

$$E(\mathbf{f}) = S(\mathbf{f})\cdot(1 - A(\mathbf{f})\mathrm{e}^{j2\pi\mathbf{k}_\varepsilon^{\mathrm{T}}\mathbf{f}}), \tag{7.11}$$

such that with constant $\mathbf{k}_\varepsilon$,

$$\begin{aligned}\phi_{ee}(\mathbf{f}) = \mathcal{E}\left\{|E(\mathbf{f})|^2\right\} = \phi_{ss}(\mathbf{f})\Big[&\left[1 - \mathrm{Re}\{A(\mathbf{f})\}\cdot\cos(2\pi\mathbf{k}_\varepsilon^{\mathrm{T}}\mathbf{f}) + \mathrm{Im}\{A(\mathbf{f})\}\cdot\sin(2\pi\mathbf{k}_\varepsilon^{\mathrm{T}}\mathbf{f})\right]^2 \\ &+\left[\mathrm{Re}\{A(\mathbf{f})\}\cdot\sin(2\pi\mathbf{k}_\varepsilon^{\mathrm{T}}\mathbf{f}) + \mathrm{Im}\{A(\mathbf{f})\}\cdot\cos(2\pi\mathbf{k}_\varepsilon^{\mathrm{T}}\mathbf{f})\right]^2\Big] \\ = \phi_{ss}(\mathbf{f})\left[1 + |A(\mathbf{f})|^2 - 2\,\mathrm{Re}\{A(\mathbf{f})\}\cdot\cos(2\pi\mathbf{k}_\varepsilon^{\mathrm{T}}\mathbf{f}) + 2\,\mathrm{Im}\{A(\mathbf{f})\}\cdot\sin(2\pi\mathbf{k}_\varepsilon^{\mathrm{T}}\mathbf{f})\right]&.\end{aligned} \tag{7.12}$$

Taking the same approach as in (7.9), based on the PDF of the MC mismatch, gives

$$\phi_{ee}(\mathbf{f}) = \phi_{ss}(\mathbf{f})\left[1 + |A(\mathbf{f})|^2 - 2\,\mathrm{Re}\left\{A^*(\mathbf{f})\mathfrak{F}\{p_{\mathbf{k}_\varepsilon}(\boldsymbol{\alpha})\}\right\}\right]. \tag{7.13}$$

The overall power spectrum of the motion-compensated prediction error is then

$$\phi_{ee}(\mathbf{f}) = \underbrace{\phi_{ss}(\mathbf{f})\left[1 + |A(\mathbf{f})|^2 - 2\,\mathrm{Re}\left\{A^*(\mathbf{f})\mathfrak{F}\{p_{\mathbf{k}_\varepsilon}(\boldsymbol{\alpha})\}\right\}\right]}_{\mathrm{I}} + \underbrace{\min\left[\Theta, \phi_{ss}(\mathbf{f})\right]|A(\mathbf{f})|^2}_{\mathrm{II}} + \underbrace{\phi_{rr}(\mathbf{f})}_{\mathrm{III}}. \tag{7.14}$$

The main factors producing non-zero components in the prediction error signal are errors in motion compensation (I: due to erroneous motion vectors or insufficient sub-sample interpolation), quantization distortion of the reference picture (II) and unpredictable content (III: noise, uncovered areas). For the case of stationary processes, the in-loop filter $A(\mathbf{f})$ could be designed for global minimization of prediction error variance. One intuitive approach would be frequency-selective switching between intra-picture and inter-picture coding,

$$|A(\mathbf{f})| = \begin{cases} 0 & \text{if} \quad \phi_{ss}(\mathbf{f}) < \phi_{ee}(\mathbf{f}) \\ 1 & \text{if} \quad \phi_{ss}(\mathbf{f}) \geq \phi_{ee}(\mathbf{f}). \end{cases} \tag{7.15}$$

As this would require an ideal filter, eventually with several pass and stop bands, it cannot practically be realized. Furthermore, moving pictures are not stationary, such that a globally optimized filter would not be optimum, and the error $\mathbf{k}_\varepsilon$ typi-

---

[11] In practical lossy coding, this effect may even be worse, as the spectrum $\phi_{qq}(\mathbf{f})$ also includes non-linear and spatially variant coding artifacts (e.g. blocking, ringing) superimposed in $e(\mathbf{n})$.

[12] Different filters may be operated in the loop for these purposes. In particular to reduce coding noise/artifacts, non-linear and shift variant filters can be beneficial, whereas for simplicity only linear filters are considered here. Still, for a non-linear filter the optimization goals would be identical: Reducing the prediction error, and improving the reconstruction.

cally also varies with the picture content. Therefore, minimizing the variance of the residual signal requires filter adaptation. Linear optimization methods can be based on a Wiener filter design, minimizing the prediction error variance for an ensemble of samples under the given motion parameter and reference picture constraints [GIROD 1993]. More practical methods of filter designs for the purpose of loop filtering were presented in [PANG, TAN 1994] [WEDI, MUSMANN 2001] [CHEN ET AL. 2012].

Wiener-based optimization principles for adaptive loop filters are equivalent to the post filter approach explained in Sec. 6.9. The optimization target can be improved prediction or reconstruction, depending on the position of the filter (before input or after output of the motion compensation stage, cf. Fig. 7.2). In the first case, motion estimation and a first pass of motion compensation needs to be performed before the filter parameters can be determined, whereas the residual error computation would need to include the filtering; due to the mutual influence, an optimization should be performed iteratively, which adds additional encoder complexity. Furthermore, in this case it may be useful to subsume the interpolation filter for sub-sample motion compensation, which additionally means that the adaptive filter needs to be defined for various sub-sample phase positions.

If the motion estimation is expected to be systematically inaccurate (e.g. no sub-sample accuracy), simple switchable lowpass filters can be operated in the loop to achieve a similar effect. ITU-T Rec. H.261 defines a separable binomial loop filter for this purpose. It should be observed that interpolation filters for sub-sample accurate motion compensation, as used in newer hybrid video coding methods, have a lowpass effect as well, which becomes less of an advantage when the motion compensation is more accurate, such that high precision of motion compensation also requires high quality of interpolation (cut-off by lowpass close to half sampling rate).

In case of more precise motion compensation, the feedback of coding errors (component II in (7.14)) becomes more prevalent, in particular at low data rates. As coding errors are caused by non-linear and spatially/temporally variant operations, usage of non-linear filters can often better fulfill the given purpose, even though it is not possible to characterize their behaviour by a frequency-domain description. An example for this are deblocking filters (see Sec. 6.9).

The motion estimation and subsequent compensation are highly accurate, i.e. $\mathcal{E}\{\mathbf{k}_\varepsilon\}\rightarrow\mathbf{0}$, and if coding errors are low, a strong loop filter could turn out to be disadvantageous. Following (7.13), this leads to a power spectrum of the prediction error signal

$$\phi_{ee}(\mathbf{f}) = \phi_{ss}(\mathbf{f})\Big[1-\left|A(\mathbf{f})\right|^2\Big], \tag{7.16}$$

which would mean that a lowpass filter with strong roll-off characteristics (including interpolation filters with overly strong lowpass characteristics) could discard a considerable part of the high frequency components in the prediction, which would then appear in the prediction error signal and would need to be newly coded, even if they had already been available in the decoded reference picture. This shows the need for high-quality sub-sample interpolation filters (when MC precision is high), which consequently have been defined in the most recent generations of video coding standards.

If *channel losses* occur, the power spectrum of the reconstructed signal is distorted by a component $K(\mathbf{f})/(1-H(\mathbf{f}))$ according to (5.16). Observe that $H(\mathbf{f})$ now consists of a motion-dependent shift and a spatial filter $A(\mathbf{f})$ in the prediction loop. The loss is propagated until intra-only coding is performed for some picture, or the motion compensation does not use the affected area for reference any more. The spatial filter may have some smoothing effect which increases over time as it is iteratively applied[13]. As motion compensation is typically shift variant (different motion vectors used depending on location), the original picture structure can become severely geometrically distorted; this effect is also increasing over time. Fig. 7.4 shows an example of an error occurring (a) and the evolution after a time span of 10 subsequent MC predictions.

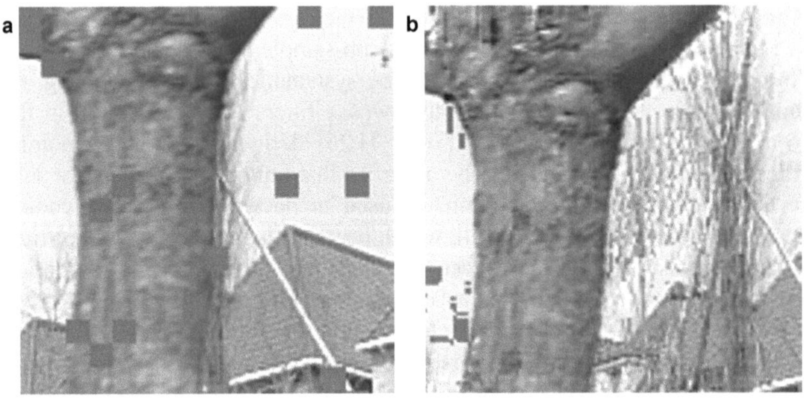

Fig. 7.4. Error propagation in case of motion-compensated prediction
a Time position where a block loss occurs **b** Propagation after 10 pictures

Here (as well as in the example in Fig. 7.4), it is assumed that motion vectors are decoded without loss. As the motion vectors influence the most relevant characteristics of the predictor filter's impulse response, wrong motion vectors would cause even more unnatural and shift variant geometric distortions in the reconstruction. The same applies to other prediction control parameters such as reference picture indices. The effect of visibility after a loss is clearly dependent on the importance of the lost component with regard to subsequent predictions. The inherent problem is again the misalignment of prediction between encoder and decoder after an error, i.e. the *drift problem* in predictive coding (Sec. 5.2.1. and

[13] In this context, stronger lowpass filtering in the loop could be beneficial for less visibility of error propagation, but it is less efficient in terms of compression performance (as discussed above).

6.3.4). Systematic provisions, e.g. stronger error protection for more important information parts can prevent the occurrence of overly large reconstruction error in case of losses. To guarantee recovery from errors, systematic insertion of intra-coded pictures (I pictures), refresh of smaller areas or other mechanisms which reset the prediction loop into a controllable state should be used (see Sec. 7.5.1 and 5.4). Furthermore, error control and concealment mechanisms (see Sec. 9.3.2) should be employed.

7.2.4 Reference pictures in motion-compensated prediction

In the context of motion estimation (Sec. 2.6.2), the definitions of *current picture* and *reference picture* were introduced. In hybrid video coding as used in current standards, the coordinate system of the motion vector field usually refers to the current (encoded/decoded) picture, and the best match is searched as a displaced area in one or several reference picture(s). In the special case of block matching, a uniform or variable-size block grid is defined over the current picture. In closed loop prediction, the reference picture must have been decoded prior to the current picture, which does not necessarily mean that it has to be a previous picture regarding capturing and display order. However, the usage of a future captured/displayed picture as reference introduces delay and the need for additional memory, both at the encoder and decoder.

The overall number of reference pictures is limited by the memory size that is available for storage at the encoder and decoder, by the need for signalling the choice and to some extent by the processing power of the encoder which has to perform the selection. Due to such limitations, early compression standards restricted the choice to only one picture (e.g. the immediately preceding picture in H.261) or two pictures (e.g. one previous and one future picture by display order in MPEG 1 and MPEG-2), see Sec. 7.8. The first scheme is often referred as *P prediction* (previous), the latter scheme as *B prediction* (bi-directional). In B prediction [PURI ET AL. 1990], the two possible motion compensation directions are denoted as *backward* and *forward oriented.* In MPEG-1 and MPEG-2, these schemes are restricted in the sense that a P picture can only be predicted from the previous I|P picture in capture/display order; a B picture can only use the previous I|P for backward oriented and the following I|P for forward oriented prediction. This requires re-ordering the pictures for coding/decoding, but only within the range of the chosen maximum distance between I|P pictures (see discussion below in context of Fig. 7.6).

More generic schemes as implemented in the AVC and HEVC standards more radically decouple coding and display order, using list indexing for signalling the prediction reference pictures. This is advantageous in terms of compression benefit [FLIERL, GIROD 2003B]. In case of inter-picture prediction, the following choices are given (where the list of references includes pictures stored in the decoded picture buffer):

- Uni(-modal) prediction: Only one reference picture is used. This is traditionally still referred to as P prediction, even though it is more general than the

H.261/MPEG-1/MPEG-2 scheme in that the reference can be a past or future picture in capture/display order.

- Bi(-modal) prediction: Two reference pictures are used, the prediction samples are averaged (weighting factors 0.5). This is still referred to as B prediction traditionally, even though being more general than the bi-directional MPEG-1/-2 scheme as it is also possible that both references come from past or from future in capture/display order. Furthermore, B pictures can also be used as reference pictures for other B or P pictures, when preceding in decoding order.
- Weighted prediction: Similar as B prediction, but with weighted averages (individual weighting factors $\neq 0.5$ for the two references).

The more generic case, where individual weights may be applied for the various reference pictures, is referred to as *multi-hypothesis prediction*, further discussed below. In any of the current video compression standards, usually no more than two different reference pictures can be used for prediction at the same time, which is intended to limit signalling overhead as well as memory accesses.

As stated above, motion vectors and computation of displaced picture differences are typically aligned with the current picture's coordinate system. In this case, it is irrelevant whether samples from a reference picture are used multiple times for prediction or whether they are not used at all; in any case, a unique prediction exists for every sample of the current picture. Discontinuities in the displacement vectors at spatially adjacent positions often indicate the presence of occlusions or new content, where the prediction would naturally fail. Fig. 7.5a shows for the case of the coordinate system referring to the current picture, that certain areas from the reference picture may not be used for prediction, and other areas may have double (or even multiple) reference usage. Whereas the first case may indicate that the area is occluded in the current picture, at least some of the double references should be ambiguous and may indicate newly uncovered areas that cannot be predicted. It should however be observed that unreliable motion estimation is more probable in case of occlusions, which can be detected by large variance of the residual and resolved by switching to intra-picture coding.

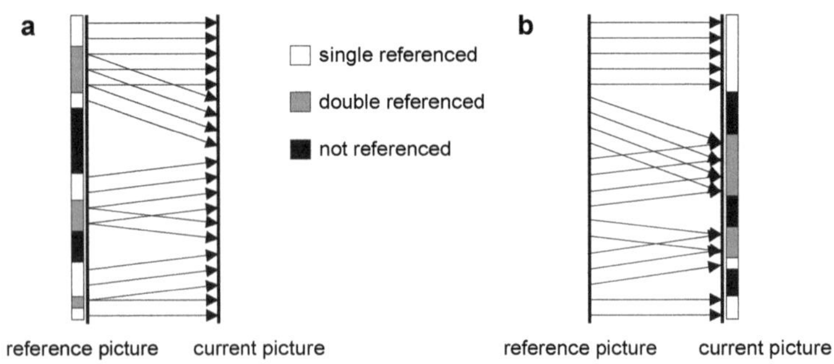

Fig. 7.5. Motion compensation using current picture (**a**) and reference picture (**b**) for the coordinate system

Using the coordinate system of the reference picture as anchor for the displacement vector field is possible as well (but not used in current video compression standards). The prediction $\hat{\mathbf{S}}$ of the current picture then has to be generated by *projection-based motion compensation* from the reference picture, as illustrated in Fig. 7.5b. In this case, *holes* without inter-picture prediction reference may occur in the current picture. These could typically relate to dis-occluded areas, and be subject to intra-picture coding, where spatial prediction could be based on 'hole filling', i.e. interpolating or extrapolating from the hole's boundaries. In case of double references, prediction is ambiguous, such that additional signalling would be required to indicate the correct projection (if any). Projection-based motion compensation usually would require additional memory for storing the motion data, as the ordered sequence of displacement vectors does not concur with the ordered sequence of the current picture[14]. Though having some advantages in identifying occlusions and newly uncovered areas, the disadvantages mentioned here have prevented wider usage of projection-based motion compensation in video compression so far. It is however a widely used method in motion-compensated synthesis from reference picture(s), e.g. in frame rate upsampling and in depth-based view rendering, where the displacements are expressing disparity shifts between different camera views.

To analyze the coding gain which can be achieved by *B prediction*, we consider a case where the current picture is predicted from the two closest (previous and subsequent on time axis) reference pictures using motion vectors $\mathbf{k}_{-1}$ and $\mathbf{k}_{+1}$, respectively, averaging the predictions[15]:

$$\hat{s}_{\mathrm{b}}(\mathbf{n},n_3)=0.5\left[s(\mathbf{n}+\mathbf{k}_{-1},n_3-1)+s(\mathbf{n}+\mathbf{k}_{+1},n_3+1)\right]. \tag{7.17}$$

Both partial motion compensations will likely not be perfect. Assuming that deviations $\mathbf{k}_{\varepsilon,-1}$ and $\mathbf{k}_{\varepsilon,+1}$ from the true displacement occur during estimation, the motion-compensated residual is

$$\begin{aligned} e_{\mathrm{b}}(\mathbf{n},n_3)&=s(\mathbf{n},n_3)-0.5\left[s(\mathbf{n}+\mathbf{k}_{-1},n_3-1)+s(\mathbf{n}+\mathbf{k}_{+1},n_3+1)\right] \\ &=s(\mathbf{n},n_3)-0.5\left[s(\mathbf{n}+\mathbf{k}_{\varepsilon,-1},n_3)+s(\mathbf{n}+\mathbf{k}_{\varepsilon,+1},n_3)\right]+r(\mathbf{n},n_3), \end{aligned} \tag{7.18}$$

where $r(\mathbf{n},n_3)$ is the component that cannot be predicted even with perfect motion compensation. In analogy with (7.6) and (7.7), the prediction error 2D spectrum is

---

[14] This can be avoided when the *collocated* vector, i.e. from identical coordinate positions in reference picture and current picture is used, which may however be wrong particularly close to discontinuities and in case of large displacements.

[15] The reason for improvements through more general weighted, bi-predictive and multi-hypothesis schemes is similar. The specific case considered here simplifies, as almost identical statistics of the two combined prediction errors, of error in motion estimation etc. can be assumed.

$$\begin{aligned} E_{\mathrm{b}}(\mathbf{f}) &= S(\mathbf{f})\left[1-0.5\,\mathrm{e}^{-\mathrm{j}2\pi\mathbf{f}^{\mathrm{T}}\mathbf{k}_{-1}}-0.5\,\mathrm{e}^{-\mathrm{j}2\pi\mathbf{f}^{\mathrm{T}}\mathbf{k}_{+1}}\right]+R(\mathbf{f}) \\ &= S(\mathbf{f})\left[\mathrm{j}\mathrm{e}^{-\mathrm{j}\pi\mathbf{f}^{\mathrm{T}}\mathbf{k}_{\varepsilon,-1}}\sin\left(\pi\mathbf{f}^{\mathrm{T}}\mathbf{k}_{\varepsilon,-1}\right)+\mathrm{j}\mathrm{e}^{-\mathrm{j}\pi\mathbf{f}^{\mathrm{T}}\mathbf{k}_{\varepsilon,+1}}\sin\left(\pi\mathbf{f}^{\mathrm{T}}\mathbf{k}_{\varepsilon,+1}\right)\right]+R(\mathbf{f}). \end{aligned} \tag{7.19}$$

The resulting power spectrum is in case of statistical independence of the two compensation deviations (see problem 7.5)

$$\phi_{e_{\mathrm{b}}e_{\mathrm{b}}}(\mathbf{f}) = \mathcal{E}\left\{\left|E_{\mathrm{b}}(\mathbf{f})\right|^2\right\} = \phi_{ss}(\mathbf{f})\left[1-\frac{1}{2}\cos\left(2\pi\mathbf{f}^{\mathrm{T}}\mathbf{k}_{\varepsilon,+1}\right)-\frac{1}{2}\cos\left(2\pi\mathbf{f}^{\mathrm{T}}\mathbf{k}_{\varepsilon,-1}\right)\right]+\phi_{rr}(\mathbf{f}). \tag{7.20}$$

If $\mathbf{k}_{\varepsilon,-1} = \mathbf{k}_{\varepsilon,+1}$, the result will be identical with the power density spectrum for inaccurate motion compensation in uni-directional prediction (7.5). Otherwise, for statistically independent processes of the two motion compensation deviations, a similar consideration as (7.7)-(7.9) gives

$$\phi_{e_{\mathrm{b}}e_{\mathrm{b}}}(\mathbf{f}) = \phi_{ss}(\mathbf{f})\left[1-\frac{1}{2}\mathrm{Re}\left\{\mathfrak{F}\left\{p_{\mathbf{k}_{\varepsilon,-1}}(\boldsymbol{\alpha})\right\}\right\}-\frac{1}{2}\mathrm{Re}\left\{\mathfrak{F}\left\{p_{\mathbf{k}_{\varepsilon,+1}}(\boldsymbol{\alpha})\right\}\right\}\right]+\phi_{rr}(\mathbf{f}), \tag{7.21}$$

or assuming identity of the PDFs

$$\phi_{e_{\mathrm{b}}e_{\mathrm{b}}}(\mathbf{f}) = \phi_{ss}(\mathbf{f})\left[1-\mathrm{Re}\left\{\mathfrak{F}\left\{p_{\mathbf{k}_{\varepsilon}}(\boldsymbol{\alpha})\right\}\right\}\right]+\phi_{rr}(\mathbf{f}). \tag{7.22}$$

Compared to (7.9), the spectral power of the residual is only half. The actual gain through B prediction however depends on the spectral properties of the pictures and the PDF of the motion compensation errors (i.e. the errors caused by erroneous motion estimation). The result presented here would be strictly valid for the case of independent motion estimation w.r.t. the two reference pictures. Note that in case of joint estimation and/or joint coding of motion parameters, the two estimation errors could become dependent and hypothetically even compensate each other in further reducing the variance of the residual (see related considerations in problem 7.5).

Bi-directional motion-compensated prediction as implemented in the MPEG-1 and MPEG-2 standards is illustrated in Fig. 7.6a. To guarantee a causal processing sequence at the decoder, it is necessary to first process (encode/decode) P pictures which will be used for uni- or bi-directional prediction of other pictures (P pictures are shaded in Fig. 7.6a). The sequence of P pictures can be interpreted as being a video of lower frame rate, and P prediction is restricted to a time-consecutive sequential processing here[16]. Intermediate B pictures can be predicted from each one preceding and one subsequent decoded I|P reference picture. In the example shown, a causal sequence of picture encoder processing is ...1-...-4-2-3-7-5-6-... This imposes an encoding delay, as the B pictures have to be stored until both adjacent reference are available in decoding order. Additional picture memory is also necessary at the decoder side, as both reference pictures have to be

[16] Occasionally, an I picture can also appear in that sequence (e.g. for random access or recovery from errors), which then can also be used to predict the subsequent P picture, or adjacent B pictures.

available for the prediction of the B pictures in between, and those have also to be stored until they can be output in correct temporal display order[17].

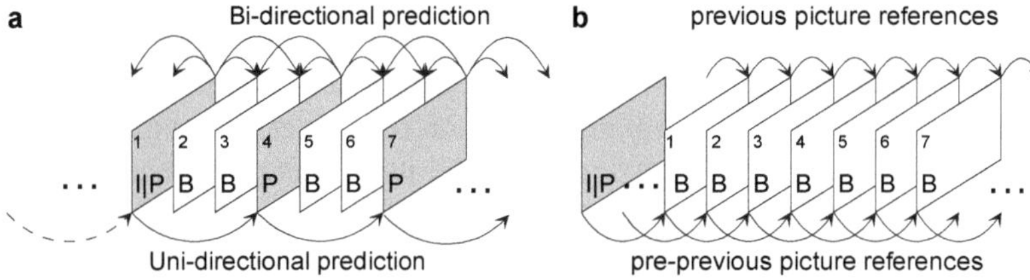

Fig. 7.6. Causal processing sequences **a** by combination of uni-directional and bi-directional prediction **b** by generalized bi-prediction with two references from previous pictures

In case of occlusions, an area to be predicted for the current picture may only be visible in one of the two pictures used for prediction. Therefore, switchable modes should be available to use either of them, or the average. This may further decrease the percentage of unpredictable areas[18] which appear as component ϕ_{rr} in (7.20). Therefore, need for intra-picture coding is usually lower in B pictures.

With more general bi-prediction concepts as they are defined by the AVC/H.264 and HEVC standards, it is also possible that *two preceding* or *two subsequent* pictures are simultaneously used as references, and B pictures can also be used as reference for subsequently decoded pictures. The advantages in terms of improving the prediction are similar, however, if two preceding pictures are used, no additional delay is introduced (Fig. 7.6b). The underlying principle of choosing prediction from multiple reference pictures is shown in Fig. 7.7. In addition to the motion vectors, the reference picture indices (k_3) have to be encoded as side information.

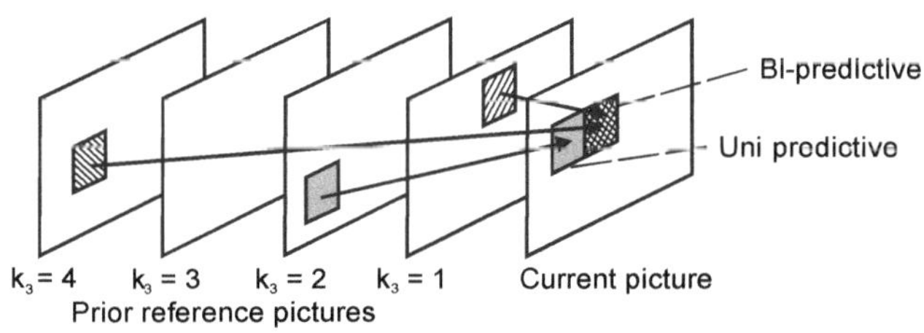

Fig. 7.7. Multi reference picture motion-compensated prediction

[17] In principle, B pictures in Fig. 7.6a are not used as prediction reference by the decoder, therefore do not need to be stored in the reference picture buffer and could be output as soon as they are decoded. Additional picture storage at the decoder output may however be necessary, when decoding and output are not fully synchronized.

[18] This may be the more important component in the case where high-quality motion estimation and compensation are performed, such that motion vector inaccuracies are less severe.

With more flexible reference picture indexing, usage of *hierarchical picture prediction structures* also is possible. Specifically, B pictures can be used as references for other B pictures in a sub-ordinate level. Starting with level L_0 which usually only consists of a subsampled sequence of I (intra-coded) and P pictures, T more levels may exist within the hierarchy, each of which inserts more pictures and provides a higher frame rate. Taking prediction references from higher levels is not possible, pictures from lower levels can be used if they were previously decoded[19]. It is also possible to implement hierarchical B structures (Figure 7.8a), or hierarchical P structures (Fig. 7.8b). The examples in Fig. 7.8 only show dyadic ratios of frame rates between the levels, but other ratios are possible. Except for the base level L_0, picture prediction chains are terminated after T steps at latest, such that propagation of errors can be better controlled than in long picture-recursive structures. Therefore, these hierarchical inter-picture prediction structures also allow efficient implementation of unequal error protection and scalable video coding (see Sec. 7.5.1).

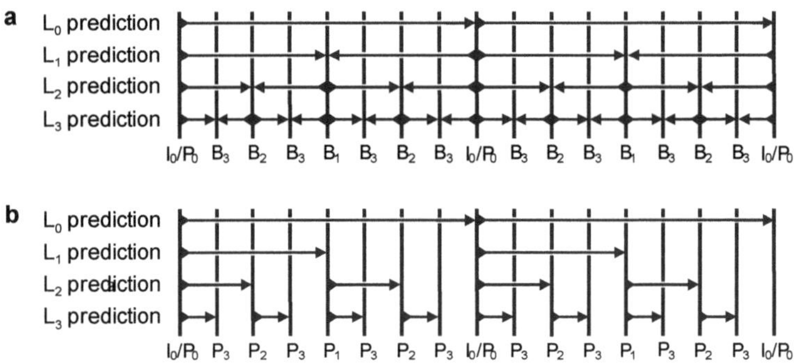

Fig. 7.8. Hierarchical picture structures with 4 levels (T=3)
a bi-directional **b** uni-directional

The effect of *quantization error feedback* according to (7.10) affects the prediction of B pictures as well. Quantization error components from the two superimposed pictures may however be correlated, particularly when both are mutually connected in the prediction process. If no correlation exists, the averaging in the prediction effects the energy of the error feedback to decrease by $0.5^2+0.5^2=0.5$, which means that the standard deviation of the fed-back quantization error component decreases by a factor $\sqrt{2}$. These effects are particularly noteworthy in the case of hierarchical B pictures (Fig. 7.8a).

The quantization error energy occurring in a picture at the lowest pyramid level L_0 influences not only the quality of the picture itself, but propagates into pic-

[19] In principle, pictures from the same level could be used as well, if they are preceding in decoding order; such a case is not shown in the examples of Fig. 7.8, as it is not commonly used in practice.

tures of the next-higher levels as follows under the assumption of statistical independence of the different paths (see Fig. 7.9)[20]:

$$\sigma_q^2 \approx \underbrace{1}_{L_0} + \underbrace{2\cdot 0.5^2}_{L_1} + \underbrace{2\cdot 0.5^2 + 4\cdot 0.25^2}_{L_2} + \underbrace{2\cdot 0.5^2 + 8\cdot 0.25^2 + 8\cdot 0.125^2}_{L_3} + \ldots \leq 1.5^T \quad (7.23)$$

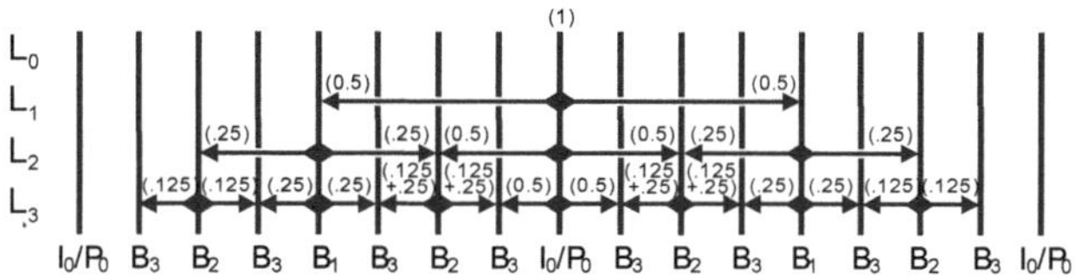

Fig. 7.9. Error propagation from an I or P picture into a hierarchical B picture structure, T=3

As a result, the quantization error energy propagated from L_0 increases by a factor of 1.5 per additional level. This would lead to the conclusion that the quantization step size should be increased by $\sqrt{1.5} \approx 1.225$ for each next higher level to keep roughly equal balance of error propagation effects over all levels. Note that this is only an approximation, as

- the quantization errors arriving at a picture (e.g. L_2 directly from L_0 and indirectly via L_1) need not necessarily to be uncorrelated such that the energy of the propagation could in fact even be higher;
- According to (5.8) the quantization error propagation assumed here only occurs for open-loop prediction, whereas in a closed-loop system, the quantization in the next level is intended to compensate the previous quantization error that is included in the prediction, see (5.11). However, when zero quantization is chosen, which is often the case due to excellent prediction particularly in B pictures, the behaviour of a closed-loop prediction system will become similar to open-loop, where the quantization error is propagated from picture to picture. This can also be interpreted as continuing quantization error feedback over the prediction loop (5.14). As prediction becomes inferior when performed from a reference coded by low quality, a simple solution is to improve the quality of references if those are expected to persist over a larger number of prediction steps.

Practically, the optimum choice of variable quantization step sizes over the different levels of a hierarchical prediction structure depends on sequence properties and on the rate-distortion operation point. For example, in sequences where the next-higher levels can be perfectly predicted due to smooth changes, the lowest level should be quantized even more accurately and the higher levels more coarse-

---

[20] Independence could e.g. be assumed when different spatial shift occurs over the different paths due to different motion compensation.

ly, where the first aspect increases the prediction quality and the second aspect enforces choosing zero quantization at higher levels[21].

In case of uni-directional hierarchical structures (Fig. 7.8b), the quantization error propagation from low to high levels is even stronger. For the dyadic structure given in the figure, quantization and prediction quality originating from a picture at level L_0 influences 2^T-1 additional pictures within the other hierarchy levels. Therefore, larger gaps in changing quantization step sizes are typically applied in a pyramid structure consisting of P pictures, as it is sometimes used in low-delay applications.

In summary, as the influence of pictures for subsequent predictions and possible propagation of quantization errors decreases towards the higher levels, they can be quantized more coarsely. On the other hand, it can be observed in hierarchical prediction structures that due to the larger temporal distance the quality of the prediction becomes inferior for the lower levels. Nevertheless, when variable quantization over the levels is applied, hierarchical picture structures typically provide significant benefit in terms of compression efficiency, as compared to picture-recursive structures. In scenes with low or moderate changes, the number of levels could be further increased for providing even more gain, whereas in cases of fast and irregular motion, the gain is often lower and the number of hierarchical levels should be limited or adapted appropriately. If no adaptation is made, setting of T=2..3 provides reasonable gain on average for typical video sequences.

As a general disadvantage, a larger number of levels increases the encoding and decoding delay in particular for the hierarchical B structure when subsequent pictures have to be accessed. However, hierarchical B picture pyramids could also be constructed using only temporally preceding pictures for reference in low-delay applications; as the temporal distance between the current picture and the second reference picture is further increased in that case, the prediction quality may however suffer.

Generally, the data rate allocations to the pictures of the different levels show significant variations in hierarchical structures (regardless if P or B), as the higher levels are quantized coarser and MC usually performs better for them, having prediction references with shorter temporal distance. In case of fixed-rate transmission, this can impose additional delay due to the need of coded bit-stream buffering.

Further disadvantages of B picture schemes are the increased number of motion parameters, higher complexity of motion estimation and compensation, and larger amount of memory accesses necessary.

---

[21] When operating an AVC codec in a hierarchical B picture structure, a step size change by increasing the *QP* value by 1 for each next-higher level is often chosen, which would correspond to a quantization step size increase of roughly $\sqrt[6]{2} \approx 1.112$. It can be found that more aggressive change sometimes gives better PSNR values on average, but quality fluctuation between the pyramid levels also increases, which may not necessarily be desirable considering subjective quality.

Fig. 7.10 shows prediction error pictures for cases of motion-compensated prediction from previous pictures, including cases of multi-frame and bi-prediction. Fig. 7.11 illustrates examples of bi-directional and hierarchical prediction structures. Generally, lower prediction error can be observed with bi-directional schemes due to the fact that dis-occluded areas from future reference pictures can be used; however, the prediction error increases with higher temporal distance within the hierarchy structure (contrast of residual pictures enhanced 1.5x).

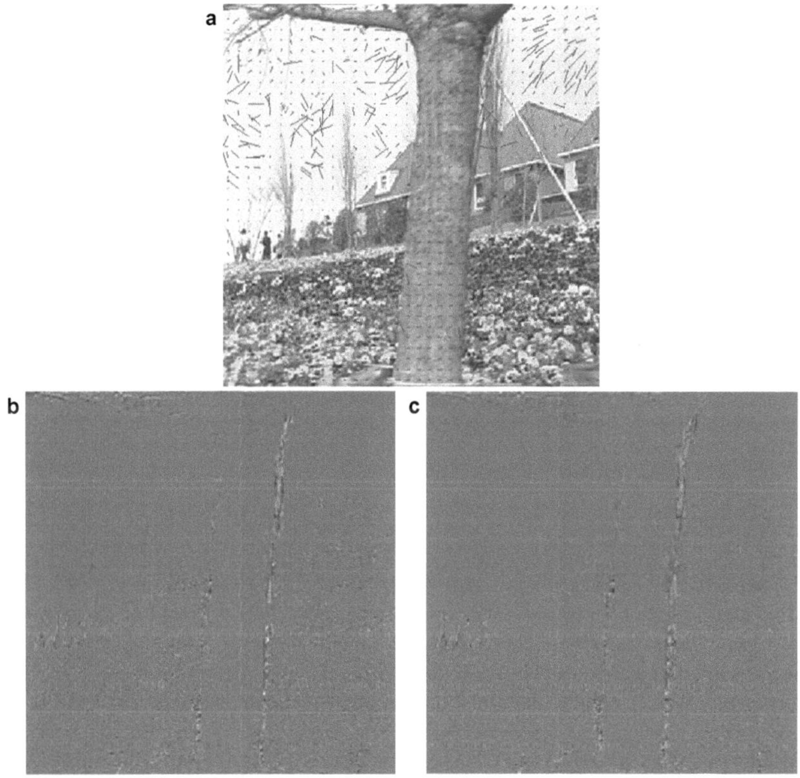

Fig. 7.10. Motion compensated prediction with 1/4 sample accuracy, 8-tap interpolation, block size 8x8. **a** Original picture (with MV overlay), and prediction error pictures: **b** uni-directional (P) prediction from previous reference picture **c** bi-prediction from two previous reference pictures

In cases of changed lighting conditions e.g. by flashlights or reflections, or trick modes such as fades or dissolves, *weighted prediction* can be applied, where the sample amplitudes of either the entire picture or a local area are adjusted by a gain α and an offset β[22],

[22] Weighted prediction can be applied both for uni- and bi-prediction cases; in the latter case, individual values α and β could be used for the two reference pictures. For simplicity,

$$\hat{s}(\mathbf{n}) = \alpha(\mathbf{n})s(\mathbf{n}-\mathbf{k}) + \beta(\mathbf{n}) . \tag{7.24}$$

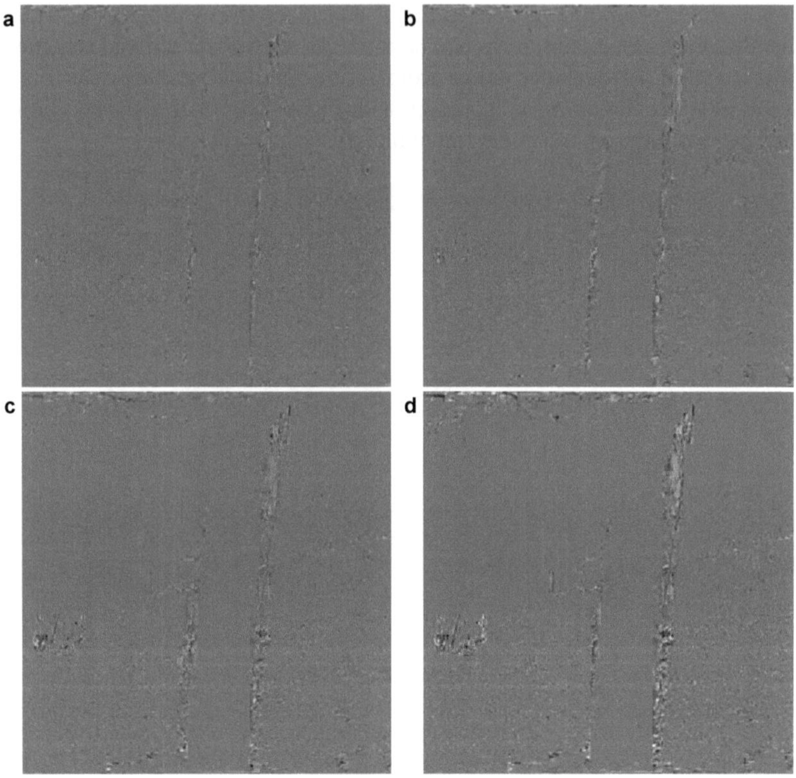

Fig. 7.11. Hierarchical prediction with 1/4 sample accuracy (same picture as in Fig. 7.10a), 8-tap interpolation, block size 8x8. **a** Bi-directional, distance 1 (L3 of Fig. 7.8a) **b** ditto, distance 2 (L2) **c** ditto, distance 4 (L1) **d** uni-directional, distance 4 (L1 of Fig. 7.8b)

Multi-hypothesis prediction. If several references are available, it is either possible to select one of them or construct a superposition. In principle, the selection of the best suitable motion vector could be interpreted as picking one out of different possible prediction hypotheses. Sub-sample accurate motion compensation could be interpreted as a weighted superposition (by interpolation filtering) of several full-sample hypotheses, where however the choice of weighting factors is limited. If more than one reference picture is available for selection (Fig. 7.7), bi-prediction is an average-based superposition of two hypothesis choices. In principle, generalized bi-prediction could also superimpose two predictions with different displacement shifts from the same reference picture, combination with weighted prediction is another example case. A general formulation of multi-

---

sub-sample interpolation and loop filters are not shown in the prediction equation. Observe the identity to (6.13) where the same approach is used to synthesize illumination changes.

hypothesis prediction, subsuming all of these cases and allowing a total of P hypotheses, is given as

$$\hat{s}(\mathbf{n})=\sum_{p=0}^{P-1} w_p(\mathbf{n})\tilde{s}(\mathbf{n}-\mathbf{k}_p)\,, \tag{7.25}$$

Here, $\mathbf{k}_p$ includes the reference picture position and the corresponding spatial displacement. By proper superposition of different hypotheses it should be possible to improve the prediction quality. In order to limit the number of possible combinations (which would increase the encoder search complexity and the amount of side information) a set of pre-defined combinations should be defined from which the weighting factors are determined. For example, in case of bi-prediction, the weighting factors correspond to the filter coefficients in sub-sample interpolation, multiplied by 0.5 for each of the references. Overlapping-block motion compensation described in the next paragraph is another example which implicitly gives individual weights for single samples.

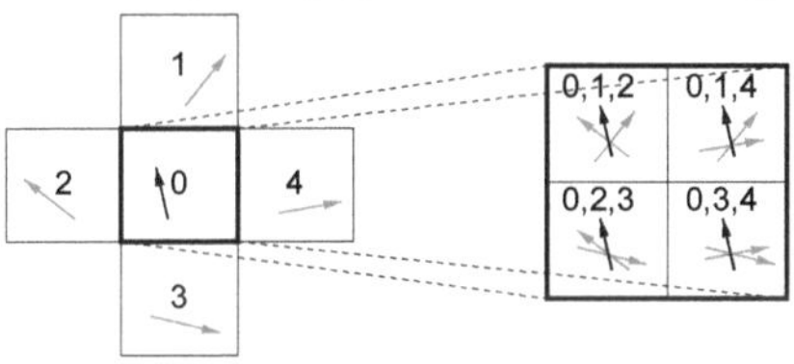

Fig. 7.12. OBMC: usage of motion vectors from neighbored block2 '1'…'4' for weighted superposition of different prediction errors in quarter tiles of the current block '0', as described by the weighting functions (7.26)

Overlapping-block MC. At moving objects' boundaries, discontinuities between adjacent block motion vectors typically occur. As it is unlikely that object boundaries coincide with block boundaries, unpredictable structures or structures caused by compromising motion vectors would typically appear in the prediction residual. A possible solution to this problem is *overlapping block motion compensation* (OBMC) [ORCHARD, SULLIVAN 1994], where a weighted superposition of several motion-compensated predictions (or equivalently, of residuals) is performed. For this purpose, neighbor blocks' motion vectors are used to compute alternative predictions/residuals for the current block, and finally all the different versions are combined. This has the effect of smoothing prospectively erroneous components in prediction. An example is given in Fig. 7.12. The prediction in the current block (0) is generated using its own motion vector $\mathbf{k}_0$ and the four motion vectors $\mathbf{k}_p$ of its horizontal and vertical neighbors (p=1…4), using (7.25) with P=5 hypotheses and weighting functions which implement a suitable fade-in transition over the block boundaries. Example weighting matrices for block size of 8x8 samples are given as[23]:

[23]These weighting functions are similar to those defined in ITU-T Rec. H.263 and MPEG-4 part 2. Other weighting functions are possible, e.g. including all eight neighbored blocks. Usually, at each position of the block, the weights should have unity sum.

$$\mathbf{W}_0 = \frac{1}{8}\begin{bmatrix} 4&5&5&5&5&5&5&4 \\ 5&5&5&5&5&5&5&5 \\ 5&5&6&6&6&6&5&5 \\ 5&5&6&6&6&6&5&5 \\ 5&5&6&6&6&6&5&5 \\ 5&5&6&6&6&6&5&5 \\ 5&5&5&5&5&5&5&5 \\ 4&5&5&5&5&5&5&4 \end{bmatrix} \quad \mathbf{W}_1 = \frac{1}{8}\begin{bmatrix} 2&2&2&2&2&2&2&2 \\ 1&1&2&2&2&2&1&1 \\ 1&1&1&1&1&1&1&1 \\ 1&1&1&1&1&1&1&1 \\ 0&0&0&0&0&0&0&0 \\ 0&0&0&0&0&0&0&0 \\ 0&0&0&0&0&0&0&0 \\ 0&0&0&0&0&0&0&0 \end{bmatrix} \quad \mathbf{W}_2 = \frac{1}{8}\begin{bmatrix} 2&1&1&1&0&0&0&0 \\ 2&1&1&1&0&0&0&0 \\ 2&2&1&1&0&0&0&0 \\ 2&2&1&1&0&0&0&0 \\ 2&2&1&1&0&0&0&0 \\ 2&2&1&1&0&0&0&0 \\ 2&1&1&1&0&0&0&0 \\ 2&1&1&1&0&0&0&0 \end{bmatrix}$$

$$\mathbf{W}_3 = \frac{1}{8}\begin{bmatrix} 0&0&0&0&0&0&0&0 \\ 0&0&0&0&0&0&0&0 \\ 0&0&0&0&0&0&0&0 \\ 0&0&0&0&0&0&0&0 \\ 1&1&1&1&1&1&1&1 \\ 1&1&1&1&1&1&1&1 \\ 1&1&2&2&2&2&1&1 \\ 2&2&2&2&2&2&2&2 \end{bmatrix} \quad \mathbf{W}_4 = \frac{1}{8}\begin{bmatrix} 0&0&0&0&1&1&1&2 \\ 0&0&0&0&1&1&1&2 \\ 0&0&0&0&1&1&2&2 \\ 0&0&0&0&1&1&2&2 \\ 0&0&0&0&1&1&2&2 \\ 0&0&0&0&1&1&2&2 \\ 0&0&0&0&1&1&1&2 \\ 0&0&0&0&1&1&1&2 \end{bmatrix}. \tag{7.26}$$

This specific method of OBMC requires weighted superposition of at most three different prediction signals, but computation is simplified if neighbors have identical displacements. For optimum performance, OBMC should already be considered during motion estimation, as the best choice of vectors is no longer independent for the current block and its neighbors [HEISING ET AL. 2001]. The OBMC method is not fully consistent for the case of *smooth variations* in the motion vector field as occurring e.g. in cases of zoom or rotation. It will then usually cause a spatial blurring or smearing of the prediction and the residual. An alternative for such cases is warping prediction (Sec. 7.2.5).

Decoder-side motion vector derivation (DMVD). Additional side information data is necessary in multi-hypothesis prediction, when additional reference pictures are used, specifically for motion vectors, reference picture indices and weighting factors. A scheme to derive such data at the decoder was presented in [KAMP, WIEN 2012]. The method is shown in Fig. 7.13. An L-shaped template is constructed from samples in already decoded areas which are beyond the left and top boundaries of the current block. Motion estimation is performed by matching the template against corresponding positions from one or several reference pictures, assuming that the motion of the current block will approximately equal the motion of the template. The best-found matches can then be used for multi-hypothesis prediction, where weights can be either uniform or related to the quality of the prediction (derived from the cost function of the template matching).

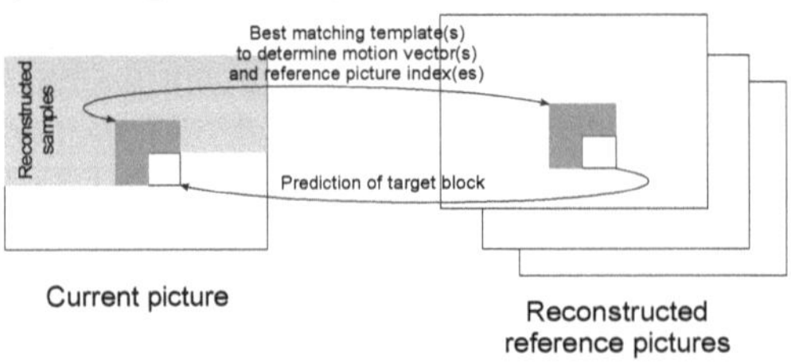

Fig. 7.13. Decoder-side motion vector derivation (DMVD) with several reference pictures

References in case of stream switching. The need of stream switching e.g. occurs in

- broadcast applications, when a user switches to another video programme;
- streaming applications, when a user decides to start viewing a live stream or uses fast forward/backward modes on a stored stream;
- replay of stored video, when a user is using fast forward/backward;
- lossy video transmission, when in case of data losses another more reliable stream shall provide error concealment.

When inter-picture prediction is used, switching at the position of a picture that relies on one or several other decoded reference picture(s) which are not available (as they were sent before the switching was made) is not reasonable. If such a functionality is intended, it is necessary to systematically define points in the stream, with reference to pictures where a clean decoding process can start. A typical way of implementing this is by definition of random access points, starting with a picture that is intra-only coded. This type of picture is called I picture or instantaneous decoder refresh (IDR) picture. Beyond random access, recovery after transmission errors can also start from these points.

In case of bi-directional (and possibly hierarchical) prediction structures, which are often used in broadcast and streaming applications due to their good compression performance, a problem occurs due to the fact that decoding order and display order are not identical. This has the consequence that pictures that precede the random accessible picture in display order may appear in the stream afterwards. On the other hand, for good prediction performance, these pictures might not only refer to the new random accessible picture, but also to other pictures that are not known to a decoder which has just started decoding the stream.

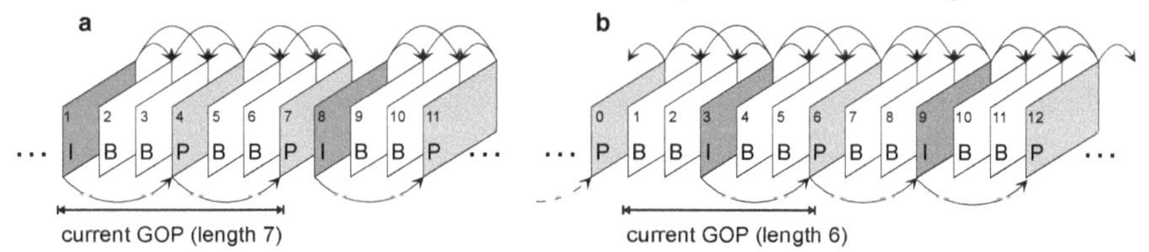

Fig. 7.14. Concepts of **a** closed GOP, and **b** open GOP for the example of an 'IBBP' picture coding structure as typically used in MPEG-2

In MPEG-2, a *group of pictures* (GOP) structure is defined, where each GOP shall have one I picture as random access point[24] in the beginning. In the so-called *closed GOP* concept (Fig. 7.14a), no pictures of the GOP shall refer to pictures of a previous or subsequent GOP, which means in the shown example, that the GOP starts with a random access point (I picture in case of MPEG-2), and cannot end with a B picture. This firstly deteriorates the compression performance (due to the lack of bi-directional prediction at the end), and also may make the GOP bounda-

---

[24] Random access is required in broadcast (when switching channels), live streaming and file access (when jumping to later point in a video). Similar problems occur in *bit-stream splicing*, when parts of two videos shall be concatenated without re-encoding.

ries more visible as kind of 'switching' artifacts in normal operation (i.e. decoding the entire movie without stream switching). As a solution, the *open GOP* structure (Fig. 7.14b) can be used, where however in case of stream switching at the given entry point of the GOP, it is not possible to decode the first two B pictures in the given example. It is therefore reasonable to discard those pictures, which may be done based on a (non-normative) decision at the decoder side.
The problem of inefficiency of closed GOP structures is becoming even worse in case of hierarchical B picture structures, as definable with the AVC and HEVC standards. Nevertheless, the definition of an entirely intra coded, 'instantaneous decoder refresh' (IDR) picture as random access point in AVC/HEVC requires that any pictures that follow the IDR (in decoding order) shall not take reference to any picture preceding the IDR, which is equivalent to the definition of the closed GOP structure[25].

Neither MPEG-2 nor AVC define a normative decoder behavior for 'leading' pictures that are found in the bit stream after a random access point, but cannot be correctly decoded because they refer to some earlier reference that is not available (such as the first two B pictures in Fig. 7.14b). HEVC additionally defines another fully intra-coded type as *clean random access* (CRA) picture, where un-decodable leading pictures may follow in the bit stream but are identified as such and may or may not be decoded (same as open GOP), and *broken link access* (BLA) pictures, where it is obligatory not to decode (which is appropriate for splicing, where otherwise reference to a picture from a different scene might be made).

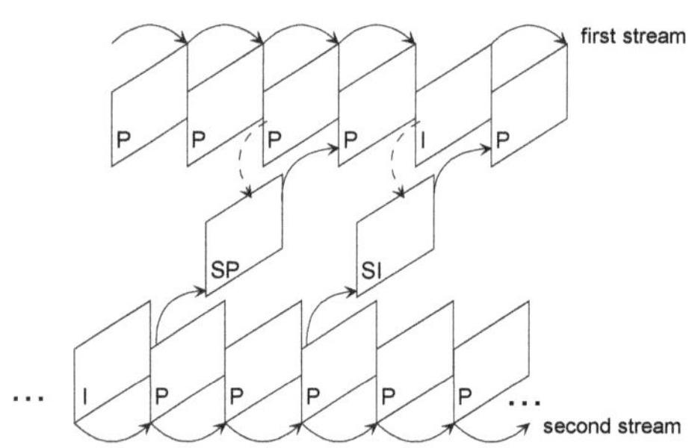

Fig. 7.15. Placement of switching pictures in a multiple-stream representation of a video sequence

AVC further defines *switching pictures* (*S* pictures) which allow transition between prediction processes of multiple streams that represent the same video signal, but possibly at different resolution or rate/quality [KARCZEWICZ, KURCEREN 2003]. The concept was originally proposed in [GIROD, FÄRBER, HORN 1997]. It allows seam-

[25] Unlike MPEG-2, the AVC standard does not normatively specify a GOP structure. Nevertheless, the distance between pictures at the lowest level of a hierarchical picture structure is often denoted as 'GOP length', e.g. length 8 in the example of Fig. 7.8a. This may however not be identical with the period for random access.

less switching (and splicing) of bit streams, however requires access to both versions at the encoder, and pre-definition of the switching positions. The 'S' picture contains the differential information that is necessary to change from the decoder state (which means the state of the reference picture memory in the recursive loop) of one stream to that of the other stream. The principle is shown in Fig. 7.15. If the S picture is complementing a P or I picture from the other stream, it is denoted as SP or SI picture, respectively. The information related to both types of S pictures adds overhead (additional data rate) as compared to a single-stream coder.

7.2.5 Accuracy of motion compensation

Regarding motion estimation and compensation, the following modalities of accuracy exist w.r.t. their representation:

- Degree of (sub-)sample precision of displacement vectors; in case of sub-sample precision also the quality of the interpolation filter (in terms of approaching the accuracy of ideal sub-sample interpolation);
- Maximum range of displacement vectors;
- Maximum time distance between current picture and reference pictures;
- Resolution of the displacement vector field representation (i.e. density of vectors over space and time);
- Expressiveness of the underlying motion model, in terms of representing translational, rotational, affine etc. vector field mapping, and eventually describing discontinuities at object boundaries.

All these accuracy dimensions are inter-related with methods for encoding motion data (see Sec. 7.4), as the rate for representation (as well as the complexity) may increase by the increased accuracy. The necessary accuracy also depends on statistical properties of a given video.

The necessary *sub-sample precision* and the *interpolation accuracy* also depend on the amount of detail and motion in the given picture. High-frequency areas showing edges, detailed texture etc. require higher accuracy than smooth areas. A more detailed analysis can be performed similar to (7.9) and (7.13), where the occurrence of high frequencies in the prediction error signal firstly depends on the spectrum of the picture, on the precision of motion compensation and on the frequency attenuation of the interpolation filter. If the same content is available in different picture resolutions,

- the maximum range as well as the displacement vectors themselves scale with the spatial picture resolution, and with the temporal distance (or reciprocally with frame rate);
- the necessary displacement vector field resolution may be scaled reciprocally with the picture resolution, i.e. it could usually be expected that lower resolutions require smaller block sizes (if the same content is shown with fewer samples); this could however depend on the amount of additional high-frequency detail that appears in the higher-resolution picture;

- the displacement vector precision (sample/sub-sample) should approximately be constant, if the downsampled pictures are generated with suitable lowpass filters for alias suppression[26].

The necessary accuracy in terms of *range*, *resolution* and *precision* is highly sequence dependent, in particular the expected motion range (e.g. velocity of camera and object movements), the granularity of motion (e.g. sizes of objects, amount of global versus local motion), and the amount of spatial detail. The *expressiveness* is mostly related to the method of motion representation. For example, a block-based representation assuming one identical motion vector for horizontal/vertical shift of an entire block cannot truly represent scaling or rotation of content, unless the resolution is increased (small block sizes). Models such as affine mapping (7.29) can describe continuous changes over the displacement vector field efficiently by a low number of parameters. Discontinuities of the motion vector field, as they occur at moving object boundaries, are again either requiring higher resolution, or precise definition of the discontinuity positions by external means such as contour/shape coding.

Sub-sample luma interpolation in HEVC. The properties of interpolation for sub-sample precision in motion compensation play an important role for the quality of the prediction. The fractional sample interpolation for luma samples in HEVC is one of the best schemes available so far. It uses separable application of an 8-tap FIR filter for the half-sample positions and a 7-tap FIR filter for the quarter-sample positions. Using only 7 taps rather than the 8 for the quarter-sample interpolation positions is appropriate since those are relatively close to an integer-sample position. The actual filter coefficient values of the FIR interpolation filters were partially derived from DCT basis function equations [UGUR ET AL. 2013]. They are represented as 6-bit integer values as shown in Tab. 7.1. Herein, the interpolated sample would be exactly between n=3 and n=4 for the phase position d=1/2, and close to n=3 or n=4 for the phase positions d=1/4 or d=3/4, respectively.

Tab. 7.1. Filter Coefficients for Luma Fractional Sample Interpolation in HEVC (values to be multiplied by 2^{-6})

Phase position	n=							
d, $h(n)$=	0	1	2	3	4	5	6	7
d=1/4	−1	4	−10	58	17	−5	1	(0)
d=1/2	−1	4	−11	40	40	−11	4	1
d=3/4	(0)	1	−5	17	58	−10	4	−1

[26] It can however be observed that mostly for video sequences containing spatial aliasing due to improper (sub-)sampling, higher sub-sample precision of motion compensation such as 1/8 sample or beyond provides improved prediction.

Block-based MC with fixed block size. Block-based motion compensation with fixed block size and translational shift does neither support continuous variations nor discontinuities in the motion vector field properly, if the block sizes are too large, and would tend to require excessive side information if the sizes are too small[27]. Even though, a fixed-size grid of blocks, with only one displacement vector defined per 16x16 sample unit, was used in early video compression standards until MPEG-2.

Variable block size MC. An example for variable block size motion compensation [CHAN, YU 1990] is shown in Fig. 7.16. In general, the blocks do not even need to have square shape (the case shown in the figure). A simple approach is deciding during estimation about splitting a larger block into two or four equal sub-partitions whenever justified by an improved prediction and rate cost. The example in the figure expresses a sub-partitioning by quad-tree codes (Sec. 6.1.2) for each of the largest block units A, B, C. Instead of signalling displacement vectors separately for each small prediction block, it is also possible to merge adjacent block partitions that have identical motion vectors for a more efficient representation (methods of signalling and reference to standards where this is applied are further discussed in Sec. 7.4).

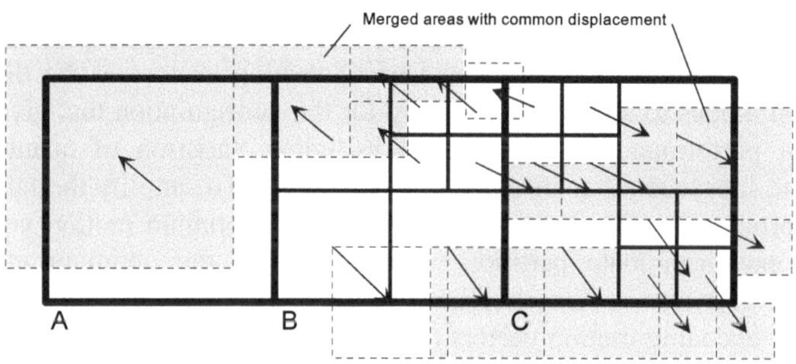

Largest block unit	A	B	C
Quad-tree code	1	0(1110(1111))	0(0(1111)10(1111)1)
Number of MVs	1	7	10

Fig. 7.16. Motion compensation using variable block sizes

Beyond regular (e.g. quad-tree) block-based splitting and merging, motion partitions can be established at even finer granularity for better alignment with true

[27] Beyond that, small block sizes such as 4x4 are more vulnerable to noise during estimation. Generally, this could be resolved by imposing constraints based on homogeneity with the vectors of adjacent blocks, which however can also be implemented by allowing variable block sizes, i.e. use small blocks only when necessary.

motion boundaries, e.g. by the method of *geometric partitioning* as presented in Fig. 6.5.

Rate-distortion optimization in motion estimation. The rate to be spent for the displacement vectors, the reference picture index and for variable-size partitioning to support finer resolution of the motion vector field has to pay off such that it is more than compensated by the achieved savings in bit rate for encoding of the motion-compensated residual in the hybrid coder. Therefore, the selection of optimum motion compensation (including the choice of displacement and partitioning parameters) can be expected to be rate dependent. Simple minimization of picture difference is no longer an appropriate criterion with regard to rate-distortion performance, as another motion compensation configuration may be similarly good in performance, but less expensive in terms of side information.

The decision about the best motion compensation constellation should include knowledge about the rate to be spent for the displacement vectors and the reference index, for variable block partition modes and for residual signal encoding. Methods for rate-distortion optimized decisions (see Sec. 4.3 and further discussion in Sec. 7.2.7) should be employed, taking into account the rate/quality trade-off. This can be implemented by applying Lagrangian multipliers in a cost function similar to (4.43). In an extreme case of encoder optimization, motion estimation and coding mode selection could perform test-wise encoding of multiple prediction error candidate blocks, using various motion vectors in the different block partition modes to search exhaustively for the configuration that gives best rate-distortion performance, possibly even including variation of quantization settings for the subsequent transform coefficient coding. To simplify the selection, a two-stage process is often applied, where first one optimum motion vector is selected for each candidate partition mode, and then the optimum mode is searched (see e.g. [WIEGAND, GIROD 2001]), where R_{motion}, R_{mode} and R_{coeff} are the rates to be spent for encoding motion vectors (and reference picture indices), partitioning mode information and transform coefficients in residual coding, respectively. This two-step procedure is

$$\mathbf{k}_{\text{opt}}(\text{mode}) = \arg\min_{\mathbf{k}} \left\{ D(\mathbf{k}) + \lambda_{\text{motion}} R_{\text{motion}}(\mathbf{k}) \right\}, \tag{7.27}$$

$$\text{mode}_{\text{opt}} = \arg\min_{\text{mode}} \left\{ D(\text{mode}) + \lambda_{\text{mode}} \left[R_{\text{motion}}(\mathbf{k}_{\text{opt}}(\text{mode})) + R_{\text{mode}} + R_{\text{coeff}}(\text{mode}) \right] \right\}. \tag{7.28}$$

In (7.27), $D(\mathbf{k})$ is the cost function applied in motion estimation (see Sec. 2.6.2). In (7.28), the Lagrangian multiplier weights the overall rate to be spent by a candidate partition mode, including the rates for the motion vector, signalling of the mode and the residual coding by transform coefficients. D(mode) is the actual distortion occurring in reconstruction.

Investigation of the actual coding rate for transform coefficients is necessary, as the motion-compensated residual signal does *not* have a white spectrum, as was shown in (7.14), or if occluded areas appear in the residual. Therefore, a residual with more energy but better compaction to few coefficients can be a better choice than another with less energy, but being spectrally more flat (see Problem 7.2).

Instead of determining the rate by true encoding, an approximate optimum solution would be to select the motion vector which gives the minimum sum of *logarithmic* amplitude of coefficients (and giving some weight regarding the number of coefficients that fall below the quantization threshold). This would implicitly consider the transform coding gain of the given prediction error residual (cf. (4.34) and (5.38)), but it might be undesirable to perform a transform and logarithmic mapping of coefficient amplitudes as part of the motion estimation. A less complex method with similar effect is using the *sum of absolute transformed differences* (SATD) as criterion in motion estimation. For this, it is necessary to apply a 2D transform to the residual before taking the decision on motion estimation. This can be restricted to a subset of best candidates, and a simplified transform such as the Walsh basis (2.250) is often used instead of a DCT. Unlike the picture domain sum of absolute differences (SAD), the SATD favors residuals which concentrate energy in fewer coefficients, which can provide some benefit for the two-stage approach described above, if used as $D(\mathbf{k})$ in (7.27), it is likely to select a motion vector candidate that would better minimize the R_{coeff} part of the criterion in (7.28).

Global MC. In *Global Motion Compensation* (GMC), a warping function describing mapping to coordinates **n** in the current picture from **n**' in the reference picture (or vice versa) is defined by a parametric model, e.g. the affine transformation,

$$\underbrace{\begin{bmatrix} n_1' \\ n_2' \end{bmatrix}}_{\mathbf{n}'} = \underbrace{\begin{bmatrix} a_1 & a_2 \\ a_3 & a_4 \end{bmatrix}}_{\mathbf{A}} \cdot \underbrace{\begin{bmatrix} n_1 \\ n_2 \end{bmatrix}}_{\mathbf{n}} + \underbrace{\begin{bmatrix} k_1 \\ k_2 \end{bmatrix}}_{\mathbf{k}}, \tag{7.29}$$

or the bilinear transformation

$$\begin{aligned} n_1' &= b_1 n_1 n_2 + a_1 n_1 + a_2 n_2 + k_1 \\ n_2' &= b_2 n_1 n_2 + a_3 n_1 + a_4 n_2 + k_2 . \end{aligned} \tag{7.30}$$

Beyond translation shift **k**, other appearance modifications such as rotation, scaling or shear can be described by the additional parameters. In cases where a global (e.g. camera) motion is present, provided that the given parametric model can describe such mapping, encoding of additional local motion compensation parameters may not be necessary, or they may just express small deviations from the global model, which can be expected to reduce the amount of side information. The mapping **n**→**n**' requires sub-sample shifts with high-quality spatial interpolation to provide sufficient quality of the prediction[28]. However, even models such as (7.29) and (7.30) may not provide a perfect compensation of camera motion,

---

[28] GMC is defined in MPEG-4 part 2, but as only bilinear interpolation (2.231) is used for generating the warped reference, which then can be significantly blurred, it does not provide significant compression benefit. An overview over various mapping methods and the physical background of camera projection is given in [MCA, SEC. 4.5].

particularly when optical distortions are present or when the camera is moving its position and occlusions occur.

Warping prediction. Warping based on parametric coordinate mapping such as (7.29) and (7.30) can also be applied locally, where due to the smaller range it can be expected to provide a more accurate representation of the true motion than either global warping or block-based translational motion. Translational block-matching motion estimation could be interpreted as a *sub-sampled* approximation of a dense motion vector field, where sub-sampling factors are equal to the block width and height. The dense field is then reconstructed by a hold element, such that a constant motion vector is applied for the entire block. Alternatively, the motion vector field can be interpreted as a grid of values, from which the horizontal and vertical displacement shifts can be densely interpolated at any position. Methods of this type have been introduced as *triangle-based motion compensation* (TMC) using affine warping (7.29) [BRUSEWITZ 1990] and *control grid interpolation* (CGI) [SULLIVAN, BAKER 1991] where bilinear interpolation (2.231) is used component-wise to generate the dense sample-related displacements from the control grid's displacement vector samples, which is equivalent to the parametric model (7.30). Hierarchical grid structure are also possible [HUANG, HSU 1994]

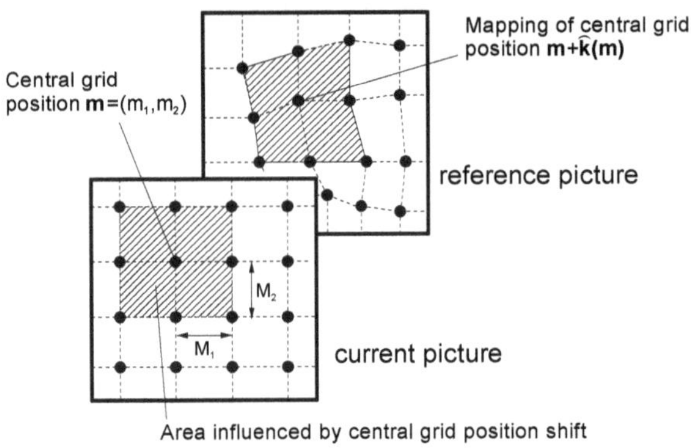

Fig. 7.17. Displacement field interpolation based on bilinear warping from control positions

Assume that a regularly sampled grid of control values with distances M_1 and M_2 horizontally and vertically is defined, as shown in Fig. 7.17. In this example, bilinear interpolation computes sample-wise motion vectors within the rectangular area bounded by 4 control points with their associated displacements $\hat{\mathbf{k}}(m_1, m_2)$

$$\begin{aligned}\mathbf{k}(n_1,n_2) = \hat{\mathbf{k}}(m_1,m_2)\big(1-(n_1/M_1-m_1)\big)\big(1-(n_2/M_2-m_2)\big)\\ +\hat{\mathbf{k}}(m_1+1,m_2)(n_1/M_1-m_1)\big(1-(n_2/M_2-m_2)\big)\\ +\hat{\mathbf{k}}(m_1,m_2+1)\big(1-(n_1/M_1-m_1)\big)(n_2/M_2-m_2)\\ +\hat{\mathbf{k}}(m_1+1,m_2+1)(n_1/M_1-m_1)(n_2/M_2-m_2)\end{aligned} \tag{7.31}$$

$$\text{with } m_iM_i \le n_i < (m_i+1)M_i \text{ and } m_i = \left\lfloor \frac{n_i}{M_i} \right\rfloor$$

Bilinear interpolation is complexity-wise the most simple approach of warping MC, since the individual motion shifts of samples from a rectangular grid can be computed efficiently by adding a constant difference relative to the last position[29]. The value of each control grid point influences interpolation within a rectangular area of size $(2M_1-1)(2M_2-1)$, as also shown in Fig. 7.17. Within each quarter of that area, the values of the three other neighbor control grid points take influence as well (eight in total), which means that the optimum displacement grid parameters $\hat{\mathbf{k}}(m_1,m_2)$ mutually influence each other. However, the influence is highest in close vicinity, which can be used to reduce the computational complexity. As an exhaustive joint estimation would be impractical, displacements are better estimated one by one, but in an iterative fashion. Either the result of conventional block matching with blocks centered at the control points can be used as starting point of such an iteration, or feature-based matching based on movement of characteristic local key points can be applied.

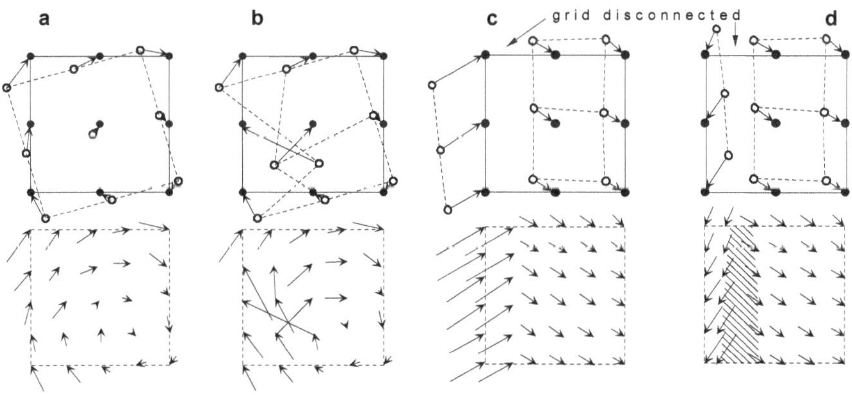

Fig. 7.18. Examples of warping grids. *Top*: Support grid in current picture (●) and reference picture (o); *Bottom*: Interpolated displacement vector field
a Case of rotational motion **b** Unreasonable configuration with overlapping vectors. Motion discontinuities in cases of an occluded area (**c**) and a dis-occluded area (**d**)

---

29 In case of the affine transformation (7.29) with 6 parameters, the shifts of 3 grid points uniquely define the mapping to be used within triangular grid partitions.

Fig. 7.18a illustrates how rotational motion can be captured by a warping motion model based on control points. Warping compensation however has severe disadvantages when the assumption about the underlying motion model is wrong. This is particularly the case at moving object boundaries, where the displacement vector field is discontinuous and occlusions may be present.

As the estimation may be unreliable in such cases, it can also produce unreasonable results that can however easily be detected and classified as cases where warping should not be applied. As an example, *overlapping* displacement vectors from adjacent grid points appear unreasonable, as they would lead to inverted coordinate relationships within partial areas (Fig. 7.18b). Overlaps could however occur, when a foreground object is moving much faster than the background; in that case, an occluded area should be expected. A possible solution to solve this problem is shown in Fig. 7.18c/d. When large differences between adjacent grid vectors are observed, the warping grid is artificially disconnected, and extrapolation of displacement vectors can be performed towards the position of the object boundary. The determination of this boundary can also be part of the estimation process [OHM 1994B], [OHM 1996] [HEISING 2002], employing a similar method as geometric partitioning (Fig. 6.5) between the diverging or converging grid positions. Overlapping motion vectors as in Fig. 7.18b, which are in fact reasonable in the case of fast moving objects, are then fully supported as well. In [HEISING 2002], also a hybrid switching between warping interpolation and an overlapping block matching method has been investigated. Though warping MC appears to be a better model for the underlying physical motion process, the estimation of the local warping parameters as well as the motion compensation itself are significantly more complex than in block-based approaches.

7.2.6 Hybrid coding of interlaced video signals

Interlaced formats (see Sec. 1.3.1 and 2.3.3) have been dominating in analog video and still exist in digital video. In fact, with interlaced sampling only half number of samples need to be processed as compared to a progressive video of same spatial resolution and with a frame rate equalling the interlaced field rate. Due to the interrelation of vertical and temporal sampling and the possibility of mixed alias occurring, motion-compensated prediction of interlaced video requires special attention and also has motivated the design of dedicated coding tools for that purpose. In hybrid coding, this concerns both the motion compensation and the 2D transform coding of the residual. Methods for interlaced coding were developed in the context of the MPEG-2 video compression standard, establishing one of its major differences compared to MPEG-1. Similar methods since then have been included in newer video coding standards MPEG-4 part 2 and AVC.

Generally, any video compression algorithm should be able to code interlaced video by deciding at the sequence level whether field pictures (separate) or frame pictures (two fields merged) shall be input to the encoder and output by the decoder. Except for the algorithm that needs to make the correct decision at the encoder side, the mechanism itself is a simple pre-/post-processing, and some signalling mechanism is needed to inform the receiver how to interpret and output the data.

In the next simplest approach, which already requires a modification of the compression algorithm, a decision is made at the frame level whether two field pictures shall be encoded independently ('field mode') or jointly, combined into one frame ('frame mode'). In AVC, this choice is denoted as *picture adaptive frame/field* (PAFF) mode selection. In field mode, each field is separately processed as standalone picture; the frame mode merges two fields by alternating lines and processes the entire picture. No re-design of core parts of the codec is necessary, the scheme can be implemented by accessing the picture samples in the decoded picture buffer as needed by the specific mode. It should however be observed that a field picture cannot consistently be predicted from a frame picture (and vice versa); therefore, on-the-fly rearrangement of picture lines from the frame buffer as well as mode dependent vertical scaling of motion vectors is required. Frequent changing between the modes does not appear too useful anyway, as the characteristics of a sequence (in terms of which mode is better suitable) would not be expected to change too rapidly.

Indeed, pictures of a sequence will likely have some areas where it is more appropriate to encode in frame mode (e.g. when content is standing still or constantly moving horizontally) and other areas where field mode gives better performance (e.g. with fast and inhomogeneous, in particular vertical motion). Therefore, adaptive switching between frame and field mode can additionally be performed at the level of macroblocks in MPEG-2, MPEG-4 part 2 and AVC (in the latter denoted as *macroblock adaptive frame/field*, MBAFF). Three prediction modes for interlaced video as defined in MPEG-2 are shown in Fig. 7.19[30], where the first two can be invoked also in AVC, either globally at picture level (PAFF), or macroblock adaptive (MBAFF), whereas the third is only defined in MPEG-2 [PURI, ARAVIND, HASKELL 1993]:

- In *frame prediction*, one common motion vector is used for both fields. Depending on the vertical motion shift, prediction of an area in a current field could be performed either using the field of same or opposite parity of the reference picture, depending on whether the (frame-related) vertical motion shift is by an even or odd number of samples. As evident from Fig. 7.19a, this mode would require different scaling of the motion vector for the lines belonging to the even and odd fields, as the sampling time distances between even/odd and odd/even fields of current/reference picture are different. Beyond that, vertical interpolation for sub-sample motion compensation could cause problems. Out of these reasons, it is unlikely that frame prediction would ever be chosen in certain cases of vertical motion.
- In *field prediction*, independent motion vectors can be used for the two fields. Each field can be predicted from the field of either same or opposite parity in the reference picture. Vertical sub-sample interpolation can be applied as usual in the selected reference field which could however fail in cases where

---

[30] These are modes defined by the MPEG-2 standard.

alias is present due to the field subsampling. In addition to the configuration illustrated in Fig. 7.19b, it is possible that the second field of a frame uses the first field for prediction reference when fields are encoded independently (in picture adaptive field mode). This will often improve the compression performance in case of pictures or partitions that are intra-coded.

– *Dual prime prediction* performs sort of de-interlacing of the reference picture, that can be interpreted as on-the-fly generation of a progressive reference picture from the two reference fields (for which the motion vector can be used to take into account the motion shift between the even and odd lines). Actual reference samples can be computed on the fly as needed. Common motion vectors are used, but scaled according to the effective temporal distance between the respective current field and the virtual reference picture. The method also solves the inconsistency problems of the frame mode mentioned above, but generation of reference samples (necessary at encoder and decoder) as well as the motion estimation at the encoder are significantly more complex.

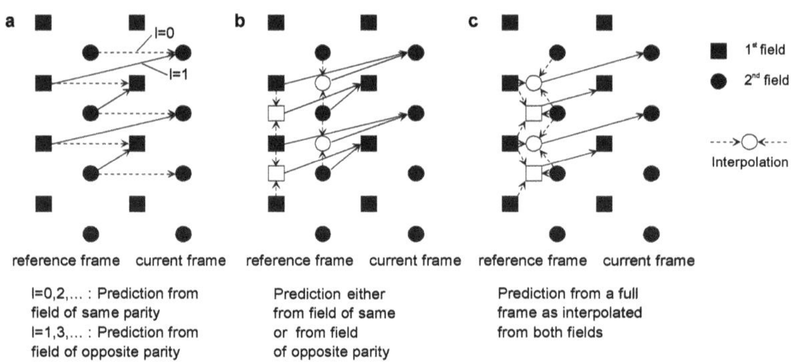

Fig. 7.19. Motion-compensated prediction schemes for fields in interlaced video: **a** frame prediction with inconsistency for odd vertical shifts **b** field prediction **c** 'dual prime' prediction

In AVC, the frame/field encoding decision is made for each vertical pair of macroblocks (each 16×32 luma region) in a frame, whereas in MPEG-2, the frame/field decision is made at the macroblock level, where it is necessary to handle half-height motion compensation blocks in field mode, which can cost some overhead in motion rate. AVC does not define a method like dual-prime, which would have been overly complicated in the combination with variable block size MC, intra prediction etc. Since in AVC a mixture of field and frame macroblock pairs may occur within an picture, the methods that are used for zig-zag scanning of transform coefficients, prediction of motion vectors, prediction of intra prediction modes, intra-picture sample prediction, deblocking filtering, and context modeling in entropy coding are modified to account for such a mixture. In AVC's MBAFF, one field cannot use the macroblocks in the other field of the same picture as a reference for motion prediction. Thus, sometimes PAFF coding can be more efficient than MBAFF coding (particularly in the case of rapid global motion, scene change, or intra picture refresh).

After the prediction residual is computed, the blocks for the computation of the block transform are either constructed from frames or from fields, depending on whether the frame mode or the field mode is selected (see Fig. 7.20). A transform mode is not necessarily coupled with the MBAFF mode selection; however, in PAFF, the transform operation is implicitly coupled to the mode due to the separate or combined processing of samples.

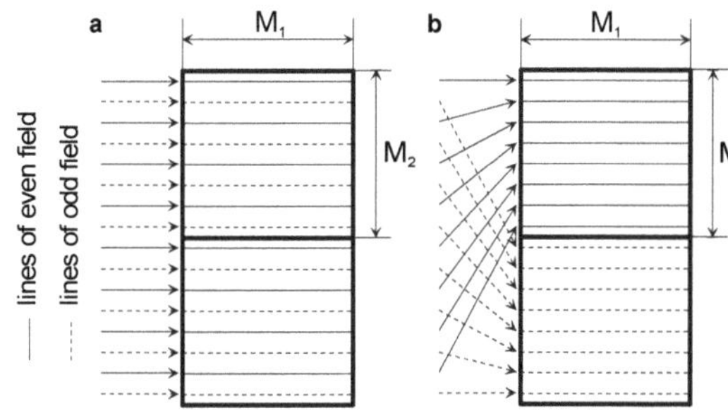

Fig. 7.20. **a** Block transform in frame mode (lines interleaved) **b** separate block transforms in the even and odd fields

7.2.7 Optimization of hybrid encoders

Standards for video compression typically define only the bit stream syntax and semantics and the operation of decoders (cf. Sec. 7.8). Hence, optimization of encoders offers freedom for improved compression and quality of decoded video, still being compatible with any standard-compliant decoding devices. Key issues are the optimization of motion compensation, mode decision and quantization of transform coefficients.

Compression performance can be improved by *rate distortion optimization.* The typical solution for this is made by Lagrangian optimization (cf. Sec. 4.3 and 7.2.5) [Sullivan, Wiegand 2005].

Rate distortion optimization as discussed so far usually assumed instantaneous decisions for each local area. Using a pre-defined Lagrangian multiplier λ for that purpose implicitly assumes a constant slope of the rate distortion function, which is usually not given if a picture consists of areas with different amount of spatial detail and different amount of change over time. In fact, variation of the quantizer step size depending on such local properties can improve the average performance, for which it is however necessary to test different quantization settings for each local area (e.g. block), investigate the benefit they give in terms of rate-distortion performance and choose those settings which give the highest benefit in minimizing the distortion for a given rate budget (i.e. prefer areas by better quantization where the slope of the $D(R)$ function is largest). This is entitled as *picture-*

level rate distortion optimization which usually requires multiple coding passes before the final setting can be decided. This can be beneficial firstly due to the fact that the amount of detail changes locally, and the operational point on the $D(R)$ function therefore is different for different areas, e.g. depending on the amount of zero transform coefficients. Furthermore, due to context-dependent coding of coefficients, it may be beneficial to enforce isolated non-zero coefficients into the zero value, since the amount of bits to be spent in case of non-zero coding would not be justified by the decrease of distortion. The contrary is true in higher detailed areas. Concepts like this are denoted as block-level *rate distortion optimized quantization* (RDOQ).

Next, it should be observed that various *recursive dependencies* exist in the (motion compensated or intra) prediction loop, as well as in coding of side information parameters such as motion and mode information. It was already pointed out in Sec. 7.2.4 that not all picture types (I, P, B) are equally relevant by their contribution to the overall distortion. It is usually a good strategy to give more relevance to pictures that have a large effect on predicting other pictures, e.g. an I picture in the beginning of a sequence that is followed by other pictures with low amount of changes; a similar consideration should apply to P and B pictures at the lower levels of hierarchical prediction structures. In this context, optimum settings highly depend on the amount of change in the sequence; with fast changes, the influence of encoding quality to subsequent pictures typically decays faster. Methods that utilize such dependency chains in the context of rate-distortion optimization bear high potential in hybrid encoder improvement, but again require multiple encoding passes, and due to the necessity of performing look-ahead over several pictures, causing additional latency. Manageable solutions with restriction to selected cases are described in [Ortega, Ramchandran, Vetterli 1994] [Ramchandran, Ortega, Vetterli 1994] and [Beermann, Wien, Ohm 2002]. For optimum decision in cross-picture look-ahead, the error feedback of the prediction loop (5.14) has to be analyzed to decide about the optimum way of jointly minimizing rate and distortion. Solutions which try to describe the behaviour of a video codec, taking into account recursive influences over multiple pictures (based on motion and mode decisions generated from a first pass) have been described in [Schumitsch et al. 2003] [Winken et al. 2007] and [Rusert 2007]. Such strategies are again not suitable for low-delay applications.

Another aspect of encoder optimization relates to *rate control*, in cases when a usage scenario has the requirement of constant-rate encoding, e.g. for transmission over networks with rate constraints. Rate regulation mechanisms are usually based on varying the quantizer step size, but the degrees of freedom depend on the allowable delay and buffer sizes (see Sec. 9.3). The interdependency between rate control and rate-distortion optimization mechanisms, in particular best strategies to hold variations in the quantizer step size as close as possible to the R-D optimum case, should be considered in this context.

Finally, even though formally not being part of the encoder loop, *preprocessing* can have a large effect on compression performance. With a rate given to encode a sequence of certain properties (amount of detail, motion changes etc.), it has to be decided which is the best spatio-temporal resolution by which the sequence should be encoded. Possibly, smoothing out some details or reducing resolution can be a better strategy in terms of perceptual reconstruction quality, rather than introducing compression artifacts due to rate constraints. Semantic criteria can be taken into account by identifying *regions of interest* to be encoded by higher quality, e.g. human faces or letters in a videoconferencing scenario. Similarly, blurring (or resolution reduction) can be applied to less relevant regions. In the context of quantizer optimization, region of interest definition and perceptual weighting functions can also be taken into account, similar as in still image coding (see Sec. 6.4.2). Here, video offers considerable degrees of freedom to optimize encoding for better subjective quality at a given rate. For example, human observers are hardly able to detect detailed structures in the case of fast motion, which means that effects such as motion blur could be mimicked by pre-processing as well, which could also improve the performance of motion-compensated prediction by producing less prediction errors or allowing representation of motion parameters with less precision.

7.2.8 Hybrid coding using subband/wavelet transforms

Replacing the block transform by a subband or wavelet transform in the encoding and prediction loop of a hybrid coder facing the following problems:

- For motion compensation and mode switching in hybrid coders the block-based approach is often used for its simplicity and effectiveness. This can produce discontinuities and local high spatial frequencies in the residual at positions where adjacent blocks' motion vectors diverge, or where switching between two modes (intra/inter) or picture reference indices occurs. Whereas boundaries of block-wise motion compensation and block-based transform can be aligned, subband and wavelet transforms have a block-overlapping effect by their longer impulse responses, which can cause interferences (e.g. ringing artifacts) at the residual's discontinuities.
- The advantage of spatial scalability of the 2D wavelet transform cannot be transferred in a hybrid coder structure. Due to the fact that the transform is not shift invariant (i.e. the result depends on the subsampling phase), subsampling in the motion compensation loop (by discarding higher-frequency bands from the residual) may not allow an artifact-free reconstruction after inverse motion compensation at lower resolution.

Solutions to the first problem include usage of warping-based or overlapping block MC (Sec. 7.2.5); smooth blending functions reduce the problem of amplitude discontinuities in case of intra-mode switching [MARTUCCI ET AL. 1997] [HEISING ET AL. 2001]. Performance gain over block-transform based hybrid coding (under comparable optimization conditions) is reported, however the complexity is also significantly increased.

At first sight, it appears that the second problem could be solved by *subband domain motion compensation.* Here, the frequency transform over the spatial coordinates is performed *first* (outside the prediction loop), and then motion-compensated prediction coding is applied in parallel to the spatial images of sub-sampled transform coefficients. Different methods have been proposed both for motion compensation in the wavelet transform domain and in the DCT domain e.g. [BOSVELD ET AL. 1992] [YANG, RAMCHANDRAN 1997] [SHEN, DELP 1999] [BENZLER 2000] [VAN DER AUWERA ET AL. 2002]. None of such methods has been shown to be competitive in compression performance, compared to hybrid coding with one motion compensation at full resolution. Reasons are the *shift variance* of the transforms, which prevents a reasonably precise prediction, and the alias inherent in the band-filtered and sub-sampled images, which again are highly shift-variant and unpredictable between different sub-sampling phases (especially in high-frequency bands where even frequency reversion could be occurring in sub-sampling, see Sec. 2.8.1)[31].

A possible solution to these problems is the application of an *over-complete* wavelet transform (ODWT), which is a representation of the lowpass and highpass filtered signals without sub-sampling. It is however not necessary to encode and transmit the over-complete information, as it can be reconstructed at the decoder side from the *critically sampled* signal [ZACIU ET AL. 1996]. A block diagram of an over-complete domain in-band motion-compensated wavelet video codec is shown in Fig. 7.21. For the prediction at each spatial resolution scale, only information from this scale and all lower scales is needed, such that spatial scalability is guaranteed without drift between the predictions of encoder and decoder. The ODWT representation as needed for MC prediction is generated locally by an inverse DWT up to the required resolution level. As the motion shift is known, it can even be decided a priori which phases of the ODWT need to be reconstructed, such that the complexity can be reduced [AUWERA ET AL. 2002]. Likewise, it is possible to combine the steps of IDWT and ODWT into just one interpolation filtering operation [LI, KEROFSKY 2002].

In general, besides additional complexity, one problem of ODWT hybrid coding is the difficulty to determine motion parameters that are suitable across spatial resolutions. Rate-distortion optimization is more difficult than in 2D wavelet coding and in conventional hybrid coding due to the tight integration between transform and motion compensation processes.

Still, the problem exists that alias-free subsampling is not possible with non-ideal subband filters. Therefore, alias will typically be present in the scaled-down pictures even with the ODWT approach. It should further be noted that the motion compensation should be aligned between all subbands, in order to avoid possible phase distortion artifacts in the reconstruction. Finally, in terms of storage requirements and motion compensation operations over all scales, there is no benefit of the ODWT method, compared to a differential pyramid that so far has shown to be a better choice for spatially scalable video coding with hybrid coders (see Sec. 7.3).

---

[31] If content is moving, it is likely that it appears in the reference picture with a different subsampling phase.

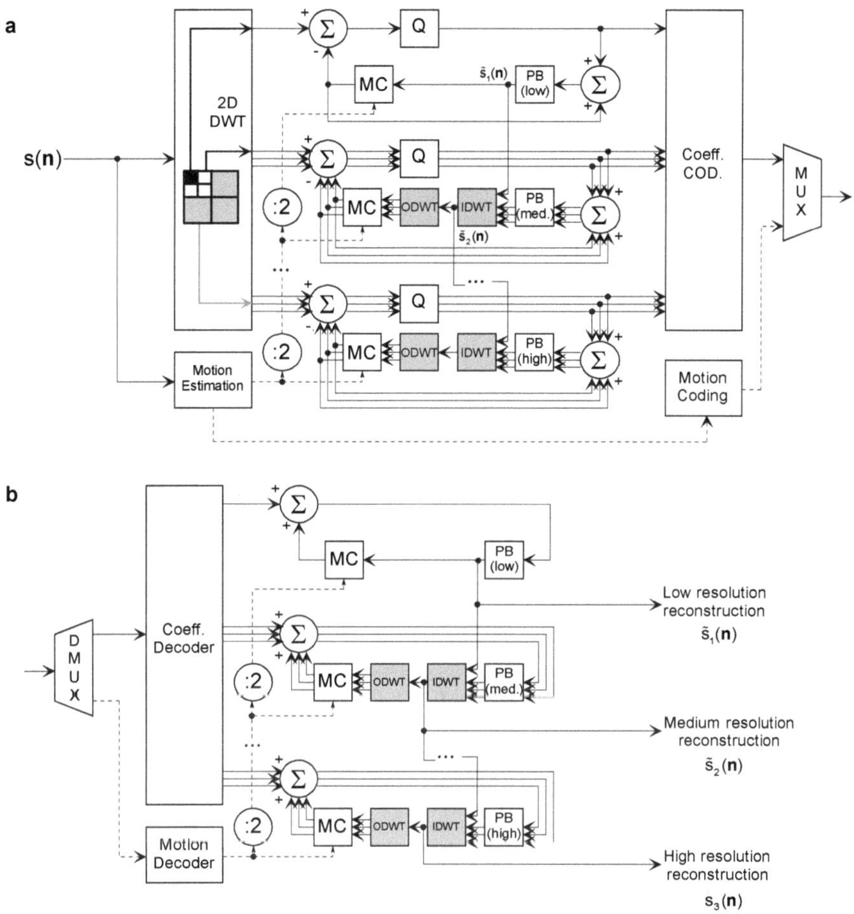

Fig. 7.21. Video codec using an over-complete wavelet transform in the prediction loop: **a** Encoder **b** Decoder [DWT: Discrete Wavelet Transform; IDWT: Inverse DWT; ODWT: Over-complete DWT; MC: Motion Compensation; Q: Quantization; PB: Picture buffer]

7.3 Spatio-temporal transform coding

7.3.1 Inter-picture transform and subband coding

Transform coding can also be applied over three (two spatial and one temporal) dimensions. Approaches using block-based [NATARJAN, AHMED 1977] and sub-band/wavelet transforms [KARLSSON, VETTERLI 1987] had already been proposed in the early history of video coding research. The discrete block transform over a group of M_3 pictures would covers a 'cube' of side lengths M_1, M_2 and M_3 in the two spatial and one temporal dimensions, respectively (Fig. 7.22a).

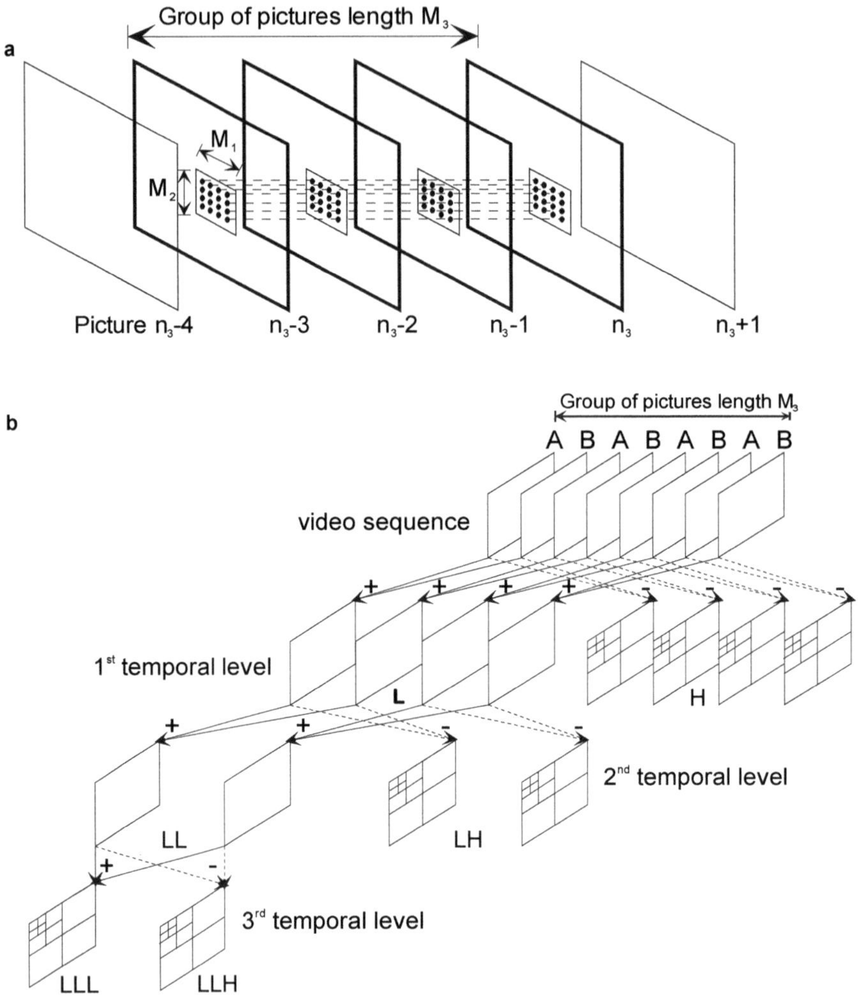

Fig. 7.22. **a** Extension of a block transform into the temporal dimension by definition of a spatio-temporal 'cube' **b** Spatio-temporal wavelet decomposition using T=3 levels of a temporal wavelet tree based on Haar functions

Fig. 7.22b illustrates a 3D Wavelet transform using a Haar filter basis for processing along the temporal axis[32]. This can be interpreted as decomposition of picture pairs into average (lowpass) and difference (highpass) coefficient pictures

$$\begin{aligned}\underbrace{s_u(n_1,n_2,n_3)}_{s_\mathrm{L}(\mathbf{n})} &= \frac{1}{2}\left[\underbrace{s_{u+1}(n_1,n_2,2n_3)}_{s_\mathrm{A}(\mathbf{n})}+\underbrace{s_{u+1}(n_1,n_2,2n_3+1)}_{s_\mathrm{B}(\mathbf{n})}\right] \\ \underbrace{c_u(n_1,n_2,n_3)}_{c_\mathrm{H}(\mathbf{n})} &= \underbrace{s_{u+1}(n_1,n_2,2n_3)}_{s_\mathrm{A}(\mathbf{n})}-\underbrace{s_{u+1}(n_1,n_2,2n_3+1)}_{s_\mathrm{B}(\mathbf{n})}.\end{aligned} \tag{7.32}$$

For simplicity, the even and odd pictures (polyphase components of temporal decomposition) are denoted as 'A' and 'B'[33], whereas the lowpass and highpass outputs of the corresponding level are 'L' and 'H', respectively; in a subsequent level of the wavelet tree, two consecutive 'L' pictures of the current level will become 'A' and 'B'. In an iterated decomposition this gives 'LL', 'LLL', 'LLH' etc. A 2D spatial wavelet transform can be applied to the decomposed pictures of the terminating temporal bands. The localization property of the wavelet transform provides energy compaction such that instantaneous temporal changes only influence single collocated highpass pictures. With T temporal wavelet tree levels, the resulting group of pictures length (or highest subsampling factor in the temporal-axis transform) is $M_3=2^T$. The following interpretations can be made:

- In case of no motion, all relevant information will be concentrated in the lowest-frequency band pictures, which then establish an average over M_3 pictures each; if the picture content is stable over time, this discards noisy variations.
- The 3D (spatio-temporal) spectrum of a moving scene is sheared but still sparse (cf. Sec. 2.2.2). It can however be expected that the placement of higher-energy components is spread over several bands, when the filters are non-ideal with overlapping Fourier transfer functions. This effect even becomes worse with large motion, since then higher spatial frequencies may be folded into lower temporal frequencies due to alias. Therefore, the actual sparseness of the spectrum may be difficult to exploit.

Strategies for encoding 3D frequency coefficients require careful consideration of these effects. 3D wavelet schemes could allow utilization of contexts which implicitly relate to the shear effects in the 3D spectrum, however such methods would fail in case of fast motion with alias folding. The 3D wavelet decomposition of Fig. 7.22b is interpreted as a *wavelet transform cube* in Fig. 7.23. A similar 3D zero-tree method extended from SPIHT (Sec. 6.4.4) has been introduced in

---

[32] A non-orthonormal version of the Haar basis is used in the following, which allows to interpret the temporal lowpass and highpass filter outputs as average and difference processes, respectively.

[33] Note that the meaning of polyphase position 'B' is different from the 'B pictures' commonly used in hybrid coding. Actually, the 'H' pictures as output from the process have some similarity with residual (P,B) pictures from motion compensated prediction

[KIM, XIONG, PEARLMAN 2000]. A 3D extension of EBCOT was proposed in [XU, LI, ZHANG 2000].

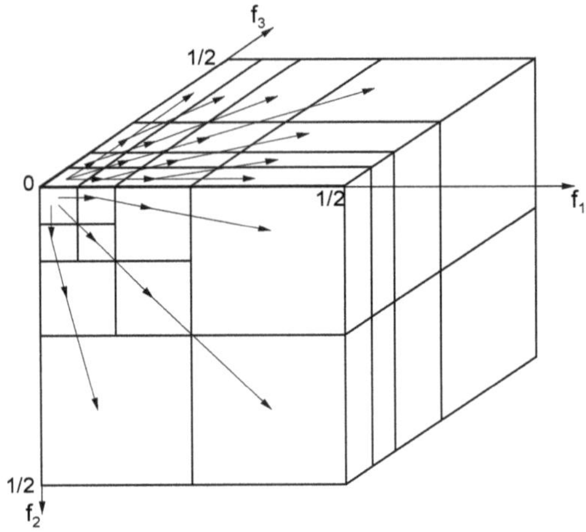

Fig. 7.23. 3D Wavelet transform cube with example zero-tree correspondences

As in replenishment coding, as a first remedy it is reasonable to perform a *change detection* and not use temporal-axis correspondences in case of dissimilarity. The temporal-axis filtering should be switched off when no compression advantage would be achievable.

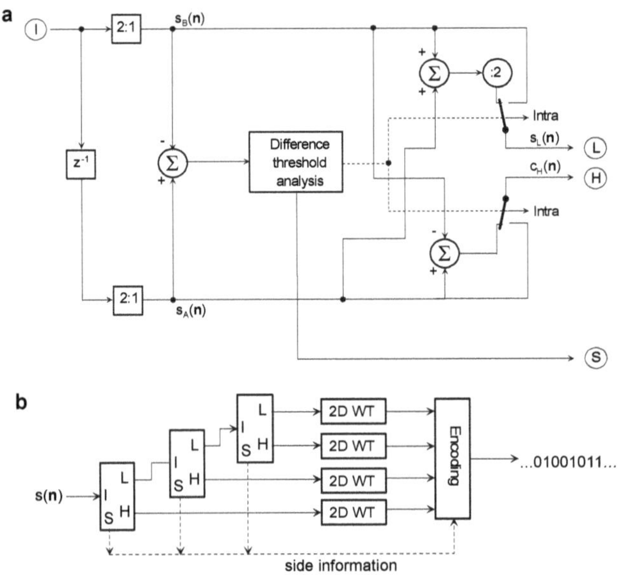

Fig. 7.24. 3D Wavelet decomposition with temporal-change adaptation
a Temporal 2-band split element **b** Cascaded structure

As significant changes may occur only for parts of the picture (e.g. in case of fast object motion), *local change* analysis is most useful [PODILCHUK, JAYANT, NOLL 1990] [QUELUZ 1992], where intra coding is applied instead of temporal-axis Haar filtering[34]. A block diagram of such a motion-adaptive method is shown in Fig. 7.24. If the picture difference goes beyond a threshold, the temporal-axis filtering is switched off in position I ('intra-picture'). A non-orthonormal set of Haar filters using different norms for the lowpass and highpass basis functions as in (7.32) is used here in the 2-band split element, which is implemented by a polyphase structure (cf. Sec. 2.8.3).

The approaches of block transform and Haar-based wavelet transform may produce visible 'switching' at temporal block boundaries in case of higher compression (discarding of non-zero higher-frequency coefficients of the temporal component). To avoid this, longer filter kernels could be used in the case of wavelet transform, which however have the disadvantage that enabling/disabling switching of the temporal-axis decomposition cannot be implemented as simple as in the case of Haar filters. Furthermore, if stronger motion is present, longer filters could produce even more severe artifacts such as ghosting in the lower temporal-axis frequency decomposition pictures. As a reasonable solution, the biorthogonal 5/3 filter pair (2.325) could be used. If implemented in a lifting structure (Fig. 2.55b), it is also possible to implement local adaptation such as switching or motion compensation, which will be further discussed in the subsequent section.

7.3.2 Motion-compensated temporal filtering

Application of motion compensation (MC) to align the transform operation along the temporal axis with a motion trajectory was introduced in [KRONANDER 1989], which however resulted in an over-complete representation in case of inhomogeneous motion displacement. Whereas the combination of MC with temporal-axis block transforms requires multi-picture correspondences which may cause inconsistencies in case of locally varying motion and occlusion, subband/wavelet transforms constructed from cascaded two-band structures can straightforwardly be implemented as *motion-compensated temporal filtering* (MCTF), using only correspondences between very few pictures in each step when short kernel filters or lifting implementations are used[35]. It should be emphasized that due to the fact that temporal and spatial dimensions are independent, combination of MCTF with 2D wavelet transform is not inherently necessary; moreover, a cascaded MCTF structure could likewise be interpreted as an extension of hierarchical picture concepts in open-loop hybrid coding.

---

[34] Note that in case of fast motion, the averaging applied by the Haar lowpass filter may cause ghosting artefacts, which will appear in the reconstruction as well when the corresponding highpass information is discarded.

[35] Indeed, when motion compensation is applied within the context of fast transform algorithms this should likewise be possible with block transforms.

In simplest case, Haar basis functions (2.312) are used, applying motion shift $\mathbf{k}=[\ k_1\ \ k_2\ ]^{\mathrm{T}}$ between pairs of pictures, with z-transfer functions for lowpass and highpass, respectively,

$$H_0(z)=\frac{\sqrt{2}}{2}+\frac{\sqrt{2}}{2}z_1^{-k_1}z_2^{-k_2}z_3^{-1} \text{ and } H_1(z)=-\frac{\sqrt{2}}{2}z_1^{k_1}z_2^{k_2}+\frac{\sqrt{2}}{2}z_3^{-1}. \qquad (7.33)$$

In a similar notation as in (7.32), and considering a non-orthonormal modification of the filters where H_0 and H_1 are scaled by $\sqrt{2}/2$ and $\sqrt{2}$, respectively, this can be expressed as follows. Let 'A'/'B', 'L' and 'H' express even-/odd-indexed input, lowpass and highpass output pictures, respectively,

$$s_{\mathrm{L}}(\mathbf{n})=s_{\mathbf{A}}(\mathbf{n}+\tilde{\mathbf{k}})+s_{\mathrm{B}}(\mathbf{n}) \text{ and } c_{\mathrm{H}}(\mathbf{n})=s_{\mathbf{A}}(\mathbf{n})-s_{\mathrm{B}}(\mathbf{n}+\mathbf{k}), \qquad (7.34)$$

where typically $\tilde{\mathbf{k}}=-\mathbf{k}$, and the coordinate systems of A|H and B|L are pair-wise aligned. If k_1 and k_2 are integer numbers which are constant over the entire picture pair[36], a perfectly invertible analysis/synthesis relationship exists except for boundary samples where correspondences might refer to samples outside of the other picture. In the latter case, periodic extension of pictures could be used; alternatively, intra switching as shown in Fig. 7.24a can be applied, which is even more consistent in case of missing correspondences (denoted as 'unconnected samples' here).

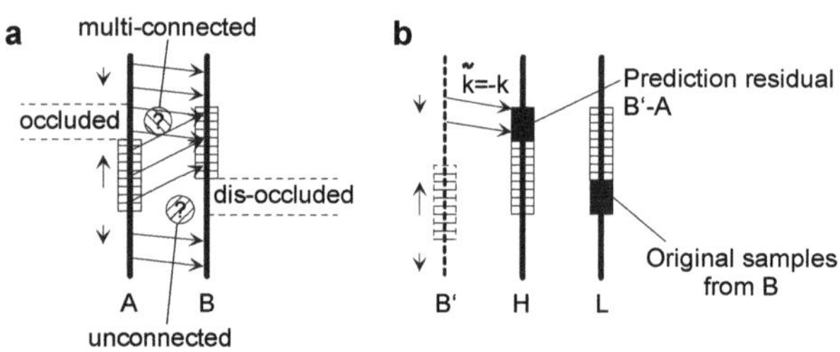

Fig. 7.25. **a** occurrence of 'unconnected' and 'multiple connected' areas **b** insertion of original samples from B in L, prediction errors from A in H

If motion vectors are *spatially varying* over the picture, unconnected samples could be present anywhere, and beyond that, multiple connections may occur in cases of motion vector overlaps (Fig. 7.25a). All uniquely connected samples can be reconstructed by synthesis filtering, which needs to include the inverse MC (sample accurate) mapping. In the multi-connected case, information would be duplicated in different transform coefficients of (7.34), while in the unconnected case, information would be missing and reconstruction would be impossible. In [OHM 1993], a first method was suggested to substitute unconnected samples in L by original samples from picture B, and multiple connected samples in H by plain picture differences (prediction errors), such that unique reconstruction is possible.

[36] e.g. in case of global translational motion

This leads to three classes of samples, 'connected' where the normal Haar lowpass and highpass filter operations are performed, 'picture B intra' and 'picture A predicted'. This method guarantees perfect reconstruction only for the case of sample-accurate MC; the prediction of the samples in picture *A* can either be done from reconstructed B samples, or from another picture B' (if it is a preceding picture, the motion vector should be inverted, see Fig. 7.25b). In the time sequence, the multi-connected case can also be interpreted as areas becoming occluded, whereas the unconnected case would refer to dis-occluded areas.

Motion-compensated lifting filters. Implementation of non-orthonormal Haar filters by the lifting structure (2.343) inherently enforces the spatial coordinate relationships as defined in (7.33) and (7.34), where the coordinate system of B coincides with L, and A with H. In the lifting implementation of Fig. 2.55, the prediction and update filters are replaced by 2D filters integrating the motion shift, such that $P(\mathbf{z}) = -z_1^{k_1} z_2^{k_2}$ and $U(\mathbf{z}) = \frac{1}{2} z_1^{\tilde{k}_1} z_2^{\tilde{k}_2}$,

$$c_{\mathrm{H}}(\mathbf{n}) = s_{\mathrm{A}}(\mathbf{n}) - s_{\mathrm{B}}(\mathbf{n}+\mathbf{k}) \text{ and } s_{\mathrm{L}}(\mathbf{n}) = s_{\mathrm{B}}(\mathbf{n}) + \frac{1}{2} c_{\mathrm{H}}(\mathbf{n}+\tilde{\mathbf{k}}) = \frac{1}{2}\left[s_{\mathrm{B}}(\mathbf{n}) + s_{\mathrm{A}}(\mathbf{n}+\tilde{\mathbf{k}})\right], \quad (7.35)$$

Except for normalization, this is equivalent to (7.34) for the case $\tilde{\mathbf{k}} = -\mathbf{k}$ and full-sample shift. Re-defining the motion-compensated Haar filters by a lifting structure however makes it possible to release the restriction of full-sample shifts and achieve perfect reconstruction for arbitrary motion vector fields, by simply including additional sub-sample interpolation filters in $P(\mathbf{z})$ and $U(\mathbf{z})$, where perfect reconstruction can always be achieved by the reverse lifting in the synthesis operation. Even the case $\tilde{\mathbf{k}} \neq -\mathbf{k}$ becomes possible, however this could produce ghosting artifacts in the *L* picture. The interpretation by lifting filters was first provided in [PESQUET, BOTTREAU 2001], [LUO ET AL. 2001] and [SECKER, TAUBMAN 2001]. A special case had previously been developed in [OHM, RÜMMLER 1997], where it was shown that the polyphase kernels of 1D or 2D biorthogonal filter pairs can be used as perfect-reconstructing interpolation filters in the case of a half-sample accurate (bilinear interpolated) motion compensation with temporal-axis Haar filters; an operational MCTF coding system based on this approach was later reported in [CHOI, WOODS 1999]. An analysis about theoretical performance bounds is given in [FLIERL, GIROD 2003A].

As in MC prediction of hybrid coding, usage of high-quality interpolation filters in $P(\mathbf{z})$ and $U(\mathbf{z})$ provides better compression gain, because improved motion compensation reduces the energy in H and concentrates information in L. The lifting flow allows referencing (sub-)sample shifts against any sampling position, guaranteeing perfect reconstruction [SECKER, TAUBMAN 2001].

Extended temporal axis lifting filters. One single analysis level of the temporal-axis wavelet tree, again by view of a pair-wise picture decomposition, is illustrated in Fig. 7.26a, giving yet another interpretation of the motion-compensated Haar filters. The MC prediction step in the lifting filter structure (resulting in the H picture) is practically identical to unidirectional MC prediction in hybrid coding, when operating in a hierarchical picture structure (see Fig. 7.8b). The spatial MC operations (optionally including sub-sample interpolation) are now directly included in the lifting structure of the temporal filter.

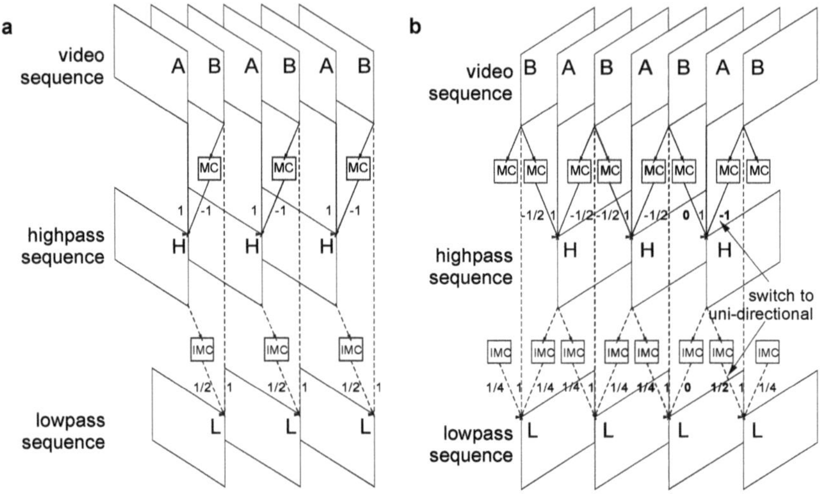

Fig. 7.26. MC wavelet transformation step A/B→H/L in lifting structure **a** Haar filter with uni-directional prediction and update **b** 5/3 filter with bi-directional prediction and update, giving effectively 5-tap lowpass filter

It is now also straightforward to extend this scheme to bi-directional picture prediction concepts. The prediction step then is identical to the *bi-predictive* pyramid (Fig. 7.8a). In the definition used here, odd pictures B expose the lower (temporally subsampled) level of the pyramid. The entire principle is shown in Fig. 7.26b, which can be derived from the lifting implementation of the bi-orthogonal 5/3 filter (2.342) in Fig. 2.55b, where the temporal prediction and update filters are including spatial shift and interpolation for MC. Also the update step is performed bi-directionally, where still the reverse correspondence between MC and IMC should be observed to avoid artifacts in the L picture[37]. Similar to the case of MC prediction coders, it is also possible to switch dynamically between forward, backward and bi-directional prediction, or implement intra mode switching. If for example an H picture shall only be computed by the prediction of a picture A from the subsequent picture B, the left weight of the prediction step can be set to 0, and the right weight to −1. The subsequent update operations should be similarly aligned. An example is shown in the data flow of the rightmost H picture in Fig. 7.26b.

[37] The symmetry of prediction and update steps, both by the relationship of MC/IMC and with regard to the lifting-filter weights is important for optimum encoding performance, cf. Sec. 7.3.3.

An important difference between hierarchical prediction schemes in hybrid coding and temporal wavelet decomposition by MCTF structures is the impossibility to operate MCTF in a closed-loop structure (i.e. the encoder being aware of the prediction quality that the decoder has available). This is due to the fact that in the decomposition the update step *follows* the prediction step, whereas during the synthesis at the decoder, the inverse update step must be performed *before* the inverse prediction is applied for reconstruction (i.e. the decoder reconstructs the adjacent picture(s) B first before reconstructing the picture A, whereas the encoder first has to predict A from B and can therefore only use the originals, not the decoded pictures). On the other hand, with good MC prediction at low data rates, it could be expected that major parts of the H pictures are quantized into zero, which would have the consequence that the L pictures are merely copied from the underlying pictures B during the update step. This would mean that in extreme cases of low-rate closed-loop coding, the update step could become by tendency irrelevant at the decoder. If it is however executed at the encoder from original pictures A and B, the L picture that is used for prediction of A is similar to the underlying picture B, but the temporal lowpass filtering operating can have a denoising effect which could be beneficial when the motion compensation in prediction succeeds. Otherwise, due to the divergence of prediction and update references of encoder and decoder, MCTF based hierarchical schemes tend to have more quality fluctuations than closed-loop hybrid coders with hierarchical picture structure, which would be most noticeable at the pictures A of the least relevant level. They however provide better advantage in the range of low-distortion encoding (high rates). Though closed-loop coding is not applicable in the context of prediction/update lifting, the drift becomes clearly less severe than with open-loop hybrid coders due to the update step. The effects of quantization and error propagation are further discussed in the next section.

Alternative MC schemes. When MCTF is combined with block transform coding, block-based motion compensation is an appropriate choice. In that case, deblocking post processing can further enhance the quality at low rates. When the lowpass and highpass pictures resulting from MCTF are fed into a subband or wavelet transform, artificial block structures due to MC could lead to interference with the transform basis functions, and cause not only blocking artifacts in cases when high frequency coefficients are quantized into zero, but also additional ringing effects. To avoid this, combination of 3D subband and wavelet coders with warping MC was described in [OHM 1994B] and [SECKER, TAUBMAN 2001]. Alternatively, overlapping-block methods can be used [HANKE, RUSERT, OHM 2003]. The block overlap tends to blur prediction differences in the H picture in the vicinity of motion boundaries, but will also produce more blurred areas in the L picture where the motion is inconsistent. This is beneficial for higher compression efficiency as well as improved subjective quality, when L picture sequences shall be displayed directly with lower frame rate (see quality improvement in Fig. 7.27).

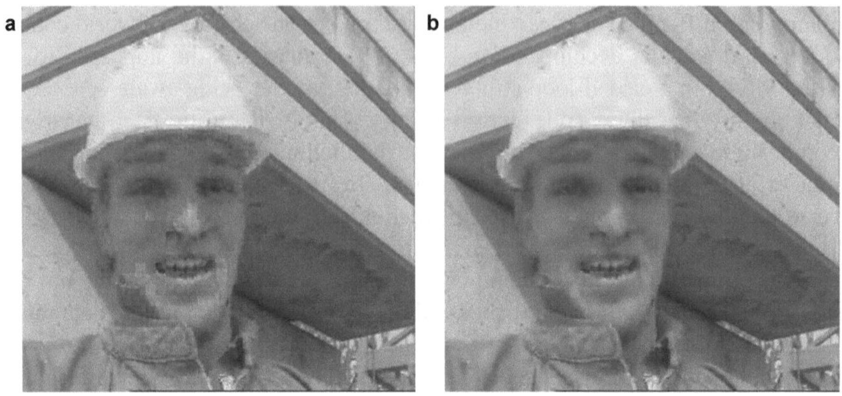

Fig. 7.27. Reconstruction of video pictures from an MCTF wavelet coded representation with block-based MC (*a*) and with OBMC/deblocking (*b*)
[pictures courtesy HANKE/RUSERT/THULL]

7.3.3 Quantization and encoding of MCTF-based representations

The transforms used in MCTF (in the versions introduced here) are not orthonormal[38]. For quantization, it is therefore necessary to scale the mapping of the transform domain quantization error to the decoding error accordingly. For simplicity, we assume that an orthonormal spatial transform is used, such that the aspect of optimum quantization can be analyzed separately for the temporal transform; otherwise, scaling factors for spatial and temporal transform have to be multiplied.

Firstly, the case of Haar filters is regarded in detail. To obtain an orthonormal representation from (7.35), H pictures would need to be multiplied by $a_{\mathrm{H}} = 1/\sqrt{2}$, L pictures by $a_{\mathrm{L}} = \sqrt{2}$, according to (2.343). Practically, instead of scaling the amplitudes, quantization step sizes can be scaled by the reciprocal factors, which avoids non-integer factors due to the fact that the ratio $a_{\mathrm{L}}/a_{\mathrm{H}}=2$, which means that L pictures should be quantized by half step size. Compared to hybrid coding, this is a clearly larger ratio than typically applied in adjusting the step sizes of P versus I pictures in hybrid coding, which otherwise have similar characteristics as L and H, respectively (see Sec. 7.2.4). This can also be interpreted through the signal flow in the lifting filter structure in Fig. 7.28a. A quantization error q_{H} of the H component is first weighted by a factor ½ and then subtracted from the quantization error q_{L} of the L component. The error injected into the reconstructed B is then $\Delta_{\mathrm{B}}=q_{\mathrm{L}}-\frac{1}{2}q_{\mathrm{H}}$. Now, in the synthesis of A, the prediction is generated from the reconstructed B and added to H. The reconstruction error in A is $\Delta_{\mathrm{A}}=q_{\mathrm{L}}+\frac{1}{2}q_{\mathrm{H}}$, which shows that in both reconstructed pictures the quantization errors in H only contribute by half amplitude. Under the assumption

---

[38] Beyond the purpose of simpler computation, this is necessary to avoid discontinuity artifacts between 'normal' and 'unconnected/multi-connected' samples.

of independent quantization errors in L and H, and assuming that a non-normalized quantization error q is scaled into the values $q_L=q/a_L$ and $q_H=q/a_H$, the reconstruction error variance is:

$$\sigma^2_{\Delta_B} = \mathcal{E}\left\{\left(\frac{\sqrt{2}}{2}q\right)^2\right\} + \mathcal{E}\left\{\left(-\sqrt{2}\frac{1}{2}q\right)^2\right\} = \mathcal{E}\{q^2\}$$
$$\sigma^2_{\Delta_A} = \mathcal{E}\left\{\left(\frac{\sqrt{2}}{2}q\right)^2\right\} + \mathcal{E}\left\{\left(\sqrt{2}\frac{1}{2}q\right)^2\right\} = \mathcal{E}\{q^2\}. \tag{7.36}$$

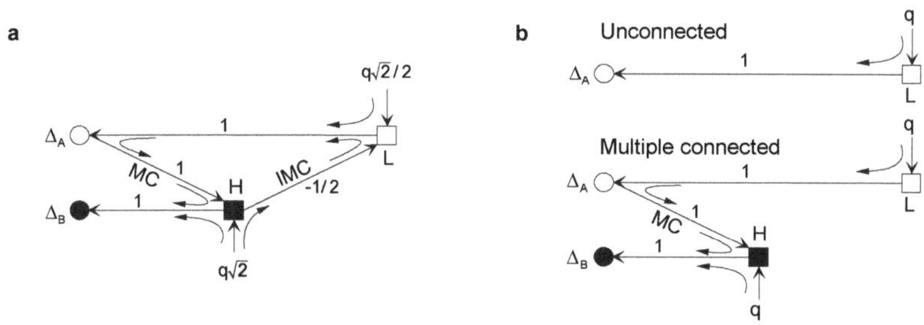

Fig. 7.28. Synthesis signal flow of quantization errors: **a** in the Haar lifting structure **b** in case of unconnected / multiple-connected samples

From (7.36), $\sigma^2_{q_L}$ and $\sigma^2_{q_H}$ are equally spread into the reconstruction error variances of pictures A and B, respectively. This is however only true if the path via IMC/L/H/MC compensates half of the quantization error, i.e. IMC has to be an *exact inverse* of MC. If this was not the case, the second path propagates a quantization error ½q_H' into another sample of H, and the reconstruction error in A would become $\Delta_A=q_L+q_H-½q_H'$. If q_H and q_H' are correlated, which can be assumed for small misalignments, there would still be an almost perfect compensation. Strictly spoken, only for the cases of full-sample shifts or ideal sub-sample interpolation, MC and IMC can perfectly match. Otherwise, the choice of the interpolation filters has a direct influence on the reconstruction error[39].

Similarly, scaling of quantization step sizes for other types of filters can be derived from (5.56). For example, in the case of MCTF with 5/3 filters as in Fig. 7.26b, the lowpass component should be quantized more accurately by a factor which is the ratio of the Euclidean norm of the filter coefficients,

$$\sqrt{\frac{\left(-\frac{1}{2}\right)^2+1^2+\left(-\frac{1}{2}\right)^2}{\left(-\frac{1}{8}\right)^2+\left(\frac{1}{4}\right)^2+\left(-\frac{3}{4}\right)^2+\left(\frac{1}{4}\right)^2+\left(-\frac{1}{8}\right)^2}} = \sqrt{\frac{48}{23}} \tag{7.37}$$

[39] Due to this, it can be observed that a sub-sample accuracy of 1/8 sample in motion compensation still provides gain in MCTF coding, while for hybrid coding usually a saturation is reached for about 1/4 sample accuracy.

as compared to the highpass component, which means that the quantizer step size for the lowpass signal should be smaller by roughly 1.45. It should be noted that, compared to the factor 1.225 derived for the case of hierarchical bi-prediction in hybrid coding under (7.23), this ratio is still higher and would theoretically impose a coding gain of approximately 0.7 dB per additional level at high rates[40], and again assuming perfect match of MC and IMC. A theoretical analysis of the coding gain achievable by MCTF is performed in [FLIERL, GIROD 2003], who came to the conclusion that up to 40% reduction in rate, as compared to a hybrid coder with same methods of motion compensation, could be achieved at high rates.

Quantization of unconnected and multiple-connected samples. The synthesis flows in cases of unconnected and multiple-connected samples are illustrated in Fig. 7.28b. For the unconnected case, $\Delta_B=q$ should be chosen if a normalization factor $a_L=1$ is applied, such that $\sigma_{\Delta_B}^2=\mathcal{E}(q^2)$. For the multi-connected case, plain prediction error samples are embedded into the *H* picture, which need to be quantized with a normalization factor $a_H=1$ causing an error variance $\mathcal{E}(q^2)$ in the reconstructed picture A. If additionally the prediction reference (e.g. in the current picture B or in a previously decoded picture) had been affected by another $\sigma_{\Delta_B}^2=\mathcal{E}(q^2)$, the total error under assumption of statistical independence becomes $\sigma_{\Delta_A}^2=2\mathcal{E}(q^2)$.[41] If the latter effect occurs, multiple-connected positions suffer both from the error that is intruding via the reference sample, and from the quantization error of the prediction residual. As such errors can propagate and be superimposed over several levels of an MCTF pyramid, quality fluctuations could become more severe. This is particularly the case when non-orthonormal Haar filters are used, where the ratio of quantization step sizes for 'connected' and 'multi-connected' samples is even larger than for the case of 5/3 filters.

Due to the relevant difference in optimum quantizer setting for *L*/*H* connected, unconnected and multi-connected positions, it is necessary to adjust the quantization weights (or the normalization factors) locally. However, in a full (spatio-temporal) 3D coder, transform coefficients are quantized instead of samples. In the simplest case, the number of sample types can be counted in a local area to adjust the quantization statistically. Beyond that, the direct synthesis effect of a given transform's basis functions onto the spatial sample position can be analyzed for even better local adjustment. With known motion parameters and mode set-

---

[40] Practically, such a coding gain cannot be realized in cases where the residual (respectively the samples in the H picture are very small due to motion compensation, and therefore quantized into the zero value

[41] This increase of errors at multiple-connected positions only occurs if the encoder performs prediction from original samples, which is similar to the error propagation in open-loop prediction systems. It should however be noted that this can be resolved by two-pass encoding, if sufficient computing resources are available, and the encoder generates decoded references solely for the purpose of better matching the prediction references of multiple-connected samples.

tings, this can be realized by an additional spatial transform applied to the sample pattern of optimum quantizer step size settings, see [OHM 1994A].

Encoding the motion-compensated 3D wavelet coefficients can be performed similar as in 2D wavelet coding (Sec. 6.4.4) or 3D wavelet coding without MC (Sec. 7.3.1). Due to the usage of MC, conventional 2D wavelet coders can more straightforwardly be applied to the subband L/H pictures resulting from MCTF processing without specific context-dependent coding over the time axis; this is particularly true in configurations where the entire temporal transform is performed first. This case shall be denoted as 't+2D' transform, corresponding to the scheme shown in Fig. 7.22b where temporal processing is performed first. '2D+t' approaches where the spatial wavelet transform is performed first, followed by MCTF processing over sequences from spatial subbands have also been investigated, but so far have been found to be less efficient in terms of overall compression. This is most likely due to the fact that alias residing in the 2D subbands after spatial subsampling is not shift invariant, making motion compensation difficult.

7.4 Coding of side information (motion, modes, partitioning)

Beyond efficient entropy coding of the transform coefficients (same as still picture coding, as described in Sec. 6.4.1), the main target in video coding is keeping the number of relevant coefficients as low as possible through efficient prediction. On the other hand, this requires signalling of side information which should not exceed the amount of bits saved in coding the transform coefficients. The most important components of side information in video compression are:

- Partitioning information (jointly or separate for the transform size/depth and partitions where identical motion parameters are used);
- Prediction reference information (including intra mode signalling, i.e. not using a different reference picture, as special case);
- Prediction parameter information, such as motion vectors for inter prediction, or directional modes for intra prediction;
- Information for adaptation and switching of loop or interpolation filter parameters;
- Parameters which turn on/off specific coding tools.

Whereas early video compression standards used fixed entities (16x16 macroblocks) as units for prediction, the H.263 and MPEG-4 part 2 standards already allow usage of different motion vectors at sub-macroblock level, where in case of MPEG-4 the partitions can be of arbitrary shape (see Sec. 7.8). For efficient compression, variable block-size partitioning with signalling that is equivalent to quad-tree coding (Sec. 6.1.2), is most widely used. For this purpose, a constant (typically square) size partition grid is defined as top level, which can further be split into sub-partitions. In the case of the AVC/H.264 standard, a 16x16 macroblock can further be sub-divided into 16x8, 8x16 or 8x8 sub-blocks, where for the case of 8x8, a second level for sub-division into 8x4, 4x8 or 4x4 can follow. This means that at each of the two levels, one out of 4 choices (no split, two bi-splits, one quad-split) can be selected (see Fig. 7.29). In HEVC, the following, even more flexible approach is used:

- Coding Tree Block structure: HEVC splits a picture into a grid of coding tree blocks (CTBs). The size NxN of a luma CTB can be chosen as N=16, 32, or 64 samples, with the latter being found to usually provide the best compression. Chroma CTBs have corresponding sizes, depending on the color sub-sampling format. The luma and the chroma CTBs together with the associated syntax form the coding tree unit (CTU).
- Coding Blocks and Units: The quad-tree syntax attached to the CTU specifies the size and position of all contained luma and chroma coding blocks (CBs), where again the chroma sizes are determined from the luma dependent on the chroma sub-sampling scheme. The root of the quad tree is associated with the CTU. For the case of no split, the size of the luma CTB is also the largest possible size of the luma CB. Luma and Chroma CBs together with the associated syntax form a Coding Unit (CU). A CTU contains one or multiple CUs, and below the CU, further partitioning into prediction units (PUs) and transform units (TUs) is done, the latter by optionally extending signalling by a second quad tree below the CU level. An example CTB partitioning into CBs and transform blocks (TBs), along with the associated quad tree structure, is shown in Fig. 7.30.
- Prediction Units and Prediction Blocks: The decision whether to code a picture area using the intra or inter prediction mode is made at the CU level. Depending on the prediction mode decision, the luma and chroma CBs can be further split into luma and chroma prediction blocks (PBs). HEVC supports variable PB sizes ranging from 64x64 down to 4x4 samples for intra prediction, and down to 4x8 or 8x4 samples for motion compensated prediction. The CB can be chosen not to be split, to be split into one of six different bi-split modes, or to be quad-split into prediction blocks (PB), see Fig. 7.31[42].
- Tree-structured Transform Blocks and Units: The prediction residual is coded using a block transform. The luma/chroma CB can be identical to the respective transform block (TB) or it can be further split into smaller TBs. In terms of the transform tree, it is possible in inter-picture prediction that the transform spans across prediction block boundaries[43].

[42] Intra prediction supports only square PB sizes, whereas the six different non-square sizes can only be used for inter prediction.

[43] In intra prediction, the block prediction loop follows the transform block sizes, e.g. when a PB of size 32x32 uses a certain intra prediction mode, it could still be split into 4x4 transform blocks, and the prediction process is performed over those 4x4 blocks, whereas all blocks use the same mode.

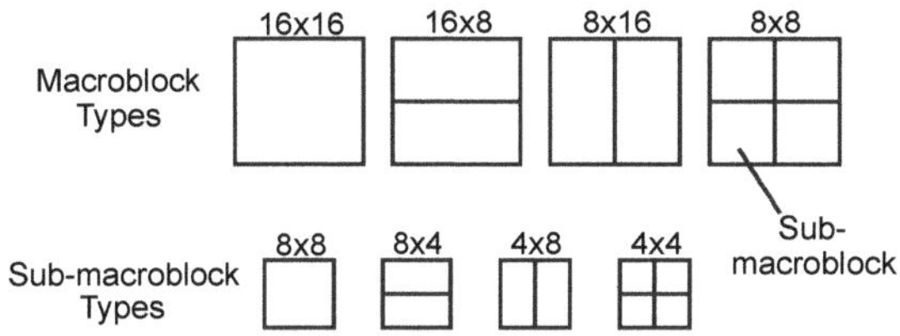

Fig. 7.29. Partitioning of a macroblock (top) and a sub-macroblock (bottom) for motion-compensated prediction, as defined in H.264/AVC

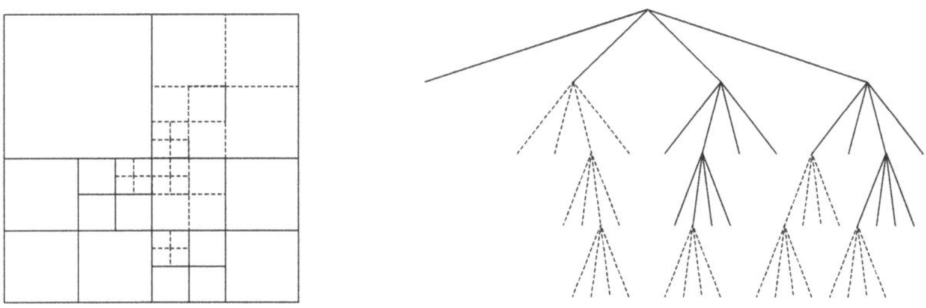

Fig. 7.30. Partitioning in HEVC: Subdivision of a CTB into CBs and TBs. Solid lines indicate CB boundaries and dotted lines indicate TB boundaries. Left: the CTB with its partitioning, right: the corresponding quad tree.

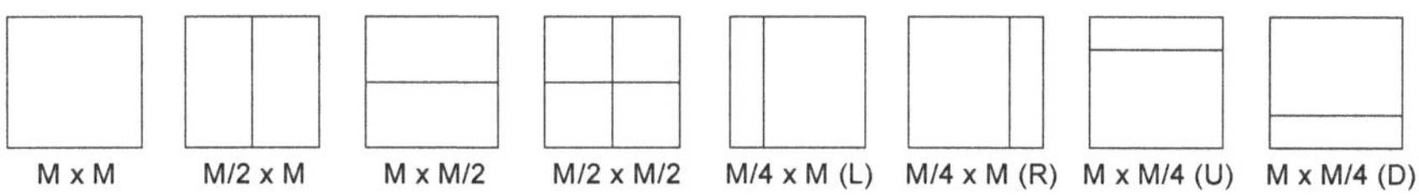

Fig. 7.31. Partitioning of a CB of size *M*x*M* into one of 8 different PB shapes, as defined in HEVC

Motion information is to be encoded/signaled for each of the sub-partitions (prediction unit in HEVC or each sub-partition of sub-macroblocks in AVC). Before further discussing specific methods for the encoding of motion vectors, some properties of motion vector fields and motion trajectories should be considered.

If an object moves through a scene or the camera is moved, the spatio-temporal coherence in a motion vector field is high, which can be utilized in motion estimation (Sec. 2.6.2), but also for more compact encoding of motion parameters. For both purposes, a prediction of the expected motion vector can be performed from the spatial neighborhood in the same picture, and/or from a motion vector in a temporally adjacent picture, which is most likely on the same motion trajectory. In estimation, it is then merely necessary to investigate a relatively small search range around the predicted value, for encoding it becomes more likely that the deviation from the prediction is small, such that entropy coding becomes more efficient.

Once estimated, motion parameters are usually encoded lossless. However it should be noted that this does not mean a lossless representation of the underlying motion vector field. In fact, most methods for encoding of locally varying motion vector fields are based on sub-sampling, e.g. through the size of partitions where samples are compensated by the same motion vector or same motion model.

In fact, the underlying motion vector field itself can be interpreted as a correlated signal, for which an approximation is generated from the encoded representation through a reconstruction. This approximation can even by purpose be made less accurate, e.g. using rate distortion optimization during estimation (cf. Sec. 7.2.7). In this case, the method of motion parameter coding has to be analyzed by the estimation procedure. If the spatio-temporal coherence is exploited in motion vector coding, a method which preferably estimates a spatio-temporally continuous motion vector field usually also requires a lower rate for encoding, due to better predictability of local motion vectors.

If predictive coding is applied, due to the fact that encoding of parameters should be lossless, no difference exits here between open-loop and closed-loop (DPCM) coding of motion vectors. In the whole context, it should be observed that the chosen sub-sample precision of motion vectors has to be retained in coding. With same search/encoding range, increasing the sub-sample precision usually requires more bits on the side of motion vector coding.

Predictive coding. Encoding prediction differences for motion vectors is more efficient than coding of original values, since the value can be expected to be smaller and to have a more consistent probability distribution, such that entropy coding is likely to become more efficient. Fig. 7.32 illustrates for the example of horizontal motion displacement that values of motion vector prediction differences Δk_1 (computed with reference to the left neighbor) are better concentrated around the value zero, and have smaller variance than the original k_1 values. The statistical distribution of motion vector differences is often assumed to follow a symmetric and exponentially decreasing (e.g. Laplacian) PDF.

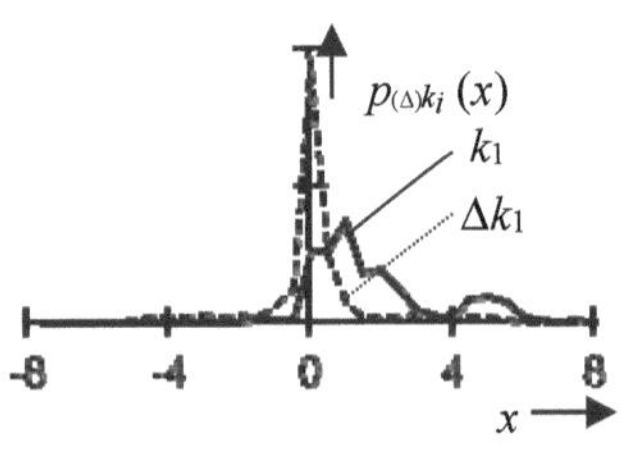

Fig. 7.32. Probability distributions of the horizontal motion vector component k_1 in a video sequence, and of the associated prediction difference component Δk_1

Only motion vectors from previously decoded partitions may be used for prediction. Linear prediction (e.g. averaging from a set of neighbor values) could be applied, but in cases where the neighbors' motion vectors are different, this might lead to an unreasonable result, as it is more likely that one of the neighbor candidates also applies for the current partition. *Median prediction* could be used, providing a 'majority decision'. Component-wise median prediction of MVs is one possible prediction method in the AVC standard for MV prediction at the macroblock and sub-macroblock levels, where already decoded vectors from three adjacent prediction blocks are used as input to median computation[44]. Instead of median prediction, only one predictor candidate is used in cases of 8x16 and 16x8 partitions, and when only one of the adjacent prediction blocks uses the same reference picture as the current block.

Alternatively, several predictors can be defined from nearest-neighbor candidates, where the index of the best predictor is encoded directly; this is referred as *motion vector competition* or *advanced motion vector prediction*, used in HEVC. Herein, depending on a rule-based consistency check among spatially neighbored prediction blocks and one vector from the collocated area in the current block's reference picture (the latter denoted as *temporal motion vector prediction*, TMVP), two candidates are selected and a flag is sent for signalling the one that is finally chosen. Again, the difference (prediction error) between that prediction and the actual value of the current block's motion vector is finally coded.

Reference picture index coding. If motion-compensated prediction can choose one of several reference pictures (as in case of AVC and HEVC), it is further necessary to signal the reference picture index (or two indices in case of bi-prediction). For this purpose, coding via reference picture lists or via the time difference can be used. Efficient coding of the reference picture index is also based on coherence with adjacent, already decoded blocks, which is achieved by defining prediction rules; explicit signalling is necessary only in cases where the prediction fails. In competition-based schemes, it is also useful to determine the reference picture index prediction along with the motion vector prediction to save signalling overhead. In cases where the reference picture index of the current block is different from the reference picture index of a neighbor block from which the motion vector prediction is computed, it is further useful to *scale* the motion vector prediction value w.r.t. the actual time distance between the current and the reference picture.

Spatial hierarchical coding. Multi-resolution or hierarchical motion estimation algorithms (cf. Sec. 2.6.2) are intended to support smoother and spatially more consistent motion vector fields. During the hierarchical process of estimation, an increasingly higher-accuracy representation of the motion vector field[45] is usually achieved.

---

[44] Actually, applying median computation independently to horizontal and vertical displacement components can again lead to unreasonable combinations; indeed, another choice would be usage of a vector median predictor [ASTOLA, HAAVISTO, NEUVO 1990], where the motion vector candidates are sorted by length.

[45] Accuracy can relate to precision (full-sample or sub-sample, i.e. quantization of the vectors), and to the spatial resolution (i.e. sampling) of the vector field. Note that, if the picture itself is scaled, both of these accuracies scale as well: A sample shift becomes a

This can be directly used for hierarchical encoding of the motion vector field by differential pyramid coding (Sec. 2.8.6). Motion vectors within a quad-tree like hierarchy can be encoded differentially (Fig. 7.33b), such that four vectors at the finer resolution are predicted from one parent at the lower resolution, or are signaled as identical (i.e. no split). Advantages are as follows:

- The differential pyramid has a decorrelating effect by itself, which can be more efficient than recursive one-step prediction in some cases;
- Motion parameters attached to the different levels of a pyramid can be applied to video pictures of different spatial resolution levels, such that *scalable coding* (see Sec. 7.2.4) implicitly includes scalability of side information;
- The latter bullet also implies that motion parameters from different hierarchy levels have different relevance, which can be exploited for error protection in case of transmission over lossy channels;
- Quad-tree coding (Sec. 6.1.2) allows signalling the extension of (block shaped) areas for which identical parameters are valid.

The aforementioned method has the disadvantage that the motion vectors of finer resolution are only determined from the directly collocated position at coarser resolution, whereas in cases of discontinuity, dependencies with neighbored positions can exist. An alternative is applying median prediction within hierarchical prediction. An example method for 3-tap median prediction from motion vectors of the next coarser hierarchy level is shown in Fig. 7.34[46]. The dots indicate the centers of square blocks, where the motion vectors are virtually positioned. For each dot, the three spatially closest values of the next-coarser resolution (e.g. predicting a value in the 4x4 grid from three values in the 8x8 grid) are used as input for the median computation.

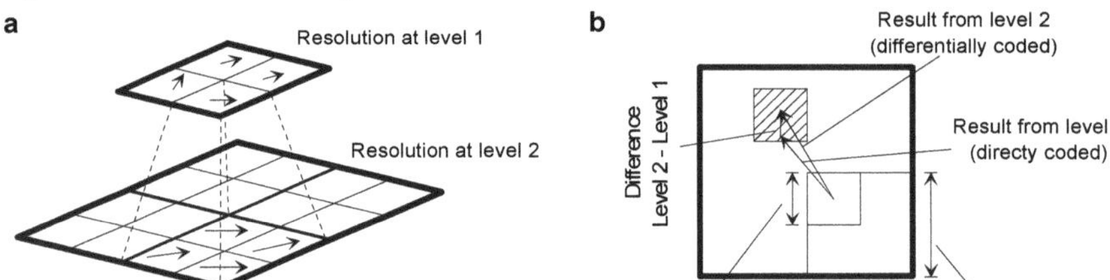

Fig. 7.33. Hierarchical coding of a block-based motion vector field:
a resolution accuracy within the hierarchy levels **b** differential encoding

Direct, skip and merge modes. In many cases, a perfect prediction of the motion vector (zero difference from predicted value) is possible. To achieve an even more compact signalling, combination of zero motion vector difference coding with inference of modes can be used in the following special cases:

---

half-sample shift in the 2:1 subsampled image, and a vector expressing motion within a 16x16 block maps to an 8x8 block after 2:1 subsampling.

[46] The median values can be computed separately for the horizontal/vertical vector components, or by establishing an ordered list by vector length (Euclidean vector norm).

- In direct mode, the motion vector as well as the prediction reference information, i.e. indexing of reference picture(s) and selection of uni- or bi-prediction can be derived from decoded values of neighboring regions, but a non-zero residual could still be present
- In skip mode, the same information as in direct mode can be derived, but additionally the residual is implicitly encoded as zero[47]
- In merge mode (defined in the HEVC standard), a prediction block is associated with the same motion vector and prediction reference information as a neighboring area, but a non-zero residual may need to be encoded.

In candidate based approaches, rules and mechanisms have to be defined to establish a valid (finite) list of candidates, where the signalling is then an index information referring to a position in that list.

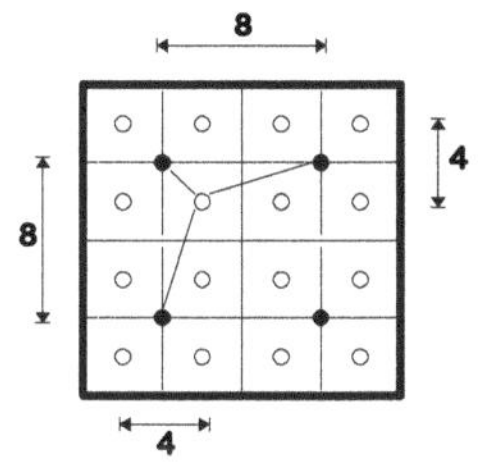

Fig. 7.34. Hierarchical 3-tap median prediction of block motion vectors (example predicting 4x4 block vectors placed at block center positions from spatially closest 8x8 blocks)

The motion vector field typically shows coherence in both spatially adjacent and temporally adjacent locations. If motion information from another picture shall be used for prediction, there is an inconsistency since the motion parameters themselves would describe the *trajectory* along which largest temporal coherence can be expected, and this is not known before decoding. To resolve this in a simple way, the motion vector of the *collocated area* of the reference picture is usually used for prediction. Fig. 7.35a/b show how this is implemented in the *temporal direct mode* of AVC. The first case (a) is using the motion vector from the collocated block of a temporally preceding reference picture. The second case (b) is for a hierarchical B picture structure where the motion vector from the collocated block in a reference picture at lower hierarchy level is used. As however the temporal distance between the B picture and the respective pictures from the lower level is different, it is necessary to scale the motion vector accordingly[48]. Fig. 7.35c shows the case of derivation of motion parameters in spatial direct and skip modes of the AVC standard, where the component-wise median value from three adjacent blocks is used, similar to computation in motion vector prediction.

---

[47] Note that in MPEG-4 part 2, skip mode is only invoked in case of *zero* motion vector. Therefore, the skip mode of newer standards (AVC, HEVC) where it can also be applied in case of perfect MV *prediction*, is more efficient.

[48] Scaling is also necessary in cases temporal direct (a) and spatial direct (c) when the motion vector taken from the reference picture or the neighbored block refers to a different temporal distance than applicable for the current block and its reference.

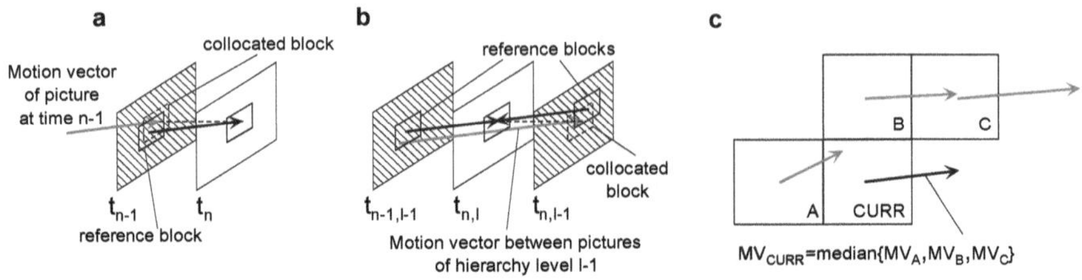

Fig. 7.35. a/b Usage of motion vector of the reference picture's collocated area in temporal direct mode of AVC **c** Derivation of motion vector as median value from neighbored blocks in spatial direct and skip modes

As the collocated position may deliver a wrong candidate in particular when the motion is large and/or the vector field is discontinuous, usage of motion vectors that are truly on the correct trajectory would be preferable, but it would be necessary to project vectors from their positions in the reference pictures to the current picture, which is difficult to determine when the coordinate system of the current picture is the anchor. In this context, it may also happen that either none or several vectors are projected into the same position, such that contradicting values may be found. As a remedy, either all available vectors may be used as candidates (which would require additional signalling), median computation could be performed or additional rules would be required, e.g. choosing the vector with smallest magnitude or least deviation from the neighbors. Even though such methods are not used in current standards, it can be expected that better analysis of temporal motion trajectories could provide further advantages in motion data coding; this would however require more computational complexity as well as storage of predicted candidate motion vectors with higher precision.

Even when the collocated position is used, in accessing the temporal candidate, it is necessary to store the motion information of reference pictures. To save memory, the granularity for storing the temporal motion candidates is restricted to an equivalent of a 16×16 block grid even in the HEVC standard, although smaller PB structures are used at the corresponding location in the reference picture.

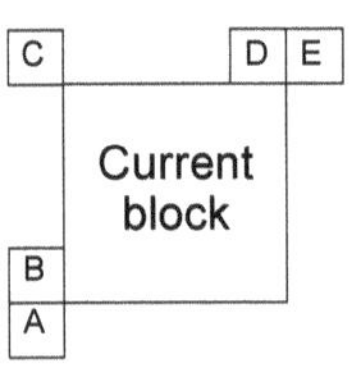

Fig. 7.36. Positions of five spatial neighbor block candidates A…E investigated for the current prediction block in HEVC (same arrangement of candidates used in merge and predictive MV coding)

The merge mode as implemented in HEVC is conceptually similar to the direct and skip modes in AVC; however, one out of several available candidates is signaled explicitly, similar to motion vector competition in prediction. The set of possible candidates in the merge mode consists of spatial neighbor candidates, a temporal candidate and additional generated candidates or zero values (if necessary). The number is fixed, and the list is filled until it is complete with candidates of different value. Fig. 7.36 shows the positions of five spatial candidates, where an extreme case is shown that a large prediction block would only have neighbors of smaller sizes. Similarly, the merge mode is often used to allocate the same vector of a larger adjacent block to a smaller block.

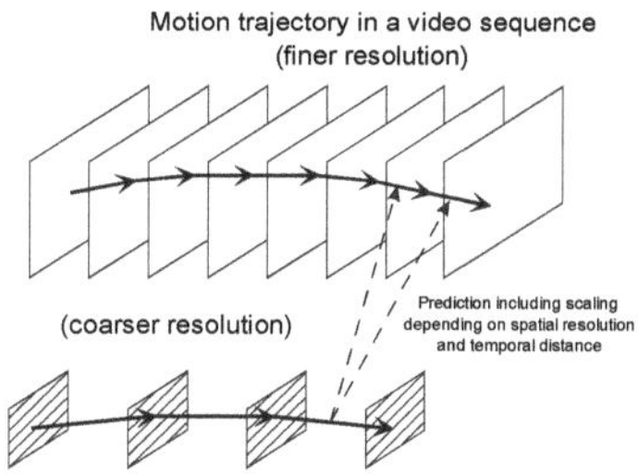

Fig. 7.37. Hierarchical coding of motion trajectories

In case of spatial or temporal hierarchical picture structures, usage of predictor candidates (for differential as well as direct/skip/merge coding) over the hierarchy implicitly gives a coarse-to-fine coding of a multi-picture motion trajectory (see Fig. 7.37). Therefore, it can be expected that the hierarchical structure is beneficial both in terms of lower rate for the motion-compensated residual as well as motion vector coding, provided that the same content appears (even though moved) over a longer sequence of pictures).

In case of B pictures, if the motion trajectory is coherent over time, it can further be expected that forward and backward motion vectors would typically not be much different with exception of the sign. Hence, joint coding as e.g. used in skip and direct modes is becoming even more efficient and saves rate for duplicate motion vectors.

Coding of intra prediction modes. Switching to intra coding could be interpreted as a special case of self-referencing a picture. Due to the special properties of intra coding, it may however not be useful to include intra-mode switching as a special case of reference index, as usually a decoding process with different parameters needs to be invoked. As an example, refer to the coding of directional intra prediction modes in HEVC (see section 6.3.5).

Entropy coding of side information parameters. Efficient prediction and differential coding of side information (such as motion vectors) allows reaching lower rates in entropy coding. This is usually true for non-adaptive and adaptive entropy coding, where however adaptive coding allows better consideration of specific sequence properties. For motion vector differences, often codes with exponential distribution of code-word length (e.g. EG codes, see Sec. 4.4.3) are used due to the expected Laplacian distribution of difference values (cf. Fig. 7.32). Nevertheless, even when prediction is applied, non-linear dependencies may exist, and usage of adaptive context-based coding can provide additional benefit. For example, it is typically sequence dependent how often the prediction of motion vectors is successful, how often direct/skip/merge are used, to which extent already the first candidate in motion vector competition is the best choice etc.

Side information for multi-component video. Sharing of side information parameters is usually applied if the video consists of multiple components such as luma and chroma, RGB, auxiliary channels like alpha, shape or depth maps (cf. Sec. 7.6). If components are aligned by their coordinate systems, they should not be treated differently in coding (e.g. it would not make sense to impose a different motion shift to different color components, as the underlying physical motion should be identical, and artifacts might become visible otherwise). In cases where certain components are subsampled, it is further necessary to scale motion vectors accordingly.

Derivation of side information parameters. Due to the fact that side information in general can require a significant amount of the video data rate (30% or even higher percentages are often found in particular at low rates). Therefore, prediction or more sophisticated methods of derivation by the decoder, such as proper selection of candidate motion vectors or modes, are crucial for efficient coding. With proper adaptive entropy coding, even unlikely cases or modes which may only be used for some types of video sequences, can beneficially be included. In this context, it is important to make the choice under rate-distortion criteria (see section 7.2.7). Furthermore, in terms of keeping the data rate as low as possible, exploitation of dependencies by conditional decoding is highly beneficial. Examples for this are methods of decoder-side motion vector derivation (Sec. 7.2.4) or other kinds of parameter deduction from neighbored, already decoded areas. In this context, the increased complexity that is caused through checking additional conditions or performing additional processing should also be justified by the performance improvement.

7.5 Scalable video coding

Scalable video coding allows decoding a useful signal from a partial bit stream, either with reduced frame rate (*temporal scalability*), picture size (*spatial scalability*) or quality (*fidelity scalability*, also referred to as *SNR scalability*).

The simplest approach of scalable video coding would not use motion compensation, i.e. perform intra-only coding over a sequence of pictures. As however efficient video coding can largely benefit by utilizing temporal coherence via motion, this does not achieve attractive compression. On the other hand, in motion-compensated coding, motion information should be scalable along with video texture. However, scalable coding in a recursive prediction loop is difficult to implement, as scaling a bit stream could be interpreted like a data loss, which would lead to drift (divergence of encoder and decoder side prediction processes). Therefore, scalable versions of motion compensated hybrid prediction/transform coding are typically using the layered coding approach, i.e. adding another option of prediction from low-to-high resolution/quality in order to circumvent this problem. As an alternative, 3D transform coding could implicitly provide a scalable representation in an embedded fashion (see Sec. 5.4.2), which however is also difficult to combine with motion compensation in particular in case of spatial scalability, where the shift variance of the transform is problematic. The latter methods are not further discussed in the current section.

7.5.1 Scalability in hybrid video coding

The picture-recursive processing of hybrid video coding imposes drift (5.8) when different predictions were used at encoder and decoder sides. Layered coding can guarantee that each lower layer in the scalable representation is self-contained, if prediction does not use decoded information from higher layers. In closed-loop systems, the prediction of lower layers will consequently be worse than it could be if all higher enhancement layer information was usable in the prediction, as expressed in (5.15). This however does not penalize the operational point of the base (lowest) layer, which will implicitly perform like a single layer coder at the same rate; as however the lower layer information is used for prediction of the higher layers, and this part of the prediction is worse than it could be from the viewpoint of the higher layer itself, it is unlikely that the rate-distortion performance of scalable hybrid coding could always be as good as in a single-layer coder, i.e. it cannot be expected that a scalable coder will achieve at multiple rate points the same low distortion performance as single layer coders operated independently and optimized for those rate points. Such general presumptions must nevertheless be regarded carefully, as video is a multi-dimensional signal and hybrid video coding has significant degrees of freedom to choose the best reference for prediction. It has furthermore to be observed that side information (e.g. motion and mode) establishes an important part of the entire bit rate in hybrid coding, which should also be considered to be re-used/predicted across the layers as much as possible.

From an application point of view, the performance of scalable coding is often compared against *simulcast* (cf. sec. 5.4.1), which is offering independent coded streams for the different resolutions or qualities. In simulcast, a decoder would simply disregard those sub-streams that are not representing the intended target operation point. Even though typically the bitrate can be significantly larger in simulcast due to the redundancy between the sub-streams, the decoder design is also simpler.

In the following sub-sections, different principles of scalable hybrid coding are discussed in detail, which also will include solutions trying to find a trade-off between the base and enhancement layer quality. Usually, two-layer configurations are considered, however extension to more layers as in Fig. 5.13 is straightforward by using multiple enhancement layer coder/decoder building blocks.

MC prediction loops in layered coding. To support a self-decodable base layer, it is necessary to implement an MC prediction loop for it at the encoder and decoder. Operations of spatial decimation and interpolation for variation of picture size between the layers should be applied outside of the prediction loop[49]. In simplest case, the encoding of the residual error from the decoded base layer could be performed directly as intra-picture coding (Fig. 7.38a), without running MC prediction at the same enhancement layer. This approach would be suboptimum in any case where information relevant for prediction could be drawn from previously decoded pictures at the enhancement layer. For example, if the base layer has lower spatial resolution, it would be necessary to encode the differential information towards higher resolution for each picture again. If the base layer is quantized with poor quality, the residual error towards higher quality would be similar for subsequent enhancement layer pictures. Therefore, temporal correlation would exist between the base-to enhancement residuals of subsequent pictures in such a case, particularly when the resolution/quality gap between the layers is large[50]. To resolve this, MC prediction with a separate loop can be applied in the enhancement layer coding (Fig. 7.38b). If detail information is available from previously decoded enhancement layer pictures, it is likely that the residual error will become lower.

With separate prediction loops for base and enhancement layers, the encoder complexity obviously increases, but this also depends on the mode of scalability. If the enhancement layer operates at higher spatial resolution (i.e. spatial scalability) different (or at least refined) motion vectors should be used to provide good prediction quality in the enhancement layer. For fidelity scalability, identical motion parameters could better be used in both prediction loops[51]. In the latter case, it is possible to combine both prediction loops at the decoder. To illustrate this, Figs. 7.39a/b show similar structures as Figs. 7.38a/b, and Fig. 7.39c is completely equivalent with Fig. 7.39b, i.e. each coder/decoder structure of Fig. Fig. 7.39b/c can be combined with the other under the assumption that the two motion compensation operations are identical[52]. It is still necessary to run two separate loops

---

[49] For pure fidelity scalability, the decimation and interpolation elements can be omitted.

[50] This correlation between residual errors will be higher for cases of no or small change of the video scene, where all finer detail structures must again and again be encoded for each enhancement layer picture. In contrary, for cases of larger changes (in particular when motion compensation fails, e.g. in case of occluded areas), prediction from the collocated base layer picture can be more efficient than prediction from previously decoded enhancement layer pictures.

[51] This may however not be optimum under rate-distortion criteria, where motion parameters are usually selected depending on rate.

[52] See problem 7.3.

at the encoder to guarantee base-layer stability; however, in Fig. 7.39c only one decoder prediction loop is needed, if only one of the layers shall be reconstructed, which is the normal case in many applications. The single-loop decoder is possible when the inverse prediction loops are identical systems (in principle LSI, but identically adapted locally), such that it is irrelevant whether the available residual signals are added before or after the loop. Another interesting conclusion that can be drawn from the two structures is the equivalence of predicting the original picture of the enhancement layer, or the residual in the loop. One major difference is that in case of Fig. 7.39c, the prediction loop of the enhancement layer encoder processes reconstructed pictures, whereas in the case of Fig. 7.39b (as well in Fig. 7.38b), it processes residual signals which are later on added to the reconstructed base layer. Fig. 7.39c can be further modified in a way that only one inverse transform is performed at the decoder, i.e. the residual computation between base and enhancement layers is performed in the transform domain. Still, this would be equivalent in terms of the coded representation (if small deviations e.g. by rounding are ignored).

The identity of Fig. 7.39b/c unveils that it is completely equivalent to either predict the base-to-enhancement residual by motion compensation in the enhancement layer (7.39b), or the enhancement-layer motion compensated residual from the base layer residual (7.39c). At the encoder side of the method of Fig. 7.39c, the prediction residual from the base layer is subtracted from the prediction residual of the enhancement layer, which is denoted as *residual prediction.* When however the motion compensation in base and enhancement layers is different, these two predictions are no longer exchangeable. This is typically the case

– in spatial scalability, where base and enhancement layer have different spatial resolution;
– in fidelity scalability, when different motion vectors were used in the enhancement layer loop, which may be useful due to the different prediction reference.

In case of diverging motion compensation, shift variant components such as ghosting edges may appear in the final residual. To circumvent this problem, methods have been developed to re-compute a residual from the base layer using the motion parameters of the enhancement layer [LI ET AL. 2013]. This would however require an additional motion compensation loop both at the encoder and decoder, which may be undesirable out of complexity reasons, unless justified by sufficient compression benefit.

Furthermore, it is beneficial to allow switching between different prediction modes (base to enhancement, enhancement only, residual). For example, without significant change of the picture contents over time, cases will typically occur where using *only* the previous picture of the (higher quality) enhancement layer reconstruction allows best prediction, without taking reference to the collocated base layer picture. A more flexible structure which enables this mode additionally is depicted in Fig. 7.40. Observe that, similar to the scheme of Fig. 7.38c, entire reconstructed pictures are stored in the scheme of Fig. 7.40. The current enhancement layer picture can either be predicted entirely from the upsampled base layer, from a previously decoded enhancement layer reconstruction[53], or from the average (bi-prediction) of both. Additionally, a *residual prediction* mode is implemented, where the motion-compensated residual of the enhancement layer is predicted from the collocated residual of the base layer[54]. All cases can be justified by specific (local or global) sequence properties, and should be enabled for efficient scalable hybrid coding.

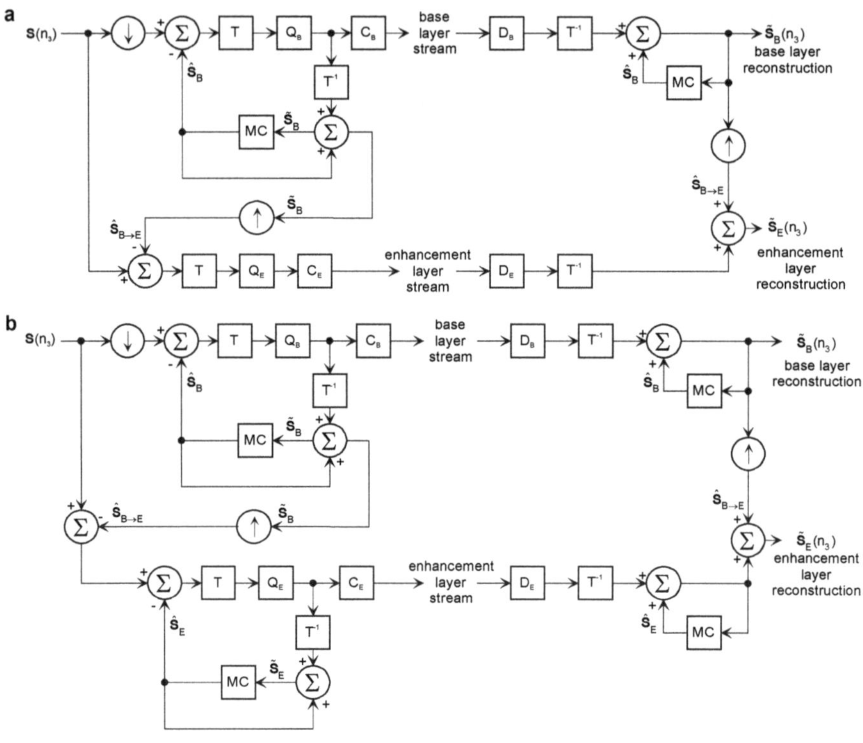

Fig. 7.38. Spatial scalability combined with quantizer scalability in a hybrid coder: **a** without MC prediction in enhancement layers **b** with double loop, supporting MC prediction in enhancement layer residual [T: transform; ↓/↑ : decimation/interpolation; Q_B: quantizer of base layer; Q_E: quantizer of enhancement layer]

---

53 The first of these cases is equivalent to the 'intra only' enhancement coding of Fig. 7.38a, whereas the second case would be equivalent to simulcast (no decoded lower layer used for higher layer prediction).

54 The method of Fig. 7.38b, where the base-to-enhancement residual is further reduced by motion compensated prediction, is not included here, even though it gives different results in case of different motion compensation in base and enhancement layers. Including it as a a fifth mode would in principle be possible, but would require a second picture memory to store residuals, which may not be justified under complexity considerations.

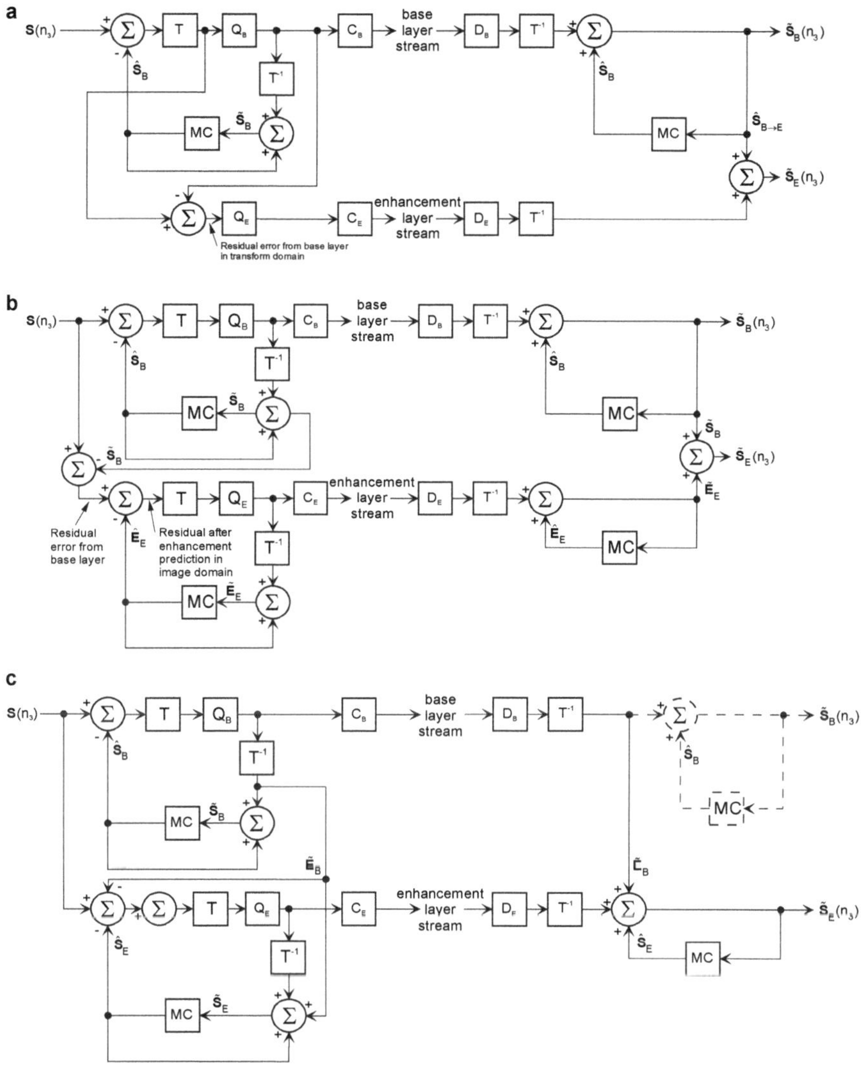

Fig. 7.39. Quantizer scalability in an MC prediction coder without (**a**) and with prediction in the enhancement layer (**b**/**c**), both latter structures have identical base and enhancement information components, respectively (**b** double-loop decoding; **c** single-loop decoding, base layer reconstruction optionally possible with second loop)

The options included in Fig. 7.40 also include configurations as they have been implemented in scalable versions of different video compression standards:

– The scalable extension of AVC (often nicknamed as Scalable Video Coding, SVC) supports configurations 1 and 4, as well as configuration 2, however the latter restricted to the case where the base layer is intra coded. This way, a decoder can be implemented running only one loop of motion compensation when the highest layer is decoded, it however requires a dedicated design of the enhancement layer decoder.

– The scalable version of HEVC supports (nicknamed Scalable HEVC, SHVC) configurations 1, 2 and 3. This requires running multiple motion compensations loops (one for each layer) when the highest layer is decoded. It allows re-using single-layer decoder(s) without any further modification for decoding the higher layer(s), by just providing the collocated (and up-sampled, in case of spatial scalability) decoded picture of the lower layer in the reference picture buffer of the higher layer.

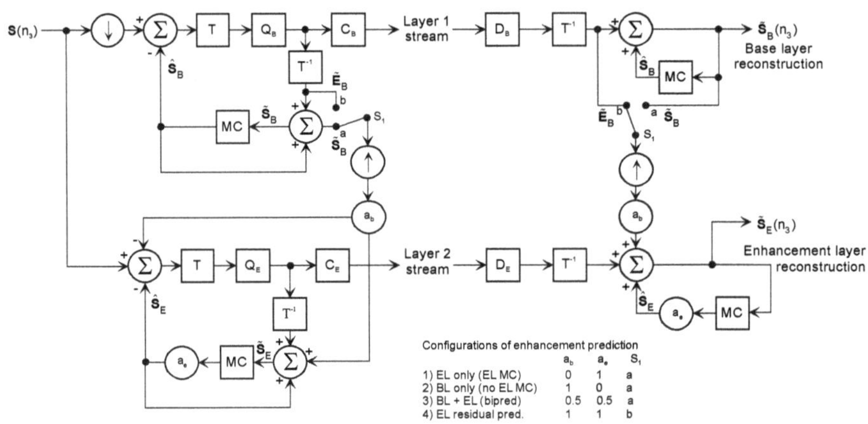

Fig. 7.40. Double-loop spatial scalability with different switchable adaptation mechanisms for enhancement layer prediction

Control of drift. To enforce better compression performance at the highest layer's rate point, prediction at the encoder side could *always* be performed from the highest enhancement-layer decoded result as reference, irrespective of what information the decoder would have available (Fig. 7.41). This specific method of 'SNR scalability' was defined in the MPEG-2 standard. While this approach does not penalize the enhancement layer, it decreases the lower layer quality by (potentially severe) drift. This is basically similar to the case of data losses, caused by the fact that different predictions are used at encoder and decoder when only partial information is received, with the consequence of error propagation as in (5.16).

A practical method of drift control, which is a variant of the double-loop method of Fig. 7.39 and guarantees unconditional base layer stability by penalizing the compression performance of the enhancement layer, was suggested in [Wu, Li, Zhang 2000], entitled as 'progressive fine granularity scalability'. Here, enhancement layer information of a bit-plane representation is partially used for prediction by a sophisticated prediction mechanism which terminates error propagation after a fixed number of pictures in cases where not the full enhancement information may be available at the decoder.

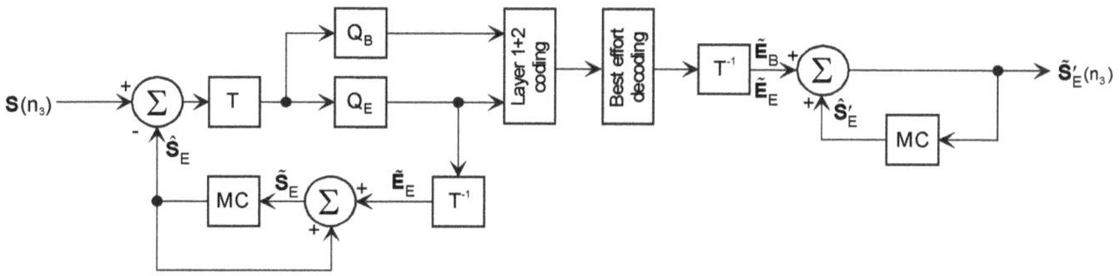

Fig. 7.41. Hybrid SNR scalability structures with drift at the base layer

Temporal scalability. In temporal scalability, lower layers shall enable decoding with lower frame rates. If the base layer shall be predicted self-contained, i.e. use only its own set of reference pictures, this can be achieved by constructing a lower frame rate sequence using frame skipping. Enhancement layers supplementing for higher frame rates consist of pictures from the previously skipped positions. They can be predicted from reference pictures of any lower layers or the same layer.

Hierarchical picture structures as introduced in Sec. 7.2.4 (Fig. 7.8) are therefore an almost ideal solution for temporal scalability, implicitly providing embedded layers, with no rate penalty compared to the single-layer scheme[55]. By defining the picture sequence of the lowest frame-rate base layer as 'key pictures', stability for decoding that frame rate is guaranteed whenever the base layer is received. Beyond that, the availability of key pictures allows to achieve an adjustable trade-off between coding efficiency and drift even in other scalability modes such as spatial and fidelity scalability. The base layer consisting of key pictures is a stable and drift-free point, from which decoding of enhancement layer(s) can be re-started whenever losses of higher layers would have occurred. For base layer pictures other than the key pictures, strategies may be applied where a certain amount of drift is admitted for the benefit of higher compression performance. An example is given in Fig. 7.42, where, except for the key pictures, all other pictures can be predicted from references with higher resolution or quality[56]. The configuration shown here would be drift-free for cases when the layered structure is defined as 'spatial/fidelity scalability first' (i.e. whenever a new temporal layer is invoked, the following layers should first enhance it to full resolution and quality before going to the next-higher frame rate). Starting from the same key pictures, but decoding the sequence of layers in a 'temporal scalability first' approach (i.e. first go to highest frame rate at lowest spatial resolution/quality), a certain amount

[55] Observe however that an optimized single layer coder may not decide using a larger number of hierarchical layers for sequences with large amount of changes, whereas a scalable coder has to use a certain structure as imposed by the scalability requirement.

[56] Note that the scalable version of AVC does not allow using such a structure with higher resolution in the enhancement layer, as it does not define downsampling as part of the prediction process.

of drift would occur, but it can be expected to not dramatically affect the quality since the decoding chains are relatively short picture sequences in any case, and averaging two predictions limits the propagation effect, cf. (7.18)ff.

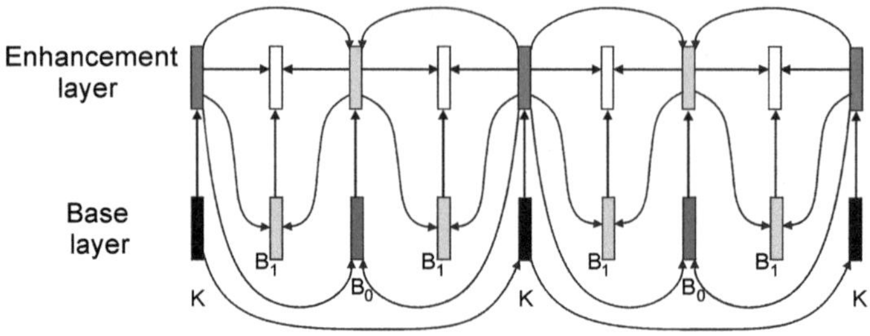

Fig. 7.42. Dependency example with limited drift in case of enhancement layer loss in a hierarchical B picture structure (after SCHWARZ)

Scalability of motion and mode information. Side information such as motion, partitioning and mode parameters can contribute a significant percentage of the total rate, in particular when more precision (e.g. usage of small block sizes) is enabled. In particular for the case of spatial and fidelity scalability, it can be expected that such parameters are highly coherent between layers. Specifically,

- if selection of motion, partitioning and mode parameters is performed using rate distortion criteria, they can be different (typically more precise) for the higher rate, but the parameters determined for the lower rate/layer can be used as prediction for the higher rate/layer parameters.
- if the higher layer has higher spatial resolution (spatial scalability), the lower layer parameters can also be used as predictors, however it will be necessary to scale the partition sizes and motion vectors according to the upsampling factor.

It is not guaranteed that the prediction of side information parameters from the lower layer is the optimum choice. Since it is common to exploit spatial and/or temporal redundancy in the encoding of motion and mode parameters within the partitions, it could well be that the predictors from spatially or temporally collocated partitions at higher resolution provide better prediction than the lower-resolution layer.

7.5.2 Standardization in scalable video coding

Scalable video coding (SVC) part of the AVC standard. Scalability functionality had been included in various video coding standards since MPEG-2. However, until the advent of the Scalable Video Coding (SVC) extension of the AVC standard[57], a significant penalty in compression performance usually occurred compared to single-layer coding. The main reasons for this were:

---

[57] Defined per Annex G of ISO/IEC 14496-10 | ITU-T Rec. H.264.

- Early approaches of fidelity scalability came with base layer drift (MPEG-2) or were inefficient due to pure intra-picture enhancement (MPEG-4 part 2 Fine Granularity Scalability, FGS[58]).
- The approaches for spatial scalability used the layered coding approach, but mainly targeted residual coding of picture information, as such performing duplicate encoding of side information parameters (motion, mode switching etc.) for base and enhancement layers; this duplication was even more costly, since the coding of side information in general lacked efficiency e.g. in terms of predicting motion parameters. Beyond that, prediction mode switching also lacked some flexibility (e.g. substituting bi-prediction in the enhancement layer, which can be efficient in certain cases, by also allowing bi-prediction between upsampled base and one enhancement layer picture), and the entropy coding of the scalable coding modes was also rather inefficient in MPEG-2 and MPEG-4 part 2 spatial scalability.
- Temporal scalability in MPEG-2 and MPEG-4 part 2 is efficient as far as B pictures are used, however due to the lack of hierarchical picture structures in standards prior to AVC, this would not be applicable for more than two layers of temporal scalability.

The most important properties of SVC that provide efficiency in scalability are described in the subsequent paragraphs.

Temporal scalability is most efficient in SVC by usage of hierarchical picture prediction structures (see examples of a B picture pyramid and a P picture pyramid with 4 temporal layers each in Fig. 7.8). In general, layers with higher temporal resolution must not be used for prediction of layers with lower temporal resolution. Furthermore, pictures at the lowest layer of the hierarchy (I and P in Fig. 7.8) are denoted as *key pictures* and play a role as re-synchronization or stream switching points.

For spatial and SNR scalability, relevant information from lower layers is utilized as far as possible, including motion, mode and partitioning information. Fig. 7.43 shows how the block partitioning and the modes are inherited from the base layer to the enhancement layer in case of dyadic upsampling in spatial scalability. Motion vectors are scaled up by a factor of two, where optionally a difference can be encoded. Special modes such as skip and direct can also be inherited from the lower to the higher layer. Prediction block partitioning is upsampled, which is straightforward in the case of dyadic upsampling factors, except for the case of base layer blocks with prediction block sizes of 16, where four 16x16 blocks have to be established at the enhancement layer (center of Fig. 7.43). In case of non-dyadic upsampling factors, motion vectors are scaled by that factor, and rounding to the next possible prediction block boundary (at each 4th sample position) is performed.

Prediction from a lower to a higher layer is implemented as an additional prediction mode, whereas any other modes (inter motion compensation, intra prediction) are still available for prediction from a reference of the previous decoded

58 In the meantime, FGS has been removed from the MPEG-4 Visual (part 2) standard.

pictures from the layer itself. For prediction from the lower to the higher layers, only reconstructed blocks decoded in intra mode (denoted as Intra-BL)[59], or the reconstructed residual signal of motion compensated blocks can be used. This avoids implementing a full reconstruction of motion compensated blocks over all lower layers, and therefore only a single motion compensation loop needs to be operated at the decoder[60]. However, multiple encoder loops have to be run in a closed-loop fashion at different rate points in the overall layered structure (which may consist of various combinations of temporal, spatial and SNR scalability).

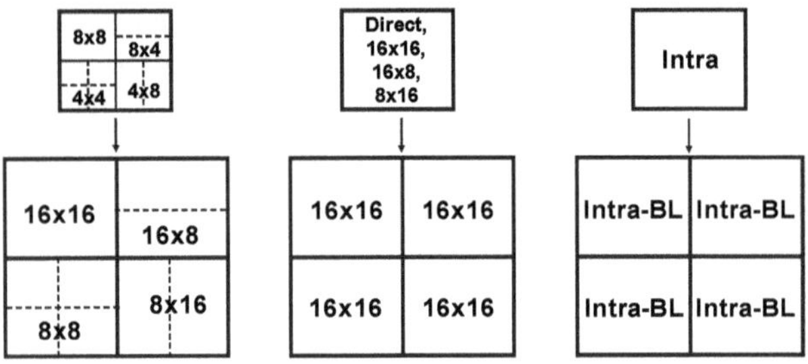

Fig. 7.43. Prediction of partitioning and mode information in SVC spatial scalability.

A typical configuration of spatial scalability is shown in Fig. 7.44. Prediction structures of all layers are spatially aligned in a way that intra-coded macroblocks, MC residual data, motion, mode and partitioning information from layer l are upsampled and used as optional additional prediction in the corresponding areas of layer l+1. For spatial scalability, often dyadic (powers of two) factors are used in terms of picture sizes of the layers, but non-dyadic factors are also possible.

Fidelity Scalability is implemented with two basic operation principles: Coarse-granular SNR scalability (CGS) and medium-granular SNR scalability (MGS).

– CGS is designed to support a few selected points with usually at least 50% bit-rate increase from one layer to the next. It works like spatial scalability without spatial resolution change between the layers, i.e. no upsampling filter is necessary and only the quantizer step size is usually changed from layer to layer. From CGS, SVC bit streams can be converted to single-layer AVC bit streams by a method called *bit-stream rewriting*. Even though multiple MC

[59] Observe that for intra coded pictures, it is necessary to reconstruct the base layer signal in a spatial prediction loop. However, it is not necessary to rely on prediction information from other time instances.

[60] With identical motion compensation loops in base and enhancement layers, it would not make a difference whether the original signal or the residual signal is predicted from the base layer (see Fig. 7.39). This is however not the case here, firstly due to usage of different motion and mode (e.g. block partitioning) information, second due to usage of additional components such as the deblocking filter in the loop. Therefore, the benefit achievable by predicting the residual signal component is typically lower in SVC than for intra and motion/mode information.

loops are run at the encoder, at the decoder only one IDCT and one MC loop are necessary (corresponding to the encoder loop of highest available layer), as explained in the context of Fig. 7.39c.

- MGS is a combination of simple CGS syntax with a more advanced motion-compensated prediction structure, allowing the use of adaptive prediction from either base layer or enhancement layer, and assigning additional intermediate layers established by sub-groups of transform coefficients, which can be interpreted as a kind of *frequency partitioning*. MGS NAL units can be partially discarded with relatively small amount of drift occurring, such that layers between two CGS layers can additionally be defined.

SVC uses layered coding (bottom-up prediction) for both picture information and side information. Beyond that, the entropy coding methods of AVC (CABAC and CAVLC) are extended by a few context models (in particular referring contexts from lower layers for conditions in the entropy coding of higher layers), which makes the compression more efficient.

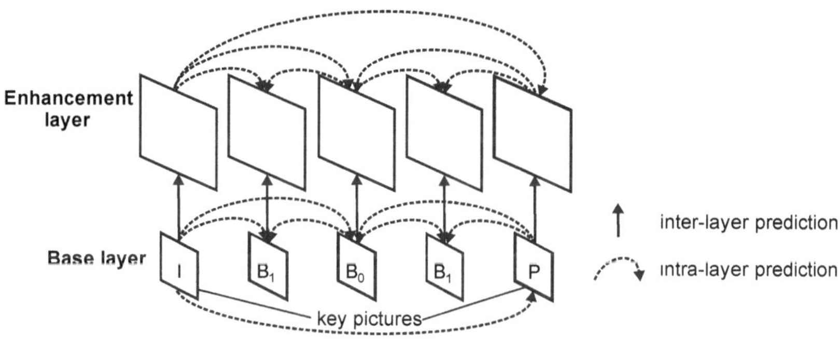

Fig. 7.44. Two layers of spatial scalability in AVC scalable coding

Drift-free operation is only guaranteed at those stable rate point(s) where the encoding is running its prediction loop(s). However, if a sufficient number of stable points are available, stability is implicitly achieved over a broader range of rates, even though between the stable rate points some drift may occur. This will in particular be guaranteed when the enhancement-layer prediction chains are short, as it is the case in hierarchical temporal prediction structures (where however the key pictures need always to be kept stable).

The sequence of layers is not uniquely defined in SVC, but rather specified by three parameters *d*, *t* and *q* where

- *d* stands for dependency_id which indicates a layer with separate sequence parameter set, which is either a spatial or CGS layer
- *q* stands for quality_id which indicates a quality refinement layer (CGS or MGS)
- *t* stands for temporal_id which indicates temporal resolution

Decoding of a part of the bit stream with higher (*d*,*t*,*q*) parameter is not useful, if a part of lower layer information is missing. These parameters are included in the

NAL unit headers of SVC packets (see Sec. 9.2) and could be evaluated within a network without touching the actual video stream. Furthermore, the NAL unit type of base layer packets can be interpreted by conventional single-layer SVC decoders of the respective profile. This way, a legacy AVC decoder is able to receive an SVC stream and decode the base layer, but would ignore any data packets belonging to the enhancement layer.

For the specific case of video coding, the *over-completeness* of SVC according to (2.372) does not necessarily mean that the encoded bit rate is significantly increased if the scalability functionality is enabled. In worst case, meaning that no lower layer information could be useful for predicting the higher layer, the performance would be identical to the case of independent coding, i.e. simulcast. This is however most likely to happen for video sequences with low amount of changes, which are uncritical for compression anyway. In more general cases, the performance can be close to a single layer codec, in particular when the encoder is designed to provide a trade-off between lower and higher compression jointly. The SVC approach of layered prediction in spatial scalability implicitly offers additional degrees of freedom beyond single layer coding in the enhancement layer, particularly the prediction of motion vectors beyond macroblock boundaries and virtual extension of the base layer block sizes due to upsampling. However, to achieve such benefit, SVC may require careful optimization of encoding decisions, which can significantly increase of encoder complexity.

Scalable extension of HEVC. The scalability extension to HEVC enables spatial and coarse grain SNR scalability, and is referred to as SHVC. Temporal scalability as supported in the HEVC base specification may also be combined with spatial and SNR scalability modes. SHVC uses multi-loop coding, where both base and enhancement layer pictures have to be fully decoded to make them available as prediction reference. In spatial scalability, the decoded base layer picture is resampled by a normatively defined interpolation filter; it is then used as additional reference picture for enhancement layer prediction, which enables both inter-layer texture and motion parameter prediction. Otherwise, the coding tools in SHVC do not include low-level changes, such that existing decoder devices could perform the enhancement layer processing. The base layer bit stream can anyway be interpreted by legacy decoders.

The interpolation filter is a 7- or 8-tap FIR filter (depending on upsampling phase position) for luma, and a 4-tap FIR filter for chroma re-sampling, where filters are defined with an accuracy of 1/16 sample precision (rounding for the respective position may be necessary in case of non-dyadic upsampling ratio). The filter design is similar (for ¼ and ½ sample phase positions identical) as for sub-sample interpolation in motion compensation (cf. Sec. 7.2.5), i.e. high quality interpolation with low frequency cut-off close to half of the sampling rate. This enables projection of base layer reconstructed samples values for prediction in the enhancement layer, inserting the upsampled base layer picture into the enhancement layer reference picture list, where usage for reference can then be signaled in the same manner as inter-frame prediction. By using bi-prediction, an encoder can enforce averaging information from the base layer and previously decoded enhancement layer pictures, or also employ weighted prediction. SHVC further supports the prediction of motion information from the decoded base layer by re-using the temporal motion vector prediction (TMVP) process of HEVC (cf. Sec. 7.4), which inherits the scaled motion vector as well as the prediction mode from the collocated partition of the reference picture, again without changing the core part of the HEVC process when decoding enhancement layer information.

7.6 Multi-view video coding

3D displays[61] (stereo and multi-view) require two or more views of the same scene, which have overlapping visible areas. When coded independently, the number of samples and therefore the data rate would increase approximately linearly by the number of views, compared to a conventional (monoscopic, single view) display. Some measures can be taken to avoid this. For example, it is common to encode only one of the views by highest quality or resolution, assuming that the human visual system would still detect the multi-view cues necessary for the 3D illusion when the remaining view(s) are of lower quality.

A special case of representing stereo video is achieved by so-called *frame compatible formats*, where two pictures are reduced in terms of number of samples and placed into one. Typical methods are[62]

- side-by-side packing, where the horizontal resolution is reduced by a factor of two;
- top-bottom packing, where the vertical resolution is reduced by a factor of two.

Frame compatible formats have the disadvantage that it is not possible to exploit the inter-view similarities, as conventional hybrid video coding does not allow prediction (e.g. by motion compensation) between distinct parts of the same picture. Therefore, joint compression of multiple views showing the same scene has become a relevant topic in video coding.

A 3D illusion is achieved by feeding the observer's eyes by different pictures, which approximately reflect the *parallax shift* that occurs depending on the distance between an observer and an observed object. Normally, the parallax shift would be zero when an observer focuses an object with both eyes. For a display situation, a rendered parallax of zero would be interpreted such that the object would be on the screen, and it may be confusing when the observer tries to focus

---

[61] It is beyond the scope of this book to discuss details about 3D display technology. The interested reader is referred to [GENG 2013].

[62] Both methods can also be combined with quincunx subsampling, where each second line/column needs additionally to be shifted before executing the packing. This could produce additional artificial high frequency components which may be undesirable for the subsequent compression.

an object behind, or in front of the screen (see Sec. 3.1.4). Typically, the pictures are *rectified* for output on stereo displays, i.e. equivalent to camera capture with parallel optical axes where the parallax shift occurs only horizontally. This also diminishes undesirable effects regarding the deviation between maximum focus plane and parallax shift. With rectified pictures, corresponding positions of different camera views can only be displaced horizontally[63].

The *disparity shift* (parallax counted by number of samples) is equivalent to motion shift within a temporal sequence, assuming that the camera would have been moved when capturing a static scene. Therefore, exploitation of *inter-view* dependencies (between pictures captured from different cameras at same time instance) can be performed very similar to *inter-picture* motion compensated coding. For example, in hybrid coding, it is sufficient to define picture(s) from other camera view(s) as additional references, performing disparity compensated prediction as if it were motion compensation[64].

As in motion compensated compression, multi-view compression is facing the problem that a foreground object may occlude part of the background in the other camera view, such that those areas cannot even be predicted by disparity compensated approaches. Inter-view prediction can simply be defined as an additional mode (co-existing and competing with inter-picture motion compensated prediction), however it may not be used frequently in scenes with no motion or low amount of motion, where the similarity over time can be expected to be larger than the similarity between views. Exploitation of inter-view redundancy is however efficient in cases where no temporal reference pictures can be used, e.g. in picture positions that shall be intra coded for the purpose of random access etc.[65] Furthermore, the gain achieved generally by inter-view compression relies on the similarity between the different camera captures, which is mainly dependent on the baseline distance between the cameras and the scene depth structure (e.g. less similarity for objects which are close to the cameras).

A method of hybrid multi-view video coding (MVC) was defined as an amendment and has been included as annex H in the AVC standard [VETRO ET AL. 2011], and a similar approach has recently been defined for HEVC [SULLIVAN ET AL. 2013]. In its basic design concepts, both implement inter-view prediction by taking benefit from the flexible multiple reference picture management capabilities in AVC and HEVC. The key elements of this concept are:

---

63 Also for the case of non-rectified views, limitations exist on positions where correspondences can reasonably be found, due to the properties of *epipolar geometry* (see MCA, SEC. 4.8). This however requires exact knowledge about camera parameters.

64 Likewise, it is possible to add one more 'view dimension' to 3D transform/wavelet coding, and perform 4D coding with motion compensation in the temporal, and disparity compensation in the view axis [YANG ET AL. 2006].

65 In case of scenes with low amount of temporal change, the bitrate to be spent for cyclic intra coding (when required for random access, error robustness etc.) can take a significant amount of the overall bitrate. Therefore, employing inter-view references can provide large compression gain in such cases.

- No changes to the core-level design of the previous standard definition (slice level and below); this way, existing implementations of monoscopic video decoders can be re-purposed, as they do not need to be aware whether a reference picture is coming from a different time instance or a different view, as long as it is available in the reference picture buffer. Therefore, changes are only necessary to high-level syntax, e.g., to specify view dependency and random access points.
- A single view denoted as *base view* does not depend on other views and can be extracted from a multi-view bit stream (identified by NAL unit type byte, similar to SVC base layer). Non-MVC decoders can ignore the enhancement data and decode the base view for monoscopic display.

An example prediction structure for the case of 5 camera views is shown in Fig. 7.45. Inter-view prediction is usually limited to use references only from the same time instance.

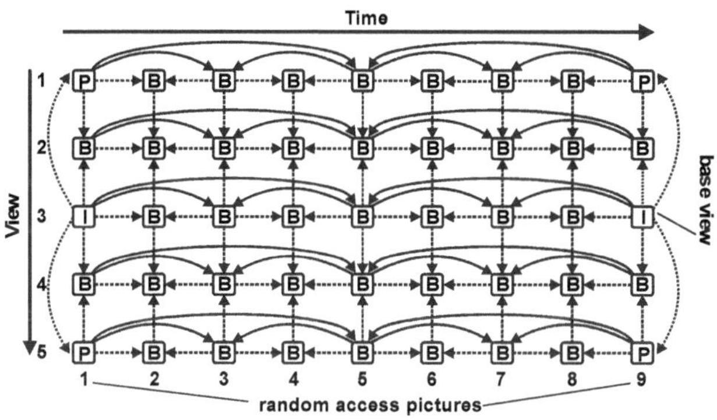

Fig. 7.45. Hierarchical B prediction across dimensions of time and 5 views in MVC (base view on top)

Essentially, the decoded pictures from other views are inserted into the reference picture lists of the current view for use in prediction processing. The decoder receives a displacement vector and a reference picture index, and performs either 'disparity' or 'motion' displacement compensated prediction, depending on whether the reference picture is an inter-view reference or an inter-frame (from the same view at different time instance) reference; actually, the decoder does not even need to be aware which type it is. As a result, the reference picture lists of a current picture can include temporally distant reference pictures of the current view as well as inter-view reference pictures from other views of the same time instance. The prediction is adaptive, such that the best choice among temporal and inter-view references (or an average from bi-prediction or weighted prediction) can be selected.

Data rate savings compared to simulcast (independent coding of views) are typically 20-30% for same quality (e.g. measured by PSNR) for stereo, and can go to 50% and higher for multi-camera captures[66].

The actual compression advantage however highly depends on the amount of redundancy between the views (where the similarity depends on the camera baseline and the distance between camera and captured scene) and on the amount of temporal change in the scene. Therefore, it can be beneficial to implement additional compression tools that target exploitation of redundancy that is specifically present in multi-view video, as described in the subsequent paragraphs.

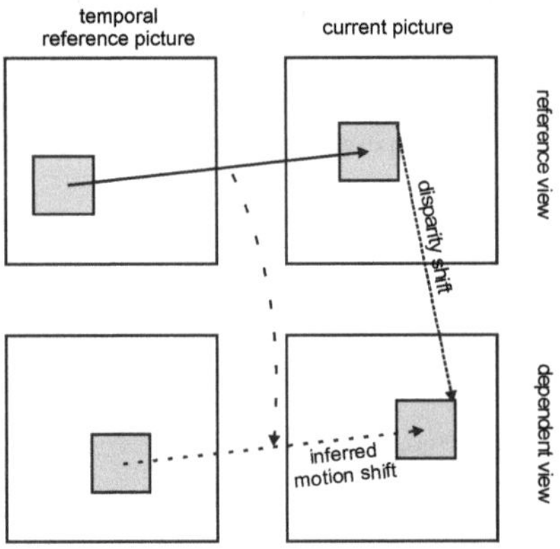

Fig. 7.46. Illustration of motion vector inference, where the motion parameters of a dependent view are derived from the reference view, taking into account the disparity shift

Disparity vector derivation and improvement of motion/disparity compensated coding: It can be beneficial to use a disparity mapping between views, even if the current block is not coded using inter-view prediction. In this case, if either a spatial or temporal neighbor has been coded in inter-view mode, a disparity vector can be derived which is likely suitable for the current block as well (due to the fact that typically larger areas have constant depth value and depth is also consistent over time). The derived disparity vector can be used

- to predict a motion vector to be used in motion compensated prediction of the second view from the corresponding motion vector of the first view, which is expected at a displaced position found via the derived disparity vector (Fig. 7.46);
- to use the derived disparity vector as an additional predictor candidate in disparity compensated prediction;

[66] It should be noted that also the 'frame compatible' formats de facto perform simulcast, as the left and right views packed into the same picture cannot reference each other by displacement prediction. In terms of complexity, the multi-view scheme would be equal to the frame-compatible approach, if the total number of samples is identical (i.e. using half-horizontal or half-vertical sizes for the stereo pictures), but better compression performance can be expected.

– in the case that motion compensated prediction is used in both, predict the motion compensated residual of the second view from that of the first view, using the location correspondence via the derived disparity vector.

Illumination compensation: Prediction may fail when cameras capturing the same scene are not calibrated in color transfer, or if specular lighting effects caused by surface reflections are present. Illumination compensation can be used in these cases, which however only should be applied to parts which are predicted from an inter-view reference picture. Simple methods use a scaling factor α and an offset β to compensate for illumination changes (6.13), it should however be avoided to spend too much side information for this. Alternatively, parameters can be derived from information that the decoder has available; for this, decoded samples from the top neighboring row and left neighboring column of the current block can be used (see Fig. 7.47). Those have to be compared against the corresponding neighboring samples of the reference block (using the disparity vector which is available anyway in case of inter-view prediction), to determine the scale and offset parameters by solving an equation system.

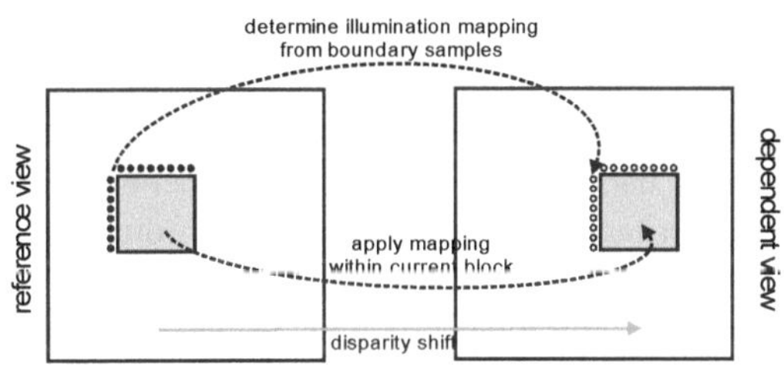

Fig. 7.47. Neighboring samples for the derivation of illumination compensation parameters

In stereoscopic display, good 3D quality is still perceived subjectively when one of the views is encoded with larger error or less spatial resolution (see Sec. 3.1.4). In stereo compression, it is generally reasonable to encode the base view with higher quality, since this also improves the prediction of dependent view(s). If criteria of sample fidelity are loosened, an even stronger compression can be achieved by performing view synthesis of additional views. For this purpose, disparity or depth information can be used to project picture information from the reference view to the positions where they would be visible in the (current) target view. A review of view synthesis methods can be found in [Izquierdo, Ohm 1999]. Though it would be possible to generate the depth map at the receiver end, this might not have sufficient quality at positions where depth estimation is difficult (e.g. at depth discontinuities and occlusions occurring at object boundaries). Therefore, generating high-quality depth maps during the production of 3D video, and encoding them as part of the 3D video representation is reasonable. Such depth information can be used to generate a continuum of views, as e.g. needed for autostereoscopic displays, or to allow adapting the depth range for different types of stereoscopic displays, for the given view situation or by user preference.

Depth map compression can be achieved by operating a second (monochrome) video codec independently for the depth data. In this case, the 'video texture plus depth' compression does not utilize any correspondences between texture and depth. The depth data are conveyed as additional samples as an auxiliary component plane, where some additional syntax is necessary to explain their meaning. An early approach has been defined in MPEG-4 Visual (ISO/IEC 14496-2). The aforementioned MVC extension of AVC has also been recently extended into a video-plus-depth format [CHEN ET AL. 2013], which maintains MVC stereo compatibility, while encapsulating coded texture and depth maps into a single bit stream. Texture and depth map data can also be coded at different spatial resolution. Depth data are typically relatively smooth within the area of objects or background in the scene, but have sharp discontinuities at object boundaries. Though reduction of resolution could be disadvantageous for conveying the correct position of boundaries, it significantly reduces the encoder and decoder complexity (in terms of number of samples that need to be processed).

Typical depth maps have a simple structure of large areas with few changes (flat or gradually increasing) and sharp edges (depth discontinuities). Therefore, they can be compressed to a much lower rate than the video itself. In depth-based view synthesis, the depth map could also be interpreted as a kind of side information; in this context, joint compression of texture and depth data can further be considered, e.g. using the same motion parameters for both, extract clues about depth discontinuities from the texture data, or use depth maps for better compression of dependent texture views. Inheritance of motion parameters from texture to depth (or vice versa) can be achieved by either sharing them, or by inserting the other component's collocated vector as additional predictor in the vector coding process.

The preservation of edges in depth maps is important since inaccurate edge reconstruction may lead to significant objective distortion and perceivable artifacts in synthesized views. Due to this fact, filters in the prediction loop that are optimized for compression of texture, particularly interpolation filters which could produce ringing around sharp edges, deblocking filters etc. should be switched off. Alternatively, they could be replaced by morphological filters, rank order filters or other types of non-linear, edge preserving filters (see MCA, SEC. 2.1.3). Another interesting characteristic of depth images is the precise object boundary information which is inherent, corresponding with depth discontinuities in the scene. This often matches with amplitude edges in the corresponding texture component, and can therefore be predicted from them. Further methods to take benefit of joint texture and depth coding are:

- Re-using (temporal) motion parameters of the texture for the depth map (or vice versa);
- Using depth maps (which can eventually be expected to be more precise than the block-wise disparity vectors) to improve the inter-view prediction in a view-synthesis mode [HANNUKSELA ET AL. 2013];

- Using the depth discontinuities to improve motion compensation at moving object boundaries by more precise partitioning of the areas where motion vectors are valid.

Not all of the methods in the last three bullets can be used simultaneously, depending on whether the texture information is exploited for depth coding (and needs to be coded first) or vice versa.

Partition-based depth coding. Sharp edges (with a unique value at each side) are expected at depth discontinuities. Therefore, specific coding modes can be used for this purpose, including coding of the edge position/shape and of the depth values at both sides. For coding the edge values, the following methods can be employed:

- Geometric partitioning (separation by a straight line) as shown in Fig. 6.5;
- Precise coding of the contour by chain coding (Fig. 6.6) or context-based binary coding (Fig. 6.2b);
- Derivation from collocated texture edge positions (when available).

The method under the latter bullet can also be used as a predictor for more efficient explicit compression in the first two bullets. In case of depth discontinuities, it can be beneficial to skip a subsequent transform coding step. Further efficiency can be achieved when a lookup table is used to code the distinct depth values at the two sides of the edge, instead of directly signalling the depth amplitudes.

View synthesis prediction (VSP). In VSP, information from an available depth map is utilized to warp texture data from a reference view to the target view for the purpose of prediction. When depth maps are used for image-based rendering, view synthesis is typically implemented as forward warping (similar to Fig. 7.5b), using the coordinate system and the depth map associated with the reference view. Though this is in principle possible also for the purpose of prediction (where it also requires specification of a hole-filling procedure filling the gaps that remain after warping), this is not practical. In cases of unforeseeable large disparity values it might be required to first generate an entire synthetic picture and store it in the reference picture buffer before proceeding to encode or decode the current picture. This would lead to memory / complexity increase and processing delay at the decoder. On-the-fly processing is enabled by block-based backward VSP (BVSP) schemes, where the depth information of the current block is inferred to determine the corresponding samples in the inter-view reference picture. Further, when texture is coded prior to depth, a derived disparity vector as described earlier can be used, and if a homogeneous disparity shift is used over small block areas (e.g. size 4x4 instead of shifts that can vary for each sample), BVSP can be completely integrated into the conventional motion compensation process e.g. of AVC or HEVC.

3D AVC. This extension of AVC [HANNUKSELA ET AL. 2013], defined in annex J of the standard, uses some of the aforementioned approaches. It allows compression of video and depth data, but more explicitly exploits inter-component dependencies between texture and depth and introduces joint coding of texture and depth data. Particularly the second (non-base) texture view is coded more efficiently by utilizing motion vector correspondences and more efficient coding of displacement vectors, derived disparity vectors, illumination compensation and BVSP. Generally, the additional gain achievable by more dedicated coding tools for depth and dependent texture, compared to the class of MVC-like codecs that simply re-use existing motion compensation for inter-view prediction, is in the range of reducing the bit rate by almost 20% on average over several sequences, but highly depending on the similarity between the views. Further benefit can be achieved by employing encoder optimization, such that the trade-off between rate and synthesis quality is optimized.

3D HEVC. To achieve higher compression efficiency, 3D HEVC (currently under development) applies specific tools for exploiting the correlation of motion and texture, as well as residual data between views. However, compression of the base view is still compatible with baseline HEVC, such that monoscopic video can be decoded by version 1 legacy decoders. The most important elements are conceptually close to the methods described above:

- Disparity vector derivation based on neighboring blocks, which can then be used for improved motion and/or disparity vector inheritance of the dependent view(s) and depth map(s);
- Inter-view residual prediction, which comes by the assumption that the residual signal after motion compensation would be similar in different views;
- Illumination compensation (only applied in case of inter-view references) with parameters derived from neighboring blocks;
- Specific 'depth modeling modes' exploiting the unique characteristics of depth maps such as edge characteristics, flat surfaces, limited amount of values stored in a lookup table;
- Modifications of motion compensation (avoidance of interpolation and of loop filtering elements) for depth map prediction, utilization of depth information for better region partitioning in texture motion of dependent views;
- View synthesis prediction, which utilizes depth information for texture coding, and can be used as an alternative prediction mode in case of inter-view references.

When the number of required views increases, e.g. for free view adaptation in interactive navigation or 3D displays with extended view range, multiple camera inputs are usually required, but even then both depth estimation and artifact-free synthesis remain difficult for content with small structures, transparent (non-opaque) objects, reflections and specular effects on object surfaces etc. Visible artifacts could be due to incorrect depth data, missing temporal consistency (e.g., fluctuation) of depth data and holes that occur in the synthesized picture during projection. In wide-range multi-view coding, dense views may be required to be encoded for high reproduction quality, and the trade-off w.r.t compressed data rate, complexity of capturing systems and processing complexity becomes critical.

This could become a critical burden for compressing video output from dense light field arrays or holographic capturing systems, imposing new challenges to utilize various dimensions of redundancy appropriately.

7.7 Synthesis based video coding

Video compression methods using the hybrid coding approach have been highly optimized in terms of accuracy of motion description, exploitation of intra and inter picture redundancies, entropy coding with adaptation to source statistics and context information already available at the decoder, etc. One of the main driving forces in improving hybrid video compression has been the increased accuracy of motion compensation, along with the more efficient coding of motion parameters. If however squared-error sample fidelity is used as sole criterion for optimizing rate-distortion performance, limits of compression could be approached soon. Tendencies that might provide further improvement of compression in the future are

- More closely modelling of human perception would be a key to increase coding efficiency of image and video coding. This should include visibility of errors depending on combinations of spatial detail and motion in particular. Criteria to judge the fidelity of texture and edges, as well as quantifying the subjective annoyance of compression artifacts would simplify such a task. An inherent difficulty may however be the existence of individual differences in the perception of human subjects (cf. Sec. 3.3.1)
- Certain structures of texture and motion can be subjectively irrelevant in terms of sample reproduction, e.g. granular movements of water surfaces, movements of leaves of trees, crowds of people, smoke. These are structures which could cost extremely high data rate in conventional (hybrid) coding, eventually having unpredictable motion components and/or non-sparse 2D spectra. In these cases, *synthesis of visually similar* content could help to largely reduce the data rate without or with only minor effect on subjective quality.
- Current approaches of motion compensation are usually based on compensating between picture pairs, coherences over longer-duration time spans (e.g. motion trajectory over several pictures) are hardly taken into account. Optimization of both motion and MC residual coding over groups of pictures (e.g. taking into account how relevant the encoding of the information is for the entire evolution of the video scene) could further improve the subjective quality. In this context, inherent properties of motion trajectories, such as constant or accelerated velocity, should be better exploited.
- Selectively improved encoding of *semantically relevant objects* or regions of interest, along with simplification or suppression of less relevant content, could also provide subjectively improved appearance.

Note that the approaches presented in this section are not suitable for arbitrary types of scenes (e.g. due to absence of content where the given method could be applied or due to lack of real-time encoding capability). Therefore, usage of switchable solutions, where synthesis based coding approaches can be operated as additional modes (in simplest case as an alternative reference provided in the decoded picture buffer) of a conventional (e.g. hybrid) coding scheme would be straightforward. Further, some of the aspects stated above could also be achieved by encoder optimization in existing (hybrid) video coding schemes, specifically by using perceptual distortion criteria and multi-pass coding investigating the dependency relations over multiple pictures.

7.7.1 Region-based video coding

Object-based coding. The general flaws of block-based motion compensation were described above; also none of the techniques described so far provides a consistent solution to the problem of *occlusions* which necessarily occur when objects are moving differently within a video scene. The ultimate goal of *object-based coding* is to describe the objects contained in a sequence *separately* by as low number of parameters (or information bits) as possible. The nature of these objects would not be relevant for the purpose of encoding, but could be useful in the context of their automatic extraction. The background or other large areas, which are either static or subject to consistent (e.g. global) motion, are treated like any other object in the scene. In particular, if a part of the background is temporarily occluded, it is then very efficient to recall it from a background memory when it appears again [THOMA, BIERLING 1989]. It should however be considered that state of the art video coding schemes use multiple reference pictures, where by appropriate encoder decision content that is foreseen to appear again later could be kept in the picture buffer for a longer time (denoted as long term prediction reference). Another type of background memory is denoted as *sprite* or *mosaic*, which is usually constructed by combining several pictures from a moving camera into a single larger picture, and projecting (warping) them into a target background (see [MCA, SEC. 4.9]); a sprite coding method has indeed been included in the MPEG-4 Visual (ISO/IEC 14496-2) video compression standard.

For any object within a scene, it is necessary to provide the *shape/position* and the *color/texture* information for each time instance. The *motion* information should be the key to keep the representation as compact as possible, which means that changes in the other parameters need only be described if they cannot be determined by the motion. A block diagram of object-based video coding as originally proposed in [MUSMANN, HÖTTER, OSTERMANN 1989] is shown in Fig. 7.48. This method is by concept more similar to replenishment methods than to hybrid coding, as it is assumed that the object's changes can either be modeled accurately enough or the model fails entirely. The color/texture information of the object is transmitted once and retained, until the deviation from the real appearance would grow so large that it needs to be newly transmitted, this case is entitled as *model failure*. In the other case of *model compliance*, the new picture can be described accurately enough mainly by the model parameters of motion and change of shape. Both 2D [HÖTTER 1990] and 3D motion models [OSTERMANN 1994] [KAMPMANN 2002] have been investigated. Unlike hybrid coding, no prediction error is foreseen to be encoded in the original methods, which implicitly incurs the tendency that quality deteriorates until the next model failure occurs.

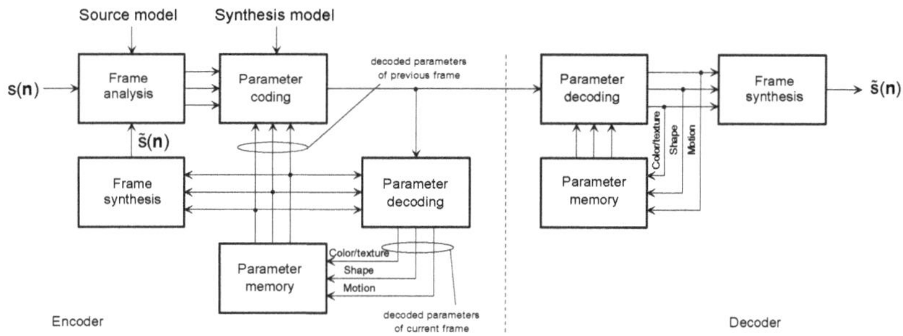

Fig. 7.48. Block diagram of an object-based video encoder and decoder as proposed in [MUSMANN, HÖTTER, OSTERMANN 1989]

As hybrid coding, the method in Fig. 7.48 is recursive, since both analysis and synthesis refer to a series of previously reconstructed pictures. Motion and shape parameters are encoded as differential information as well; in [HÖTTER 1994], the method is interpreted as a generalization of conventional hybrid coding, the latter one using block-based shape and motion models, and a block transform for texture coding.

The restriction imposed previously, postulating that texture changes shall exclusively be encoded by the intra-picture 'model failure' information, performs often worse than encoding of prediction errors. This fact was considered when defining MPEG-4 Visual (ISO/IEC 14496-2), which is the only standard so far that has included elements of object-based video coding. The MPEG-4 decoding process for a scene consisting of *video objects* is shown in Fig. 7.49. The scene description information defines the positions of the objects (where the background is one of them), and also the information which objects shall be visible in cases of spatial overlaps. For each single object, information components of *motion*, *shape* and *texture* have to be decoded. The motion information is used to exploit the temporal redundancy in both shape and texture information. Binary shape coding is performed by context-based arithmetic coding (CAE), as introduced in Sec. 6.1. Intra or residual texture coding is using DCT transform coding of block size 8x8, with *block padding* or *shape adaptive* methods for the boundary blocks, as explained in Sec. 6.1.3.

In addition to the binary shape, MPEG-4 also supports encoding of *gray-level shape* which allows to define *transparency* of objects by an *alpha map* which describes a blending factor for each sample. The alpha map is encoded by the same method that is used for the texture, re-using the texture's motion parameters within a separate hybrid coding loop[67].

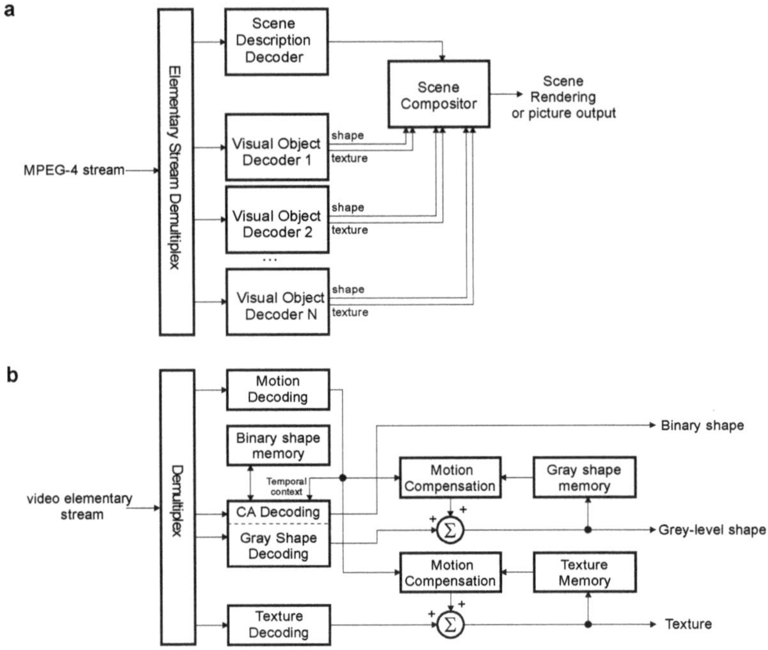

Fig. 7.49. Object-based decoding in MPEG-4 **a** Composition of N video objects into a scene or a picture **b** Structure of one single video object decoder

Object-based coding so far has not practically been shown to be applicable for improved compression of arbitrary video scenes; most reports of significant improvements from the early publications were related to simple videoconferencing sequences with one foreground object and static background. Scene analysis and exact segmentation is necessary before the coder can be operated. Beyond that, optimum allocation of rate to the components of motion, shape and texture is non-trivial (where however similar approaches are used in rate-distortion optimization of standard video codecs, decision for variable block size partitioning being a simplified method of segmentation and shape coding). The combination of a conventional (block-based) hybrid video coding with an object-based hybrid coding can therefore be beneficial, where both consist of common building blocks, but additional complexity would be necessary for implementing the elements of more precise shape coding, intended to be invoked whenever beneficial.

---

[67] Also AVC and HEVC allow adding alpha maps as 'auxiliary picture' channels. However, these are coded independent of the texture, and are not used for the purpose of data rate saving as in object-based coding.

7.7.2 Super-resolution synthesis

From camera shots with insufficient resolution, higher-resolution images can be generated from a series of pictures showing the same content but with slightly different sampling positions. This could be caused by camera or object motion; for an overview, see [PARK, PARK AND KANG 2003]. The aggregation of information from the samples requires a registration (practically done by motion estimation), combination into a non-uniform grid of samples, and transformation into a uniform grid of higher resolution. Simple methods use direct interpolation from the non-uniform grid, or estimation based methods such as constrained least squares, maximum a posteriori or projection onto convex sets, where the latter typically provide better quality. The quality of motion estimation is crucial for the reliability of any of the methods, and care should be taken that super-resolution information is only generated in cases where the identical area is safely available in multiple pictures. In terms of video coding, an interesting aspect of super-resolution comes by the fact that unlike sub-sample accurate motion compensation, which is performed by interpolation using only one picture, it provides a way to estimate samples by combining multiple pictures. However, trials of applying super-resolution for motion-compensated prediction in the context of video coding so far have shown that it is mainly beneficial when the original pictures are inappropriately sampled, i.e. if they show aliasing, such that conventional interpolation cannot be employed to generate a proper sub-sample prediction.

7.7.3 Moving texture synthesis

Dynamic textures are structures in a video which are changing randomly over time, but can be identified by spatio-temporal stationarity. Examples are water surfaces, flames, moving crowds, plants moving in the wind, smoke and clouds. There is good reason to assume that dynamic textures captured by a camera can be replaced by moving structures which are not identical, but perceptually similar.

In [NDJIKI ET AL. 2007], dynamic texture synthesis is performed within sequences of pictures entitled as 'synthesis bursts', which replicate texture from 'reference bursts' that have previously been transmitted to the decoder, typically with high quality (see Fig. 7.50). The synthesis itself is performed in this work based on the graph-cut algorithm proposed by [KWATRA ET AL. 2005], where spatio-temporal patches are selected based on similarity by a template-matching approach, and 'glued together' in a way such that the differences at the transition seams are minimized. To decide whether a certain spatio-temporal region is suitable for synthesis, a metric derived from VQM [ONG ET AL. 2004] is used. If a region is moving entirely e.g. due to camera or object motion, it is necessary to pre-compensate the motion and synthesize it accordingly, which can be done based on side information using an affine model (7.29). Furthermore, better quality is achieved if 'forward synthesis' from a previous reference burst and 'backward synthesis' from a subsequent reference burst are combined, which however would cause significant encoding delay.

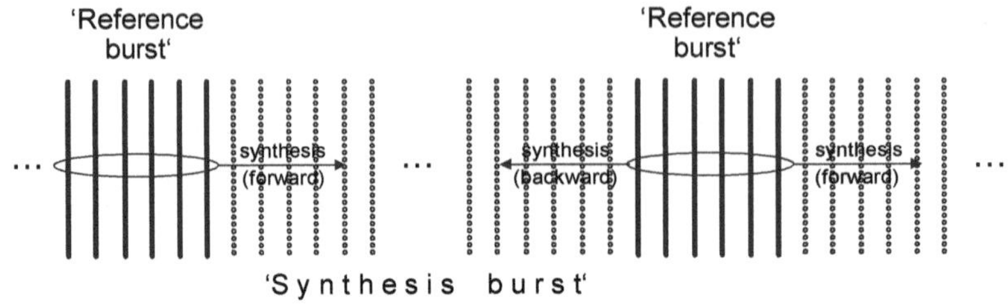

Fig. 7.50. Generation of dynamic texture within a group of sythesized pictures, trained from one or several groups of adjacent reference pictures (after NDJIKI)

Whereas the graph cut approach is an ad-hoc method derived and extended from the theory of Markov random fields (Sec. 2.5.4 and 6.6.2), another method for dynamic texture synthesis which is based on a state transition model with stationarity assumption about the underlying physical process was proposed in [DORETTO ET AL. 2003]. A dynamic texture sequence shall consist of pictures $\mathbf{s}(n)$, where all samples of the picture at time position n are written sequentially by rows into the vector $\mathbf{s}(n)$. The dynamic texture shall be modeled via a vector ARMA-process (autoregressive moving average) of a low-dimensional state vector $\mathbf{x}(n)$ and an innovation vector $\mathbf{v}(n)$; using this low-dimensional random vector, a linear mapping into the samples of a synthesized picture vector $\tilde{\mathbf{s}}(n)$ is performed, which has the same (high) dimensionality as $\mathbf{s}(n)$. Additionally, a mean image[68] $\overline{\mathbf{s}}$ as computed by time averaging from the original sequence of texture pictures $\mathbf{s}(n)$, and a noise vector $\mathbf{w}(n)$ are superimposed:

$$\mathbf{x}(n) = \mathbf{A}\mathbf{x}(n-1) + \mathbf{B}\mathbf{v}(n) \quad ; \quad \tilde{\mathbf{s}}(n) = \mathbf{C}\mathbf{x}(n) + \overline{\mathbf{s}} + \mathbf{w}(n)\,. \tag{7.38}$$

Besides the statistical characterizations of the noise components $\mathbf{v}$ and $\mathbf{w}$, in which the samples are assumed to be independent and identically distributed (i.i.d.), and the initial state vector $\mathbf{x}(0)$, the properties of the model are mainly ruled by the three matrices $\mathbf{A}$, $\mathbf{B}$ and $\mathbf{C}$ which need to be determined from the original data. For this purpose, a sequence of N pictures is analysed. This is done by arranging N vectors of each entity column-wise into a matrix, such that (7.38) can be re-written as

$$\begin{aligned}&\mathbf{X}(n) = \mathbf{A}\mathbf{X}(n-1) + \mathbf{B}\mathbf{V}(n)\\ &\qquad \text{with} \quad \mathbf{X}(n) = \left[\mathbf{x}(n-N+1) \quad \mathbf{x}(n-N+2) \quad \cdots \quad \mathbf{x}(n)\right],\end{aligned} \tag{7.39}$$

and the computation of N synthesized output pictures up to time instance n can then be defined as[69]

---

68 $\tilde{\mathbf{s}}(n)$, $\hat{\mathbf{s}}(n)$ and $\overline{\mathbf{s}}$ have the same number and arrangement of samples as the original picture in $\mathbf{s}(n)$.

69 The matrices consist of N columns from the current time positions and its N–1 predecessors. $\overline{\mathbf{S}}$ simply contains identical vectors $\overline{\mathbf{s}}$ in each of its N columns.

$$\tilde{\mathbf{S}}(n) = \mathbf{C}\mathbf{X}(n) + \overline{\mathbf{S}} + \mathbf{W}(n). \tag{7.40}$$

As nothing is known about $\mathbf{w}(n)$, but it is expected to have no systematic influence on the dynamic behaviour, a modified version of the picture, $\hat{\mathbf{s}}(n) = \mathbf{s}(n) - \overline{\mathbf{s}}$ is computed here by subtracting the average picture. The sequence of pictures $\hat{\mathbf{s}}(n)$ is combined in a matrix $\hat{\mathbf{S}}(n)$, similar to (7.40). A possible approximation of the matrix $\mathbf{C}$ from the given observation, which shall be synthesized as similar as possible in $\tilde{\mathbf{s}}(n)$, as well as the set of underlying state vectors in $\mathbf{X}(n)$ can be obtained by singular value decomposition of $\hat{\mathbf{S}}(n)$,[70]

$$\hat{\mathbf{S}}(n) = \underbrace{\mathbf{U}}_{\mathbf{C}}\underbrace{\mathbf{\Lambda}^{(1/2)}\mathbf{V}^{\mathrm{T}}}_{\mathbf{X}(n)}. \tag{7.41}$$

[Doretto et al. 2003] assume the order of the state model should be smaller than the number of training pictures N, which can be achieved by keeping only $P<N$ singular values in the computation of $\mathbf{X}(n)$[71]. As a second step, a least squares approximation of $\mathbf{A}$ can be found by comparing the results that the observation gives at two subsequent time instances, e.g. going back to (7.38),

$$\mathbf{A} = \arg\min \left\| \mathbf{X}(n) - \mathbf{A}\mathbf{X}(n-1) \right\|^2. \tag{7.42}$$

Similarly, $\mathbf{B}$ can be determined in a second step by modeling the residual error between $\mathbf{X}(n)$ and $\mathbf{A}\mathbf{X}(n-1)$ from an i.i.d. process $\mathbf{v}(n)$ (i.e. in principle 'whitening' the residual error by a matrix $\mathbf{B}^{-1}$).

This dynamic texture model can be used in various ways. As originally suggested in [Doretto et al. 2003], assuming stationarity, once the model parameters $\mathbf{A}$, $\mathbf{B}$ and $\mathbf{C}$ as well as the statistics of the innovation $\mathbf{v}$ and additive noise $\mathbf{w}$ are determined, synthesis can be run ad infinitum. Direct usage in coding was presented in [Ballé, Stojanovic, Ohm 2011], where the model is trained from a relatively small number of past pictures N (e.g. N=5) at each time instance, and a state model of dimension N is used to predict the subsequent picture of the dynamic texture. Here, as the reconstruction in (7.40) is perfect, it is not necessary to consider $\mathbf{w}(n)$, nor is any assumption about $\mathbf{v}(n)$ necessary, as it would be equivalent to the prediction error of the model, which cannot be compensated. Nevertheless, it has been shown that by using this dynamic texture model it is possible in the context of an AVC codec to reduce the prediction error for natural dynamic textures such as water surfaces, compared to conventional motion-compensated prediction.

It should however be observed that the described algorithms for dynamic texture synthesis usually fail in case of other than 'dynamic texture' type of motion.

---

[70] Herein, $\mathbf{U}$ contains the eigenvectors of $\hat{\mathbf{S}}(n)\hat{\mathbf{S}}(n)^{\mathrm{T}}$, $\mathbf{V}$ the eigenvectors of $\hat{\mathbf{S}}(n)^{\mathrm{T}}\hat{\mathbf{S}}(n)$, and $\mathbf{\Lambda}^{(1/2)}$ the singular values (square roots of related eigenvalues). The number of independent singular values is equal to the rank of $\hat{\mathbf{S}}(n)$, which usually is identical to N (provided that the number of pictures observed is smaller than the number of samples in each picture, and no identical pictures are observed).

[71] It is then also possible to determine the statistics of $\mathbf{w}(n)$ by comparing the observation against the reconstruction given by the model.

For example, in the case of the SVD-based approach, and assuming a dominant translation motion in addition to the dynamic texture motion due to camera or object movements, the vector $\bar{\mathbf{s}}$ would reproduce a mean image with strong motion blur, such that $\hat{\mathbf{s}}(n) = \mathbf{s}(n) - \bar{\mathbf{s}}$ would not any more allow a reasonable model generation via the SVD. As a countermeasure, as in [NDJIKI ET AL. 2007] a pre-compensation of 'deterministic' motion can be applied, e.g. as captured by global motion models (see Sec. 7.2.5). In that case, it is necessary to combine the dynamic synthesis with reverse warping according to the deterministic motion component. An example is shown in Fig. 7.51, where the case without global pre-compensation (Fig. 7.51b) shows heavy blurring of the synthesis result, which is avoided with global motion compensation in Fig. 7.51c.

Fig. 7.51. **a** Original picture from a video sequence containing zoom **b/c** dynamic texture synthesis without/with global motion compensation (source: STOJANOVIC)

7.8 *Video coding standards*

Open standards have played an important role in the development of video compression algorithms, and have been referred as examples of practical implementation of compression tools at various places in this chapter. The most important series of recommendations and standards for video compression were defined by working groups of the International Telecommunication Union (with video compression mainly defined in the ITU-T H.26x series for application domains of telecommunications and in the ITU-R BT.x series for television broadcast), and by the ISO/IEC *Moving Picture Experts Group* (MPEG), where the standards MPEG-1/-2/-4/-H include video compression parts. There are commonalities between the standards defined by the both groups, and significant work has been done jointly and published as twin text.

Principles of video coding standardization. Standards are about describing interoperability between systems which should be able 'understanding' each other by following rules of generating, transmitting and interpreting the data. In principle, the goal is to specify how a bit stream is uniquely converted into a video sequence of reconstructed pictures. To achieve this, an *interface layer* allows a high-level interpretation and access to the encoded video data, and a *video coding layer* (VCL) allows reconstruction of the video from the data. This chapter is mostly related to the VCL, whereas interfacing (systems) concepts are further discussed in Ch. 9. For the VCL, all video compression standards up to now have largely followed the principle to define the *bit-stream syntax*, the *bit-stream semantics* and the *decoding process*. The syntax relates to the structure of a bit stream, such that it can be parsed and interpreted as a sequence of codewords, which is typically the first step of the decoder operation. The semantics specifies the way by which these codewords are to be interpreted during the subsequent decoding process, e.g. the mapping of VLC codewords into reconstructed values, interpretation of side information parameters etc. The decoding process finally generates the output signal by a well-defined concatenation of processing steps.

The approach it is not necessarily giving an exact specification of encoder designs, it is left to the responsibility of the implementer to build an encoding device which produces a bit stream of valid syntax that can be transformed into a usable output signal by a compliant decoding device. This enables further *improving encoders* without losing compatibility, such that different encoder manufacturers can compete for the best technology during the lifetime of a standard, without need to change existing (legacy) decoding devices. It also includes an option to build low-cost encoders e.g. for low power consumption constraints, which might not show the optimum compression possible within the framework of the standard.

Consequently, conformance with video compression standards is typically defined for decoders. Unlike still image compression, where lack of processing complexity can easily be compensated by additional latency, video comes with a certain number of pictures per second or samples per second, such that a rigid real time requirement applies to decoders. Therefore, reasonable limitations of decoder complexity within the framework of a more generic standard are usually imposed by definition of *profiles* and *levels*, where a certain profile/level combination describes a *decoder conformance point*[72]:

- A *profile* specifies a collection of algorithmic elements (tools)[73] which has to be supported in the decoder to produce the expected output;
- A *level* is related to a maximum resolution of the signal which shall be supported by the decoder at a given conformance point, e.g. expressed in numbers of samples that have to be processed per second, or in terms of maximum bit rates.[74]

The profile/level indication is usually conveyed in a header of the bit stream, such that a decoding device can immediately decide if it will be able to process this stream without even starting to decode it.

When new standards or extensions to existing standards are defined, *compatibility* with previous solutions could be desirable [OKUBO 1992]. In *forward compatibility*, the bit stream generated according to an older version of a standard shall also be interpretable by decoders of a newer generation. Backward compatibility is the reverse principle, where a new standard's bit stream can at least for partial or low-quality reconstruction be interpreted by a decoding device which operates according to an older version. The former case can easily be achieved if the new standard or the extension defines a *superset* in terms of algorithmic elements of the old standard. Backward compatibility is more difficult to realize, and often will sacrifice the compression efficiency of a newer generation of standards, or will enforce higher complexity of decoding devices. As backward compatibility is often asserted to be important for acceptance in the market, this is a trade-off to be considered.

It is advantageous if the bit stream follows a hierarchical layer structure allowing to (re-) start decoding at certain positions, e.g. for switching of channels or resynchronization after data losses. Each of the layers could be coming with some header which in addition to the positioning itself gives other information that is necessary for decoding of the data below. Such layers are:

- the sequence *layer*, which subsumes a concatenated sequence of pictures[75],

---

[72] This approach has been taken systematically in all video standards since MPEG-2/H.262. It is beyond the scope of this book to give a detailed description of all profiles and levels of existing standards, as their number is large and frequently updated.

[73] In the MPEG-4 standard, another entity of *object type* is introduced, which allows a more systematic grouping of tools. Profiles are then defined as combinations of object types.

[74] It is preferable in this context to have levels nested, i.e. a decoder being able to operate on a higher level should be able to interpret a bit stream which is subject to a lower level's constraints. HEVC defines additionally *tiers*, which within on level define two or more different operation points in terms of maximum bit rate; also here, the higher tier is always capable to decode a stream of the lower tier as well.

[75] The sequence layer can also be used for *groups of pictures*, where a picture at the beginning should be instantaneously decodable (by not using prediction references to other pictures. The MPEG-2 standard defines a specific group of pictures layer for that purpose.

- the *picture layer*, which contains the information relevant for one picture,
- the *sub-picture layer(s)*[76], in various standards denoted as *slice* or *video object plane* or *group of blocks* layer, intended e.g. to support re-starting of entropy decoding if only part of a picture was lost (cf. Sec. 4.4.6),
- the *local decoding layer*, where the core decoding process is defined. This is entitled as *macroblock* in many standards, or *coding unit* in the recent HEVC development. This usually further includes smaller (block) partitions for prediction and transform of the luma and chroma components.

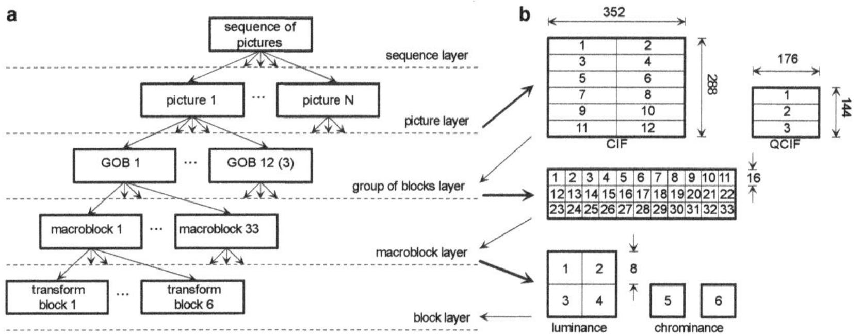

Fig. 7.52. **a** Hierarchy definition in the ITU-T Rec. H.261 syntax **b** Partitioning of a picture into GOBs in the cases of CIF and QCIF picture sizes, MBs within one GOB, DCT coding blocks within one MB

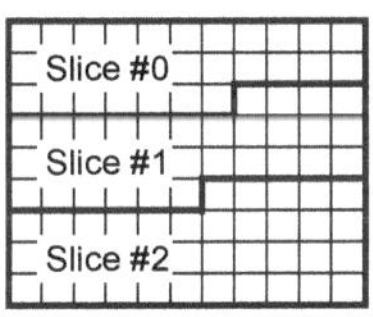

Fig. 7.53. Example of a slice structure as defined by the AVC standard.

Each of the layers typically associates some high-level parameters e.g. for enabling/disabling specific coding tools, provide information about picture sizes, timing etc. These are shared throughout all layers below. For the example of AVC and HEVC, these are the *sequence parameter set* (SPS) for the sequence layer, the *picture parameter set* (PPS) for the picture layer, the *slice header* for the slice layer. The bulk of video payload data comes at the sub-picture (slice) layer. Higher level parameter sets require a much lower amount of information, but when lost, they significantly impact the interpretation of lower layer data and may make reconstruction impossible. Therefore, they are often transported with higher priority or stronger error protection mechanisms. Fig. 7.52 shows another example of layer hierarchy for the case of the H.261 standard, which is however relatively static in terms of the number of next lower layer entities that are attached to a higher layer. Newer standards are much more flexible in terms of enabling varia-

[76] The sub-picture can also cover the entire picture, but formally this layer would still exist.

ble picture sizes, arbitrary picture sub-partitions in slices (see Fig. 7.53), more flexible prediction referencing between pictures etc. Definitions about parameter sets, sequence and picture structures, picture referencing, timing and buffering are also denoted as *high-level syntax* of a video compression standard. AVC and HEVC further include *supplemental enhancement information* (SEI) and *video usability information* (VUI) elements which give additional hints about the nature and arrangement of the pictures or the bit stream, without being normally needed for running the core decoding process.

Digital video source format specifications. Analog television was characterized by having a fixed number of active lines (e.g. 480 in US/Japan, 576 in Europe), as necessary to be interpreted by TV sets. Even though digital video could be much more flexible, video cameras, displays, projectors, and even graphics adapters of PCs and smartphones follow a certain set of source resolution conventions, since this is beneficial for interoperability (see also Sec. 1.3.1). In case of natural video, the most relevant source format specifications are ITU-R BT.601 (for standard TV resolution), ITU-R BT.709 (for HD resolutions) and ITU-R BT.2020 (for ultra HD resolutions).

First-generation hybrid coding standards. All relevant standards that were defined from approximately 1990 until today are based on the principle of hybrid video coding, combining motion-compensated DPCM, 2D DCT-based transform coding and various elements of spatial prediction. In the beginning, the concept was kept quite simple, using motion-compensated prediction with strictly limited maximum displacement range, avoiding sophisticated sub-sample interpolation, and using fixed block sizes for motion compensation (typically 16x16 samples). Further, restrictions are often made in terms of source formats supported (no arbitrary picture sizes, frame rates etc.). Under this, we categorize[77]

- ITU-T Rec. H.261 '*Video Codec for Audiovisual Services at p x 64 kbit/s*', first released in 1989[78], which was primarily designed for usage in conversational (videophone and video conferencing) applications, where a short encoding and decoding delay is of primary importance. For motion compensation, it used full-sample accuracy and a switchable spatial lowpass filter in the loop.
- MPEG-1 Video (ISO/IEC 11172-2 *'Coding of Moving Pictures and associated Audio for Digital Storage Media at up to about 1.5 Mbit/s, Part 2: Video*'), released in 1991 [LEGALL 1992], was originally designed for storage applications of the CIF source format on CDs, and marks a transition between

[77] ITU-T Rec. H.120 '*Codecs for Videoconferencing using Primary Digital Group Transmission*', first released in 1984, could additionally be listed here. It uses a replenishment method based on spatial 2D DPCM, intended for a transmission rate of 2 *Mb/s*; the second version of 1988 also included motion compensation for improved compression, but this has hardly ever been used as it was superseded by H.261 and other successors.

[78] The first release was still entitled '... at n x 384 bit/s'.

first and second generations. The main novelties compared to H.261 are extension of the motion compensation range, usage of half-sample accurate MC and usage of bi-directional prediction by defining the type of B pictures (cf. Fig. 7.6a).

- ITU-T Rec. J.81 (formerly ITU-R CMTT.723) '*Transmission of Component Coded Digital Television Signals for Contribution-Quality Applications at the Third Hierarchical Level of ITU-T Rec. G.702*', released in 1993, which was intended for encoded rates around 34 *Mb/s*, using intra-field, inter-field or motion-compensated hybrid coding with half-sample accuracy. It did not find wide use, since H.262 | MPEG-2 became available in the same application domain, providing better compression quality due to usage of B pictures.

Second-generation hybrid coding standards. The second generation of hybrid coding improved the compression performance more extensively by using half-sample accuracy and enlarged ranges of motion compensation, as well as bi-(directional) prediction. Further, designs were made more flexible w.r.t. supporting various picture sizes and frame rates. DCT of fixed size 8x8 is generally used. Some of these standards also allow using variable block size of motion compensation (down to 8x8), improved methods of entropy coding are employed, and tools for extended functionalities (interlaced coding, scalable and multi-view coding). Compared to the early first generation, this could achieve 30-50% bit rate reduction while retaining the quality identical. This category includes the following standards:

- MPEG-2 Video | H.262 (ISO/IEC 13818-2 '*Generic Coding of Moving Pictures and Associated Audio Information – Part 2: Video*', twin text with ITU-T Recommendation H.262), was first released in 1995. This covers a wide range of applications and data rates; larger picture sizes up to HD resolution are also supported. MPEG-2 decoders are forward compatible (i.e. able to decode bit streams) w.r.t. MPEG-1, but as one major change, the standard additionally includes dedicated tools for interlaced video (specific field processing). MPEG-2 further defines tools for scalable coding (spatial and SNR) and multi-view coding in specific profiles, but neither of the related profiles gives a reasonable trade-off between compression performance, complexity and functional benefit.
- ITU-T Rec. H.263, '*Video Coding for Low Bit Rate Communication*', first released in 1996, was targeted to achieve significantly better compression than H.261, mainly in the same application domain of video communication. Besides various measures for improving the motion compensation and its accuracy, improved prediction mechanisms such as overlapping block MC (Sec. 7.2.5), multi-picture prediction, and a loop filter with adaptation to the strength of quantization are included. Additional methods for error-resilient transmission and modes for temporal, spatial and SNR scalability are also defined. These enhancements are often subsumed under the names of 'H.263+' and 'H.263++' [CÔTÉ ET AL. 1998], but are rarely used in full combination.

H.263 also defines sets of tools as profiles, specifically designed for usage in communication applications.

- MPEG-4 Visual (ISO/IEC 14496-2: '*Coding of Audiovisual Objects – Part 2: Visual*'), was released in its first version in 1999. It includes a video coding specification, which is mainly different from MPEG-2 and H.263 by specifying tools for new functionalities related to encoding of *video objects* which can have either rectangular or arbitrary shape[79]. Video scenes can be composed from several objects which may change by position, appearance, size etc. independent of each other, but it is also possible to embed video elements into 3D scenes (cf. Sec. 9.4). The tool of 'sprite coding' denotes a static background picture that can be warped into a scene, and filled by foreground objects by means of scene composition. MPEG-4 part 2 further includes tools for scalable coding, which however lack efficiency in compression. Other coding tools worthwhile mentioning were partially added in later versions. Among them are MC with quarter-sample accuracy, global MC and shape-adaptive DCT. Specific tools for the lossless and near lossless quality range are defined in the *studio profiles*.

From some of their aspects, particularly higher precision of motion compensation and usage of loop filtering, the extensions of both H.263 and MPEG-4 part 2 mark the transition to the third generation.

Third-generation hybrid coding standards. The third generation of hybrid coding further improves the compression performance by even better motion compensation using quarter-sample accuracy, variable block sizes, multiple reference pictures and temporal hierarchy structures; more sophisticated spatial filters are operated in the prediction loop, enhanced methods of spatial prediction are used. Transforms are operated with variable block sizes, and concepts of context-based and adaptive entropy coding are heavily used. Transforms are specified through integer precision approximations, which solves implementation specific deviations between encoder and decoder, known to potentially cause drift in the first and second generation standards. Compared to the second generation, the combination of various tools again gives approximately 50% (significantly more for HEVC) of bit rate reduction. Due to more flexible referencing and entropy coding, additional functionality such as scalable coding and multi-view coding also became more efficient in terms of compression performance than with second generation standards. The category includes

- *Advanced Video Coding* (AVC, specified as ITU-T Rec. H.264 and ISO/IEC 14496-10). The first version of AVC was released in 2003 (for and overview, see [WIEGAND ET AL. 2003]); the development was performed by the *Joint Video Team* (JVT), which consisted of members of ISO/IEC WG11 (MPEG) and ITU-T SG 6/16 Video Coding Experts Group (VCEG). AVC allows predic-

---

[79] A decoder designed to support the simple profile of MPEG-4 part 2 is also able to decode an H.263 baseline bit stream, i.e. the latter establishes a subset in terms of syntax, semantics and decoding tools.

tion block partitioning from 16x16 macroblocks (MB) down to 4x4 prediction block sizes, 4x4 integer-precision DCT (plus 8x8 in most newer profiles), motion compensated prediction with quarter-sample accuracy, uni- or bi- or weighted prediction, directional intra prediction with 8 prediction directions plus a flat (DC) mode, an in-loop deblocking filter controlled by the local coding state and sample differences across the block boundary (cf. Sec. 6.9) is operated in the loop. Two different entropy coding methods are defined (depending on profile), *context-adaptive VLC* (CAVLC) and *context-adaptive binary arithmetic coding* (CABAC), which are universally applied to all elements of the code syntax. Picture referencing is very flexible, allowing hierarchical picture referencing structures. Important extensions of AVC were the series of High Profiles defined in the range extensions of 2004; the scalable extension of 2006 which implements a layered coding approach (cf. Sec. 7.5.1), but is more efficient than any previous scalable video coding due to more intense re-use of information from the base layer; and the multi-view extension of 2008. The latter is using the flexible referencing structures for inter-view prediction in stereo/multi-view (cf. Sec. 7.6), whereas being backward compatible to single-view (monoscopic) decoding of the base view. More recent extensions defined in 2012 and 2013 provided the capability to encode depth maps and more efficient combined compression of multi-view video plus depth.

- *High-efficiency Video Coding* (HEVC, released in 2013 as ISO/IEC 23008-2 and ITU-T H.265); for and overview, see [SULLIVAN ET AL. 2012]. The HEVC development is performed by the *Joint Collaborative Team on Video Coding* (JCT-VC) of MPEG and VCEG. One of the main targets of specifying HEVC was a significant improvement in compression (compared to AVC), particularly for video formats of high and ultra-high resolution, which is reported to become as large that for the same quality the bit rate can be reduced by approx. 50% in relevant cases [OHM ET AL. 2012]. HEVC is also based on the hybrid video coding principle, but allows even more flexibility in adaptation than AVC, and particularly further improves motion compensation and intra prediction. Some of the most important novelties are: usage of larger *coding tree units* (CTU) of typical size 64x64 for luma blocks, further partitioned into square-shaped coding units (CU) of equal or smaller sizes down to 8x8; more flexible partitioning of motion compensation areas (down to 4x8/8x4 samples) within *prediction units* (PU) below the CU level, with efficient signalling of motion vectors through new advanced motion vector prediction (AMVP) and merge methods; defining, below the CU level, *transform units* (TU) with sizes from 4x4 up to 32x32; intra prediction (operated in a prediction loop according to TU block sizes) with 33 directional modes, additional flat (DC) and plane (amplitude surface) prediction from boundary samples. Improved interpolation filters for motion compensation are employed, and in addition to the deblocking filter which is similar to AVC's, a non-linear sample processing denoted as sample adaptive offset (SAO) is operated in the prediction loop (see Sec. 6.9). Entropy coding is based on CABAC, similar to

AVC, but simplified by reducing the number of context-coded bins, and lower number of contexts overall. Special signalling allows skipping transform and loop filtering, which provides better compression performance for computer graphics type of content. If additionally the quantization is omitted, lossless compression is possible as well. Recently, range extensions for color sampling formats beyond 4:2:0 and extended bit depth, as well as multi-view and scalable extensions were defined; a 3D extension for more efficient combined texture/depth coding, and an extension for more efficient coding of graphics/screen content are also under development [SULLIVAN ET AL. 2013].

Robustness to data errors/losses and flexibility for operation over a variety of network environments is enabled by a number of design aspects in AVC and with some improved variations in HEVC as well. For this, the high-level syntax of the latter defines:

- Parameter set structures: A parameter set is supposed to contain information that is expected to rarely change and is usually associated with a large number of bits of the video payload. The parameter set structure provides a secure mechanism for conveying data that are essential to the decoding process. The concepts of sequence and picture parameter sets known from AVC are extended by one higher hierarchy level defining the *video parameter set* (for highest-level information, e.g. describing the structure of an overall bit stream and its relevant sub-streams).
- NAL unit structure: Each piece of bit stream syntax structure is placed into a logical data packet called a network abstraction layer (NAL) unit. Depending on the NAL unit type and other header information, it is possible to identify the purpose of the payload contained, e.g. parameter sets, video coding layer data, etc.
- Slices, tiles and wavefront processing: A slice is an entity which can be decoded independently from other slices in terms of entropy coding and intra prediction. It can either be an entire picture or a region of a picture. Slices are often used for re-synchronization after data losses. In case of packetized transmission, it would be undesirable to have a slice spanning over several packets due to expectation of losses. Therefore, the maximum payload size (in number of bits) within a slice is restricted, such that the number of MBs or CTBs may be variable. Slices typically contain MBs or CTBs in row-wise scanning order[80], but additional provisions can be taken to reduce cascading dependencies by allowing *wavefront parallel processing* (Fig. 7.54c). Additionally, HEVC defines tiles, which are also independently decodable, but only come with a minimum amount of header information. A *tile* consists of a rectangular arranged group of CTBs, which is also useful to enable parallel

[80] AVC also allows different ordering of macroblocks in slices, entitled as *flexible macroblock ordering* (FMO). This e.g. allows defining two slices interleaved as black/white checkerboard patterns, which in case where one of them is lost would allow a certain degree of reconstruction of samples from the other.

processing and allow random local access to individual regions of a picture. Examples of partitioning a picture into a slice structure and a tile structure are shown in Fig. 7.54a/b.

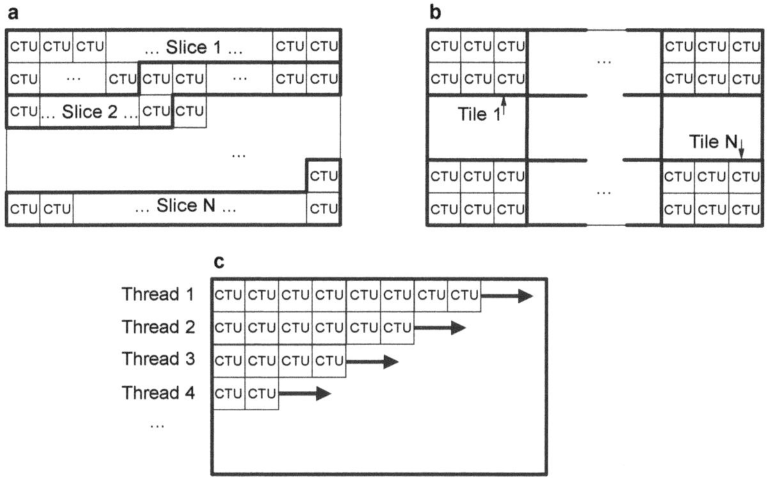

Fig. 7.54. Example partitioning of a picture into slices (a) and tiles (b), and illustration of wavefront parallel processing (c)

Reference picture buffering and signalling: Compared to AVC, the mechanisms to signal storage, removal and addressing of previously decoded pictures in the reference picture buffer have been slightly simplified and became more robust due to direct usage of the *picture order count* (POC) values in HEVC. For each slice, two lists (called reference picture list 0 and list 1) are defined, for which at the slice level a unique mapping exists to the pictures (and their POC values) that are currently in the reference picture buffer. For reference picture indexing during prediction, a simple pointer to the list is sufficient, where uni-prediction generally uses list 0, whereas bi-prediction uses both lists.

7.9 Problems

Problem 7.1.

A motion-compensated prediction of a video signal is performed. In our toy example, the misalignment of motion compensation follows a uniform discrete distribution into values $k_{\varepsilon,i} \in \{-1,0,1\}$, and shall be statistically independent between horizontal and vertical dimensions ($i=1,2$).

a) Determine the power spectrum $\phi_{ee}(f_1,f_2)$ of the prediction error picture, if the corresponding original picture from the sequence has a power spectrum $\phi_{ss}(f_1,f_2)$.
b) Which would be the optimum cut-off frequency of a lowpass filter in the prediction loop, such that the variance of the prediction error signal is minimized?
c) The power spectrum of the picture shall be described in horizontal direction as $\phi_{ss}(f)=A|1-f|$. Compute the coding gain that is achievable by motion-compensated prediction, both without and with the lowpass filter from b).

Problem 7.2.

In motion estimation, it shall be decided which of two motion vectors $\mathbf{k}_1$ and $\mathbf{k}_2$ shall be chosen. The following prediction error blocks have been found:

$$\mathbf{E}(\mathbf{k}_1)=\begin{bmatrix}10 & 10\\ 0 & -10\end{bmatrix} \quad ; \quad \mathbf{E}(\mathbf{k}_2)=\begin{bmatrix}20 & 20\\ 20 & 20\end{bmatrix}$$

a) Determine the best vector $\mathbf{k}_{1|2}$ following the criterion of minimum error variance.
b) Apply a 2D Walsh transform of block size 2x2. Determine the coded bit rate, using the table from problem 6.3c. Based on a criterion to minimize the rate, which of the vectors would now be the preferable choice? Give an interpretation.

Problem 7.3.

In the block schematics for scalable hybrid coders as given in Fig. 7.39b/c, reconstructed signals would be identical if the entire information (base and enhancement layer) is used and motion compensation (MC) is identical in all MC blocks of base and enhancement layers. Show that also the prediction errors at the respective quantizer inputs (QE und QB) are then identical in both schemes.

[Hint: Use the relationships between the respective signals in the frequency domain, similar to (5.7) and (5.8) to simplify the proof. Use $H(\mathbf{f})$ for the entire block of motion compensation without explicitly expressing the sizes of the motion vectors.]

Problem 7.4.

The properties of the first picture from a video sequence of 10 pictures are modeled by a separable 2D AR(1) process of parameters $\rho_1=\rho_2=\sqrt{3}/2$ and $\sigma_s^2=16$. By a camera pan, a horizontal translation movement of exactly 10 samples/picture leftwards occurs. The picture size shall be $N_1 \mathrm{x} N_2=90\mathrm{x}50$ samples.

a) The signal is encoded using $R=6$ bit/sample, without utilizing any correlation between samples. Determine the minimum possible distortion D.

In the following points, the signal coded according to a) is regarded as the original reference for subsequent compression.

b) By which amount [bit/sample] can the rate be reduced without introducing additional distortion, if *i)* 1D and *ii)* 2D spatial correlation in the signal is utilized?
c) The exact motion shift shall be known to the decoder, such that perfect motion-compensated prediction can be performed (without using side information). Sketch those areas in the second picture which have to be newly coded and those which can

be perfectly predicted. By this, determine the possible reduction in bit rate due to MC prediction.

d) Compute the necessary total bit rate for encoding of the sequence with distortion D, if all known spatial and temporal correlation properties are used.

Due to a finite length of a transmission buffer used for rate regulation, the maximum number of bits to be used for encoding of any single picture is limited to 9000.

e) Compute the minimum additional distortion D_1 (compared to the encoding cases of a/b), which is now introduced for the first picture of the sequence.

f) How will the distorted representation of the first picture affect the prediction of the second picture? Which power spectrum of the motion-compensated prediction error will be found within those areas of the second picture which could otherwise have been perfectly predicted?

Problem 7.5.

Compute the power density spectrum of the prediction residual in bi-directional motion compensation (7.20), under the assumption that the deviations from true displacement $\mathbf{k}_{\boldsymbol{\varepsilon},-1}$ and $\mathbf{k}_{\boldsymbol{\varepsilon},+1}$ are statistically independent. How would the result change if the two deviations were identical, and if they would have opposite sign?

8 Speech and Audio Coding

Typical appearance models of speech and audio signals can be determined from physical models of the human vocal tract and of musical instruments, respectively. These are often expressed as filters excited by a source (source-filter model). The importance of periodic components is related to the harmonic properties of many audio sources; harmonic spectra appear in cases of voiced speech and many types of musical instruments, which can directly be related to the physical process of sound generation, e.g. excitation of a resonance body by a string. Beyond harmonics, also transient and noise-like components are important. Whereas harmonic signals tend to be sparse in frequency, transient signals are sparse in time. Stationary noise is not sparse in either of the domains, but bears a large amount of perceptual irrelevance and therefore it may not be necessary to reproduce its exact time and frequency characteristics. Exploitation of perceptual irrelevance aspects for saving data rate in compressed speech and audio is more advanced than it is in the state of the art compression of image and video signals. Most speech coding schemes are based on linear prediction methods. Synthesis models for speech generation have reached an advanced status and are employed for parametric coding at low rates with only few parameters. In speech coding, understandability of the reconstructed signal is an important design criterion, whereas in coding of audio signals, psycho-acoustic properties of hearing are used to achieve transparent quality, even when the reconstruction is lossy compared to a digital (PCM) original. Transform coding based on block-overlapping or subband transforms is most frequently used for waveform coding of audio signals. Sound synthesis, enabling synthetic generation of natural or newly-created sounds, is a very mature technology as well. It is however mostly based on nonlinear models, such that universal usage in analysis/synthesis audio coding with automatic computation of synthesis parameters is still an area that requires further development. Other fields of intense development in audio coding are targeting efficient compression of combined speech/general audio signals, and of multi-channel or 3D audio captures.

© Springer-Verlag Berlin Heidelberg 2015

J.-R. Ohm, *Multimedia Signal Coding and Transmission*,
Signals and Communication Technology, DOI 10.1007/978-3-662-46691-9_8

8.1 Coding of speech signals

8.1.1 Linear predictive coding

Most speech codecs are based on *Linear Predictive Coding* (LPC) (Sec. 2.6.2), which stands in tight relationship with autoregressive models (Sec. 2.6.1). The speech generation in the human vocal tract can be modeled by acoustic excitation of a tube having varying perimeters, which then by its properties of reflection and resonance is mapped into parameters of an autoregressive synthesis filter [RABINER, SCHAFER 1978]; the z-domain singularities (poles) of such a filter correspond to the resonance frequencies of the tube, which give a reason for the dominance of harmonic components in voiced speech. The excitation is performed via the vocal cords, which can additionally modulate the air that is flowing through the tube in a harmonic fashion. For unvoiced speech, the tube behaves less resonant, and the contribution of throat, jaws and tongue to the generation of the sound is larger.

To determine optimum filter parameters, the Wiener-Hopf equation (2.208) can be used[1], based on autocorrelation values as measured from segments of signal samples. Typical filter lengths used in this context are $P=12\ldots16$. However, in case of voiced speech, this may not even cover the duration of the pitch period, such that even some periodic structures can be found in the residual signal. Therefore, a secondary stage of *long term prediction* is additionally applied, where the time lag is usually determined by pitch analysis rather than the Wiener-Hopf equation. One or very few coefficients applied on top of a variable delay line are usually sufficient for this purpose.

If an AR filter is excited by a signal of flat spectral properties (white noise or impulse-like signals), the spectral properties of the output are close to the properties of the original signal as modeled by the filter; however, with white noise excitation as in an original AR model, the output signal would rather appear like colored noise with spectral peaks around the resonance frequencies of the filter. In general for quality of audible perception, adequacy of the amplitude spectrum is of highest importance. For voiced speech however, the phase should at least be consistent between the formants (harmonics), such that arbitrary phase variations of the excitation signal, as caused by noise excitation, would lead to an unnatural impression. With a regular or mixed pulse train excitation, the output comes closer to a harmonic signal. When the true prediction error is coded as excitation (usage of closed or open loop prediction) the reproduction of the true signal would be guaranteed at least within an expected margin of error. However, sample-wise

[1] There are alternative ways of solving the Wiener-Hopf equation through inversion of the covariance matrix, in particular Levinson-Durban recursion and Cholesky decomposition are relevant in this context. In the recursive computation, alternative structures for LPC filters, such as ladder filter structure related to partial correlation (PARCOR) coefficients can directly be determined (see [RABINER, SCHAFER 1978] and the further notes by the end of the current section).

predictive coding is not suitable for low data rates. Therefore, LPC schemes with excitation strategies for the synthesis filter, being consistent with the aforementioned properties of speech generation models, have been developed.

DPCM and adaptive DPCM methods. By these schemes, the true waveform of the speech signal is approximated as close as possible. As scalar quantization has to be used in closed-loop systems with sample-wise prediction, the reduction of bit rate is limited to factors of 2-3 compared to 8-bit PCM (i.e. 3-4 bits/sample) when good, understandable speech quality shall be achieved. The basic principle was explained in Sec. 5.2.1. In *Adaptive DPCM* (ADPCM), methods with either forward or backward adaptation of predictor filters are used, where forward adaptation typically uses filter lengths of 12-16 taps, whereas in combination with backward adaptation shorter filters of 4-8 taps are used. In case of forward adaptation, it is necessary to encode the filter parameters as side information.

For the quantization, either forward adaptive (quantization step size to be transmitted as side information), or backward adaptive methods are used. For the latter case, the step size Δ is increased in cases where several subsequent quantization decisions have the same sign (indicating high quantization error feedback as from (5.14)), and it is decreased when permanent toggling of positive/negative signs is observed. Such an adaptation is also consistent with the properties of the human auditory sense, where the perception of noise depends on the amplitude of the signal.

For speech coding, application of open-loop DPCM schemes has also been suggested (entitled as D*PCM in [JAYANT, NOLL 1984]). Firstly, the effect of drift propagation is usually much less severe for speech than it is in image and video coding, as the synthesis filters are strictly stable with pole locations farther away from the unit circle of the z transform. Furthermore, it is argued that due to the filtering of the quantization error by the synthesis filter in (5.7), with the input assumed as being spectrally white, a kind of 'noise shaping' occurs, which puts most of the quantization noise into those spectral components that are emphasized by the synthesis filter, which could be advantageous under psycho-acoustic criteria. As however this is hardly controllable in an open-loop system with direct quantization of the error, the more systematic approach of noise shaping as often applied in the excitation-based LPC synthesis methods (see below) appears to be a better choice.

Excitation-based LPC synthesis. In this class of speech coders (Fig. 8.1a), the LPC synthesis filter is fed by vectors of innovation signal samples (excitation sequences) selected from a codebook. Different from the vector prediction methods presented in Sec. 5.2.4, the synthesis filter is a one-step recursive filter as in usual AR modeling, which means that prediction relies on a number of subsequent previous samples, including the directly preceding sample. Unlike closed-loop DPCM, the quantization cannot be implemented in the domain of prediction errors, as the first samples in the excitation vector recursively influence subsequent samples. The methodology to find the best excitation signal is an analysis-by-synthesis approach, where the candidate excitation vectors are fed into the filter and the reconstruction sequences are compared against the original. In the process of comparison, signal-dependent spectral weighting is often applied. This introduces a distortion criterion based on the *spectral noise shaping* mentioned above, which is also asserted to give best understandability of the synthesized speech signal. This is often done by re-using the (all pole) LPC synthesis filter, but modify the z polynomial w.r.t. smaller radial position of poles, and/or by introducing additional zero locations with the same angular orientations as the poles, but at smaller radial position. This way, the frequency weighting function gives a flattened approximation of the assumed original signal spectrum, supporting the view that the error would be least perceivable at positions where the spectral energy of the signal itself is high (e.g. the resonance frequencies of the vocal tract).

Excitation-based synthesis methods are most widely used for speech coding at low bit rates, where still a natural impression of the speech (retaining the individual characteristics of a speaker as well) shall be achieved. For unvoiced speech, generation of a good synthesis result is also uncritical from white-noise type (random) innovation vectors. Different excitation-based methods can better be characterized by their effect on voiced speech segments, where a unique pitch property in the signal can only be obtained by feeding a train of excitation pulses into the synthesis filter[2]. The following methods are typically employed (for more detail, the reader is referred to [HANZO ET AL. 2007], [VARY, MARTIN 2006]):

- *Multi-pulse excitation* (MPE), which is based on optimum placement of non-equidistant pulses, where the number of pulses shall be as low as possible;
- *Regular pulse excitation* (RPE), similar to MPE, but with equidistant pulses; in MPE and RPE, the set of different excitation configurations (of which only a finite number exists) establishes the excitation codebook, which does not need to be stored explicitly;
- *Code excited linear prediction* (CELP), in which excitation signals are selected from a codebook without specifically differentiating between voiced and unvoiced speech segments (in fact, the codebook is usually designed such that both cases are appropriately supported, and generated by putting randomly generated samples rather than using a training sequence as in conventional VQ); a variant of CELP is *vector sum excited linear prediction* (VSELP).
- *Harmonic vector excitation coding* (HVXC), where the excitation signal particularly for voiced speech segments can be efficiently presented by few parameters such as pitch and gain. In the MPEG-4 standard, a speech codec is defined where HVXC is combined with CELP, the latter used for unvoiced segments.

---

[2] For fast search in the code space of excitation signals, it is also possible to analyse the pitch of the original signal beforehand, and investigate only those candidates which would incur a similar pitch.

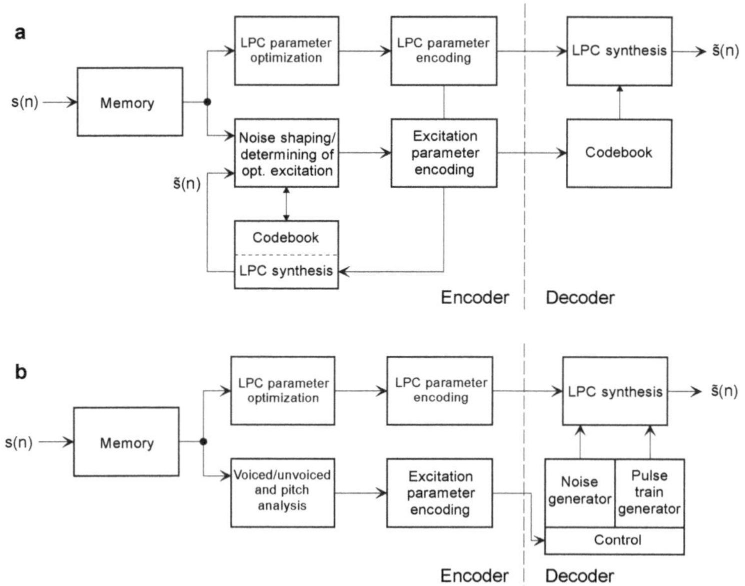

Fig. 8.1. Speech coding: **a** Analysis by synthesis method **b** LPC vocoder method

In particular for encoding at very low bit rates (3 *kb/s* and below) also mixtures of these methods are used, e.g. *mixed excitation linear prediction* (MELP), which is a hybrid between CELP and MPE.

If forward adaptation of LPC parameters is used, the percentage of bits to be spent for the side information can become quite significant at low rates. Different representations of LPC parameters can be used for the purpose of encoding, e.g. *PARCOR coefficients* [ITAKURA, SAITO 1972], *log area coefficients* [MAKHOUL ET AL. 1985] or *line spectrum pairs* [ITAKURA, SUGAMURA 1979], which can in principle be interpreted as alternative forms of the filter coefficients directly computed from the Wiener-Hopf equation, with identical residual error (in case of mathematically precise mapping). It should however be noted that direct quantization of the filter coefficients modifies their properties, such that the original design goal of the Wiener-Hopf equation (minimizing the variance of the prediction error) might no longer be fulfilled. Unlike the LPC coefficients representing the direct implementation as a recursive filter, the alternative representations usually are less sensitive versus direct quantization when employing Euclidean distance criteria in their respective representation domain, such that the resulting spectral model has less deviation from original spectral properties [VISWANATHAN, MAKHOUL 1975].

Efficient representation of LPC parameters is crucial in case of low-rate speech coding. An example is the OPUS-SILK speech codec [VOS ET AL. 2013], which uses a combination of long and short term prediction, line spectrum frequency (LSF) representation by vector quantization, combined with a secondary scalar quantization stage and prediction of LSF coefficients over time. All indices are furthermore entropy coded. Unlike the above mentioned approaches, the excitation signal is compressed by trellis coded uniform scalar quantization (Sec. 4.6) with variable offset values. The offset is determined by a pseudo random number generator which is recursively controlled from previous quantization decisions (i.e. backward adaptive). The optimum trellis path is determined by the Viterbi algorithm. The SILK algorithm can also efficiently compress stereo speech signals, where in advance of the compression, switchable mid/side coding (similar to Fig. 8.6) is used, with subsequent prediction of the side channel from the mid channel for compensation of run-time stereo effects.

8.1.2 Parametric (synthesis) coding

The vocoder principle (Fig. 8.1b) was originally designed for speech synthesis, e.g. to be used for outputting a written text as a synthetic speech signal. In the LPC synthesis model, compact parameters are used to describe the filter and the excitation. For representing the filter parameters, the methods described in the previous paragraph can be used; the excitation signal is either synthetic noise or a pulse train of a given period (pitch frequency of the speech signal). In addition, by systematic modification of the filter parameters and the excitation signal, tone height, spectral shape, amplitude etc. can be varied, such that e.g. either male-like or female-like voices can be output. Vocoders in general lack naturalness, but still provide a good understandability of speech.

For extreme-low bit rate coding, vocoders can directly be applied, which requires in first place determination of the model parameters, i.e. LPC parameters and excitation signal parameters, i.e. voiced/unvoiced classification, the pitch period for the voiced segments, and the amplitude (variance) of the excitation signal. Filter parameters can be obtained by the Wiener-Hopf equation (2.206) or equivalent, the excitation signal parameters can be gained by analysing the residual error of the prediction.

Hybrid solutions between vocoder-based coding and the previously-discussed excitation-based coding (the latter targeting minimization of some objective error criterion when determining the residual) are possible as well. This is reasonable, since in terms of decoder building blocks, a large amount of similarity exists. In extreme case, synthesis-based coding could be designed such that speech recognition is applied, and the sequence of words/letters spoken is conveyed as text or as phonemes, along with timing parameters. A *text to speech* approach, employing any available speech synthesis method, could then be applied at the decoder end; however, any individual speaker properties and prosody would be lost for the sake of extreme compression, and methods to recover them would increase the rate again.

8.1.3 Speech coding standards

The approach usually taken in speech coding standardization is different from the common approach used in defining image and video coding standards (Sec. 7.8).

Table 8.1. Overview of relevant standards for encoding of speech signals[3]

Standard	Codec type	Bit rate [*kb/s*]
ITU-T G.711	PCM	64
ITU-T G.711.1	Wideband embedded extension of G.711 (MDCT / weighted VQ for high band)	64, 80, 96
ITU-T G.718	Frame error robust narrow/wideband embedded variable bit-rate	8-32
ITU-T G.722.2	Adaptive multi-rate wideband (AMR-WB)	~16
ITU-T G.726	ADPCM	16, 24, 32, 40
ITU-T G.727	Embedded ADPCM	16, 24, 32, 40
ITU-T G.728	Low-delay CELP	16
ITU-T G.729	Algebraic CELP (ACELP)	8
ITU-T G.729.1	Embedded variable bit rate extension of G.729	8-32
ITU-T G.723.1	ACELP / MP-LPC	5.3 / 6.3
GSM 06.10	RPE-LTP	13
GSM Enhanced Full Rate	ACELP	12.2
GSM Half Rate 06.20	VSELP	5.6
MPEG-4 Audio	CELP scalable narrow band	3.85-12.2 (steps ~ 2 *kb/s*)
MPEG-4 Audio (ISO/IEC 14496-3)	CELP scalable wide band	10.9-23.8
OPUS-SILK (IETF RFC 6716)	LPC/LTP with trellis coded excitation	10-32
MPEG-4 Audio (ISO/IEC 14496-3)	HVXC scalable	2 / 4
FS 1015 (US Fed. Secure)	LPC-10E	2.4
FS 1016	CELP	4.8
FS 1017	MELP	2.4

[3] PCM: *pulse code modulation*, ADPCM: *adaptive differential PCM*, CELP: *code excited linear prediction*, LPC: *linear predictive coding*, MP-LPC: *multi-pulse LPC*, RPE-LTP: *regular pulse excitation / long term prediction*, VSELP: *vector sum excited linear prediction*, HVXC: *harmonic vector excitation coding*, MELP: *mixed excitation linear prediction*

As the intelligibility of speech is an important criterion in speech transmission services, also encoding algorithms are commonly specified, thus guaranteeing as far as possible a foreseen perceptual quality.

Table 8.1 gives an overview about some of common speech coding standards available today, most of which are based on the principle of linear prediction. These standards are used for speech encoding and transmission either standalone or combined with other media types in multimedia communication systems.

8.2 Generic audio, music and sound coding

8.2.1 Transform coding of audio signals

Generic audio signals are a mixed composition from different sound sources, which could be musical instruments, voice/speech/singing, synthetic or natural (by microphone) sound captures. Due to the large variety, it is difficult if not impossible to characterize all musical instruments by a unique physical model. Nevertheless, an excitation-plus-filter model (as in speech coding) is appropriate for various kinds of instruments, particularly those which are based on periodic oscillation (e.g. string) amplified by a resonance body, as it is typically the case for string instruments. There are also cases where the excitation is basically noise, but a harmonic sound is produced due to steerable resonances, as for woodwind and brass instruments. In other cases, the excitation could be a transient applied to an oscillating string, as for guitar or piano, causing a decaying harmonic without explicit sustain period; or sustain extended by interaction of the string with a resonance body. Along with various percussive instruments which usually have a broad spectrum but short temporal envelope, the possible multitude of instruments is complemented by the fact that various synthetic sounds are being generated in music, and various non-music natural sounds (such as clapping) need to be encoded along with the music.

Based on such properties of musical and non-musical sound signals, methods of *transform coding* establish the most widely used audio compression technology of today, as they allow reconstruction by good quality for arbitrary audio signals, if appropriate adaptation to the time-varying properties of the spectrum is made[4]. For spectral-domain coding, it is also straightforward to employ mechanisms which take into account the psycho-acoustic properties of hearing (Sec. 3.2.2),

[4] In contrast to that, e.g. music signals are often perceived strangely distorted when encoded by an LPC method which is originally designed for speech signals. This is partially due to the fact that AR models (even of higher order) are not fully appropriate for audio signals; they do not optimally represent spectrally sparse (quasi periodic), nor temporally sparse (transient) signals, which are rather typical for the important classes of harmonic and percussive musical instruments, respectively.

mainly related to frequency masking, by using spectral-dependent quantization. A generic method for audio transform coding is illustrated in Fig. 8.2. The psycho-acoustic model is assumed to be based mainly on frequency-dependent masking thresholds; in addition, temporal masking can be used. For signal coding, block-overlapping transforms and subband filter banks are suitable to avoid perceivable artifacts at block boundaries. To better adapt for local signal properties, the window length of the transform can be variable; cases of transients are better captured by short analysis windows (better time resolution), whereas long windows provide much better compression (and concentration of energy to only few coefficients) in the case of stationary, sustained and harmonic-like sounds, which are sparse in the frequency domain. For determining the frequency profile and applying masking thresholds, which refer to quantization step sizes to be chosen, a separate windowed DFT is often run additionally for time-frequency analysis. Analysis for perception-related frequency groups requires specific partitioning of the frequency axis, where the other transform's basis functions used for the purpose of encoding may not be suitable[5]. The masking thresholds themselves are signal dependent and change over time, to fully exploit the fact that dominant spectral components can mask coding noise within other frequency bands (cf. Sec. 3.2.2). Once the masking function is known, the frequency coefficients are quantized and encoded. The perceptual model itself does not need to be unique in terms of the compression technology, which gives headroom for encoder optimization. It is however necessary to convey control parameters (e.g. frequency dependent quantization step sizes) as side information to the decoder. Further compression can be achieved by utilizing redundancy over time (e.g. predicting relevance of frequency components), or between different channels of a recording. Another option is synthesis of certain components, e.g. high frequencies or entire additional channels in case of multi-channel audio. For this purpose, dedicated synthesis models need to be employed, which also may require transmission of parameters as side information.

In audio coders which make extensive use of such weighting functions, up to 1024 frequency bands are normally used (e.g. in the block-overlapping modified DCT defined in the Advanced Audio Coding standard). However, a better resolution of the time axis is required when transients occur, or if temporal masking functions shall be employed additionally. In such cases, it is beneficial to switch to a transform with less frequency bands, thus allowing shorter basis functions.

---

[5] For a good approximation of the frequency sensitivity properties of human listeners by psycho-acoustic weighting functions (see Sec. 3.2.2), a sufficiently fine frequency resolution of the transform is required. While transforms used for compression typically often use either equal-bandwidth frequency channels, the perceptually adjusted Bark scale shows an approximately linear increase of bandwidth for frequency groups above 500 Hz. Therefore, in the psycho-acoustic optimization of encoding, varying numbers of uniform-width frequency bands have to be combined.

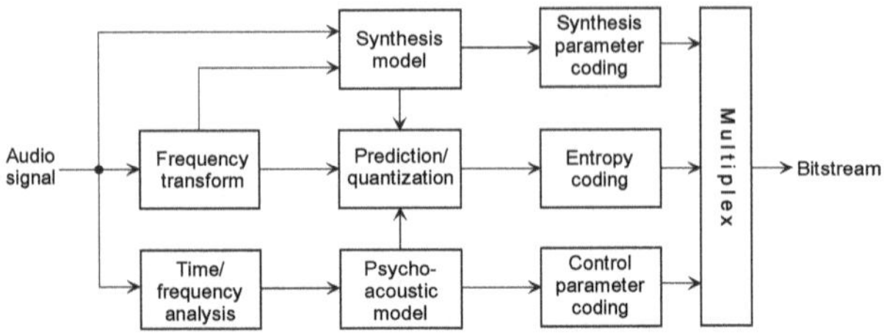

Fig. 8.2. Generic structure of an audio encoder using transform coding combined with psycho-acoustic model and synthesis model

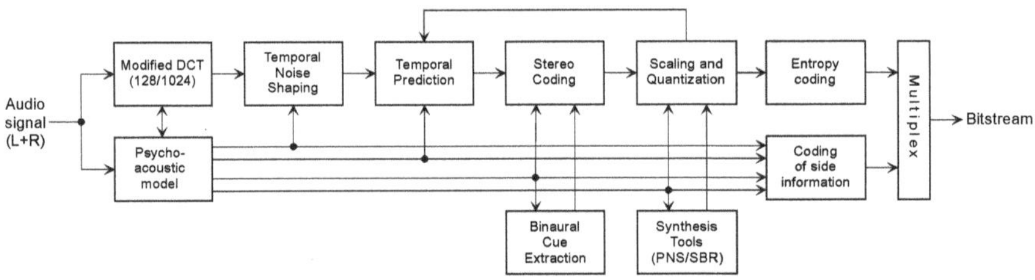

Fig. 8.3. MPEG-2/-4 Advanced Audio Coding (AAC)

For the principle of transform coding (Sec. 5.3) in the way specifically applied to audio signals, the combination of compression tools of the Advanced Audio Coding (AAC) standard[6] gives a good example to understand what methodologies are used in state-of-the-art audio transform coding. A block diagram is shown in Fig. 8.3. The lower part of the figure contains elements where the transform coefficients are to some extent synthesized, which is more specifically discussed in Sec. 8.2.2 for perceptual noise substitution (PNS) and spectral band replication (SBR), and in Sec. 8.2.3 for binaural cue coding (BCC).

Transform with switchable resolution. As a monolithic transform with block overlap, a *Modified DCT* (MDCT) similar to (2.289)-(2.292) is used for frequency decomposition[7]. The window size and hence the number of transform coefficients is switchable between 128 and 1024 typically, where the short size is better suita-

[6] AAC was originally defined as part of MPEG-2 (ISO/IEC 13818-7), and later extended in MPEG-4 (ISO/IEC 14496-3) and MPEG-D (ISO/IEC 23003). As compared to the 'MP3' codec (MPEG-1 audio layer 3, see Sec. 8.2.4), AAC provides further reduction of the data rate by approximately a factor of two without loss of perceptual quality. The design of MP3 is somehow inconsistent, as it uses two concatenated transforms for compatibility with MPEG-1 audio layers 1 and 2.

[7] The Scalable Sampling Rate profile of AAC defines a pseudo QMF filterbank in order to achieve variations of the sampling rate by completely discarding higher frequency bands. This was originally intended for the purpose of rate scalability on variable-bitrate channels, but so far has hardly been used in practical implementations.

ble for transients and exploitation of temporal masking effects, the long size for more periodic or stationary segments. Further, the larger number of coefficients also increases the frequency resolution, therefore allowing a more precise adaptation for frequency masking (see Sec. 3.2.2). It is necessary to employ special *transition windows* at the positions where the size is switched. Transition windows are not symmetric, i.e. differently shaped on both ends, but the overall constancy condition (2.290) still holds. However, in case of the MDCT, this can be compromising with regard to orthogonality, which may however not be critical when transients (e.g. at the beginning of new notes) are present anyway.

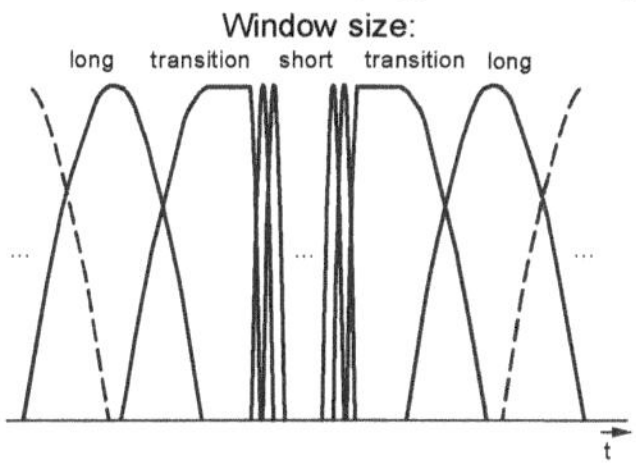

Fig. 8.4. Window switching in Advanced Audio Coding (AAC) by using transition windows between long and short length MDCT bases

Temporal noise shaping. TNS is useful for signals that have temporally concentrated (impulse-like or transitional) behaviour, and is targeting exploitation of temporal masking effects. The basic idea is based on the observation that signals that are temporally sparse (impulse-like) have a broad spectrum, where however neighbored coefficients tend to be dependent (predictable) and vice versa, signals with a sparse (e.g. harmonic) spectrum are temporally extended, but predictable in the time domain [HERRE 1999]. Going back to the latter case and considering an analogy with open-loop (D*PCM) prediction as discussed for the case of speech signals (Sec. 8.1.1), the quantization error was shaped by the spectrum of the signal. For the dual case of a temporally sparse signal, if open-loop prediction is applied over the series of *transform coefficients*, the effect will be that a higher reconstruction error applies at those time positions where the signal changes, such that temporal masking likely applies. This way, it is possible to exploit temporal masking by an across-frequency open-loop prediction, which in some cases also could give a compression advantage, since spectra of transients are non-sparse, but have some coherence in amplitude and phase between neighbored spectral coefficients[8].

Temporal prediction of transform coefficients. In the case of harmonic signals, a significant statistical dependency also exists between the transform coefficients of subsequent blocks. This is exploited by invoking a (closed loop) DPCM scheme, i.e. prediction over time is performed for coefficients of the same fre-

---

[8] An extreme case of this would be the spectrum of a Dirac impulse, which is perfectly predictable over frequency.

quency, using the coefficient from the previous transform window as prediction reference. Predictor coefficients do not need to be transmitted, since backward adaptation using the LMS algorithm (2.221)-(2.223) is employed. In case of switching between short and long windows, prediction is considered to be meaningless and therefore disabled by default; otherwise, the prediction can be disabled by encoder control. The temporal prediction, when used, has the side effect that temporal window-dependent fluctuations of the coding noise (by frequently switching on and off low-amplitude coefficients) are minimized, such that modulation of errors by the 'window frequency' is avoided. The newer AAC version defined in MPEG-4 additionally contains the *long term prediction* (LTP) tool, which is a forward adaptive scheme (requiring transmission of side information to convey the coefficients) operating in the time domain. This is not explicitly shown in Fig. 8.3, as it requires an additional inverse transform synthesis from the quantized coefficients, the prediction is followed by forward transform and TNS in this case. LTP is specifically beneficial in case of harmonic signals, where the misalignment between the harmonics and the actual analysis frequencies of the MDCT may otherwise cause high prediction errors.

Scaling, quantization and entropy coding. Non-uniform quantizers (power-law, i.e. exponential characteristic with offset) are used for the representation of the frequency coefficients. The psychoacoustic model determines the allowable quantization errors within the different frequency components. For this, the coefficients are grouped into non-uniform scale factor bands, which can e.g. be derived from the Bark scale (see Sec. 3.2.2)[9]. Besides a global scale factor (adapting the quantization step size over all frequency bands), individual scale factors can be chosen for each band. The individual scale factors are encoded differentially, starting from the lowest band. Huffman code tables are used for encoding; scale factors are skipped for those bands where all coefficients are zero.

The common entropy coder is based on a Huffman (table) approach with a systematic fixed-length part. Quantization levels larger than 16 are signaled by an escape symbol. The sign is binary coded and appended to the respective codeword. 11 Huffman tables optimized for different statistics are available and can be selected individually for each band; all-zero frequency bands are also explicitly signaled.

For extremely low rates (<40 *kb/s*) a method of *Transform Domain Weighted Interleaved Vector Quantization* (TWIN-VQ) is additionally defined, which completely replaces the separate quantization and encoding of transform coefficients. The other building blocks of AAC are retained without changes. The selection of the vectors is also controlled by the perceptual model. The notion *interleaving* reflects the method of ordering the spectral coefficients into vectors, which are preferably assembled from frequency subbands of similar properties.

---

[9] In fact, the actual grouping of bands which are subject to equal quantization is made by encoder choice.

Scalable and error resilient AAC coding. As an alternative to the quantization and entropy coding described above, a scalable coding mode is defined, which is based on embedded quantization and *Bit Slice Arithmetic Coding* (BSAC), an arithmetic coding method applied in combination with bit-plane coding, similar to the methods described in Sec. 4.3. The arithmetic coding uses contexts over bit-planes; different context models are provided, and the best is selected based on statistics observed so far. The temporal prediction loop is closed at the highest bit plane, which incurs drift when enhancement layers are discarded. Therefore, even though the scalable BSAC method has been reported to be equivalent in performance with single-layer AAC at high rate points, its compression usually becomes clearly inferior at low rates [YANG ET AL. 2006]. It should however be noted that the main purpose is to achieve graceful degradation under transmission rate variations and channel losses.

BSAC can also be employed in combination with TWIN-VQ. Here, the information encoded by TWIN-VQ establishes the base layer, and a residual error is encoded by BSAC as enhancement layer. Therefore, even though the lowest rate point may be better due to the efficient base layer compression, the performance at higher rates typically is worse compared to the method using only BSAC.
As the main application area for AAC with BSAC is in the domain of lossy transmission scenarios, additional mechanisms are defined for error-resilient transmission of AAC streams. These are, more specifically, Huffman code-word reordering (HCR) avoiding error propagation across spectral components, virtual codebooks (VCB) which assigns more robust codewords for amplitude coefficients of large amplitude, and reversible variable length coding (RVLC), see Sec. 4.4.3.

Lossless coding of audio signals. Transform coding schemes such as AAC are not able to achieve lossless compression at a reasonable rate, due to the fact that larger bit-depths are required for lossless (LS) representation of transform coefficients than for the original PCM representation; transform coefficients are also often computed with floating point precision in audio compression. Methods for lossless coding of audio are therefore rather based on time-domain predictive coding (using optimized predictors with forward adaptation, i.e. solving the Wiener-Hopf equation), which e.g. in the scheme proposed in [LIEBCHEN 2003] is used in combination with systematic entropy coding such as Rice codes. This can also be used in combination with lossy coding, where a conventional AAC encoder is run to represent a base layer, and lossy to lossless (LLS) coding is operated on the residual signal. This approach may however show some high amplitudes in the residual, when perceptual masking is used in the encoding of the base layer.

8.2.2 Synthesis based coding of audio and sound signals

The transition between waveform coding (trying to reproduce the true shape of the signal waveform by signal approximation methods, based on PSNR or perceptional distortion criteria) and synthesis based coding (trying to reproduce a signal that is also perceptually similar to the original, but represented by a model and a possibly small set of parameters) is seamless, or at least, both approaches can be combined. Newer versions of AAC therefore contain compression tools that can be categorized as synthesis based.

Analyzing psycho-acoustic properties of human hearing, certain noise-like (non-harmonic) signal components exist, for which traditional waveform coding, which includes preservation of the phase relationships, is not necessary. These components can be substituted within their respective frequency bands by noise signals of same variance, which is denoted as *perceptual noise substitution* (PNS). This principle of synthetic generation of signal components has been further developed towards more general mechanisms of *spectral band replication* (SBR), which also allows substituting higher-frequency harmonics by synthetic signal components. Typically then, only a narrowband signal of 8 *kHz* bandwidth is encoded, and all higher frequency components are synthetically reproduced, which is done differently for harmonic and non-harmonic components. Generally, the assumption is that both the harmonic series as well as the spectral roll-off characteristics continue in the high-frequency range in a similar way as they can be analyzed from the lower frequencies. Only in cases where a significant deviation from this prediction occurs, additional parameters need to be encoded. The frequency analysis and synthesis is performed using a modulated (QMF) filterbank with complex basis functions, which provides a more precise frequency separation and also more consistent phase alignment between the (actually encoded) low frequency and the (synthesized) high frequency components. More details about the method can be found in [DIETZ ET AL 2002].

AAC with bandwidth extension allows transparent (without audible impairment) encoding of most stereo audio signals at lowest bit rates around 48 *kb/s*. Perceptually adapted audio coding schemes can eventually impose high *measurable* distortions (when computed as SNR criteria), while the effect may hardly be *audible*, exploiting the principle of irrelevance. Artifacts can however become audible through occurrence of

- Non-harmonic distortions[10];
- Frequency-selective and non-stationary noise;
- 'Roughness' of the signal, which becomes in particular perceivable due to temporal fluctuations between transform blocks;
- Loss of high frequencies.

Audio signals as they typically have to be encoded could stem from natural sound recordings, but synthetically generated sounds have become more important in music production during the last decades. Synthesis methods exist which are capable to imitate natural sounds such that they are almost indistinguishable. These synthesis methods can often be characterized by a small set of parameters (e.g.

[10] Observe that clipping of harmonic signals as usually occurring in analog audio systems and amplifiers would mainly cause harmonic distortions, which are more convenient to the human listener.

filters, fed from oscillators or noise sources; associated properties, coupling and modulation of different sources; temporal envelope characteristics). Audio synthesis description formats such as MIDI are supported by many devices like PC sound cards, mobile phones etc. This enables the possibility to encode both natural and synthetic signals by the characterization of a suitable synthesis model and a small set of synthesis parameters. However, direct mapping of a sound available as a waveform into a parametric synthesis model is a largely unresolved problem. For example, frequency modulation (FM) is able to generate very natural instrument imitations or new synthetic sounds; however, FM is a nonlinear modulation scheme, such that it is not straightforward to automatically determine the parameters which are necessary to obtain a desired synthesis result.

A widely used principle to find best settings in parametric coding is analysis-by-synthesis optimization. This can be highly complex, in particular if the parameters are sensitive, if the parameter space is multi-dimensional and/or has local minima of the cost function, and if optimization includes the selection of the synthesis method as well. When recording is not made in a reverberation-free environment, room modeling or canceling out room influences or artificially added effects becomes an important part of the analysis problem. If the distortion is of LSI nature, de-convolution (e.g. compensation of reverberation/echo) could be used.

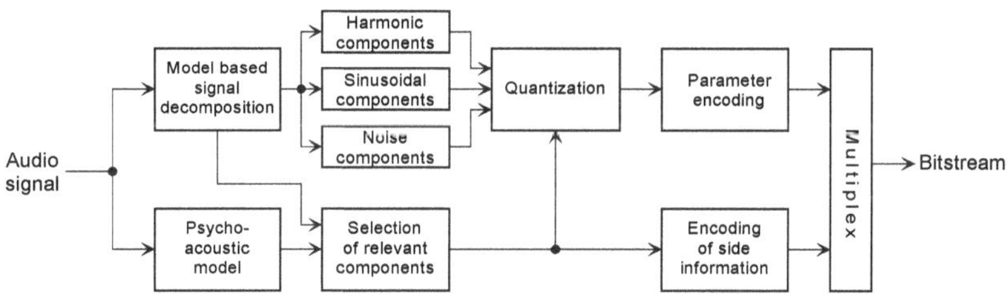

Fig. 8.5. Block diagram of the parametric audio coding scheme HILN as defined by the MPEG-4 standard

Possible applications of parametric audio coding can be identified in the range of extremely low bit rates, where waveform coding will fail or produce unacceptable coding distortions. A method for parametric coding of audio signals is the *Harmonic and Individual Lines plus Noise* (HILN) scheme defined in the audio part (14496-3) of the MPEG-4 standard, which gives good synthesis quality for cases of monophonic sources (single musical instruments) or polyphonic sounds of low complexity. A block schematic of HILN is shown in Fig. 8.5. The first step is a decomposition of the signal into harmonic components, sinusoidal components and noise components, all of which are parametrically encoded based on a perception model using relevance detection and weighting. At the decoder end, the reverse principle of compositing the synthesized signal from these three components is performed. The method has been developed for compressed rates between 4 and 16 *kb/s*. For monophonic signals, reconstruction from a 6 *kb/s* stream still delivers roughly acceptable quality, but is by far not transparent when compared to the original. Most relevant properties of the signal are retained, which would at such low rates never be possible with one of the plain waveform coding methods described in Sec. 8.1.3. Further extensions to parametric coding were proposed which include *transients* as an additional synthesis element [SCHUIJERS ET AL. 2003], being better capable to model temporal-envelope characteristics such as attack properties, and reportedly improving the perceived synthesis quality also in case of more complex polyphonic sounds.

Another important aspect of natural audio synthesis is related to the room environment in which signals are typically generated. Even though this can mostly be seen as a problem of signal composition, it could also be used for parametric audio compression, when properties of rooms can be modeled by a low number of parameters, such as early echo positions and reverberation time. This however requires specific audio scene composition after decoding, consisting of mixing audio objects and superimposing effects, followed by rendering according to the given reproduction system (loudspeakers, headphones).

The MPEG-4 standard defines tools for supporting such an entirely *synthetic audio* representation. Beyond the low-level waveform representation methods discussed so far, this includes

- A flexible music synthesis language denoted as *structured audio orchestra language* (SAOL);
- An efficient wavetable synthesis based on the *structured audio sample bank format* (SASBF);
- An adoption of general MIDI (musical instrument digital interface) formats to support sound synthesis devices;
- Capability for flexible mixing of audio objects, insertion of audio effects and modeling of sound propagation in 3D rooms;

For more information on the aspects of mixing and rendering, the reader is referred to [MCA, SEC. 7.6].

8.2.3 Coding of stereo and multi-channel audio signals

Stereo redundancy. For exploitation of *stereo redundancy*, it has to be observed that directional identification in stereophonic signals or in binaural perception is not only based on amplitude differences between the left and right channels ('intensity stereo'). Moreover, the delay between the signals arriving at the left and right ears ('time-of-arrival stereo') and the frequency attenuation (caused by the occlusion through the head) play important roles as well. The latter aspect, which also effects that one of the ears does not receive sound waves directly and therefore has interaction with (room) environment properties, is particularly important in *artificial head stereophony*. In cases where the stereophonic signal is recorded using microphones at distant positions, delays between the sounds arriving through left and right channels are occurring. If different sound sources with different delays are present (the usual case), this prevents direct similarity matching of the left and right signals, making exploitation of redundancy difficult. A simple method in stereo signal encoding, as it is implemented in the baseline version of the AAC standard, is shown in Fig. 8.6. The system is switchable between the following modes:

- Separate encoding of left and right channels;
- 'Mid/side' (M/S) coding, i.e. encoding of a (monophonic) sum signal and a difference signal from the two stereo channels;
- Intensity coding, i.e. coding of one channel, and use of an intensity factor to synthesize the second channel as an amplitude-modified copy of the first. Intensity coding is mainly useful for synthetically mixed sources where both stereo channels are originating from a monophonic source and have been placed on the stereo basis by amplitude balance adjustment.

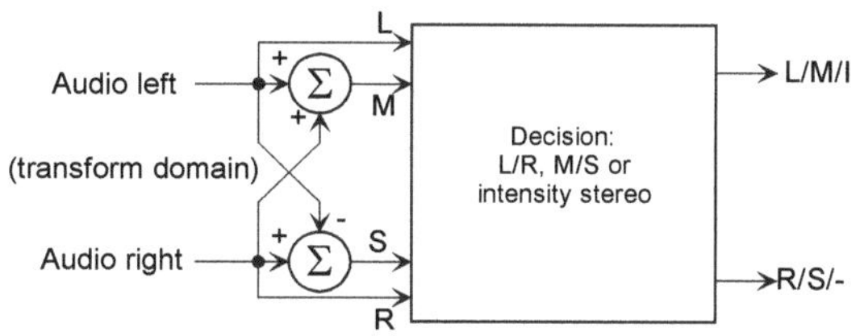

Fig. 8.6. Stereo signal coding modes in Advanced Audio Coding (AAC)

M/S coding can be rather inefficient in case of time-of-arrival stereo, as duplications (echoes) would appear in both sum and difference signals. Therefore, more sophisticated methods of stereo coding resolve the problem of variable delays occurring in stereo mixes of multiple sources by frequency-band dependent handling of the redundancy. The basic principle is generation of a monophonic down-mix, which includes all sound information from both stereo channels, and uses *cue parameters* as side information to express the differences between left and right channels. These can be used to re-generate the spectral and temporal envelopes of both channels from the monophonic down-mix [BAUMGARTE, FALLER 2003] [FALLER, BAUMGARTE 2003]. A block diagram is shown in Fig. 8.7.

The cue parameters are frequency dependent. Processing is usually done within frequency groups, where the center frequencies and bandwidths are determined from the perceptually motivated Bark scale (see Sec. 3.2.2). The most important parameters are

- Inter-channel level differences (ICLD), best represented on a logarithmic scale,
- Inter-channel time differences (ICTD), which are computed by searching the maximum of cross correlation between the left and right channel from the time signals of the respective frequency band[11],

[11] Alternatively, usage of inter-channel phase differences (ICPD) has been proposed which are computed by averaging the phase shifts between the complex frequency coefficients in the corresponding band [BREEBART ET AL. 2005]. It should however be observed that the phase of periodic signals is ambiguous and therefore cannot be directly mapped into time lag.

- Inter-channel correlation (ICC), i.e. the actual maximum value of the cross correlation function in the given frequency band.

The handling by separate frequency bands is motivated by the fact that firstly it is more likely that each band is dominated by only one of the sources such that it is better possible to find matching correspondences between the left and right channels, and second, even if several sources appear in the band, the variable delays and corresponding temporal hulls can be synthesized more precisely than for the temporal signal. For the purpose of side parameter derivation, complex-exponential modulated filterbanks with non-uniform, psycho-acoustically motivated frequency partitioning and often under-critical sampling (i.e. over-complete information in the frequency domain) provide good performance.

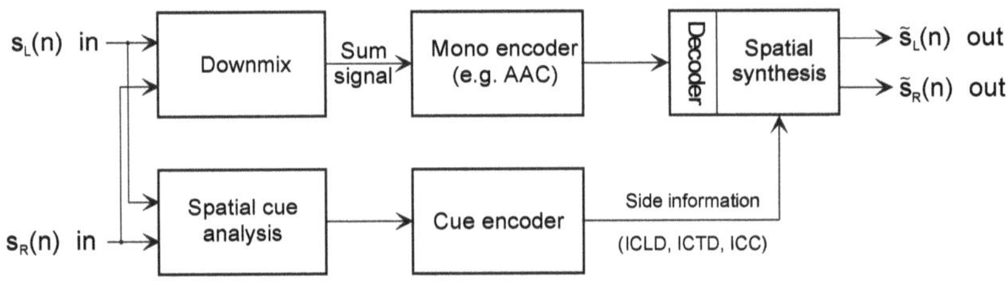

Fig. 8.7. General principle of binaural cue coding

This basic scheme of parametric stereo coding (PSC), as it is used in the high efficiency version of AAC, can be extended into more generic parametric coding of multi-channel sources, where an N-channel source is mapped into an M-channel down-mix plus N sets of cue parameters at the encoder, and the down-mix signal is encoded conventionally. The reconstruction is performed by decoding the down-mix and using the cue parameters for synthesis of all N channels. As a side effect, such a scheme has the advantage that it is compatible e.g. with conventional stereo coding, if a legacy scheme is used for the encoding of the down-mix, and the cue parameters can simply be ignored by a legacy decoder. Such a concept forms the basis of *Spatial Audio Coding* (SAC), as described in the subsequent paragraph.

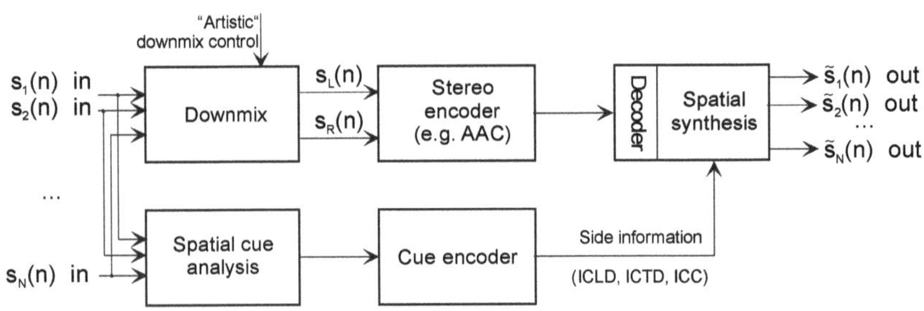

Fig. 8.8. Spatial audio coding (SAC) block diagram, with N input channels and stereo compatible down-mix

Spatial audio coding (SAC). The general principle of SAC for the case of a two-channel (stereo compatible) down-mix is shown in Fig. 8.8. In the basic SAC scheme introduced in [HERRE ET AL. 2005], spatial parameter estimation is performed in the domain of a bandpass representation, more specifically generated by a 64-channel complex 'pseudo QMF' filterbank, with few lowest frequency bands subdivided into finer frequency resolution by feeding the output of the first filter bank into a secondary bank with 4 or 8 channels typically. The spatial cue parameters include ICLD and ICC as described above, but additionally, a more accurate synthesis can be achieved through usage of

- Channel Prediction Coefficients (CPC), used to predict output channels (or combinations thereof) from input channels (or combinations);
- Prediction errors, i.e. waveform differences between the parametric description and the actual waveforms.

The down-mix is generated by applying a complementary synthesis filterbank, combining information from the available input channels in an appropriate way[12].

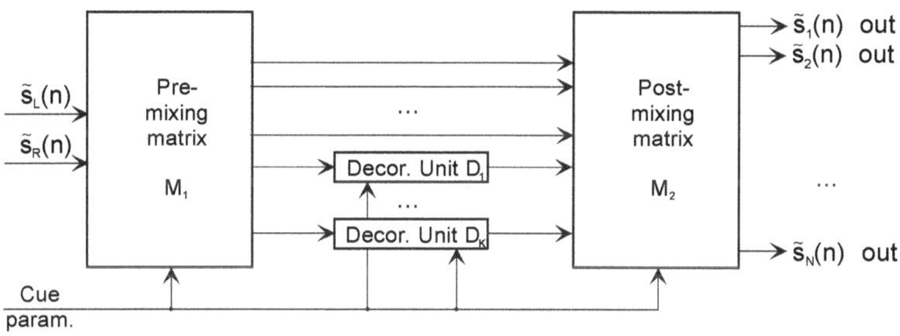

Fig. 8.9. Generic 2-channel into N-channel synthesis circuit using matrix units M_i and decorrelation units D_j

CLD, ICC and CPC parameters are determined per frequency band. For synthesis, CLD and CPC are basically employed to determine factors of *matrix units* that are used to reconstruct the N channels from the M-channel down-mix (CLD for M=1, mono, and CPC for M=2, stereo down-mix) by a linear mapping, as shown in Fig. 8.8. This is however only sufficient if the correspondence between the channels can indeed be described as a linear mix. Otherwise, particularly to model delays and diffuse sound elements, ICC parameters are used by *decorrelation units* (denoted as D_j in Fig. 8.9) between the two matrix units, with the goal to generate output signals with a variable degree of correlation. The decorrelation units basically consist of an allpass filter implemented in a lattice structure, augmented by

[12] It is beyond the scope of this book to describe methods for optimization of the downmix. Practical approaches include automatic downmix by exploiting the cue information e.g. to avoid comb filter effects / erasure of frequency bands. On the other hand, an 'artistic' downmix is generated semi-automatically, tuning the parameters by human interaction, are also used.

additional tools to better reconstruct the temporal and frequency hulls of the given channels. One of these tools is *temporal envelope shaping* (TES), which is very similar to the TNS tool of AAC as described in Sec. 8.2.1.
SAC can be operated over a wide range of data rates, where the rate is mainly influenced by the temporal and spectral resolution of the spatial parameter coding, their quantization and on the decision on whether a residual between the synthesis result and the original is encoded additionally

Multi-channel feeds such as 5+1 and 7+1 can be represented in a very compact way by using SAC, generated synthetically from stereo by the spatial parameters. To simplify the analysis and synthesis, it is possible to employ cascades of basic building blocks. For this purpose, it is sufficient to define two types of simple analysis and synthesis units[13], namely

- Two-to-one and one-to-two (2-1, 1-2) elements with M=1, N=2;
- Three-to-two and two-to-three (3-2, 2-3) elements with M=2, N=3.

A configuration where these elements are appropriately combined for the encoding of a 5+1 signal into a stereo down-mix plus associated spatial cue parameters is shown in Fig. 8.10.

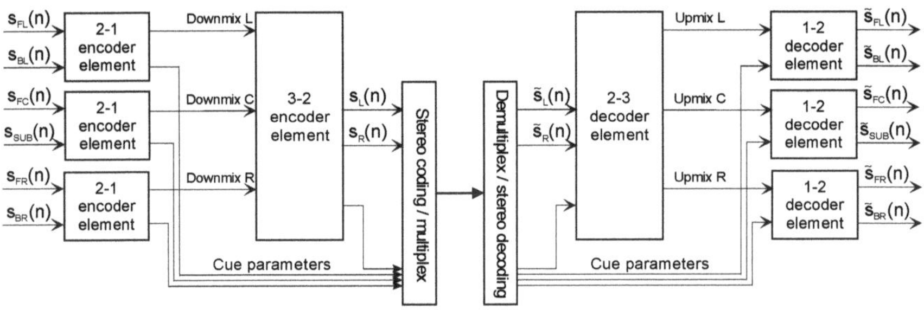

Fig. 8.10. Spatial audio coding (SAC) of a 5+1 multi-channel signal representation with compatible stereo base layer

Spatial Audio Object Coding (SAOC) is a combination of SAC with the idea of sound-object based synthesis coding [BREEBAART ET AL. 2008]. Its main advantage is to allow preserving the possibility for the listener to adjust the mix based on personal preference, and to perform an optimization for the given rendering configuration (stereo or surround) independent of the coded representation. A synthesis parameter representation allows very efficient parametric coding of individual audio objects (tracks), as discussed in Sec. 8.2.2. Additionally, a rendering functionality is included to interactively render the audio objects into an acoustical scene for several types of reproduction systems (various numbers of loudspeakers or binaural for headphones). For this purpose, SAOC encodes Object Level Differences (OLD), Inter-Object Cross Coherences (IOC) and Down-mix Channel Level Differences (DCLD) into a parameter bit stream.

---

[13] Restriction to these simple types significantly reduces the number of CLD and CPC parameters, as compared to the more generic topology of Fig. 8.8.

The basic approach of SAOC is still backward compatible with SAC, i.e. the main information is sent by a mono or stereo down-mix. Additional parametric side information (such as level or time differences) is then encoded per object, allowing to modify or remove the object in the down-mix, or enable modification of its appearance in the multi-channel rendering output.

It should be noted that most existing spatial audio mixtures for surround experience (such as 5+1, 7+1 and beyond) are artificially generated (not resulting from actual microphone recording in a 3D space). A general problem of audio rendering by loudspeakers comes by the fact that usually a 'sweet spot' exists at which the spatial impression is optimum, and multiple users can only be served with certain limitations. To resolve this, the number and density of loudspeakers can be increased (e.g. the front center speaker in case of 5+1 allows better localization of sound sources at that position). Nevertheless, even then sound sources that are closer than the loudspeakers cannot be rendered with good spatial localization. In *sound field synthesis* [Boone, Verheijen 1993], an extremely large number of loudspeakers (>200) e.g. arranged as a ring along walls (and optionally elevated) are used to actually generate a field of acoustic waves as they would be present in a natural environment with sound sources present at certain locations and in a given room. The idea is based on Huygen's principle, which states that a primary sound source can be replaced by a number of secondary at different positions, which just generate the behaviour of the acoustic waves from the primary source as they would appear at the respective position. Whereas it is relatively straightforward to simulate sources synthetically within such an environment, capturing of natural acoustic scenes by microphones exposes many unsolved problems, as it is hardly possible to sample a natural sound field by a sufficiently dense-packed array of microphones. Furthermore, the simulation of elevated sound sources is rather limited when only a ring of loudspeakers at the same height are used.

Generally, it can be foreseen that compression of composite audio signals feeding large arrangements of loudspeakers is challenging in terms of data rate, processing needs and fidelity requirements. In this context, it can be expected that the signals can be encoded by equal or at least similar mechanisms as SAC. Frequency dependent level and time differences, as well as correlation between the channels are the main mechanisms in simulating the sound source position in the sound field, and in case of dense packing of speakers, these parameters can be expected to be even more compact. Further, the number of loudspeakers can vary per installation. On the other hand, it is unlikely that hierarchical arrangement of the above-mentioned 2-1 and 3-2 elements would create a reasonable stereo or mono down-mix, when the number of input channels is extremely large, such that the processing effort is huge.

An alternative method is *ambisonics* [Fellgett 1975] [Gerzon 1977] [Hope 1979], an equivalent approach of the original concept was entitled as *eidophony* [Scherer 1977] during early years of development. The idea is based on description of the sound pressure field at the origin of a spherical coordinate system (Fig. 8.11a), such that the directional arrival of any sound or acoustic wave can be localized by a description from two angles. In the case of the original first order ambisonics,

The signal allowing a directional localization of a 3D sound source can be encoded into only three or four channels, which are equivalent to one microphone with spherical characteristics and two or three microphones with eight characteristics, the latter orientated by the three orthogonal directions of the $[W_1\ W_2\ W_3]$ 3D coordinate space[14] (Fig. 8.11b). Each of them has dipol (+/–) characteristic, i.e. when the sound wave intrudes from the opposite end, the amplitude becomes negative. The spherical channel W_0 is equivalent to intensity (mono down-mix), whereas the three other channels allow coding of the directional position. Specifically, a sound source $s(t)$ which is positioned at azimuth angle ϕ and elevation angle θ (relative to the W_1 frontal axis) is mapped as follows:

$$s_{W_0}(t) = \frac{\sqrt{2}}{2} s(t) \quad ; \quad s_{W_1}(t) = s(t)\cos\phi\cos\theta$$
$$s_{W_2}(t) = s(t)\sin\phi\cos\theta \quad ; \quad s_{W_3}(t) = s(t)\sin\theta \qquad (8.1)$$

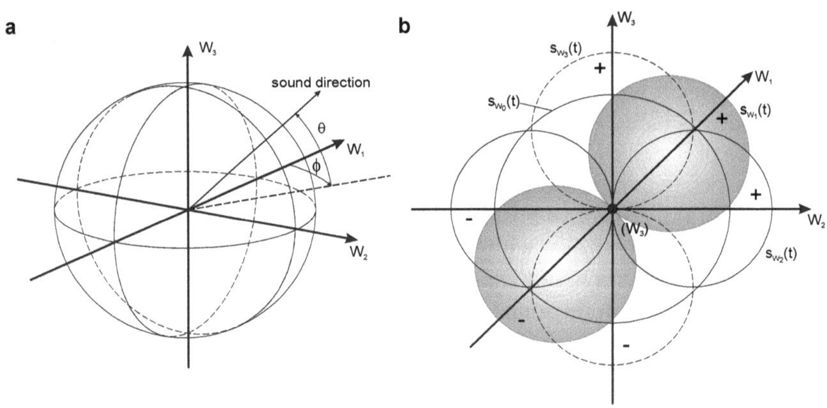

Fig. 8.11. **a** Spherical coordinate system used in ambisonics **b** Illustration of the directional characteristics of four channels (equivalent to one spherical and three orthogonal eight-characteristic microphones) in first-order ambisonics

Note that this approach is also a straightforward extension of 'mid/side' stereophony, where the left and right channels recorded by microphones with cardioid characteristic (providing level maxima of right / left channels at angles $\phi = 0^0$ and $\phi = 180^0$) can be generated by adding / subtracting the signals coming from a spherical and an eight-characteristic microphone (the latter with orientation along the left-to-right W_2 axis).

In a similar way, the signals from the four channels can be decoded to feed an arrangement of arbitrary many loudspeakers, which should then be positioned on a sphere with approximately uniform azimuth and elevation angles. Assuming that a loudspeaker shall be positioned at azimuth angle ϕ and elevation angle θ (rela-

[14] The third non-spherical microphone and associated channel are only necessary when an elevation angle shall be supported.

tive to the W_1 axis), it has to be driven by the following mixture of the four channels:

$$s_{\phi,\theta}(t) = \frac{\sqrt{2}}{2} s_{W_0}(t) + s_{W_1}(t)\cos\phi\cos\theta + s_{W_2}(t)\sin\phi\cos\theta + s_{W_3}(t)\sin\theta . \qquad (8.2)$$

This approach reconstructs the sound field perfectly at the sweet spot, which is the center of the sphere; however, at positions where the listener comes closer to one of the loudspeakers, directional orientation is lost due to the fact that the sound pressure amplitude from the closest loudspeaker becomes predominant.

The problem of first order ambisonics, only giving optimum rendering of the sound field around the 'sweet spot' can also be interpreted by the fact that level differences between adjacent speakers are not sufficiently diverse (e.g. only different by a factor of $\sqrt{2}$ for two speakers at same elevation angle with azimuth change by 45 degrees), such that a wrong direction can be perceived when the listener comes closer to certain speakers. As a possible solution to this problem, *higher-order ambisonics* (HOA) has been developed [Poletti 2005], where additional higher-order spherical harmonics (such as sine and cosine basis functions dependent on multiples of ϕ and θ are used. In terms of directional characteristics, these possess more and more narrow (shotgun-characteristic like) maxima, such that increasingly better directional separation becomes possible when more channels are used. However, whereas it is straightforward to generate the related multiple channels synthetically from monophonic sources, building actual microphones to record a natural sound-field requires significant effort (including signal processing for calibration and compensation of delays due to the necessary placement of microphone capsules off the coordinate system center).

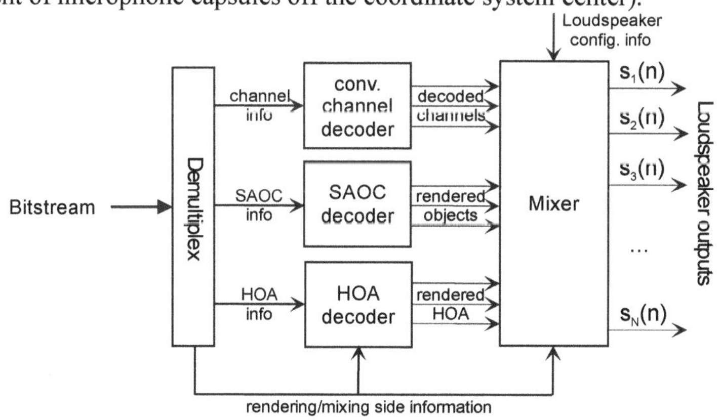

Fig. 8.12. 3D audio coding architecture

In terms of compression, the ambisonics concept is attractive, as due to the orthogonality of the different basis functions of spherical harmonics over the angular dimensions, it inherently removes redundancy between the channels.

HOA in combination with SAOC are core elements of the 3D audio representation in the recently defined standard MPEG-H audio (ISO/IEC 23008-3). The general architecture is shown in Fig. 8.12. Depending on the output (loudspeaker configuration), the rendered HOA channel information is mixed together with conventional (composite) channels and rendered (spatial) audio objects. This mixing process allows flexible adaptation for various output configurations.

Table 8.2. Overview of common formats and algorithms used in waveform coding of audio signals

Standard	Compression technology	Typical stereo data rate (CD quality) [kb/s]	Typical stereo data rate (FM quality) [kb/s]
MPEG-1 Layer 1 (ISO/IEC 11172-3 & 13818-3)	Polyphase filterbank, 32 bands	384	256
MPEG-1 Layer 2	As layer 1, improved support of psychoacoustic models	256	192
MPEG-1 Layer 3 ('MP3')	Tandem polyphase filterbank + MDCT, 192 / 576 bands switchable	180	128
Ogg Vorbis	MDCT, VQ	160	120
Windows Media Audio	MDCT	160	120
MPEG-2 AAC (ISO/IEC 13818-7)	MDCT, 128 / 1024 bands switch., TNS, stereo coding	128	64
MPEG-4 AAC-LC (ISO/IEC 14496-3)	As above, PNS	64-96	40-48
MPEG-4 High Efficiency AAC	As above, additional synthesis tools SBR, parametric stereo	≥48	≥24
MPEG-D USAC (ISO/IEC 23003-3)	Switchable ACELP/MDCT, SBR, SAC for stereo/multi-channel	≥40	≥20
OPUS (IETF RFC 6716)	Switchable LPC (SILK) and MDCT (CELT), low delay	≥40	≥16

8.2.4 Music and sound coding standards

In audio coding standardization, a similar paradigm is mostly used as in image/video compression: An open standard may define only the bit stream syntax and semantics, as well as the related decoder operation, but some freedom exists for the design of the encoder. With such an approach, standard-compatible encoders[15] may provide different degrees of quality for the decoded signal, typically balancing performance versus complexity. For example, if a pre-recorded audio signal is targeted for multiple consumption, encoding needs to be performed only once, which justifies operating a higher complexity encoder; in case of real time operation on a battery powered mobile device, low complexity may be more relevant. In *encoder optimization* for audio signals, appropriate adjustment for perception models is one of the key aspects, but rate-distortion criteria can be used as well (see Sec. 3.2 and 4.3).

For various common audio waveform coding algorithms, an overview of configurations and typical stereo data rates is given in Table 8.2. Both cases of CD quality (close to perceptually lossless with typical material) and FM radio quality (lossy) are shown. These numbers are mainly intended to demonstrate the progress that was made in audio compression. They were collected from reports about typical operational ranges, but should not be regarded as an absolute evidence about the benefits of different codecs. More details can be found in the sections below.

The audio part of the MPEG-1 standard (ISO/IEC 11172-3 *'Coding of Moving Pictures and associated Audio for Digital Storage Media at up to about 1.5 Mbit/s, Part 3: Audio'*) is probably the key representative of the first generation of dedicated digital audio compression. It defines the syntax and semantics of three different audio codecs, which follow a forward-compatible principle over 'layers'. Layer 1 and 2 codecs both use the same polyphase (subband) filterbank, transforming the signal into 32 uniform-width frequency bands. Signal adaptive bit allocation in quantization (see Sec. 5.3.2) is applied, no entropy coding is employed. In this context, MPEG-1 Audio Layer 2 was the first codec making more extensive use of perceptual weighting functions in encoding. This is rather limited here, due to the usage of only 32 equal-width frequency bands, which cannot be mapped directly with the nonlinear Bark scale of 24 frequency groups (cf. Sec. 3.2). In the improved scheme of MPEG-1 Audio Layer 3 (often nicknamed *MP3*), a second filter bank is supplemented which is based on a block-overlapping transform (Modified DCT). Sub-division of each of the initial 32 bands into 6 or 18 narrower bands can be chosen adaptively; this allows a spectral decomposition into a total of either 6x32=192 or 18x32=576 frequency bands. With a sampling rate of 44.1 *kHz*, this gives a minimum bandwidth resolution of about 22/576 *kHz*=38.2 *Hz*, which is sufficient for the purpose of weighting based on the Bark scale. The number of frequency bands can be selected depending on signal properties (transient/harmonic). As the 32-band polyphase filterbank is still used, a certain level of forward compatibility with Layers 1 and 2 is retained, even though the core quantization and encoding steps are completely different in Layer 3; for decoding of Layer 1 or Layer 2 streams, a Layer 3 decoder has to switch to a different coefficient decoding algorithm, but can at least partially use the same filter bank. A typical block diagram of an 'MP3' encoder is shown in Fig. 8.13.

[15] 'Compatible' means that an encoder is producing standard compliant bit streams.

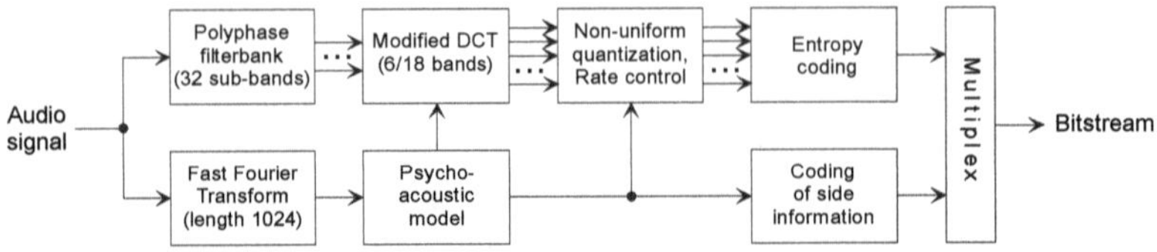

Fig. 8.13. Typical MPEG-1 Layer 3 ('MP3') Audio encoder

The codecs of MPEG-1 Layers 1-3 are also identically used in MPEG-2 a.k.a. ISO/IEC 13818-3 '*Generic Coding of Moving Pictures and Associated Audio Information - Part 3: Audio*'. After finalization of the first edition of the MPEG-2 standard, it became however evident that significantly better compression performance could be achieved when the tandem configuration of transforms would be replaced by a monolithic transform, and various other compression tools would be supplemented (cf. Sec. 8.1.3). This initiated the definition of a new part of MPEG-2 (ISO/IEC 13818-7 '*Generic Coding of Moving Pictures and Associated Audio Information - Part 7: Advanced Audio Coding*'). AAC then also became the primary technology for audio waveform coding in the MPEG-4 standard (ISO/IEC 14496-2: '*Coding of Audiovisual Objects – Part 3: Audio*'), which in its basic configuration is again identical to MPEG-2 part 7. Since then, AAC has been extended by various tools, where for plain mono or stereo compression 'High Efficiency AAC' (HE-AAC) is currently one of the best technologies available[16]. Compared to MP3, a reduction of the data rate by roughly a factor of 2 could be achieved, retaining the same quality; in general, this gain is even more noticeable for lower data rates (in the range of lossy compression).

MPEG-4 AAC with parametric synthesis tools such as *spectral band replication* (SBR), *perceptual noise substitution* (PNS) and *spatial audio coding* (SAC) is reported to provide transparent (without perceptible impairment) encoding of stereo audio signals at rates of *96 kb/s* and below.

In the MPEG-1 and MPEG-2 part 3 standards, the layers define combinations of tools, but no conformance points (profiles/levels) beyond that, e.g. defining requirements in terms of maximum bit rate, number of channels to be decoded in real time etc. had been defined[17]. MPEG-2 AAC defines three different *profiles*, which take the role of defining tool combinations, as well as bit rate limitations and real-time decoding requirements. These are *AAC LC* (low complexity), *AAC Main* (extension of LC by the temporal prediction tool) and *AAC SSR* (extension of LC by the scalable sample rate tool). In MPEG-4 audio, the entity of *object types* is introduced additionally to define combinations of coding tools. A profile

[16] An encoder block diagram is shown in Fig. 8.3.

[17] Originally, in MPEG-1 it had been targeted that a decoding device should be able to process stereo audio in real time at 44.1 KHz sampling rate and with a maximum encoded rate of 384 kb/s, which was the target for audio when stored along with a compressed video on a CD.

is then supporting one or several object types. Forward-compatible relationships with the MPEG-2 AAC profiles are retained by defining them as corresponding MPEG-4 AAC object types, where MPEG-4 AAC profiles are then defined including them, but may also include other new object types. Some of these profiles would require implementation of very specific tools and are hardly used practically. Those which are currently most relevant are

- AAC LC (also often denoted as 'AAC Profile'), which mainly adds perceptual noise substitution (PNS) into the previous MPEG-2 profile;
- High Efficiency AAC (HE-AAC) which includes spectral band replication (SBR) on top of AAC LC
- HE-AAC version 2 which further includes parametric stereo (PS).

Other tools (some of which were described in Sec. 8.2.1) have not been widely used in products. For a more complete overview of object types and profiles, the reader is referred to [QUACKENBUSH 2012].

Elements of HE-AAC are also used in the MPEG-D *Unified Speech and Audio Coding* (USAC) standard, ISO/IEC 23003-3. This is a switchable solution, which alternatively employs a speech or an audio coding algorithm, depending on the current signal property. For speech coding, an ACELP codec (see Sec. 8.1) is used. The audio coding algorithm is similar to HE-AAC version 2, but the more generic (beyond-stereo) method of spatial audio coding (SAC) is used for better compression in case of multiple channels. USAC is mainly intended for low-rate application areas.

Another approach of switchable speech/audio compression is the OPUS codec (IETF RFC 6716), which employs the SILK algorithm (see Sec. 8.1.1) in combination with a transform codec for generic audio, denoted as *constrained energy lapped transform* (CELT). The latter is a successor of the VORBIS audio codec. It is capable for low delay operation by using short window lengths of the MDCT. To compensate for the disadvantage this would give in case of non-transient (particularly harmonic) signals, it is combined with a pitch analysis, which can be used to predict transform coefficients across blocks. The transform coefficients (or their residuals) are further compressed by lattice VQ with pyramid shape (Sec. 4.5.1 and 4.5.2). For this purpose, coefficients are grouped within perceptually optimized frequency bands, and the bit allocation (depending on energy) is determined per band.

Compression methods for parametric synthesis of dedicated instruments have not found wide acceptance yet, except for the case where the control parameters are known (e.g. in studio sound production). Firstly, identification and separation of single instruments from a polyphonic audio still requires significant development in order to achieve acceptable quality (see MCA, SEC. 6.3). Further, even if the sound source is separately available, determining suitable synthesis parameters is not straightforward. Another area that has raised significant attention during the last decade is multi-channel, spatial and 3D audio coding. In these cases, particularly when 'sound fields' shall be generated from a large number of loudspeakers, the data rate might grow tremendously unless the dependency between the channels is appropriately exploited. First approaches of the aforementioned algorithms for Spatial Audio Coding (SAC) and Spatial Audio Object Coding (SAOC) have been specified in MPEG-D part 1 (ISO/IEC 23003-1, also nicknamed 'MPEG Surround') and part 2 (ISO/IEC 23003-2), respectively. Standardization for more dense representation of sound fields, entitled as '3D Audio', is ongoing and likely to be specified in MPEG-H part 3 (ISO/IEC 23008-3).

9 Transmission and Storage of Multimedia Data

Transmission and storage are core parts of multimedia communication systems. Before multimedia content can be consumed, it has to be made available to the user either via a transmission network or locally from a storage medium. Digital techniques firstly are advantageous compared to traditional (analog) solutions, as they can provide better quality, consuming less transmission channel bandwidth. In contrast to analog technology, where transmission channels and storage media were designed for monolithic media types such as video or speech signals, digital representation enables conveying multimedia content by any transmission and storage medium, provided that sufficient channel capacity is made available. This chapter introduces solutions for digital media transmission and storage in scenarios where traditional services are being replaced, and also where new services are evolving, like internet streaming and mobile access to media sources. Traditional monolithic-media systems are more and more superseded by heterogeneous network environments and diversified presentation devices (terminals) for different situations. This includes mobile access, home access etc., such that requirements for flexible media adaptation are becoming ever more important.

9.1 Digital multimedia services

Digital representation, transmission and storage has enabled a mass dissemination of multimedia signals. A driving force of this development is the convergence of traditionally separate sectors of *telecommunications*, *computers*, audiovisual *entertainment*, and personal *photography/video*. Mobile devices have come into everyday use, allowing access to multimedia information everywhere, or to capture and send videos, photos and audio recordings to other people. Digital techniques using compressed formats for storage and transmission have a number of advantages over traditional analog media[1]:

---

[1] The term 'traditional analog' covers recording and transmission of audiovisual signals on canvas, paper, film, magnetic tapes, records etc. in a very wide sense.

© Springer-Verlag Berlin Heidelberg 2015
J.-R. Ohm, *Multimedia Signal Coding and Transmission*,
Signals and Communication Technology, DOI 10.1007/978-3-662-46691-9_9

- Copying of bit streams / files without quality loss, no physical/chemical ageing of media representation on backup copies;
- Less bandwidth consumption in transmission when efficient compression schemes are used; by which e.g. personalized transmission of multimedia signals over wireless channels is enabled;
- Simple integration of different media for usage in multi-functional devices;
- Simple transformation from one format into another;
- Lower susceptibility against losses in storage and transmission, when optimized accordingly;
- Augmentation by additional information (metadata) simplifies management and search; digital signal processing also allows analysis for content recognition and characterization;
- Easier usage of random-access storage, recording and playback media; by tendency, digital storage (as well as the transmission) is becoming increasingly cheaper.

On contrary, disadvantages are:

- If the channel quality largely degrades, digital transmission often shows a 'threshold breakdown' of quality; achieving 'graceful degradation' in case of severe losses requires more effort (sophisticated combination of source and channel coding) than for some analog transmission methods;
- Due to continuous innovations, digital technology appears to outdate quickly, 'ageing' of formats and devices can be observed. In this context, also *transcoding* into new formats can become necessary in order to make them replayed by a new type of device. Beyond the computational effort, transcoding is often a lossy process that could degrade the quality unless designed carefully.
- Due to the simplicity of acquiring and managing digital media data, unauthorized distribution, copy and access have become easier than for analog media. Even though it may be argued whether this is a real disadvantage, traditional business models which were associated with media like movies, video, audio have become obsolete and needed to be replaced by new models eventually including access restrictions which are not always user friendly.

In short, the following main reasons can be identified for digital multimedia representations to take over the previous analog-dominated domains: They are cheaper and they provide more flexibility and functionality. However, audiovisual media are intended for business, where the simplified universal access to multimedia sources makes dissemination harder to be controlled. Beyond that, no security mechanism is prone against cracking by the long term.

Mobility, *personalization*, *interactivity* and *spatialization* are central new functionality aspects, which elevate digital media over preceding analog techniques. Mobile networks enable ubiquitous multimedia access and dissemination, providing usage for personalized purposes from anywhere. In the context of mobility, further integration of traffic systems (vehicles, intelligent traffic analysis and routing systems) with multimedia systems can be expected. On the other hand, digital multimedia systems enable *virtual mobility*, where by interactivity and spatialization (e.g. 3D displays or 3D navigation), the illusion of being in a different place is provided. This is often claimed to decrease the need for physical travel, which however is questionable. Virtual mobility just supplements different ways to experience the world and to communicate.

For recording and transmission of digital media signals, various storage media and transmission channels which are used are quite different w.r.t. *capacity* (storage, access and transmission data rate), *latency* (access and transmission delay), and *transmission quality* (error characteristics, fluctuations etc.). This will directly or indirectly map into the level of user satisfaction. From the perspective of the user, it is irrelevant where the content is residing, provided that the service is sufficiently fast and has acceptable quality. One can stick to this abstract point of view, and, when designing an overall system, just care about the impact that a compression algorithm operated over a channel has on the quality perceived by the user.

It is not necessary here to make any difference between transmission and storage; both cases can be interpreted as abstract channels with specific characteristics. In fact, communication networks today are a mixture of transmission paths and storage devices, e.g. providing temporary storage at proxy servers or anywhere else on the route, which eventually provides lower latency and better quality delivered to the user. Emerging services for cloud storage and processing are even more resolving any distinction between local and remote operation. In case of broadcast or other live streaming, temporary local storage at the receiver allows new functionality like pause, rewind and slow-motion replay modes at least over a certain range of the timeline.

Aspects related to the adaptation of digital multimedia representations for transmission over channels of different characteristics are discussed in Sec. 9.2 and Sec. 9.3. In Sec. 9.4, more specific methods of transmission for the cases of broadcast and internet/mobile streaming are introduced.

9.2 Network interfaces

By digital multimedia signal transmission, traditional ways of transmitting *one type of data over one network* specifically designed for that purpose, as it was the case for analog radio or TV broadcast, is increasingly broken up. Likely, only digital technology will survive, as it is more effective, economical, and increases availability. Traditional analog services like radio and TV broadcast use(d) monolithic systems, where the entire chain from the acquisition device's output, all parts of the transmission system and the receiver are based on a unique signal resolution and representation. To some extent, the first generations of digital media broadcast (*Digital Audio Broadcast*, DAB; *Digital Video Broadcast*, DVB – see Sec. 9.4) are following a similar paradigm: For the given domain of applications, the encoded data formats, transport mechanisms etc. are specified in a 'monolithic' fashion, expecting that broadcast service providers would install an infrastructure dedicated for this purpose anyway, and customers would buy the equipment required for receiving and decoding.

The aforementioned paradigm is changing through the fact that for digital encoding and decoding the hardware is cheap and programmable, which is likely to make life cycles of devices much shorter than they were during the analog age, but also allows updating. Furthermore, the Internet and mobile networks are much less homogeneous than a traditional broadcast domain, therefore also entitled as *network of networks*. Even though, the original design of the Internet protocol stack was not intended for real-time streaming of media data, both transmission bandwidth and latency are variable. The internet transmission mechanism is based on *packet switching*, where the available capacity is dynamically allocated to different users, depending on the frequency by which they release data packets into the network. Particularly in case of network overload, latency increases due to necessity of buffering, and packets may anyway be discarded if they would arrive late[2].

Digital wireless and mobile transmission networks of today (3G, 4G/LTE, WLAN, WIMAX and beyond) also use packet switching primarily, where however better control mechanisms with regard to packet losses exist at the air interface, compared to those available for the Internet at large. Packet units are small, and re-transmission is used in case of data losses, which can be executed fast, with minor observable latency, at the air interface.

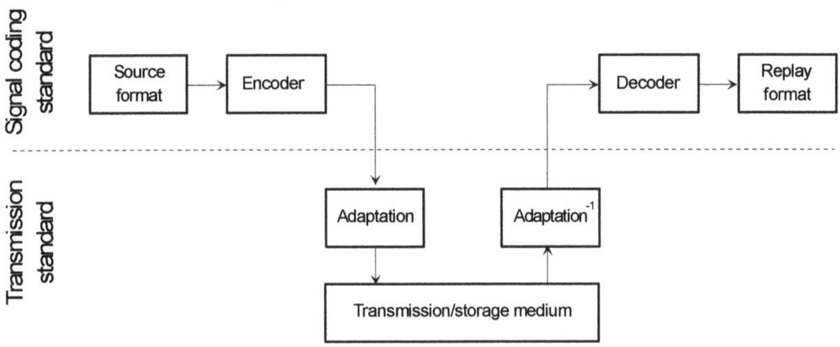

Fig. 9.1. Adaptation of a generic multimedia signal encoder/decoder to a transmission or storage medium

Due to the high variability and flexibility of digital transmission, it is necessary to achieve interoperability and adaptation of transmission mechanisms with any kind of channels. A high-level view of this problem is shown in Fig. 9.1. In the ideal case, it should be possible to define adaptation mechanisms *completely independ-*

---

2 This situation has somewhat changed at least when the network is not too close to overload. Routers are fast, wide-area connections using fiber optic links are providing large capacity, such that voice or video over IP services today have less latency in intercontinental connections, than they used to have by analog satellite links in the past.

ent of the source coding/decoding algorithm, such that a flexible adaptation can be made *after encoding is finalized*, depending on the type of network and instantaneous network conditions. This could further take into account the *replay format* of the terminal, such that no unnecessary information is transmitted which is not used at the receiver end. This ideal case cannot be achieved for all source coding algorithms. If networks are operated in a chain, this chain will always be as weak as its weakest link, e.g. if a narrow-bandwidth and error-prone channel path is present anywhere in the transmission chain, this will constrain the overall quality.

NAL of AVC and HEVC. As a more concrete example, in the Advanced Video Coding (AVC) and High Efficiency Video Coding (HEVC) standards, a *network abstraction layer* (NAL) encapsulates the coded representation of the video itself as well as supplemental header parameters, which enables conveying data of the video coding layer (VCL) as well as the high level syntax (HLS) by a variety of transport systems or storage media (see Fig. 9.2a). Control data is carried in the *parameter sets* – the *sequence parameter set* (SPS) which applies to a series of coded pictures, and the *picture parameter set* (PPS) that applies to one or more pictures within such a series[3]. A parameter set such as SPS and PPS contains highly-important information that is expected to change less frequently. Each VCL NAL unit contains an identifier that refers to the content of the corresponding PPS, and each PPS contains an identifier that refers to the content of the corresponding SPS. In AVC, *data partitioning* additionally allows rearranging parts of the VCL information by priority, e.g. giving motion/mode information more robustness against data losses. The same is possible when scalable representations are used (Sec. 7.3), which naturally allow separate transport of the different layers within classes of NAL packets.
The coded video data is organized into *NAL units*, which are data packets consisting of a number of bytes. The first byte is a header byte containing an indication of the NAL unit type, for some types followed by additional bytes of header data (e.g. in case of scalable coding: the information about the layer contained). The remaining bytes contain payload data of the type indicated by the header. The NAL unit structure is generic for usage in both packet-oriented and bit-stream oriented transport. A NAL unit stream consists of a series of NAL units.

For continuous streaming, each NAL unit is prefixed by a start code. *Emulation prevention bytes* are inserted as necessary within the payload to guarantee that the unique start code prefixes are not appearing in other parts of the stream. For alignment of bit streams with byte boundaries, 'stuffing' data are added as necessary. In certain packet-oriented systems (e.g., IP systems), identification of the beginning of NAL units is implicit such that no start code prefix would be needed. However, as the formatting of the payload data should remain the same as for stream transmission (e.g. for allowing transmission of the same stream over different network types), the emulation prevention is performed as well.

---

[3] HEVC also defines a *video parameter set* (VPS) which provides information about the high level structure of entire bit streams.

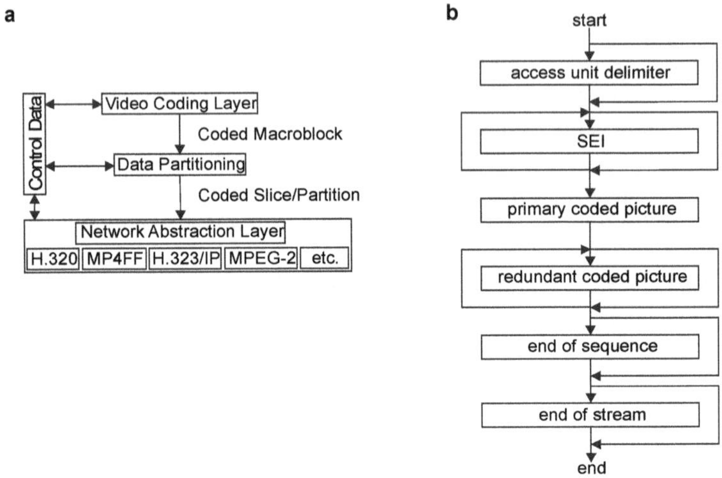

Fig. 9.2. **a** Layer structure of AVC **b** structure of access units

NAL units come as VCL and non-VCL types. Non-VCL NAL units contain any associated additional information such as the aforementioned parameter sets *video usability information* (VUI) and *supplemental enhancement information* (SEI)[4].

A set of NAL units forms an *access unit*, which typically relates to decoded picture information at a given time instance (see Fig. 9.2b). Each access unit contains a set of VCL NAL units that together compose a *primary coded picture*. An optional *access unit delimiter* aids in locating the start of the access unit, some SEI data may also precede the primary coded picture. Subsequent VCL NAL units in the same access unit may contain information for redundant pictures (e.g. for the purpose of error resilience), auxiliary pictures (e.g. alpha maps), enhancement layer(s) in scalable coding or other views in multi-view coding. An independently decodable sequence of pictures that uses a single SPS is denoted as *coded video sequence* (CVS). An *end of sequence* NAL unit may indicate that the coded picture is the last picture of a CVS; an *end of stream* NAL unit may indicate that the stream is ending. Each CVS starts with an *instantaneous decoding refresh* (IDR) access unit. An IDR access unit contains an intra-coded picture that is decodable without referencing any previous pictures (cf. Sec. 7.8).

[4] VUI and SEI data in AVC and HEVC are intended to enhance usability of the decoded video signal but are not necessary for decoding the values of the samples in the video pictures

9.3 Adaptation for channel characteristics

Transmission channels can be characterized by their *capacity*, which can be interpreted as the rate (number of bits per second) which can be conveyed error-free. It can be fixed (in case of static channels) or time variant. Channels are typically used for transporting multiple streams of data simultaneously, which is achieved by *multiplexing*. A multiplex can be arranged in a way that all participants are allocated identical data rates, whereas multiplexes with dynamic allocation of rates are possible as well, more economically using the overall available rate. Therefore, both from the aspect of channel properties as well as multiplexing, with regard to multimedia transmission, classification into *Variable Bit Rate* (VBR) or *Constant Bit Rate* (CBR) schemes is relevant. VBR transmission and storage is better suitable for multimedia sources, many compression algorithms by their nature produce variable-rate output. In case of a multiplex system, an advantage of VBR can be achieved when it can statistically be expected that some sources at a given time instance produce high rate, while at the same time some other sources produce medium or low rate[5]. Variable-rate channels are often based on packet switching, where delays and packet losses may occur when the capacity of the underlying physical channel(s) is superseded by the multiplex. This is in particular critical in real-time transmission and streaming. Therefore, even in statistical multiplex VBR channels it is common to put some restriction on the data rate variation which the sources are allowed to produce. For CBR channels, the rate generated by the encoder has to be limited anyway; *rate control* mechanisms have to be employed in this context.

A second important criterion in usage of multimedia codecs is the *error characteristic* of the channel. For synchronous stream transmission over CBR networks (e.g. in satellite, terrestrial or cable broadcast), bit losses may occur at the physical layer, but are often no longer observable at the application layer, as error correction by channel coding is typically applied [BOSSERT 1999]. If errors are uncorrectable, entire packets of data have to be discarded anyway, such that the resulting effect would be similar to packet losses in case of packet switching. Whereas *random errors* are most likely to be correctable, *burst errors* (as occurring in mobile networks) are more severe.

A distinction criterion for best strategies of error control relates to the potential existence of a *back channel*, which can be used to inform the transmitting end about problems in transmission. For *point-to-point* connections, usage of back channels allows implementing efficient error control by re-transmission, provided that it does not cause unacceptable latency. For point-to-multipoint (e.g. broadcast, multicast), where the channel condition varies between different users, it is more suitable to include error protection for typical or worst case conditions beforehand in the system design.

---

5 The beneficial effect is denoted as *statistical multiplexing*, and the capacity saving can become not larger than the single source's ratio of peak rate over mean rate.

Further, in optimizing transmission strategies it is important to distinguish between live streaming, streaming of stored media and file transfer applications. The case of point-to-point *file transfer* is uncritical regarding transmission errors, as re-transmissions can be requested repeatedly, until the file is finally available at the receiver without errors. On the other hand, this has the highest latency as the data cannot be used before the entire file has been transferred. In *streaming of stored content*, replay starts after a sufficient amount of data has been received, expecting that the remaining data flow will not be slowing down, and a sufficient amount of new data would arrive while previously received data are decoded and replayed for output. The following aspects have to be considered here:

- As network conditions by the time of transmission are not foreseeable in advance during encoding, different streams supporting various expected conditions can be pre-computed, or real-time *transcoding* can be employed at the time of transmission. The latter however increases the server complexity and most likely decreases the quality (cf. Sec. 5.4.1). Scalable coding is an alternative allowing flexible adaptation of a once encoded stream for transmission.
- Latency in case of streaming is mainly noticed by the user through the initial delay from the time of request until the signal starts replaying. Structural encoding delay can be ignored in case of pre-computed streams, the relevant latency adds up from transmission, buffering at the receiver end and decoding latency. Increasing the latency can typically be utilized to improve the quality of transmission.

For *live streaming*, real-time encoding is necessary. This however also has the advantage that any adaptation of the encoding process as required by network conditions can be made instantaneously. In the case of multicast live streaming, individual adaptation for different transmission chains can further become necessary. Due to limitation of computational power, real-time encoders often generate streams from which a lower reconstructed quality is output at the decoder than it would be possible by using highly optimized off-line encoders, which can take benefit by testing a larger variety of possible coding modes, perform look-ahead decision etc.

9.3.1 Rate and transmission control

The bit rate resulting by the data compression methods described in Ch. 4-8 highly depends on the variability[6] of the sources; for example, if the same quantizer setting (quantization step size) is used, signals with less changes over time or space will cause less bit rate output with typical compression algorithms. This means if the quantizer setting (distortion) is fixed, variable bit rate (VBR) will be the normal case for multimedia sources. If a transmission or storage medium allocates *fixed transmission rate*, constant bit rate (CBR) encoding is necessary which

---

[6] Resolution, amount of detail, strength of motion etc.

implies that the distortion varies depending on source properties. Typical examples where this is necessary are modem channels (ISDN, xDSL), broadcast channels (DVB, DAB) and storage media with a constant replay rate (tape recorders, CD players). For fixed bit rate output of an encoder, the bits are not output instantaneously but written into a first-in-first-out (FIFO) buffer, which is filled by variable rate from the encoder output, and streamed out by fixed rate to the channel input (see Fig. 9.3). For the purpose of rate control, it is then necessary to adjust the distortion, which is usually done by changing the quantizer step size. A model for this is the *leaky bucket* [REIBMAN, HASKELL 1992], assuming a volumetric container being filled with bits ('water drops') in a discontinuous flow, a constant quantity per time unit is running out. The task is then to regulate the input flow by a margin such that the bucket never overflows, but that also at least a minimum amount of information always is retained within the bucket, as otherwise channel capacity would be wasted.

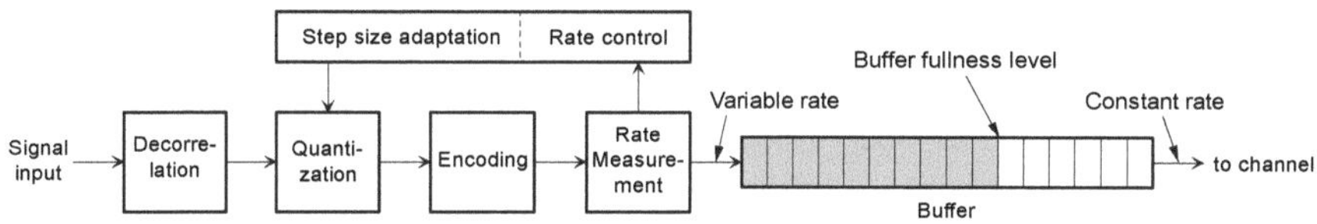

Fig. 9.3. Rate control by quantizer step size adaptation

It is not sufficient to only serve the channel by a fixed rate without taking into regard the schedule of the decoder. If for example a video picture is scheduled to be readily decoded for display at a certain point in time, the information being related to this picture must have been available for decoding a certain time span in advance, which is at minimum the time required for decoder processing. This can only be achieved in the given scenario, when a complementary buffer is implemented at the decoder, which works symmetrically with the leaky bucket: It is filled by a constant amount of bits per time unit, but is flushed by regular distances, where the flushing stops when all bits necessary to decode one access unit (e.g. a picture of video) have been released from the buffer[7]. The buffer model can be described by three parameters R, B and F, where R is the (constant output) rate transmitted over the channel, B is the buffer size and F is an initial buffer fullness which shall be reached before decoding starts [RIBAS-CORBERA ET AL. 2003]. An example timing diagram related to the decoder buffer fullness is shown in Fig. 9.4. An initial delay that can be quantified as F/R occurs before decoding can start. Both cases of buffer overflow and buffer underflow have to be prevented[8]. In

---

[7] The example of video is regarded here, as it establishes one of the most challenging cases, due to the high fluctuations in data rate that are inherent by statistical properties of video signals.

[8] A *decoder buffer overflow* could occur if *stuffing bits* were sent over the channel in case of *encoder buffer underflow*. In that case, another mechanism would need to be implemented to detect these stuffing bits as not to be written into the decoder buffer.

principle, the rate control algorithm should track the (hypothetical) state of the decoder buffer for this purpose. Typically, the quantizer step size is changed when the buffer occupancy either approaches a certain margin towards 0 or B. Algorithms for buffer control can become quite complex, e.g. taking into account expected bit budgets for I-, P- and B-type pictures[9], to avoid over-fluctuation. Simple regulation of the quantization step size Δ for the purpose of rate control may also be sub-optimum in the sense of rate-distortion optimization (RDO). Combinations of rate control with RDO are however straightforward [CHOU, MIAO 2001], as increasing or decreasing the Lagrangian multiplier λ in RDO algorithms is directly related to a decrease or increase of rate, respectively. Generally, it is then necessary to adapt the quantizer accordingly.

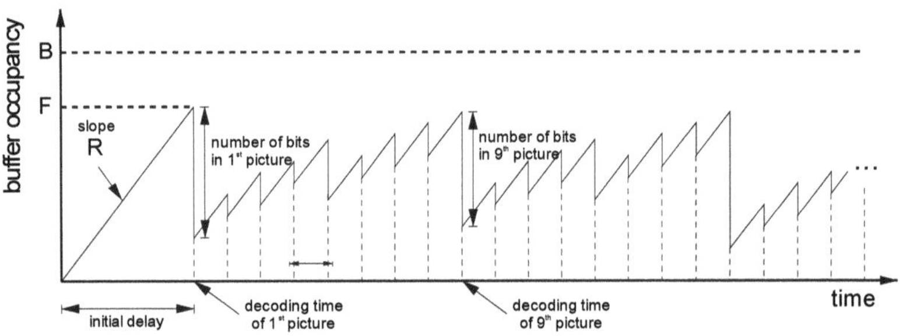

Fig. 9.4. Decoder buffer timing diagram, case of synchronous picture decoding and display

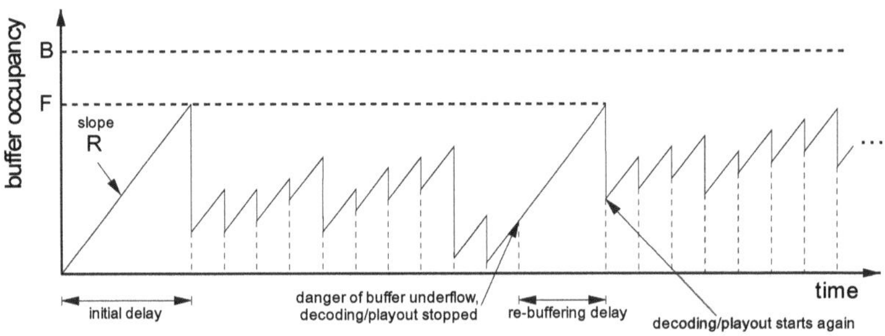

Fig. 9.5. Decoder buffer timing diagram with re-buffering

Typically, if the decoder buffer runs into underflow during readout, losses occur because the information required for decoding is not available. This is in particular critical for real-time applications with scheduled timing of outputs, e.g. synchronous frame rates of video or audio signals. In cases where a low delay is

---

[9] In Fig. 9.4, an example is shown where each 8th picture could be an intra-coded (IDR) picture consuming more data rate. In the time between two such pictures, the buffer fills up gradually, such that sufficient headroom is available when the bits for the next I-type picture are flushed. In principle, this means that the transmission of bits related to the I-type pictures takes significantly longer time than a normal picture sampling period.

required, but synchronous picture play-out is not important (e.g. low-cost videoconferencing), the decoder could perform decoding whenever all bits related to a new picture are available (unless the delay constraint is violated), but the appearance may become jerky. In case of streaming from storage servers, *re-buffering* can be performed where decoding is stopped for a certain period until the buffer is sufficiently full again. This causes additional delay and has the consequence that the replay time e.g. of a movie is extended by the time of the re-buffering period(s)[10]. An example is shown in Fig. 9.5.

The method of re-buffering is not useful for constant-rate/constant-delay networks, where instead a sufficiently stable rate control should be implemented at the encoder such that decoder buffer underflow does never occur. Re-buffering is a decoder-side emergency mechanism to prevent loss of control in cases of transmission over variable-rate or variable-delay networks, as the time by which the information arrives at the decoder cannot fully be controlled by the encoder side. It can nevertheless be tried at the encoder to implement transmission strategies where the need for re-buffering is kept to a minimum. For this case, it is necessary that the transmitter end has some knowledge about *network parameters* and considers them in its scheduling decisions for transmission. In the case of VBR, the rate-related slope of the decoder buffer timing diagrams in Figs. 9.4 and 9.5 becomes variable. In this case, the fullness level B can also be approached arbitrarily close, and the transmitter simply would stop sending information when this point is reached. This corresponds to the fact that in VBR transmission it is allowed to send with a *peak rate* $R_{\max}$ for a certain period. The advantage is that the initial delay becomes shorter, and that the buffer resources can be managed more flexibly. Unfortunately, networks which support VBR transmission typically also have variable delay. This requires the implementation of more intelligent *scheduling mechanisms* in the decision about optimum timing for transmitting and decoding the information. In the case of packet-based networks, the schedules for packetization at the transmitter and de-packetization at the receiver, taking into account expected delays and delay jitter, data losses and re-transmission (see subsequent section) are a crucial part of such strategies, and receiver/decoder buffers have to be extended such that the overall transmission chain can cope with more extreme situations.

Hypothetical reference decoder (HRD). For definition of a standard-conforming decoder, it is important in a real-time system to specify how (and when) bits need to be available for decoding and how the decoded pictures are stored and output. Video compression standards such as MPEG-2, AVC and HEVC achieve this by specifying input (coded stream) and output (decoded picture) buffer models and developing an implementation-independent model of a receiver, entitled as HRD. An encoder is not allowed to create a bit stream that (in terms of timing and computational resources) cannot be decoded by the HRD. If a decoder implementation adequately mimics the behavior of the HRD (perhaps adding some extra buffering

---

[10] In case of audio, slowing down the play-out speed is more complicated.

to account for practical variation of picture decoding speeds), it fulfills a necessary condition for the ability of decoding conformant bit streams. In AVC, the HRD specifies the operation of two buffers: the *coded picture buffer* (CPB) and the *decoded picture buffer* (DPB). The CPB models the arrival and removal timing of the coded bits. The HRD design of AVC and HEVC is similar in spirit to buffer concepts in MPEG-2 video, but is more flexible in support of transmitting the video at a variety of bit rates without excessive delay. Multiple pictures can be stored for use as references for decoding later pictures in bit-stream order, and any reference picture can be located either in the past or future (regarding display order). The decoder replicates the multi-picture buffer of the encoder in its DPB according to the reference picture buffering type. For this purpose, *memory management control operations* (MMCO) are specified in the AVC bit stream[11]. To this extent, the HRD also specifies a model of the DPB to ensure that excess memory is not needed in a decoder to store the pictures used as references and to store pictures for reordering between bit-stream order and display order.

Post encoder rate control. Embedded and scalable coding can also be recognized as a viable solution to achieve rate and error control. Scalable streams are implicitly organized as multiple-rate streams with clear priority which information has to go first when only a certain bit-rate budget can be transmitted. Implementation of constant-rate or variable-rate streaming becomes extremely simple, as no transcoding or full-stream switching has to be invoked any more (only partial streams of enhancement layers have to be switched on or off). In principle, more flexible solutions could be envisaged, where decisions about transmission rate control are done by media aware network elements (MANE) anywhere in the transmission path, when it turns out that the assumptions made initially at the transmitter side are in conflict with the actual delay or with the instantaneous data rate that can be transmitted.

9.3.2 Error control

Some source coding algorithms introduced in Ch. 4-8 provide inherent tools for resilience against transmission losses[12]. *Resynchronization* provides insertion of entry points to media streams where the decoding process can restart in a case when an unrecoverable error has occurred previously, or when random access is required e.g. to allow switching between different media sources. Typically, error resilience and re-synchronization penalize efficiency in terms of compression, but provide better quality in a lossy transmission situation [WANG ET AL. 2000]. As an

[11] HEVC does not define MMCO commands, since the timing of decoded picture buffer operations can directly be determined from the reference picture list structure and associated picture order count (POC).

[12] Some of these methods, e.g. data partitioning and scalable coding, assume that a part of the information can be conveyed by 'first class' priority, where less transmission errors shall occur; this aspect will be further discussed below.

example, when all pictures of a video sequence are intra coded, no propagation of errors over time can occur, but worse compression will be achieved than by using inter-picture (e.g. motion compensated) coding. On the other extreme, if only the first picture is intra-picture coded, and all subsequent pictures are coded by inter-picture prediction (which could make sense provided that no scene change occurs and no random access is required), compression will be much better, but in principle infinite propagation of errors could occur. As a compromise between both extremes, *cyclic intra-picture coding* is often used, preventing infinite propagation of errors[13] without too large penalty on compression performance. Two different methods are shown in Fig. 9.6. The partial refresh over selected areas which are changed from picture to picture (Fig. 9.6b) is advantageous if data rate fluctuations shall be kept low e.g. for a simpler buffer control and lower overall delay. The full-picture refresh as a by-product gives well-defined access or entry points for decoding, and is therefore more appropriate for broadcast or storage applications. The partial refresh may have the disadvantage that the systematic positions and move of refresh areas becomes visible as additional artifacts. Generally, since any refresh requires spending more data rate, the optimum refresh frequency depends on the actual probability and other characteristics (e.g. burst behavior) of transmission errors.

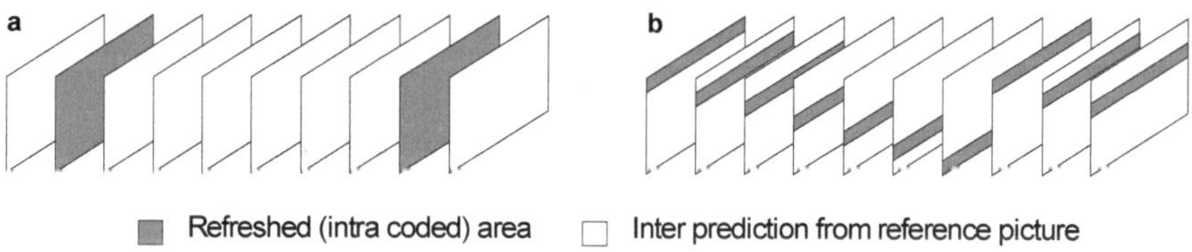

Fig. 9.6. Cyclic intra-picture coding to avoid infinite propagation of transmission error effects **a** performed over entire pictures **b** performed over parts of pictures

Better error resilience of a coding method does not necessarily imply a decrease in compression performance. E.g., bi-prediction and hierarchical B pictures in MC prediction often achieve better error resilience without sacrificing (typically even increasing) compression performance. Effects of errors are usually increased by *dependencies* in the decoding process. In prediction and in any schemes with backward adaptation (such as backward adaptation in entropy coding), dependency chains exist which cause new information being useless if it relies on previous information that was received and interpreted erroneously. With self-contained units of data (e.g. slices in AVC, group of blocks in H.261), dependencies are finite, and entry points for recovery from errors can be defined.

Best strategies for error control are largely depending on network characteristics. Networks providing *Quality of Service* (QoS) typically give guarantees about mean error rate and error variations like burst duration in addition to the other QoS parameters already discussed in the previous section. Another important

---

[13] Alternatives are methodologies which perform a non-intra update of the decoded picture memory, e.g. by usage of switching pictures (cf. Sec. 7.2.4).

distinction criterion is to be made for networks *with* or *without back channels* providing the transmitter with information about the state of the received data.

QoS-supporting networks often allow to define different QoS classes (also entitled as *differentiated services*). This can be exploited in cases where different priority levels can be assigned to different parts of the bit stream, e.g. in scalable coding or with data partitioning. In case of best-effort networks without QoS guarantee, it is usually still possible to achieve a similar effect by implementing *application layer QoS*, which could be achieved by *unequal error protection* (UEP) applied to different parts of the stream. *Forward Error Correction* (FEC) as described in the following subsection is one of the most appropriate methods for error protection. In case of networks providing a back channel, *automatic repeat request* (ARQ) can also be applied. It has also been pointed out that hybrid solutions between FEC and ARQ provide additional advantages [CHANDE ET AL. 1999] [ZHANG, KASSAM 1999].

Forward error correction (FEC). FEC is a common method of channel coding[14], where redundancy is supplemented to the information, such that detection or correction of errors becomes feasible. In block-based channel coding, this is often implemented as *Cyclic Redundancy Check* (CRC) coding. Herein, N information bits are protected by K redundancy bits, which typically allows detecting of up to K errors occurring in the coded block of length $N+K$, or correct up to $K/2$ errors (in case of even K). The code rate is $N/(N+K)$. CRC codes which are often applied in multimedia signal transmission are *Bose-Chaudhury-Hocquenghem* (BCH) codes [HOCQUENGHEM 1959] [BOSE, RAY-CHAUDHURY 1960] and *Reed-Solomon* (RS) codes [REED, SOLOMON 1960]. The other large class are the *convolutional codes*, whose state transitions define valid sequences of bits that would arrive at the decoder provided that no bit error would have occurred. The *Viterbi algorithm* [VITERBI 1967] [FORNEY 1973] is then used[15] to find the valid bit sequence with closest distance to the sequence received. Convolutional codes are characterized by the *constraint length* and the *code rate r*, which specifies the percentage of the information bits carried within the total rate sent over the channel.

One of the most efficient classes of convolutional codes existing today are the turbo codes [BERROU, GLAVIEUX 1993], which are based on iterative convergence of two decoders operating on interleaved versions of the same data block; either the bits are recovered correctly or in case of failure, all contributing bits need to be regarded as lost. *Rate-Compatible Punctured Convolutional Codes* (RCPC) [HAGENAUER 1988] further allow adjusting the amount of redundancy by a post-encoding operation[16].

---

[14] For a general reference on channel coding, see e.g. [BOSSERT 1999].

[15] As a variant for source coding, the Viterbi algorithm was described in Sec. 4.6.

[16] 'Puncturing' removes redundancy bits at designated positions, such that the code rate becomes higher, the error correction capability decreases.

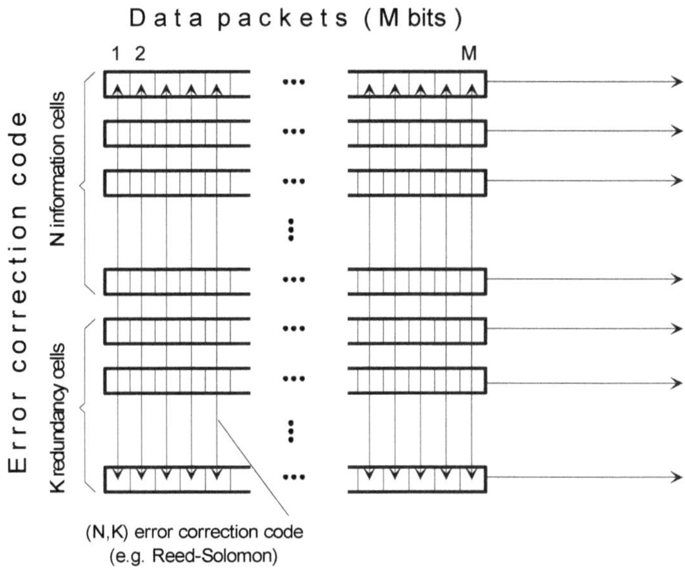

Fig. 9.7. Interleaving combined with error correction coding for error protection in case of burst errors

For burst error cases, the effect of error correction can only be obtained if the redundancy of the code exceeds the error burst length. As this can hardly be guaranteed, error bursts on the channel can be dispersed (regarding their effect to the bit stream) by *interleaving*. For this, a sequence of information bits is arranged in a matrix that is loaded row-wise, where FEC is then applied column-wise across all rows (or vice versa). When the bits in the matrix are transmitted row by row, it is likely that the burst only affects one or a few rows, which means that the erroneous bits can be recovered by the column-wise FEC. Fig. 9.7 shows an example how interleaving can be applied in the packet-based transmission, where the 'bursts' occur as complete data packets losses. This can again be solved by applying an FEC column-wise *across packets*. An additional row-wise FEC could then be applied to correct single bit errors within the packets, if necessary. In the specific case of packet losses, the locations of losses are usually known e.g. from the sequence numbers in the packet headers, which makes the situation different from a random bit-error scenario, where it is unknown which bits were wrong. In the case of known error/loss position it is possible to recover the complete information perfectly when no more than K losses occur within the block of N+K information packets sent, i.e. the maximum number of recoverable packet losses is equal to the number of redundant packets.

Exploitation of latency and ARQ. If a back channel exists between receiver and transmitter, and sufficient latency is allowed by the application, error control can also invoke re-transmission of lost information. In [CHANDE ET AL. 1999] it is shown that this can significantly increase the error resilience as compared to interleaving FEC techniques, when the same amount of redundant information is used[17]. In [Chou, Miao 2001], rate-distortion optimal streaming over a channel with packet losses and a complementary back channel is studied. For a given set of packets and a time delivery deadline, it has to be decided which packet to send by the next available transmission slot. The goal would be to achieve minimum distortion under the given rate constraint, which in particular for the case of video coding must also include analysis of signal dependencies (such as picture dependencies discussed above). In any case, only packets with high priority should be re-sent, to avoid additional network overload. For the case of ARQ applications, the scheduling mechanisms discussed in the previous section under pure transmission delay aspects have to be modified by including the ARQ round-trip delay and the probability of remaining errors after ARQ re-transmission. ARQ approaches are important in Internet streaming, for 3G (and beyond) mobile networks, and for wireless LAN.

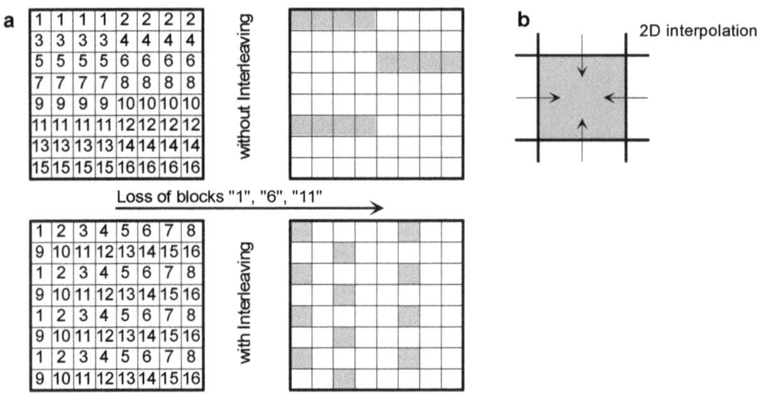

Fig. 9.8. a Spatial interleaving of block-wise picture coding, where signal blocks are combined into packets 1-16 **b** combined with interpolation for error concealment

Error concealment. Error concealment reduces the effect of errors by post-decoding signal recovery processing [Wang, Zhu 1998]. Concealment is not part of the normative decoder process and therefore does not need to be standardized. However, in an encoded representation format, certain methods of concealment can be supported by encoder actions, e.g. by using error resilient mechanisms such as inserting resynchronization points more or less frequently, using data partitioning, etc. Furthermore, even when the decoder is not able to operate properly due to data losses, some benefit could still be drawn from partial information received. Specific tools for error resilient encoding may also require addi-

[17] Combination with FEC is also possible, e.g. by using *rate constrained punctured convolutional* (RCPC) coding, where a weaker code (less redundancy) is a sub-set of the stronger code, and additional redundancy is sent upon ARQ instead of re-sending the full information.

tional data rate, as by purpose some redundancy is left in the encoded source to allow concealment of errors. An example is utilization of *spatial signal interleaving*[18]. If data packets are assembled from coded information of locally neighbored signal areas, larger concatenated areas may be un-decodable in case of packet losses, as it is shown for the example of an image signal in Fig. 9.8a, top. By interleaving, the errors can be spread over different areas of the signal (Fig. 9.8a, bottom). A possible concealment approach in this case would be interpolation of information from correctly received neighbored areas (Fig. 9.8b). However, it is then no longer possible to utilize prediction from direct neighbored blocks, such that compression performance is suffering (e.g. by not being able to perform intra prediction, prediction of transform coefficients or context-adaptive entropy coding from the closest neighborhood).

For video transmission, methods of temporally consistent concealment are important. The most simple concealment method is *picture freezing*, where the decoding process is stopped up to the next resynchronization point. Even though the subjective perception in case of picture freezing may be less annoying than error propagation which would occur otherwise (cf. Fig. 7.4), it is not very convenient for subjective impression if applied frequently. An alternative are techniques of motion-compensated concealment [KAISER, FAZEL 1999], where it is tried to get as much as possible advantage from motion vectors correctly received to map previously correctly decoded areas into subsequent pictures. In cases where motion vectors are lost as well, techniques of motion vector field interpolation can be applied as part of the concealment; temporally and spatially adjacent, correctly received motion vectors can be used for this purpose. Nevertheless, there is no guarantee that motion-compensated concealment would always lead to improved quality; in particular in cases where the motion vector field is inhomogeneous, the threat of introducing unnatural artifacts is considerably high, and it is difficult to identify such artifacts without reference to the original picture, which is not available at the decoder. A better optimized solution could likely be between the simple freeze-picture concealment and improved methods, where the latter ones should only be applied when motion information is classified as being consistent and reliable. For a review of various methods, refer to [WANG, ZHOU 1998] and [WANG ET AL. 2000].

More advanced approaches for concealment are based on *synthesis* of lost information. This has successfully been applied in case of transmission losses in speech transmission, where vocoder methods are used to substitute missing samples. In case of audio, particularly for harmonic sounds, phase consistent substitution of a lost segment is important. For image and video signals, color and texture synthesis are promising methods in this context, which were also discussed as advanced compression methods (Sec. 6.6.2 and 7.7.3); areas which are recognized to be homogeneous by color can usually be filled by linear interpolation or extrapolation. In case of texture synthesis, parameters of a model have to be derived

---

[18] AVC defines a slice type with *flexible macroblock ordering* (FMO), where interleaving by a checkerboard structure similar to Fig. 9.8 can be implemented.

from available samples, and it has then to be assumed that the model would be homogeneously valid over the area to be concealed. Methods that are useful in this context have to take into account that the boundary between the area with available samples and the concealed area does not show as artificial seam. This can e.g. be achieved by conditional AR (CAR) synthesis, or synthesis based on template matching from the available samples.

In this context, it can also be considered to transmit synthesis parameters as side information, which are not intended to be used in the conventional decoding process, but only for concealment after error-prone decoding. Conceptually, this still establishes a concept of adding redundancy for recovery of the source, which would however be quite different from conventional channel coding approaches which operate at the level of coded bits.

9.4 Media transport, storage and reproduction

To realize an interoperable system for multimedia data compression, transmission, decoding and reproduction[19], further aspects have to be observed which go beyond plain interfacing of encoding algorithms and networks as discussed in Sec. 9.2. Particularly, unique identification of the data streams to be decoded, multiplexing, synchronization of streams and decoded content (e.g. synchronous presentation of video and audio) are necessary. Hence, specifications are needed for systems-level syntax and related mechanisms, which includes identification, access, multiplexing and synchronization of data streams, aspects of user interaction and presentation of the decoded content via output devices.

In addition, mechanisms have to be provided which allow combining different sub-streams (video, audio, supplemental information) for transport.

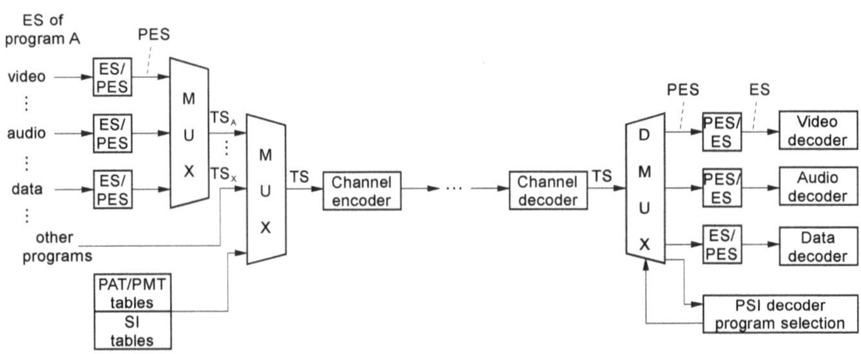

Fig. 9.9. Transport stream multiplex of MPEG-2

[19] Other aspects in this context are user interaction and rights management.

9.4.1 Broadcast applications

In broadcast, the *transport stream hierarchy* of MPEG-2 is widely used. As shown in Fig. 9.9, each encoding device (e.g. video and audio encoders) generates *elementary streams* (ES), which can directly be decoded by the complementary decoding device. To make the ESs piecewise accessible, decomposition into *packetized elementary streams* (PES) is performed. The *transport stream* (TS) is a concatenation of PES, as actually to be transmitted over the network[20]. Additionally, higher-level multiplexes of several TS can be constructed, which are e.g. related to different programs. Supplemental information at this level are the *program association tables* (PAT), containing a list of all TS contained in the multiplex, including indices for the *program map tables* (PMT) which are associated with the particular program units and used to manage the TS packets. In addition, *service information* (SI) tables can be provided, which e.g. can be used for conveying *electronic program guides* (EPG).

In the PES, packet sizes are variable with the length indicated in the header. Each packet also contains the ES identification which is needed to re-assemble the ES. Other optional parts of the PES header contain information about priority, copyright, scrambling for access protection, additional error detection for the previous packet and synchronization information.

MPEG-2 TS packets have a fixed length of 188 bytes[21], the payload data are filled from PES packets. A transport stream contains (in principle in arbitrary sequence) PES data from all associated elementary streams, which belong to a program of a given *program identifier* (PID). In addition, the TS-packet header also contains information about priority, scrambling etc. *program clock recovery* (PCR) information is used for synchronization between different transport streams in the TS multiplex. Private data can also be conveyed within the header or in ES sections. This can e.g. be used for the purpose of encryption and to identify access rights in Pay TV.

Synchronization. For time-accurate presentation of the decoded data, like synchronous replay of video and audio, it is necessary to provide synchronization mechanisms at the ES level. To support synchronization, the ESs are supplemented by *time stamps*. These can be defined based on a relative reference (e.g. the start of a movie) or based on absolute time. Different types of time stamps include:

---

[20] In Internet applications, an MPEG-2 TS could again be transported via RTP packets, where duplicate packaging is necessary due to different packet payload sizes; in such cases however, it is tried to keep the overhead to a minimum. For other networks, such as DVB, the MPEG-2 TS is the primary transport mechanism which directly interfaces with the network layer (cf. Sec. 9.2).

[21] The number of 188 bytes was adopted from packet payload size in asynchronous transfer mode (ATM) as originally planned for broadband ISDN.

- *decoding time stamps* (DTS) indicate the time by which the decoding of the related picture shall be finished;
- *composition time stamps* (CTS, only available in MPEG-4, where several objects can be composed into a scene before rendering) indicate the time by which the composition result shall be available;
- *presentation time stamps* (PTS) indicate the time by which the presentation (e.g. display) shall be performed.

Additional requirements related to synchronization, in particular in MPEG-4, can refer to user interaction, or activating certain parts of the ESs upon request.

Digital Video Broadcast (DVB). As a typical application example for transmission methodologies in broadcast services defined for digital media, the Digital Video Broadcast (DVB) series of standards is reviewed in some detail here, where the bit stream transmission is built around the systems parts of the MPEG-2 standard. DVB was designed to re-use previous analog TV channels and infrastructure for digital services. With the same frequency bandwidth that was originally reserved for one analog program, the introduction of digital video compression allowed increasing the number of programs, increasing the quality or resolution (e.g. expand from SD to HD), or provide a combination of these improvements. Originally defined as a European standard by the European Telecommunication Standards Institute (ETSI), DVB has also been adopted by various countries outside Europe. Beside some extensions to the MPEG-2 TS *by program specific information* (PSI) which allows e.g. to embed EPGs, and by definition of conditional access mechanisms as necessary for pay TV services, DVB specifications are mainly defining physical transmission over satellite TV channels (DVB-S), cable TV channels (DVB-C) and terrestrial TV channels (DVB-T).

For each of these configurations, modulation and physical-layer error protection mechanisms are specified. The first building blocks are identically used in DVB-S, DVB-C and DVB-T as shown in Fig. 9.10. Following the base-band interface, which defines the access to the packetized bytes residing in an MPEG-2 transport stream, an ‘energy dispersal’ is performed which has the goal of suppressing the carriers of the subsequent modulation as far as possible for avoidance of inter-channel interference; in principle, this is a scrambling process which is particularly effective if longer runs of ‘0’ or ‘1’ bits would be present. Channel coding is a combination of an outer encoder based on a Reed-Solomon [204,188] CRC code, an interleaver and an ‘inner’ encoder, which is a punctured convolutional encoder with rates selectable as r=1/2, 2/3, 3/4, 5/6 or 7/8, thus being flexible in terms of the amount of redundancy added.

The modulation schemes are different for the three types of physical channels. DVB-S uses *quadrature phase shift keying* (QPSK), which allows transmission of 2 bits per modulation symbol. If the bandwidth of a satellite transponder is 36 *MHz*, the source payload rate which can be conveyed by these methods is variable between 26.1 *Mb/s* (for convolutional code rate 1/2) and 45.6 *Mb/s* (for convolutional code rate 7/8). Cable channels provide much better transmission quality (in terms of channel bit error rate) than satellite channels. Hence, DVB-C uses *quad-*

rature amplitude modulation (QAM), which carries a higher number of bits per modulation symbol. It can be selected as either 16-QAM (4 bits/symbol), 32-QAM (5 bits/symbol) or 64-QAM (6 bits/symbol). Differential encoding enables receiver synchronization of the phase. DVB-T is the most complex method; terrestrial TV channels exhibit frequency-selective loss characteristics, which is even worse for the case of mobile reception. Due to this reason, a spreading of the coded binary information over multiple carriers is performed. This is implemented through *orthogonal frequency division multiplex* (OFDM) [VAN NEE, PRASAD 2000], with error correction across carriers, such that losses of some of them can be compensated. OFDM is also combined with quadrature modulation, where QPSK (2 bits/symbol), 16-QAM (4 bits/symbol) and 64-QAM (6 bits/symbol) can be used as modulation methods, to be applied depending on expected channel error characteristics.

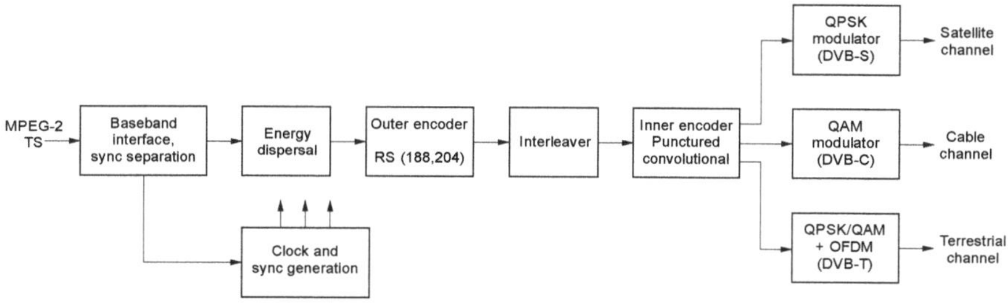

Fig. 9.10. Block diagram of DVB transmitters

In a hierarchical variant of DVB-T modulation, different quadrature modulation methods can also be mixed, such that different parts of the bit stream can be transmitted with variable protection (as e.g. QPSK is more resilient against transmission errors than 64-QAM). This provides more graceful degradation when the channel quality deteriorates. In the OFDM signal, a 'guard interval' has to be inserted after each carrier symbol, which allows suppressing interference from delayed components of the modulated signal, which are here expected to arrive from neighbored transmission stations (TV towers). The necessary length of the guard interval depends on the possible maximum time lag between signals from different neighbored TV towers that could be received by one receiver, and can be set to a maximum of approx. 220 μs in DVB-T[22]. By this, neighbored terrestrial transmitters can use the same carrier frequencies when sending the same signal in DVB-T, which allows a much more economic bandwidth utilization than in analog TV where such a case would have caused interference problems and the second signal would have been displayed as 'ghost picture'. Examples of usable data rates which can be transmitted over one previous analog TV channel are given in Table 9.1 for different combinations of modulation methods and convolutional code rates. If hierarchical modulation is used, combinations in between are possi-

[22] This value considers the typical height of TV towers and the earth surface curvature.

ble, such that e.g. using QPSK (code rate 1/2) embedded into 16-QAM (code rate 5/6), a rate can be achieved which adds up from (R_{QPSK}+$R_{16\text{-}QAM}$). This way, for example joint transmission of data with high priority at 4.7 *Mb/s* and data with low priority at 15.67 *Mb/s* is possible.

Table 9.1. Usable data rates (*Mb/s*) per terrestrial TV channel for different combinations of code rates and modulation types in DVB-T (after REIMERS)

Code rate *r*=	QPSK	16-QAM	64-QAM
1/2	4.7	9.4	14.10
2/3	6.27	12.53	18.8
3/4	7.05	14.10	21.15
5/6	7.83	15.67	24.68
7/8	8.23	16.45	24.68

In DVB, typically up to 5 digital standard definition (SD) programs are transported within the bandwidth of one previous analog TV channel, when the MPEG-2 standard is used for video and audio encoding.

The newest generation of DVB, namely DVB-x2 (x=C,S,T), still uses MPEG-2 transport stream infrastructure, but at the coding layer replaces the MPEG-2 video codec by AVC[23]. This allowed the introduction of HD broadcast services, in combination with improved modulation and error correction schemes. All three schemes use BCH codes for outer coding and LDPC[24] codes for inner coding in the interleaved packet structure. In combination with higher number of symbols in the modulation schemes, this allows reaching an increase of the useful bit rate. For example, up to 4096-QAM is used in DVB-C2, and up to 256-QAM in DVB-T2. DVB-C2 uses QAM in combination with multi-carrier transmission (coded OFDM), which enables frequency interleaving as additional dimension for error correction. The maximum number of subcarriers in the OFDM scheme of DVB-T2 is increased by a factor of 4 in comparison to DVB-T, and overall more flexibility is added, which can be used to either increase the data rate or the robustness of transmission. This way, the channel capacity (useful bit rate that can be transmitted lossless under typical operation conditions) is increased by more than 30% in DVB-S2 and almost 50% in DVB-T2, compared to the first generation of DVB. Also in DVB-C2, the capacity is significantly increased (in particular for low channel S/N, where more than doubling of the rate could be achieved), and more robustness against pulse-like disturbances is also gained. Along with the usage of AVC, which typically requires less than half of the transmission rate of MPEG-2 video, this means that in most of the DVB transmission scenarios, SD pro-

[23] Support for HEVC is currently also under development in DVB-x2, for prospective usage in high and ultra-high definition broadcast. HEVC-based services are planned to be started from 2016

[24] See sec. 9.3.2.

grammes can be replaced by same number (or even increased number) of HD programmes with clearly higher quality, using the same physical channel bandwidth.

Digital Audio Broadcast (DAB) uses basically equivalent transmission methods as the first generation of DVB-T, but data rates are lower (as useful for plain audio transmission).

Regardless of the significant advances that the DVB schemes have brought in digital media transmission, a tendency can be observed that traditional broadcast services (with pre-defined time-line for a multitude of users, without significant amount of interactivity) are replaced by more individual on-demand services, or that internet transmission also takes over TV-like live transmission in 'IP-TV'. In this context, physical transmission channels (in particular terrestrial frequency bands) that were previously reserved for TV broadcast are being allocated to mobile networks.

9.4.2 Communication services

Linear broadcast services require more precise timing behaviour than on-demand retrieval of media. Such real-time requirements are even more sensitive in communication services, where bi-directional conversation would become difficult when large delays were occurring during coding, transmission and decoding[25]. Basically, the same media codecs can be used for such services, but certain coding/decoding tools may need to be disabled when they would inherently impose a delay, such as

- large block length in transform coding of audio signals;
- look-ahead coding for improved performance;
- extensive re-ordering of capture/display picture sequence in video compression, e.g. when using B or hierarchical B group of picture structures.

On the other hand, communication applications are often point-to-point, or even when multiple participants communicate (e.g. in a conference), their number is significantly smaller, and unlike broadcast scenarios, each participant is usually known, such that compression and transmission methods (including error control) can individually be negotiated.

Interoperability for communication services has primarily been defined by recommendations of the ITU-T[26]. In particular, the following number ranges of specifications are relevant in this context:

- Transmission, multiplexing and synchronization H.220–H.229
- Systems aspects H.230–H.239

---

[25] The same applies in control, maintenance and surveillance applications, when real-time reaction is required based on audiovisual media presentation to a remote observer.

[26] Except for some proprietary applications which are not intended to be interoperable with other vendors' products, such as closed environments for conferencing or social networking.

- Communication procedures H.240–H.259
- Coding of moving video H.260–H.279
- Related systems aspects H.280–H.299
- Systems and terminal equipment for audiovisual services H.300–H.349
- Mobile applications and service aspects H.500ff.
- Broadband applications and service aspects H.600ff.
- IPTV applications and service aspects H.700ff.

The need for multiplexing and synchronizing various media, and adapting their transmission to various types of networks also arises in communication applications. In the H.22x series, H.222 is identical with the respective part of MPEG-2 Systems; H.223 is defining multiplexes, synchronization and error control for transmission for various networks in low-rate applications. The corresponding control procedures are specified in the H.24x series, where some are specific for negotiating control parameters of certain codecs (e.g. H.241 relates to negotiation procedures when using H.264 in video communication services). H.245 is defining more generic systems-level control for audiovisual and multimedia communication applications, such as conference setup, negotiation of codecs and their operational settings, bit rates etc.

9.4.3 File storage

Further important aspects that must be defined at the systems level are *file formats* for storage of media and associated data. Basically, an entire video with associated audio could just be written from the sequence of bits in an MPEG-2 transport stream into a binary file, and uniquely (at least as entire sequence) retrieved, decoded and replayed from there. By invoking stream scanning for random access points, mechanisms such as fast forward and rewind can be implemented. This however requires initial download, i.e. availability of the entire file, before decoding and replay can be started. A file format should however be a more flexible container format to support extracting only parts of the media stored in the file, replay in re-arranged order, efficient skipping of certain parts, editing, re-writing etc. In case of scalable formats, it is furthermore desirable to enable extraction of media with different resolution. For these functionalities, it is necessary to structure the data into separately accessible parts, and to define pointers that enable access to these parts. Furthermore, like in case of broadcast streams, the relevant timing and synchronization information for data output has to be associated with the data. Additional metadata needs to be associated as well, if available.

As an open standard (non-proprietary formats), the *ISO base media file format* (IBMFF), which is defined by identical text in ISO/IEC 14496-12 (MPEG-4 part 12) and ISO/IEC 15444-12 (JPEG2000 part 12), has been designed to achieve this. It establishes the common ground (in terms of unique signalling and mechanisms) for other more media specific formats, such as the *MP4 file format* (14496-14) the *Advanced Video Coding (AVC) file format* (14496-15), which specifically aligns with the concept of NAL unit access, and JPEG2000's *Compound image file format* (15444-6).

The IBMFF contains timing, structure and media information for a 'presentation', which e.g. consists of one or several video sequences, possibly along with audio. The purpose is to allow interchange, management, editing, and presentation of the media. In this context, it is not relevant where the file is stored; it could be local on the system where the presentation shall be output, or is may be remote, still to be delivered via a network by some streaming mechanism; in the latter case, independence from any particular network protocol, but efficient handling in general with any network is intended. The file structure is object-oriented; decomposition into constituent objects shall be simple, with the structure of these objects inferred directly from their type. The following definitions are relevant in this context:

- *Sample:* All data associated with a single timestamp. Special types of samples are foreseen for random access and sync;
- *Chunk:* A contiguous set of samples, allocated to one track;
- *Track:* A timed sequence of related samples, which could be further structured as chunks (belonging to one level), or sub-tracks (associated with different levels). The assignment of sub-tracks with levels relates to a specific feature, e.g. a scalability layer;
- As pointers to samples and tracks, the *sample table* is a kind of packed directory for the timing and physical layout of the samples in a track, and the *hint track* is a special track containing instructions for packaging one or more tracks into a streaming sequence;
- *Box:* An object-oriented building block defined by a unique type identifier (which signals the data type contained) and length of the data. There are numerous types of boxes for specific data types. Type and size are both signaled in the header of the box. The header can use compact or extended size (32 or 64 bits) and compact or extended types Typically only the media data box(es), i.e. the bulk payload, need the 64-bit size. The size is the entire size of the box, including the header fields, and all contained data. This is mainly used for efficient parsing, to skip data of boxes that shall not be used. A *container box* may be used to group a set of related boxes. A *segment* consists of one (or more) movie (or movie fragment) boxes, with other associated media or metadata boxes

The file may also contain hints for streaming, e.g. when the origin of the data was recorded from a transport stream (TS), or it is intended to be output to a TS with a foreseen structure. The elementary streams of the TS are stored sequentially 'as is' within the corresponding media data boxes.

Basically, similar concepts are used in the Matroska format defined by an open source project [WWW.MATROSKA.ORG]. Other relevant (proprietary) multimedia container formats are Apple's QuickTime (of which an earlier version had founded the basis for developing MP4 and IBMFF), Microsoft's Advanced Streaming Format (ASF) and Audio Video Interleave (AVI), Adobe's Flash. Most of these formats are able to support storage of proprietary media formats of the respective technology owners, as well as carrying other media data such as those from open international standards.

9.4.4 Internet and mobile media streaming

In *streaming* applications, decoding and replay of media is instantaneous, i.e. it can start as soon as a sufficient timeline of data is available, whereas transmission of data to be replayed in the future is ongoing.

Internet streaming. The *internet protocol* (IP) abstracts from the underlying physical layer. Though the Internet was originally designed for file transfer, today real-time media streaming (particularly video) produces the majority of Internet traffic, with tendency of significant further growth. On top of IP, the *transport control protocol* (TCP) is widely used, which guarantees reliable transmission by acknowledgement of IP packet reception over each link in the transmission chain (between switches, routers). As this was originally considered of being too slow for real time streaming applications (due to its link-to-link hand shaking, re-transmission mechanisms and other protocol overhead such as congestion-dependent flow control), an alternative lower-layer transport protocol, the *unified datagram protocol* (UDP) was defined. UDP does not employ acknowledgement and retransmission mechanisms, thus encountering the danger that packets would arrive late or become lost in case of network congestion. Losses typically occur, when overflow of input or output buffers happens along the network path, e.g. in switches. Both TCP and UDP also employ simple CRC based error detection and would discard packets which are affected by bit errors. As a higher-layer transport protocol, the *real-time transport protocol* (RTP) is primarily used on top of UDP for Internet media streaming. RTP provides timing and synchronization information associated with the data and the header information can be used to re-order packets by their original sequence at the receiver end. Timing information conveyed in RTP can also be used for synchronization of different media, or to decide whether a packet is outdated and may be discarded. Payload type identification is conveyed for certain common types such as video and audio streams according to ISO/IEC standards or ITU-T recommendations. Examples for this are[27]

- RFC 2250 RTP payload format for MPEG1/MPEG2 video (including capabilities to encapsulate transport streams and program streams)
- RFC 4629 RTP payload format for ITU-T Rec. H.263 video (superseding RFC2429/RFC3555)
- RFC 6184 RTP payload format for AVC/H.264 video (superseding 3984)
- RFC 6190 RTP payload format for AVC scalable video coding
- RTP 6416 (superseding 3016) RTP payload format for MPEG-4 audio/visual streams

Preliminary drafts also exist for RTP transport of HEVC and the multi-view (MVC) extension of AVC.

[27] RFC = request for comments, usually issued with a deadline until which comments or modifications are to be made; otherwise, the recommended specification would be regarded as approved.

RTP can be used in conjunction with the *real-time control protocol* (RTCP) which is primarily intended to provide feedback about the instantaneous network QoS (delay, packet loss rate etc.), e.g. to be used for optimizing transmission w.r.t. better error resilience.

Newer versions of IP such as IP v6 also foresee implementation of QoS and classified (differentiated) services. On the other hand, as the Internet is a 'network of networks' in which older-generation routers and switches may exist, the functionality of QoS mechanisms might not be guaranteed for any link that could be established. Alternatively, QoS could be achieved by *overlay networks* which are operated as application-specific services at a higher layer of the protocol stack.

By assigning different network QoS levels or priority classes for sub-streams of different relevance, full advantage can be drawn from scalable (layered) media representations or data partitioning methods. To guarantee QoS, it is necessary by the network side to perform reservation of transmission capacity, and *call admission control* (CAC) to prevent network overload.

For transmission capacity reservation over internet connections, the *reservation protocol* (RSVP) was defined. The transmitting device is then informed about the capacity granted for the connection. Beyond high priority packets (e.g. header or parameter information, base layer of a scalable stream) that are transmitted more safely, additional packets (e.g. enhancement layer) can be transported by the network on a best-effort basis, if available capacity allows. This principle can only work satisfactorily, provided that each transmitting device observes the negotiated parameters, for which the simplest way is the implementation of a leaky bucket (see Sec. 9.3.1) counting the number of packets that is produced by the source within a certain time interval. The duration of this interval corresponds to the negotiated *peak duration*, during which the source is allowed to send by a rate higher than the negotiated mean. If the buffer runs full, the rate has to be reduced by rate control mechanisms.

Even in cases where no reservation guarantee is provided by the network, a transmitter can estimate the currently available network capacity by receiving feedback from the receiver side about the transmission quality (e.g., in terms of packet loss, error burst characteristics, delay and its variation). Then, a useful approach is preventive reduction of the transmission rate on a voluntary basis, or usage of application-layer error protection.

Optimum adaptation of rate, considering the available capacity of the network requires rate control. However, as media streaming often uses *pre-encoded data* to be transported from a server to a client, rate control on-the-fly cannot be used. Possible alternatives are:

- Storage of multiple streams, supporting different rates, and applying *stream switching* depending on available transmission rate;
- Transcoding, i.e. on-the-fly decoding and re-encoding by the target rate (cf. Sec. 9.3.1);
- Usage of scalable streams, which saves storage capacity compared to 1. and saves computational complexity compared to 2.

In spite of provisions taken as described, packet losses frequently occur in Internet streaming, but their effect can be minimized if error resilience and error concealment tools as previously discussed are employed.

Both in communication and streaming applications, it is desirable to allow consumption of media within web browsing interfaces. Though web browsing usually establishes hypertext transfer protocol (HTTP) based connection, with HTTP linking to the more secure TCP transmission, it is possible to embed plugins that are based on RTP. Such an approach is for example taken by the open source *Web RTC* initiative ('web real time communication'), where conversational services can be run as plugin within a web browsing environment [WWW.WEBRTC.ORG]. For streaming, where the timing constraints are less rigid than for conversational applications, even direct usage of the TCP/HTTP stack for transmission of media has become possible due to the improvement of the internet infrastructure and capacity. However, due to the fact that useful transmission rate can vary significantly in TCP, efficient adaptation has to be supported.

One technology using such an approach is *dynamic adaptive streaming over HTTP* (DASH)[28], which has been defined as international standard in ISO/IEC 23009 [SODAGAR 2009]. DASH includes a mechanism for *media presentation description* (MPD) which for a given media data expresses the structure in terms of data chunks (each of which is associated with a hyperlink), their timing, duration and synchronization, as well as different representations (in terms of data amount/rate) available in the adaptation set. On the server, the data chunks associated with a media item can be stored in one or multiple files, along with the MPD file(s). The knowledge about the timed structure and its different available representations can be used for dynamic rate adaptation (i.e. switching to lower/higher rate depending on instantaneous network conditions), and also or for user interaction such as rewind / fast forward and partial replay. For the multi-rate representations, either simulcast style (i.e. independent coding for the various rates) or scalable style (layered coding, when supported by the codec) can be used. The data chunks can simply be retrieved by http get commands from the side of a DASH client, and once available, can be managed/re-ordered in the correct replay sequence by just using the MPD information.

To some extent, solutions available for internet streaming of media are inefficient (in terms of necessary overhead for signalling, as well as insufficient mechanisms of identifying data or partial data in caching for more efficient distribution) and lack flexibility (in terms of personalized access and adaptation between instantaneous media and network properties), due to the fact that the internet was originally designed for the purpose of non-real-time text transfer. Basically, it should be possible to achieve similar functionality transparent over the network, as if the media data were stored locally. This should also include remote or distributed processing of media data. One step in this direction may be a new standard defined as systems layer of MPEG-H (23008-1), entitled as *'MPEG media*

[28] Sometimes also denoted as 'MPEG DASH'.

transport' (MMT), which exactly tries to target the problems above. As it is still under development, it would be premature going into technical details here.

Wireless networks. Wireless transmission for mobile multimedia communication is getting more important by the increasing usage of wearable computers (laptops, tablets, smart phones or other mobile devices with multimedia capabilities, e.g. built-in cameras, color LCD screens etc.). Unlike the situation 10 years ago, none of these is restricted to low resolution and/or low quality any more. Whereas wireless networks such as LTE/4G are becoming available which provide higher transmission capacity, multimedia data are also establishing the most relevant traffic over wireless links. For digital wireless networks, it is necessary to distinguish between the case of stationary (in terms of position) transmitter and receiver, and the case where at least one of them is moving, i.e. the mobile transmit/receive scenario.

In the first case, the transmission capacity which can be made available highly depends on the stationarity of network conditions which would allow reliable adaptation of modulation and error protection methods, interference cancellation, etc. This is significantly simpler with fixed-position devices, where it is possible to establish optimum settings by measuring the wireless channel during initialization, making updates when necessary. Typical wireless LAN (WLAN) technology as available today, e.g. designed in compliance to the standards IEEE 802.11 or 802.16 (WIMAX), provide peak transmission rates of 100 *Mb/s* and above; however, when several users need to share this capacity, this can lead to similar error characteristics based on congestion as in wired packet-switching networks.

In the second case of 'true' mobile communication, the device operated by a user is moving, such that transmission characteristics could change permanently. This leads to more severe errors already at the physical layer, and less capability (or need for more complex algorithms) for adaptation of the transmission. Nevertheless, the mobile network infrastructure available today can provide transmission capacity that typically allows to use streaming of media such as compressed video in real-time in the downlink (depending on current location, speed and network load). Whereas this was hardly possible with second generation ('2G') mobile networks such as GSM (*global standard for mobile communication*), and its extensions GPRS (*general packet radio service*) and EDGE (*enhanced data rates for global evolution*), it already became more realistic with the advent of the third generation ('3G') *universal mobile telecommunication system* (UMTS)[29]. Whereas GSM serves the different users within a base station's network cell via time-

[29] UMTS is a framework which was developed under the roof of the ITU-T. An alternative denotation is IMT-2000 (International Mobile Telecommunications 2000). In fact, UMTS from the beginning was not homogeneous, but comprised an entire group of standards, some of which were newly defined radio transmission methods and protocol mechanisms, but also integrating existing networks like GSM-based GPRS and EDGE, as long as they are available. The initially foreseen data rate of UMTS/3G is up to 2 *Mb/s*, which however would barely be available to individual users.

division multiple access (TDMA), UMTS uses a code-division (CDMA) method which allows making more efficient use of the available mobile bandwidth resources and interference cancellation. UMTS has further progressing into the *long term evolution* (LTE), meanwhile rather denoted as '4G'. LTE uses even more advanced allocation of frequency resources by combining CDMA with *orthogonal frequency division multiplex* (OFDM), and further supports *multiple input multiple output* (MIMO) transmission, which also allows better location-dependent transmission based on antenna diversity. Starting from 3G, packet switching is used exclusively in mobile networks, where packets are however much smaller in size (below 100 bytes in both UMTS and LTE typically) than ordinary IP packets. Packet losses may frequently occur, which can also be caused by bit errors or burst errors occurring at the physical layer. The usual method to compensate for packet losses is re-transmission (see below). Quality of Service mechanisms are defined as well. The available bandwidth and the probability of transmission losses depend on the number of users which share the same channel, but are also continuously improved by implementing more advanced transmission methods.

If an IP connection is established via UMTS or LTE (which is the case for the bulk of media transmission), the IP packets are chunked into sub-packets of smaller size for the mobile network error control, using a specific protocol stack for the network link layer. Both FEC and ARQ are employed for this purpose. When ARQ is performed between the mobile device and the base station, the additional round-trip delay is extremely short (unless the transmission fails repeatedly due to a bad physical link). On the other hand, it is still necessary to carefully decide in a mobile streaming application which packets are worthwhile to re-transmit, because re-transmission increases the network load, which again decreases the transmission quality etc. This can however be managed efficiently due to the availability of QoS negotiation mechanisms.

Today's mobile devices usually support multiple transmission protocols, and select the best alternative path available in the given environment. For example, 'hot spots' like airports, trains, fairs are equipped by WLAN, WIMAX or successors thereof, which provide cheap transmission and high bandwidth, but are based on a pico-cellular topology. Each base station covers a radius of 10-50 meters only, such that access is extremely limited to one location. New technology will also evolve in the medium-range radio networks. From this background, future generations of mobile networks will integrate all these approaches and additionally support new error-prone modulation, coding and multiplexing methods as wellas advanced physical layer processing to achieve an even higher bandwidth. Another potential for further improvements is established by *adhoc networks*, where any mobile device can take over tasks such as routing towards other mobile devices or towards a base station. These technologies might bring mobile multimedia transmission even further away from traditional centralized broadcast services, becoming more and more de-centralized.

A Vector and Matrix Algebra

Vector and matrix notations are frequently used in this book, as these allow expressing operations applied to groups of samples in a very efficient way. Linear mathematical operations can straightforwardly be expressed by vector and matrix algebra as well. The underlying conventions are summarized in this section.

A vector is a one-dimensional structure of K scalar values. We typically use column vectors, i.e. vertical structures. A $K\text{x}L$ matrix is a two-dimensional structure with L rows and K columns, which can also be written as a collection of L row vectors,

$$\mathbf{A} = \begin{bmatrix} a_{11} & \cdots & a_{1K} \\ \vdots & \ddots & \vdots \\ a_{L1} & \cdots & a_{LK} \end{bmatrix} = \begin{bmatrix} \mathbf{a}_1^{\mathrm{T}} \\ \vdots \\ \mathbf{a}_L^{\mathrm{T}} \end{bmatrix}. \tag{A.1}$$

Higher-dimensional structures are called *tensors*. Matrices and vectors are indicated by bold letters, where in principle the denomination of value types is retained (see Appendix C). For example, **s** and **S** are a vector and a matrix consisting of samples $s(\mathbf{n})$ or $s_{\mathbf{k}}$. The most important rules used in vector and matrix operations are summarized in the remaining part of this section.

The transpose of a vector is

$$\mathbf{a}^{\mathrm{T}} = \begin{bmatrix} a_1 \\ \vdots \\ a_K \end{bmatrix}^{\mathrm{T}} = \begin{bmatrix} a_1 & \cdots & a_K \end{bmatrix}. \tag{A.2}$$

The conjugate $\mathbf{A}^*$ contains the conjugate entries of $\mathbf{A}$. The transpose of a matrix is performed by exchange of rows and columns,

$$\mathbf{A}^{\mathrm{T}} = \begin{bmatrix} a_{11} & \cdots & a_{1K} \\ \vdots & \ddots & \vdots \\ a_{L1} & \cdots & a_{LK} \end{bmatrix}^{\mathrm{T}} = \begin{bmatrix} a_{11} & \cdots & a_{L1} \\ \vdots & \ddots & \vdots \\ a_{1K} & \cdots & a_{LK} \end{bmatrix}. \tag{A.3}$$

Similarly, the Hermitian of a matrix with complex entries uses conjugate entries in the transpose, $\mathbf{A}^{\mathrm{H}}=[\,\mathbf{A}^*]^{\mathrm{T}}$. The inner product (also dot product) of two vectors of same length K is a scalar value

© Springer-Verlag Berlin Heidelberg 2015

J.-R. Ohm, *Multimedia Signal Coding and Transmission*,
Signals and Communication Technology, DOI 10.1007/978-3-662-46691-9

$$\mathbf{a}\cdot\mathbf{b}=\mathbf{a}^{\mathrm{T}}\mathbf{b}=\begin{bmatrix}a_1 & \cdots & a_K\end{bmatrix}\cdot\begin{bmatrix}b_1\\ \vdots\\ b_K\end{bmatrix}=a_1\cdot b_1+a_2\cdot b_2+\ldots+a_K\cdot b_K\,. \tag{A.4}$$

The outer product of two vectors of length K and L is a $K\mathrm{x}L$ matrix

$$\mathbf{a}\mathbf{b}^{\mathrm{T}}=\begin{bmatrix}a_1\\ \vdots\\ a_L\end{bmatrix}\cdot\begin{bmatrix}b_1 & \cdots & b_K\end{bmatrix}=\begin{bmatrix}a_1b_1 & a_1b_2 & \cdots & a_1b_K\\ a_2b_1 & a_2b_2 & & \vdots\\ \vdots & & \ddots & \\ a_Lb_1 & \cdots & & a_Lb_K\end{bmatrix}. \tag{A.5}$$

The product of a vector of length K and a $K\mathrm{x}L$ matrix is a vector of length L

$$\mathbf{A}\mathbf{x}=\begin{bmatrix}a_{11} & \cdots & a_{1K}\\ \vdots & \ddots & \vdots\\ a_{L1} & \cdots & a_{LK}\end{bmatrix}\cdot\begin{bmatrix}x_1\\ \vdots\\ x_K\end{bmatrix}=\begin{bmatrix}a_{11}x_1+\ldots+a_{1K}x_K\\ \vdots\\ a_{L1}x_1+\ldots+a_{LK}x_K\end{bmatrix}. \tag{A.6}$$

The inner matrix product (first matrix is $K\mathrm{x}L$, second matrix $M\mathrm{x}K$) is an $M\mathrm{x}L$ matrix

$$\begin{aligned}\mathbf{AB}&=\begin{bmatrix}a_{11} & \cdots & a_{1K}\\ \vdots & \ddots & \vdots\\ a_{L1} & \cdots & a_{LK}\end{bmatrix}\cdot\begin{bmatrix}b_{11} & \cdots & b_{1M}\\ \vdots & \ddots & \vdots\\ b_{K1} & \cdots & b_{KM}\end{bmatrix}\\ &=\begin{bmatrix}a_{11}b_{11}+\ldots+a_{1K}b_{K1} & \cdots & a_{11}b_{1M}+\ldots+a_{1K}b_{KM}\\ \vdots & \ddots & \vdots\\ a_{L1}b_{11}+\ldots+a_{LK}b_{K1} & \cdots & a_{L1}b_{1M}+\ldots+a_{LK}b_{KM}\end{bmatrix}.\end{aligned} \tag{A.7}$$

If two matrices of equal size are multiplied, one of them has to be transposed. Here, the following relation holds:

$$\mathbf{A}^{\mathrm{T}}\mathbf{B}=\left[\mathbf{B}^{\mathrm{T}}\mathbf{A}\right]^{\mathrm{T}}. \tag{A.8}$$

A matrix generated by element-wise multiplication of two equal-size matrices is the Hadamard product

$$\mathbf{A}\circ\mathbf{B}=\begin{bmatrix}a_{11} & \cdots & a_{1K}\\ \vdots & \ddots & \vdots\\ a_{L1} & \cdots & a_{LK}\end{bmatrix}\circ\begin{bmatrix}b_{11} & \cdots & b_{1K}\\ \vdots & \ddots & \vdots\\ b_{L1} & \cdots & b_{LK}\end{bmatrix}=\begin{bmatrix}a_{11}b_{11} & \cdots & a_{1K}b_{1K}\\ \vdots & \ddots & \vdots\\ a_{L1}b_{L1} & \cdots & a_{LK}b_{LK}\end{bmatrix}. \tag{A.9}$$

The Frobenius product is a scalar value which extends the concept of dot product (A.4) to matrices and tensors,

$$\mathbf{A}:\mathbf{B}=\sum_{l=1}^{L}\sum_{k=1}^{K}a_{lk}b_{lk}=\mathrm{tr}\left(\mathbf{A}^{\mathrm{T}}\mathbf{B}\right)=\mathrm{tr}\left(\mathbf{B}^{\mathrm{T}}\mathbf{A}\right). \tag{A.10}$$

The Kronecker product of two matrices (sizes $K\mathrm{x}L$ and $M\mathrm{x}N$) is performed such that each element of the first matrix is multiplied by each element of the second. The result is a matrix of size $KM\mathrm{x}LN$, which can be partitioned into KL sub-matrices, each of size $M\mathrm{x}N$:

$$\mathbf{A}\otimes\mathbf{B}=\begin{bmatrix} a_{11} & \cdots & a_{1K} \\ \vdots & \ddots & \vdots \\ a_{L1} & \cdots & a_{LK} \end{bmatrix} \otimes \begin{bmatrix} b_{11} & \cdots & b_{1M} \\ \vdots & \ddots & \vdots \\ b_{N1} & \cdots & b_{NM} \end{bmatrix} = \begin{bmatrix} a_{11}\mathbf{B} & \cdots & a_{1K}\mathbf{B} \\ \vdots & \ddots & \vdots \\ a_{L1}\mathbf{B} & \cdots & a_{LK}\mathbf{B} \end{bmatrix}. \tag{A.11}$$

The cross product of two vectors (defined in a 3D coordinate space with orthogonal axes) is a vector that is perpendicular with the plane spanned by the two vectors:

$$\mathbf{a}\times\mathbf{b}=\begin{bmatrix} a_1 \\ a_2 \\ a_3 \end{bmatrix} \times \begin{bmatrix} b_1 \\ b_2 \\ b_3 \end{bmatrix} = \begin{bmatrix} a_2 b_3 - a_3 b_2 \\ a_3 b_1 - a_1 b_3 \\ a_1 b_2 - a_2 b_1 \end{bmatrix} = \begin{bmatrix} 0 & -a_3 & a_2 \\ a_3 & 0 & -a_1 \\ -a_2 & a_1 & 0 \end{bmatrix} \begin{bmatrix} b_1 \\ b_2 \\ b_3 \end{bmatrix} = -\mathbf{b}\times\mathbf{a}. \tag{A.12}$$

The determinant of a square matrix of size KxK is the sum over $K!$ possible permutations $(\alpha,\beta,..,\omega)$ of numbers $(1,2,..,K)$, where k is the number of inversions within a permutation (sequence $a_{1,\alpha}a_{1,\beta}$, $\alpha>\beta$):

$$\det(\mathbf{A})=\begin{vmatrix} a_{11} & a_{12} & \cdots & a_{1K} \\ a_{21} & a_{22} & & a_{2K} \\ \vdots & \vdots & \ddots & \vdots \\ a_{K1} & a_{K2} & \cdots & a_{KK} \end{vmatrix} = \sum_{(\alpha,\beta,..,\omega)} (-1)^k a_{1\alpha} a_{2\beta} \ldots a_{K\omega}\,. \tag{A.13}$$

This can better be interpreted as computing the sum of 'products over diagonals' within the periodically extended matrix. All products of diagonals oriented in parallel with the primary (trace) axis contribute by positive sign, all secondary-oriented diagonals (top right to bottom left and parallel) by negative sign; e.g. for cases K=2 and K=3:

$$\det(\mathbf{A})=\begin{vmatrix} a_{11} & a_{12} \\ a_{21} & a_{22} \end{vmatrix} = a_{11}a_{22} - a_{12}a_{21}$$

$$\det(\mathbf{A})=\begin{vmatrix} a_{11} & a_{12} & a_{13} \\ a_{21} & a_{22} & a_{23} \\ a_{31} & a_{32} & a_{33} \end{vmatrix} = a_{11}a_{22}a_{33} + a_{12}a_{23}a_{31} + a_{13}a_{21}a_{32} \tag{A.14}$$

$$-a_{11}a_{23}a_{32} - a_{12}a_{21}a_{33} - a_{13}a_{22}a_{31}.$$

Furthermore, the expression for the absolute value of the determinant $|\mathbf{A}|=|\det(\mathbf{A})|$ is used for various purposes. Inverting a matrix, $\mathbf{A}^{-1}$ is useful in many places, e.g. to solve linear equation systems $\mathbf{A}\mathbf{x}=\mathbf{b} \Rightarrow \mathbf{x}=\mathbf{A}^{-1}\mathbf{b}$. Additional conditions are $[\mathbf{A}^{-1}]^{-1}=\mathbf{A}$ and $\mathbf{A}^{-1}\mathbf{A}=\mathbf{A}\mathbf{A}^{-1}=\mathbf{I}$, i.e. the matrix multiplied by its inverse results in the *identity matrix*

$$\mathbf{I}=\begin{bmatrix}1 & 0 & \cdots & & 0\\ 0 & 1 & 0 & & \vdots\\ \vdots & 0 & 1 & \ddots & \\ & & \ddots & \ddots & 0\\ 0 & \cdots & & 0 & 1\end{bmatrix}. \tag{A.15}$$

To be invertible, the matrix **A** has to be of square shape. If the determinant of the matrix and all determinants of sub-matrices are unequal to zero (i.e. if the matrix has full rank), it is invertible, otherwise it is called to be singular. The inversion of matrices of sizes 2x2 and 3x3 is performed as follows:

$$\mathbf{A}^{-1}=\begin{bmatrix}a_{11} & a_{12}\\ a_{21} & a_{22}\end{bmatrix}^{-1}=\frac{1}{\det(\mathbf{A})}\cdot\begin{bmatrix}a_{22} & -a_{12}\\ -a_{21} & a_{11}\end{bmatrix} \tag{A.16}$$

$$\begin{aligned}\mathbf{A}^{-1}&=\begin{bmatrix}a_{11} & a_{12} & a_{13}\\ a_{21} & a_{22} & a_{23}\\ a_{31} & a_{32} & a_{33}\end{bmatrix}^{-1}\\ &=\frac{1}{\det(\mathbf{A})}\cdot\begin{bmatrix}a_{22}a_{33}-a_{23}a_{32} & a_{13}a_{32}-a_{12}a_{33} & a_{12}a_{23}-a_{13}a_{22}\\ a_{31}a_{23}-a_{21}a_{33} & a_{11}a_{33}-a_{13}a_{31} & a_{13}a_{21}-a_{11}a_{23}\\ a_{21}a_{32}-a_{31}a_{22} & a_{12}a_{31}-a_{11}a_{32} & a_{11}a_{22}-a_{12}a_{21}\end{bmatrix}.\end{aligned} \tag{A.17}$$

Inversion of larger size matrices can be reduced to recursive inversion of sub-matrices by the following formula, where any $\mathbf{A}_{11}$ and $\mathbf{A}_{22}$ again shall be square matrices:

$$\begin{aligned}&\mathbf{A}=\begin{bmatrix}\mathbf{A}_{11} & \mathbf{A}_{12}\\ \mathbf{A}_{21} & \mathbf{A}_{22}\end{bmatrix}\quad\Rightarrow\quad\mathbf{A}^{-1}\\ &=\begin{bmatrix}\left[\mathbf{A}_{11}-\mathbf{A}_{12}\mathbf{A}_{22}^{-1}\mathbf{A}_{21}\right]^{-1} & -\mathbf{A}_{11}^{-1}\mathbf{A}_{12}\left[\mathbf{A}_{22}-\mathbf{A}_{21}\mathbf{A}_{11}^{-1}\mathbf{A}_{12}\right]^{-1}\\ -\mathbf{A}_{22}^{-1}\mathbf{A}_{21}\left[\mathbf{A}_{11}-\mathbf{A}_{12}\mathbf{A}_{22}^{-1}\mathbf{A}_{21}\right]^{-1} & \left[\mathbf{A}_{22}-\mathbf{A}_{21}\mathbf{A}_{11}^{-1}\mathbf{A}_{12}\right]^{-1}\end{bmatrix}.\end{aligned} \tag{A.18}$$

Further,

$$[\mathbf{AB}]^{-1}=\mathbf{A}^{-1}\mathbf{B}^{-1}\quad;\quad[c\mathbf{A}]^{-1}=\frac{1}{c}\mathbf{A}^{-1}. \tag{A.19}$$

The eigenvector of a square matrix is a vector which, when multiplied by the matrix, gives a scaled version of itself. The scaling factor is the associated eigenvalue λ. A non-singular (full rank) KxK matrix has K different eigenvectors $\mathbf{\Phi}_k$ and K eigenvalues λ_k:

$$\mathbf{A}\cdot\mathbf{\Phi}_k=\lambda_k\cdot\mathbf{\Phi}_k,\ 1\le k\le K. \tag{A.20}$$

In principle, the eigenvalues can be computed by solving the linear equation system $[\mathbf{A}-\lambda_k\mathbf{I}]\cdot\mathbf{\Phi}_k=0$, for which solutions must exist provided that

$\det\left[\mathbf{A}-\lambda_k\mathbf{I}\right]=0$. Computing the determinant gives the coefficients α_i of the *characteristic polynomial* $\alpha_K\lambda_k{}^K+\alpha_{K-1}\lambda_k{}^{K-1}+\ldots+\alpha_1\lambda_k+\alpha_0=0$, the solutions of which are the K eigenvalues λ_k. Substituting those into (A.20) again gives the conditions for obtaining the eigenvectors, where it is further necessary to impose a condition about their norm. For our purpose, it is reasonable to use the orthonormality constraint, $\mathbf{\Phi}_k{}^{\mathrm{T}}\mathbf{\Phi}_k=1$. The eigenvectors generally fulfill the orthogonality principle (A.24). The trace of a $K\mathrm{x}K$ matrix is the sum of the elements along its primary diagonal axis:

$$\mathrm{tr}\left[\mathbf{A}\right]=\sum_{k=1}^{K}a_{k,k}\,. \tag{A.21}$$

Additionally,

$$\mathrm{tr}\left[\mathbf{AB}\right]=\mathrm{tr}\left[\mathbf{BA}\right]\,;\quad \mathrm{tr}\left[\mathbf{A}\times\mathbf{B}\right]=\mathrm{tr}\left[\mathbf{A}\right]\cdot\mathrm{tr}\left[\mathbf{B}\right]. \tag{A.22}$$

The *Euclidean norm* of a vector is its scalar product (A.4) with itself, subsequently square-rooted

$$\sqrt{\mathbf{a}^{\mathrm{T}}\mathbf{a}}=\sqrt{\begin{bmatrix}a_1 & \cdots & a_K\end{bmatrix}\begin{bmatrix}a_1\\ \vdots\\ a_K\end{bmatrix}}=\sqrt{\sum_{k=1}^{K}a_k{}^2}\,. \tag{A.23}$$

Orthogonality means that the scalar product of any two different vectors from a set is zero. A stronger criterion is *orthonormality*, where additionally, the Euclidean norm of all vectors shall be one (unity),

$$\mathbf{t}_i{}^{\mathrm{H}}\mathbf{t}_j=\mathbf{t}_j{}^{\mathrm{H}}\mathbf{t}_i=0\ \text{ for } i\neq j\,;\ \mathbf{t}_i{}^{\mathrm{H}}\mathbf{t}_i=1\ \text{ for all } i. \tag{A.24}$$

When a set of orthonormal vectors is interpreted as rows (or columns) of a matrix $\mathbf{T}$, (A.24) implies that $\mathbf{T}^{\mathrm{H}}\mathbf{T}=\mathbf{I}$ (or $\mathbf{T}\mathbf{T}^{\mathrm{H}}=\mathbf{I}$). In case of square orthonormal matrices (length of vectors identical with the number of vectors in the set), furthermore $\mathbf{T}^{-1}=\mathbf{T}^{\mathrm{H}}$. Orthogonality as a relationship between vectors is furthermore a special case of *bi-orthogonality*. Assume that a set of linear independent (but not necessarily orthogonal) vectors $\mathbf{t}_i$ establishes a *basis system*; these are arranged as rows (or columns) of a matrix $\mathbf{T}$. Then, a *dual basis* $\tilde{\mathbf{T}}$ shall exist with the property that its row vectors $\tilde{\mathbf{t}}_j$ are orthogonal with $\mathbf{t}_i$ when $i\neq j$. This gives

$$\mathbf{T}\tilde{\mathbf{T}}^{\mathrm{T}}=\mathbf{I}\Leftrightarrow \mathbf{t}_i{}^{\mathrm{T}}\tilde{\mathbf{t}}_j=\tilde{\mathbf{t}}_j{}^{\mathrm{T}}\mathbf{t}_i=0\ \text{ for } i\neq j\,;\ \mathbf{t}_i{}^{\mathrm{T}}\tilde{\mathbf{t}}_i=\tilde{\mathbf{t}}_i{}^{\mathrm{T}}\mathbf{t}_i=1\ \text{ for all } i. \tag{A.25}$$

The orthonormal basis system is a special case thereof, where

$$\tilde{\mathbf{t}}_i=\mathbf{t}_i{}^*\Leftrightarrow \tilde{\mathbf{T}}=\mathbf{T}^*\Rightarrow \mathbf{T}^{-1}=\mathbf{T}^{\mathrm{H}}. \tag{A.26}$$

In matrix notation, it is also possible to express equation systems $\mathbf{T}\mathbf{x}=\mathbf{c}$ where the number of unknowns in $\mathbf{x}$ is not equal to the number of equations. Assume that K is the number of elements in $\mathbf{x}$, L is the dimension of $\mathbf{c}$ (or the number of equations), such that $\mathbf{T}$ is of size $K\mathrm{x}L$, non-square and not invertible. A solution to this problem has to impose additional conditions, e.g. the minimum least-squares fit:

$$\|\mathbf{e}\|^2 = \|\mathbf{c}-\mathbf{Tx}\|^2 = [\mathbf{c}-\mathbf{Tx}]^{\mathrm{H}}[\mathbf{c}-\mathbf{Tx}] \overset{!}{=} \min . \tag{A.27}$$

A solution, being optimum in the case that the conditions in the equation system are not disturbed by noise, is obtained by the *pseudo inverse matrix* $\mathbf{T}^{\mathrm{P}}$:

$$\mathbf{x} = \mathbf{T}^{\mathrm{P}}\mathbf{c} , \tag{A.28}$$

where the following cases are distinguished:

$$\begin{aligned} K<L &: \mathbf{T}^{\mathrm{P}} = (\mathbf{T}^{\mathrm{H}}\mathbf{T})^{-1}\mathbf{T}^{\mathrm{H}} \quad ; \quad \mathbf{T}^{\mathrm{P}}\mathbf{T} = \mathbf{I}^{(K\mathrm{x}K)} \\ K=L &: \mathbf{T}^{\mathrm{P}} = \mathbf{T}^{-1} \quad ; \quad \mathbf{T}^{\mathrm{P}}\mathbf{T} = \mathbf{T}\mathbf{T}^{\mathrm{P}} = \mathbf{I}^{(K\mathrm{x}K)} \\ K>L &: \mathbf{T}^{\mathrm{P}} = \mathbf{T}^{\mathrm{H}}(\mathbf{T}\mathbf{T}^{\mathrm{H}})^{-1} \quad ; \quad \mathbf{T}\mathbf{T}^{\mathrm{P}} = \mathbf{I}^{(L\mathrm{x}L)} . \end{aligned} \tag{A.29}$$

The pseudo inverse matrix has size LxK, and for the case of $K=L$ and full rank of $\mathbf{T}$ is identical to the conventional inverse of $\mathbf{T}$, where also the solution of the equation system is unique. For $K<L$, the problem is *over-determined*, which means that more conditions exist than unknowns shall be computed. It is typically not possible in this case to force the right hand side of (A.27) into the value zero, because the conditions may be contradicting. For $K>L$, the equation system is *under-determined*, which means that less conditions than unknowns exist; then, pseudo inversion would find at least one set of values $\mathbf{x}$ that fulfills (A.27) perfectly, but different sets will likely exist as alternative solutions. An example for this is the case when $\mathbf{x}$ is a signal vector, and $\mathbf{c}$ a corresponding vector of transform coefficients with less values; here, it is no possible to recover arbitrary signal vectors from $\mathbf{c}$; moreover, the recovered one will be the one that has best correlation with the incomplete set of reconstruction basis vectors in $\mathbf{T}^{\mathrm{P}}$. In the latter case when furthermore $\mathbf{T}$ represents an orthogonal basis, $\mathbf{TT}^{\mathrm{H}}=\mathbf{I}$ and therefore $\mathbf{T}^{\mathrm{P}}=\mathbf{T}^{\mathrm{H}}$. The most compact representation (least squared reconstruction error from a finite number of elements in $\mathbf{c}$) is then given by the eigenvector basis $\mathbf{\Phi}$ from (A.20), provided that $\mathbf{A}$ expresses the corresponding expectation (i.e. being the covariance matrix) of $\mathbf{x}$.

For non-square matrices or square matrices with non-full rank, *singular value decomposition* (SVD) can be applied with a similar purpose as eigenvector decomposition. $\mathbf{A}$ is an KxL matrix with rank $R \le \min(K,L)$. Then, a KxK-Matrix $\mathbf{\Psi}$ and an LxL-Matrix $\mathbf{\Phi}$ are defined such that

$$\mathbf{\Phi}^{\mathrm{T}}\mathbf{A}\mathbf{\Psi} = \mathbf{\Lambda}^{(1/2)} = \underbrace{\left[\begin{array}{cccc|c} \sqrt{\lambda(1)} & 0 & \cdots & 0 & \vdots \\ 0 & \ddots & \ddots & \vdots & \mathbf{0} \\ \vdots & \ddots & \ddots & 0 & \vdots \\ 0 & \cdots & 0 & \sqrt{\lambda(R)} & \\ \cdots & \mathbf{0} & \cdots & & \mathbf{0} \end{array}\right]}_{R \quad\quad K-R} \begin{array}{l} \left.\vphantom{\begin{array}{c}1\\1\\1\\1\end{array}}\right\} R \\ \} \; L-R \end{array} . \tag{A.30}$$

The elements in $\mathbf{\Lambda}^{(1/2)}$ are the R *singular values* $\lambda^{1/2}(r)$ of $\mathbf{H}$. These are square roots of the R non-zero eigenvalues, which are identical from the LxL matrix $\mathbf{AA}^{\mathrm{T}}$ and the MxM-Matrix $\mathbf{A}^{\mathrm{T}}\mathbf{A}$ (for min(K,L)>R, the remaining eigenvalues of $\mathbf{AA}^{\mathrm{T}}$ or $\mathbf{A}^{\mathrm{T}}\mathbf{A}$ are zero). The columns of $\mathbf{\Phi}$ are the eigenvectors $\boldsymbol{\phi}_r$ of $\mathbf{AA}^{\mathrm{T}}$, the columns of $\mathbf{\Psi}$ the eigenvectors $\boldsymbol{\psi}_r$ of $\mathbf{A}^{\mathrm{T}}\mathbf{A}$, where in correspondence with (A.20) the following conditions apply:

$$\mathbf{\Psi}^{\mathrm{T}}\left[\mathbf{A}^{\mathrm{T}}\mathbf{A}\right]\mathbf{\Psi} = \mathbf{\Lambda}^{(K)} \text{ and } \mathbf{\Phi}^{\mathrm{T}}\left[\mathbf{AA}^{\mathrm{T}}\right]\mathbf{\Phi} = \mathbf{\Lambda}^{(L)}. \tag{A.31}$$

By multiplying both sides of (A.30) with inverses of $\mathbf{\Phi}$ and $\mathbf{\Psi}$, it is possible to express $\mathbf{A}$ as a weighted superposition from R orthogonal component matrices,

$$\mathbf{A} = \mathbf{\Phi}\mathbf{\Lambda}^{(1/2)}\mathbf{\Psi}^{\mathrm{T}} = \sum_{r=1}^{R}\lambda^{1/2}(r)\cdot\boldsymbol{\phi}_r\boldsymbol{\psi}_r^{\mathrm{T}}. \tag{A.32}$$

The *generalized inverse* $\mathbf{A}^{\mathrm{G}}$ of (A.32) is the corresponding weighted superposition of inverted component matrices, which is for the full-rank case R=min(K,L) identical to the pseudo inverse (A.29):

$$\mathbf{A}^{\mathrm{G}} = \mathbf{\Psi}\mathbf{\Lambda}^{(-1/2)}\mathbf{\Phi}^{\mathrm{T}} = \sum_{r=1}^{R}\lambda^{-1/2}(r)\cdot\boldsymbol{\phi}_r\boldsymbol{\psi}_r^{\mathrm{T}}. \tag{A.33}$$

SVD allows solutions selecting or omitting specific singular values and corresponding components, e.g. for the purpose of more compact or less noise sensitive approximation.

B Symbols and Variables

Unique symbols are used in equations whenever possible. As far as duplicate symbols are used, usually no ambiguity will occur in a given context.

Table B.1. Index variables and signal sizes over different dimensions in signal and spectral domains

	horizontal	vertical	depth (distance)	temporal
World coordinates **W**	W_1	W_2	W_3	t *)
Speed **V** in 3D world space	V_1	V_2	V_3	-
Size in 3D world space	S_1	S_2	S_3	-
Continuous coordinate **t**	t_1	t_2	-	t_3
Speed **u** in image plane	u_1	u_2	-	-
discrete coordinate **n**	n_1	n_2	-	n_3
Sampling interval	T_1	T_2	-	T_3
Sub-sampled interval	m_1	m_2	-	m_3
Size of finite discrete signal	N_1	N_2	-	N_3
Size of a sub-sampled signal or block	M_1	M_2	-	M_3
Frequency	f_1	f_2	-	f_3
z transform	z_1	z_2	-	z_3
Discrete frequency	k_1	k_2	-	k_3
Number of discrete spectral coefficients	U_1	U_1	-	U_3
Index of filter coefficient, filter order	p_1, q_1; P_1,Q_1	p_2, q_2; P_2,Q_2	-	p_3, q_3; P_3,Q_3
Index of impulse response, coordinate shift	k_1	k_2	-	k_3

*) also used for 1D signals, typically with temporal dependency

Multimedia signals are often *multi-dimensional* (e.g. row, column and temporal directions in video). Table B.1 lists the *running variables*, *sizes* or *limits* and spectral denotations for each of the up to four dimensions. Table B.2 lists the *signal*

© Springer-Verlag Berlin Heidelberg 2015

J.-R. Ohm, *Multimedia Signal Coding and Transmission*,
Signals and Communication Technology, DOI 10.1007/978-3-662-46691-9

types and their representation in the signal domain and in the spectral domain, where for the vector index variables (**n**, **z**, **f** etc.) the respective index variables from the Table B.1 have to be supplemented, depending on the number of dimensions.

Table B.2. Signal types in signal and spectral domains

	Signal domain	Spectral domain
Original signal, continuous	$s(\mathbf{t})$	$S(\mathbf{f})$
Original signal, sampled	$s(\mathbf{n})$	$S_{(\delta)}(\mathbf{f})$, $S(\mathbf{z})$
Signal estimate	$\hat{s}(\mathbf{n})$	$\hat{S}_{(\delta)}(\mathbf{f})$
Signal reconstruction	$\tilde{s}(\mathbf{n})$	$\tilde{S}_{(\delta)}(\mathbf{f})$
Binary signal	$b(\mathbf{n})$	--
Noise signal	$v(\mathbf{n})$	$V_a(\mathbf{f})$
Filter output	$g(\mathbf{n})$	$G_{(\delta)}(\mathbf{f})$, $G(\mathbf{z})$
Prediction error residual	$e(\mathbf{n})$, $r(\mathbf{n})$	$E_{(\delta)}(\mathbf{f})$, $R_{(\delta)}(\mathbf{f})$
Window function	$w(\mathbf{n})$	$W(\mathbf{f})$
Quantization error signal	$q(\mathbf{n})$	$Q_{(\delta)}(\mathbf{f})$
Interference signal	$k(\mathbf{n})$	$K_{(\delta)}(\mathbf{f})$
Filter coefficient (non-	$a(\mathbf{p})$	$A(\mathbf{z})$
recursive, recursive)	$b(\mathbf{p})$	$B(\mathbf{z})$
Filter impulse response	$h(\mathbf{n})$	$H_{(\delta)}(\mathbf{f})$, $H(\mathbf{z})$)
Transform coefficient	$c_k(\mathbf{m})$	$C_k(\mathbf{z})$

ϕ_{ss}, ϕ_{gg}, ...	Power spectra of s, g, ...
φ_{ss}, φ_{gg}, ...	Autocorrelation functions of s, g, ...
ϕ_{sg}, φ_{sg}	Cross power spectra, cross correlation
μ_{ss}, μ_{sg}	Auto covariance, cross covariance
ρ_{ss}, ρ_{sg}	Correlation coefficient (normalized)
$\delta(\cdot)$	Dirac impulse, unit impulse (Delta function)
$\varepsilon(\cdot)$	Unit step function
$\gamma(\cdot)$	Geometric transformation mapping
λ_k	Eigenvalue
$\boldsymbol{\phi}_k$	Eigenvector
$\mathbf{t}_k$	Basis vector
$\mathbf{T}$	Transform matrix
$\mathbf{H}$, $\mathbf{G}$	Filter matrix
$\mathbf{I}$	Identity matrix
Δ	Quantizer step size, distance function
$I(S_j)$	Self information of discrete state S_j
$p_s(x)$	Probability density function of process s over random variable x
$P_s(x)$	Cumulative distribution function

$\Pr(S_j)$, $\Pr(j)$	Probability (of discrete state S_j)
$f(\cdot)$	Amplitude mapping function
$p_{sg}(x,y)$	Joint probability density function of processes s and g over random variables x and y
$P_{sg}(x,y)$	Joint cumulative probability
$\Pr(S_i,S_j)$	Joint Probability (of discrete states S_i and S_j)
$p_{s\|g}(x\|y)$	Conditional probability density function, process s conditioned by g
$p_{\mathbf{s}}(\mathbf{x})$	Vector probability density of vector process $\mathbf{s}$ over random vector $\mathbf{x}$
σ_s^2,σ_g^2	Variance of s, g, ...
m_s, m_g	Mean value of s, g, ...
Q_s, Q_g	Quadratic mean value (energy, power) of s, g, ...
$m_s^{(P)}$, $c_s^{(P)}$	P^{th} order moment, central moment
$\mathcal{E}\{\cdot\}$	Expected value
$L\{\cdot\}$	System transfer function
$\mathrm{Im}\{\cdot\}$	Imaginary part of function
$\mathrm{Re}\{\cdot\}$	Real part of function
$H(S)$	Entropy over set S
$H(S_1,S_2)$	Joint entropy over sets S_1,S_2
$H(S_1\|S_2)$	Conditional entropy over sets S_1,S_2
$I(S_1;S_2)$	Mutual information over sets S_1,S_2
$\mathrm{d}(\cdot,\cdot)$	Difference function, distance function, distortion
i,j	Code symbol index
r	Iteration number
A, $A_{\mathrm{max/min}}$	Amplitude value, maximum and minimum amplitude
B	Number of bits
F	Focal length
I, J	Number of codewords, code symbols or letters in discrete alphabet
K	Vector length
L	Constraint length
R	Number of iterations
T	Number of levels
$\mathbf{\Phi}_{ss}$, $\boldsymbol{\phi}_{ss}$	Autocorrelation matrix, vector
$\mathbf{C}_{ss}$, $\mathbf{c}_{ss}$	Autocovariance matrix, vector
$\mathbf{s}$, $\mathbf{g}$, ...	Vector from an ordered sequence of scalars s, g, ...
$\mathbf{S}$, $\mathbf{G}$, ...	Matrix from an ordered field of scalars s, g, ...
$\mathbf{A}$, $\mathbf{T}$, $\mathbf{F}$	Coordinate mapping or sampling matrices in signal and frequency domains
$\mathbf{R}$, $\boldsymbol{\tau}$	Rotation matrix, translation vector
HSV	Hue, saturation, value color components
RGB	Red, green, blue (primary colors)
YC_bC_r	Luminance and color difference components
$\mathcal{A}$, $\mathcal{B}$	Discrete alphabets consisting of letters $\mathcal{A}_i$, $\mathcal{B}_i$
$\mathcal{C}$	Code, codebook
$\mathcal{F}\{\cdot\}$	Fourier transform

$\mathcal{C}t$	Context
$\mathcal{S}$	Set, state
$\mathcal{V}$	Scale space
$\mathcal{W}$	Wavelet space
$\mathcal{R}^K$	Vector space of K dimensions
ε	Step size factor, small incremental value
η	Efficiency parameter
κ	Number of dimensions in signal
λ	Wavelength, Lagrangian multiplier
ϕ,θ	Angles (e.g. of rotation)
τ	Period, shift (continuous)
$\varphi(\tau)$	Scaling function, interpolation function
$\psi(\tau)$	Wavelet function
F_i	Frequency of static value (e.g. for sampling, modulation)
T_i	Signal-domain unit of static value (e.g. time or spatial distance)
Θ	Threshold, scaling factor
$\mathbf{\Lambda}$	Eigenvalue matrix, shape of an area
$\mathbf{\Pi}$	Search range, shape of filter mask, projection profile
Φ, Ψ	Transform matrices of optimized linear transforms

C Glossary and Acronyms

1D	One Dimensional
1080i	1080 (lines) interlaced
1080p	1080 (lines) progressive
2D	Two Dimensional
3D	Three Dimensional
4K	4000 samples per line (approx.)
8K	8000 samples per line (approx.)
720p	720 (lines) progressive
AAC	Advanced Audio Coding
AC	Alternating Component, or Arithmetic Coding
ACF	Auto Correlation Function
ADPCM	Adaptive DPCM
ALF	Adaptive Loop Filter
ALS	Audio Lossless Coding
AMVP	Advanced Motion Vector Prediction
ANN	Artificial Neural Network
AR	Autoregressive (Model)
ARQ	Automatic Repeat Request
AVC	Advanced Video Coding
BCC	Binaural Cue Coding
BCH	Bose-Chaudhury-Hocquengem (code)
BDPSNR	Bjøntegaard Delta PSNR
BSAC	Bit Slice Arithmetic Coding
BT	Broadcast Technology (series of ITU-R standards)
CABAC	Context Adaptive Binary Arithmetic Coding
CAC	Call Admission Control
CAE	Context-dependent Arithmetic Coding
CAR	Conditional AR
CAVLC	Context Adaptive VLC
CB	Code book, or Coding Block
CBP	Coded Block Pattern
CBF	Coded Block Flag
CBR	Constant Bit Rate
CCD	Charge Coupled Device
CCF	Cross Correlation Function
CD	Compact Disc
CDF	Cumulative Distribution Function
CELP	Code Excited Linear Prediction
CG	Coding gain
CGI	Control Grid Interpolation
CGS	Coarse Granularity Scalability
CIE	Commission International d'Eclairage
CIF	Common Intermediate Format
CMOS	Complementary Metal Oxide Silicon
CPB	Coded Picture Buffer
CPC	Channel Prediction Coefficient
CQE	Continuous Quality Evaluation
CRA	Conditional Random Access
CRC	Cyclic Redundancy Check
CRT	Cathode Ray Tube
CTB	Coding Tree Block
CTU	Coding Tree Unit (in HEVC)
CU	Coding Unit (in HEVC)
DAB	Digital Audio Broadcast
DASH	Dynamic Adaptive Streaming over HTTP
DC	Direct (flat) Component
DCP	Disparity Compensated Prediction
DCT	Discrete Cosine Transform
DEC	Decoder
DFT	Discrete Fourier Transform
DM	Delta Modulation, or Direct Mode
DMVD	Decoder-side Motion Vector Derivation
DPB	Decoded Picture Buffer
DPCM	Differential PCM
DRF	Distortion Rate Function
DSC	Distributed Source Coding
DSCQS	Double Stimulus Constant Quality Scale
DSIS	Double Stimulus Impairment Scale
DSL	Digital Subscriber Loop
DST	Discrete Sine Transform
DTS	Decoding Time Stamp
DV	Disparity Vector

© Springer-Verlag Berlin Heidelberg 2015

J.-R. Ohm, *Multimedia Signal Coding and Transmission*,
Signals and Communication Technology, DOI 10.1007/978-3-662-46691-9

DVB	Digital Video Broadcast
DVD	Digital Versatile Disc
DWT	Discrete Wavelet Transform
EBCOT	Embedded Block Coding with Optimum Truncation
EC	Entropy Coding
ECQ	Entropy Constrained Quantization
EDGE	Enhanced Data Rates for Global Evolution
EG	Exponential Golomb (code)
ENC	Encoder
EOB	End of Block
EPG	Electronic Program Guide
ES	Elementary Stream
ETSI	European Telecommunication Standardization Institute
EZBC	Embedded Zero-tree Block Coding
EZW	Embedded Zero-tree Wavelet Coding
FEC	Forward Error Correction
FFT	Fast Fourier Transform
FGS	Fine Granularity Scalability
FIR	Finite Impulse Response (filter)
FLC	Fixed Length Coding
FM	Frequency Modulation
FSVQ	Finite State VQ
GGD	Generalized Gaussian distribution
GLA	Generalized Lloyd Algorithm
GMC	Global MC
GMRF	Gauss Markov Random Field
GOB	Group of Blocks
GPRS	General Packet Radio Service
GOP	Group of Pictures
GR	Golomb Rice (code)
GSVQ	Gain Shape VQ
HDR	High Dynamic Range
HD(TV)	High Definition (TV)
HEVC	High Efficiency Video Coding
HRD	Hypothetical Reference Decoder
HT	Haar Transform
HTTP	Hyper Text Transfer Protocol
HVS	Human Visual System
IBC	Intra Block Copy
IBMFF	ISO Base Media File Format
IC	Illumination Compensation
ICC	Inter Channel Correlation
ICLD	Inter Channel Level Difference
ICTD	Inter Channel Time Difference
IDCT	Inverse DCT
IDFT	Inverse DFT
IDR	Instantaneous Decoder Refresh
IDWT	Inverse DFT
IID	Independent identically distributed
IIR	Infinite Impulse Response (filter)
IMC	Inverse MC
IP	Internet Protocol
ISO/IEC	International Standardization Organisation / International Electrotechnical Commission
ITU-R	International Telecommunication Union – Radiocommunication Sector
ITU-T	International Telecommunication Union - Telecommunication Sector
JCT-VC	Joint Collaborative Team on Video Coding
JCT-3V	Joint Collaborative Team on 3D Video Extensions Development
JVT	Joint Video Team
KLT	Karhunen Loève Transform
LCD	Liquid Crystal Display
LDPC	Low Density Parity Check (code)
LLS	Lossy to lossless
LMS	Least Mean Square
LOT	Lapped Orthogonal Transform
LPB	Less probable Bin
LPC	Linear Predictive Coding
LS	Lossless
LSB	Least Significant Bit
LSI	Linear Shift Invariant (system)
LSP	Line Spectrum Pair
LTE	Long Term Evolution (of UMTS, a.k.a. 4G)
LTI	Linear Time Invariant (system)
LVQ	Lattice Vector Quantization
LZ(W)	Lempel-Ziv(-Welsh) coding
MA	Moving Average
MAD	Minimum Absolute Difference
MB	Macro Block
MBAFF	MB Adaptive Frame/Field (coding)
MC	Motion Compensation
MCP	MC Prediction
MCTF	MC Temporal Filtering
MDC	Multiple Description Coding
MDCT	Modified DCT
MDDT	Mode Dependent Directional Transform
ME	Motion Estimation
MELP	Mixed Excitation Linear Prediction
MGS	Medium Granularity Scalability
MIDI	Musical Instruments Digital Interface
MIMO	Multiple Input Multiple Output (transmission)
MMCO	Memory Management Control Operation
MOS	Mean Opinion Score
MPB	More Probable Bin

MPE	Multi Pulse Excitation
MPEG	Moving Picture Experts Group
MRF	Markov Random Field
M/S	Mid/Side coding
MSB	Most Significant Bit
MSE	Mean Square Error
MV	Motion Vector
MVC	Multi-view Video Coding
NAL	Network Abstraction Layer
NTSC	National Television Standards Committee
OFDM	Orthogonal Frequency Division Multiplex
ORDF	Operational Rate Distortion Function
OBMC	Overlapping Block MC
ODWT	Over-complete DWT
PAFF	Picture Adaptive Frame/Field (coding)
PAL	Phase Alternating Line (analog TV format)
PARCOR	Partial Correlation (coefficients)
PB	Prediction Block
PCM	Pulse Code Modulation
PDF	Probability Density Function
PDS	Power Density Spectrum
PEAQ	Perceptual Audio Quality
PES	Packetized Elementary Stream
PESQ	Perceptual Speech Quality
PMF	Probability Mass Function
PNS	Perceptual Noise Substitution
POC	Picture Order Count
PRF	Perfect Reconstruction Filter
PSC	Parametric Stereo Coding
PSF	Point Spread Function
PSK	Phase Shift Keying
PSNR	Peak Signal to Noise Ratio
PTS	Presentation Time Stamp
PU	Prediction Unit (in HEVC)
Q	Quantizer
QAM	Quadrature Amplitude Modulation
QCIF	Quarter CIF
QMF	Quadrature Mirror Filter
QoS	Quality of Service
QPSK	Quadrature PSK
QT	Quad tree
QVGA	Quarter VGA
RAC	Relative Address Coding
RAP	Random Access Picture
RC	Rate control
RDF	Rate Distortion Function
RDO	Rate Distortion Optimization
RDOQ	RD Optimized Quantization
READ	Relative Address Element Designate
RFC	Request for Comments
RLC	Run Length Coding
RPE	Regular Pulse Excitation
RPL	Reference picture list
RPM	Reference picture memory
RQT	Residual Quad-Tree (in HEVC)
RS	Reed-Solomon (code)
RSVP	Reservation Protocol
RTC	Real Time Communication
RTCP	Real Time Control Protocol
RTP	Real Time Transfer Protocol
RVLC	Reversible VLC
SAC	Scalable Audio Coding
SAD	Sum of Absolute Differences
SAO	Signal Adaptive Offset
SAOC	Spatial Audio Object Coding
SAOL	Structured Audio Orchestra Language
SATD	Sum of Absolute Transform-domain Differences
SBC	Subband Coding
SBR	Spectral Band Replication
SC	Stimulus comparison
SD(TV)	Standard Definition TV
SECAM	Séquentiel couleur à mémoire (analog TV format)
SEI	Supplemental Enhancement Information
SFM	Spectral Flatness Measure
SMPTE	Society of Motion Picture and Television Engineers
SNR	Signal to Noise Ratio
SNRSEG	Segmental SNR
SPIHT	Set Partitioning in Hierarchical Trees
SQ	Scalar Quantization
SS	Single stimulus
SSIM	Structure Similarity Measure
SSR	Scalable Sample Rate
STFT	Short Time Fourier Transform
SVC	Scalable Video Coding
SVD	Singular Value Decomposition
TB	Transform Block
TC	Transform Coding
TCP	Transport Control Protocol
TCQ	Trellis Coded Quantization
TDAC	Time Domain Aliasing Cancellation
TES	Temporal Envelope Shaping
TMVP	Temporal Motion Vector Prediction
TNS	Temporal Noise Shaping
TS	Transport Stream
TU	Transform Unit (in HEVC)
TV	Television
UDP	Unified Datagram Protocol
UEP	Unequal Error Protection

UHD(TV) Ultra High Definition (TV)
UMTS Universal Mobile Telecommunication System
UVLC Universal VLC
VBR Variable Bit Rate
VBST Variable Block Size Transform
VCEG Video Coding Experts Group
VCL Video Coding Layer
VGA Video Graphics Array
VLC Variable Length Coding
VQ Vector Quantization
VQM Video Quality Measure
VSELP Vector Sum Excited Linear Prediction
VSP View Synthesis Prediction
VUI Video Usability Information
WHT Walsh Hadamard Transform
WLAN Wireless LAN
WP Weighted prediction
WSNR Weighted SNR
WT Wavelet Transform
WVGA Widescreen VGA

D References

Ahmed, N. , Natarjan, T. , Rao, K. R.: Discrete cosine transform. *IEEE Trans. Comp.* 23 (1974), pp. 90-93

Alexander, S. T. , Rajala, S. A.: Image compression results using the LMS adaptive algorithm. *IEEE Trans. Acoust., Speech, Signal Process.* 33 (1985), pp. 712-714

Anderson, J. B. , Mohan, S.: Sequential coding algorithms : A survey and cost analysis. *IEEE Trans. Commun.* 32 (1984), pp. 169-176

Anderson, J. B. , Mohan, S.: Source and Channel Coding. Dordrecht: Kluwer 1991

Antonini, M. , Barlaud, M. , Mathieu, P. , Daubechies, I.: Image coding using wavelet transform. *IEEE Trans. Image Process.* 1 (1992), pp. 205-220

Apostolopoulos, J.: Error-resilient video compression through the use of multiple states. *Proc. IEEE ICIP* (2000), pp. 2585-2588

Aravind, R. , Gersho, A.: Low rate image coding with finite-state vector quantization. *Proc. IEEE ICASSP* (1986), pp. 4.3.1-4.3.4

Van der Auwera, G. , Munteanu, A. , Schelkens, P. , Cornelis, J.: Bottom-up motion compensated prediction in the wavelet domain for spatially scalable video coding. *IEE Electron. Lett.* 38 (2002), pp. 1251-1253

Ayanoglu, E. , Gray, R. M.: The design of predictive trellis waveform coders using the generalized Lloyd algorithm. *IEEE Trans. Commun.* 34 (1986), pp. 1073-1080

Ballard, D. H. , Brown, C. M.: Computer Vision. Englewood Cliffs: Prentice Hall, 1985

Ballé, J.: Image Compression by Microtexture Synthesis. *Aachen Series on Multimedia and Communications Engineering* no. 11, Aachen: Shaker, 2012

Ballé, J. , Stojanovic, A. , Ohm, J.-R.: Models for static and dynamic texture synthesis in image and video compression. *IEEE J. Sel. Top. Sig. Proc.* 5 (2011), pp. 1353–1365

Barnsley, M.: Fractals everywhere. San Diego: Academic Press 1988

Baumgarte, F. , Faller, C.: Binaural cue coding – Part I: Psychoacoustic fundamentals and design principles. *IEEE Trans. Speech Audio Proc.* 11 (2003), pp. 509-519

Beermann, M. , Wien, M. , Ohm, J.-R.: Look-ahead coding considering rate/distortion-optimization. *Proc. IEEE ICIP* (2002), vol. I, pp. 93-95

Bellifemine, F. , Capellino, A. , Chimienti, A. , Picco, R. , Ponti, R.: Statistical analysis of the 2D-DCT coefficients of the differential signal for images. *Signal Process. : Image Commun.* 4 (1992), pp. 477-488

© Springer-Verlag Berlin Heidelberg 2015

J.-R. Ohm, *Multimedia Signal Coding and Transmission*,
Signals and Communication Technology, DOI 10.1007/978-3-662-46691-9

Benzler, U.: Spatial scalable video coding using a combined subband-DCT approach. *IEEE Trans. Circ. Syst. Video Tech.* 10 (2000), pp. 1080-1087

Berger, T.: Rate Distortion Theory. Englewood Cliffs: Prentice-Hall 1971

Besag, J.: On the statistical analysis of dirty pictures. *J. Roy. Stat. Soc. B* 48 (1986), pp. 259-302

Bhaskaran, V.: Predictive VQ schemes for grayscale image compression. *Proc. IEEE GLOBECOM* (1987), pp. 436-441

Bjøntegaard, G.: Calculation of Average PSNR Differences between RD curves. *ITU-T SG16/Q6, 13th VCEG Meeting*, Austin, Texas, USA, April 2001, Doc. VCEG-M33 (available from http://wftp3.itu.int/av-arch/video-site/0104_Aus/)

Bjøntegaard, G.: Improvements of the BD-PSNR model. *ITU-T SG16/Q6, 35th VCEG Meeting*, Berlin, Germany, 16th - 18th July, 2008, Doc.VCEG-AI11 (available from http://wftp3.itu.int/av-arch/video-site/0807_Ber/)

Blahut, R. E.: Computation of channel capacity and rate-distortion functions. *IEEE Trans. Inf. Theor.* 18 (1972), pp, 460-473

Blahut, R. E.: Principles and Practice of Information Theory. Reading: Addison-Wesley 1987

Boone, M.M. , Verheijen, E.N.G.: Multi-channel sound reproduction based on wave field synthesis, *Proc. 95th AES Conv.*, New York 1993, preprint 3719

Bose, N. K.: Applied Multidimensional Systems Theory. New York: Van Nostrand Reinhold, 1983

Bose, R. C. , Ray-Chaudhuri, D. K.: On A Class of Error Correcting Binary Group Codes. *Information and Control* 3 (1960), pp. 68–79

Bossert, M.: Channel Coding for Telecommunications. New York: Wiley, 1999

Bostelmann, G.: A simple high quality DPCM-codec for video telephony using 8 Mbit/s. *Nachrichtentechnische Zeitschrift* 28 (1974), Heft 3

Bosveld, F. , Lagendijk, R. L. , Biemond, J.: Compatible spatio-temporal subband encoding of HDTV. *Signal Process.* 28 (1992), pp. 271-289

Breebaart, J. , Engdegård, J. , Falch, C. , Hellmuth, O. , Hilpert, J. , Hoelzer, A. , Koppens, J. , Oomen, W. , Resch, B. , Schuijers, E. , Terentiev, L.: Spatial Audio Object Coding (SAOC) - The **Upcoming** MPEG Standard on Parametric Object Based Audio Coding. *Proc. 124th AES Conv.*, Amsterdam 2008, preprint 7377

Britanak, V. , Yip, P. C. , Rao, K. R.: Discrete Cosine and Sine Transforms: General Properties, Fast Algorithms and Integer Approximations. New York: Academic Press, 2010

Brünig, M. , Niehsen, W.: Fast full-search block matching. *IEEE Trans. Circ. Syst. Video Tech.* 11 (2001), p. 241-247

Brusewitz, H.: Motion compensation with triangles, *Proc. 3rd Intern. Conf. on 64 kbit Coding of moving Video*, Rotterdam, Netherlands, Sept. 1990.

Budagavi, M. , Fuldseth, A. , Bjøntegaard, G. , Sze, V. , Sadafale, M.: Core transform design in the High Efficiency Video Coding (HEVC) standard. *IEEE J. Sel. Top. Sig. Proc.* 7 (2013), pp. 1029-1041

Burt, P. J. , Adelson, E. H.: The Laplacian pyramid as a compact image code. *IEEE Trans. Commun.* 31 (1983), pp. 532-540

Cao, X. , Lai, C. , Wang, Y. , Liu, L. , Zheng, J. , He, Y.: Short Distance Intra Coding Scheme for High Efficiency Video Coding. *IEEE Trans. Image Proc.* 22 (2013), pp. 790-801

Chai, B. B. , Vass, J. , Zhuang, X.: Statistically adaptive wavelet image coding. In *Visual Information Representation, Communication, and Image Processing*, pp. 73-95, Marcel Dekker, 1999

Chan, M.-H. , Yu, Y.-B.: Variable size block matching motion compensation and application to video coding. *Proc. IEE* 137 (1990), pp. 202-212

Chelappa, R. , Kashyap, R.L.: Texture synthesis using 2-D noncausal autoregressive models. *IEEE Trans. Acoust., Speech, Signal Process.* 33 (1985), pp. 194-203

Chellappa, R. , Jain, A. (eds.): *Markov Random Fields: Theory and Applications*. Academic Press, 1996

Chen, W.-H. , Pratt, W. K.: Scene adaptive coder. *IEEE Trans. Commun.* 32 (1984), pp. 225-232, extended version of a contribution to *IEEE ICC* (1981), Philadelphia, PA, June 1981

Chen, W.-H. , Smith, C. H.: Adaptive coding of monochrome and color images. *IEEE Trans. Commun.* 25 (1977), pp. 1285-1292

Chen, W.-H. , Smith, C. H. , Fralick, S.: A fast computational algorithm for the discrete cosine transform. *IEEE Trans. Commun.* 25 (1977), pp. 1004-1009

Tsai, C.-Y. , Chen, C.-Y. , Yamakage, T. , Chong, I. S. , Huang, Y.-W. , Fu, C.-M. , Itoh, T. , Watanabe, T. , Chujoh, T. , Karczewicz, M. , Lei, S.-M.: Adaptive loop filtering for video coding, *IEEE J. Sel. Top. Sig. Proc.* 7 (2013), pp. 934-945

Chen, Y. , Hannuksela, M. M. , Suzuki, T. , Hattori, S.: Overview of the MVC+D 3D video coding standard. *J. Vis. Commun. Image Represent.* 25 (2014), pp. 679-688

Choi, S.-J. , Woods, J. W.: Motion-compensated 3-D subband coding of video. *IEEE Trans. Image Process.* 8 (1999), pp. 155-167

Chou, P. A. , Lookabaugh, T. , Gray, R. M. (1989A): Entropy-constrained vector quantization. *IEEE Trans. Acoust., Speech, Signal Process.* 37 (1989), pp. 31-42

Chou, P. A. , Lookabaugh, T. , Gray, R. M. (1989B): Optimal pruning with applications to tree-structured source coding and modeling. *IEEE Trans. Inf. Theor.* 35 (1989), pp. 299-315

Chou, P. A. , Miao, Z.: Rate-distortion optimized streaming of packetized media. *Microsoft Research Tech. Rep.* MSR-TR-2001-35, February 2001

Chuang, G. C.-H. , Kuo, C. C. J.: Wavelet descriptor of planar curves: Theory and applications. *IEEE Trans. Image Proc.* 5 (1996), pp. 56-70

Commission Internationale d'Éclairage (CIE): *Commission internationale de l'Eclairage proceedings*, 1931; also included in ISO 11664-1:2007

Clarke, R. J.: Transform Coding of Images. London: Academic Press 1985

Claypoole, R. L. , Davis, G. , Sweldens, W. , Baraniuk, R. G.: Nonlinear Wavelet Transforms for Image Coding, *Asilomar Conf. Sig., Syst., Comp.*, pp. 662-667, Nov. 1997

Conway, J. H. , Sloane, N. J. A.: Sphere Packings, Lattices and Groups. New York: Springer 1988

Côté, G. , Erol, B. , Gallant, M. , Kossentini, F.: H.263+: Video Coding at Low Bit Rates. *IEEE Trans. Circ. Syst. Video Tech.* 8 (1998) pp. 849-866

Cover, T. M. , Thomas, J. A.: Elements of Information Theory. New York: Wiley 1991

Crochiere, R. E. , Rabiner, L. R.: Multirate Digital Signal Processing. Englewood Cliffs: Prentice-Hall 1983

Cuperman, V. , Gersho, A.: Vector predictive coding of speech at 16 kbits/s. *IEEE Trans. Commun.* 33 (1985), pp. 685-696

Daubechies, I.: Orthonormal basis of compactly supported wavelets. *Comm. Pure Applied Math.* 41 (1988), pp. 909-996

Daubechies, I.: The wavelet transform, time-frequency localization and signal analysis. *IEEE Trans. Inf. Theor.* 36 (1990), pp. 961-1005

Daubechies, I. , Sweldens, W.: Factoring wavelet transforms into lifting steps. *J. Fourier Anal. Appl.* 4 (1998), pp. 247-269

Doretto, G. , Chiuso, A. , Wu, Y. N. , Soatto, S.: Dynamic textures. *Intl. J. Comp. Vision* 51 (2003), pp. 91–109

Dietz, M. , Liljeryd, L. , Kjörling, K. , Kunz, O.: Spectral band replication, a novel approach in audio coding. *Proc. 112th AES Conv.*, Munich, 2002, preprint 5553

Dudgeon, D. E. , Mersereau, R. M.: Multidimensional Digital Signal Processing. Englewood Cliffs: Prentice-Hall 1984

Dufaux, F. , Ebrahimi, T. , Sullivan, G. J.: The JPEG XR Image Coding Standard. *IEEE Sig. Proc. Mag.* 26 (2009), pp. 195-199,204

Dunham, M. O. , Gray, R. M.: An algorithm for the design of labeled-transition finite-state vector quantizers. *IEEE Trans. Commun.* 33 (1985), pp. 83-89

Elias, P.: Universal codeword sets and representations of the integers, *IEEE Trans. Inf. Theor.* 21 (1975), pp. 194-203

Faller, C. , Baumgarte, F.: Binaural cue coding – Part II: Schemes and Applications. *IEEE Trans. Speech Audio Proc.* 11 (2003), pp. 509-531

Fellgett. P.: Ambisonics. Part One: General system description. Studio Sound, August 1975, pp. 20-40.

Fischer, T. R.: A pyramid vector quantizer. *IEEE Trans. Inf. Theor.* 32 (1986), pp. 568-583

Fischer, T. R. , Wang, M.: Entropy-constrained trellis-coded quantization. *IEEE Trans. Inf. Theor.* 38 (1992), pp. 415-426

Flierl, M. , Girod, B.: Investigation of motion-compensated lifting wavelet transform. *Proc. Picture Coding Symposium* (2003), pp. 59-62, Saint-Malo, April 2003 (A)

Flierl, M. , Girod, B.: Generalized B pictures and the draft H.264/AVC video-compression standard. *IEEE Trans. Circ. Syst. Video Tech.* 13 (2003), pp. 587- 597 (B)

Forney, G. D., jr.: The Viterbi algorithm. *Proc. IEEE* 61 (1973), pp. 268-278

Foster, J. , Gray, R. M. , Dunham, M. O.: Finite-state vector quantization for waveform coding. *IEEE Trans. Inf. Theor.* 31 (1985), pp. 348-359

Freeman, H.: Boundary encoding and processing. *Picture Processing and Psychopictorics*, B. S. Lipkin and A. Rosenfeld (eds.), New York: Academic Press 1970

Fu, C.-M. , Alshina, E. , Alshin, A. , Huang, Y.-W. , Chen, C.-Y. , Tsai, C.-Y. , Hsu, C.-W. , Lei, S.-M. , Park, J.-H. , Han, W.-J.: Sample Adaptive Offset in the HEVC Standard. IEEE Trans. *Circ. Syst. Video Tech.* 22 (2012), pp.1755-1764

Gabor, D.: Theory of communication. *J. Inst. Elect. Eng.* 93 (1946), pp. 429-457

Geng, J.: Three-dimensional display technologies, *Advances in Optics and Photonics* 5 (2013), pp. 456-535

Gersho, A. , Gray, R. M.: Vector Quantization and Signal Compression. Dordrecht: Kluwer 1992

Gerzon, M.A.: Design of ambisonic decoders for multi speaker surround sound. *Proc. 58th AES Convention*, New York 1977

Gharavi, H. , Tabatabai, A.: Subband coding of monochrome and color images. *IEEE Trans. Circ. Syst.* 35 (1988), pp. 207-214

Gilge, M. , Engelhardt, T. , Mehlan, R.: Coding of arbitrarily shaped image segments based on a generalized orthogonal transform. *Signal Process. : Image Commun.* 1 (1989), pp. 153-180

Gimlett, J. I.: Use of 'activity' classes in adaptive transform image coding. *IEEE Trans. Commun.* 23 (1975), pp. 785-786

Girod, B.: The efficiency of motion-compensating prediction for hybrid coding of video sequences. *IEEE J. Sel. Areas Commun.* 5 (1987), pp. 1140-1154

Girod, B.: Psychovisual aspects of image communication. *Signal Process.* 28 (1992), pp. 239-251

Girod, B.: Motion compensation: Visual aspects, accuracy and fundamental limits. *Motion Analysis and Image Sequence Processing*, M. I. Sezan and R. L. Lagendijk (eds.), Dordrecht: Kluwer 1993

Girod, B.: Rate-constrained motion estimation. *Proc. Visual Commun. Image Process.* (1994), SPIE vol. 2308

Girod, B.: Bidirectionally decodable streams of prefix code-words. *IEEE Comm. Lett.* 3 (1999), pp. 245-247

Girod, B. , Aaron, A. M. , Rane, S. , Rebollo-Monedero, D.: Distributed video coding. *Proc. IEEE* 93 (2005), pp. 71-83

Girod, B. , Färber, N. , Horn, U.: Scalable codec architectures for Internet video-on-demand. *Proc. 1997 Asilomar Conf. Signals and Syst.*, Pacific Grove, 1997

Goyal, V.K. , Kovacevic, J. , Arean, R. , Vetterli, M.: Multiple description transform coding of images. Proc. IEEE ICIP (1998), vol. 1, pp 674 – 678

Gray, R. M.: Sliding block source coding. *IEEE Trans. Inf. Theor.* 21 (1975), pp. 357-368

Gray, R. M.: Vector quantization. *IEEE ASSP Mag.* 1 (1984), no. 2, pp. 4-29

Gray, R. M.: Source Coding Theory. Dordrecht: Kluwer 1990

de Haan, G. , Biezen, P. W. A. C. , Huijgen, H. , Ojo, O. A.: True-motion estimation with 3-D recursive search block matching. *IEEE Trans. Circ. Syst. Video Tech.* 3 (1993), pp. 368-379

Habibi, A.: Hybrid coding of pictorial data, *IEEE Trans. Comm.* 22 (1974), pp. 614-626

Hagenauer, J.: Rate-compatible punctured convolutional codes (RCPC codes) and their applications. *IEEE Trans. Commun.* 36 (1988), pp. 389-400

Hampel, H. , Arps, R. , Chamzas, D. D. , Dellert, D. , Duttweiler, D. L. , Endoh, T. , Equitz, W. , Ono, F. , Pasco, R. , Sebestyen, I. , Starkey, C. J. , Ulban, S. J. , Yamazaki, Y. , Yoshida, T.: Technical features of the JBIG standard for progressive bi-level image compression. *Signal Process.: Image Commun.* 4 (1992), pp. 103-111

Hanke, K. , Rusert, T. , Ohm, J.-R.: Motion-compensated 3D video coding using smooth transitions. Proc. SPIE VCIP (2003), vol. 5022, pp. 933-940

Hannuksela, M. M. , Rusanovskyy, D. , Su, W. , Chen, L. , Li, R. , Aflaki, P. , Lan, D. , Joachimiak, M. , Li, H. , Gabbouj, M.: Multiview-video-plus-depth coding based on the Advanced Video Coding standard. *IEEE Trans. Image Proc.* 22 (2013), pp. 3449-3458

Hanzo, L. J. , Somerville, C. , Woodard, J.: Voice and Audio Compression for Wireless Communications. New York: Wiley, 2007

Heising, G. , Marpe, D. , Cycon, H. L. , Petukhov, A. P.: Wavelet-based very low bit-rate video coding using image warping and overlapped block motion compensation, *IEE Proceedings - Vision, Image and Signal Processing* 148 (2001), no. 2, pp. 93-101

Heising, G.: Efficient and robust motion estimation in grid-based hybrid video coding schemes. *Proc. IEEE ICIP* (2002), Vol. I, pp. 697-700

Herre, J.: Temporal noise shaping, quantization and coding methods in perceptual audio coding: A tutorial introduction. *Proc. AES 17th Intl. Conf. High Quality Audio Coding*, Florence, Sept. 1999

Herre, J. , Purnhagen, H. , Breebaart, J. , Faller, C. , Disch, S. , Kjörling, K. , Schuijers, E. , Hilpert, J. , Myburg, F.: The reference model architecture for MPEG spatial audio coding. *Proc. 118th AES Conv.*, Barcelona 2005, preprint 6447

Hilbert, D.: Über die stetige Abbildung einer Line auf ein Flächenstück. *Mathematische Annalen* 38 (1891), pp. 459-460

Ho, Y.-S. , Gersho, A.: Variable-rate multi-stage vector quantization for image coding. *Proc. IEEE ICASSP* (1988), pp. 1156-1159

Hope, A.: Ambisonics – The theory and patents. *Studio Sound*, Oct 1979, pp. 36-44

Hocquenghem, A.: Codes correcteurs d'erreurs. *Chiffres* 2 (1959): pp. 147–156

Hötter, M.: Object-oriented analysis-synthesis coding based on moving two-dimensional objects. *Signal Process. : Image Commun.* 2 (1990), pp. 409-428

Hötter, M.: Optimization and efficiency of an object-oriented analysis-synthesis coder. *IEEE Trans. Circ. Syst. Video Tech.* 4 (1994), pp. 181-194

Hsiang, S.-T. , Woods, J. W.: Embedded image coding using zeroblocks of subband/ wavelet coefficients and context modeling. *Proc. IEEE ISCAS* (2000), vol. 3, pp. 662-665

Huang, C.-L. , Hsu, C.-Y.: A new motion compensation method for image sequence coding using hierarchical grid interpolation. *IEEE Trans. Circ. Syst. Video Tech.* 4 (1994), pp. 42-51

Hubel, D. H. , Wiesel, T. N.: Receptive fields, binocular interaction and functional architecture in the cat's visual cortex. *J. Physiol. (London)* 160 (1962), pp. 106-154

Huffman, D. A.: A method for the construction of minimum redundancy codes. *Proc. IRE* 40 (1952), pp. 1098-1101

Itakura, F.: *Minimum* prediction residual principle applied to speech recognition. *IEEE Trans. Acoust., Speech, Signal Process.* 23 (1975), pp. 52-72

Itakura, F. , Saito, S.: On the optimum quantization of feature parameters in the PARCOR speech synthesizer. *Proc. 1992 Conf. Speech Commun.*, pp. 434-437

Itakura, F. , Sugamura, M.: LSP speech synthesizer, its principle and implementation. *Trans. Committee on Speech Res.*, ASJ, vol. S79-46, Nov. 1979

Izquierdo M. E. , Ohm, J.-R.: Image-based rendering and 3D modeling : A complete framework. *Signal Proc.: Image Commun.* 15 (2000), pp. 817-858

Jacquin, A. E.: Image coding based on a fractal theory of iterated contractive image transformations. *IEEE Trans. Image Process.* 1 (1992), pp. 18-30

Jähne, B.: Digital Image Processing. New York/Berlin: Springer 2005 (6th edition)

Jain, A. K.: A fast Karhunen-Loève transform for a class of random processes. *IEEE Trans. Commun.* 24 (1976), pp. 1023-1029

Jain, A. K.: Fundamentals of Digital Image Processing. Englewood Cliffs: Prentice-Hall 1989

Jayant, N. S. , Noll, P.: Digital Coding of Waveforms. Englewood Cliffs: Prentice-Hall 1984

Johnston, J. D.: A filter family designed for use in quadrature mirror filter banks. *Proc. IEEE ICASSP* (1980), pp. 291-294

Kamp, S. , Wien, M.: Decoder-Side Motion Vector Derivation for Block-Based Video Coding, *IEEE Trans. Circ. Syst. Video Tech.* 22 (2012), pp. 1732,1745

Kampmann, M.: Automatic 3-D face model adaptation for model-based coding of videophone sequences. *IEEE Trans. Circ. Syst. Video Tech.* 12 (2002), pp. 172-182

Kaneko, T. , Okudaira, M.: Encoding of arbitrary curves based on the chain code representation. *IEEE Trans. Commun.* 33 (1985), pp. 697-707

Karczewicz, M. , Kurceren, R.: The SP- and SI-frames design for H.264/AVC. *IEEE Trans. Circ. Syst. Video Tech.* 13 (2003), pp. 637-644

Karlsson, G. , Vetterli, M.: Subband coding of video signals for packet switched networks. *Proc. Visual Commun. Image Process.* (1987), SPIE vol. 845, pp. 446-456

Katto, J. , Ohki, J. , Nogaki, S. , Ohta, M.: A wavelet codec with overlapped motion compensation for very low bit-rate environment. *IEEE Trans. Circ. Syst. Video Tech.* 4 (1994), pp. 328-338

Kiang, S.-Z. , Baker, R. L. , Sullivan, G. J. , Chiu, C.-Y.: Recursive optimal pruning with applications to tree structured vector quantizers. *IEEE Trans. Image Process.* 1 (1992), pp. 162-169

Kim, B.-J. , Xiong, Z. , Pearlman, W. A.: Low bit-rate scalable video coding with 3D SPIHT. *IEEE Trans. Circ. Syst. Video Tech.* 10 (2000), pp. 1374-1387

Kim, T.: Side match and overlap match vector quantizers for images. *IEEE Trans. Image Process.* 1 (1992), pp. 170-185

Koga, T. , Iinuma, K. , Hirano, A. , Iijima, Y. , Ishiguro, T.:Motion compensated interframe coding for video conferencing. Proc. Nat. Telecomm. Conf. (1981), New York, pp. G.5.3.1-G.5.3.5

Kondo, S. Sasai, H. , Kadono, S.: Tree-structured hybrid intra prediction. *Proc. IEEE ICIP (2004)*, vol. 1, pp. 473-476

Kovacevic, J. , Vetterli, M.: Nonseparable perfect reconstruction filter banks and wavelet bases for $\mathcal{R}^n$. *IEEE Trans. Inf. Theor.* 38 (1992), pp. 533-555

Kronander, T.: Some aspects of perception based image coding. *PhD Thesis*, Linköping University, 1989

Krüger, H. , Geiser, B. , Vary, P. , Li, H. T. , Zhang, D.: Gosset Lattice Spherical Vector Quantization with Low Complexity. *Proc. IEEE ICASSP* (2011), pp. 485-488

Kunt, M. , Ikonomopoulos, A. , Kocher, M.: Second-generation image coding techniques. *Proc. IEEE* 73 (1985), pp. 549-574

Kwatra, V. , Schödl, A. , Essa, I. , Turk, G. , Bobick, A.: Graph-cut textures: Image and video synthesis using graph cuts. *Proc. Intl. Conf. Comp. Graph. and Interact. Techn.* (SIGGRAPH 2003), pp. 277–286.

Kwatra, V. , Essa, I. , Bobick, A. , Kwatra, N.: Texture optimization for example-based synthesis. *Proc. Intl. Conf. Comp. Graph. and Interact. Techn.* (SIGGRAPH 2005), pp. 795–802.

Lam, E. Y. , Goodman, J. W.: A mathematical analysis of the DCT coefficient distributions for images. *IEEE Trans. Image Proc.* 9 (2000), pp. 1661-1666

Le Gall, D.: The MPEG video compression algorithm. *Signal Process.: Image Commun.* 4 (1992), pp. 129-140

Le Gall, D. , Tabatabai, A.: Subband coding of digital images using symmetric short kernel filters and arithmetic coding techniques. *Proc. IEEE ICASSP* (1988), pp. 761-764

Lewis, A. S. , Knowles, G.: Image compression using the 2D wavelet transform. *IEEE Trans. Image Proc.* 1 (1992), pp. 244-250

Li, J. , Lei, S.: An embedded still image coder with rate distortion optimization. *IEEE Trans. Image Proc.* 8 (1999), pp 913-924

Li, X. , Chen, J. , Rapaka, K. , Karczewicz. M.: Generalized interlayer residual prediction for scalable extension of HEVC. *Proc. IEEE ICIP* (2013), pp. 1559-1562

Li, X. , Kerofsky, L.: High performance resolution scalable video coding via all-phase motion compensated prediction of wavelet coefficients. *Proc. SPIE VCIP* (2002), vol. 4671, pp. 1080-1089

Liebchen, T.: MPEG-4 Lossless Coding for High-Definition Audio. *Proc. 115th AES Conv.*, New York 2003, preprint 5872

Linde, Y. , Buzo, A. , Gray, R. M.: An algorithm for vector quantizer design. *IEEE Trans. Commun.* 28 (1980), pp. 84-95

List, P. , Joch, A. , Lainema, J. , Bjontegaard, G. , Karczewicz, M.: Adaptive deblocking filter. *IEEE Trans Circ. Syst. Video Tech.* 13 (2003), pp. 614-619

Lloyd, S. P.: Least squares quantization in PCM. (1957) Reprint: *IEEE Trans. Commun.* 30 (1982), pp. 129-137

Luo, L. , Li, J. , Li, S. , Zhuang, Z. , Zhang, Y.-Q.: Motion compensated lifting wavelet and its application in video coding. *Proc. ICME 2001*, July 2001

Ma, J. , Plonka, G.: The Curvelet Transform. *IEEE Sig. Proc. Mag.* 27 (2010), pp. 118-133

Mallat, S. G.: A Wavelet Tour of Signal Processing, Third Edition: The Sparse Way. Amsterdam/Boston/New York: Elsevier/Academic Press, 2008

Mallat, S. G. , Zhang, Z.: Matching pursuits with time-frequency dictionaries, *IEEE Trans. Signal Proc.* 41 (1993), pp. 3397-3415

Malvar, H. S.: Signal Processing with Lapped Transforms. Norwood: Artech House 1991

Malvar, H. S. , Staelin, D. H.: The LOT : Transform coding without blocking effects. *IEEE Trans. Acoust., Speech, Signal Process.* 37 (1989), pp. 553-559

Malvar, H. S. , Hallapuro, A. , Karczewicz, M. , Kerofsky, L.: Low-complexity transform and quantization in H.264/AVC. *IEEE Trans. Circ. Syst. Video Tech.* 13 (2003), pp. 598- 603

Maragos, P. A. , Schafer, R. W. , Mersereau, R. M.: Two-dimensional linear prediction and its application to adaptive predictive coding of images. *IEEE Trans. Acoust., Speech, Signal Process.* 32 (1984), pp. 1213-1229

Marcellin, M. W. , Fischer, T. R.: Trellis coded quantization of memoryless and Gauss-Markov sources. *IEEE Trans. Commun.* 38 (1990), pp. 82-93

Marpe, D. , Cycon, H. L.: Very low bit-rate video coding using wavelet-based techniques. *IEEE Trans. Circ. Syst. Video Tech.* 9 (1999), pp. 85–94

Marpe, D. , Schwarz, H. , Wiegand, T.: Context-based adaptive binary arithmetic coding in the H.264/AVC video compression standard. *IEEE Trans. Circ. Syst. Video Tech.* 13 (2003), pp. 620-636

Marr, D.: Vision: A Computational Investigation into the Human Representation and Processing of Visual Information. New York: Freeman 1982

Martucci, S. A. , Sodagar, L. , Chiang, T. H. , Zhang, Y.-Q.: A zerotree wavelet coder. *IEEE Trans. Circ. Syst. Video Tech.* 7 (1997), pp. 109-118

Marzetta, T. L.: Two-Dimensional Linear Prediction: Autocorrelation Arrays, Minimum-Phase Prediction Error Filters, and Reflection Coefficient Arrays. *IEEE Trans. Acoust., Speech, Signal Process.* 28 (1980), pp. 725-733.

Max, T.: Quantizing for minimum distortion. *IRE Trans. Inf. Theor.* 6 (1960), pp. 7-12

Mayer, C. , Höynck, M.: Linear phase perfect reconstruction filter banks with alias-reduced lowpass filters for spatially scalable video coding. *Proc. 10th Aachen Symp. Signal Theory* (2001), pp. 397-402, Aachen, September 2001

McMillan, L. , Bishop, G.: Plenoptic modeling: An image-based rendering system. *Proc. SIGGRAPH* (1995), Los Angeles, Aug. 1995

Meagher, D.: Octree Encoding: A New Technique for the Representation, Manipulation and Display of Arbitrary 3-D Objects by Computer. *Rensselaer Polytechnic Institute*, Technical Report IPL-TR-80-111, 1980

Meiri, A. Z. , Yudilevich, E.: A pinned sine transform image coder. *IEEE Trans. Commun.* 29 (1981), pp. 1728-1735

Meşe, M. , Vaidyanathan, P. P.: Optimized Halftoning using dot diffusion and methods for inverse halftoning. *IEEE Trans. Image Process.* 9 (2000), pp. 691-709

Modestino, J. W. , Bhaskaran, V.: Robust two-dimensional tree encoding of images. *IEEE Trans. Commun.* 29 (1981), pp. 1786-1791

Mohan, S. , Sood, A. K.: A multiprocessor architecture for the (M,L)-algorithm suitable for VLSI implementation. *IEEE Trans. Commun.* 34 (1986), pp. 1218-1224

Musmann, H. G. , Pirsch, P. , Grallert, H. J.: Advances in picture coding. *Proc. IEEE* 73 (1985), pp. 523-548

Musmann, H. G. , Hötter, M. , Ostermann, J.: Object-oriented analysis-synthesis coding of moving images. *Signal Process. : Image Commun.* 1 (1989), pp. 117-138

Nam, K. M. , Kim, J.-S. , Park, R.-H.: A fast hierarchical motion vector estimation algorithm using mean pyramid, *IEEE Trans. Circ. Syst. Video Tech.* 5 (1995), pp. 341-351

Natarajan, T. , Ahmed, N.: On Interframe Transform Coding. *IEEE Trans. Comm.* 25 (1977), pp. 1323-1329

Ndjiki-Nya, P. , Hinz, T. , Wiegand, T.: Generic and robust video coding with texture analysis and synthesis. *Proc. IEEE Intl. Conf. Multim. and Expo* (ICME 2007), pp. 1447–1450.

Ndjiki-Nya, P. , Doshkov, D. , Kaprykowsky, H. , Zhang, F. , Bull, D. , Wiegand, T.: Perception-oriented video coding based on image analysis and completion: A review. *Signal Proc.: Image Comm.* 27 (2012), pp. 579-594

van Nee, R. , Prasad, R.: OFDM for Wireless Multimedia Communications. Boston: Artech House, 2000

Neff, R. , Zakhor, A.: Very low bit rate coding based on matching pursuit, *IEEE Trans. Circ. Syst. Video Tech.* 7 (1997), pp. 158-171

Netravali, A.N. , Robbins, J. D.: Motion compensated television coding, Part I. *Bell Syst. Tech. J.* 58 (1979), pp. 631-670

Netravali, A. N. , Robbins, J. D.: Motion-compensated coding : Some new results. *Bell Syst. Tech. J.* 59 (1980), pp. 1735-1745

Nguyen, T. , Marpe, D.: Performance Analysis of HEVC-Based Intra Coding for Still Image Compression. *Proc. Picture Coding Symposium* (2012), Krakow, May 2012

Norkin, A. , Bjontegaard, G. , Fuldseth, A. , Narroschke, M. , Ikeda, M. , Andersson, K. , Zhou, M. , van der Auwera, G.: HEVC Deblocking Filter. *IEEE Trans. Circ. Syst. Video Tech.* 22 (2012), pp.1746-1754

Ohm, J.-R.: Advanced packet-video coding based on layered VQ and SBC techniques. *IEEE Trans. Circ. Syst. Video Tech.* 3 (1993), pp. 208-221

Ohm, J.-R. (1994A): Three-dimensional subband coding with motion compensation. *IEEE Trans. Image Process.* 3 (1994), pp. 559-571

Ohm, J.-R. (1994B): Motion-compensated 3-D subband coding with multiresolution representation of motion parameters. *Proc. IEEE ICIP* (1994), vol. III, pp. 250-254

Ohm, J.-R.: Multimedia Communication Technology. New York/Berlin: Springer, 2004

Ohm, J.-R.: Multimedia Content Analysis. New York/Berlin: Springer, 2015 (to appear)

Ohm, J.-R. , Sullivan, G. J. , Schwarz, H. , Tan, T. K. , Wiegand, T.: Comparison of the coding efficiency of video coding standards - including high efficiency video coding (HEVC). *IEEE Trans.Circ. Syst. Video Tech.* 22 (2012), pp. 1669–1684

Okubo, S.: Requirements for high quality video coding standards. *Signal Process. : Image Commun.* 4 (1992), pp. 141-151

Ong, E. P. , Yang, X. , Lin, W. , Lu, Z. , Yao, S.: Video quality metric for low bitrate compressed video. *IEEE Trans. Image Proc.* 3 (1994), pp. 693-699

Orchard, M. T. , Sullivan, G. J.: Overlapped block motion compensation : An estimation-theoretic approach. *Proc. IEEE ICIP* (2004), pp. 3531-3534

Ortega, A. , Ramchandran, K. , Vetterli, M.: Optimal trellis-based buffered compression and fast approximations. *IEEE Trans. Image Process.* 3 (1994), pp. 26-40

Ostermann, J.: Object-based analysis-synthesis coding (OBASC) based on the source model of moving flexible 3-D objects. *IEEE Trans. Image Proc.* 3 (1994), pp. 705-710

Pang, K. K. , Tan, T. K.: Optimum loop filters in hybrid coders. *IEEE Trans. Circ. Syst. Video Tech.* 4 (1994), pp. 158-167

Park, S. C. , Park, M. K. , Lang, M. G.: Super-resolution image reconstruction: a technical overview. *IEEE Sig. Proc. Mag.* 20 (2003), pp. 21-36

Peano, G.: Sur une courbe, qui remplit toute une aire plane. *Mathematische Annalen* 36 (1890), pp. 157–160

Pennebaker, W. B. , Mitchell, J. L.: JPEG Still Image Compression Standard. New York: Van Nostrand Reinhold 1993

Pennebaker, W. B. , Mitchell, J. L. , Langdon, G. G. , Arps, R. B.: An overview of the basic principles of the Q-coder adaptive binary arithmetic coder. *IBM J. Res. Develop.* 32 (1988), pp. 717-752

Pesquet-Popescu, B. , Bottreau, V.: Three-dimensional lifting schemes for motion-compensated video compression. *Proc. IEEE ICASSP* (2001), pp. 1793-1796

Plataniotis, K. N. , Venetsanopoulos, A. N.: Color Image Processing and Applications. New York/Berlin: Springer, 2000

Podilchuk, C. I. , Jayant, N. S. , Noll, P.: Sparse codebooks for the quantization of non-dominant subbands in image coding. *Proc. IEEE ICASSP* (1990), pp. 2837-2840

Poletti, M. A.: Three-dimensional surround sound systems based on spherical harmonics. *Journ. AES* 53 (2005), pp. 1004-1025

Proakis, G. , Salehi, M.: Fundamentals of Communication Systems. London: Pearson Education, 2007

Princen, J. P. , Bradley, A. B.: Analysis/synthesis filter bank design based on time domain aliasing cancellation. *IEEE Trans. Acoust., Speech, Signal Process.* 34 (1986), pp. 1153-1161

Puri, A. , Aravind, R. , Haskell, B. G.: Adaptive frame/field motion compensated video coding. *Signal Process. : Image Commun.* 5 (1993), pp. 39-58

Puri, A. , Aravind, R. , Haskell, B. G. , Leonardi, R.: Video coding with motion-compensated interpolation for CD-ROM applications. *Signal Process. : Image Commun.* 2 (1990), pp. 127-144

Purves, D. , Augustine, G. J. , Fitzpatrick, D. , Hall, W. C. , LaMantia, A.-S. , McNamara, J. O. , White, L. E.: Neuroscience, 4th edition. Sunderland: Sinauer Associates, 2012

Quackenbush, S.: MPEG Audio Coding Advances. In *The MPEG Representation of Digital Media*, L. Chiariglione (ed.), pp. 125-139, New York: Springer 2012

Queluz, M. P.: A 3-dimensional subband coding scheme with motion-adaptive subband selection. *Proc. EUSIPCO* (1992), *Signal Process. VI : Theor. and Appl.*, pp. 1263-1266, Amsterdam: Elsevier 1992

Rabiner, L. R. , Schafer, R. W.: Digital Processing of Speech Signals. Englewood Cliffs: Prentice-Hall 1978

Ramamurthy, B. , Gersho, A.: Classified vector quantization of images. *IEEE Trans. Commun.* 34 (1986), pp. 1105-1115

Ramchandran, K. , Vetterli, M.: Best wavelet packet bases in a rate-distortion sense. *IEEE Trans. Image Process.* 2 (1993), pp. 160-175

Ramchandran, K. , Ortega, A. , Vetterli, M.: Bit Allocation for dependent quantization with applications to multiresolution and MPEG video coders. *IEEE Trans. Image Process.* 3 (1994), pp. 533-545

Ranganath, S. , Jain, A. K.: Two-dimensional linear prediction models - Part I : Spectral factorization and realization. *IEEE Trans. Acoust., Speech and Signal Process.* 33 (1985), pp. 280-299

Reed, I. S. , Solomon, G.: Polynomial codes over certain finite fields. *SIAM J. Appl. Math.* 8 (1960), pp. 300-304

Reibman, A. R. , Haskell, B. G.: Constraints on variable bit-rate video for ATM networks. *IEEE Trans. Circ. Syst. Video Tech.* 2 (1992), pp. 361-372

Reimers, U.: DVB - The Family of International Standards for Digital Video Broadcasting. Heidelberg/New York: Springer, 2004

Reininger, R. C. , Gibson, J. D.: Distribution of two-dimensional DCT coefficients for images. *IEEE Trans. Commun.* 31 (1983), pp. 835-839

Ribas-Corbera, J. , Chou, P. A. , Regunathan, S. L.: A generalized hypothetical reference decoder for H.264/AVC. *IEEE Trans. Circ. Syst. Video Tech.* 13 (2003), pp. 674-687

Richard, C. , Benveniste, A. , Kretz, F.: Recursive estimation of local characteristics of edges as applied to ADPCM coding. *IEEE Trans. Commun.* 32 (1984), pp. 718-728

Rioul, O. , Vetterli, M.: Wavelets and signal processing. *IEEE Signal Process. Mag.* 8 (1991), no. 4, pp. 14-38

Riskin, E. A. , Gray, R. M.: A greedy tree growing algorithm for the design of variable rate vector quantizers, *IEEE Trans. Signal Process.* 39 (1991), pp. 2500-2507

Robson, J. G.: Spatial and temporal contrast sensitivity functions of the visual system. *J. Opt. Soc. Amer. A* 56 (1966), pp, 1141-1142

Roese, J. , Pratt, W. , Robinson, G.: Interframe Cosine Transform Image Coding. *IEEE Trans. Commun.* 25 (1977), pp. 1329- 1339

Rue, H. , Held, L.: Gaussian Markov Random Fields. *Monographs Stat. Appl. Prob.* no. 104, Boca Raton: Chapman & Hall/CRC, 2005

Rusert, T.: Distortion Modeling for Rate-Constrained Optimization of Scalable Video Coding. *Aachen Series on Multimedia and Communications Engineering* no. 5, Aachen: Shaker, 2007

Rusert, T. , Hanke, K. , Ohm, J.-R.: Transition filtering and optimized quantization in interframe wavelet video coding, Proc. SPIE VCIP 2003, vol. 5150, pp. 682-694

Sabin, M. , Gray, R. M.: Product code vector quantizers for waveform and voice coding. *IEEE Trans. Acoust., Speech, Signal Process.* 32 (1984), pp. 474-488

Said, A. , Pearlman, W.: A new fast and efficient image codec based on set partitioning in hierarchical trees. *IEEE Trans. Circ. Syst. Video Tech.* 6 (1996), pp. 243-250

Samet, H. : The quadtree and related hierarchical data structures. *Comput. Surv.* 16 (1984), pp. 187-260

Scherer, P.: Ein neues Verfahren der raumbezogenen Stereophonie mit verbesserter Übertragung der Rauminformation. *Rundfunktechnische Mitteilungen* 21 (1977), S. 196-204

Schumitsch, B. , Schwarz, H. , Wiegand, T.: Inter-frame optimization of transform coefficient selection in hybrid video coding. *Proc. Picture Coding Symposium* (2004), pp. 59-64

Secker, A. , Taubman, D.: Motion-compensated highly-scalable video compression using an adaptive 3D wavelet transform based on lifting. *Proc. IEEE ICIP* (2001), pp. 1029-1032

Schuijers, E. , Oomen, W. , den Brinker, B. , Breebaart, J.: Advances in parametric coding for high-quality audio. *Proc. 114th AES Conv.*, Amsterdam, 2003, preprint 5852

Shannon, C. E.: A mathematical theory of communication. *Bell Syst. Tech. J.* (1948), reprint : University of Illinois press 1949

Shannon, C. E.: Coding theorems for a discrete source with a fidelity criterion. *IRE Nat. Conv. Rec.* (1959), reprint : *Information and Decision Processes*, R. E. Machol (ed.), New York : McGraw Hill 1960

Shannon, C. E.: The Collected Papers. Piscataway : IEEE Press 1992

Shapiro, J. M.: Embedded image coding using zerotrees of wavelets coefficients. *IEEE Trans. Signal Process.* 41 (1993), pp. 3445-3462

Shen, K. , Delp, E.: Wavelet based rate scalable video compression. IEEE Trans. Circ. Syst. Video Tech. 9 (1999), pp. 109-122

Shishikui, Y.: A study on modeling of the motion compensation prediction error signal. *IEICE Trans. Commun.* 75-B (1992), pp. 368-376

Sikora, T. , Makai, B.: Shape-adaptive DCT for generic coding of video. *IEEE Trans. Circ. Syst. Video Tech.* 5, pp. 59-62, 1995

Simon, S. F.: On suboptimal multidimensional companding. *Proc. IEEE Data Compress. Conf.* (1998), pp. 438-447

Slepian, D. , Wolf, J. K.: Noiseless Coding of Correlated Information Sources. *IEEE Trans. Inf. Theo.* 19 (1973), pp. 471-480

Smith, M. J. T.: IIR analysis/synthesis systems. *Subband Image Coding*, J. W. Woods (ed.), Dordrecht : Kluwer 1991

Sodagar, I.: The MPEG-DASH standard for multimedia streaming over the Internet. *IEEE Multimedia* 18 (2011), pp. 62-67

Spearman, C.: General intelligence, objectively determined and measured. *Am. J. Psychol.* 15 (1904), pp. 201-293

Stevens, S. S. , Volkmann, J.: The relation of pitch to frequency: A revised scale. *Am. J. Psychol.* 53 (1940), pp. 329-353.

Stewart, L. C. , Gray, R. M. , Linde, Y.: The design of trellis waveform coders. *IEEE Trans. Commun.* 30 (1982), pp. 702-710

Sullivan, G. J.: Efficient Scalar Quantization of Exponential and Laplacian Random Variables. *IEEE Trans. Inf. Theo.* 42 (1996), pp. 1365-1374,

Sullivan, G. J. , Baker, R. L.: Motion compensation for video compression using control grid interpolation. *Proc. IEEE ICASSP* (1991), pp. 2713-1716

Sullivan, G. J. , Ohm, J.-R. , Han, W.-J. , Wiegand, T.: Overview of the High Efficiency Video Coding (HEVC) Standard. *IEEE Trans. Circ. Syst. Video Tech.* 22 (2012), pp.1649-1668

Sullivan, G. J. , Wiegand, T.: Rate-Distortion Optimization for Video Compression. *IEEE Signal Proc. Mag.* 15 (2005), pp. 74-90

Sullivan, G. J. , Boyce, J. , Chen, Y. , Ohm, J.-R.: Standardized extensions of High Efficiency Video Coding (HEVC). *IEEE J. Sel. Top. Signal Proc.* 7 (2013), pp. 1001–1016

Sze, V. , Budagavi, M. , Sullivan, G. J.: High Efficiency Video Coding: Algorithms and Architectures. New York: Springer 2014

Takashima, Y. , Wada, M. , Murakami, H.: Reversible variable length codes. *IEEE Trans. Commun.* 43 (1995), pp. 158-162

Tan, T. K. , Boon, C. , Suzuki, Y.: Intra prediction by template matching. *Proc. IEEE ICIP* (2006), pp. 1693-1696

Taubman, D. : High performance scalable image compression with EBCOT. *IEEE Trans. Image Proc.* 9 (2000), pp. 1158-1170

Taubman, D. S. , Marcellin, M. W.: JPEG 2000: Image Compression Fundamentals, Standards and Practice. Dordrecht: Kluwer Academic Publishers, 2001

Thoma, R. , Bierling, M.: Motion compensating interpolation considering covered and uncovered background. *Signal Process. : Image Commun.* 1 (1989), pp. 191-212

Traunmüller, H.: Analytical expressions for the tonotopic sensory scale. *J. Acoust. Soc. Am.* 88 (1990), pp. 97-100

Ugur, K. , Alshin, A. , Alshina, E. , Bossen, F. , Han, W.-J. , Park, J.-H. , Lainema, J.: Motion compensated prediction and interpolation filter design in the H.265/HEVC. *IEEE J. Sel. Top. Sig. Proc.* 7 (2013), pp. 946-956

Urey, H. , Chellappan, K. V. , Erden, E. , Surman, P.: State of the art in stereoscopic and autostereoscopic displays, *Proc. IEEE* 99 (2011), pp. 540-555

Vaidyanathan, P. P.: Multirate Systems and Filter Banks. Englewood Cliffs: Prentice Hall 1993

Vaisey, J. , Gersho, A.: Image compression with variable block size segmentation. *IEEE Trans. Signal Process.* 40 (1992), pp. 2040-2060

Vaishampayan, V. A. : Design of multiple description scalar quantizers. *IEEE Trans. Inf. Theor.* 39 (1993), pp. 821-834

Vary, P. , Martin, R.: Digital Speech Transmission. New York: Wiley 2006

i Ventura, R. M. F. , Vandergheynst, P. , Frossard, P.: Low-rate and flexible image coding with redundant representations, *IEEE Trans. Image Proc.* 15 (2006), pp. 726–739

Vetro, A. , Wiegand, T. , Sullivan, G. J.: Overview of the stereo and multiview video coding extensions of the H.264/MPEG-4 AVC standard, *Proc. IEEE* 99 (2011), pp. 626-642

Vetterli, M. , Kovacevic, J.: Wavelets and Subband Coding. Englewood Cliffs: Prentice Hall, 1995

Vetterli, M. , LeGall, D.: Perfect reconstruction FIR filter banks: Some properties and factorizations. *IEEE Trans. Acoust., Speech, Signal Proc.* 37 (1989), pp. 1057-1071

Viswanathan, R. , Makhoul, J.: Quantization properties of transmission parameters in linear predictive systems. *IEEE Trans. Acoust., Speech, Signal Proc.* 23 (1975), pp. 309-321

Viterbi, A. J.: Error bounds for convolutional codes and an asymptotically optimum decoding algorithm. *IEEE Trans. Inf. Theor.* 13 (1967), pp. 260-269

Viterbi, A. J. , Omura, J. K.: Principles of Digital Communication and Coding. New York: McGraw Hill 1985

Vos, K. , Sørensen, K. V. , Jensen, S. S. , Valin, J.-M.: Voice coding with OPUS. *Proc. 135th AES Conv.*, New York 2013, preprint 8941

Wang, Y. , Orchard, M. , Vaishampayan, V. , Reibman, A. R. : Multiple description coding using pairwise decorrelating transform. *IEEE Trans. Image Proc.* 10 (2001), pp. 351-366

Wang, Y. , Wenger, S. , Wen, J. , Katsaggelos, A. G.: Error resilient video coding techniques. *IEEE Signal Process. Mag.* 17(4) (2000), pp. 61-82

Wang, Y. , Zhu, Q.-F.: Error control and concealment for video communication: A review. *Proc. IEEE* 86 (1998), pp. 974-997

Wang, Z. , Bovik, A.C.: Mean squared error: Love it or leave it? A new look at Signal Fidelity Measures. *IEEE Signal Proc. Mag.* 26 (2009), pp. 98-117

Wang, Z. , Simoncelli, E. P.: An adaptive linear system framework for image distortion analysis. *Proc. IEEE ICIP* (2005), vol. 3, pp. 1160–1163

Weber, E. H.: De pulsu, resorptione, audita et tactu. Annotationes anatomicae et physiologicae. Leipzig: Köhler, 1834

Wedi, T. , Musmann, H.G.: Motion- and aliasing-compensated prediction for hybrid video coding. *IEEE Trans. Circ. Syst. Video Tech.* 13 (2003), pp. 577-586

Welch, T. A.: A Technique for High Performance Data Compression. IEEE Comput. 17 (1984), pp. 8-19

Wen, J. , Villasenor, J.: A class of reversible variable length codes for robust image and video coding. *Proc. IEEE ICIP* (1997), vol. II, pp. 65-68

Westerink, P. H. , Biemond, J. , Boekee, D. E. , Woods, J. W.: Subband coding of images using vector quantization. *IEEE Trans. Commun.* 36 (1988), pp. 713-719

Wiegand, T. , Girod, B.: Multi-Frame Motion-Compensated Prediction for Video Transmission. Dordrecht: Kluwer, 2001

Wiegand, T. , Sullivan, G. J. , Bjontegaard, G. , Luthra, A.: Overview of the H.264/AVC video coding standard. *IEEE Trans. Circ. Syst. Video Tech.* 13 (2003), pp. 560-576

Wien, M.: Variable block-size transforms for H.264/AVC. *IEEE Trans. Circ. Syst. Video Tech.* 13 (2003), pp. 604-613

Wien, M.: High Efficiency Video Coding: Coding Tools and Specification. New York, Heidelberg: Springer 2015

Winken, M. , Schwarz, H. , Marpe, D. , Wiegand, T.: Joint Optimization of Transform Coefficients for Hierarchical B Picture Coding in H.264/AVC. *Proc. IEEE ICIP* (2007), vol. 4, pp. 89-92

Witten, I. H. , Neal, R. M. , Cleary, J. G.: Arithmetic coding for data compression. *Comm. ACM* 30 (1987), pp. 520-540

Wittmann, S. , Wedi, T.: Transmission of post-filter hints for video coding schemes. *Proc. IEEE ICIP* (2007), pp. I-81–I-84

Woods, J. W. , O'Neill, S. D.: Subband coding of images. *IEEE Trans. Acoust., Speech, Signal Process.* 34 (1986), pp. 1278-1288

Wu, F. , Li, S. , Zhang, Y.-Q.: DCT-prediction based progressive fine granularity scalable coding. *Proc. IEEE ICIP* (2000), pp. 1903-1906

Wyner, A. , Ziv, J.: The rate-distortion function for source coding with side information at the decoder. *IEEE Trans. Inf. Theo.* 22 (1976), pp. 1-10

Xu, J. , Li, S. , Zhang, Y.-Q.: A wavelet codec using 3-D ESCOT. *Proc. IEEE-PCM2000*, Dec. 2000

Xu, J. , Zeng, B. , Wu, F.: An Overview of Directional Transforms in Image Coding. *Proc. IEEE ISCAS* (2010), pp. 3036-3039

Yang, D. T. , Kyriakakis, C. , Kuo, C.-C. J.: High-fidelity Multichannel Audio Coding. *EURASIP Book Series on Signal. Proc. and Comm.*, New York: Hindawi 2006

Yang, W. , Lu, Y. , Wu, F. , Cai, J. , Ngan, K. N. , Li, S.: 4D Wavelet-based Multi-view Video Coding. *IEEE Trans. Circ. Syst. Video Tech.* 16 (2006), pp.1385-1396

Yang, X. , Ramchandran, K.: Hierarchical backward motion compensation for wavelet video coding using optimized interpolation filters. *Proc. IEEE ICIP* (1997), vol. 1, pp. 85-88

Yang, X. , Ramchandran, K.: Scalable wavelet video coding using alias-reduced hierarchical motion compensation. *IEEE Trans. Image* Process. 9 (2000), pp. 778-791

Zaciu, R. et al.: Motion estimation and motion compensation using the overcomplete discrete wavelet transform. *Proc. IEEE ICIP* (1996), vol. I, pp. 973-976

Zhang, Q. , Kassam, S. A.: Hybrid ARQ with selective combining for fading channels. *IEEE J. Select. Areas Commun.* 17 (1999), pp. 867-880

Ziv, J. , Lempel, A.: A universal algorithm for sequential data compression, *IEEE Trans. Inf. Theor.* 23 (1977), pp. 337-343

Ziv, J. , Lempel, A.: Compression of individual sequences via variable-rate coding. *IEEE Trans. Inf. Theor.* 24 (1978), pp. 530-536

Zurmühl, R.: Matrizen. Berlin: Springer 1964

Zwicker, E.: Psychoakustik. Berlin: Springer 1982

E Index

© Springer-Verlag Berlin Heidelberg 2015
J.-R. Ohm, *Multimedia Signal Coding and Transmission*,
Signals and Communication Technology, DOI 10.1007/978-3-662-46691-9

MIX
Papier aus verantwortungsvollen Quellen
Paper from responsible sources
FSC® C105338

If you have any concerns about our products,
you can contact us on
ProductSafety@springernature.com

In case Publisher is established outside the EU,
the EU authorized representative is:
Springer Nature Customer Service Center GmbH
Europaplatz 3, 69115 Heidelberg, Germany

Printed by Libri Plureos GmbH
in Hamburg, Germany